Computational Methods for Multiphase Flows in Porous Media

COMPUTATIONAL SCIENCE & ENGINEERING

Computational Science and Engineering (CS&E) is widely accepted, along with theory and experiment, as a crucial third mode of scientific investigation and engineering design. This series publishes research monographs, advanced undergraduate- and graduate-level textbooks, and other volumes of interest to a wide segment of the community of computational scientists and engineers. The series also includes volumes addressed to users of CS&E methods by targeting specific groups of professionals whose work relies extensively on computational science and engineering.

Series Volumes

Layton, William, *Introduction to the Numerical Analysis of Incompressible Viscous Flows*

Ascher, Uri M., *Numerical Methods for Evolutionary Differential Equations*

Zohdi, T. I., *An Introduction to Modeling and Simulation of Particulate Flows*

Biegler, Lorenz T., Omar Ghattas, Matthias Heinkenschloss, David Keyes, and Bart van Bloemen Waanders, Editors, *Real-Time PDE-Constrained Optimization*

Chen, Zhangxin, Guanren Huan, and Yuanle Ma, *Computational Methods for Multiphase Flows in Porous Media*

Shapira, Yair, *Solving PDEs in C++: Numerical Methods in a Unified Object-Oriented Approach*

Computational Methods for Multiphase Flows in Porous Media

Zhangxin Chen
Guanren Huan
Yuanle Ma

Southern Methodist University
Dallas, Texas

siam
Society for Industrial and Applied Mathematics
Philadelphia

10 9 8 7 6 5 4 3 2

Library of Congress Cataloging-in-Publication Data

Chen, Zhangxin
Computational methods for multiphase flows in porous media / Zhangxin Chen, Guanren Huan, Yuanle Ma.
p. cm.
Includes bibliographical references and index.
ISBN 978-0-89871-606-1 (pbk.)
1. Multiphase flow–Mathematical models. 2. Porous materials–Mathematical models. 3. Petroleum reserves–Mathematical models. 4. Finite element method. I. Huan, Guanren. II. Ma, Yuanle. III. Title.

TA357.5.M84C45 2006
533.2'8015118–dc22

2005056416

The cover was produced from images created by and used with permission of the Scientific Computing and Imaging (SCI) Institute, University of Utah; J. Bielak, D. O'Hallaron, L. Ramirez-Guzman, and T. Tu, Carnegie Mellon University; O. Ghattas, University of Texas at Austin; K. Ma and H. Yu, University of California, Davis; and Mark R. Petersen, Los Alamos National Laboratory. More information about the images is available at *http://www.siam.org/books/series/csecover.php*.

Partial royalties from the sale of this book are placed in a fund to help students attend SIAM meetings and other SIAM-related activities. This fund is administered by SIAM, and qualified individuals are encouraged to write directly to SIAM for guidelines.

This book is dedicated
to Aijie, Christina, Paul,
and William.

Contents

List of Figures

List of Tables

Preface

Mathematical models of petroleum reservoirs have been utilized since the late 1800s (Darcy, 1856). A mathematical model consists of a set of differential equations that describe the flow of fluids in the petroleum reservoirs, together with an appropriate set of boundary and/or initial conditions. The reliability of predictions from a reservoir model depends on how well the model describes a field. To develop a model, in general simplifying assumptions need be made because the field is too complicated to be described exactly. The assumptions needed to solve a model analytically are very restrictive; many analytical solutions require that the reservoir be homogeneous and isotropic, for example. It is usually necessary to solve a mathematical model approximately using numerical methods. Since the 1950s, when digital computers became widely available, numerical models have been used to predict, understand, and optimize complex physical fluid flow processes in petroleum reservoirs. Moreover, the emergence of complex enhanced recovery techniques in the field of oil production has emphasized the need for sophisticated mathematical and computational tools, capable of modeling intricate physical phenomena and sharply changing fluid interfaces. The objective of this book is to provide researchers in the area of porous media flow, especially in petroleum reservoirs, with an overview of various multiphase flow and transport models and the current, state-of-the-art computational methods used in the solution of these models.

This book offers a fundamental and practical introduction to the use of computational methods, particularly finite element methods, in the simulation of fluid flow in petroleum reservoirs. In the presentation, we have attempted to introduce every concept in the simplest possible setting and to maintain a level of treatment that is as rigorous as possible without being unnecessarily abstract. In developing numerical methods, a brief discussion of the basic concepts has been given in the text as needed, and the reader is referred to appropriate references for more details. We have not attempted to give any mathematical proofs, but rather we review multiphase flow equations and computational methods to introduce the basic terminologies and notation. We have attempted to present a thorough discussion of practical aspects of these subjects in a consistent manner, and to focus on the mathematical formulations of these equations and on applications of the computational methods to their solution.

This book covers four major topics (flow and transport differential equations and their numerical solutions, rock and fluid properties, numerical methods, and linear system solvers), eight applications (single phase flow, two-phase flow, flow of black oil type, volatile oil flow, compositional flow, nonisothermal flow, chemical compositional flow, and flows in fractured porous media), and six special topics (welling modeling, upscaling, history matching, parallel computing, oil recovery optimization, and surface network systems). Each chapter ends with bibliographical information and exercises.

In Chapter 2, after an introduction in Section 2.1, we review the basic governing equations for single phase flow (Section 2.2), two-phase flow (Section 2.3), transport of a component in a fluid phase (Section 2.4), transport of multicomponents in a fluid phase (Section 2.5), flow of black oil type (Section 2.6), flow of volatile oil type (Section 2.7), compositional flow (Section 2.8), nonisothermal flow (Section 2.9), chemical compositional flow (Section 2.10), and flows in fractured porous media (Section 2.11). As an example, deformable media, non-Darcy's law, and other effects are discussed for single phase flow in Section 2.2. Also, alternative differential equations are developed for two-phase flow in Section 2.3; these alternative formulations can be extended to flows of other types. Relationships among all these flows are mentioned in Section 2.12.

In Chapter 3, we consider rock and fluid properties. In particular, capillary pressures and relative permeabilities are discussed for the rock properties in Section 3.1, and oil, gas, and water properties and equations of state are studied for the fluid properties in Section 3.2. In Section 3.3, temperature-dependent rock and fluid properties are described.

In Chapter 4, numerical methods are developed; an emphasis is placed on the development of finite element methods. After an introduction to the classical finite difference methods in Section 4.1, six major types of finite element methods are reviewed: standard (Section 4.2), control volume (Section 4.3), discontinuous (Section 4.4), mixed (Section 4.5), characteristic (Section 4.6), and adaptive (Section 4.7). All these finite element methods have been employed in petroleum reservoir simulation. For each method, a brief introduction, the notation, basic terminology, and necessary concepts are given.

In Chapter 5, solution techniques for solving the linear systems arising in numerical reservoir simulation are considered; both direct and iterative algorithms are introduced. In Sections 5.1 and 5.2, we discuss Gaussian elimination or Cholesky's method for tridiagonal and general banded matrices, respectively. Because the structure of a matrix depends on the ordering of nodes, Section 5.3 is devoted to this topic. Then Krylov subspace algorithms are described: conjugate gradient (CG), generalized minimum residual (GMRES), orthogonal minimum residual (ORTHOMIN), and biconjugate gradient stabilized (BiCGSTAB) iterative algorithms, respectively, in Sections 5.4–5.7. Preconditioned versions of these algorithms and the choice of preconditioners are studied in Sections 5.8 and 5.9. Practical considerations for the choice of preconditioners in reservoir simulation are given in Section 5.10. Finally, comparisons of direct and iterative algorithms are presented in Section 5.11.

In Chapters 6–12, numerical and computational methods for single phase flow, two-phase flow, flow of black oil type, compositional flow, nonisothermal flow, chemical compositional flow, and flows in fractured porous media respectively, are studied. For single and two-phase flows, numerical and analytic solutions are compared. For two-phase flow, a comparison between different numerical methods is also presented. For the flow of black oil type, different solution schemes (e.g., fully implicit, sequential, and IMPES—implicit in pressure and explicit in saturation) are assessed. The numerical and experimental examples given in Chapters 6–12 are based on the benchmark problems of the first nine comparative solution projects organized by the society of petroleum engineers and real field data analysis.

In Chapter 13, vertical and horizontal well modeling using finite difference and finite element methods is discussed. Finally, in Chapter 14 special topics on upscaling, history matching, parallel computing, oil recovery optimization, and surface network systems are briefly touched on.

This book can serve as a textbook for graduate (even advanced undergraduate) students in geology, petroleum engineering, and applied mathematics. It can be also used as a handbook for employees in the oil industry who need a basic grasp of modeling and computational method concepts. It can also serve as a reference book for geologists, petroleum engineers, applied mathematicians, and scientists in the area of petroleum reservoir simulation. Calculus, basic physics, and some acquaintance with partial differential equations and simple matrix algebra are necessary prerequisites.

Chapters 2 through 5 form the essential material for a course. Because each of Chapters 6 through 13 is essentially self-contained and independent, different course paths can be chosen. The exercise section in each chapter plays a role in the presentation, and the reader should spend the time to solve the problems.

We take this opportunity to thank many people who have helped, in different ways, in the preparation of this book. We have had incredible support from Professor Jim Douglas, Jr., and Professor Richard E. Ewing. We would like to thank Professor Ian Gladwell for reading the whole manuscript and making invaluable suggestions. The book title was suggested by Professor Roland Glowinski. Many students have made invaluable comments about the early drafts of this book. In particular, we thank Dr. Baoyan Li and Dr. Wenjun Li for carrying out some numerical experiments for us.

Zhangxin Chen, Guanren Huan, and Yuanle Ma

Zhangxin Chen: NSERC/AERI/Foundation CMG Chair in Reservoir Simulation and iCORE Industrial Chair in Reservoir Modeling

Dallas, Texas, USA

October 15, 2005

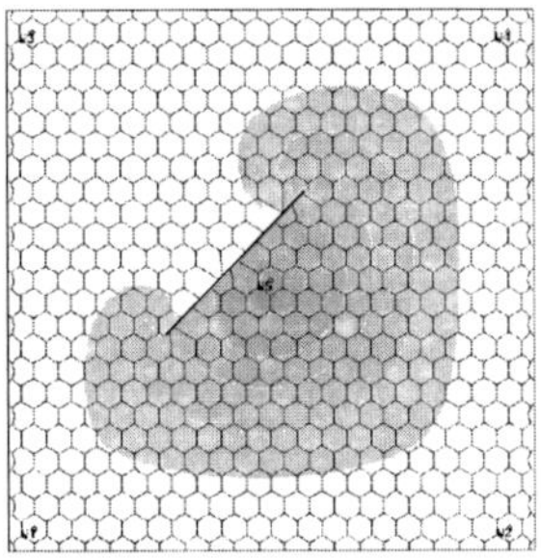

Chapter 1
Introduction

1.1 Petroleum Reservoir Simulation

In mathematical terminology, a *porous medium* is the closure of a subset of the Euclidean space $\mathbb{R}^d$ ($d = 1, 2$, or 3). A *petroleum reservoir* is a porous medium that contains hydrocarbons. The primary goal of *reservoir simulation* is to predict future performance of a reservoir and find ways and means of optimizing the recovery of some of the hydrocarbons.

The two important characteristics of a petroleum reservoir are the natures of the rock and of the fluids filling it. A reservoir is usually *heterogeneous*; its properties heavily depend on the space location. A *fractured reservoir* is heterogeneous, for example. It consists of a set of blocks of porous media (the *matrix*) and a net of fractures. The rock properties in such a reservoir dramatically change; its permeability may vary from one millidarcy (md) in the matrix to thousands md in the fractures. While the governing equations for the fractured reservoir are similar to those for an ordinary reservoir, they have additional difficulties that must be overcome. The mathematical models presented in this book take into account the heterogeneity of a porous medium, and computational methods are presented for both ordinary and fractured media.

The nature of the fluids filling a petroleum reservoir strongly depends on the stage of oil recovery. In the very early stage, the reservoir essentially contains a single fluid such as gas or oil (the presence of water can be usually neglected). Often the pressure at this stage is so high that the gas or oil is produced by simple natural decompression without any pumping effort at the wells. This stage is referred to as *primary recovery*, and it ends when a pressure equilibrium between the oil field and the atmosphere occurs. Primary recovery usually leaves 70%–85% of hydrocarbons in the reservoir.

To recover part of the remaining oil, a fluid (usually water) is injected into some wells (*injection wells*) while oil is produced through other wells (*production wells*). This process serves to maintain high reservoir pressure and flow rates. It also displaces some of the oil and pushes it toward the production wells. This stage of oil recovery is called *secondary recovery* (or *water flooding*).

In the secondary recovery, if the reservoir pressure is above the bubble point pressure of the oil phase, there is *two-phase immiscible flow*, one phase being water and the other being oil, without mass transfer between the phases. If the reservoir pressure drops below the bubble point pressure, then the oil (more precisely, the hydrocarbon phase) is split into a liquid phase and a gaseous phase in thermodynamic equilibrium. In this case, the flow is of *black oil type*; the water phase does not exchange mass with the other phases, but the liquid and gaseous phases exchange mass.

Water flooding is not very effective, and after this stage 50% or more of hydrocarbons often remain in the reservoir. Due to strong surface tension, a large amount of oil is trapped in small pores and cannot be washed out using this technique. Also, when the oil is heavy and viscous, the water is extremely mobile. If the flow rate is sufficiently high, instead of producing oil, the production wells primarily produce water.

To recover more of the hydrocarbons, several enhanced recovery techniques have been developed. These techniques involve complex chemical and thermal effects and are termed *tertiary recovery* or *enhanced recovery*. Enhanced oil recovery is oil recovery by injecting materials that are not normally present in a petroleum reservoir. There are many different versions of enhanced recovery techniques, but one of the main objectives of these techniques is to achieve miscibility and thus eliminate the residual oil saturation. The miscibility is achieved by increasing temperature (e.g., in situ combustion) or by injecting other chemical species like CO_2. One typical flow in enhanced recovery is the *compositional flow*, where only the number of chemical species is given a priori, and the number of phases and the composition of each phase in terms of the given species depend on the thermodynamic conditions and the overall concentration of each species. Flows of other types involve *thermal methods*, particularly steam drive and soak, and *chemical flooding*, such as alkaline, surfactant, polymer, and foam (ASP+foam) flooding. All flows of these types in petroleum reservoir applications are considered in this book.

1.2 Numerical Methods

In general, the equations governing a mathematical model of a reservoir cannot be solved by analytical methods. Instead, a numerical model can be produced in a form that is amenable to solution by digital computers. Since the 1950s, when digital computers became widely available, numerical models have been used to predict, understand, and optimize complex physical fluid flow processes in petroleum reservoirs. Recent advances in computational capabilities (particularly with the advent of new parallel architectures) have greatly expanded the potential for solving larger problems and hence permitting the incorporation of more physics into the differential equations. While several books are available on finite difference methods as applied to the area of porous media flow (Peaceman, 1977B; Aziz and Settari, 1979), there does not appear to be available a book that examines the application of finite element methods in this area. The purpose of this book is to attempt to provide researchers in this area, especially in petroleum reservoirs, with the current, state-of-the-art finite element methods.

Compared with *finite difference methods*, the introduction of *finite element methods* is relatively recent. The advantages of the finite element methods over the finite differences are that general boundary conditions, complex geometry, and variable material properties can be relatively easily handled. Also, the clear structure and versatility of the finite elements makes it possible to develop general purpose software for applications. Furthermore, there is a solid theoretical foundation that gives added confidence, and in many cases it is possible to obtain concrete error estimates for the finite element solutions. Finite element methods were first introduced by Courant (1943). From the 1950s to the 1970s, they were developed by engineers and mathematicians into a general method for the numerical solution of partial differential equations.

Driven by the needs for designing technologies for exploration, production, and recovery of oil and gas, the petroleum industry has developed and implemented a variety of numerical reservoir simulators using finite element methods (e.g., see the biannual SPE numerical simulation proceedings published by the society of petroleum engineers since 1968). In addition to the advantages mentioned above, finite element methods have some peculiar features when applied to reservoir simulation, such as in the reduction of grid orientation effects; in the treatment of local grid refinement, horizontal and slanted wells, and corner point techniques; in the simulation of faults and fractures; in the design of streamlines, and in the requirement of high-order accuracy of numerical solutions. These topics will be studied in detail.

The standard finite element methods and two closely related methods, *control volume* and *discontinuous* finite element methods, are covered here. Control volume finite element methods possess a local mass conservation property on each control volume, while discontinuous methods are closely related to the finite volume methods that have been utilized in reservoir simulation. Two nonstandard methods, the *mixed* and *characteristic* finite element methods, are also discussed. The reason for the development of mixed methods is that in many applications a vector variable (e.g., a velocity field in petroleum reservoir simulation) is the primary variable in which one is interested, and then the mixed methods are designed to approximate both this variable and a scalar variable (e.g., pressure) simultaneously and give a high-order approximation for both variables. The characteristic finite element methods are suitable for advection-dominated (or convection-dominated) problems. They take reasonably large time steps, capture sharp solution fronts, and conserve mass. Finally, *adaptive* finite element methods are described. These methods adjust themselves to improve approximate solutions that have important local and transient features.

1.3 Linear System Solvers

For a petroleum reservoir simulator with a number of gridblocks of order 100,000, about 80%–90% of the total simulation time is spent on the solution of linear systems. Thus the choice of a fast linear solver is crucial in reservoir simulation. In general, a system matrix arising in numerical reservoir simulation is sparse, highly nonsymmetric, and ill-conditioned. While sparse, its natural banded structure is usually spoiled by wells that perforate into many gridblocks and/or by irregular gridblock structure. Furthermore, the matrix dimension M often ranges from hundreds to millions. For the solution of such systems, *Krylov subspace algorithms* are the sole option.

Over a dozen parameter-free Krylov subspace algorithms have been proposed for solving nonsymmetric systems of linear equations. Three such leading iterative algorithms are the CGN (the conjugate gradient iteration applied to the normal equations), GMRES (residual minimization in a Krylov space), and BiCGSTAB (a biorthogonalization method adapted from the biconjugate gradient iteration). These three algorithms differ fundamentally in their capabilities. Examples of matrices can be constructed to show that each type of iteration can outperform the others by a factor on the order of $\sqrt{M}$ or M (Nachtigal et al., 1992). Moreover, these algorithms are often useless without *preconditioning*. The Krylov subspace algorithms and their preconditioned versions are discussed. The discussion of these algorithms and of their preconditioners is for algorithms of general applicability. Some guidelines are also provided about the choice of a suitable algorithm for a given problem.

1.4 Solution Schemes

Since the fluid flow models in porous media involve large, coupled systems of nonlinear, time-dependent partial differential equations, an important problem in the numerical simulation is to develop stable, efficient, robust, accurate, and self-adaptive time stepping techniques. *Explicit methods* like forward Euler methods require that a Courant–Friedrichs–Lewy (CFL) time step constraint be satisfied, while *implicit methods* such as backward Euler and Crank–Nicolson methods are reasonably stable. On the other hand, the explicit methods are computationally efficient, and the implicit methods require the solution of large systems of nonlinear equations at each time step. Explicit methods, together with linearization by some Newton-like iteration, have been frequently used in reservoir simulation. Due to the CFL condition, enormously long computations are needed to simulate a long time period (e.g., over ten years) problem in a field-scale model, and thus fully explicit methods cannot be efficiently exploited, especially for problems with strong nonlinearities.

A variation to achieve better stability without suffering too much in computation is the IMPES (implicit in pressure and explicit in saturation) scheme. This scheme works well for problems of intermediate difficulty and nonlinearity (e.g., for two-phase incompressible flow) and is still widely used in the petroleum industry. However, it is not efficient for problems with strong nonlinearities, particularly for problems involving more than two fluid phases.

Another basic scheme for solving multiphase flow equations is the *simultaneous solution* (SS) method, which solves all of the coupled nonlinear equations simultaneously and implicitly. This technique is stable and can take very large time steps while stability is maintained. For the black oil and thermal models (with a few components) considered in this book, the SS scheme is a good choice. However, for complex problems that involve many chemical components (e.g., the compositional and chemical compositional flow problems), the size of system matrices to be solved is too large, even with today's computing power.

A variety of *sequential* methods for solving equations in an implicit fashion without a full coupling have been developed. They are less stable but more computationally efficient than the SS scheme, and more stable but less efficient than the IMPES scheme. The sequential schemes are very suitable for the compositional and chemical compositional flow problems that involve many chemical components.

Finally, an *adaptive implicit scheme* can be employed in reservoir simulation. The principal idea of this technique is to seek an efficient middle ground between the IMPES (or sequential) and SS schemes. That is, at a given time step, the expensive SS scheme is confined to those gridblocks that require it, while on the remaining gridblocks the IMPES scheme is implemented. The majority of research in the solution schemes has concentrated on the stability of time stepping methods, and the efficient linearization and iterative solution of the resulting equations. The accuracy of these schemes must be also addressed. All the solution schemes mentioned are covered and compared in this book.

1.5 Numerical Examples

Many numerical examples are presented to test and compare different numerical methods, linear system solvers, and solution schemes. These examples are based on the benchmark problems of the first nine comparative solution projects organized by the Society of Petroleum Engineers. Typically, about ten organizations participated in each project. The numerical examples presented include three-dimensional black oil reservoir simulations, a coning problem study, gas cycling analysis of retrograde condensate reservoirs, steam injection simulations, dual porosity model simulations, gridding techniques, horizontal well modeling, and large-scale reservoir simulations. A couple of numerical examples are based on real field data analysis.

1.6 Ground Water Flow Modeling

There are many modeling and simulation processes that use technologies and techniques similar to those in petroleum reservoir simulation; one example is *ground water flow modeling*. Ground water is one of the most widely distributed and important resources on the earth. Over half of the population in the USA depends on ground water for its water supply, for example. Also, ground water is an important source of irrigation and industrial process water. In a large part of the USA, available sources of ground water are a fundamental constraint on development and economic activity. Ground water quality is endangered by organic, inorganic, and radioactive contaminants introduced into the ground by improper disposal or accidental spill. Protecting this quality is a problem of broad economic and societal importance.

Water movement in the subsurface has been studied for many decades by soil scientists and agricultural engineers. This research dates back to the classical work of Richards (1931). The subsurface is a multiphase system. It consists of at least three phases: the solid phase of the soil matrix, the water phase, and a gaseous phase. Other phases like a separate organic liquid phase or an ice phase may exist. The traditional approach of studying a subsurface system has concentrated exclusively on water. Over the past few decades, interest has grown in problems where other phases can be important. These include the evaluation of remediation technologies such as soil venting where the gas phase plays an important role. *Soil venting* is a technology that attempts to remove contaminants from the soil before they can seriously pollute ground water supplies. It works by pumping air through a part of the subsurface contaminated by a volatile contaminant and inducing it to volatilize so that it can be removed by the gas phase flow. Previous evaluation of this technology has indicated that it is economical and efficient in contaminant cleanup. For such an application, coupled nonlinear equations for an air-water system must be solved. While ground water modeling has become increasingly important, it is beyond the scope of this book to study it. However, we emphasize that technologies and techniques similar to those used in petroleum reservoirs apply also to ground water flow (Chen and Ewing, 1997A; Helmig, 1997).

1.7 Basin Modeling

Basin modeling is a term often used to describe three factors: the burial history of sediments, the thermal history of these sediments, and the generation, migration, and preservation of hydrocarbons. The burial history of sedimentary units is driven by sediment supply, chemical and mechanical compaction, tectonic forces, erosional and intrusive events, and sea-level changes. An understanding of this dynamical evolution of sediments is critical to basin modeling since paleostructures, porosity, sedimentary thermal conductivity, solubility, faulting, and fluid flow all depend on the sedimentary patterns of behavior. When the burial history of the sediments is known, one needs to determine their thermal history. There are two approaches to this. The first approach assumes a priori models for heat flux evolution, and the determination is carried out by fiat. The second approach uses present-day data that contain some cumulative measure of thermal history and attempts to utilize these data to reconstruct the thermal history of the sediments. After determining the sedimentary thermal history, one needs to determine the generation, migration, and preservation of hydrocarbons. In this step, one needs to figure out the ways and means of providing thermokinetic models of hydrocarbon generation from organic material and to assess their accuracy. All these factors constitute crucial parts in attempts at basin modeling. Basin modeling is a very important and complex process (Allen and Allen, 1990; Lerche, 1990; Chen et al., 2002B). However, due to the scope of this book, this topic will not be discussed further.

1.8 Units

British units are used almost exclusively in reservoir engineering in the USA. However, the use of metric systems, particularly the SI (Sisteme International) unit system, has been increasing. Hence we state the SI base units and some common derived units adapted from Campbell and Campbell (1985) and Lake (1989). The SI base quantities and units are given in Table 1.1. When the mole is used, the elementary entities must be specified; they can be atoms, molecules, ions, electrons, other particles, or specified groups of such particles in petroleum engineering.

Some SI derived units are shown in Table 1.2, and a list of useful conversions are stated in Table 1.3. Two troublesome conversions are between pressure (1 MPa $\approx$ 147 psia) and temperature (1 K = 1.8 R, Rankine). Neither the Fahrenheit nor the Celsius scale is absolute, so an additional conversion is required:

$$^\circ\mathrm{F} = \mathrm{R} - 459.67, \qquad ^\circ\mathrm{C} = \mathrm{K} - 273.16.$$

Table 1.1. *SI base quantities and units.*

Base quantity	SI unit	SI unit symbol	SPE symbol
Time	Second	s	t
Length	Meter	m	L
Mass	Kilogram	kg	M
Thermodynamic temperature	Kelvin	K	T
Amount of substance	Mole	mol	

Table 1.2. *Some common SI derived units.*

Quantity	Unit	SI unit symbol	Formula
Pressure	Pascal	Pa	N/m^2
Velocity	Meter per second		m/s
Acceleration	Meter per second squared		m/s^2
Area	Square meter		m^2
Volume	Cubic meter		m^3
Density	Kilogram per cubic meter		kg/m^3
Energy (work)	Joule	J	N·m
Force	Newton	N	kg·m/s^2
Viscosity (dynamic)	Pascal second		Pa·s
Viscosity (kinematic)	Square meter per second		m^2/s

Table 1.3. *Selected conversion factors.*

To convert from	To	Multiply by
Day (mean solar)	Second (s)	8.640000E + 04
Darcy	Meter^2 (m^2)	9.869232E − 13
Mile (U.S. survey)	Meter (m)	1.609347E + 03
Acre (U.S. survey)	Meter^2 (m^2)	4.046872E + 03
Acres	Feet^2 (ft^2)	4.356000E + 04
Atmosphere (standard)	Pascal (Pa)	1.013250E + 05
Bar	Pascal (Pa)	1.000000E + 05
Barrel	Feet^3 (ft^3)	5.615000E + 00
Barrel (petroleum 42 gal)	Meter^3 (m^3)	1.589873E − 01
British thermal unit	Joule (J)	1.055232E + 03
Dyne	Newton (N)	1.000000E − 05
Gallon (U.S. liquid)	Meter^3 (m^3)	3.785412E − 03
Hectare	Meter^2 (m^2)	1.000000E + 04
Gram	Kilogram (kg)	1.000000E − 03
Pound (lbm avoirdupois)	Kilogram (kg)	4.535924E − 01
Ton (short, 2000 lbm)	Kilogram (kg)	9.071847E + 02

The superscript ° is not used for the absolute temperature scales K and R. The volume conversions are also troublesome due to the interchangeable use of mass and standard volumes:

$$1 \text{ reservoir barrel (or bbl)} = 0.159 \text{ m}^3,$$
$$1 \text{ standard barrel (or STB)} = 0.159 \text{ SCM}.$$

The symbol SCM (standard cubic meter) is not a standard SI unit; it indicates the amount of mass contained in one cubic meter calculated at standard pressure and temperature.

The use of unit prefixes is sometimes convenient (cf. Table 1.4), but it does require care. If a prefixed unit is exponentiated, the exponent applies to the prefix as well as the unit. For example, $1 \text{ km}^2 = 1 \text{ (km)}^2 = 1 \ (10^3 \text{ m})^2 = 1 \times 10^6 \text{ m}^2$.

Table 1.4. *SI unit prefixes.*

Factor	SI prefix	Symbol	Meaning (U.S.)
10^{-9}	nano	n	One billionth of
10^{-6}	micro	μ	One millionth of
10^{-3}	milli	m	One thousandth of
10^{-2}	centi	c	One hundredth of
10^{-1}	deci	d	One tenth of
10	deka	da	Ten times
10^2	hecto	h	One hundred times
10^3	kilo	k	One thousand times
10^6	mega	M	One million times
10^9	giga	G	One billion times
10^{12}	tera	T	One trillion times

There are several quantities that have the exact same or approximate numerical value between the SI and practical units:

$$1 \text{ cp} = 1 \text{ mPa·s}, \quad 1 \text{ dyne/cm} = 1 \text{ mN/m},$$
$$1 \text{ Btu} \approx 1 \text{ kJ}, \quad 1 \text{ darcy} \approx 1\,\mu\text{m}^2, \quad 1 \text{ ppm} \approx 1 \text{ g/m}^3.$$

There are several more useful unit conversions:

$$1 \text{ atm} = 14.7 \text{ psia}, \quad 1 \text{ day} = 24 \text{ hrs}, \quad 1 \text{ ft} = 30.48 \text{ cm},$$
$$1 \text{ bbl} = 5.615 \text{ ft}^3, \quad 1 \text{ darcy} = 1{,}000 \text{ md}, \quad 1 \text{ hr} = 3{,}600 \text{ sec}.$$

More unit conversions will be stated in Chapter 16.

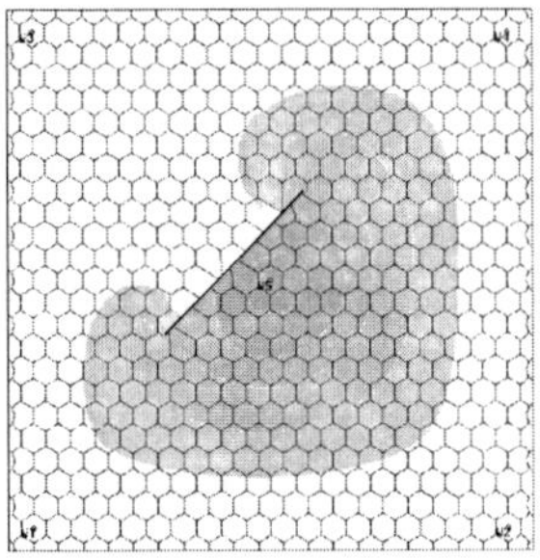

Chapter 2
Flow and Transport Equations

2.1 Introduction

Mathematical models of petroleum reservoirs have been utilized since the late 1800s. A mathematical model consists of a set of equations that describe the flow of fluids in a petroleum reservoir, together with an appropriate set of boundary and/or initial conditions. This chapter is devoted to the development of such a model.

Fluid motion in a petroleum reservoir is governed by the conservation of mass, momentum, and energy. In the simulation of flow in the reservoir, the momentum equation is given in the form of Darcy's law (Darcy, 1856). Derived empirically, this law indicates a linear relationship between the fluid velocity relative to the solid and the pressure head gradient. Its theoretical basis was provided by, e.g., Whitaker (1966); also see the books by Bear (1972) and Scheidegger (1974). The present chapter reviews some models that are known to be of practical importance.

There are several books available on fluid flow in porous media. The books by Muskat (1937; 1949) deal with the mechanics of fluid flow, the one by Collins (1961) is concerned with the practical and theoretical bases of petroleum reservoir engineering, and the one by Bear (1972) treats the dynamics and statics of fluids. The books by Peaceman (1977) and Aziz and Settari (1979) (also see Mattax and Dalton, 1990) present the application of finite difference methods to fluid flow in porous media. While the book by Chavent and Jaffré (1986) discusses finite element methods, the discussion is very brief, and most of their book is devoted to the mathematical formulation of models. The proceedings edited by Ewing (1983), Wheeler (1995), and Chen et al. (2000A) contain papers on finite elements for flow and transport problems. There are also books available on ground water hydrology; see Polubarinova-Kochina (1962), Wang and Anderson (1982), and Helmig (1997), for example.

The material presented in this chapter is very condensed. We do not attempt to derive differential equations that govern the flow and transport of fluids in porous media, but rather we review these equations to introduce the terminology and notation used throughout this book. The chapter is organized as follows. We consider the single phase flow of a fluid in a porous medium in Section 2.2. While this book concentrates on an ordinary porous medium, deformable and fractured porous media for single phase flow are also studied as an example. Furthermore, flow equations that include non-Darcy effects are described, and boundary and initial conditions are also presented. We develop the governing equations for two-phase immiscible flow in a porous medium in Section 2.3; attention is paid to the development of alternative differential equations for such a flow. Boundary and initial conditions associated with these alternative equations are established. We consider flow and transport of a component in a fluid phase and the problem of miscible displacement of one fluid by another in Section 2.4; diffusion and dispersion effects are discussed. We deal with transport of multicomponents in a fluid phase in Section 2.5; reactive flow problems are presented. We present the black oil model for three-phase flow in Section 2.6. A volatile oil model is defined in Section 2.7; this model includes the oil volatility effect. We construct differential equations for multicomponent, multiphase compositional flow, which involves mass transfer between phases in a general fashion, in Section 2.8. Although most mathematical models presented deal with isothermal flow, we also present a section on nonisothermal flow in Section 2.9. In Section 2.10, we consider chemical compositional flooding, where ASP+foam (alkaline, surfactant, and polymer) flooding is described. In Section 2.11, flows in fractured porous media are studied in more detail. Section 2.12 is devoted to discussing the relationship among all the flow models presented in this chapter. Finally, bibliographical information is given in Section 2.13. The mathematical models are briefly described in this chapter; more details on the governing differential equations and constitutive relations will be given in each of the subsequent chapters where a specific model is treated.

The term *phase* stands for matter that has a homogeneous chemical composition and physical state. Solid, liquid, and gaseous phases can be distinguished. Although there may be several liquid phases present in a porous medium, only a gaseous phase can exist. The phases are separate from each other. The term *component* is associated with a unique chemical species, and components constitute the phases.

2.2 Single Phase Flow

In this section, we consider the transport of a *Newtonian fluid* that occupies the entire void space in a porous medium under the isothermal condition.

2.2.1 Single phase flow in a porous medium

The governing equations for the *single phase flow* of a fluid (a single component or a homogeneous mixture) in a porous medium are given by the *conservation of mass*, *Darcy's law*, and an *equation of state*. We make the assumptions that the mass fluxes due to dispersion and diffusion are so small (relative to the advective mass flux) that they are negligible and that the fluid-solid interface is a material surface with respect to the fluid mass so that no mass of this fluid can cross it.

The spatial and temporal variables will be represented by $\mathbf{x} = (x_1, x_2, x_3)$ and t, respectively. Denote by ϕ the *porosity* of the porous medium (the fraction of a representative elementary volume available for the fluid), by ρ the density of the fluid per unit volume, by $\mathbf{u} = (u_1, u_2, u_3)$ the superficial *Darcy velocity*, and by q the external sources and sinks. Con-

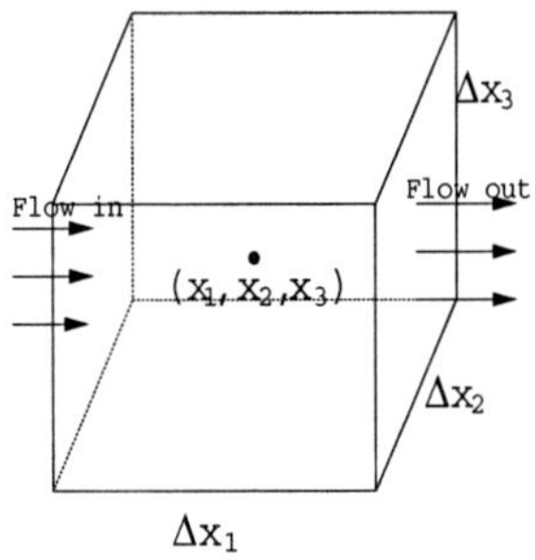

Figure 2.1. *A differential volume.*

sider a rectangular cube such that its faces are parallel to the coordinate axes (cf. Figure 2.1). The centroid of this cube is denoted (x_1, x_2, x_3), and its length in the x_i-coordinate direction is Δx_i, $i = 1, 2, 3$. The x_i-component of the *mass flux* (mass flow per unit area per unit time) of the fluid is ρu_i. Referring to Figure 2.1, the mass inflow across the surface at $x_1 - \frac{\Delta x_1}{2}$ per unit time is

$$(\rho u_1)_{x_1-\frac{\Delta x_1}{2},x_2,x_3} \Delta x_2 \Delta x_3,$$

and the mass outflow at $x_1 + \frac{\Delta x_1}{2}$ is

$$(\rho u_1)_{x_1+\frac{\Delta x_1}{2},x_2,x_3} \Delta x_2 \Delta x_3.$$

Similarly, in the x_2- and x_3-coordinate directions, the mass inflows and outflows across the surfaces are, respectively,

$$(\rho u_2)_{x_1,x_2-\frac{\Delta x_2}{2},x_3} \Delta x_1 \Delta x_3, \quad (\rho u_2)_{x_1,x_2+\frac{\Delta x_2}{2},x_3} \Delta x_1 \Delta x_3$$

and

$$(\rho u_3)_{x_1,x_2,x_3-\frac{\Delta x_3}{2}} \Delta x_1 \Delta x_2, \quad (\rho u_3)_{x_1,x_2,x_3+\frac{\Delta x_3}{2}} \Delta x_1 \Delta x_2.$$

With $\partial/\partial t$ being the time differentiation, *mass accumulation* due to compressibility per unit time is

$$\frac{\partial(\phi\rho)}{\partial t} \Delta x_1 \Delta x_2 \Delta x_3,$$

and the removal of mass from the cube, i.e., the mass decrement (accumulation) due to a sink of strength q (mass per unit volume per unit time) is

$$-q \Delta x_1 \Delta x_2 \Delta x_3.$$

The difference between the mass inflow and outflow equals the sum of mass accumulation within this cube:

$$\left[(\rho u_1)_{x_1-\frac{\Delta x_1}{2},x_2,x_3} - (\rho u_1)_{x_1+\frac{\Delta x_1}{2},x_2,x_3}\right]\Delta x_2\Delta x_3$$
$$+\left[(\rho u_2)_{x_1,x_2-\frac{\Delta x_2}{2},x_3} - (\rho u_2)_{x_1,x_2+\frac{\Delta x_2}{2},x_3}\right]\Delta x_1\Delta x_3$$
$$+\left[(\rho u_3)_{x_1,x_2,x_3-\frac{\Delta x_3}{2}} - (\rho u_3)_{x_1,x_2,x_3+\frac{\Delta x_3}{2}}\right]\Delta x_1\Delta x_2$$
$$=\left(\frac{\partial(\phi\rho)}{\partial t} - q\right)\Delta x_1\Delta x_2\Delta x_3.$$

Divide this equation by $\Delta x_1\Delta x_2\Delta x_3$ to see that

$$-\frac{(\rho u_1)_{x_1+\frac{\Delta x_1}{2},x_2,x_3} - (\rho u_1)_{x_1-\frac{\Delta x_1}{2},x_2,x_3}}{\Delta x_1}$$
$$-\frac{(\rho u_2)_{x_1,x_2+\frac{\Delta x_2}{2},x_3} - (\rho u_2)_{x_1,x_2-\frac{\Delta x_2}{2},x_3}}{\Delta x_2}$$
$$-\frac{(\rho u_3)_{x_1,x_2,x_3+\frac{\Delta x_3}{2}} - (\rho u_3)_{x_1,x_2,x_3-\frac{\Delta x_3}{2}}}{\Delta x_3} = \frac{\partial(\phi\rho)}{\partial t} - q.$$

Letting $\Delta x_i \to 0$, $i = 1, 2, 3$, we obtain the *mass conservation equation*

$$\frac{\partial(\phi\rho)}{\partial t} = -\nabla\cdot(\rho\mathbf{u}) + q, \tag{2.1}$$

where $\nabla\cdot$ is the *divergence* operator:

$$\nabla\cdot\mathbf{u} = \frac{\partial u_1}{\partial x_1} + \frac{\partial u_2}{\partial x_2} + \frac{\partial u_3}{\partial x_3}.$$

Note that q is negative for sinks and positive for sources.

Equation (2.1) is established for three space dimensions. It also applies to the one-dimensional (in the x_1-direction) or two-dimensional (in the x_1x_2-plane) flow if we introduce the factor

$$\begin{aligned} \bar{\alpha}(\mathbf{x}) &= \Delta x_2(\mathbf{x})\Delta x_3(\mathbf{x}) && \text{in one dimension,}\\ \bar{\alpha}(\mathbf{x}) &= \Delta x_3(\mathbf{x}) && \text{in two dimensions,}\\ \bar{\alpha}(\mathbf{x}) &= 1 && \text{in three dimensions.} \end{aligned}$$

For these three cases, (2.1) becomes

$$\bar{\alpha}\frac{\partial(\phi\rho)}{\partial t} = -\nabla\cdot(\bar{\alpha}\rho\mathbf{u}) + \bar{\alpha}q. \tag{2.2}$$

The *formation volume factor*, B, is defined as the ratio of the volume of the fluid measured at reservoir conditions to the volume of the same fluid measured at standard conditions:

$$B(p,T) = \frac{V(p,T)}{V_s},$$

where s denotes the standard conditions and p and T are the fluid pressure and temperature (at reservoir conditions), respectively. Let W be the weight of the fluid. Because $V = W/\rho$ and $V_s = W/\rho_s$, where ρ_s is the density at standard conditions, we see that

$$\rho = \frac{\rho_s}{B}.$$

Substituting ρ into (2.2), we have

$$\bar{\alpha}\frac{\partial}{\partial t}\left(\frac{\phi}{B}\right) = -\nabla \cdot \left(\frac{\bar{\alpha}}{B}\mathbf{u}\right) + \frac{\bar{\alpha}q}{\rho_s}. \tag{2.3}$$

While (2.1) and (2.3) are equivalent, the former will be utilized in this book except for the black oil and volatile oil models.

In addition to (2.1), we state the *momentum conservation* in the form of Darcy's law (Darcy, 1856). This law indicates a linear relationship between the fluid velocity and the *pressure head gradient*:

$$\mathbf{u} = -\frac{1}{\mu}\mathbf{k}\,(\nabla p - \rho \wp \nabla z), \tag{2.4}$$

where $\mathbf{k}$ is the *absolute permeability tensor* of the porous medium, μ is the fluid viscosity, $\wp$ is the magnitude of the gravitational acceleration, z is the depth, and ∇ is the *gradient operator*:

$$\nabla p = \left(\frac{\partial p}{\partial x_1}, \frac{\partial p}{\partial x_2}, \frac{\partial p}{\partial x_3}\right).$$

The x_3-coordinate in (2.4) is in the vertical downward direction. The *permeability* is an average medium property that measures the ability of the porous medium to transmit fluid. In some cases, it is possible to assume that $\mathbf{k}$ is a diagonal tensor

$$\mathbf{k} = \begin{pmatrix} k_{11} & & \\ & k_{22} & \\ & & k_{33} \end{pmatrix} = \operatorname{diag}(k_{11}, k_{22}, k_{33}).$$

If $k_{11} = k_{22} = k_{33}$, the porous medium is called *isotropic*; otherwise, it is *anisotropic*.

2.2.2 General equations for single phase flow

Substituting (2.4) into (2.1) yields

$$\frac{\partial(\phi\rho)}{\partial t} = \nabla \cdot \left(\frac{\rho}{\mu}\mathbf{k}\,(\nabla p - \rho\wp\nabla z)\right) + q. \tag{2.5}$$

An equation of state is expressed in terms of the *fluid compressibility* c_f:

$$c_f = -\frac{1}{V}\frac{\partial V}{\partial p}\bigg|_T = \frac{1}{\rho}\frac{\partial \rho}{\partial p}\bigg|_T, \tag{2.6}$$

at a fixed temperature T, where V stands for the volume occupied by the fluid at reservoir conditions. Combining (2.5) and (2.6) gives a closed system for the main unknown p

or ρ. Simplified expressions such as a linear relationship between p and ρ for a *slightly compressible fluid* can be used; see the next subsection.

It is sometimes convenient in mathematical analysis to write (2.5) in a form without the explicit appearance of gravity, by the introduction of a *pseudopotential* (Hubbert, 1956):

$$\Phi' = \int_{p^o}^{p} \frac{1}{\rho(\xi)\wp}\, d\xi - z, \tag{2.7}$$

where p^o is a reference pressure. Using (2.7), equation (2.5) reduces to

$$\frac{\partial(\phi\rho)}{\partial t} = \nabla \cdot \left(\frac{\rho^2\wp}{\mu}\mathbf{k}\nabla\Phi'\right) + q. \tag{2.8}$$

In numerical computations, more often we use the usual *potential* (piezometric head)

$$\Phi = p - \rho\wp z,$$

which is related to Φ' (with, e.g., $p^o = 0$ and constant ρ) by

$$\Phi = \rho\wp\Phi'.$$

If we neglect the term $\wp z \nabla\rho$, in terms of Φ, (2.5) becomes

$$\frac{\partial(\phi\rho)}{\partial t} = \nabla \cdot \left(\frac{\rho}{\mu}\mathbf{k}\nabla\Phi\right) + q. \tag{2.9}$$

In general, there is not a distributed mass source or sink in single phase flow in a three-dimensional medium. However, as an approximation, we may consider the case where sources and sinks of a fluid are located at isolated points $\mathbf{x}^{(i)}$. Then these point sources and sinks can be surrounded by small spheres that are excluded from the medium. The surfaces of these spheres can be treated as part of the boundary of the medium, and the mass flow rate per unit volume of each source or sink specifies the total flux through its surface.

Another approach to handling *point sources* and *sinks* is to insert them in the mass conservation equation. That is, for point sinks, we define q in (2.5) by

$$q = -\sum_i \rho q^{(i)} \delta(\mathbf{x} - \mathbf{x}^{(i)}), \tag{2.10}$$

where $q^{(i)}$ indicates the volume of the fluid produced per unit time at $\mathbf{x}^{(i)}$ and δ is the Dirac delta function. For point sources, q is given by

$$q = \sum_i \rho^{(i)} q^{(i)} \delta(\mathbf{x} - \mathbf{x}^{(i)}), \tag{2.11}$$

where $q^{(i)}$ and $\rho^{(i)}$ denote the volume of the fluid injected per unit time and its density (which is known) at $\mathbf{x}^{(i)}$, respectively. The treatment of sources and sinks will be discussed in more detail in later chapters (cf. Chapter 13).

2.2.3 Equations for slightly compressible flow and rock

It is sometimes possible to assume that the fluid compressibility c_f is constant over a certain range of pressures. Then, after integration (cf. Exercise 2.1), we write (2.6) as

$$\rho = \rho^o e^{c_f(p-p^o)}, \tag{2.12}$$

where ρ^o is the density at the reference pressure p^o. Using a Taylor series expansion, we see that

$$\rho = \rho^o \left\{ 1 + c_f(p - p^o) + \frac{1}{2!}c_f^2(p - p^o)^2 + \cdots \right\},$$

so an approximation results:

$$\rho \approx \rho^o\big(1 + c_f(p - p^o)\big). \tag{2.13}$$

The *rock compressibility* is defined by

$$c_R = \frac{1}{\phi}\frac{d\phi}{dp}. \tag{2.14}$$

After integration, it is given by

$$\phi = \phi^o e^{c_R(p-p^o)}, \tag{2.15}$$

where ϕ^o is the porosity at p^o. Similarly, it is approximated by

$$\phi \approx \phi^o\big(1 + c_R(p - p^o)\big). \tag{2.16}$$

Then it follows that

$$\frac{d\phi}{dp} = \phi^o c_R. \tag{2.17}$$

After carrying out the time differentiation in the left-hand side of (2.5), the equation becomes

$$\left(\phi\frac{\partial\rho}{\partial p} + \rho\frac{d\phi}{dp}\right)\frac{\partial p}{\partial t} = \nabla\cdot\left(\frac{\rho}{\mu}\mathbf{k}\left(\nabla p - \rho\wp\nabla z\right)\right) + q. \tag{2.18}$$

Substituting (2.6) and (2.17) into (2.18) gives

$$\rho\left(\phi c_f + \phi^o c_R\right)\frac{\partial p}{\partial t} = \nabla\cdot\left(\frac{\rho}{\mu}\mathbf{k}\left(\nabla p - \rho\wp\nabla z\right)\right) + q.$$

Defining the *total compressibility*

$$c_t = c_f + \frac{\phi^o}{\phi}c_R, \tag{2.19}$$

we see that

$$\phi\rho c_t\frac{\partial p}{\partial t} = \nabla\cdot\left(\frac{\rho}{\mu}\mathbf{k}\left(\nabla p - \rho\wp\nabla z\right)\right) + q, \tag{2.20}$$

which is a *parabolic equation* in p (cf. Section 2.3.2), with ρ given by (2.12).

2.2.4 Equations for gas flow

For gas flow, the compressibility c_g of gas is usually not assumed to be constant. In such a case, the general equation (2.18) applies; i.e.,

$$c(p)\frac{\partial p}{\partial t} = \nabla \cdot \left(\frac{\rho}{\mu} \mathbf{k} \left(\nabla p - \rho \wp \nabla z \right) \right) + q, \tag{2.21}$$

where

$$c(p) = \phi \frac{\partial \rho}{\partial p} + \rho \frac{d\phi}{dp}.$$

A different form of (2.21) can be derived if we use the *gas law* (the pressure-volume-temperature (PVT) relation)

$$\rho = \frac{pW}{ZRT}, \tag{2.22}$$

where W is the molecular weight, Z is the gas compressibility factor, and R is the *universal gas constant*. If pressure, temperature, and density are in atm, K, and g/cm^3, respectively, the value of R is 82.057. For a pure gas reservoir, the gravitational constant is usually small and neglected. We assume that the porous medium is isotropic; i.e., $\mathbf{k} = k\mathbf{I}$, where $\mathbf{I}$ is the identity tensor. Furthermore, we assume that ϕ and μ are constants. Then, substituting (2.22) into (2.5), we see that

$$\frac{\phi}{k}\frac{\partial}{\partial t}\left(\frac{p}{Z}\right) = \nabla \cdot \left(\frac{p}{\mu Z} \nabla p \right) + \frac{RT}{Wk} q. \tag{2.23}$$

Note that $2p\nabla p = \nabla p^2$, so (2.23) becomes

$$\frac{2\phi\mu Z}{k}\frac{\partial}{\partial t}\left(\frac{p}{Z}\right) = \Delta p^2 + 2pZ\frac{d}{dp}\left(\frac{1}{Z}\right)|\nabla p|^2 + \frac{2\mu ZRT}{Wk} q, \tag{2.24}$$

where Δ is the *Laplacian operator*:

$$\Delta p = \frac{\partial^2 p}{\partial x_1^2} + \frac{\partial^2 p}{\partial x_2^2} + \frac{\partial^2 p}{\partial x_2^3}.$$

Because

$$c_g = \frac{1}{\rho}\frac{d\rho}{dp}\bigg|_T = \frac{1}{p} - \frac{1}{Z}\frac{dZ}{dp},$$

we have

$$\frac{\partial}{\partial t}\left(\frac{p}{Z}\right) = \frac{pc_g}{Z}\frac{\partial p}{\partial t}.$$

Inserting this equation into (2.24) and neglecting the term involving $|\nabla p|^2$ (often smaller than other terms in (2.24)), we obtain

$$\frac{\phi\mu c_g}{k}\frac{\partial p^2}{\partial t} = \Delta p^2 + \frac{2ZRT\mu}{Wk} q, \tag{2.25}$$

which is a parabolic equation in p^2.

There is another way to derive an equation similar to (2.25). Define a *pseudopressure* by

$$\psi = 2\int_{p^o}^{p} \frac{p}{Z\mu}\, dp.$$

Note that

$$\nabla\psi = \frac{2p}{Z\mu}\nabla p, \qquad \frac{\partial\psi}{\partial t} = \frac{2p}{Z\mu}\frac{\partial p}{\partial t}.$$

Equation (2.23) becomes

$$\frac{\phi\mu c_g}{k}\frac{\partial\psi}{\partial t} = \Delta\psi + \frac{2RT}{Wk}q. \tag{2.26}$$

The derivation of (2.26) does not require us to neglect the second term in the right-hand side of (2.24).

2.2.5 Single phase flow in a deformable medium

Consider a *deformable porous medium* whose solid skeleton has compressibility and shearing rigidity. The medium is assumed to be composed of a *linear elastic material*, and its deformation to be small.

Let $\mathbf{w_s}$ and $\mathbf{w}$ be the displacements of the solid and fluid, respectively. For a deformable medium, Darcy's law in (2.4) is generalized as follows (Biot, 1955; Chen et al., 2004B):

$$\dot{\mathbf{w}} - \dot{\mathbf{w}}_s = -\frac{1}{\mu}\mathbf{k}\left(\nabla p - \rho\wp\nabla z\right), \tag{2.27}$$

where $\dot{\mathbf{w}} = \partial\mathbf{w}/\partial t$. Note that $\mathbf{u} = \dot{\mathbf{w}}$, so (2.27) just introduces a new dependent variable $\mathbf{w_s}$. Additional equations are needed for a closed system.

Let $\mathbf{I}$ be the identity matrix. The total *stress tensor* of the bulk material is

$$\boldsymbol{\sigma} + \sigma\mathbf{I} \equiv \begin{pmatrix} \sigma_{11}+\sigma & \sigma_{12} & \sigma_{13} \\ \sigma_{21} & \sigma_{22}+\sigma & \sigma_{23} \\ \sigma_{31} & \sigma_{32} & \sigma_{33}+\sigma \end{pmatrix}$$

with the symmetry property $\sigma_{ij} = \sigma_{ji}$. To understand the meaning of this tensor, consider a cube of the bulk material with unit size. Then σ represents the total normal tension force applied to the fluid part of the faces of the cube, while the remaining components σ_{ij} are the forces applied to the portion of the cube faces occupied by the solid. The stress tensor satisfies the *equilibrium relation*

$$\nabla\cdot\left(\boldsymbol{\sigma} + \sigma\mathbf{I}\right) + \rho_t\wp\nabla z = 0, \tag{2.28}$$

where $\rho_t = \phi\rho + (1-\phi)\rho_s$ is the mass density of the bulk material and ρ_s is the solid density. To relate $\boldsymbol{\sigma}$ to $\mathbf{w_s}$, we need a constitutive relationship between the stress and strain tensors.

Denote the *strain tensors* of the solid and fluid by $\boldsymbol{\epsilon}_s$ and $\boldsymbol{\epsilon}$, respectively, defined by

$$\epsilon_{s,ij} = \frac{1}{2}\left(\frac{\partial w_{s,i}}{\partial x_j} + \frac{\partial w_{s,j}}{\partial x_i}\right), \qquad \epsilon_{ij} = \frac{1}{2}\left(\frac{\partial w_i}{\partial x_j} + \frac{\partial w_j}{\partial x_i}\right), \qquad i,\ j = 1,2,3.$$

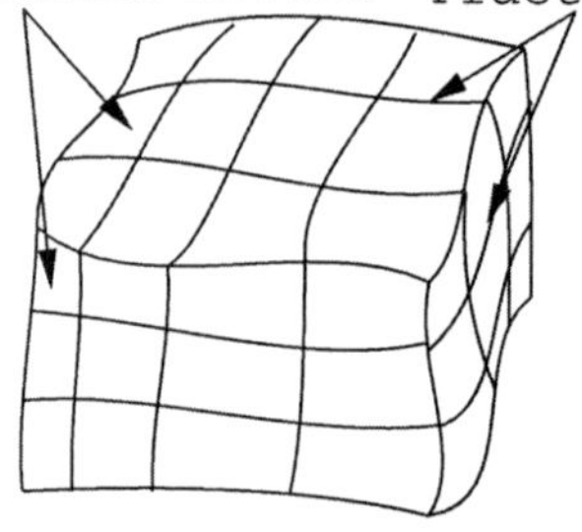

Figure 2.2. *A fractured porous medium.*

Also, define $\epsilon = \epsilon_{11} + \epsilon_{22} + \epsilon_{33}$. The stress-strain relationship is

$$\begin{pmatrix} \sigma_{11} \\ \sigma_{22} \\ \sigma_{33} \\ \sigma_{23} \\ \sigma_{31} \\ \sigma_{12} \\ \sigma \end{pmatrix} = \begin{pmatrix} c_{11} & c_{12} & c_{13} & c_{14} & c_{15} & c_{16} & c_{17} \\ \cdot & c_{22} & c_{23} & c_{24} & c_{25} & c_{26} & c_{27} \\ \cdot & \cdot & c_{33} & c_{34} & c_{35} & c_{36} & c_{37} \\ \cdot & \cdot & \cdot & c_{44} & c_{45} & c_{46} & c_{47} \\ \cdot & \cdot & \cdot & \cdot & c_{55} & c_{56} & c_{57} \\ \cdot & \cdot & \cdot & \cdot & \cdot & c_{66} & c_{67} \\ \cdot & \cdot & \cdot & \cdot & \cdot & \cdot & c_{77} \end{pmatrix} \begin{pmatrix} \epsilon_{s,11} \\ \epsilon_{s,22} \\ \epsilon_{s,33} \\ \epsilon_{s,23} \\ \epsilon_{s,31} \\ \epsilon_{s,12} \\ \epsilon \end{pmatrix},$$

where $c_{ij} = c_{ji}$ (i.e., the coefficient matrix is symmetric). Now, substitute this relationship into (2.28) to give three equations for the three unknowns $w_{s,1}$, $w_{s,2}$, and $w_{s,3}$.

As an example of the stress-strain relationship, we consider the case where the solid matrix is isotropic. In this case, with $\epsilon_s = \epsilon_{s,11} + \epsilon_{s,22} + \epsilon_{s,33}$, the relationship is given by

$$\begin{aligned} \sigma_{ii} &= 2G\left(\epsilon_{s,ii} + \frac{\nu\epsilon_s}{1-2\nu}\right) - Hp, \qquad && i = 1, 2, 3, \\ \sigma_{ij} &= 2G\epsilon_{s,ij}, && i, j = 1, 2, 3, \ i \neq j, \end{aligned}$$

where G and ν are the *Young modulus* and the *Poisson ratio* for the solid skeleton, and H is a physical constant whose value must be determined by experiments or by numerical methods (Biot, 1955; Chen et al., 2004B).

2.2.6 Single phase flow in a fractured medium

A *fractured porous medium* is a medium that is intersected by a network of interconnected *fractures*, or solution channels (cf. Figure 2.2). Such a medium could be modeled by allowing the porosity and permeability to vary rapidly and discontinuously over the whole domain. Both these quantities are much larger in the fractures than in the blocks of porous rock (called *matrix blocks*). However, the data requirement and computational cost for simulating such a single porosity model would be too great to approximate the flow in the entire medium. Instead, it is more convenient to regard the fluid in the void space as made up of two parts, one part in the fractures and the other in the matrix, and to treat each part as a continuum that occupies the entire domain. These two overlapping continua are allowed to coexist and interact with each other. There are two distinct *dual concepts*: *dual porosity* (and single permeability) and *dual porosity/permeability*. The former is considered in this section, while the latter will be studied in Section 2.11.

Since fluid flows more rapidly in the fractures than in the matrix, we assume that it does not flow directly from one block to another. Rather, it first flows into the fractures, and then it flows into another block or remains in the fractures (Douglas and Arbogast, 1990). Also, the equations that describe the flow in the fracture continuum contain a source term that represents the flow of fluid from the matrix to the fractures; this term is assumed to be distributed over the entire medium. Finally, we assume that the external sources and sinks interact only with the fracture system, which is reasonable since flow is much faster in this system than in the matrix blocks. Based on these assumptions, flow through *each block* in a fractured porous medium is given by

$$\frac{\partial(\phi\rho)}{\partial t} = -\nabla \cdot (\rho \mathbf{u}). \tag{2.29}$$

The flow in the fractures is described by

$$\frac{\partial(\phi_f \rho_f)}{\partial t} = -\nabla \cdot (\rho_f \mathbf{u}_f) + q_{mf} + q_{ext}, \tag{2.30}$$

where the subscript f represents the fracture quantities, q_{mf} denotes the flow from the matrix to the fractures, and q_{ext} indicates the external sources and sinks. The velocities $\mathbf{u}$ and $\mathbf{u}_f$ are determined by Darcy's law as in (2.4).

The *matrix-fracture transfer term* q_{mf} can be defined by two different approaches: one approach using matrix *shape factors* (Warren and Root, 1963; Kazemi, 1969) and the other based on boundary conditions imposed explicitly on matrix blocks (Pirson, 1953; Barenblatt et al., 1960). The latter approach is presented here; the former will be described in Section 2.11 and Chapter 12. The total mass of fluid leaving the ith matrix block Ω_i per unit time is

$$\int_{\partial\Omega_i} \rho \mathbf{u} \cdot \boldsymbol{\nu} \, d\ell,$$

where $\boldsymbol{\nu}$ is the outward unit normal to the surface $\partial\Omega_i$ of Ω_i and the dot product $\mathbf{u} \cdot \boldsymbol{\nu}$ is defined by

$$\mathbf{u} \cdot \boldsymbol{\nu} = u_1\nu_1 + u_2\nu_2 + u_3\nu_3.$$

The divergence theorem and (2.29) imply

$$\int_{\partial\Omega_i} \rho \mathbf{u} \cdot \boldsymbol{\nu} \, d\ell = \int_{\Omega_i} \nabla \cdot (\rho \mathbf{u}) d\mathbf{x} = -\int_{\Omega_i} \frac{\partial(\phi\rho)}{\partial t} d\mathbf{x}.$$

Now, define q_{mf} by

$$q_{mf} = -\sum_i \chi_i(\mathbf{x}) \frac{1}{|\Omega_i|} \int_{\Omega_i} \frac{\partial(\phi\rho)}{\partial t} d\mathbf{x}, \tag{2.31}$$

where $|\Omega_i|$ denotes the volume of Ω_i and $\chi_i(\mathbf{x})$ is its characteristic function, i.e.,

$$\chi_i(\mathbf{x}) = \begin{cases} 1 & \text{if } \mathbf{x} \in \Omega_i, \\ 0 & \text{otherwise.} \end{cases}$$

With the definition of q_{mf}, we now establish a boundary condition on the surface of each matrix block in a general fashion. Gravitational forces have a special effect on this condition. Moreover, pressure gradient effects must be treated on the same footing as the gravitational effects. To that end, following Arbogast (1993), we employ the pseudopotential Φ' defined in (2.7) to impose a condition on the surface of each matrix block by

$$\Phi' = \Phi'_f - \Phi^o \quad \text{on } \partial\Omega_i, \tag{2.32}$$

where, for a given Φ'_f, Φ^o is a pseudopotential reference value on each block Ω_i determined by

$$\frac{1}{|\Omega_i|}\int_{\Omega_i} (\phi\rho)\left(\psi'(\Phi'_f - \Phi^o + x_3)\right) d\mathbf{x} = (\phi\rho)(p_f), \tag{2.33}$$

with the function ψ' equal to the inverse of the integral in (2.7) as a function of p. Monotonicity of $\phi\rho$ insures a unique solution to (2.33) unless the rock and fluid are incompressible. In that case, set $\Phi^o = 0$.

For the model described, the highly permeable fracture system rapidly comes into equilibrium on the fracture spacing scale locally. This equilibrium is defined in terms of the pseudopotential, and is reflected in the matrix equations through the boundary condition (2.32).

2.2.7 Non-Darcy's law

Strictly speaking, Darcy's law holds only for a Newtonian fluid over a certain range of flow rates. As the flow rate increases, a deviation from this law has been noticed (Dupuit, 1863; Forchheimer, 1901). It has been experimentally and mathematically observed that this deviation is due to *inertia*, *turbulence*, and other high-velocity effects (Fancher and Lewis, 1933; Hubbert, 1956; Mei and Auriault, 1991; Chen et al., 2000B). Hubbert (1956) observed a deviation from the usual Darcy law at a Reynolds' number of flow of about one (based on the grain diameter of an unconsolidated medium), whereas turbulence was not noticed until the Reynolds' number approached 600 (Aziz and Settari, 1979).

A correction to Darcy's law for high flow rates can be described by a quadratic term (Forchheimer, 1901; Ward, 1964; Chen et al., 2000B):

$$\left(\mu\mathbf{I} + \beta\rho|\mathbf{u}|\mathbf{k}\right)\mathbf{u} = -\mathbf{k}\left(\nabla p - \rho\wp\nabla z\right),$$

where β indicates the inertial or turbulence factor and

$$|\mathbf{u}| = \sqrt{u_1^2 + u_2^2 + u_3^2}.$$

This equation is generally called *Forchheimer's law* and incorporates laminar, inertial, and turbulence effects. It has been the subject of many experimental and theoretical investigations. These investigations have centered on the issue of providing a physical or theoretical basis for the derivation of Forchheimer's law. Many approaches have been developed and analyzed for this purpose such as empiricism fortified with dimensional analysis (Ward, 1964), experimental study (MacDonald et al., 1979), averaging methods (Chen et al., 2000B), and variational principles (Knupp and Lage, 1995).

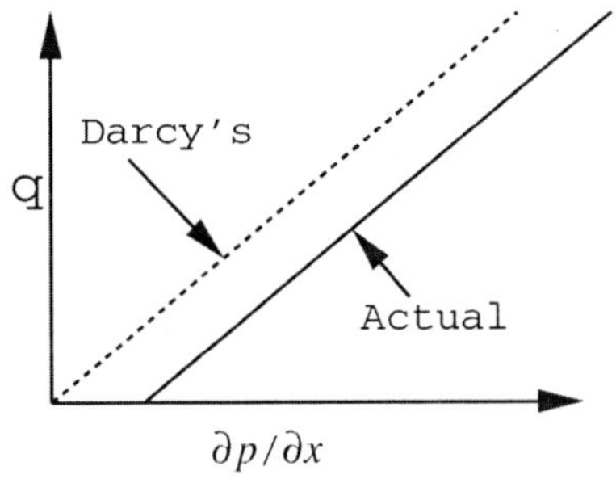

Figure 2.3. *Threshold phenomenon.*

2.2.8 Other effects

There exist several effects that introduce additional complexity in the basic flow equations. Some fluids (e.g., polymer solutions; cf. Section 2.10 and Chapter 11) exhibit *non-Newtonian phenomena,* characterized by nonlinear dependence of shear stress on shear rate. The study of non-Newtonian fluids is beyond the scope of this book, but can be found in the literature on rheology. In practice, the resistance to flow in a porous medium can be represented by Darcy's law with viscosity μ depending on flow velocity; i.e.,

$$\mathbf{u} = -\frac{1}{\mu(\mathbf{u})}\mathbf{k}\left(\nabla p - \rho\wp\nabla z\right).$$

Over a certain range of the velocity (the pseudoplastic region of flow), the viscosity can be approximated by a power law (Bird et al., 1960):

$$\mu(\mathbf{u}) = \mu_o|\mathbf{u}|^{m-1},$$

where the constants μ_o and m are empirically determined.

Other effects are related to *threshold* and *slip phenomena.* It has been experimentally observed that a certain nonzero pressure gradient is required to initiate flow. The threshold phenomenon can be seen in the relationship between q and $\partial p/\partial x$ for low rates, as shown in Figure 2.3. The slip (or Klinkenberg) phenomenon occurs in gas flow at low pressures and results in an increase of effective permeability compared to that measured for liquids. These two phenomena are relatively unimportant, and can be incorporated with a modification of Darcy's law (Bear, 1972).

2.2.9 Boundary conditions

The mathematical model described so far for single phase flow is not complete unless necessary *boundary* and *initial conditions* are specified. Below we present boundary conditions of three kinds that are relevant to (2.5). A similar discussion can be given for (2.28), which defines the displacement of the solid. Also, similar boundary conditions can be described for the dual porosity model. We denote by Γ the external boundary or a boundary segment of the porous medium domain Ω under consideration.

Prescribed pressure

When the pressure is specified as a known function of position and time on Γ, the boundary condition is

$$p = g_1 \quad \text{on } \Gamma.$$

In the theory of partial differential equations, such a condition is termed a boundary condition of the *first kind*, or a *Dirichlet boundary condition.*

Prescribed mass flux

When the total mass flux is known on Γ, the boundary condition is

$$\rho \mathbf{u} \cdot \boldsymbol{\nu} = g_2 \quad \text{on } \Gamma,$$

where $\boldsymbol{\nu}$ indicates the outward unit normal to Γ. This condition is called a boundary condition of the *second kind*, or a *Neumann boundary condition.* For an *impervious boundary*, $g_2 = 0$.

Mixed boundary condition

A boundary condition of *mixed kind* (or *third kind*) takes the form

$$g_p p + g_u \rho \mathbf{u} \cdot \boldsymbol{\nu} = g_3 \quad \text{on } \Gamma,$$

where g_p, g_u, and g_3 are given functions. This condition is referred to as a *Robin* or *Dankwerts boundary condition.* Such a condition occurs when Γ is a semipervious boundary. Finally, the initial condition can be defined in terms of p:

$$p(\mathbf{x}, 0) = p_0(\mathbf{x}), \qquad \mathbf{x} \in \Omega.$$

2.3 Two-Phase Immiscible Flow

In reservoir simulation, we are often interested in the *simultaneous flow* of two or more fluid phases within a porous medium. We now develop basic equations for multiphase flow in a porous medium. In this section, we consider two-phase flow where the fluids are *immiscible* and there is no mass transfer between the phases. One phase (e.g., water) wets the porous medium more than the other (e.g., oil), and is called the *wetting phase* and indicated by a subscript w. The other phase is termed the *nonwetting phase* and indicated by o. In general, water is the wetting fluid relative to oil and gas, while oil is the wetting fluid relative to gas.

2.3.1 Basic equations

Several new quantities peculiar to multiphase flow, such as *saturation*, *capillary pressure*, and *relative permeability*, must be introduced. The saturation of a fluid phase is defined as the fraction of the void volume of a porous medium filled by this phase. The fact that the two fluids jointly fill the voids implies the relation

$$S_w + S_o = 1, \tag{2.34}$$

where S_w and S_o are the saturations of the wetting and nonwetting phases, respectively. Also, due to the *curvature* and *surface tension* of the interface between the two phases, the pressure in the wetting fluid is less than that in the nonwetting fluid. The pressure difference is given by the capillary pressure

$$p_c = p_o - p_w. \tag{2.35}$$

Empirically, the capillary pressure is a function of saturation S_w.

Except for the accumulation term, the same derivation that led to (2.1) also applies to the mass conservation equation for each fluid phase (cf. Exercise 2.2). Mass accumulation in a differential volume per unit time is

$$\frac{\partial(\phi\rho_\alpha S_\alpha)}{\partial t}\Delta x_1 \Delta x_2 \Delta x_3.$$

Taking into account this and the assumption that there is no mass transfer between phases in the immiscible flow, mass is conserved within each phase:

$$\frac{\partial(\phi\rho_\alpha S_\alpha)}{\partial t} = -\nabla\cdot(\rho_\alpha \mathbf{u}_\alpha) + q_\alpha, \qquad \alpha = w,\ o, \tag{2.36}$$

where each phase has its own density ρ_α, Darcy velocity $\mathbf{u}_\alpha$, and mass flow rate q_α. Darcy's law for single phase flow can be directly extended to multiphase flow:

$$\mathbf{u}_\alpha = -\frac{1}{\mu_\alpha}\mathbf{k}_\alpha\left(\nabla p_\alpha - \rho_\alpha \wp \nabla z\right), \qquad \alpha = w,\ o, \tag{2.37}$$

where $\mathbf{k}_\alpha$, p_α, and μ_α are the *effective permeability*, pressure, and viscosity for phase α. Since the simultaneous flow of two fluids causes each to interfere with the other, the effective permeabilities are not greater than the absolute permeability $\mathbf{k}$ of the porous medium. The *relative permeabilities* $k_{r\alpha}$ are widely used in reservoir simulation:

$$\mathbf{k}_\alpha = k_{r\alpha}\mathbf{k}, \qquad \alpha = w,\ o. \tag{2.38}$$

The function $k_{r\alpha}$ indicates the tendency of phase α to wet the porous medium.

Typical functions of p_c and $k_{r\alpha}$ will be described in the next chapter. When q_w and q_o represent a finite number of point sources or sinks, they can be defined as in (2.10) or (2.11). Also, the densities ρ_w and ρ_o are functions of their respective pressures. Thus, after substituting (2.37) into (2.36) and using (2.34) and (2.35), we have a complete system of two equations for two of the four main unknowns p_α and S_α, $\alpha = w, o$. Other mathematical formulations will be discussed in this section. The development of single phase flow in deformable and fractured porous media is applicable to two-phase flow. We do not pursue this similar development.

2.3.2 Alternative differential equations

In this section, we derive several alternative formulations of the differential equations in (2.34)–(2.37).

Formulation in phase pressures

Assume that the capillary pressure p_c has a unique inverse function:

$$S_w = p_c^{-1}(p_o - p_w).$$

We use p_w and p_o as the main unknowns. Then it follows from (2.34)–(2.37) that

$$\begin{aligned} \nabla \cdot \left(\frac{\rho_w}{\mu_w} \mathbf{k}_w \left(\nabla p_w - \rho_w \wp \nabla z \right) \right) &= \frac{\partial (\phi \rho_w p_c^{-1})}{\partial t} - q_w, \\ \nabla \cdot \left(\frac{\rho_o}{\mu_o} \mathbf{k}_o \left(\nabla p_o - \rho_o \wp \nabla z \right) \right) &= \frac{\partial \left(\phi \rho_o (1 - p_c^{-1}) \right)}{\partial t} - q_o. \end{aligned} \tag{2.39}$$

This system was employed in the *simultaneous solution* (SS) scheme in petroleum reservoirs (Douglas et al., 1959). The equations in this system are strongly nonlinear and coupled. More details will be given in Chapter 7.

Formulation in phase pressure and saturation

We use p_o and S_w as the main variables. Applying (2.34), (2.35), and (2.37), equation (2.36) can be rewritten as

$$\begin{aligned} \nabla \cdot \left(\frac{\rho_w}{\mu_w} \mathbf{k}_w \left(\nabla p_o - \frac{dp_c}{dS_w} \nabla S_w - \rho_w \wp \nabla z \right) \right) &= \frac{\partial (\phi \rho_w S_w)}{\partial t} - q_w, \\ \nabla \cdot \left(\frac{\rho_o}{\mu_o} \mathbf{k}_o \left(\nabla p_o - \rho_o \wp \nabla z \right) \right) &= \frac{\partial \left(\phi \rho_o (1 - S_w) \right)}{\partial t} - q_o. \end{aligned} \tag{2.40}$$

Carrying out the time differentiation in (2.40), dividing the first and second equations by ρ_w and ρ_o, respectively, and adding the resulting equations, we obtain

$$\begin{aligned} &\frac{1}{\rho_w} \nabla \cdot \left(\frac{\rho_w}{\mu_w} \mathbf{k}_w \left(\nabla p_o - \frac{dp_c}{dS_w} \nabla S_w - \rho_w \wp \nabla z \right) \right) \\ &\quad + \frac{1}{\rho_o} \nabla \cdot \left(\frac{\rho_o}{\mu_o} \mathbf{k}_o \left(\nabla p_o - \rho_o \wp \nabla z \right) \right) \\ &\quad = \frac{S_w}{\rho_w} \frac{\partial (\phi \rho_w)}{\partial t} + \frac{1 - S_w}{\rho_o} \frac{\partial (\phi \rho_o)}{\partial t} - \frac{q_w}{\rho_w} - \frac{q_o}{\rho_o}. \end{aligned} \tag{2.41}$$

Note that if the saturation S_w in (2.41) is explicitly evaluated, we can use this equation to solve for p_o. After computing this pressure, the second equation in (2.40) can be used to calculate S_w. This is the *implicit pressure-explicit saturation* (IMPES) scheme and has been widely exploited for two-phase flow in petroleum reservoirs (cf. Chapter 7).

Formulation in a global pressure

The equations in (2.39) and (2.40) are strongly coupled, as noted. To reduce the coupling, we now write them in a different formulation, where a *global pressure* is used. For simplicity, we assume that the densities are constant; the formulation does extend to variable densities (Chen et al., 1995; Chen et al., 1997A). Introduce the *phase mobilities*

$$\lambda_\alpha = \frac{k_{r\alpha}}{\mu_\alpha}, \qquad \alpha = w, o,$$

and the *total mobility*

$$\lambda = \lambda_w + \lambda_o.$$

Also, define the *fractional flow* functions

$$f_\alpha = \frac{\lambda_\alpha}{\lambda}, \qquad \alpha = w, o.$$

With $S = S_w$, define the *global pressure* (Antoncev, 1972; Chavent and Jaffré, 1986)

$$p = p_o - \int^{p_c(S)} f_w\big(p_c^{-1}(\xi)\big)d\xi, \tag{2.42}$$

and the total velocity

$$\mathbf{u} = \mathbf{u}_w + \mathbf{u}_o. \tag{2.43}$$

It follows from (2.35), (2.37), and (2.42) that the total velocity is

$$\mathbf{u} = -\mathbf{k}\lambda\big(\nabla p - (\rho_w f_w + \rho_o f_o)\wp\nabla z\big). \tag{2.44}$$

Also, carrying out the differentiation in (2.36), dividing by ρ_α, adding the resulting equations with $\alpha = w$ and o, and applying (2.42), we obtain

$$\nabla \cdot \mathbf{u} = -\frac{\partial \phi}{\partial t} + \frac{q_w}{\rho_w} + \frac{q_o}{\rho_o}. \tag{2.45}$$

Substituting (2.44) into (2.45) gives a pressure equation for p:

$$-\nabla \cdot \big(\mathbf{k}\lambda\big(\nabla p - (\rho_w f_w + \rho_o f_o)\wp\nabla z\big)\big) = -\frac{\partial \phi}{\partial t} + \frac{q_w}{\rho_w} + \frac{q_o}{\rho_o}. \tag{2.46}$$

The phase velocities are related to the total velocity by (cf. Exercise 2.3)

$$\begin{aligned} \mathbf{u}_w &= f_w\mathbf{u} + \mathbf{k}\lambda_o f_w \nabla p_c + \mathbf{k}\lambda_o f_w(\rho_w - \rho_o)\wp\nabla z, \\ \mathbf{u}_o &= f_o\mathbf{u} - \mathbf{k}\lambda_w f_o \nabla p_c + \mathbf{k}\lambda_w f_o(\rho_o - \rho_w)\wp\nabla z. \end{aligned} \tag{2.47}$$

From the first equation of (2.47) and (2.36) with $\alpha = w$, we have a saturation equation for $S = S_w$:

$$\begin{aligned} \phi\frac{\partial S}{\partial t} + \nabla \cdot \left(\mathbf{k}\lambda_o f_w\left(\frac{dp_c}{dS}\nabla S - (\rho_o - \rho_w)\wp\nabla z\right) + f_w\mathbf{u}\right) \\ = -S\frac{\partial \phi}{\partial t} + \frac{q_w}{\rho_w}. \end{aligned} \tag{2.48}$$

Classification of differential equations

There are basically three types of second-order partial differential equations: elliptic, parabolic, and hyperbolic. We must be able to distinguish among these types when numerical methods for their solution are devised.

If two independent variables (either (x_1, x_2) or (x_1, t)) are considered, then second-order partial differential equations have the form, with $x = x_1$,

$$a\frac{\partial^2 p}{\partial x^2} + b\frac{\partial^2 p}{\partial t^2} = f\left(\frac{\partial p}{\partial x}, \frac{\partial p}{\partial t}, p\right).$$

This equation is (1) elliptic if $ab > 0$, (2) parabolic if $ab = 0$, or (3) hyperbolic if $ab < 0$.

The simplest elliptic equation is the *Poisson equation*

$$\frac{\partial^2 p}{\partial x_1^2} + \frac{\partial^2 p}{\partial x_2^2} = f(x_1, x_2).$$

A typical parabolic equation is the *heat conduction equation*

$$\phi\frac{\partial p}{\partial t} = \frac{\partial^2 p}{\partial x_1^2} + \frac{\partial^2 p}{\partial x_2^2}.$$

Finally, the prototype hyperbolic equation is the *wave equation*

$$\frac{1}{v^2}\frac{\partial^2 p}{\partial t^2} = \frac{\partial^2 p}{\partial x_1^2} + \frac{\partial^2 p}{\partial x_2^2}.$$

In the one-dimensional case, this equation can be "factorized" into two first-order parts:

$$\left(\frac{1}{v}\frac{\partial}{\partial t} - \frac{\partial}{\partial x}\right)\left(\frac{1}{v}\frac{\partial}{\partial t} + \frac{\partial}{\partial x}\right)p = 0.$$

The second part gives the first-order hyperbolic equation

$$\frac{\partial p}{\partial t} + v\frac{\partial p}{\partial x} = 0.$$

We now turn to the two-phase flow equations. While the phase mobilities λ_α can be zero (cf. Chapter 3), the total mobility λ is always positive, so the pressure equation (2.46) is elliptic. If one of the densities varies, this equation becomes parabolic. In general, $-\mathbf{k}\lambda_o f_w dp_c/dS$ is semipositive definite, so the saturation equation (2.48) is a parabolic equation, which is *degenerate* in the sense that the diffusion can be zero. This equation becomes hyperbolic if the capillary pressure is ignored. The total velocity is used in the global pressure formulation. This velocity is smoother than the phase velocities. It can also be used in the phase formulations (2.39) and (2.40) (Chen and Ewing, 1997B). We remark that the coupling between (2.46) and (2.48) is much less strong than between the equations in (2.39) and (2.40). Finally, with $p_c = 0$, (2.48) becomes the known *Buckley–Leverett equation* whose flux function f_w is generally nonconvex over the range of saturation values where this function is nonzero, as illustrated in Figure 2.4; see the next subsection for the formulation in hyperbolic form.

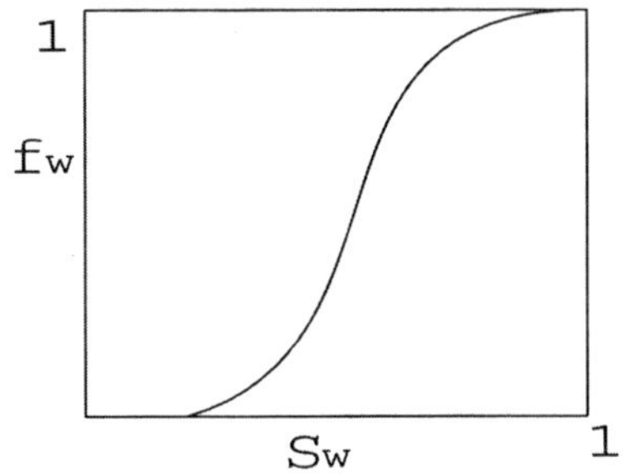

Figure 2.4. *A flux function* f_w.

Formulation in hyperbolic form

Assume that $p_c = 0$ and that rock compressibility is neglected. Then (2.48) becomes

$$\phi \frac{\partial S}{\partial t} + \nabla \cdot (f_w \mathbf{u} - \lambda_o f_w (\rho_o - \rho_w) \wp \mathbf{k} \nabla z) = \frac{q_w}{\rho_w}. \tag{2.49}$$

Using (2.45) and the fact that $f_w + f_o = 1$, this equation can be manipulated into

$$\phi \frac{\partial S}{\partial t} + \left(\frac{df_w}{dS} \mathbf{u} - \frac{d(\lambda_o f_w)}{dS} (\rho_o - \rho_w) \wp \mathbf{k} \nabla z \right) \cdot \nabla S = \frac{f_o q_w}{\rho_w} - \frac{f_w q_o}{\rho_o}, \tag{2.50}$$

which is a hyperbolic equation in S. Finally, if we neglect the gravitational term, we obtain

$$\phi \frac{\partial S}{\partial t} + \frac{df_w}{dS} \mathbf{u} \cdot \nabla S = \frac{f_o q_w}{\rho_w} - \frac{f_w q_o}{\rho_o}, \tag{2.51}$$

which is the familiar form of waterflooding equation, i.e., the *Buckley–Leverett equation*. The source term in (2.51) is zero for production since

$$\frac{q_w}{\rho_w} = f_w \left(\frac{q_w}{\rho_w} + \frac{q_o}{\rho_o} \right),$$

by Darcy's law. For injection, this term may not be zero since it equals $(1 - f_w) q_w / \rho_w \neq 0$ in this case.

2.3.3 Boundary conditions

As for single phase flow, the mathematical model described so far for two-phase flow is not complete unless necessary boundary and initial conditions are specified. Below we present boundary conditions of three kinds that are relevant to systems (2.39), (2.40), (2.46), and (2.48). We denote by Γ the external boundary or a boundary segment of the porous medium domain Ω under consideration.

Boundary conditions for system (2.39)

The symbol α, as a subscript, with $\alpha = w, o$, is used to indicate a considered phase. When a phase pressure is specified as a known function of position and time on Γ, the boundary condition reads

$$p_\alpha = g_{\alpha,1} \quad \text{on } \Gamma. \tag{2.52}$$

When the mass flux of phase α is known on Γ, the boundary condition is

$$\rho_\alpha \mathbf{u}_\alpha \cdot \boldsymbol{\nu} = g_{\alpha,2} \quad \text{on } \Gamma, \tag{2.53}$$

where $\boldsymbol{\nu}$ indicates the outward unit normal to Γ and $g_{\alpha,2}$ is given. For an impervious boundary for the α-phase, $g_{\alpha,2} = 0$.

When Γ is a semipervious boundary for the α-phase, a boundary condition of mixed kind occurs:

$$g_{\alpha,p} p_\alpha + g_{\alpha,u} \rho_\alpha \mathbf{u}_\alpha \cdot \boldsymbol{\nu} = g_{\alpha,3} \quad \text{on } \Gamma, \tag{2.54}$$

where $g_{\alpha,p}$, $g_{\alpha,u}$, and $g_{\alpha,3}$ are given functions.

Initial conditions specify the values of the main unknowns p_w and p_o over the entire domain at some initial time, usually taken at $t = 0$:

$$p_\alpha(\mathbf{x}, 0) = p_{\alpha,0}(\mathbf{x}), \qquad \alpha = w, o,$$

where $p_{\alpha,0}(\mathbf{x})$ are known functions.

Boundary conditions for system (2.40)

Boundary conditions for system (2.40) can be imposed as for system (2.39); i.e., (2.52)–(2.54) are applicable to system (2.40). The only difference between the boundary conditions for these two systems is that a prescribed saturation is sometimes given on Γ for system (2.40):

$$S_w = g_4 \quad \text{on } \Gamma.$$

In practice, this prescribed saturation boundary condition seldom occurs. However, a condition $g_4 = 1$ does occur when a medium is in contact with a body of this wetting phase. The condition $S_w = 1$ can be exploited on the bottom of a water pond on the ground surface, for example. An initial saturation is also specified:

$$S_w(\mathbf{x}, 0) = S_{w,0}(\mathbf{x}),$$

where $S_{w,0}(\mathbf{x})$ is given.

Boundary conditions for (2.46) and (2.48)

Boundary conditions are usually specified in terms of phase quantities like those in (2.52)–(2.54). These conditions can be transformed into those in terms of the global quantities introduced in (2.42) and (2.43). For the prescribed pressure boundary condition in (2.52), for example, the corresponding boundary condition is given by

$$p = g_1 \quad \text{on } \Gamma,$$

where p is defined by (2.42) and g_1 is determined by

$$g_1 = g_{o,1} - \int^{g_{o,1}-g_{w,1}} f_w\big(p_c^{-1}(\xi)\big)d\xi.$$

Also, when the total mass flux is known on Γ, it follows from (2.53) that

$$\mathbf{u} \cdot \nu = g_2 \quad \text{on } \Gamma,$$

where

$$g_2 = \frac{g_{o,2}}{\rho_o} + \frac{g_{w,2}}{\rho_w}.$$

For an impervious boundary for the total flow, $g_2 = 0$.

2.4 Transport of a Component in a Fluid Phase

Now, we consider the *transport of a component* (e.g., a solute) in a fluid phase that occupies the entire void space in a porous medium. We do not consider the effects of chemical reactions between the components in the fluid phase, radioactive decay, biodegradation, or growth due to bacterial activities that cause the quantity of this component to increase or decrease. Conservation of mass of the component in the fluid phase is given by

$$\begin{aligned}\frac{\partial(\phi c\rho)}{\partial t} &= -\nabla \cdot (c\rho\mathbf{u} - \rho\mathbf{D}\nabla c) \\ &\quad - \sum_i q_1^{(i)}(\mathbf{x}^{(i)}, t)\delta(\mathbf{x} - \mathbf{x}^{(i)})(\rho c)(\mathbf{x}, t) \\ &\quad + \sum_j q_2^{(j)}(\mathbf{x}^{(j)}, t)\delta(\mathbf{x} - \mathbf{x}^{(j)})(\rho^{(j)}c^{(j)})(\mathbf{x}, t),\end{aligned} \tag{2.55}$$

where c is the *concentration* (volumetric fraction in the fluid phase) of the component, $\mathbf{D}$ is the *diffusion-dispersion tensor*, $q_1^{(i)}$ and $q_2^{(j)}$ are the rates of production and injection (in terms of volume per unit time) at points $\mathbf{x}^{(i)}$ and $\mathbf{x}^{(j)}$, respectively, and $c^{(j)}$ is the specified concentration at source points.

Darcy's law for the fluid is expressed as in (2.4); namely,

$$\mathbf{u} = -\frac{1}{\mu}\mathbf{k}\left(\nabla p - \rho\wp\nabla z\right). \tag{2.56}$$

The mass balance of the fluid is written as

$$\begin{aligned}\frac{\partial(\phi\rho)}{\partial t} + \nabla \cdot (\rho\mathbf{u}) &= -\sum_i \rho q_1^{(i)}(\mathbf{x}^{(i)}, t)\delta(\mathbf{x} - \mathbf{x}^{(i)}) \\ &\quad + \sum_j \rho^{(j)} q_2^{(j)}(\mathbf{x}^{(j)}, t)\delta(\mathbf{x} - \mathbf{x}^{(j)}).\end{aligned} \tag{2.57}$$

The diffusion-dispersion tensor $\mathbf{D}$ in (2.55) in three space dimensions is defined by

$$\mathbf{D}(\mathbf{u}) = \phi\left\{d_m\mathbf{I} + |\mathbf{u}|\left(d_l\mathbf{E}(\mathbf{u}) + d_t\mathbf{E}^{\perp}(\mathbf{u})\right)\right\}, \tag{2.58}$$

where d_m is the *molecular diffusion* coefficient; d_l and d_t are, respectively, the *longitudinal* and *transverse dispersion* coefficients; $|\mathbf{u}|$ is the Euclidean norm of $\mathbf{u} = (u_1, u_2, u_3)$, $|\mathbf{u}| = \sqrt{u_1^2+u_2^2+u_3^2}$; $\mathbf{E}(\mathbf{u})$ is the orthogonal projection along the velocity,

$$\mathbf{E}(\mathbf{u}) = \frac{1}{|\mathbf{u}|^2} \begin{pmatrix} u_1^2 & u_1u_2 & u_1u_3 \\ u_2u_1 & u_2^2 & u_2u_3 \\ u_3u_1 & u_3u_2 & u_3^2 \end{pmatrix};$$

and $\mathbf{E}^{\perp}(\mathbf{u}) = \mathbf{I} - \mathbf{E}(\mathbf{u})$.

Physically, the tensor dispersion is more significant than the molecular diffusion; also, d_l is usually considerably larger than d_t. The density and viscosity are known functions of p and c:

$$\rho = \rho(p, c), \qquad \mu = \mu(p, c).$$

After the substitution of (2.56) into (2.55) and (2.57), we have a coupled system of two equations in c and p. Boundary and initial conditions for this system can be developed as in the earlier sections. Note that the equations described here apply to the problem of miscible displacement of one fluid by another in a porous medium. Various simplifications discussed in Section 2.2 apply to (2.56) and (2.57).

2.5 Transport of Multicomponents in a Fluid Phase

The equation used to model the transport of multicomponents in a fluid phase in a porous medium is similar to (2.55); i.e.,

$$\frac{\partial(\phi c_i \rho)}{\partial t} = -\nabla \cdot (c_i \rho \mathbf{u} - \rho \mathbf{D}_i \nabla c_i) + q_i, \qquad i = 1, 2, \ldots, N_c, \tag{2.59}$$

where c_i, q_i, and $\mathbf{D}_i$ are the (volumetric) concentration, the source/sink term, and the diffusion-dispersion tensor of the ith component, respectively, and N_c is the number of the components in the fluid. The constraint for the concentrations is

$$\sum_{i=1}^{N_c} c_i = 1.$$

Sources and sinks of a component can result from injection and production of this component by external means. They can also stem from various processes within the fluid phase, such as *chemical reactions* among components, radioactive decay, biodegradation, and growth due to bacterial activities, that cause the quantity of this component to increase or decrease, as noted earlier. In this section, we focus only on chemical reactions, i.e., a reactive flow problem.

When a component participates in chemical reactions that cause its concentration to increase or decrease, q_i can be expressed as

$$q_i = Q_i - L_i c_i, \tag{2.60}$$

where Q_i and L_i represent the *chemical production* and *loss rates*, respectively, of the ith component. To see their expressions in terms of concentrations, we consider *unimolecular*, *bimolecular*, and *trimolecular reactions* among the chemical components. These cases can be generally written as

$$s_1 \rightleftharpoons s_2 + s_3,$$

$$s_1 + s_2 \rightleftharpoons s_3 + s_4,$$

$$s_1 + s_2 + s_3 \rightleftharpoons s_4 + s_5,$$

where the s_i's denote generic chemical components. Corresponding to these reactions, Q_i and L_i can be expressed as

$$Q_i = \sum_{j=1}^{N_c} k^f_{i,j} c_j + \sum_{j,l=1}^{N_c} k^f_{i,jl} c_j c_l + \sum_{j,l,m=1}^{N_c} k^f_{i,jlm} c_j c_l c_m,$$

$$L_i = k^r_i + \sum_{j=1}^{N_c} k^r_{i,j} c_j + \sum_{j,l=1}^{N_c} k^r_{i,jl} c_j c_l,$$

where k^f and k^r are forward and reverse chemical rates, respectively. These rates are functions of pressure and temperature (Oran and Boris, 2001).

Darcy's law (2.56) and the overall mass balance equation (2.57) hold for the transport of multicomponents. Again, after Darcy's velocity is eliminated, we have a coupled system of $N_c + 1$ equations for c_i and p, $i = 1, 2, \ldots, N_c$ (cf. Exercise 2.4).

2.6 The Black Oil Model

We now develop basic equations for the simultaneous flow of three phases (e.g., water, oil, and gas) through a porous medium. Previously, we assumed that mass does not transfer between phases. The *black oil model* relaxes this assumption. It is now assumed that the hydrocarbon components are divided into a gas component and an oil component in a stock tank at standard pressure and temperature, and that no mass transfer occurs between the water phase and the other two phases (oil and gas). The gas component mainly consists of methane and ethane.

To reduce confusion, we carefully distinguish between phases and components. We use lowercase and uppercase letter subscripts to denote the phases and components, respectively. Note that the water phase is just the water component. The subscript s indicates standard conditions. The mass conservation equations stated in (2.36) apply here. However, because of mass interchange between the oil and gas phases, mass is not conserved within each phase, but rather the total mass of each component must be conserved:

$$\frac{\partial(\phi \rho_w S_w)}{\partial t} = -\nabla \cdot (\rho_w \mathbf{u}_w) + q_W \tag{2.61}$$

for the water component,

$$\frac{\partial(\phi \rho_{Oo} S_o)}{\partial t} = -\nabla \cdot (\rho_{Oo} \mathbf{u}_o) + q_O \tag{2.62}$$

for the oil component, and

$$\frac{\partial}{\partial t}\big(\phi(\rho_{Go} S_o + \rho_g S_g)\big) = -\nabla \cdot (\rho_{Go} \mathbf{u}_o + \rho_g \mathbf{u}_g) + q_G \tag{2.63}$$

for the gas component, where ρ_{Oo} and ρ_{Go} indicate the *partial densities* of the oil and gas components in the oil phase, respectively. Equation (2.63) implies that the gas component may exist in both the oil and gas phases.

Darcy's law for each phase is written in the usual form

$$\mathbf{u}_\alpha = -\frac{1}{\mu_\alpha}\mathbf{k}_\alpha\left(\nabla p_\alpha - \rho_\alpha \wp \nabla z\right), \qquad \alpha = w,\, o,\, g. \tag{2.64}$$

The fact that the three phases jointly fill the void space is given by the equation

$$S_w + S_o + S_g = 1. \tag{2.65}$$

Finally, the phase pressures are related by capillary pressures

$$p_{cow} = p_o - p_w, \qquad p_{cgo} = p_g - p_o. \tag{2.66}$$

It is not necessary to define a third capillary pressure since it can be defined in terms of p_{cow} and p_{cgo}.

The alternative differential equations developed for two phases can be adapted for the three-phase black oil model in a similar fashion (Chen, 2000). That is, (2.61)–(2.66) can be rewritten in the three-pressure formulation (cf. Exercise 2.5), in a pressure and two-saturation formulation (cf. Exercise 2.6), or in a global pressure and two-saturation formulation (cf. Exercise 2.7). In the global formulation, the pressure equation is elliptic or parabolic depending on the effects of densities. The two saturation equations are parabolic if the capillary pressure effects exist; otherwise, they are hyperbolic (Chen, 2000).

For the black oil model, it is often convenient to work with the conservation equations on "standard volumes," instead of the conservation equations on "mass" (2.61)–(2.63). The mass fractions of the oil and gas components in the oil phase can be determined by *gas solubility*, R_{so} (also called *dissolved gas-oil ratio*), which is the volume of gas (measured at standard conditions) dissolved at a given pressure and reservoir temperature in a unit volume of stock-tank oil:

$$R_{so}(p, T) = \frac{V_{Gs}}{V_{Os}}. \tag{2.67}$$

Note that

$$V_{Os} = \frac{W_O}{\rho_{Os}}, \qquad V_{Gs} = \frac{W_G}{\rho_{Gs}}, \tag{2.68}$$

where W_O and W_G are the weights of the oil and gas components, respectively. Then (2.67) becomes

$$R_{so} = \frac{W_G \rho_{Os}}{W_O \rho_{Gs}}. \tag{2.69}$$

The *oil formation volume factor* B_o is the ratio of the volume V_o of the oil phase measured at reservoir conditions to the volume V_{Os} of the oil component measured at standard conditions:

$$B_o(p, T) = \frac{V_o(p, T)}{V_{Os}}, \tag{2.70}$$

where

$$V_o = \frac{W_O + W_G}{\rho_o}. \tag{2.71}$$

Consequently, combining (2.68), (2.70), and (2.71), we have

$$B_o = \frac{(W_O + W_G)\rho_{Os}}{W_O \rho_o}. \tag{2.72}$$

Now, using (2.69) and (2.72), the *mass fractions* of the oil and gas components in the oil phase are, respectively,

$$C_{Oo} = \frac{W_O}{W_O + W_G} = \frac{\rho_{Os}}{B_o \rho_o},$$

$$C_{Go} = \frac{W_G}{W_O + W_G} = \frac{R_{so}\rho_{Gs}}{B_o \rho_o},$$

which, together with $C_{Oo} + C_{Go} = 1$, yield

$$\rho_o = \frac{R_{so}\rho_{Gs} + \rho_{Os}}{B_o}. \tag{2.73}$$

The *gas formation volume factor* B_g is the ratio of the volume of the gas phase measured at reservoir conditions to the volume of the gas component measured at standard conditions:

$$B_g(p, T) = \frac{V_g(p, T)}{V_{Gs}}.$$

Let $W_g = W_G$ be the weight of free gas. Because $V_g = W_G/\rho_g$ and $V_{Gs} = W_G/\rho_{Gs}$, we see that

$$\rho_g = \frac{\rho_{Gs}}{B_g}. \tag{2.74}$$

For completeness, the *water formation volume factor*, B_w, is defined by

$$\rho_w = \frac{\rho_{Ws}}{B_w}. \tag{2.75}$$

Finally, substituting (2.73)–(2.75) into (2.61)–(2.63) yields the conservation equations on standard volumes:

$$\frac{\partial}{\partial t}\left(\frac{\phi \rho_{Ws}}{B_w} S_w\right) = -\nabla \cdot \left(\frac{\rho_{Ws}}{B_w}\mathbf{u}_w\right) + q_W \tag{2.76}$$

for the water component,

$$\frac{\partial}{\partial t}\left(\frac{\phi \rho_{Os}}{B_o} S_o\right) = -\nabla \cdot \left(\frac{\rho_{Os}}{B_o}\mathbf{u}_o\right) + q_O \tag{2.77}$$

for the oil component, and

$$\begin{aligned} &\frac{\partial}{\partial t}\left[\phi\left(\frac{\rho_{Gs}}{B_g} S_g + \frac{R_{so}\rho_{Gs}}{B_o} S_o\right)\right] \\ &\quad = -\nabla \cdot \left(\frac{\rho_{Gs}}{B_g}\mathbf{u}_g + \frac{R_{so}\rho_{Gs}}{B_o}\mathbf{u}_o\right) + q_G \end{aligned} \tag{2.78}$$

for the gas component. Equations (2.76)–(2.78) represent balances on *standard volumes*. The volumetric rates at standard conditions are

$$q_W = \frac{q_{Ws}\rho_{Ws}}{B_w}, \quad q_O = \frac{q_{Os}\rho_{Os}}{B_o},$$
$$q_G = \frac{q_{Gs}\rho_{Gs}}{B_g} + \frac{q_{Os}R_{so}\rho_{Gs}}{B_o}. \tag{2.79}$$

Since ρ_{Ws}, ρ_{Os}, and ρ_{Gs} are constant, they can be eliminated after (2.79) is substituted into (2.76)–(2.78).

The basic equations for the black oil model consist of (2.64)–(2.66) and (2.76)–(2.78). The choice of main unknowns depends on the state of the reservoir, i.e., the *saturated* or *undersaturated state*, which will be discussed in Chapter 8.

2.7 A Volatile Oil Model

The black oil model developed above is not suitable for handling a *volatile oil reservoir*. A reservoir of volatile oil type is one that contains relatively large proportions of ethane through decane at a reservoir temperature near or above 250° F with a high formation volume factor and stock-tank oil gravity above 45° API (Jacoby and Berry, 1957). With a more elaborate two-component hydrocarbon model, a *volatile oil model*, the effect of *oil volatility* can be included. In this model, there are both oil and gas components, *solubility* of gas in both oil and gas phases is permitted, and *vaporization* of oil into the gas phase is allowed. Therefore, the two hydrocarbon components can exist in both oil and gas phases.

Oil volatility in the gas phase is

$$R_v = \frac{V_{Os}}{V_{Gs}}.$$

Using a similar approach as for the black oil model, the conservation equations on standard volumes are

$$\frac{\partial}{\partial t}\left(\frac{\phi\rho_{Ws}}{B_w}S_w\right) = -\nabla\cdot\left(\frac{\rho_{Ws}}{B_w}\mathbf{u}_w\right) + q_W \tag{2.80}$$

for the water component,

$$\frac{\partial}{\partial t}\left[\phi\left(\frac{\rho_{Os}}{B_o}S_o + \frac{R_v\rho_{Os}}{B_g}S_g\right)\right] = -\nabla\cdot\left(\frac{\rho_{Os}}{B_o}\mathbf{u}_o + \frac{R_v\rho_{Os}}{B_g}\mathbf{u}_g\right) + q_O, \tag{2.81}$$

for the oil component, and

$$\frac{\partial}{\partial t}\left[\phi\left(\frac{\rho_{Gs}}{B_g}S_g + \frac{R_{so}\rho_{Gs}}{B_o}S_o\right)\right] = -\nabla\cdot\left(\frac{\rho_{Gs}}{B_g}\mathbf{u}_g + \frac{R_{so}\rho_{Gs}}{B_o}\mathbf{u}_o\right) + q_G \tag{2.82}$$

for the gas component. In general, the hydrocarbon components (i.e., oil and gas) can be defined using *pseudocomponents* obtained from the compositional flow described in the next section.

2.8 Compositional Flow

In the black oil and volatile oil models, two hydrocarbon components are involved. Here we consider *compositional flow* that involves many components and *mass transfer* between phases in a general fashion. In a *compositional model*, a finite number of hydrocarbon components are used to represent the composition of reservoir fluids. These components associate as phases in a reservoir. We describe the model under the assumptions that the flow process is isothermal (i.e., at constant temperature), the components form at most three phases (e.g., vapor, liquid, and water), and there is no mass interchange between the water phase and the hydrocarbon phases (i.e., the vapor and liquid phases). We could state a general compositional model that involves any number of phases and components, each of which may exist in any or all of these phases (cf. Section 2.10). While the governing differential equations for this type of model are easy to set up, they are extremely complex to solve. Therefore, we describe the compositional model that has been widely used in the petroleum industry.

Instead of using the concentration, it is more convenient to employ the *mole fraction* for each component in the compositional flow, since the phase equilibrium relations are usually defined in terms of mole fractions (cf. (2.91)). Let ξ_{io} and ξ_{ig} be the *molar densities* of component i in the liquid (e.g., oil) and vapor (e.g., gas) phases, respectively, $i = 1, 2, \ldots, N_c$, where N_c is the number of components. Their physical dimensions are moles per pore volume. If W_i is the *molar mass* of component i, with dimensions mass of component i/mole of component i, then $\xi_{i\alpha}$ is related to the mass density $\rho_{i\alpha}$ by $\xi_{i\alpha} = \rho_{i\alpha}/W_i$. The *molar density* of phase α is

$$\xi_\alpha = \sum_{i=1}^{N_c} \xi_{i\alpha}, \qquad \alpha = o, g. \tag{2.83}$$

The mole fraction of component i in phase α is then

$$x_{i\alpha} = \frac{\xi_{i\alpha}}{\xi_\alpha}, \qquad i = 1, 2, \ldots, N_c, \ \alpha = o, g. \tag{2.84}$$

Because of mass interchange between the phases, mass is not conserved within each phase; the total mass is conserved for each component:

$$\begin{aligned} &\frac{\partial(\phi\xi_w S_w)}{\partial t} + \nabla\cdot(\xi_w \mathbf{u}_w) = q_w, \\ &\frac{\partial(\phi[x_{io}\xi_o S_o + x_{ig}\xi_g S_g])}{\partial t} + \nabla\cdot(x_{io}\xi_o\mathbf{u}_o + x_{ig}\xi_g\mathbf{u}_g) \\ &\qquad + \nabla\cdot(\mathbf{d}_{io} + \mathbf{d}_{ig}) = q_i, \qquad i = 1, 2, \ldots, N_c, \end{aligned} \tag{2.85}$$

where ξ_w is the molar density of water, q_w and q_i are the molar flow rates of water and the ith component, respectively, and $\mathbf{d}_{i\alpha}$ denotes the *diffusive flux* of the ith component in the α-phase, $\alpha = o, g$. In (2.85), the volumetric velocity $\mathbf{u}_\alpha$ is given by Darcy's law as in (2.64):

$$\mathbf{u}_\alpha = -\frac{1}{\mu_\alpha}\mathbf{k}_\alpha(\nabla p_\alpha - \rho_\alpha\wp\nabla z), \qquad \alpha = w,\ o,\ g. \tag{2.86}$$

In addition to the differential equations (2.85) and (2.86), there are also algebraic constraints. The *mole fraction balance* implies that

$$\sum_{i=1}^{N_c} x_{io} = 1, \quad \sum_{i=1}^{N_c} x_{ig} = 1. \tag{2.87}$$

In the transport process, the porous medium is saturated with fluids:

$$S_w + S_o + S_g = 1. \tag{2.88}$$

The phase pressures are related by capillary pressures:

$$p_{cow} = p_o - p_w, \quad p_{cgo} = p_g - p_o. \tag{2.89}$$

These capillary pressures are assumed to be known functions of the saturations. The relative permeabilities $k_{r\alpha}$ are also assumed to be known in terms of the saturations, and the viscosities μ_α, molar densities ξ_α, and mass densities ρ_α are functions of their respective phase pressure and compositions, $\alpha = w, o, g$.

The least well understood term in (2.85) is that involving the diffusive fluxes $\mathbf{d}_{i\alpha}$. The precise constitutive relations for these quantities still need to be derived; however, from a practical point of view the following straightforward extension of the single phase *Fick's law* to multiphase flow is in widespread use:

$$\mathbf{d}_{i\alpha} = -\xi_\alpha \mathbf{D}_{i\alpha} \nabla x_{i\alpha}, \qquad i = 1, 2, \ldots, N_c, \ \alpha = o, \ g, \tag{2.90}$$

where $\mathbf{D}_{i\alpha}$ is the diffusion coefficient of component i in phase α (cf. (2.58) or Section 2.10). The diffusive fluxes must satisfy

$$\sum_{i=1}^{N_c} \mathbf{d}_{i\alpha} = \mathbf{0}, \qquad \alpha = o, \ g.$$

Note that there are more dependent variables than there are differential and algebraic relations combined; there are formally $2N_c + 9$ dependent variables: x_{io}, x_{ig}, $\mathbf{u}_\alpha$, p_α, and S_α, $\alpha = w, o, g$, $i = 1, 2, \ldots, N_c$. It is then necessary to have $2N_c + 9$ independent relations to determine a solution of the system. Equations (2.85)–(2.89) provide $N_c + 9$ independent relations, differential or algebraic; the additional N_c relations are provided by the *equilibrium relations* that relate the numbers of moles.

Mass interchange between phases is characterized by the variation of mass distribution of each component in the vapor and liquid phases. As usual, these two phases are assumed to be in the *phase equilibrium state*. This is physically reasonable since the mass interchange between phases occurs much faster than the flow of porous media fluids. Consequently, the distribution of each hydrocarbon component into the two phases is subject to the condition of *stable thermodynamic equilibrium*, which is given by minimizing the *Gibbs free energy* of the compositional system (Bear, 1972; Chen et al., 2000C):

$$f_{io}(p_o, x_{1o}, x_{2o}, \ldots, x_{N_c o}) = f_{ig}(p_g, x_{1g}, x_{2g}, \ldots, x_{N_c g}), \tag{2.91}$$

where f_{io} and f_{ig} are the fugacity functions of the ith component in the liquid and vapor phases, respectively, $i = 1, 2, \ldots, N_c$. More details will be given on these fugacity functions in Chapters 3 and 9.

We end with a remark on the calculation of *mass fractions* $c_{i\alpha}$ of component i in phase α from the mole fractions $x_{i\alpha}$ (cf. Exercise 2.8)

$$c_{i\alpha} = \frac{W_i x_{i\alpha}}{\sum_{j=1}^{N_c} \left(W_j x_{j\alpha}\right)}, \qquad i = 1, 2, \ldots, N_c, \ \alpha = o, \ g, \tag{2.92}$$

and the calculation of *mass densities* ρ_α from the molar densities ξ_α

$$\rho_\alpha = \xi_\alpha \sum_{i=1}^{N_c} W_i x_{i\alpha}. \tag{2.93}$$

2.9 Nonisothermal Flow

The differential equations so far have been developed under the condition that flow is isothermal. This condition can be removed by adding a *conservation of energy equation*. This equation introduces an additional dependent variable, temperature, to the system. Unlike the case of mass transport, where the solid itself is assumed impervious to mass flux, the solid matrix does conduct heat. The average temperature of the solid and fluids in a porous medium may not be the same. Furthermore, heat may be exchanged between the phases. For simplicity, we invoke the requirement of *local thermal equilibrium* that the temperature be the same in all phases.

For multicomponent, multiphase flow in a porous medium, the mass balance and other equations are presented as in (2.85)–(2.91). Under the *nonisothermal condition*, some variables such as porosity, density, and viscosity may depend on temperature (cf. Chapter 3). The conservation of energy equation can be derived as in Section 2.2 for the mass conservation. A statement of the energy balance or *first law of thermodynamics* in a differential volume V is

Net rate of energy transported into V
+ Rate of energy production in V
= Rate of energy accumulation in V.

Using this law, the overall energy balance equation is (Lake, 1989)

$$\begin{aligned} &\frac{\partial}{\partial t}\left(\rho_t U + \frac{1}{2}\sum_{\alpha=w}^{g} \rho_\alpha |\mathbf{u}_\alpha|^2\right) + \nabla \cdot \mathbf{E} \\ &\quad + \sum_{\alpha=w}^{g} \left(\nabla \cdot (p_\alpha \mathbf{u}_\alpha) - \rho_\alpha \mathbf{u}_\alpha \cdot \wp \nabla z\right) = q_H - q_L, \end{aligned} \tag{2.94}$$

where ρ_t is the *overall density*, $\rho_t U$ is the *total internal energy*, the term $\sum_{\alpha=w}^{g} \rho_\alpha |\mathbf{u}_\alpha|^2/2$ represents *kinetic energy* per unit bulk volume, $\mathbf{E}$ is the energy flux, the term

$$\sum_{\alpha=w}^{g} \left(\nabla \cdot (p_\alpha \mathbf{u}_\alpha) - \rho_\alpha \mathbf{u}_\alpha \cdot \wp \nabla z\right)$$

is the rate of work done against the pressure field and gravity, q_H indicates the *enthalpy source term* per bulk volume, and q_L is heat loss.

The total internal energy is

$$\rho_t U = \phi \sum_{\alpha=w}^{g} \rho_\alpha S_\alpha U_\alpha + (1-\phi)\rho_s C_s T, \tag{2.95}$$

where U_α and C_s are the *specific internal energy* per unit mass of phase α and the *specific heat capacity* of the solid, respectively, and ρ_s is the density of the solid. The overall density ρ_t is determined by

$$\rho_t = \phi \sum_{\alpha=w}^{g} \rho_\alpha S_\alpha + (1-\phi)\rho_s.$$

The energy flux is made up of *convective contributions* from the flowing phases, *conduction*, and *radiation* (with all other contributions being ignored):

$$\mathbf{E} = \sum_{\alpha=w}^{g} \rho_\alpha \mathbf{u}_\alpha \left(U_\alpha + \frac{1}{2}|\mathbf{u}_\alpha|^2 \right) + \mathbf{q}_c + \mathbf{q}_r, \tag{2.96}$$

where $\mathbf{q}_c$ and $\mathbf{q}_r$ are the conduction and radiation fluxes, respectively. For multiphase flow, the conductive heat flux is given by *Fourier's law*:

$$\mathbf{q}_c = -k_T \nabla T, \tag{2.97}$$

where k_T represents the total *thermal conductivity*. For brevity, we ignore radiation, though it can be important in estimating heat losses from wellbores. Inserting (2.95)–(2.97) into (2.94) and combining the first term in the right-hand side of (2.96) with the work done by pressure, we see that

$$\begin{aligned}
&\frac{\partial}{\partial t}\left(\phi \sum_{\alpha=w}^{g} \rho_\alpha S_\alpha U_\alpha + (1-\phi)\rho_s C_s T + \frac{1}{2}\sum_{\alpha=w}^{g} \rho_\alpha |\mathbf{u}_\alpha|^2 \right) \\
&+ \nabla \cdot \left(\sum_{\alpha=w}^{g} \rho_\alpha \mathbf{u}_\alpha \left(H_\alpha + \frac{1}{2}|\mathbf{u}_\alpha|^2 \right) \right) \\
&- \nabla \cdot (k_T \nabla T) + \sum_{\alpha=w}^{g} \rho_\alpha \mathbf{u}_\alpha \cdot \wp \nabla z = q_H - q_L,
\end{aligned} \tag{2.98}$$

where H_α is the *enthalpy* of the α-phase (per unit mass) given by

$$H_\alpha = U_\alpha + \frac{p_\alpha}{\rho_\alpha}, \qquad \alpha = w, o, g.$$

As usual (Lake, 1989), if we neglect the kinetic energy and the last term in the left-hand side of (2.98), we obtain the *energy equation* for temperature T

$$\begin{aligned}
&\frac{\partial}{\partial t}\left(\phi \sum_{\alpha=w}^{g} \rho_\alpha S_\alpha U_\alpha + (1-\phi)\rho_s C_s T \right) \\
&+ \nabla \cdot \sum_{\alpha=w}^{g} \rho_\alpha \mathbf{u}_\alpha H_\alpha - \nabla \cdot (k_T \nabla T) = q_H - q_L.
\end{aligned} \tag{2.99}$$

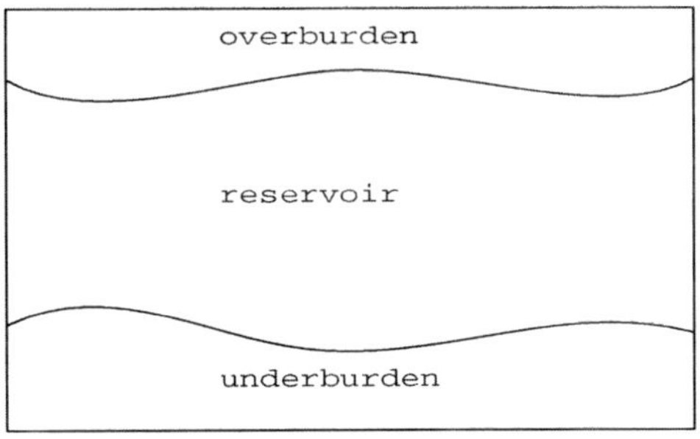

Figure 2.5. *Reservoir, overburden, and underburden.*

If desired, diffusive fluxes can be added to the left-hand side of (2.99) as in (2.85). Namely, using (2.90), the term

$$-\sum_{i=1}^{N_c}\sum_{\alpha=w}^{g} \nabla \cdot (\xi_\alpha H_{i\alpha} W_i \mathbf{D}_{i\alpha} \nabla x_{i\alpha})$$

can be inserted, where W_i is the molecular weight of component i and $H_{i\alpha}$ represents the enthalpy of component i in phase α.

In thermal methods heat is lost to the adjacent strata of a reservoir or the overburden and underburden, which is included in q_L of (2.99). We assume that the *overburden* and *underburden* extend to infinity along both the positive and negative x_3-axis (the vertical direction); see Figure 2.5. If the overburden and underburden are impermeable, heat is transferred entirely through conduction. With all fluid velocities and convective fluxes being zero, the energy conservation equation (2.99) reduces to

$$\frac{\partial}{\partial t}\left(\rho_{ob} C_{p,ob} T_{ob}\right) = \nabla \cdot (k_{ob} \nabla T_{ob}), \tag{2.100}$$

where the subscript *ob* indicates that the variables are associated with the overburden and $C_{p,ob}$ is the heat capacity at constant pressure. The initial condition is the original temperature $T_{ob,0}$ of the overburden:

$$T_{ob}(\mathbf{x}, 0) = T_{ob,0}(\mathbf{x}).$$

The boundary condition at the top of the reservoir is

$$T_{ob}(\mathbf{x}, t) = T(\mathbf{x}, t),$$

where we recall that T is the reservoir temperature. At infinity, T_{ob} is fixed:

$$T_{ob}(x_1, x_2, \infty, t) = T_\infty.$$

On other boundaries, we can use the impervious boundary condition

$$k_{ob} \nabla T_{ob} \cdot \boldsymbol{\nu} = 0,$$

where $\boldsymbol{\nu}$ represents the outward unit normal to these boundaries. Now, the rate of heat loss to the overburden is calculated by $k_{ob}\nabla T_{ob} \cdot \boldsymbol{\nu}$, where $\boldsymbol{\nu}$ is the unit normal to the interface between the overburden and reservoir (pointing to the overburden). Similar differential equations and initial and boundary conditions can be developed for the underburden.

2.10 Chemical Compositional Flow

An important method in *enhanced oil recovery* is *chemical flooding*, such as alkaline, surfactant, polymer, and foam (ASP+foam) flooding. The injection of these chemical components reduces fluid mobility to improve the sweep efficiency of a reservoir, i.e., increases the volume of the permeable medium contacted at any given time. For a chemical flooding compositional model, the governing differential equations consist of a mass conservation equation for each component, an energy equation, Darcy's law, and an overall mass conservation or continuity equation for pressure. These equations are developed under the following assumptions: local thermodynamic equilibrium, immobile solid phase, Fickian dispersion, ideal mixing, slightly compressible soil and fluids, and Darcy's law.

For this model, it is more convenient to use the concentration for each component in the mass conservation equation, as in Sections 2.4 and 2.5, since chemical reactions are involved. The mass conservation for component i is expressed in terms of the overall concentration of this component per unit pore volume:

$$\frac{\partial}{\partial t}(\phi \tilde{c}_i \rho_i) = -\nabla \cdot \left(\sum_{\alpha=1}^{N_p} \rho_i \left[c_{i\alpha} \mathbf{u}_\alpha - \mathbf{D}_{i\alpha} \nabla c_{i\alpha} \right] \right) + q_i, \qquad i = 1, 2, \ldots, N_c, \tag{2.101}$$

where the overall concentration $\tilde{c}_i$ is the sum over all phases, including the adsorbed phases,

$$\tilde{c}_i = \left(1 - \sum_{j=1}^{N_{cv}} \hat{c}_j \right) \sum_{\alpha=1}^{N_p} S_\alpha c_{i\alpha} + \hat{c}_i, \qquad i = 1, 2, \ldots, N_c; \tag{2.102}$$

N_{cv} is the total number of volume-occupying components (such as water, oil, surfactant, and air); N_p is the number of phases; $\hat{c}_i$, ρ_i, and q_i are the adsorbed concentration, mass density, and source/sink term of component i; and $c_{i\alpha}$ and $\mathbf{D}_{i\alpha}$ are the concentration and diffusion-dispersion tensor, respectively, of component i in phase α. The term $1 - \sum_{j=1}^{N_{cv}} \hat{c}_j$ represents the reduction in pore volume due to adsorption.

The density ρ_i is related to pressure by (2.6). For slightly compressible fluids, it is given by (2.12); i.e., at a reference phase pressure p_r, it equals

$$\rho_i = \rho_i^o \left(1 + C_i^o (p_r - p_r^o) \right), \tag{2.103}$$

where C_i^o is the constant compressibility and ρ_i^o is the density at the reference pressure p_r^o.

The *diffusion-dispersion tensor* $\mathbf{D}_{i\alpha}$ is an extension of (2.58) to multiphase flow:

$$\mathbf{D}_{i\alpha}(\mathbf{u}_\alpha) = \phi \left\{ S_\alpha d_{i\alpha} \mathbf{I} + |\mathbf{u}_\alpha| \left(d_{l\alpha} \mathbf{E}(\mathbf{u}_\alpha) + d_{t\alpha} \mathbf{E}^\perp(\mathbf{u}_\alpha) \right) \right\}, \tag{2.104}$$

where $d_{i\alpha}$ is the *molecular diffusion* coefficient of component i in phase α; $d_{l\alpha}$ and $d_{t\alpha}$ are, respectively, the *longitudinal* and *transverse dispersion* coefficients of phase α; $|\mathbf{u}_\alpha|$ is the Euclidean norm of $\mathbf{u}_\alpha = (u_{1\alpha}, u_{2\alpha}, u_{3\alpha})$, $|\mathbf{u}_\alpha| = \sqrt{u_{1\alpha}^2 + u_{2\alpha}^2 + u_{3\alpha}^2}$; $\mathbf{E}(\mathbf{u}_\alpha)$ is the orthogonal projection along the velocity,

$$\mathbf{E}(\mathbf{u}_\alpha) = \frac{1}{|\mathbf{u}_\alpha|^2} \begin{pmatrix} u_{1\alpha}^2 & u_{1\alpha} u_{2\alpha} & u_{1\alpha} u_{3\alpha} \\ u_{2\alpha} u_{1\alpha} & u_{2\alpha}^2 & u_{2\alpha} u_{3\alpha} \\ u_{3\alpha} u_{1\alpha} & u_{3\alpha} u_{2\alpha} & u_{3\alpha}^2 \end{pmatrix};$$

and $\mathbf{E}^{\perp}(\mathbf{u}_\alpha) = \mathbf{I} - \mathbf{E}(\mathbf{u}_\alpha)$, $i = 1, 2, \ldots, N_c$, $\alpha = 1, 2, \ldots, N_p$. The source/sink term q_i combines all rates for component i and is expressed as

$$q_i = \phi \sum_{\alpha=1}^{N_p} S_\alpha r_{i\alpha} + (1-\phi) r_{is} + \tilde{q}_i, \tag{2.105}$$

where $r_{i\alpha}$ and r_{is} are the *reaction rates* of component i in the α fluid phase and rock phase, respectively, and $\tilde{q}_i$ is the injection/production rate of the same component per bulk volume. The volumetric velocity $\mathbf{u}_\alpha$ is given by Darcy's law as in (2.86):

$$\mathbf{u}_\alpha = -\frac{1}{\mu_\alpha}\mathbf{k}_\alpha(\nabla p_\alpha - \rho_\alpha \wp \nabla z), \qquad \alpha = 1, 2, \ldots, N_p. \tag{2.106}$$

The energy conservation equation is given as in (2.99):

$$\begin{aligned} &\frac{\partial}{\partial t}\left(\phi \sum_{\alpha=1}^{N_p} \rho_\alpha S_\alpha U_\alpha + (1-\phi)\rho_s c_s T\right) \\ &\quad + \nabla \cdot \sum_{\alpha=1}^{N_p} \rho_\alpha \mathbf{u}_\alpha H_\alpha - \nabla \cdot (k_T \nabla T) = q_H - q_L. \end{aligned} \tag{2.107}$$

The heat loss to the overburden and underburden can be calculated as in Section 2.9.

In the simulation of chemical flooding, a pressure equation for the aqueous phase (e.g., phase 1) is obtained by an overall mass balance on volume-occupying components. Other phase pressures are evaluated using the capillary pressure functions, as in (2.89):

$$p_{c\alpha 1} = p_\alpha - p_1, \qquad \alpha = 1, 2, \ldots, N_p, \tag{2.108}$$

where $p_{c11} = 0$ for convenience. Introduce the phase mobility

$$\lambda_\alpha = \frac{k_{r\alpha}}{\mu_\alpha} \sum_{i=1}^{N_{cv}} \rho_i c_{i\alpha}, \qquad \alpha = 1, 2, \ldots, N_p,$$

and the total mobility

$$\lambda = \sum_{\alpha=1}^{N_p} \lambda_\alpha.$$

Note that

$$\sum_{i=1}^{N_{cv}} \rho_i \mathbf{D}_{i\alpha} \nabla c_{i\alpha} = \mathbf{0}, \quad \sum_{i=1}^{N_{cv}} r_{i\alpha} = \sum_{i=1}^{N_{cv}} r_{is} = 0, \quad \alpha = 1, 2, \ldots, N_p.$$

Now, by adding (2.101) over i, $i = 1, 2, \ldots, N_{cv}$, we obtain the pressure equation (cf. Exercise 2.9)

$$\begin{aligned} &\phi c_t \frac{\partial p_1}{\partial t} - \nabla(\lambda \mathbf{k} \nabla p_1) \\ &\quad = \nabla \cdot \sum_{\alpha=1}^{N_p} \lambda_\alpha \mathbf{k}(\nabla p_{c\alpha 1} - \rho_\alpha \wp \nabla z) + \sum_{i=1}^{N_{cv}} \tilde{q}_i, \end{aligned} \tag{2.109}$$

where the total compressibility c_t is defined by

$$c_t = \frac{1}{\phi}\frac{\partial}{\partial p_1}\sum_{i=1}^{N_{cv}} \phi \tilde{c}_i \rho_i .$$

Assume that the rock compressibility is given by (2.16); i.e., at the reference pressure p_r^0,

$$\phi = \phi^o\big(1 + c_R(p_r - p_r^o)\big). \tag{2.110}$$

With $p_r = p_1$ and using (2.103) and (2.110), we have

$$\phi\tilde{c}_i\rho_i = \phi^o\tilde{c}_i\rho_i^o\left(1 + (c_R + C_i^0)(p_1 - p_1^o) + c_R C_i^0(p_1 - p_1^o)^2\right).$$

Neglecting the higher-order term in this equation, it becomes

$$\phi\tilde{c}_i\rho_i \approx \phi^o\tilde{c}_i\rho_i^o\left(1 + (c_R + C_i^0)(p_1 - p_1^o)\right). \tag{2.111}$$

Applying (2.111), the total compressibility c_t is simplified to

$$c_t = \frac{\phi^o}{\phi}\sum_{i=1}^{N_{cv}} \tilde{c}_i\rho_i^o\left(c_R + C_i^0\right). \tag{2.112}$$

Note that there are more dependent variables than there are differential and algebraic relations; there are formally $N_c + N_{cv} + N_cN_p + 3N_p + 1$ dependent variables: $c_i, \hat{c}_j, c_{i\alpha}, T$, $\mathbf{u}_\alpha$, p_α, and S_α, $\alpha = 1, 2, \ldots, N_p$, $i = 1, 2, \ldots, N_c$, $j = 1, 2, \ldots, N_{cv}$. Equations (2.101) and (2.106)–(2.109) provide $N_c + 2N_p$ independent relations, differential or algebraic; the additional $N_{cv} + N_cN_p + N_p + 1$ relations are given by the constraints

$$\begin{aligned}
&\sum_{\alpha=1}^{N_p} S_\alpha = 1 \quad \text{(a saturation constraint)},\\
&\sum_{i=1}^{N_{cv}} c_{i\alpha} = 1 \quad (N_p \text{ phase concentration constraints}),\\
&c_i = \sum_{\alpha=1}^{N_p} S_\alpha c_{i\alpha} \quad (N_c \text{ component concentration constraints}),\\
&\hat{c}_j = \hat{c}_j(c_1, c_2, \ldots, c_{N_c}) \quad (N_{cv} \text{ adsorption constraints}),\\
&f_{i\alpha}(p_\alpha, T, c_{1\alpha}, \ldots, c_{N_c\alpha}) = f_{i\beta}(p_\beta, T, c_{1\beta}, \ldots, c_{N_c\beta})\\
&\qquad\qquad (N_c(N_p - 1) \text{ phase equilibrium relations}),
\end{aligned} \tag{2.113}$$

where $f_{i\alpha}$ is the fugacity function of the ith component in the α-phase.

2.11 Flows in Fractured Porous Media

A dual porosity model has been developed for single phase flow in Section 2.2.6. This concept can be generalized to flows of other types. As an example, we consider the compositional flow in fractured porous media. For brevity of presentation, we neglect the diffusive effects.

2.11.1 Dual porosity/permeability models

In the development of the dual porosity model for single phase flow in Section 2.2.6, the fluid was assumed to flow only from the matrix into the fractures, not vice versa. Also, there was no connection between matrix blocks. Now, we consider a more general case without these two assumptions. In this general case, the mass balance equations in the matrix also contain the matrix-fracture transfer terms, $i = 1, 2, \ldots, N_c$:

$$\begin{aligned} &\frac{\partial(\phi\xi_w S_w)}{\partial t} + \nabla\cdot(\xi_w \mathbf{u}_w) = -q_{w,mf}, \\ &\frac{\partial(\phi[x_{io}\xi_o S_o + x_{ig}\xi_g S_g])}{\partial t} + \nabla\cdot(x_{io}\xi_o\mathbf{u}_o + x_{ig}\xi_g\mathbf{u}_g) = -q_{i,mf}, \end{aligned} \tag{2.114}$$

where it is assumed that the external source/sink terms do not interact with this system. In the fracture system, the mass balance equations are

$$\begin{aligned} &\frac{\partial(\phi\xi_w S_w)_f}{\partial t} + \nabla\cdot(\xi_w \mathbf{u}_w)_f = q_{w,mf} + q_w, \\ &\frac{\partial(\phi[x_{io}\xi_o S_o + x_{ig}\xi_g S_g])_f}{\partial t} + \nabla\cdot(x_{io}\xi_o\mathbf{u}_o + x_{ig}\xi_g\mathbf{u}_g)_f \\ &\qquad = q_{i,mf} + q_i, \qquad i = 1, 2, \ldots, N_c, \end{aligned} \tag{2.115}$$

where the subscript f represents the fracture quantities. Equations (2.86)–(2.91) remain valid for both the matrix and the fractures.

The matrix-fracture transfer terms for the *dual porosity/permeability model*, $q_{w,mf}$ and $q_{i,mf}$, are defined following Warren and Root (1963) and Kazemi (1969). The transfer term for a particular component is directly related to the *matrix shape factor* σ, the fluid mobility, and the potential difference between the fracture and matrix systems. The capillary pressure, gravity, and viscous forces must be properly incorporated into this term. Furthermore, the contribution from a pressure gradient across each matrix block (and the molecular diffusion rate for each component) must be also included. For brevity of presentation, we neglect the diffusion rate.

The treatment of a pressure gradient across a block is based on the following observation: for an oil matrix block surrounded with water in the fractures, the pressure differences are

$$\Delta p_w = 0, \quad \Delta p_o = \wp(\rho_w - \rho_o).$$

Analogously, for an oil block surrounded with gas fractures and a gas block surrounded with water fractures, respectively,

$$\Delta p_g = 0, \quad \Delta p_o = \wp(\rho_o - \rho_g)$$

and

$$\Delta p_w = 0, \quad \Delta p_g = \wp(\rho_w - \rho_g).$$

We introduce the *global fluid density* in the fractures

$$\rho_f = S_{w,f}\rho_w + S_{o,f}\rho_o + S_{g,f}\rho_g,$$

and define the pressure gradient effect

$$\Delta p_\alpha = \wp \left|\rho_f - \rho_\alpha\right|, \qquad \alpha = w, o, g.$$

Now, the transfer terms that include the contributions from the capillary pressure, gravity, and viscous forces, and the pressure gradients across matrix blocks are

$$\begin{aligned} q_{w,mf} &= T_m \frac{k_{rw}\xi_w}{\mu_w}\left(\Phi_w - \Phi_{w,f} + L_c \Delta p_w\right), \\ q_{i,mf} &= T_m \left\{ \frac{k_{ro}x_{io}\xi_o}{\mu_o}\left(\Phi_o - \Phi_{o,f} + L_c\Delta p_o\right) \right. \\ &\qquad \left. + \frac{k_{rg}x_{ig}\xi_g}{\mu_g}\left(\Phi_g - \Phi_{g,f} + L_c\Delta p_g\right) \right\}, \end{aligned} \tag{2.116}$$

where Φ_α is the phase potential,

$$\Phi_\alpha = p_\alpha - \rho_\alpha \wp z, \qquad \alpha = w, o, g,$$

L_c is the *characteristic length* for the matrix-fracture flow, and

$$T_m = k\sigma\left(\frac{1}{l_{x_1}^2} + \frac{1}{l_{x_2}^2} + \frac{1}{l_{x_3}^2}\right)$$

is the *matrix-fracture transmissibility* with σ the *shape factor* and l_{x_1}, l_{x_2}, and l_{x_3} the matrix block dimensions. When the matrix permeability $\mathbf{k}$ is a tensor and different in the three coordinate directions, the matrix-fracture transmissibility is modified to

$$T_m = \sigma\left(\frac{k_{11}}{l_{x_1}^2} + \frac{k_{22}}{l_{x_2}^2} + \frac{k_{33}}{l_{x_3}^2}\right), \qquad \mathbf{k} = \mathrm{diag}(k_{11}, k_{22}, k_{33}).$$

2.11.2 Dual porosity models

For the development of a dual porosity model, the matrix blocks act as a source term to the fracture system. In this case, there are two approaches for deriving this model: the Warren–Root approach as in Section 2.11.1 and the approach based on boundary conditions imposed explicitly on matrix blocks as in Section 2.2.6.

The Warren–Root approach

In this approach, the mass balance equations in the matrix become

$$\begin{aligned} &\frac{\partial(\phi\xi_w S_w)}{\partial t} = -q_{w,mf}, \\ &\frac{\partial(\phi[x_{io}\xi_o S_o + x_{ig}\xi_g S_g])}{\partial t} = -q_{i,mf}, \qquad i = 1, 2, \ldots, N_c, \end{aligned} \tag{2.117}$$

where $q_{w,mf}$ and $q_{i,mf}$ are defined by (2.116). The balance equations (2.115) in the fractures remain unchanged.

The boundary conditions approach

For a dual porosity model of the compositional flow under consideration, the fluid flow in the matrix system can be modeled in the same way as in (2.31) for single phase flow. Let the matrix system be composed of disjoint blocks $\{\Omega_i\}$. On *each block* $\{\Omega_i\}$ the mass balance equations hold, $i = 1, 2, \ldots, N_c$:

$$\begin{aligned} &\frac{\partial(\phi \xi_w S_w)}{\partial t} + \nabla \cdot (\xi_w \mathbf{u}_w) = 0, \\ &\frac{\partial(\phi [x_{io}\xi_o S_o + x_{ig}\xi_g S_g])}{\partial t} + \nabla \cdot (x_{io}\xi_o \mathbf{u}_o + x_{ig}\xi_g \mathbf{u}_g) = 0. \end{aligned} \tag{2.118}$$

The mass balance equations in the fractures are defined as in (2.115) with $q_{w,mf}$ and $q_{i,mf}$ given by (cf. Exercise 2.10)

$$\begin{aligned} q_{w,mf} &= -\sum_j \chi_j(\mathbf{x}) \frac{1}{|\Omega_j|} \int_{\Omega_i} \frac{\partial(\phi \xi_w S_w)}{\partial t} d\mathbf{x}, \\ q_{i,mf} &= -\sum_j \chi_j(\mathbf{x}) \frac{1}{|\Omega_j|} \int_{\Omega_i} \frac{\partial(\phi [x_{io}\xi_o S_o + x_{ig}\xi_g S_g])}{\partial t} d\mathbf{x} \end{aligned} \tag{2.119}$$

for $i = 1, 2, \ldots, N_c$.

We impose boundary conditions for the matrix equations (2.118) as in Section 2.2.6. For $\xi_1, \xi_2, \ldots, \xi_N$ fixed, we define the *phase pseudopotential*

$$\Phi'_\alpha(p_\alpha, \xi_1, \xi_2, \ldots, \xi_N) = \int_{p_\alpha^o}^{p_\alpha} \frac{1}{\rho_\alpha(\xi, \xi_1, \xi_2, \ldots, \xi_N)\wp} d\xi - z, \tag{2.120}$$

where p_α^0 is some reference pressure, $\alpha = o, g$. The inverse of this integral is denoted $\psi'_\alpha(\cdot, \xi_1, \xi_2, \ldots, \xi_N)$, again for $\xi_1, \ldots, \xi_N$ fixed.

Now, the boundary conditions for (2.118) on the surface $\partial\Omega_i$ of each matrix block Ω_i are, for $i = 1, 2, \ldots, N_c$, $\alpha = o, g$,

$$\begin{aligned} &x_{i\alpha} = x_{i\alpha,f}, \\ &\Phi'_\alpha(p_\alpha, x_{1\alpha}, x_{2\alpha}, \ldots, x_{N\alpha}) = \Phi'_{\alpha,f}(p_{\alpha,f}, x_{1\alpha,f}, x_{2\alpha,f}, \ldots, x_{N\alpha,f}) - \Phi_\alpha^o, \end{aligned} \tag{2.121}$$

where, for a given $\Phi'_{\alpha,f}$, Φ_α^o is a pseudopotential reference value on each block Ω_i determined by

$$\begin{aligned} \frac{1}{|\Omega_i|} \int_{\Omega_i} (\phi\rho_\alpha)\Big(&\psi'_\alpha(\Phi'_{\alpha,f} - \Phi_\alpha^o + x_3, x_{1\alpha,f}, x_{2\alpha,f}, \ldots, x_{N\alpha,f}), \\ &x_{1\alpha,f}, x_{2\alpha,f}, \ldots, x_{N\alpha,f}\Big) d\mathbf{x} \\ = (\phi\rho_\alpha)&(p_{\alpha,f}, x_{1\alpha,f}, x_{2\alpha,f}, \ldots, x_{N\alpha,f}). \end{aligned} \tag{2.122}$$

If we assume that $\partial\rho_\alpha/\partial p_\alpha \geq 0$ (for $x_{1\alpha}, x_{2\alpha}, \ldots, x_{N\alpha}$ fixed), (2.122) is solvable for Φ_α^o (for incompressible α-phase fluid, set $\Phi_\alpha^o = 0$). The second equation in (2.121) applies to the first equation in (2.118), for the water component, the pseudopotential depends only on pressure.

This model implies that the fracture system, being highly permeable, quickly comes into chemical and mechanical equilibrium locally on the fracture spacing scale. This equilibrium is defined in terms of the mole fractions and the *chemical equilibrium pseudopotentials*, and is reflected in the matrix equations through the boundary conditions in (2.121).

2.12 Concluding Remarks

In this chapter, the basic fluid flow and transport equations have been developed for a hierarchy of models: single phase, two-phase, black oil, volatile oil, compositional, thermal, and chemical. This hierarchy of models correspond to different oil production stages. Their governing differential equations consist of the mass and energy conservation equations and Darcy's law. We have chosen to start with the simplest model for single phase flow and to end with the most complex model for chemical flooding. This approach can be reversed; that is, we can start with the chemical model, and in turn derive the thermal, compositional, volatile oil, black oil, two-phase, and single phase models.

In the chemical model, we have considered the general case where there are N_c chemical components, each of which may exist in any or all of the N_p phases. The basic equations consist of a mass conservation equation for each component (2.101), an energy equation for temperature (2.107), Darcy's law for the volumetric velocity of each fluid phase (2.106), an overall mass conservation for a phase pressure (2.109), and algebraic constraints (2.113) that describe physical and chemical phenomena peculiar to chemical flooding. The flow equations allow for compressibility of soil and fluids, dispersion and molecular diffusion, chemical reactions, and phase behavior. Even though the displacement mechanisms are different in the thermal and chemical methods, there is not much difference between the corresponding models, both of which include mass and energy conservation and Darcy's law. The mass equation is usually solved in terms of the mole fraction for each component in the thermal case (cf. (2.85)), while it is solved in terms of the volumetric concentration in the chemical case. In addition, the emphasis is placed on the solution of compositions and temperature in the former case, while it is on the solution of compositions and reactions for the components involved in the latter.

When flow is isothermal, the model equations in the chemical and thermal methods become the basic equations for compositional flow. An energy equation is not required in the compositional model, which now consists of a mass conservation equation in terms of the mole fraction for each component (2.85), Darcy's law for the phase volumetric velocity (2.86), and phase equilibrium relations for the computation of compositions (2.91). In this model, N_c components form at most three phases (e.g., vapor, liquid, and water), and mass interchanges only between the hydrocarbon phases (i.e., the vapor and liquid phases).

Instead of three fluid phases, if only a single phase is present in an entire porous medium, the mass conservation equation for each component in the compositional model becomes the transport equation of multicomponents in the fluid phase (2.59). When at most two components are involved, this equation reduces to the transport equation (2.55) for a component.

The black oil and volatile oil models can be treated as simplified, two-component compositional models. In these models, the hydrocarbon system is composed of the gas (mainly methane and ethane) and oil components at stock-tank conditions. There is no mass transfer between the water phase and the oil and gas phases. In the black oil model, the gas component can exist in the oil and gas phases. In the volatile oil model, both hydrocarbon components can exist in these two phases. The black oil model is not suitable for handling a volatile oil reservoir. The governing differential equations of these two models are generally written in terms of volumetric rates at standard conditions; see (2.76)–(2.78) and (2.80)–(2.82).

The model for two-phase immiscible flow is a special case of the black oil model; when two phases are considered and there is no mass transfer between them, the two-phase immiscible flow model results, which consists of a mass conservation equation (2.36) and Darcy's law for each phase (2.37). Finally, when only a single phase is present, the model for two-phase flow reduces to that for single phase flow (cf. (2.1) and (2.4)).

The relationship among the models is presented for ordinary porous media. For a fractured porous medium, the concept of dual porosity and dual porosity/permeability can be incorporated. Examples for single phase and compositional flows in fractured media have been discussed in Sections 2.2.6 and 2.11, respectively.

Limitations of the basic fluid flow equations for all the models presented in this chapter have not been fully discussed. Non-Newtonian fluids are not considered in subsequent chapters. Also, all considerations will be based on Darcy's law in place of the momentum balance equation. Non-Darcy's law and non-Newtonian phenomena have been briefly described in Sections 2.2.7 and 2.2.8 for single phase flow.

2.13 Bibliographical Information

The book by Aziz and Settari (1979) covered the single phase flow model through the black oil model, while the models covered in Peaceman's book (1977) included the compositional flow model. The nonisothermal and chemical compositional flow models are presented in a quite condensed fashion in this chapter. For more information on the physics of these two models, the reader should refer to the book by Lake (1989) and to the technical documentation by Delshad et al. (2000) (also see Chapters 10 and 11).

Exercises

2.1. Derive equation (2.12) from equation (2.6).

2.2. Derive the equation of mass conservation (2.36) for the simultaneous flow of two fluids in a porous medium.

2.3. Derive system (2.47) in detail.

2.4. Consider the transport equation of multicomponents in a fluid phase in a porous medium (cf. (2.59)),

$$\phi \frac{\partial (c_i \rho)}{\partial t} = -\nabla \cdot (c_i \rho \mathbf{u} - \rho \mathbf{D} \nabla c_i) + \rho q_i, \qquad i = 1, 2, \ldots, N_c, \tag{2.123}$$

and Darcy's law for the fluid

$$\mathbf{u} = -\frac{1}{\mu} \mathbf{k} \nabla p. \tag{2.124}$$

Recall the equation of state (cf. (2.6))

$$\frac{d\rho}{\rho} = c_f dp, \tag{2.125}$$

where we assume that the compressibility factor c_f is constant. Based on (2.123)–(2.125) and the concentration constraint

$$\sum_{i=1}^{N_c} c_i = 1,$$

prove that the pressure equation

$$\phi c_f \frac{\partial p}{\partial t} - \nabla \cdot \left(\frac{1}{\mu} \mathbf{k} \nabla p \right) = \sum_{i=1}^{N_c} q_i \tag{2.126}$$

holds, provided that the "higher-order" quadratic term $c_f \mathbf{u} \cdot \nabla p$ is neglected. Equation (2.126) can be utilized along with $N_c - 1$ equations of form (2.123) to describe the transport of multicomponents in a fluid or the *compressible miscible displacement process*.

2.5. Assume that the capillary pressures p_{cow} and p_{cgo} take the forms $p_{cow} = p_{cow}(S_w)$ and $p_{cgo} = p_{cgo}(S_g)$ and have respective inverse functions p_{cow}^{-1} and p_{cgo}^{-1}. Express equations (2.61)–(2.66) in a three-pressure (p_w, p_o, p_g) formulation.

2.6. Under the same assumptions as in Exercise 2.5, express equations (2.61)–(2.66) in a pressure (p_o) and two-saturation (S_w, S_g) formulation.

2.7. Consider three-phase immiscible flow

$$\begin{aligned} \frac{\partial(\phi \rho_\alpha S_\alpha)}{\partial t} &= -\nabla \cdot (\rho_\alpha \mathbf{u}_\alpha) + q_\alpha, \\ \mathbf{u}_\alpha &= -\frac{k_{r\alpha}}{\mu_\alpha} \mathbf{k} \left(\nabla p_\alpha - \rho_\alpha \wp \nabla z \right), \qquad \alpha = w, o, g, \end{aligned} \tag{2.127}$$

and the additional constraints

$$\begin{aligned} &S_w + S_o + S_g = 1, \\ &p_{cw}(S_w, S_g) = p_w - p_o, \quad p_{cg}(S_w, S_g) = p_g - p_o, \end{aligned} \tag{2.128}$$

where $p_{cw} = -p_{cow}$ and $p_{cg} = p_{cgo}$. The phase and total mobilities and the fractional flow functions are defined in the same manner as in Section 2.3:

$$\lambda_\alpha = \frac{k_{r\alpha}}{\mu_\alpha}, \quad \lambda = \sum_{\alpha=w}^{g} \lambda_\alpha, \quad f_\alpha = \frac{\lambda_\alpha}{\lambda}, \quad \alpha = w, o, g,$$

where f_α depends on the saturations S_w and S_g.

(i) Prove that there exists a function $(S_w, S_g) \longmapsto p_c(S_w, S_g)$ such that

$$\nabla p_c = f_w \nabla p_{cw} + f_g \nabla p_{cg} \tag{2.129}$$

if and only if the following equations are satisfied:

$$\frac{\partial p_c}{\partial S_w} = f_w \frac{\partial p_{cw}}{\partial S_w} + f_g \frac{\partial p_{cg}}{\partial S_w}, \quad \frac{\partial p_c}{\partial S_g} = f_w \frac{\partial p_{cw}}{\partial S_g} + f_g \frac{\partial p_{cg}}{\partial S_g}. \tag{2.130}$$

(ii) Show that a necessary and sufficient condition for existence of a function p_c satisfying (2.130) is

$$\frac{\partial f_w}{\partial S_g} \frac{\partial p_{cw}}{\partial S_w} + \frac{\partial f_g}{\partial S_g} \frac{\partial p_{cg}}{\partial S_w} = \frac{\partial f_w}{\partial S_w} \frac{\partial p_{cw}}{\partial S_g} + \frac{\partial f_g}{\partial S_w} \frac{\partial p_{cg}}{\partial S_g}. \tag{2.131}$$

This condition is referred to as the *total differential condition*.
(iii) When condition (2.131) is satisfied, the function p_c is

$$\begin{aligned} p_c(S_w, S_g) = & \int_1^{S_w} \left\{ f_w(\xi, 0) \frac{\partial p_{cw}}{\partial S_w}(\xi, 0) + f_g(\xi, 0) \frac{\partial p_{cg}}{\partial S_w}(\xi, 0) \right\} d\xi \\ & + \int_0^{S_g} \left\{ f_w(S_w, \xi) \frac{\partial p_{cw}}{\partial S_g}(S_w, \xi) + f_g(S_w, \xi) \frac{\partial p_{cg}}{\partial S_g}(S_w, \xi) \right\} d\xi, \end{aligned} \tag{2.132}$$

where we assume that the integrals are well defined. We introduce the global pressure and the total velocity

$$p = p_o + p_c, \quad \mathbf{u} = \mathbf{u}_w + \mathbf{u}_o + \mathbf{u}_g. \tag{2.133}$$

Write equations (2.127) and (2.128) in terms of the main unknowns p, S_w, and S_g.

2.8. Let the mass and molar densities, $\rho_{i\alpha}$ and $\xi_{i\alpha}$, of component i in phase α be related by $\xi_{i\alpha} = \rho_{i\alpha}/W_i$, where W_i is the molar mass of component i, $i = 1, 2, \ldots, N_c$, $\alpha = o, g$. Prove equations (2.92) and (2.93).

2.9. Derive the pressure equation (2.109) in detail.

2.10. Derive the matrix-fracture transfer terms $q_{w,mf}$ and $q_{i,mf}$ in equation (2.119) for the dual porosity model of compositional flow.

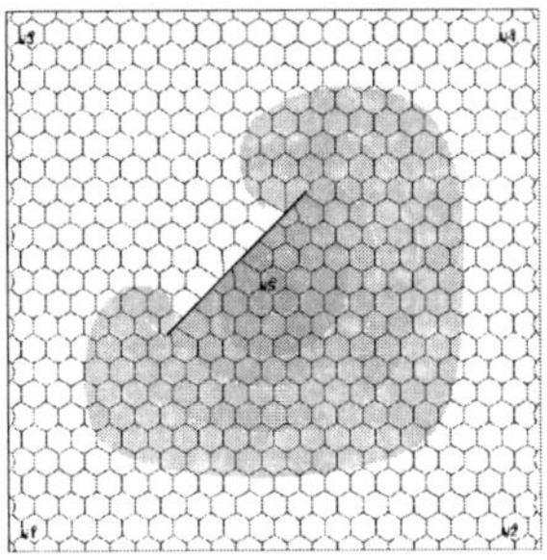

Chapter 3
Rock and Fluid Properties

The basic flow and transport equations presented in the preceding chapter and computational methods used to solve them depend on rock and fluid properties of porous media. In this chapter, we discuss these properties. In particular, capillary pressures, relative permeabilities, formation volume factors, densities, solubility, viscosities, compressibilities, and equations of state are described for the two-phase, black oil, volatile oil, and compositional models. The equations of state deal with the distribution of hydrocarbon components into phases. Temperature-dependent rock and fluid properties will be also studied for the thermal model. In chemical flooding, very complex physical and chemical phenomena occur between the reservoir rock and fluids, such as adsorption and cation exchange. For this reason, the rock and fluid properties for the chemical compositional model will be discussed in detail in Chapter 11.

This chapter is organized as follows. In Section 3.1, rock properties are given; capillary pressure and relative permeability functions for two-phase and three-phase flows are reviewed. Then, in Section 3.2, fluid properties, such as PVT (pressure-volume-temperature) data for water, oil, and gas, are stated. The equations of state for compositional flow are also examined in this section. In Section 3.3, temperature-dependent rock and fluid properties are considered. Finally, bibliographical information is given in Section 3.4.

3.1 Rock Properties

3.1.1 Capillary pressures

In two-phase flow, a discontinuity in fluid pressure occurs across an interface between any two immiscible fluids (e.g., water and oil). This is a consequence of the interfacial tension that exists at the interface. The discontinuity between the pressure in the nonwetting phase (say, oil), p_o, and that in the wetting phase (say, water), p_w, is referred to as the *capillary pressure*, p_c:

$$p_c = p_o - p_w, \tag{3.1}$$

where the phase pressures at the interface are taken from their respective sides. A typical curve of the capillary pressure is shown in Figure 3.1. The capillary pressure depends on the

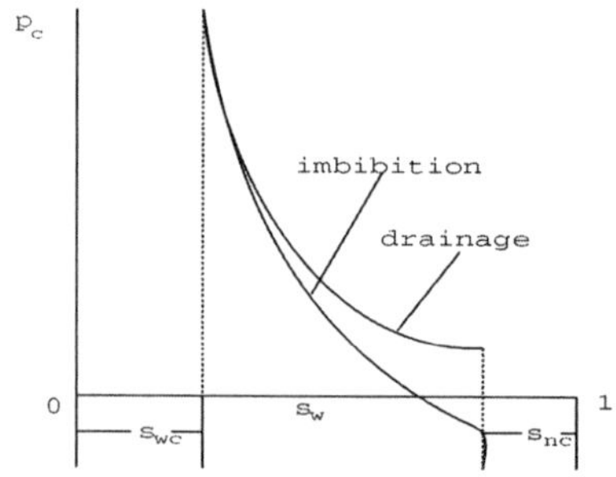

Figure 3.1. *Typical capillary pressure curve.*

wetting phase saturation S_w and the direction of saturation change (drainage or imbibition). The phenomenon of dependence of the curve on the history of saturation is called *hysteresis*. While it is possible to develop a model that takes into account the hysteresis resulting from the saturation history (Mualem, 1976; Bedrikovetsky et al., 1996), in most cases the direction of flow can be predicted, and only a set of capillary pressures are needed. Various curves describing a drainage or imbibition cycle can be found in Brooks and Corey (1964), van Genuchten (1980), and Corey (1986).

The value p_{cb} that is necessary to start displacement is termed a *threshold pressure* (Bear, 1972). The saturation value at which the wetting phase can no longer be displaced by applying a pressure gradient is referred to as *irreducible saturation*. The capillary pressure curve has an asymptote at whose value the pressure gradient remains continuous in both phases. This can be observed by considering vertical gravity equilibrium. When the value of the irreducible saturation of the nonwetting phase is approached, an analogous situation occurs at the other end of the curve during the imbibition process (Calhoun et al., 1949; Morrow, 1970).

In the discussion so far, the capillary pressure has been assumed to depend only on the saturation of the wetting phase and its history. In general, however, it also depends on the surface tension σ, porosity ϕ, permeability k, and the contact angle θ with the rock surface of the wetting phase, which, in turn, depend on the temperature and fluid compositions (Poston et al., 1970; Bear-Bachmat, 1991):

$$J(S_w) = \frac{p_c}{\sigma \cos\theta}\sqrt{\frac{k}{\phi}},$$

which is the *J-function*. If the contact angle is ignored, this function becomes

$$J = \frac{p_c}{\sigma}\sqrt{\frac{k}{\phi}}.$$

Using the J-function, typical curves for p_c can be obtained from experiments. This function is also the basis for some theoretical methods of measuring permeability k (Ashford, 1969).

For three-phase flow, two *capillary pressures* are needed:

$$p_{cow} = p_o - p_w, \qquad p_{cgo} = p_g - p_o. \tag{3.2}$$

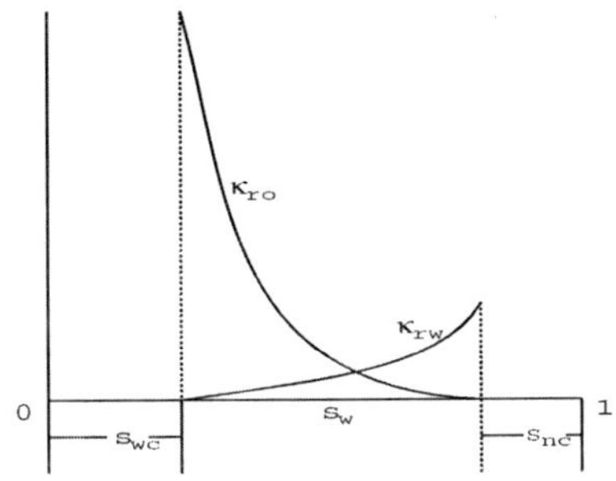

Figure 3.2. *Typical relative permeability curves.*

Note that the third capillary pressure p_{cgw} can be found using p_{cow} and p_{cgo}:

$$p_{cgw} = p_g - p_w = p_{cow} + p_{cgo}.$$

The capillary pressures p_{cow} and p_{cgo} are usually assumed to take the forms (Leverett and Lewis, 1941)

$$p_{cow} = p_{cow}(S_w), \quad p_{cgo} = p_{cgo}(S_g), \tag{3.3}$$

where S_w and S_g are the phase saturations of water and gas, respectively. These forms remain in wide use (cf. Exercises 3.1 and 3.2), though revised forms have been proposed (Shutler, 1969).

3.1.2 Relative permeabilities

Two-phase flow

Measurements on *relative permeabilities* have been made mostly for two-phase flow. Typical curves suitable for an oil-water system with water displacing oil are presented in Figure 3.2. The value of S_w at which water starts to flow is termed the *critical saturation*, S_{wc}, and the value S_{nc} at which oil ceases to flow is called the *residual saturation*. Analogously, during a drainage cycle S_{nc} and S_{wc} are referred to as the critical and residual saturations, respectively.

The slopes of capillary pressure curves at irresidual saturations must be finite in numerical simulation, so these curves themselves cannot be utilized to define the saturation value at which the displaced phase becomes immobile. This saturation value is found using the residual saturation at which the relative permeability of this phase is zero. Darcy's law implies that the phase stops flowing because the mobility becomes zero (not because the external force becomes zero). As a result, it is not necessary to distinguish the critical and residual saturations.

As for capillary pressures, relative permeabilities depend not only on the wetting phase saturation S_w, but also on the direction of saturation change (drainage or imbibition). Figure 3.3 shows the phenomenon of dependence of a relative permeability for the nonwetting phase on the history of saturation. Note that the curve of imbibition is always lower than that of drainage. For the wetting phase, the relative permeability does not depend on the history of saturation.

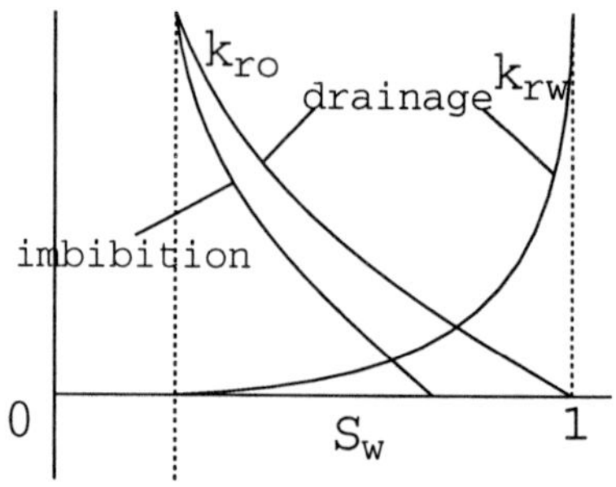

Figure 3.3. *Hysteresis in relative permeability curves.*

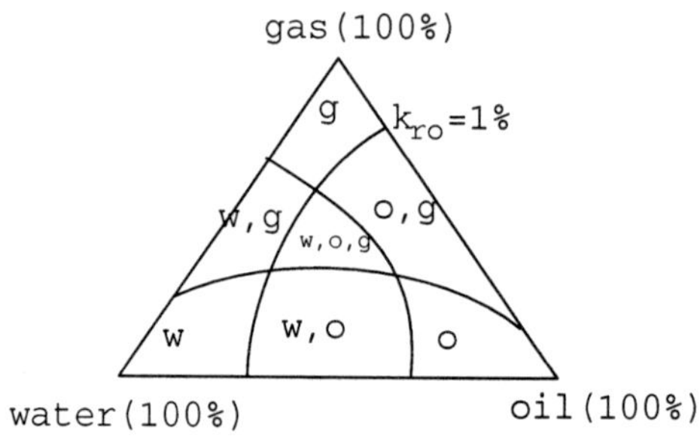

Figure 3.4. *A three-phase ternary diagram.*

Wettability of the rock also strongly influences relative permeabilities (Owens and Archer, 1971). Because of this, reservoir fluids should be employed for experiments instead of refined fluids.

Relative permeabilities must be determined empirically or experimentally for each particular porous medium of interest. However, the literature is rich on analytical expressions for the relationship between relative permeabilities and the saturation of the wetting phase (Corey, 1986). These expressions were usually obtained from simplified porous media models (e.g., bundle of capillary tubes and capillary tube networks); see Exercises 3.3 and 3.4.

Three-phase flow

In contrast, the determination of *relative permeabilities* for three-phase flow is rather difficult. From experiments, a *ternary diagram* for the relationship between the relative permeabilities and saturations can be shown as in Figure 3.4. This diagram is based on the level curve of the relative permeability being equal to 1% for each phase. From it we can figure out where single, two-, or three-phase flow occurs under different combinations of saturations. In the triangular region bounded by the three level curves, for example, three fluids flow simultaneously.

Starting from Leverett and Lewis (1941), most of the measurements on three-phase relative permeabilities have been experimental. These measurements have indicated that

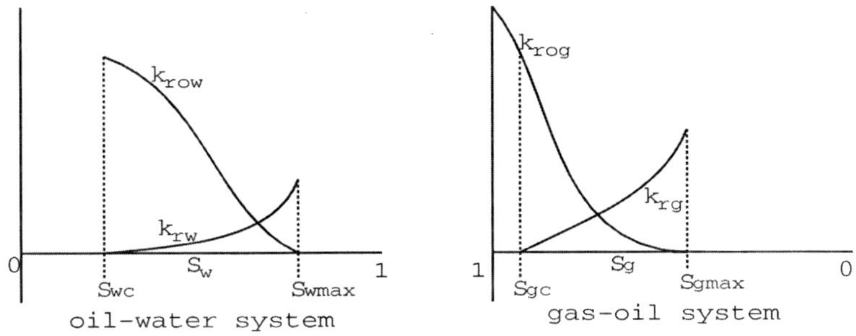

Figure 3.5. *Relative permeability curves in a three-phase system.*

the relative permeabilities for the wetting and nonwetting phases in a three-phase system are functions of their respective saturations as they are in a two-phase system (Corey et al., 1956; Snell, 1962):

$$k_{rw} = k_{rw}(S_w), \quad k_{rg} = k_{rg}(S_g). \tag{3.4}$$

The relative permeability for the intermediate wetting phase is a function of the two independent saturations:

$$k_{ro} = k_{ro}(S_w, S_g). \tag{3.5}$$

The functional form in (3.5) is rarely known. In practice, the estimation of three-phase relative permeabilities is based on two sets of two-phase data: the relative permeability in an intermediate and wetting system,

$$k_{row} = k_{row}(S_w), \tag{3.6}$$

and that in an intermediate and nonwetting system,

$$k_{rog} = k_{rog}(S_g). \tag{3.7}$$

The underlying concept is that for the wetting phase, both the intermediate and nonwetting phases act like a single nonwetting phase, while for the nonwetting phase, both the intermediate and wetting phases behave as a single wetting phase. Figure 3.5 illustrates typical relative permeability curves for a water, oil, and gas system in an isotropic porous medium. The point where $k_{row} = 0$ indicates the maximum water saturation rather than the critical oil saturation since the oil saturation can be further reduced by increasing the gas saturation. It has been experimentally observed, however, that a nonzero residual (or minimal) oil saturation S_{or} exists when oil is displaced simultaneously by water and gas. The earlier remark on hysteresis of the relative permeability for the nonwetting phase also applies to the three-phase system.

The simplest procedure to determine k_{ro} is

$$k_{ro} = k_{row} k_{rog}. \tag{3.8}$$

Other models were suggested by Stone (1970; 1973), Corey (1986), and Delshad and Pope (1989). As an example, we describe two of Stone's models, model I and model II.

Stone's model I

The saturations are normalized as follows:

$$S_{no} = \frac{S_o - S_{or}}{1 - S_{wc} - S_{or}}, \qquad S_o \geq S_{or},$$

$$S_{nw} = \frac{S_w - S_{wc}}{1 - S_{wc} - S_{or}}, \qquad S_w \geq S_{wc},$$

$$S_{ng} = \frac{S_g}{1 - S_{wc} - S_{or}}.$$

Note that

$$S_{no} + S_{nw} + S_{ng} = 1.$$

The relative permeability of oil is defined by

$$k_{ro} = S_{no}\beta_w\beta_g. \tag{3.9}$$

To determine β_w, we take $S_g = S_{ng} = 0$; i.e., the three-phase system reduces to a water-oil system. In this case, $\beta_g = 1$ and $k_{ro} = k_{row}$, which, together with (3.9), gives

$$\beta_w = \frac{k_{row}(S_w)}{1 - S_{nw}}. \tag{3.10}$$

Similarly, to determine β_g, we choose $S_w = S_{wc}$ so that $\beta_w = 1$ and $k_{ro} = k_{rog}$. Then using (3.9) yields

$$\beta_g = \frac{k_{rog}(S_g)}{1 - S_{ng}}. \tag{3.11}$$

Substituting (3.10) and (3.11) into (3.9) gives the expression of k_{ro} for *Stone's model I.*

This model reduces exactly to two-phase data only if the following condition is satisfied:

$$k_{row}(S_{wc}) = k_{rog}(S_g = 0) = 1. \tag{3.12}$$

Otherwise, the relative permeability $k_{ro}(S_w, S_g)$ provides only an approximation to the two-phase data. A model that does not have this limitation can be obtained if the oil-gas data are measured in the presence of irreducible water. In this case, a water-oil system at S_{wc} and a gas-oil system at $S_g = 0$ are physically identical; i.e., both systems satisfy $S_w = S_{wc}$ and $S_o = 1 - S_{wc}$. Hence (3.12) is equivalent to the definition of the absolute permeability being the effective permeability of oil in the presence of S_{wc}.

Set

$$k_{row}(S_{wc}) = k_{rog}(S_g = 0) = k_{rc}.$$

Then Stone's model I can be modified as follows:

$$k_{ro} = k_{rc}S_{no}\beta_w\beta_g, \tag{3.13}$$

where

$$\beta_w = \frac{k_{row}(S_w)}{(1 - S_{nw})k_{rc}}, \qquad \beta_g = \frac{k_{rog}(1 - S_g)}{(1 - S_{ng})k_{rc}}.$$

Stone's model II

In the definition of Stone's model I, S_{or} must be specified. In fact, this value can be predicted from an equation derived from *channel-flow* considerations:

$$k_{ro} = (k_{row} + k_{rw})(k_{rog} + k_{rg}) - (k_{rw} + k_{rg}), \tag{3.14}$$

where $k_{ro} \geq 0$ is required (i.e., negative values of k_{ro} mean immobile oil). As for Stone's model I, to satisfy (3.12), *model II* can be altered as follows:

$$k_{ro} = k_{rc}\left\{(k_{row}/k_{rc} + k_{rw})(k_{rog}/k_{rc} + k_{rg}) - (k_{rw} + k_{rg})\right\}. \tag{3.15}$$

3.1.3 Rock compressibility

The *rock compressibility* is defined by

$$c_R = \frac{1}{\phi}\frac{d\phi}{dp}. \tag{3.16}$$

After integration, it becomes (cf. Exercise 3.5)

$$\phi = \phi^o e^{c_R(p-p^o)}, \tag{3.17}$$

where ϕ^o is the porosity at a reference pressure p^o. Using a Taylor series expansion, we see that

$$\phi = \phi^o \left\{1 + c_R(p - p^o) + \frac{1}{2!}c_R^2(p - p^o)^2 + \cdots\right\},$$

so an approximation results:

$$\phi \approx \phi^o \left(1 + c_R(p - p^o)\right).$$

3.2 Fluid Properties

An accurate analysis of fluid properties is required before a reservoir simulator is performed. Examples of these properties include formation volume factors, densities, solubilities, viscosities, and compressibilities of fluids. In general, a representative sample of reservoir hydrocarbons is obtained and studied, and these fluid properties are then measured in the laboratory. Then the information is used to predict the phase changes that will occur both in the reservoir and in the surface separators. A *separator* is a pressure vessel (either horizontal or vertical) utilized for the purpose of separating well fluids into gaseous and liquid components.

When laboratory data are not available for the fluid properties, they can be calculated from *empirical formulas*. In this section, we state these formulas. For more details on fluid properties, the reader can refer to Carr et al. (1954), Chew and Connally (1959), Dempsey (1965), Wichert and Aziz (1972), Dranchuk et al. (1974), Beggs and Robinson (1975), Numbere et al. (1977), Standing (1977), Meehan (1980A; 1980B), Vasquez and Beggs (1980), and Craft and Hawkins (1991).

In three-phase flow (e.g., the flow of black oil type), when all gas dissolves into the oil phase there is no gas phase present; i.e., $S_g = 0$. In this case, a reservoir is in the *undersaturated state*. If all three phases coexist, the reservoir is in the *saturated state*. A *bubble point* is defined as the state in which the flow system entirely consists of liquids (water and oil), and the reservoir pressure at this point is the *bubble point pressure*. Any slight reduction in pressure (or increase in volume) at fixed temperature produces gas.

3.2.1 Water PVT properties

The black oil and volatile oil models presented in the preceding chapter require four water PVT properties for simulation:

- water density at standard conditions ρ_{Ws},
- water formation volume factor B_w,
- water compressibility c_w,
- water viscosity μ_w.

When laboratory data are not available, empirical formulas can be used to calculate them from the following given data:

- pressure p and temperature T of a reservoir,
- salinity of water S_{ALI},
- solution gas/water ratio R_{SW}.

Water density at standard conditions

We can use the following empirical correlation data in a linear interpolation for the *water density* at standard conditions against the *water salinity*:

Salinity:	0,	100,000,	200,000,	280,000 (ppm),
ρ_{Ws}:	1.0,	1.07300,	1.1370,	1.18600 (g/cm^3).

Water formation volume factor

The *water formation volume factor* B_w (RB/STB; cf. Chapter 16) can be computed by the empirical formula

$$B_w = (A + Bp + Cp^2)F_{SB}, \tag{3.18}$$

where the constants A, B, and C depend on the formation temperature (T_F) and the gas saturation status, p (psia) is the formation pressure, and F_{SB} is the *salinity correction factor* for B_w:

$$\begin{aligned} F_{SB} = \big\{&5.1 \times 10^{-8} p + (5.47 \times 10^{-6} - 1.95 \times 10^{-10} p)(T_F - 60) \\ &- (3.23 \times 10^{-8} - 8.5 \times 10^{-13} p)(T_F - 60)^2\big\} S_{ALI} + 1, \end{aligned}$$

with T_F (°F) being the formation temperature and S_{ALI} the salinity percentage (1% = 10,000 ppm). The constants A, B, and C in (3.18) can be determined from

$$A = A_1 + A_2 T_F + A_3 T_F^2,$$
$$B = B_1 + B_2 T_F + B_3 T_F^2,$$
$$C = C_1 + C_2 T_F + C_3 T_F^2,$$

where in the saturated case (the gas phase exists)

$$A_1 = 0.9911, \quad A_2 = 6.35 \times 10^{-5}, \quad A_3 = 8.5 \times 10^{-7},$$
$$B_1 = -1.093 \times 10^{-6}, \quad B_2 = -3.497 \times 10^{-9}, \quad B_3 = 4.57 \times 10^{-12},$$
$$C_1 = -5 \times 10^{-11}, \quad C_2 = 6.429 \times 10^{-13}, \quad C_3 = -1.43 \times 10^{-15},$$

and in the undersaturated case (there exists no gas phase)

$$A_1 = 0.9947, \quad A_2 = 5.8 \times 10^{-6}, \quad A_3 = 1.02 \times 10^{-6},$$
$$B_1 = -4.228 \times 10^{-6}, \quad B_2 = 1.8376 \times 10^{-8}, \quad B_3 = -6.77 \times 10^{-11},$$
$$C_1 = 1.3 \times 10^{-10}, \quad C_2 = -1.3855 \times 10^{-12}, \quad C_3 = 4.285 \times 10^{-15}.$$

The range of validity for B_w is

$$1{,}000 < p < 5{,}000 \text{ psi}, \quad 100 < T < 250° \text{ F}, \quad 0 \leq S_{ALI} < 25.$$

Water isothermal compressibility

The *water compressibility* c_w (1/psi) is calculated from salinity, temperature, and pressure by the empirical formula

$$c_w = (\hat{A} + \hat{B}\, T_F + \hat{C}\, T_F^2) 10^{-6} (1 + 0.0089\, R_{SW}) F_{SC}, \tag{3.19}$$

where the constants $\hat{A}$, $\hat{B}$, and $\hat{C}$ depend on the formation pressure:

$$\hat{A} = 3.8546 - 1.34 \times 10^{-4} p,$$
$$\hat{B} = -0.01052 + 4.77 \times 10^{-7} p,$$
$$\hat{C} = 3.9267 \times 10^{-5} - 8.8 \times 10^{-10} p.$$

The solution gas/water ratio R_{SW} (SCF/STB) is zero in the undersaturated case, while in the saturated case

$$R_{SW} = (A_{R_{SW}} + B_{R_{SW}} p + C_{R_{SW}} p^2) \cdot \left\{1 - (0.0753 - 1.73 \times 10^{-4} T_F) S_{ALI}\right\},$$

where

$$A_{R_{SW}} = 2.12 + 3.45 \times 10^{-3} T_F - 3.59 \times 10^{-5} T_F^2,$$
$$B_{R_{SW}} = 0.0107 - 5.26 \times 10^{-5} T_F + 1.48 \times 10^{-7} T_F^2,$$
$$C_{R_{SW}} = -8.75 \times 10^{-7} + 3.9 \times 10^{-9} T_F - 1.02 \times 10^{-11} T_F^2.$$

Finally, the salinity correction factor F_{SC} for c_w is defined by

$$F_{SC} = \Big\{-0.52 + 2.7 \times 10^{-4} T_F - 1.14 \times 10^{-6} T_F^2 + 1.121 \times 10^{-9} T_F^3\Big\} S_{ALI}^{0.7} + 1.$$

The range of validity for the c_w estimate is

$$1{,}000 < p < 6{,}000 \text{ psi}, \quad 80 < T < 250^\circ \text{F}, \quad 0 \le S_{ALI} < 25.$$

Water viscosity

The *water viscosity* μ_w (cp) is computed from salinity, temperature, and pressure by

$$\mu_w = 0.02414 \times 10^{247.8/(T_K - 140)} F_{SV} F_{PV}, \tag{3.20}$$

where T_K is the formation temperature in K; that is, $T_K = 273.15 + T_C$, with $T_C = (T_F - 32)/1.8^\circ$ C. The salinity correction factor F_{SV} for μ_w is

$$F_{SV} = 1 - 1.87 \times 10^{-3} S_{ALI}^{1/2} + 2.18 \times 10^{-4} S_{ALI}^{2.5} + (T_F^{1/2} - 0.0135\, T_F)(2.76 \times 10^{-3} S_{ALI} - 3.44 \times 10^{-4} S_{ALI}^{1.5}),$$

and the pressure correction factor F_{PV} for μ_w is

$$F_{PV} = 1 + 3.5 \times 10^{-12} p^2 (T_F - 40).$$

The range of validity for μ_w is

$$32 < T < 572^\circ \text{F}, \quad 0 \le S_{ALI} < 25.$$

An example of finding the water PVT properties is given in Exercise 3.6.

3.2.2 Oil PVT properties

Five quantities for the oil PVT properties with respect to the bubble point pressure (p_b) are required for the black oil and volatile oil models described in the preceding chapter:

- dissolved gas-oil ratio R_{so},
- oil formation volume factor B_o,
- oil compressibility c_o,
- oil viscosity μ_o,
- oil viscosity compressibility c_μ.

Again, when laboratory data are not available for these quantities, empirical formulas can be employed to compute them from the following given data:

- pressure p and temperature T of a reservoir,
- produced gas-oil ratio measured at separator conditions G_{OR},
- oil gravity A_{PI},
- raw gas gravity (unity for air) Y_G,
- pressure p_{sep} and temperature T_{sep} at separator conditions.

Initial bubble point pressure

An initial *bubble point pressure* p_{bi} (psia) can be obtained using the empirical formula

$$p_{bi} = \left(\frac{G_{OR}}{A_0 \, Y_{GS} \, \exp\left(C_0 \, A_{PI}/T_R\right)} \right)^{1/B_0}, \tag{3.21}$$

where G_{OR} (SCF/STB) is the observed *gas-oil ratio*, A_{PI} (°API) is the *oil gravity* defined by $A_{PI} = 141.5/D_{OB} - 131.5$, D_{OB} (g/cm^3) is the *surface oil density* at standard conditions, Y_{GS} is the *corrected gas gravity* (unity for air) defined by

$$Y_{GS} = Y_G \left(1 + 5.912 \times 10^{-5} A_{PI} \, T_{sep} \log\left(\frac{p_{sep}}{114.7} \right) \right),$$

Y_G is the *gas gravity* (unity for air), p_{sep} (psia) and T_{sep} (°F) are the pressure and temperature of a separator, and T_R (R) is the reservoir temperature ($T_R = T_F + 460$). The constants A_0, B_0, and C_0 needed in the computation of the bubble point pressure are

$$\begin{aligned} A_0 &= 0.0362, \quad B_0 = 1.0937, \quad C_0 = 25.724 \quad \text{if } A_{PI} \leq 30, \\ A_0 &= 0.0178, \quad B_0 = 1.1870, \quad C_0 = 23.931 \quad \text{if } A_{PI} > 30. \end{aligned}$$

The range of validity of the bubble point estimate is

$$30 < p_{sep} < 535 \text{ psi}, \quad 76 < T_{sep} < 150^\circ \text{ F}.$$

In addition,

$$0.511 < Y_G < 1.351 \quad \text{if } 15.3 < A_{PI} \leq 30^\circ \text{API}$$

and

$$0.53 < Y_G < 1.259 \quad \text{if } 30.6 < A_{PI} < 59.5^\circ \text{API}.$$

Dissolved gas-oil ratio

An empirical formula for the *dissolved gas-oil ratio* R_{so} (SCF/STB) is

$$R_{so} = A_0 \, Y_{GS} \, p_b^{B_0} \exp\left(\frac{C_0 \, A_{PI}}{T_R} \right). \tag{3.22}$$

This formula is exploited to find the functional relationship between R_{so} and the bubble point pressure p_b.

Oil formation volume factor

In the saturated case, the *oil formation volume factor* B_o (RB/STB) can be expressed as a function of the dissolved gas-oil ratio:

$$B_o(p_b) = 1 + \tilde{A}(T_F - 60)\frac{A_{PI}}{Y_{GS}} + \left(\tilde{B} + \tilde{C}(T_F - 60)\frac{A_{PI}}{Y_{GS}}\right) R_{so}, \tag{3.23}$$

where the constants $\tilde{A}$, $\tilde{B}$, and $\tilde{C}$ are

$$\begin{array}{lll} \tilde{A} = 1.751 \times 10^{-5}, & \tilde{B} = 4.677 \times 10^{-4}, & \tilde{C} = -1.811 \times 10^{-8} \\ & & \text{if } A_{PI} < 30, \\ \tilde{A} = 1.1 \times 10^{-5}, & \tilde{B} = 4.67 \times 10^{-4}, & \tilde{C} = 1.337 \times 10^{-9} \\ & & \text{if } A_{PI} \geq 30. \end{array}$$

When the reservoir pressure p is larger than the bubble point pressure p_b, i.e., in the undersaturated state, the formation volume factor can be evaluated from B_o at p_b, the oil compressibility c_o (1/psi), and pressure:

$$B_o(p, p_b) = B_o(p_b) \exp\big(- c_0(p - p_b)\big), \tag{3.24}$$

or approximately from

$$B_o(p, p_b) \approx B_o(p_b)\big(1 - c_0(p - p_b)\big). \tag{3.25}$$

The range of validity of the expressions for B_o is

$$30 < p_{sep} < 535 \text{ psi}, \quad 76 < T_{sep} < 150^\circ \text{F}, \quad 15.3 < A_{PI} < 59.5^\circ \text{API}.$$

In addition, above the bubble point it is required that

$$0.511 < Y_G < 1.351, \quad 111 < p < 9,485 \text{ psi};$$

below the bubble point,

$$0.511 < Y_G < 1.351, \quad 14.7 < p < 4,542 \text{ psi} \quad \text{if } A_{PI} \leq 30^\circ \text{API}$$

and

$$0.53 < Y_G < 1.259, \ 14.7 < p < 6,025 \text{ psi} \quad \text{if } 30.6 < A_{PI} < 59.5^\circ \text{API}.$$

Oil isothermal compressibility

The *oil compressibility* c_o (1/psi) can be calculated by the empirical formula

$$c_o = \frac{-1,433 + 5R_{so} + 17.2\,T_F - 1,180\,Y_{GS} + 12.61A_{PI}}{100,000\,p_b}, \tag{3.26}$$

where R_{so}, T_F, Y_{GS}, A_{PI}, and p_b are defined as earlier.

Oil viscosity

The *oil viscosity* μ_o (cp) is calculated using the Beggs–Robinson equation (Beggs and Robinson, 1975). It is computed differently in the saturated case than in the undersaturated case. In the former case, it is calculated through the "dead" oil viscosity by the formula

$$\mu_o(p_b) = \bar{A}\,\mu_{do}^{\bar{B}}, \tag{3.27}$$

where μ_{do} (cp) is the "dead" oil viscosity, $\bar{A} = 10.715(R_{so}+100)^{-0.515}$, and $\bar{B} = 5.44(R_{so}+150)^{-0.338}$. The *dead oil viscosity* μ_{do} can be found through the empirical formula

$$\mu_{do} = 10^{\bar{C}} - 1,$$

where

$$\bar{C} = 10^{C'}\,T_F^{-1.163}, \quad C' = 3.0324 - 0.02023\,A_{PI}.$$

The validity ranges for μ_o and μ_{do} are, respectively,

$$30 < p_{sep} < 535 \text{ psi}, \quad 70 < T_{sep} < 150^\circ \text{ F}$$

and

$$70 < T < 295^\circ \text{ F}, \quad 16 < A_{PI} < 58^\circ \text{ API}.$$

In the latter case, μ_o is calculated by

$$\mu_o(p, p_b) = \mu_o(p_b)\left(\frac{p}{p_b}\right)^{A'}, \tag{3.28}$$

where p (psia) is the reservoir pressure and

$$A' = 2.6\,p^{1.187}\exp\left(-8.98\times10^{-5}p - 11.513\right).$$

The validity range is

$$15.3 < A_{PI} < 59.5^\circ \text{ API}, \quad 0.511 < Y_G < 1.351, \quad 111 < p < 9,485 \text{ psi}.$$

Oil viscosity compressibility

In the black oil model, the *oil viscosity compressibility* c_μ (1/psi) is often used to evaluate the oil viscosity in the undersaturated case:

$$\mu_o(p, p_b) = \mu_o(p_b)\left(1 + c_\mu(p - p_b)\right), \tag{3.29}$$

where

$$c_\mu = \left(1 + p_b^{-1}\right)^{B'} - 1,$$
$$B' = 2.6(1 + p_b)^{1.187}\exp\left(-8.98\times10^{-5}(1 + p_b) - 11.513\right).$$

An example of calculating the oil PVT properties is given in Exercise 3.7.

3.2.3 Gas PVT properties

The black oil and volatile oil models described in the preceding chapter require two functional (parameter) relationships of gas PVT properties with respect to pressure:

- gas deviation factor Z or formation volume factor B_g,
- gas viscosity μ_g.

When laboratory data are not available, empirical formulas can be used to compute them from the following given data:

- pressure p and temperature T of a reservoir,
- raw gas gravity (unity for air) Y_G,
- content of CO_2, H_2S, and N_2: Y_{CO_2}, Y_{H_2S}, and Y_{N_2}.

Reduced pressure and temperature

Before the gas deviation factor Z is evaluated, it is necessary to compute the dimensionless, *reduced pressure* p_{red} and *temperature* T_{red}:

$$p_{red} = \frac{p}{p_{pc}}, \quad T_{red} = \frac{T_R}{T_{pc}}, \tag{3.30}$$

where p (psia) is the formation pressure, T_R (R) is the formation temperature (recall that $T_R = T_F + 460$), and p_{pc} (psia) and T_{pc} (R) are the *pseudocritical pressure* and *temperature* of gas, respectively. p_{pc} and T_{pc} are estimated from the gas gravity for both condensate and miscellaneous reservoir gas. The computed values are corrected for acid gas using the Wichert–Aziz correction. Before the Wichert–Aziz corrections were introduced, the following empirical formulas for p_{pc} and T_{pc} had been employed:

$$\begin{aligned} p_{pc0} &= A_{pc} + B_{pc}\, Y_G + C_{pc}\, Y_G^2, \\ T_{pc0} &= \hat{A}_{pc} + \hat{B}_{pc}\, Y_G + \hat{C}_{pc}\, Y_G^2, \end{aligned} \tag{3.31}$$

where Y_G is the *raw gas density* (unity for air). For the surface gas, the constants in (3.31) are calculated by

$$\begin{aligned} A_{pc} &= 677, \quad B_{pc} = 15, \quad C_{pc} = -37.5, \\ \hat{A}_{pc} &= 168, \quad \hat{B}_{pc} = 325, \quad \hat{C}_{pc} = -12.5, \end{aligned}$$

while for the condensate gas, they are given by

$$\begin{aligned} A_{pc} &= 706, \quad B_{pc} = -51.7, \quad C_{pc} = -11.1, \\ \hat{A}_{pc} &= 187, \quad \hat{B}_{pc} = 330, \quad \hat{C}_{pc} = -71.5. \end{aligned}$$

The *Wichert–Aziz corrections* for p_{pc} and T_{pc} have been used in recent years:

$$p_{pc} = \frac{p_{pc0}(T_{pc0} - W_A)}{T_{pc0} + Y_{H_2S}\,(1 - Y_{H_2S})W_A}, \qquad T_{pc} = T_{pc0} - W_A,$$

where the Wichert–Aziz correction factor W_A (°F) is given by

$$\begin{aligned} W_A = 120\left((Y_{CO_2} + Y_{H_2S})^{0.9} - (Y_{CO_2} + Y_{H_2S})^{1.6}\right) \\ -15\left(Y_{H_2S}^{0.5} - Y_{H_2S}^{4}\right), \end{aligned}$$

and Y_{H_2S} and Y_{CO_2} (decimal) are the contents of H_2S and CO_2, respectively. The ranges of validity of p_{pc} and T_{pc} are $0.36 < Y_G < 1.3$ for a condensate fluid and $0.56 < Y_G < 1.71$ and $Y_{H_2S} + Y_{CO_2} < 0.8$ for miscellaneous gas.

Gas deviation factor Z

The *gas deviation factor* Z is calculated using the method developed by Dranchuk et al. (1974) who used the Benedict–Webb–Rubin equation of state fitted to the Standing–Katz Z-factor correlation. The resulting nonlinear equation is then solved by the Newton–Raphson iteration (cf. Chapter 8):

$$Z = \frac{0.27\, p_{red}}{\rho_{gr}\, T_{red}}, \tag{3.32}$$

where ρ_{gr} is the reduced gas density and is evaluated using the Newton–Raphson iteration

$$\begin{aligned} \rho_{gr}^{i+1} &- \rho_{gr}^{i} - \mathcal{F}(\rho_{gr}^{i})/\mathcal{F}'(\rho_{gr}^{i}), \\ \mathcal{F}(\rho_{gr}^{i}) &= A_r\,(\rho_{gr}^{i})^6 + B_r\,(\rho_{gr}^{i})^3 + C_r\,(\rho_{gr}^{i})^2 + E_r\,\rho_{gr}^{i} \\ &\quad + F_r\,(\rho_{gr}^{i})^3\left(1 + G_r\,(\rho_{gr}^{i})^2\right)\exp\left(-G_r\,(\rho_{gr}^{i})^2\right) - H_r, \\ \mathcal{F}'(\rho_{gr}^{i}) &= 6A_r\,(\rho_{gr}^{i})^5 + 3B_r\,(\rho_{gr}^{i})^2 + 2C_r\,\rho_{gr}^{i} + E_r \\ &\quad + F_r\,(\rho_{gr}^{i})^2\left(3 + G_r\,(\rho_{gr}^{i})^2(3 - 2G_r\,(\rho_{gr}^{i})^2)\right)\exp\left(-G_r\,(\rho_{gr}^{i})^2\right), \end{aligned}$$

where

$$\begin{aligned} A_r &= 0.06423, & B_r &= 0.5353\,T_{red} - 0.6123, \\ C_r &= 0.3151\,T_{red} - 1.0467 - \frac{0.5783}{T_{red}^2}, & E_r &= T_{red}, \\ F_r &= \frac{0.6816}{T_{red}^2}, & G_r &= 0.6845, \\ H_r &= 0.27\,p_{red}, & \rho_{gr}^0 &= \frac{0.27\,p_{red}}{T_{red}}. \end{aligned}$$

The iteration process converges rapidly (with fewer than five iterations) with a good initial ρ_{gr}^0. The range of validity of formula (3.32) for the Z-factor is

$$0 < p_{red} < 30, \quad 1.05 \le T_{red} < 3,$$

which covers the range of possible reservoir conditions including high pressure and temperature reservoirs.

Gas formation volume factor

The *gas formation volume factor*, B_g (RB/SCF), the ratio of the volume V_g of the gas phase measured at reservoir conditions to the volume V_{Gs} of the gas component measured at standard conditions, can be calculated using the gas deviation factor Z:

$$B_g = \frac{0.00504\ Z\ T_R}{p}, \tag{3.33}$$

where p (psia) is the formation pressure.

Gas viscosity

The *gas viscosity* μ_g (cp) is evaluated based on an estimation of the gas density using the real gas law (with a Z-factor correction). The pseudocritical pressure and temperature are corrected for nonhydrocarbon components. μ_g is calculated by the Lee–Gonzalez correction (Dempsey, 1965):

$$\mu_g = \frac{\exp(F)\ \mu_c}{T_{red}}, \tag{3.34}$$

where

$$\begin{aligned}
F &= \breve{A} + \breve{B}\ T_{red} + \breve{C}\ T_{red}^2 + \breve{D}\ T_{red}^3,\\
\breve{A} &= \breve{A}_0 + \breve{A}_1\ p_{red} + \breve{A}_2\ p_{red}^2 + \breve{A}_3\ p_{red}^3,\\
\breve{B} &= \breve{B}_0 + \breve{B}_1\ p_{red} + \breve{B}_2\ p_{red}^2 + \breve{B}_3\ p_{red}^3,\\
\breve{C} &= \breve{C}_0 + \breve{C}_1\ p_{red} + \breve{C}_2\ p_{red}^2 + \breve{C}_3\ p_{red}^3,\\
\breve{D} &= \breve{D}_0 + \breve{D}_1\ p_{red} + \breve{D}_2\ p_{red}^2 + \breve{D}_3\ p_{red}^3,
\end{aligned}$$

with the constants given by

$$\begin{aligned}
\breve{A}_0 &= -2.4621182, & \breve{A}_1 &= 2.97054714,\\
\breve{A}_2 &= -0.286264054, & \breve{A}_3 &= 8.05420522 \times 10^{-3},\\
\breve{B}_0 &= 2.80860949, & \breve{B}_1 &= -3.49803305,\\
\breve{B}_2 &= 0.36037302, & \breve{B}_3 &= -1.04432413 \times 10^{-2},\\
\breve{C}_0 &= -0.793385684, & \breve{C}_1 &= 1.39643306,\\
\breve{C}_2 &= -0.149144925, & \breve{C}_3 &= 4.41015512 \times 10^{-3},\\
\breve{D}_0 &= 0.0839387178, & \breve{D}_1 &= 0.186408848,\\
\breve{D}_2 &= 0.0203367881, & \breve{D}_3 &= 6.09579263 \times 10^{-4}.
\end{aligned}$$

The *corrected gas viscosity* μ_c (cp) in formula (3.34) is defined by (Carr et al., 1954)

$$\begin{aligned}
\mu_c = {} & (1.709 \times 10^{-5} - 2.062 \times 10^{-6}\ Y_G)T_F\\
& + 8.188 \times 10^{-3} - 6.15 \times 10^{-3} \log(Y_G)\\
& + Y_{N_2}\big(9.59 \times 10^{-3} + 8.48 \times 10^{-3} \log(Y_G)\big)\\
& + Y_{CO_2}\big(6.24 \times 10^{-3} + 9.08 \times 10^{-3} \log(Y_G)\big)\\
& + Y_{H_2S}\big(3.73 \times 10^{-3} + 8.49 \times 10^{-3} \log(Y_G)\big),
\end{aligned}$$

where Y_{N_2} (decimal) is the content of N_2. μ_c is the viscosity of a gas mixture at 14.7 psia and reservoir temperature. An example of computing the gas PVT properties is given in Exercise 3.8.

3.2.4 Total compressibility

For single phase flow in a porous medium, the *total compressibility* c_t (1/psi) is

$$c_t = c_f + \frac{\phi^o}{\phi} c_R, \tag{3.35}$$

where c_f is the fluid compressibility. For multiphase flow (e.g., three-phase flow with water, oil, and gas), the total compressibility c_t is

$$c_t = S_w c_w + S_o c_o + S_g c_g + \frac{\phi^o}{\phi} c_R. \tag{3.36}$$

3.2.5 Equations of state

Several mathematical techniques to handle the hydrocarbon behavior (the distribution of chemical components among phases) are available. The most common are based on (1) the *K-value approach*, (2) *equations of state*, and (3) a variety of empirical tables from experiments. In this section, we discuss the first two techniques.

Equilibrium K-values

Let x_{io} and x_{ig} be the mole fractions of component i in the liquid (e.g., oil) and vapor (e.g., gas) phases, respectively, $i = 1, 2, \ldots, N_c$ (the number of components). The equilibrium *flash vaporization ratio* for this component is defined by

$$K_i = \frac{x_{ig}}{x_{io}}, \qquad i = 1, 2, \ldots, N_c, \tag{3.37}$$

where the quantity K_i is the *equilibrium K-value* of component i. At low pressure, these K-values are readily related to the mixture pressure and temperature (see an example in Section 3.3.2). In fact, they are easily estimated from the vapor pressure data of pure components. At high pressure, the K-values are functions of overall compositions. The introduction of the compositions into the K-value functions adds considerable complexity to the flash computation.

Equations of state

While the K-value approach is easy to set up, it lacks generality and may result in inaccurate reservoir simulation. In recent years, the *equations of state* (EOSs) have been more widely employed because they produce more consistent compositions, densities, and molar volumes. The most famous EOS is the van der Waals EOS (Reid et al., 1977). Here we discuss three more accurate EOS: Peng–Robinson, Redlich–Kwong, and Redlich–Kwong–Soave. The method of volume translation is widely used for correcting volumetric deficiencies of the original EOS (Peneloux et al., 1982).

The Peng–Robinson equation of state

The mixing principle for the Peng–Robinson EOS is

$$a_\alpha = \sum_{i=1}^{N_c}\sum_{j=1}^{N_c} x_{i\alpha}x_{j\alpha}(1-\kappa_{ij})\sqrt{a_i a_j},$$
$$b_\alpha = \sum_{i=1}^{N_c} x_{i\alpha}b_i, \qquad \alpha = o, g,$$

where κ_{ij} is a *binary interaction parameter* between components i and j, and a_i and b_i are empirical factors for pure component i. The interaction parameters account for molecular interactions between two unlike molecules. By definition, κ_{ij} is zero when i and j represent the same component, small when i and j represent components that do not differ much (e.g., when components i and j are both alkanes), and large when i and j represent components that are substantially different. Ideally, κ_{ij} depends on pressure and temperature and on the identities of components i and j (Zudkevitch and Joffe, 1970; Whitson, 1982).

The factors a_i and b_i can be computed from

$$a_i = \Omega_{ia}\alpha_i \frac{R^2T_{ic}^2}{p_{ic}}, \qquad b_i = \Omega_{ib}\frac{R\,T_{ic}}{p_{ic}},$$

where we recall that R is the *universal gas constant*, T is the temperature, T_{ic} and p_{ic} are the *critical temperature* and *pressure*, the EOS parameters Ω_{ia} and Ω_{ib} are given by

$$\Omega_{ia} = 0.45724, \quad \Omega_{ib} = 0.077796,$$
$$\alpha_i = \left(1-\lambda_i\left[1-\sqrt{T/T_{ic}}\right]\right)^2,$$
$$\lambda_i = 0.37464 + 1.5423\omega_i - 0.26992\omega_i^2,$$

and ω_i is the *acentric factor* of component i. The acentric factors roughly express the deviation of the shape of a molecule from a sphere (Reid et al., 1977). Define

$$A_\alpha = \frac{a_\alpha p_\alpha}{R^2T^2}, \qquad B_\alpha = \frac{b_\alpha p_\alpha}{RT}, \qquad \alpha = o, g,$$

where the pressure p_α is given by the *Peng–Robinson two-parameter equation of state* (Peng and Robinson, 1976)

$$p_\alpha = \frac{RT}{V_\alpha - b_\alpha} - \frac{a_\alpha(T)}{V_\alpha(V_\alpha+b_\alpha)+b_\alpha(V_\alpha-b_\alpha)} \tag{3.38}$$

with V_α being the molar volume of phase α. Introduce the compressibility factor

$$Z_\alpha = \frac{p_\alpha V_\alpha}{RT}, \qquad \alpha = o, g.$$

Equation (3.38) can be expressed as a *cubic equation* in Z_α:

$$\begin{aligned} Z_\alpha^3 - (1-B_\alpha)Z_\alpha^2 + (A_\alpha - 2B_\alpha - 3B_\alpha^2)Z_\alpha \\ - (A_\alpha B_\alpha - B_\alpha^2 - B_\alpha^3) = 0. \end{aligned} \tag{3.39}$$

This equation has three roots. When only one root is real, it is selected. If there are three real roots, the selection of the right one depends on the dominance of the liquid phase or the vapor phase (cf. Chapter 9). Now, for $i = 1, 2, \ldots, N_c$ and $\alpha = o, g$, the *fugacity coefficient* of component i in a mixture can be obtained from

$$\begin{aligned}\ln \varphi_{i\alpha} = &\frac{b_i}{b_\alpha}(Z_\alpha - 1) - \ln(Z_\alpha - B_\alpha)\\ &- \frac{A_\alpha}{2\sqrt{2}B_\alpha}\left(\frac{2}{a_\alpha}\sum_{j=1}^{N_c} x_{j\alpha}(1-\kappa_{ij})\sqrt{a_i a_j} - \frac{b_i}{b_\alpha}\right)\\ &\cdot \ln\left(\frac{Z_\alpha + (1+\sqrt{2})B_\alpha}{Z_\alpha - (1-\sqrt{2})B_\alpha}\right).\end{aligned} \tag{3.40}$$

The *fugacity* of component i is

$$f_{i\alpha} = p_\alpha x_{i\alpha}\varphi_{i\alpha}, \qquad i = 1, 2, \ldots, N_c, \ \alpha = o, g.$$

Finally, the distribution of each hydrocarbon component into the liquid and vapor phases is given by the thermodynamic equilibrium relation

$$f_{io}(p_o, x_{1o}, x_{2o}, \ldots, x_{N_c o}) = f_{ig}(p_g, x_{1g}, x_{2g}, \ldots, x_{N_c g}) \tag{3.41}$$

for $i = 1, 2, \ldots, N_c$.

The Redlich–Kwong equation of state

The *Redlich–Kwong two-parameter EOS* is given by

$$p_\alpha = \frac{RT}{V_\alpha - b_\alpha} - \frac{a_\alpha}{V_\alpha(V_\alpha + b_\alpha)}, \qquad \alpha = o, g. \tag{3.42}$$

With $Z_\alpha = p_\alpha V_\alpha/(RT)$, this equation can be written as the cubic equation

$$Z_\alpha^3 - Z_\alpha^2 + (A_\alpha - B_\alpha - B_\alpha^2)Z_\alpha - A_\alpha B_\alpha = 0, \qquad \alpha = o, g. \tag{3.43}$$

The correct choice of root can be made as in the Peng–Robinson two-parameter EOS. In the present case, the EOS parameters Ω_{ia}, Ω_{ib}, and α_i are

$$\begin{aligned}&\Omega_{ia} = 0.42748, \quad \Omega_{ib} = 0.08664,\\ &\alpha_i = T/T_{ic}.\end{aligned}$$

All other quantities A_α, B_α, a_α, b_α, a_i, and b_i have the same definitions as in the Peng–Robinson EOS, $i = 1, 2, \ldots, N_c$, $\alpha = o, g$. The fugacity coefficient of component i in a mixture can be obtained from the equation

$$\begin{aligned}\ln \varphi_{i\alpha} = &\frac{b_i}{b_\alpha}(Z_\alpha - 1) - \ln(Z_\alpha - B_\alpha)\\ &- \frac{A_\alpha}{B_\alpha}\left(\frac{2}{a_\alpha}\sum_{j=1}^{N_c} x_{j\alpha}(1-\kappa_{ij})\sqrt{a_i a_j} - \frac{b_i}{b_\alpha}\right)\ln\left(\frac{Z_\alpha + B_\alpha}{Z_\alpha}\right).\end{aligned} \tag{3.44}$$

The Redlich–Kwong–Soave equation of state

The *Soave modification* of the Redlich–Kwong EOS defines the EOS parameter α_i as

$$\alpha_i = \left(1 + \lambda_i\left[1 - \sqrt{T/T_{ic}}\right]\right)^2, \qquad i = 1, 2, \ldots, N_c,$$

where $\lambda_i = 0.48 + 1.574\omega_i - 0.176\omega_i^2$ and ω_i is the acentric factor for component i. The definitions of all other quantities and of the fugacity coefficients are the same as in the Redlich–Kwong EOS. The Peng–Robinson and Redlich–Kwong–Soave EOSs have been extensively utilized in predicting enhanced oil recovery (EOR) phase behavior.

3.3 Temperature-Dependent Properties

3.3.1 Rock properties

The *rock properties* for nonisothermal flow are similar to those for the isothermal black oil and compositional models, but now these properties depend on temperature. In particular, the capillary pressures are of the form

$$p_{cow}(S_w, T) = p_o - p_w, \quad p_{cgo}(S_g, T) = p_g - p_o. \tag{3.45}$$

Analogously, the relative permeabilities for water, oil, and gas are

$$\begin{aligned} k_{rw} &= k_{rw}(S_w, T), & k_{row} &= k_{row}(S_w, T), \\ k_{rg} &= k_{rg}(S_g, T), & k_{rog} &= k_{rog}(S_g, T), \\ k_{ro} &= k_{ro}(S_w, S_g, T). \end{aligned} \tag{3.46}$$

Stone's models I and II defined in Section 3.1.2 can be adapted for the oil relative permeability k_{ro}, for example.

As an example, the relative permeability functions k_{rw} and k_{row} for a water-oil system can be defined by

$$\begin{aligned} k_{rw} &= k_{rwro}(T)\left(\frac{S_w - S_{wir}(T)}{1 - S_{orw}(T) - S_{wir}(T)}\right)^{nw}, \\ k_{row} &= k_{rocw}(T)\left(\frac{1 - S_w - S_{orw}(T)}{1 - S_{orw}(T) - S_{wc}(T)}\right)^{now}, \end{aligned} \tag{3.47}$$

and k_{rg} and k_{rog} for a gas-oil system by

$$\begin{aligned} k_{rg} &= k_{rgro}(T)\left(\frac{S_g - S_{gr}}{1 - S_{wc}(T) - S_{oinit} - S_{gr}}\right)^{ng}, \\ k_{rog} &= k_{rocw}(T)\left(\frac{1 - S_g - S_{wc}(T) - S_{org}(T)}{1 - S_{wc}(T) - S_{org}(T)}\right)^{nog}, \end{aligned} \tag{3.48}$$

where nw, now, ng, and nog are nonnegative real numbers; S_{wc}, S_{wir}, S_{orw}, S_{org}, and S_{gr} are the *connate water saturation*, *irreducible water saturation*, *residual oil saturation* in the water-oil system, residual oil saturation in the gas-oil system, and residual gas saturation; k_{rwro}, k_{rocw}, and k_{rgro} are the water relative permeability at the residual oil saturation for the water-oil system, the oil relative permeability at the connate water saturation, and the gas relative permeability at $S_g = 1 - S_{wc}(T) - S_{oinit}$ for the gas-oil system, respectively; and S_{oinit} is the initial oil saturation in the gas-oil system. Finally, for the rock properties, one must consider the *thermal conductivity* and *heat capacity* of the reservoir, overburden, and underburden.

3.3.2 Fluid properties

Water properties

Physical properties of water and steam, such as density, internal energy, enthalpy, and viscosity, can be found from a water-steam table (Lake, 1989). Such a table is given in terms of the independent variables: pressure and temperature. In the saturated state of a reservoir, there is free gas; in this case, pressure and temperature are related, and only one of them is employed as an independent variable.

Oil properties

While any number of hydrocarbon components can be treated in the differential system describing the nonisothermal multiphase, multicomponent flow developed in the preceding chapter, computational work and time significantly increase as the number of components increases. It is often computationally convenient (or necessary) to group several similar chemical components into one mathematical component. In this way, fewer components (or *pseudocomponents*) need be simulated in practical applications.

The oil phase is a mixture of hydrocarbon components, and these components range from the lightest component, methane (CH_4), to the heaviest component, bitumen. A way to reduce the number of components is to introduce pseudocomponents, as noted. According to the composition of each pseudocomponent, one can deduce its physical properties, such as its pseudomolecular weight (which may not be a constant), critical pressure and temperature, compressibility, density, viscosity, thermal expansion coefficient, and specific heat. These properties are functions of pressure and temperature.

The most important property is the oil and gas phase *viscosity dependence on temperature*:

$$\mu_{io} = \exp\left(a_1 T^{b_1}\right) + c_1, \quad \mu_{ig} = a_2 T^{b_2},$$

where T is in absolute degrees, a_1, b_1, c_1, a_2, and b_2 are empirical parameters that can be measured in the laboratory, and μ_{io} and μ_{ig} are the viscosities of the ith component in the oil and gas phases, respectively.

Equations of state

The EOSs defined in Section 3.2.5 can be also used to define the fugacity functions $f_{i\alpha}$ for nonisothermal flow, which now depend on temperature. Because of complexity of flow of this type, however, the equilibrium K-value approach introduced in Section 3.2.5 is more often used to describe the equilibrium relations:

$$x_{iw} = K_{iw}(p, T)x_{io}, \quad x_{ig} = K_{ig}(p, T)x_{io}, \quad i = 1, 2, \ldots, N_c. \tag{3.49}$$

One example for evaluating the K-values $K_{i\alpha}$ uses the empirical formula

$$K_{i\alpha} = \left(\kappa_{i\alpha}^1 + \frac{\kappa_{i\alpha}^2}{p} + \kappa_{i\alpha}^3\, p\right) \exp\left(-\frac{\kappa_{i\alpha}^4}{T - \kappa_{i\alpha}^5}\right), \tag{3.50}$$

where the constants $\kappa_{i\alpha}^j$ are obtained in the laboratory, $i = 1, 2, \ldots, N_c$, $j = 1, 2, 3, 4, 5$, $\alpha = w, g$.

3.4 Bibliographical Information

For more information on the water PVT properties, the reader should consult Numbere et al. (1977), Meehan (1980A; 1980B), and Craft and Hawkins (1991). For the oil PVT properties, the reader should refer to Chew and Connally (1959), Beggs and Robinson (1975), Standing (1977), and Vasquez and Beggs (1980). For the gas PVT properties, the reader should see Carr et al. (1954), Dempsey (1965), Wichert and Aziz (1972), and Dranchuk et al. (1974). Finally, more details on the equations of state can be found in Peng and Robinson (1976) and Coats (1980).

Exercises

3.1. A capillary pressure for an oil-water system is computed. Given the empirical formula

$$p_{cow}(S_w) = p_{cowmin} + B \ \ln\left(\frac{S_w - S_{wc} + \epsilon}{1 - S_{wc}}\right),$$

where we recall that S_{wc} is the connate water saturation, ϵ is a small positive number, and

$$p_{cowmax} = p_{cow}(S_{wc}), \qquad B = \frac{p_{cowmax} - p_{cowmin}}{\ln\left(\epsilon/(1 - S_{wc})\right)},$$

and given the input data

$$\epsilon = 0.01, \quad S_{wc} = 0.22, \quad p_{cowmin} = 0, \quad p_{cowmax} = 6.3 \text{ (psia)},$$

find the corresponding values of p_{cow} for these values of S_w: 0.22, 0.30, 0.40, 0.50, 0.60, 0.80, 0.90, and 1.00.

3.2. A capillary pressure for a gas-oil system is determined. Given the empirical formula

$$p_{cgo}(S_g) = p_{cgomin} + B \ \ln\left(\frac{1 - S_g - S_{or} - S_{wc} + \epsilon}{1 - S_{or} - S_{wc}}\right),$$

where we recall that S_{or} is the residual oil saturation and

$$p_{cgomax} = p_{cow}(1 - S_{or} - S_{wc}), \qquad B = \frac{p_{cgomax} - p_{cgomin}}{\ln\left(\epsilon/(1 - S_{or} - S_{wc})\right)},$$

and given the input data

$$\epsilon = 0.01, \quad S_{wc} = 0.22, \quad S_{or} = 0.18, \quad p_{cgomin} = 0, \quad p_{cgomax} = 3.9 \text{ (psia)},$$

find the corresponding values of p_{cgo} for these values of S_w: 0.00 0.04, 0.10, 0.20, 0.30, 0.40, 0.50, 0.60, 0.70, and 0.78.

3.3. A relative permeability for an oil-water system is calculated. Given the empirical formulas

$$k_{rw}(S_w) = k_{rwmax}\left(\frac{S_w - S_{wc}}{1 - S_{or} - S_{wc}}\right)^{nw},$$

$$k_{row}(S_w) = \left(\frac{1 - S_w - S_{or}}{1 - S_{or} - S_{wc}}\right)^{now}, \qquad S_{wc} \le S_w \le S_{wmax},$$

where $k_{rwmax} = k_{rw}(S_{wmax})$, $S_{or} = 1 - S_{wmax}$, and nw and now are positive numbers, and given the input data

$$S_{wc} = 0.4, \quad S_{or} = 0.2, \quad k_{rwmax} = 0.2, \quad nw = now = 2,$$

find the corresponding values of k_{rw} and k_{row} for these values of S_w: 0.40, 0.42, 0.44, 0.50, 0.60, 0.70, 0.76, 0.78, 0.80, and 1.00.

3.4. A relative permeability for a gas-oil system is evaluated. Given the empirical formulas

$$k_{rg}(S_g) = \left(\frac{S_g - S_{gr}}{1 - S_{gr} - S_{or} - S_{wc}}\right)^{ng},$$

$$k_{rog}(S_g) = \left(\frac{1 - S_g - S_{or} - S_{wc}}{1 - S_{or} - S_{wc}}\right)^{nog}, \qquad S_{gr} \le S_g \le 1 - S_{or} - S_{wc},$$

where S_{gr} is the critical mobile gas saturation and ng and nog are positive numbers, and given the input data

$$S_{wc} = 0.4, \quad S_{or} = 0.2, \quad S_{gr} = 0.02, \quad ng = 0.83, \quad nog = 7.5,$$

find the corresponding values of k_{rg} and k_{rog} for these values of S_g: 0.020, 0.039, 0.058, 0.115, 0.172, 0.210, 0.286, 0.400, and 0.600.

3.5. Derive equation (3.17) from equation (3.16).

3.6. The water PVT properties for the black oil model are calculated. The given data are

Water salinity = 100,000 ppm ($S_{ALI} = 100{,}000/10{,}000 = 10$),

Gas saturation state: saturated,

Formation temperature $T_F = 250°$ F,

$T_C = (250 - 32)/1.8 = 121.11°$ C,

$T_K = 273.15 + T_C = 349.26$ K,

Formation pressure $p = 5{,}000$ psia.

Calculate (1) the water density at standard conditions ρ_{Ws} in g/cm^3 and lbm/SCF (g/cm^3 = 0.016018463 lbm/SCF); (2) the water formation volume factor B_w (RB/STB); (3) the water compressibility c_w (1/psi); and (4) the water viscosity μ_w (cp).

3.7. The oil PVT properties for the black oil model are computed. The given data are

Formation pressure $p = 6{,}000,\ 5{,}004.2,\ 3{,}000,\ 2{,}000,\ 1{,}000$ psia,
Formation temperature $T_F = 250°$ F,
$T_R = T_F + 460 = 710$ R,
Produced gas-oil ratio $G_{OR} = 1{,}000$ SCF/STB,
Oil gravity $A_{PI} = 40°$ API,
Raw gas gravity $Y_G = 0.6$,
Pressure at separator conditions $p_{sep} = 100$ psia,
Temperature at separator conditions $T_{sep} = 85°$ F.

(A) Compute the bubble point pressure.
(B) Evaluate the oil PVT properties: (1) the dissolved gas-oil ratio R_{so} (SCF/STB), (2) the oil viscosity μ_o (cp), (3) the oil compressibility c_o (1/psi), (4) the oil viscosity compressibility c_μ (1/psi), and (5) the oil formation volume factor B_o (RB/STB) vs. the given formation pressures.

3.8. The gas PVT properties for the black oil model are evaluated. The given data are

Formation pressure $p = 7{,}500$ psia,
Formation temperature $T_F = 250°$ F,
$T_R = T_F + 460 = 710$ R,
Raw gas gravity $Y_G = 0.6$,
Content of CO_2, H_2S, and $N_2 = 0.0$,
Gas is a condensate.

Calculate (1) the reduced pressure p_{red} and reduced temperature T_{red}, (2) the gas deviation factor Z, (3) the gas formation volume factor B_g (RB/SCF), and (4) the gas viscosity μ_g (cp).

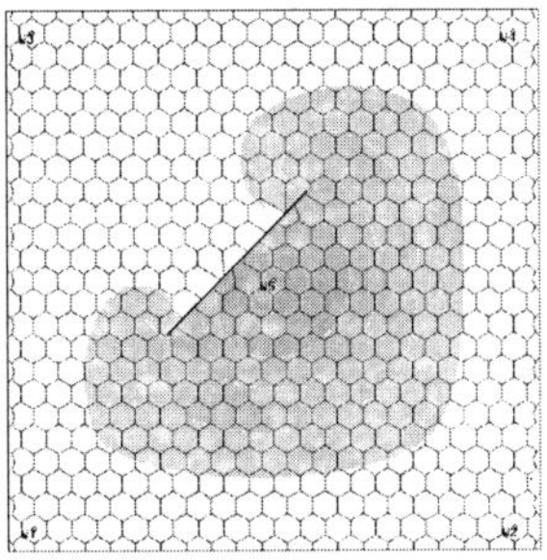

Chapter 4
Numerical Methods

A *numerical method* for solving a differential equation problem involves discretizing this problem, which has infinitely many degrees of freedom, to produce a *discrete problem*, which has finitely many degrees of freedom and can be solved using a computer. Compared with *finite difference methods*, the introduction of *finite element methods* is relatively recent. The advantages of finite elements over finite differences are that general boundary conditions, complex geometry, and variable material properties can be relatively easily handled. Also, the clear structure and versatility of the finite element methods makes it possible to develop general purpose software for applications. Furthermore, there is a solid theoretical foundation that gives added confidence, and in many cases it is possible to obtain concrete error estimates for the finite element solutions. Finite element methods were first introduced by Courant in 1943. From the 1950s to the 1970s, they were developed by engineers and mathematicians into a general method for the numerical solution of partial differential equations.

When applied to petroleum reservoir simulation, finite element methods have some peculiar features, such as in the reduction of grid orientation effects, in the treatment of local grid refinement, horizontal and slanted wells, and corner point techniques, in the simulation of faults and fractures, in the design of streamlines, and in the requirement of high-order accuracy of numerical solutions. These topics will be studied in detail in subsequent chapters.

Because we compare finite difference solutions with finite element solutions, we very briefly review finite difference methods in Section 4.1. The books by Peaceman (1977A; 1977B) and Aziz and Settari (1979) gave detailed information on the use of these methods in reservoir simulation. We concentrate on the finite element methods that have been employed in reservoir simulation in the past two decades. Six major types of finite element methods are covered: *standard* (Section 4.2), *control volume* (Section 4.3), *discontinuous* (Section 4.4), *mixed* (Section 4.5), *characteristic* (Section 4.6), and *adaptive* (Section 4.7). For each method, a brief introduction, the notation, basic terminology, and necessary concepts are given. Except for the control volume methods, these methods are taken from the book by one of the authors (Chen, 2005); for a more detailed description of the methods and their theoretical results, the reader should refer to that book. Some of more recent methods such as *multiscale*, *particle*, and *mesh-free* are not presented here. Many different gridding techniques are presented in this chapter. The eighth comparative solution project (CSP) organized by the society of petroleum engineers (SPE) is presented to compare some of these gridding techniques in Section 4.7. Finally, bibliographical information is given in Section 4.8.

4.1 Finite Difference Methods

4.1.1 First difference quotients

We describe *first* and *second difference quotients* for functions of two space variables, x_1 and x_2, and of time, t. Reduction to functions of one space variable and extension to functions of three space variables are straightforward.

Consider a function $p(x_1, x_2, t)$ of x_1, x_2, and t. The first partial derivative of p with respect to x_1 can be defined in one of the following ways:

$$\begin{aligned}
\frac{\partial p}{\partial x_1}(x_1, x_2, t) &= \lim_{h_1 \to 0} \frac{p(x_1 + h_1, x_2, t) - p(x_1, x_2, t)}{h_1}, \\
\frac{\partial p}{\partial x_1}(x_1, x_2, t) &= \lim_{h_1 \to 0} \frac{p(x_1, x_2, t) - p(x_1 - h_1, x_2, t)}{h_1}, \\
\frac{\partial p}{\partial x_1}(x_1, x_2, t) &= \lim_{h_1 \to 0} \frac{p(x_1 + h_1, x_2, t) - p(x_1 - h_1, x_2, t)}{2h_1}.
\end{aligned}$$

We replace this derivative by a difference quotient. For this, we utilize the *Taylor series expansion*

$$p(x_1 + h_1, x_2, t) = p(x_1, x_2, t) + \frac{\partial p}{\partial x_1}(x_1, x_2, t)h_1 + \frac{\partial^2 p}{\partial x_1^2}(x_1^\star, x_2, t)\frac{h_1^2}{2},$$

where $x_1 \le x_1^\star \le x_1 + h_1$ and $h_1 > 0$ is a fixed number. The last term in this equation is a *remainder* that involves a second partial derivative of p. Then $\partial p/\partial x_1$ can be obtained from

$$\frac{\partial p}{\partial x_1}(x_1, x_2, t) = \frac{p(x_1 + h_1, x_2, t) - p(x_1, x_2, t)}{h_1} - \frac{\partial^2 p}{\partial x_1^2}(x_1^\star, x_2, t)\frac{h_1}{2}. \tag{4.1}$$

The expression

$$\frac{p(x_1 + h_1, x_2, t) - p(x_1, x_2, t)}{h_1}$$

is referred to as a *forward difference quotient*, and it approximates the derivative $\partial p/\partial x_1$ with an *error* of the first order in h_1.

Similarly, we have

$$\frac{\partial p}{\partial x_1}(x_1, x_2, t) = \frac{p(x_1, x_2, t) - p(x_1 - h_1, x_2, t)}{h_1} - \frac{\partial^2 p}{\partial x_1^2}(x_1^{\star\star}, x_2, t)\frac{h_1}{2}, \tag{4.2}$$

where $x_1 - h_1 \le x_1^{\star\star} \le x_1$, and the quantity

$$\frac{p(x_1, x_2, t) - p(x_1 - h_1, x_2, t)}{h_1}$$

is called a *backward difference quotient*. This quantity also gives a first-order approximation to $\partial p/\partial x_1$.

Next, we use the Taylor series expansions with remainders involving a third partial derivative of p:

$$\begin{aligned} p(x_1+h_1,x_2,t) &= p(x_1,x_2,t) + \frac{\partial p}{\partial x_1}(x_1,x_2,t)h_1 \\ &\quad + \frac{\partial^2 p}{\partial x_1^2}(x_1,x_2,t)\frac{h_1^2}{2!} + \frac{\partial^3 p}{\partial x_1^3}(x_1^\star,x_2,t)\frac{h_1^3}{3!}, \\ p(x_1-h_1,x_2,t) &= p(x_1,x_2,t) - \frac{\partial p}{\partial x_1}(x_1,x_2,t)h_1 \\ &\quad + \frac{\partial^2 p}{\partial x_1^2}(x_1,x_2,t)\frac{h_1^2}{2!} - \frac{\partial^3 p}{\partial x_1^3}(x_1^{\star\star},x_2,t)\frac{h_1^3}{3!}, \end{aligned}$$

where $x_1 \le x_1^\star \le x_1 + h_1$ and $x_1 - h_1 \le x_1^{\star\star} \le x_1$. Subtracting these two equations and solving for $\partial p/\partial x_1$ yields

$$\begin{aligned} \frac{\partial p}{\partial x_1}(x_1,x_2,t) &= \frac{p(x_1+h_1,x_2,t) - p(x_1-h_1,x_2,t)}{2h_1} \\ &\quad - \left(\frac{\partial^3 p}{\partial x_1^3}(x_1^\star,x_2,t) + \frac{\partial^3 p}{\partial x_1^3}(x_1^{\star\star},x_2,t)\right)\frac{h_1^2}{12}. \end{aligned} \tag{4.3}$$

The quotient

$$\frac{p(x_1+h_1,x_2,t) - p(x_1-h_1,x_2,t)}{2h_1}$$

is termed a *centered difference quotient*, and it approximates $\partial p/\partial x_1$ with a higher order, i.e., second order in h_1.

From (4.1), (4.2), and (4.3), it would appear preferable to employ the centered difference approximation to $\partial p/\partial x_1$. This is not always the case. Which quotient is used depends on the particular problem (see Section 4.1.8).

It is sometimes necessary to use a difference quotient to approximate $\partial p/\partial x_1$ computed halfway between x_1 and $x_1 + h_1$. Analogously to (4.3), we can obtain

$$\begin{aligned} \frac{\partial p}{\partial x_1}\left(x_1 + \frac{h_1}{2}, x_2, t\right) &= \frac{p(x_1+h_1,x_2,t) - p(x_1,x_2,t)}{h_1} \\ &\quad - \left(\frac{\partial^3 p}{\partial x_1^3}(x_1^\star,x_2,t) + \frac{\partial^3 p}{\partial x_1^3}(x_1^{\star\star},x_2,t)\right)\frac{h_1^2}{48}, \end{aligned} \tag{4.4}$$

where $x_1 \le x_1^\star,\ x_1^{\star\star} \le x_1 + h_1$. In summary, we have defined three first difference quotients in x_1. The same quotients can be introduced in x_2 and t.

4.1.2 Second difference quotients

We exploit the Taylor series expansions with remainders involving a fourth partial derivative of p:

$$
\begin{aligned}
p(x_1 + h_1, x_2, t) &= p(x_1, x_2, t) + \frac{\partial p}{\partial x_1}(x_1, x_2, t)h_1 \\
&+ \frac{\partial^2 p}{\partial x_1^2}(x_1, x_2, t)\frac{h_1^2}{2!} + \frac{\partial^3 p}{\partial x_1^3}(x_1, x_2, t)\frac{h_1^3}{3!} + \frac{\partial^4 p}{\partial x_1^4}(x_1^\star, x_2, t)\frac{h_1^4}{4!}, \\
p(x_1 - h_1, x_2, t) &= p(x_1, x_2, t) - \frac{\partial p}{\partial x_1}(x_1, x_2, t)h_1 \\
&+ \frac{\partial^2 p}{\partial x_1^2}(x_1, x_2, t)\frac{h_1^2}{2!} - \frac{\partial^3 p}{\partial x_1^3}(x_1, x_2, t)\frac{h_1^3}{3!} + \frac{\partial^4 p}{\partial x_1^4}(x_1^{\star\star}, x_2, t)\frac{h_1^4}{4!},
\end{aligned}
$$

where $x_1 \leq x_1^\star \leq x_1 + h_1$ and $x_1 - h_1 \leq x_1^{\star\star} \leq x_1$. Adding these two equations and solving for $\partial^2 p/\partial x_1^2$ yields

$$
\begin{aligned}
\frac{\partial^2 p}{\partial x_1^2}(x_1, x_2, t) &= \frac{p(x_1 + h_1, x_2, t) - 2p(x_1, x_2, t) + p(x_1 - h_1, x_2, t)}{h_1^2} \\
&- \left(\frac{\partial^4 p}{\partial x_1^4}(x_1^\star, x_2, t) + \frac{\partial^4 p}{\partial x_1^4}(x_1^{\star\star}, x_2, t)\right)\frac{h_1^2}{24}.
\end{aligned}
\tag{4.5}
$$

The expression

$$
\Delta_{x_1}^2 p(x_1, x_2, t) = \frac{p(x_1 + h_1, x_2, t) - 2p(x_1, x_2, t) + p(x_1 - h_1, x_2, t)}{h_1^2} \tag{4.6}
$$

defines a *centered second difference quotient*, which approximates the partial derivative $\partial^2 p/\partial x_1^2$ with a second-order accuracy in h_1.

Equation (4.5) is derived with the left and right intervals at x_1 of equal length. We now consider p on the intervals $(x_1 - h_1', x_1)$ and $(x_1, x_1 + h_1'')$, where h_1' and h_1'' are not necessarily the same, and introduce a difference quotient for the second derivative

$$
\frac{\partial}{\partial x_1}\left(a(x_1, x_2, t)\frac{\partial p}{\partial x_1}\right),
$$

where a is a given function. Using Taylor series expansions as above, the following approximations hold:

$$
\begin{aligned}
&\left(a\frac{\partial p}{\partial x_1}\right)\left(x_1 - \frac{h_1'}{2}, x_2, t\right) \\
&\qquad \approx a\left(x_1 - \frac{h_1'}{2}, x_2, t\right)\frac{p(x_1, x_2, t) - p(x_1 - h_1', x_2, t)}{h_1'}, \\
&\left(a\frac{\partial p}{\partial x_1}\right)\left(x_1 + \frac{h_1''}{2}, x_2, t\right) \\
&\qquad \approx a\left(x_1 + \frac{h_1''}{2}, x_2, t\right)\frac{p(x_1 + h_1'', x_2, t) - p(x_1, x_2, t)}{h_1''}.
\end{aligned}
\tag{4.7}
$$

Note that

$$\frac{\partial}{\partial x_1}\left(a\frac{\partial p}{\partial x_1}\right)(x_1,x_2,t) \approx \left\{\left(a\frac{\partial p}{\partial x_1}\right)\left(x_1+\frac{h_1''}{2},x_2,t\right) - \left(a\frac{\partial p}{\partial x_1}\right)\left(x_1-\frac{h_1'}{2},x_2,t\right)\right\} \Big/ \left(\left(x_1+\frac{h_1''}{2}\right)-\left(x_1-\frac{h_1'}{2}\right)\right).$$

Consequently, using (4.7), we see that

$$\begin{aligned}
&\frac{\partial}{\partial x_1}\left(a\frac{\partial p}{\partial x_1}\right)(x_1,x_2,t)\\
&\approx \left\{a\left(x_1+\frac{h_1''}{2},x_2,t\right)\frac{p(x_1+h_1'',x_2,t)-p(x_1,x_2,t)}{h_1''}\right.\\
&\quad \left. -a\left(x_1-\frac{h_1'}{2},x_2,t\right)\frac{p(x_1,x_2,t)-p(x_1-h_1',x_2,t)}{h_1'}\right\}\Big/\frac{h_1'+h''}{2},
\end{aligned}$$

which we write as

$$\Delta_{x_1}(a\Delta_{x_1}p). \tag{4.8}$$

This approximation to $\frac{\partial}{\partial x_1}\left(a\frac{\partial p}{\partial x_1}\right)$ is of second order in h_1, where $h_1 = \max\{h_1', h_1''\}$. A similar definition can be given for $\Delta_{x_2}(a\Delta_{x_2}p)$.

4.1.3 Grid systems

There are two types of *grid systems* commonly employed in reservoir simulation, *block-centered* and *point-distributed grids*. Let the integer i indicate the index in the x_1-direction, and the integer j denote the index in the x_2-direction. Furthermore, let $x_{1,i}$ and $x_{2,j}$ represent the ith and jth values of x_1 and x_2, respectively. Then we set

$$p_{ij} = p(x_{1,i}, x_{2,j}).$$

Block-centered grid

A rectangular solution domain Ω is divided into rectangles, and the point $(x_{1,i}, x_{2,j})$ is at the center of the rectangle (i, j), as in Figure 4.1. The left side of the rectangle is at $x_{1,i-\frac{1}{2}}$, and the right side is at $x_{1,i+\frac{1}{2}}$. Similarly, $x_{2,j-\frac{1}{2}}$ and $x_{2,j+\frac{1}{2}}$ are the bottom and top sides of the rectangle (i, j). This type of grid is called a *block-centered grid*. It is specified by the sequences $0 = x_{1,\frac{1}{2}} < x_{1,\frac{3}{2}} < \cdots$ and $0 = x_{2,\frac{1}{2}} < x_{2,\frac{3}{2}} < \cdots$ if $\Omega = (0,1)^2$ is the unit square, for example. Also, we see that

$$\begin{aligned}
x_{1,i} &= \frac{1}{2}\left(x_{1,i-\frac{1}{2}} + x_{1,i+\frac{1}{2}}\right),\\
h_{1,i} &= x_{1,i+\frac{1}{2}} - x_{1,i-\frac{1}{2}},\\
h_{1,i-\frac{1}{2}} &= x_{1,i} - x_{1,i-1}.
\end{aligned}$$

Similar notation can be given for the x_2 variable.

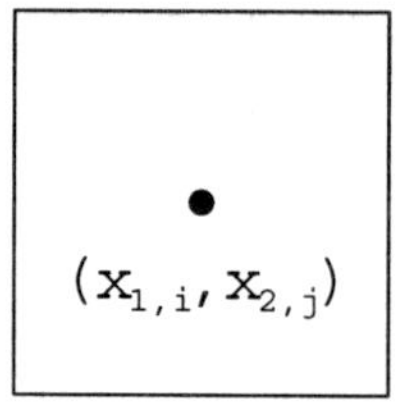

Figure 4.1. *A block-centered grid.*

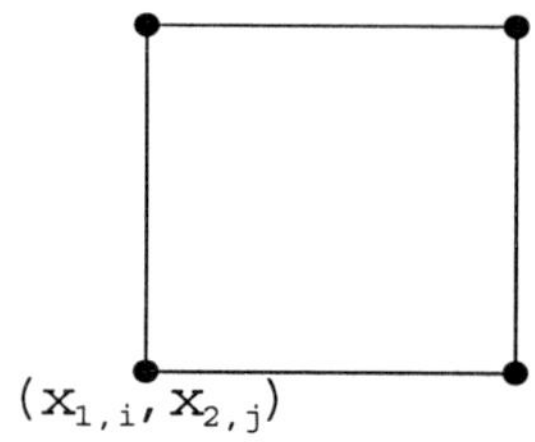

Figure 4.2. *A point-distributed grid.*

Point-distributed grid

In the other type of grid, the point $(x_{1,i}, x_{2,j})$ is now a vertex of a rectangle, as in Figure 4.2. This grid is referred to as a *point-distributed grid.* In this case, the grid is specified by the sequences $0 = x_{1,0} < x_{1,1} < \cdots$ and $0 = x_{2,0} < x_{2,1} < \cdots$ for $\Omega = (0, 1)^2$. Also, note that

$$x_{1,i-\frac{1}{2}} = \frac{1}{2}\left(x_{1,i-1} + x_{1,i}\right),$$
$$h_{1,i} = x_{1,i} - x_{1,i-1}.$$

4.1.4 Treatment of boundary conditions

As we will see, the difference equations written for the two grid systems are the same in form. There are, however, significant differences between them. Specifically, when the grids are not uniform, the locations of points and block boundaries do not coincide. Also, the treatment of boundary conditions is different. Here we introduce difference equations to approximate the boundary conditions described in Section 2.2.9.

Boundary conditions of the first kind

Suppose that we are given the following boundary condition at $x_1 = 0$:

$$p(0, x_2, t) = g(x_2, t). \tag{4.9}$$

This is a boundary condition of the *first kind*, i.e., the Dirichlet kind. In reservoir simulation, Dirichlet boundary conditions arise when pressure on the reservoir boundary or at a well is

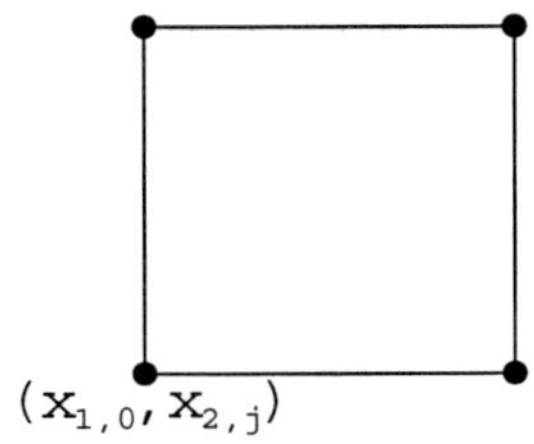

Figure 4.3. *The Dirichlet boundary condition for a point-distributed grid.*

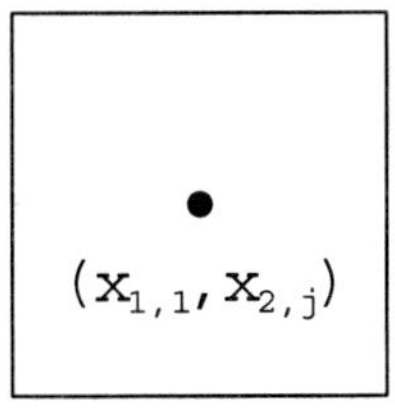

Figure 4.4. *The Dirichlet boundary condition for a block-centered grid.*

specified. For a point-distributed grid (cf. Figure 4.3), this boundary condition is given by

$$p_{0j}^n = g_j^n. \tag{4.10}$$

Equation (4.10) is utilized whenever p_{0j}^n is required in a difference equation.

For a block-centered grid, the closet point to the boundary is $(x_{1,1}, x_{2,j})$ (cf. Figure 4.4). The value of p_{1j}^n must be *extrapolated* to this point. The simplest approach is

$$p_{1j}^n = g_j^n, \tag{4.11}$$

which is only of first-order accuracy in space. A second-order approximation uses

$$\frac{1}{2}\left(3p_{1j}^n - p_{2j}^n\right) = g_j^n. \tag{4.12}$$

Note that (4.12) must be included in the system of difference equations to be solved. For this reason, the block-centered grid is sometimes modified by use of half blocks at Dirichlet boundaries (cf. Figure 4.5).

Boundary conditions of the second kind

Consider the following boundary condition at $x_1 = 0$:

$$\frac{\partial p}{\partial x_1}(0, x_2, t) = g(x_2, t). \tag{4.13}$$

This is a boundary condition of the *second kind*, i.e., the Neumann kind, and can be used to express a flow rate across a boundary or to specify an injection or production rate at a well.

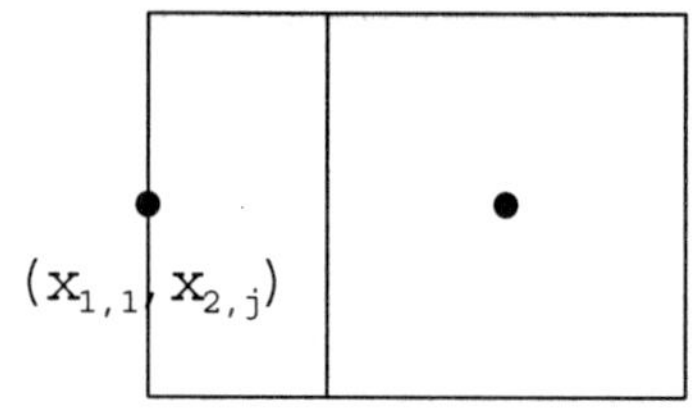

Figure 4.5. *The use of half blocks at the Dirichlet boundary.*

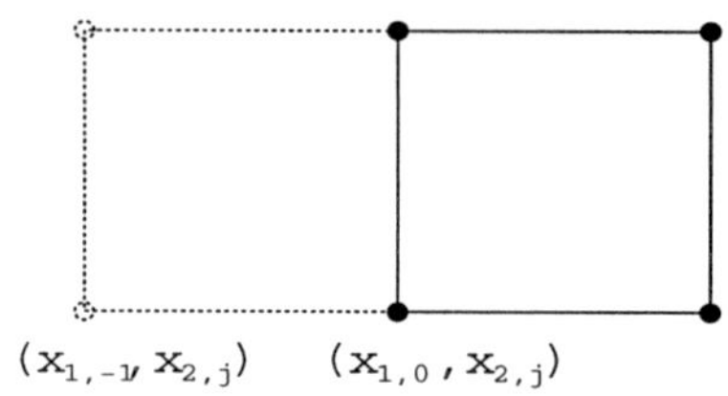

Figure 4.6. *A reflection point for a point-distributed grid.*

For a point-distributed grid, (4.13) can be approximated by

$$\frac{p_{1j}^n - p_{0j}^n}{h_{1,1}} = g_j^n, \tag{4.14}$$

which is a first-order approximation. A second-order accurate scheme uses a *reflection* (*ghost*) point; for each j, we introduce an auxiliary point $(x_{1,-1}, x_{2,j})$ (cf. Figure 4.6). The boundary condition (4.13) is discretized using the centered difference at $x_1 = 0$:

$$\frac{p_{1j}^n - p_{-1j}^n}{2h_{1,1}} = g_j^n. \tag{4.15}$$

Equation (4.15) is exploited to eliminate p_{-1j}^n from the difference equation for the differential equation at $x_1 = 0$. The first- and second-order approximations for (4.13) can be also defined for a block-centered grid, using a modification similar to that for (4.9).

Boundary conditions of the third kind

A boundary condition of the *third kind* has the form

$$\left(a\frac{\partial p}{\partial x_1} + bp\right)(0, x_2, t) = g(x_2, t), \tag{4.16}$$

where the functions a and b are given. As noted in the preceding chapter, such a condition occurs when part of the external boundary is semipervious. For a point-distributed grid, this equation can be approximated by

$$a_{0j}^n \frac{p_{1j}^n - p_{-1j}^n}{2h_{1,1}} + b_{0j}^n p_{0j}^n = g_j^n, \tag{4.17}$$

where we recall that $(x_{1,-1}, x_{2,j})$ is a reflection point. It is difficult to approximate (4.16) for a block-centered grid.

4.1.5 Finite differences for stationary problems

We consider the *stationary problem* in two space dimensions on a rectangular domain Ω:

$$-\nabla \cdot (a\nabla p) = f(x_1, x_2), \qquad (x_1, x_2) \in \Omega, \tag{4.18}$$

where the functions a and f are given. Function a is assumed to be positive on Ω. A pressure equation for incompressible flow is stationary, for example. As pointed out earlier, there are two types of grids widely used in reservoir simulation; the difference equations are the same in form for both grids. Equation (4.18) at grid point (i, j) can be approximated by

$$\begin{aligned}
&-\frac{a_{i+\frac{1}{2},j}\dfrac{p_{i+1,j}-p_{i,j}}{h_{1,i+\frac{1}{2}}} - a_{i-\frac{1}{2},j}\dfrac{p_{i,j}-p_{i-1,j}}{h_{1,i-\frac{1}{2}}}}{h_{1,i}} \\
&-\frac{a_{i,j+\frac{1}{2}}\dfrac{p_{i,j+1}-p_{i,j}}{h_{2,j+\frac{1}{2}}} - a_{i,j-\frac{1}{2}}\dfrac{p_{i,j}-p_{i,j-1}}{h_{2,j-\frac{1}{2}}}}{h_{2,j}} = f_{ij},
\end{aligned} \tag{4.19}$$

where $p_{ij} = p(x_{1,i}, x_{2,j})$ and $a_{i+\frac{1}{2},j} = a(x_{1,i+\frac{1}{2}}, x_{2,j})$. If we define

$$a_{1,i+\frac{1}{2},j} = \frac{a_{i+\frac{1}{2},j}h_{2,j}}{h_{1,i+\frac{1}{2}}},$$
$$a_{2,i,j+\frac{1}{2}} = \frac{a_{i,j+\frac{1}{2}}h_{1,i}}{h_{2,j+\frac{1}{2}}},$$

(4.19) can be then written as

$$\begin{aligned}
&-a_{1,i+\frac{1}{2},j}(p_{i+1,j}-p_{i,j}) + a_{1,i-\frac{1}{2},j}(p_{i,j}-p_{i-1,j}) \\
&\quad - a_{2,i,j+\frac{1}{2}}(p_{i,j+1}-p_{i,j}) + a_{2,i,j-\frac{1}{2}}(p_{i,j}-p_{i,j-1}) = F_{ij},
\end{aligned} \tag{4.20}$$

where $F_{ij} = f_{ij}h_{1,i}h_{2,j}$. F_{ij} may be interpreted as the integral of $f(x_1, x_2)$ over a rectangle with area $h_{1,i}h_{2,j}$. The *truncation error* is the error incurred by replacing a differential equation by a difference equation. From the discussion in Section 4.1.2, the truncation error in the approximation of the difference scheme (4.20) to (4.18) is of second order in both h_1 and h_2. This scheme is the commonly used *five-point stencil* scheme for two-dimensional problems (cf. Figure 4.7). For some points near or on the boundary of the solution domain, it involves one or two fictitious points outside the domain. The values of p at these points are eliminated, depending on which type of grid and boundary condition is employed. Equation (4.20) can be written in matrix form involving unknowns $\{p_{i,j}\}$, and must be solved via a direct or iterative algorithm; see the next chapter. An example is given in Exercise 4.1.

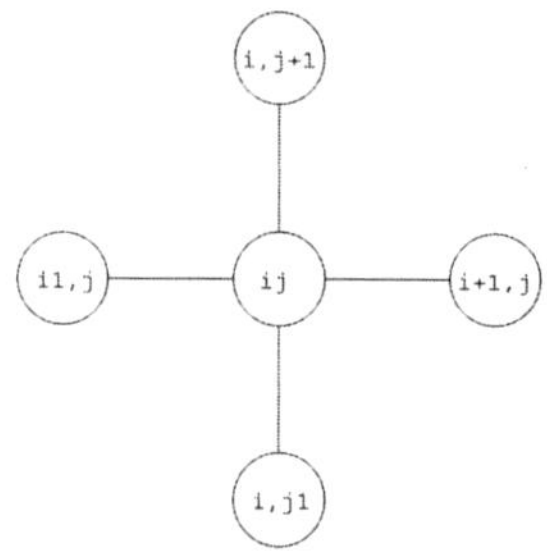

Figure 4.7. *A five-point stencil scheme.*

4.1.6 Finite differences for parabolic problems

We turn to the *transient* (*parabolic problem*) in two dimensions on a rectangular domain Ω:

$$\phi\frac{\partial p}{\partial t} - \nabla \cdot (a\nabla p) = f(x_1, x_2, t), \qquad (x_1, x_2) \in \Omega,\ t > 0, \tag{4.21}$$

where a, f, and ϕ are given functions of x_1, x_2, and t. Functions a and ϕ are assumed to be positive and nonnegative on Ω, respectively. From the preceding chapter, a pressure equation for compressible flow is parabolic. For a parabolic problem, in addition to a boundary condition, an initial condition is also needed:

$$p(x_1, x_2, 0) = p_0(x_1, x_2).$$

Let $\{t^n\}$ be a sequence of real numbers such that

$$0 = t^0 < t^1 < \cdots < t^n < t^{n+1} < \cdots.$$

For the transient problem, we proceed from the initial solution at t^0 to a solution at t^1; in general, we obtain a solution at t^{n+1} from solutions at the previous time levels. Thus the solution procedure advances through time. Set

$$\Delta t^n = t^{n+1} - t^n, \qquad n = 1, 2, \ldots,$$

and

$$p_{ij}^n = p(x_{1,i}, x_{2,j}, t^n).$$

Forward difference scheme

The simplest difference scheme for (4.21) is to replace the second partial derivatives in space by a second difference at t^n and $\partial p/\partial t$ by a forward difference. The resulting scheme is a centered second difference in space and a forward difference in time, and is called the *forward difference* scheme (or *forward Euler* scheme):

$$\begin{aligned}\phi_{ij}^n \frac{p_{i,j}^{n+1} - p_{i,j}^n}{\Delta t^n} h_{1,i}h_{2,j} - a_{1,i+\frac{1}{2},j}^n(p_{i+1,j}^n - p_{i,j}^n) + a_{1,i-\frac{1}{2},j}^n(p_{i,j}^n - p_{i-1,j}^n) \\ - a_{2,i,j+\frac{1}{2}}^n(p_{i,j+1}^n - p_{i,j}^n) + a_{2,i,j-\frac{1}{2}}^n(p_{i,j}^n - p_{i,j-1}) = F_{ij}^n\end{aligned} \tag{4.22}$$

for $n = 0, 1, 2, \ldots$. Note that this equation can be solved *explicitly* for $p_{i,j}^{n+1}$. The use of an explicit scheme brings about a *stability* problem. For $a = \phi = 1$ and $f = 0$, for example, a stability analysis (cf. Section 4.1.7) shows that the time and space step sizes must satisfy the condition

$$\Delta t \left(\frac{1}{h_1^2} + \frac{1}{h_2^2} \right) \leq \frac{1}{2} \tag{4.23}$$

to obtain stability, where $\Delta t = \max\{\Delta t^n : n = 0, 1, \ldots\}$. Hence the forward difference scheme is *conditionally stable*.

Backward difference scheme

The stability condition (4.23) on the time steps is inherent in the forward difference scheme, and can be removed by evaluating the second partial derivatives at t^{n+1}:

$$\begin{aligned} &\phi_{ij}^{n+1} \frac{p_{i,j}^{n+1} - p_{i,j}^{n}}{\Delta t^n} h_{1,i} h_{2,j} \\ &\quad - a_{1,i+\frac{1}{2},j}^{n+1} (p_{i+1,j}^{n+1} - p_{i,j}^{n+1}) + a_{1,i-\frac{1}{2},j}^{n+1} (p_{i,j}^{n+1} - p_{i-1,j}^{n+1}) \\ &\quad - a_{2,i,j+\frac{1}{2}}^{n+1} (p_{i,j+1}^{n+1} - p_{i,j}^{n+1}) + a_{2,i,j-\frac{1}{2}}^{n+1} (p_{i,j}^{n+1} - p_{i,j-1}^{n+1}) = F_{ij}^{n+1}. \end{aligned} \tag{4.24}$$

As we go from n to $n+1$, (4.24) defines $p_{i,j}^{n+1}$ *implicitly* and is termed the *backward difference* (or *backward Euler*) scheme. At each time level t^{n+1}, a linear system of algebraic equations must be solved. This system has the same form as that arising from the stationary problem. A stability analysis indicates that scheme (4.24) is *unconditionally stable*; that is, there is no restriction on the time step Δt that can be used (cf. Section 4.1.7).

The truncation errors for both the forward and backward difference schemes are of second order in h_1 and h_2 and of first order in Δt. To improve accuracy in time, the Crank–Nicholson difference scheme can be exploited, for example.

Crank–Nicholson difference scheme

Another implicit difference scheme for (4.21) is to replace the average

$$\left(\partial p(t^{n+1})/\partial t + \partial p(t^n)/\partial t \right)/2$$

by the difference quotient $(p^{n+1} - p^n)/\Delta t^n$:

$$\begin{aligned} &\phi_{ij}^{n+1} \frac{p_{i,j}^{n+1} - p_{i,j}^{n}}{\Delta t^n} h_{1,i} h_{2,j} \\ &\quad - \frac{1}{2} \Big\{ a_{1,i+\frac{1}{2},j}^{n+1} (p_{i+1,j}^{n+1} - p_{i,j}^{n+1}) - a_{1,i-\frac{1}{2},j}^{n+1} (p_{i,j}^{n+1} - p_{i-1,j}^{n+1}) \\ &\qquad + a_{2,i,j+\frac{1}{2}}^{n+1} (p_{i,j+1}^{n+1} - p_{i,j}^{n+1}) - a_{2,i,j-\frac{1}{2}}^{n+1} (p_{i,j}^{n+1} - p_{i,j-1}^{n+1}) \\ &\qquad + a_{1,i+\frac{1}{2},j}^{n} (p_{i+1,j}^{n} - p_{i,j}^{n}) - a_{1,i-\frac{1}{2},j}^{n} (p_{i,j}^{n} - p_{i-1,j}^{n}) \\ &\qquad + a_{2,i,j+\frac{1}{2}}^{n} (p_{i,j+1}^{n} - p_{i,j}^{n}) - a_{2,i,j-\frac{1}{2}}^{n} (p_{i,j}^{n} - p_{i,j-1}^{n}) \Big\} \\ &= \frac{1}{2} \left(F_{ij}^{n+1} + F_{ij}^{n} \right). \end{aligned} \tag{4.25}$$

The truncation error for this scheme is of second order in h_1, h_2, and Δt. This implicit scheme is also unconditionally stable. Moreover, it gives rise to a system of simultaneous equations that is of the same form as that arising from the backward difference scheme.

4.1.7 Consistency, stability, and convergence

We give the basic definitions of *consistency*, *stability*, and *convergence* of a finite difference scheme. We concentrate on pure initial value problems. When boundary conditions are included, the definitions must be extended to initial boundary value problems (Thomas, 1995). Furthermore, we focus on one-dimensional transient problems, and the solution domain is the entire x_1-axis; i.e., $-\infty < x_1 < \infty$. Let $x_{1,i} = ih$, $i = 0, \pm 1, \pm 2, \ldots$, and $t^n = n\Delta t$, $n = 0, 1, 2, \ldots$.

Consistency

For two real numbers ϵ and $h > 0$, we write

$$\epsilon = \mathcal{O}(h)$$

if there is a positive constant C such that

$$|\epsilon| \leq Ch.$$

A finite difference scheme $L_i^n P_i^n = G_i^n$ is (pointwise) *consistent* with the partial differential equation $\mathcal{L}p = \mathcal{F}$ at point (x, t) if for any smooth function $v = v(x, t)$,

$$R_i^n \equiv \left(\mathcal{L}v - \mathcal{F}\right)|_i^n - \left\{L_i^n v(ih, n\Delta t) - G_i^n\right\}\right) \to 0 \tag{4.26}$$

as h, $\Delta t \to 0$ and $(ih, n\Delta t) \to (x, t)$. Note that the truncation errors for the forward difference scheme (4.22) and the backward difference scheme (4.24) take the form

$$R_i^n = \mathcal{O}(h^2) + \mathcal{O}(\Delta t),$$

whereas the truncation error for the Crank–Nicholson scheme (4.25) has the form

$$R_i^n = \mathcal{O}(h^2) + \mathcal{O}\left((\Delta t)^2\right).$$

Hence these schemes are consistent with (4.21) (cf. Exercise 4.3).

Stability

A finite difference scheme is *stable* if the effect of an error (or perturbation) made in any stage of computation is not propagated into larger errors in later stages of the computation, i.e., if local errors are not magnified by further computation. A difference scheme can be examined for stability by substituting into it *perturbed values* of the solution.

We consider the one-dimensional version of (4.21) (with $x = x_1$):

$$\frac{\partial p}{\partial t} = \frac{\partial^2 p}{\partial x^2}. \tag{4.27}$$

Let P_i^n be a solution of the corresponding forward difference scheme, and let its perturbation $P_i^n + \epsilon_i^n$ satisfy the same scheme:

$$\frac{(P_i^{n+1} + \epsilon_i^{n+1}) - (P_i^n + \epsilon_i^n)}{\Delta t} = \frac{(P_{i+1}^n + \epsilon_{i+1}^n) - 2(P_i^n + \epsilon_i^n) + (P_{i-1}^n + \epsilon_{i-1}^n)}{h^2}.$$

Because of the definition of P_i^n, we see that

$$\frac{\epsilon_i^{n+1} - \epsilon_i^n}{\Delta t} = \frac{\epsilon_{i+1}^n - 2\epsilon_i^n + \epsilon_{i-1}^n}{h^2}. \tag{4.28}$$

We expand the error ϵ_i^n in a Fourier series of the form

$$\epsilon_i^n = \sum_k \gamma_k^n \exp(\bar{i}kx_i),$$

where $\bar{i} = \sqrt{-1}$. The analysis can be simplified somewhat if we assume that a solution to the error equation (4.28) has one term (dropping the subscript k in γ_k^n)

$$\epsilon_i^n = \gamma^n \exp(\bar{i}kx_i). \tag{4.29}$$

We substitute (4.29) into (4.28) and solve for the *amplification factor*

$$\gamma = \gamma^{n+1}/\gamma^n.$$

The *von Neumann criterion* for stability is that the modulus of this factor must not be greater than one (Thomas, 1995). Using (4.28) and (4.29), we see that

$$\frac{\gamma^{n+1} - \gamma^n}{\Delta t} = \frac{\gamma^n \exp(\bar{i}kh) - 2\gamma^n + \gamma^n \exp(-\bar{i}kh)}{h^2}. \tag{4.30}$$

Since

$$\exp(\bar{i}kh) - 2 + \exp(-\bar{i}kh) = 2\cos(kh) - 2 = -4\sin^2(kh/2),$$

it follows from (4.30) that

$$\gamma^{n+1} = \left(1 - \frac{4\Delta t}{h^2}\sin^2\left(\frac{kh}{2}\right)\right)\gamma^n.$$

Dividing this equation by γ^n, we obtain

$$\gamma = 1 - \frac{4\Delta t}{h^2}\sin^2\left(\frac{kh}{2}\right).$$

Thus the von Neumann criterion for stability is satisfied if

$$\left|1 - \frac{4\Delta t}{h^2}\sin^2\left(\frac{kh}{2}\right)\right| \le 1. \tag{4.31}$$

Inequality (4.31) is satisfied when the *stability condition*

$$\frac{\Delta t}{h^2} \le \frac{1}{2} \tag{4.32}$$

holds. Therefore, the forward difference scheme for (4.27) is stable under condition (4.32); i.e., this scheme is *conditionally stable*, as noted earlier.

We perform a similar von Neumann stability analysis for the backward difference scheme (4.24) for equation (4.27). In this case, the error equation takes the form

$$\frac{\epsilon_i^{n+1} - \epsilon_i^n}{\Delta t} = \frac{\epsilon_{i+1}^{n+1} - 2\epsilon_i^{n+1} + \epsilon_{i-1}^{n+1}}{h^2}. \tag{4.33}$$

Substituting (4.29) into (4.33) and performing simple algebraic calculations yields the equation for the amplification factor γ,

$$\gamma = \frac{1}{1 + (4\Delta t/h^2)\sin^2(kh/2)},$$

which is always less than or equal to one for any choice of k, Δt, and h. Hence the backward difference scheme is *unconditionally stable*. An analogous analysis shows that the Crank–Nicholson scheme is also unconditionally stable (cf. Exercise 4.4).

Convergence

Finite difference schemes are used because their solutions approximate the solutions to certain partial differential equations. What we really need is that the solutions of difference schemes can be made to approximate the solutions of the differential equations to any desired accuracy. Namely, we need *convergence* of the finite difference solutions to those of the differential equations. Specifically, a finite difference scheme $L_i^n P_i^n = G_i^n$ approximating the partial differential equation $\mathcal{L}p = \mathcal{F}$ is (pointwise) *convergent* if for any (x, t), P_i^n converges to $p(x, t)$, as h, $\Delta t \to 0$ and $(ih, n\Delta t) \to (x, t)$.

As an example, we consider the forward difference scheme (4.22) for equation (4.27):

$$\frac{P_i^{n+1} - P_i^n}{\Delta t} = \frac{P_{i+1}^n - 2P_i^n + P_{i-1}^n}{h^2}. \tag{4.34}$$

Using the analysis in Sections 4.1.1 and 4.1.2, it follows from (4.27) that

$$\frac{p_i^{n+1} - p_i^n}{\Delta t} = \frac{p_{i+1}^n - 2p_i^n + p_{i-1}^n}{h^2} + \mathcal{O}(h^2) + \mathcal{O}(\Delta t). \tag{4.35}$$

Define the error

$$z_i^n = P_i^n - p_i^n,$$

and subtract (4.35) from (4.34) to yield

$$z_i^{n+1} = (1 - 2\mathcal{R})z_i^n + \mathcal{R}(z_{i+1}^n + z_{i-1}^n) + \mathcal{O}(h^2\Delta t) + \mathcal{O}\big((\Delta t)^2\big),$$

where $\mathcal{R} = \Delta t/h^2$. If $0 < \mathcal{R} \le 1/2$, the coefficients on the right-hand side of this equation are nonnegative. Thus we see that

$$\begin{aligned} |z_i^{n+1}| &\le (1-2\mathcal{R})\,|z_i^n| + \mathcal{R}\left(|z_{i+1}^n| + |z_{i-1}^n|\right) + C\Delta t\left(h^2+\Delta t\right) \\ &\le Z^n + C\Delta t\left(h^2+\Delta t\right), \end{aligned} \tag{4.36}$$

where $Z^n = \sup_i\{|z_i^n|\}$ and the constant C is a uniform constant used to bound the "big $\mathcal{O}$" terms. Taking the supremum over i on the left-hand side of (4.36), we obtain

$$Z^{n+1} \le Z^n + C\Delta t\left(h^2+\Delta t\right). \tag{4.37}$$

Applying inequality (4.37) repeatedly implies

$$Z^{n+1} \le Z^0 + C(n+1)\Delta t\left(h^2+\Delta t\right).$$

Initially, let $Z^0 = 0$. Then we have

$$\begin{aligned} \left|P_i^{n+1} - p(ih,(n+1)\Delta t)\right| &\le Z^{n+1} \\ &\le C(n+1)\Delta t\left(h^2+\Delta t\right) \\ &\to 0 \end{aligned}$$

as $(n+1)\Delta t \to t$ and $h,\ \Delta t \to 0$. Therefore, we have proven convergence of the forward difference scheme for (4.27) under condition (4.32). Convergence of the backward and Crank–Nicholson difference schemes can be also shown (cf. Exercises 4.5 and 4.6).

There is a connection between stability and convergence. In fact, a consistent, two-level difference scheme (i.e., it involves two time levels) for a well-posed linear initial value problem is stable if and only if it is convergent. This is the *Lax equivalence theorem* (Thomas, 1995).

4.1.8 Finite differences for hyperbolic problems

For the introduction of finite differences for hyperbolic problems, we consider the model problem

$$\frac{\partial p}{\partial t} + b\frac{\partial p}{\partial x} = 0, \tag{4.38}$$

where b is a constant and $x = x_1$. This problem is a *one-way wave problem*. The one-dimensional Buckley–Leverett equation is of this form (cf. Section 2.3.2). A boundary condition for (4.38) depends on the sign of b. If this problem is imposed on a bounded interval (l_1, l_2), for example, only an *inflow boundary condition* is needed. That is, p is given at l_1 if $b > 0$, and it is given at l_2 if $b < 0$. For brevity of presentation, we consider problem (4.38) over the entire real line $\mathbb{R}$. Of course, in any case, an initial condition must be given:

$$p(x,0) = p_0(x).$$

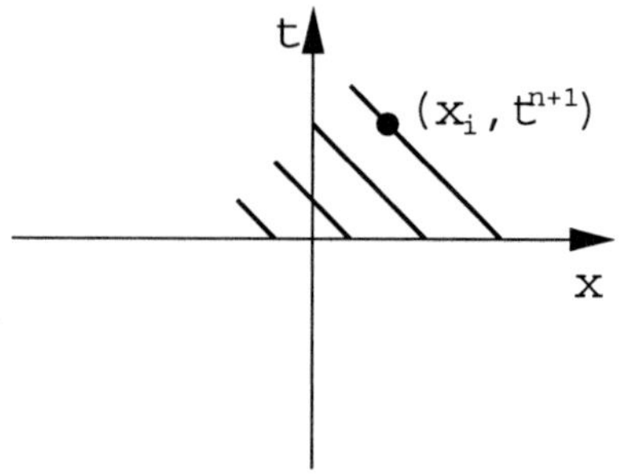

Figure 4.8. *Characteristics for problem* (4.38) *when* $b < 0$.

Explicit schemes

We consider an *explicit scheme* for problem (4.38):

$$\frac{p_i^{n+1} - p_i^n}{\Delta t} + b\frac{p_{i+1}^n - p_i^n}{h} = 0, \tag{4.39}$$

which is consistent with (4.38) (cf. Exercise 4.7). The amplification factor γ (cf. Section 4.1.7 and Exercise 4.8) for (4.39) satisfies

$$\gamma = 1 + \frac{b\Delta t}{h}(1 - \cos(kh)) - i\frac{b\Delta t}{h}\sin(kh).$$

In the case $b > 0$, $|\gamma| > 1$ (cf. Exercise 4.9). Thus, by the von Neumann criterion for stability, the difference scheme (4.39) is always *unstable*. In the case $b < 0$, it can be checked (cf. Exercise 4.10) that scheme (4.39) is stable, provided that

$$\frac{|b|\Delta t}{h} \leq 1. \tag{4.40}$$

This is the Courant–Friedrichs–Lewy (CFL) condition. That is, scheme (4.39) is conditionally stable if $b < 0$.

It is not surprising that scheme (4.39) is a good choice for problem (4.38) when $b < 0$, and a bad choice when $b > 0$. When $b < 0$, the characteristic for (4.39) through any point runs down to the right towards the x-axis (cf. Figure 4.8). Scheme (4.39) must then follow back in the same direction. For this reason, when $b > 0$, a good choice for (4.38) is

$$\frac{p_i^{n+1} - p_i^n}{\Delta t} + b\frac{p_i^n - p_{i-1}^n}{h} = 0. \tag{4.41}$$

In fact, when $b > 0$, it can be seen (cf. Exercise 4.11) that scheme (4.41) is stable under condition (4.40). (It is always unstable for $b < 0$.)

The explicit difference schemes (4.39) and (4.41) are *one-sided*. Based on the stability analysis above, only the *upwind* versions are conditionally stable.

There are other difference schemes for solving problem (4.38). The centered scheme in space is

$$\frac{p_i^{n+1} - p_i^n}{\Delta t} + b\frac{p_{i+1}^n - p_{i-1}^n}{2h} = 0. \tag{4.42}$$

This scheme yields the amplification factor γ (cf. Exercise 4.12)

$$\gamma = 1 - \bar{i}\frac{b\Delta t}{h}\sin(kh).$$

Since $|\gamma|^2 = 1 + b^2(\Delta t)^2\sin^2(kh)/h^2 \geq 1$, we see that scheme (4.42) is always unstable.

Implicit schemes

A stability analysis analogous to that in the explicit case shows that one-sided stable fully implicit difference schemes must be upwind. When $b < 0$, the *upwind implicit scheme* is

$$\frac{p_i^{n+1} - p_i^n}{\Delta t} + b\frac{p_{i+1}^{n+1} - p_i^{n+1}}{h} = 0, \tag{4.43}$$

and when $b > 0$, it is

$$\frac{p_i^{n+1} - p_i^n}{\Delta t} + b\frac{p_i^{n+1} - p_{i-1}^{n+1}}{h} = 0. \tag{4.44}$$

Scheme (4.43) has the amplification factor γ (cf. Exercise 4.13)

$$\gamma = \left(1 - \frac{b\Delta t}{h}(1 - \cos(kh)) + \bar{i}\frac{b\Delta t}{h}\sin(kh)\right)^{-1},$$

and thus

$$|\gamma|^2 = \left(1 - 4\frac{b\Delta t}{h}\sin^2\left(\frac{kh}{2}\right)\left(1 - \frac{b\Delta t}{h}\right)\right)^{-1} \leq 1 \quad \text{if } b < 0.$$

Hence scheme (4.43) is unconditionally stable when $b < 0$. A similar argument can be used to prove that scheme (4.44) has the same stability property when $b > 0$.

Now, we consider a fully implicit analogue to scheme (4.42):

$$\frac{p_i^{n+1} - p_i^n}{\Delta t} + b\frac{p_{i+1}^{n+1} - p_{i-1}^{n+1}}{2h} = 0. \tag{4.45}$$

The amplification factor γ of this scheme is (cf. Exercise 4.14)

$$\gamma = \left(1 + \bar{i}\frac{b\Delta t}{h}\sin(kh)\right)^{-1},$$

which satisfies $|\gamma| \leq 1$. Therefore, scheme (4.45) is unconditionally stable, in contrast with the always unstable scheme (4.42). A centered scheme in time (e.g., the Crank–Nicholson scheme) can be also defined for the solution of problem (4.38) (cf. Exercises 4.15–4.17).

Numerical dispersion

The local *truncation error* associated with the upwind version of the difference scheme (4.39) for problem (4.38) with $b < 0$ is (cf. Exercise 4.18)

$$R_i^n = -\frac{bh}{2}\frac{\partial^2 p}{\partial x^2}(x_i, t^n) - \frac{\Delta t}{2}\frac{\partial^2 p}{\partial t^2}(x_i, t^n) + \mathcal{O}(h^2) + \mathcal{O}\big((\Delta t)^2\big). \tag{4.46}$$

Differentiation of (4.38) with respect to t gives

$$\frac{\partial^2 p}{\partial t^2} = -b\frac{\partial^2 p}{\partial x \partial t},$$

and differentiation with respect to x yields

$$\frac{\partial^2 p}{\partial x \partial t} = -b\frac{\partial^2 p}{\partial x^2}.$$

Consequently,

$$\frac{\partial^2 p}{\partial t^2} = b^2\frac{\partial^2 p}{\partial x^2},$$

which is substituted into (4.46) to give

$$R_i^n = -\frac{bh}{2}\left(1 + \frac{b\Delta t}{h}\right)\frac{\partial^2 p}{\partial x^2}(x_i, t^n) + \mathcal{O}(h^2) + \mathcal{O}\big((\Delta t)^2\big). \tag{4.47}$$

This is the local truncation error associated with scheme (4.39).

By definition (4.26) of the local truncation error, (4.47) can be written as

$$\begin{aligned}\frac{p_i^{n+1} - p_i^n}{\Delta t} + b\frac{p_{i+1}^n - p_i^n}{h} &= \left\{\frac{\partial p}{\partial t} + b\frac{\partial p}{\partial x} + a_{num}\frac{\partial^2 p}{\partial x^2}\right\}(x_i, t^n) \\ &\quad + \mathcal{O}(h^2) + \mathcal{O}\big((\Delta t)^2\big),\end{aligned} \tag{4.48}$$

where

$$a_{num} = \frac{bh}{2}\left(1 + \frac{b\Delta t}{h}\right). \tag{4.49}$$

Therefore, we are, in fact, solving the difference equation (4.39) for the diffusion-convection problem

$$\frac{\partial p}{\partial t} + b\frac{\partial p}{\partial x} + a_{num}\frac{\partial^2 p}{\partial x^2} = 0,$$

rather than for the pure hyperbolic problem (4.38). That is, the truncation error of (4.39) includes the *numerical dispersion* term a_{num}.

If we consider the diffusion-convection problem

$$\frac{\partial p}{\partial t} + b\frac{\partial p}{\partial x} - a\frac{\partial^2 p}{\partial x^2} = 0, \qquad a > 0,$$

and develop a difference scheme similar to (4.39), then the above truncation error analysis indicates that the solution of the resulting difference equation will be associated with the problem

$$\frac{\partial p}{\partial t} + b\frac{\partial p}{\partial x} - (a - a_{num})\frac{\partial^2 p}{\partial x^2} = 0.$$

When the physical diffusion coefficient a is small, a serious problem arises. If numerical dispersion is severe (it is frequently so), a_{num} can easily dominate a. Consequently, the numerical dispersion swamps the physical dispersion, leading to a sharp front being severely smeared (cf. Exercise 4.20). The solution of hyperbolic problems using finite element methods will be discussed in Sections 4.4 and 4.6. In particular, the characteristic finite element methods introduced in Section 4.6 reduce numerical dispersion.

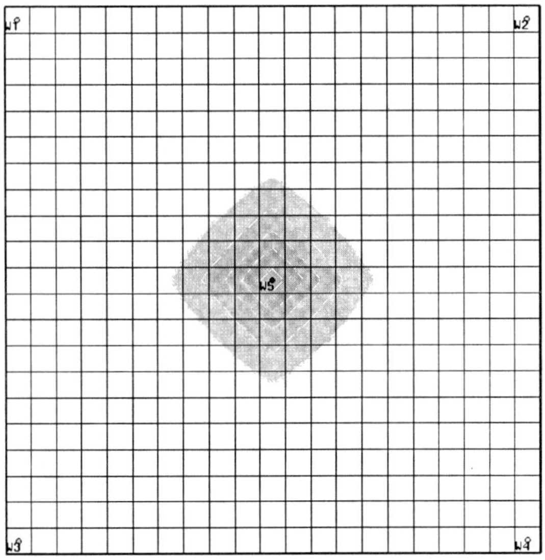

Figure 4.9. *A five-point finite difference example.*

4.1.9 Grid orientation effects

Another drawback of finite difference methods is that the solution of a partial differential problem using these methods heavily depends on spatial orientations of a computational grid, known as *grid orientation effects*. In petroleum reservoir simulation, this means that drastically different predictions from simulators can be obtained from different grid orientations.

If an upwind technique is used as in (4.39) for a two-dimensional counterpart, the resulting numerical dispersion is related to the quantity (cf. (4.49))

$$\frac{h_1}{2}\frac{\partial^2 p}{\partial x_1^2} + \frac{h_2}{2}\frac{\partial^2 p}{\partial x_2^2},$$

which is not rotationally invariant and is thus directionally dependent. When modeling multiphase flow with a high mobility ratio (mainly due to a large viscosity ratio), once a preferential flow pattern has been established, the greater mobility of the less viscous fluid causes this flow path to dominate the flow pattern. With the five-point (in two space dimensions) or seven-point (in three dimensions) finite difference stencil scheme, preferred flow paths are established along the coordinate directions (cf. Figure 4.9, where a two-phase flow example is shown; cf. Exercise 4.1 and Chapter 7). Then the use of an upwind stabilizing technique greatly enhances flow in these preferred directions. This grid orientation effect is dramatic in cases with very high mobility ratios. Therefore, different discretization methods and gridding techniques must be introduced.

4.2 Standard Finite Element Methods

4.2.1 Finite element methods for stationary problems

The exposition in this section has two purposes: to introduce the terminology and to summarize the basic ingredients that are required for the development of finite element methods.

A one-dimensional model problem

As an introduction, we consider a stationary problem in one dimension

$$\begin{aligned} -\frac{d^2p}{dx^2} &= f(x), \qquad 0 < x < 1, \\ p(0) &= p(1) = 0, \end{aligned} \tag{4.50}$$

where f is a given real-valued piecewise continuous bounded function. Note that (4.50) is a two-point boundary value problem (e.g., a one-dimensional elliptic pressure equation).

As shown in the previous section, finite difference methods for (4.50) involve replacing the second derivative by a difference quotient that involves the values of p at certain points. The discretization of (4.50) using finite element methods is different. These methods start by rewriting (4.50) in an equivalent variational formulation. For this, we introduce the *scalar product*

$$(v, w) = \int_0^1 v(x)w(x)\, dx$$

for real-valued piecewise continuous bounded functions v and w, and we define the *linear space*

$$V = \Big\{ v : v \text{ is a continuous function on } [0, 1], \ \frac{dv}{dx} \text{ is piecewise} \\ \text{continuous and bounded on } (0, 1), \text{ and } v(0) = v(1) = 0 \Big\}.$$

We also define the *functional* $F : V \to \mathbb{R}$

$$F(v) = \frac{1}{2}\left(\frac{dv}{dx}, \frac{dv}{dx}\right) - (f, v), \qquad v \in V,$$

where $\mathbb{R}$ is the set of real numbers. At the end of this subsection it will be shown that finding p for (4.50) is equivalent to the *minimization problem*

$$\text{Find } p \in V \text{ such that } F(p) \leq F(v) \qquad \forall v \in V. \tag{4.51}$$

Problem (4.51) is a *Ritz variational form* of (4.50).

In terms of computation, (4.50) can be expressed in a more useful, direct formulation. Multiplying the first equation of (4.50) by any $v \in V$, called a *test function*, and integrating over (0, 1), we see that

$$-\left(\frac{d^2p}{dx^2}, v\right) = (f, v).$$

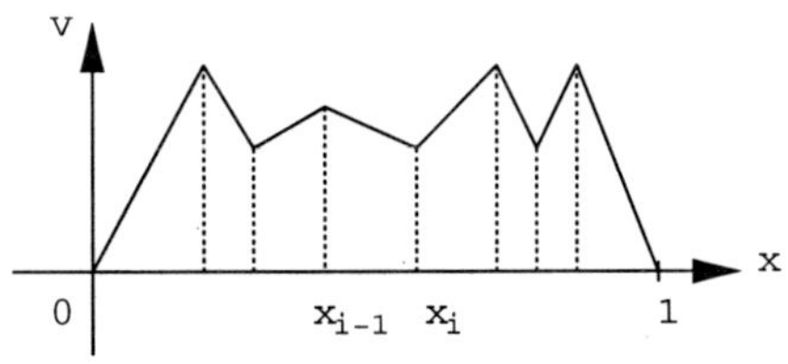

Figure 4.10. *An illustration of a function $v \in V_h$.*

Application of integration by parts to this equation yields

$$\left(\frac{dp}{dx}, \frac{dv}{dx}\right) = (f, v), \tag{4.52}$$

where we use the fact that $v(0) = v(1) = 0$ from the definition of V. Equation (4.52) is called a *Galerkin variational* or *weak form* of (4.50). If p is a solution to (4.50), then it also satisfies (4.52). The converse also holds if d^2p/dx^2 exists and is piecewise continuous and bounded in (0, 1), for example (cf. Exercise 4.21). It can be seen that (4.51) and (4.52) are equivalent (see the end of this subsection).

We now construct finite element methods for solving (4.50). Toward that end, for a positive integer M, let $0 = x_0 < x_1 < \cdots < x_M < x_{M+1} = 1$ be a *partition* of (0, 1) into a set of subintervals $I_i = (x_{i-1}, x_i)$ with length $h_i = x_i - x_{i-1}$, $i = 1, 2, \ldots, M+1$. Set $h = \max\{h_i : i = 1, 2, \ldots, M+1\}$. The *step size* h measures how fine the partition is. Define the *finite element space*

$$V_h = \{v : v \text{ is a continuous function on } [0, 1], v \text{ is linear on each subinterval } I_i, \text{ and } v(0) = v(1) = 0\}.$$

See Figure 4.10 for an illustration of a function $v \in V_h$. Note that $V_h \subset V$ (i.e., V_h is a subspace of V).

The discrete version of (4.51) is

$$\text{Find } p_h \in V_h \text{ such that } F(p_h) \le F(v) \qquad \forall v \in V_h. \tag{4.53}$$

Method (4.53) is referred to as the *Ritz finite element method*. In the same manner as for (4.52) (see the end of this subsection), (4.53) is equivalent to the problem

$$\text{Find } p_h \in V_h \text{ such that } \left(\frac{dp_h}{dx}, \frac{dv}{dx}\right) = (f, v) \qquad \forall v \in V_h. \tag{4.54}$$

This is usually termed the *Galerkin finite element method.*

It is easy to see that (4.54) has a unique solution. In fact, let $f = 0$, and take $v = p_h$ in (4.54) to give

$$\left(\frac{dp_h}{dx}, \frac{dp_h}{dx}\right) = 0,$$

so p_h is a constant. It follows from the boundary condition in V_h that $p_h = 0$.

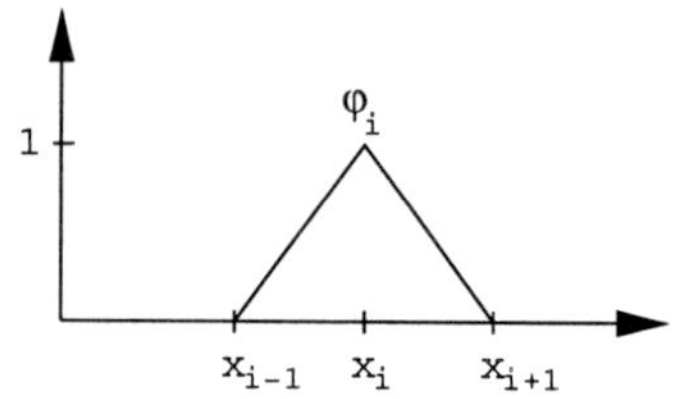

Figure 4.11. *A basis function in one dimension.*

We introduce the *basis functions* $\varphi_i \in V_h$, $i = 1, 2, \dots, M$,

$$\varphi_i(x_j) = \begin{cases} 1 & \text{if } i = j, \\ 0 & \text{if } i \neq j. \end{cases}$$

That is, φ_i is a continuous piecewise linear function on [0, 1] such that its value is one at *node* x_i and zero at other nodes (cf. Figure 4.11). It is called a *hat* or *chapeau function*. Any function $v \in V_h$ has the unique representation

$$v(x) = \sum_{i=1}^{M} v_i \varphi_i(x), \qquad 0 \leq x \leq 1,$$

where $v_i = v(x_i)$. For each j, take $v = \varphi_j$ in (4.54) to see that

$$\left(\frac{dp_h}{dx}, \frac{d\varphi_j}{dx} \right) = (f, \varphi_j), \qquad j = 1, 2, \dots, M. \tag{4.55}$$

Set

$$p_h(x) = \sum_{i=1}^{M} p_i \varphi_i(x), \qquad p_i = p_h(x_i),$$

and substitute it into (4.55) to give

$$\sum_{i=1}^{M} \left(\frac{d\varphi_i}{dx}, \frac{d\varphi_j}{dx} \right) p_i = (f, \varphi_j), \qquad j = 1, 2, \dots, M. \tag{4.56}$$

This is a linear system of M algebraic equations in the M unknowns $p_1, p_2, \dots, p_M$. It can be written in matrix form as

$$\mathbf{Ap} = \mathbf{f}, \tag{4.57}$$

where the matrix $\mathbf{A}$ and vectors $\mathbf{p}$ and $\mathbf{f}$ are given by

$$\mathbf{A} = \begin{pmatrix} a_{11} & a_{12} & \dots & a_{1M} \\ a_{21} & a_{22} & \dots & a_{2M} \\ \vdots & \vdots & \ddots & \vdots \\ a_{M1} & a_{M2} & \dots & a_{MM} \end{pmatrix}, \quad \mathbf{p} = \begin{pmatrix} p_1 \\ p_2 \\ \vdots \\ p_M \end{pmatrix}, \quad \mathbf{f} = \begin{pmatrix} f_1 \\ f_2 \\ \vdots \\ f_M \end{pmatrix}$$

with

$$a_{ij} = \left(\frac{d\varphi_i}{dx}, \frac{d\varphi_j}{dx}\right), \quad f_j = (f, \varphi_j), \quad i, j = 1, 2, \ldots, M.$$

The matrix $\mathbf{A}$ is referred to as the *stiffness matrix*, and $\mathbf{f}$ is the *source vector*.

By the definition of the basis functions,

$$\left(\frac{d\varphi_i}{dx}, \frac{d\varphi_j}{dx}\right) = 0 \quad \text{if } |i - j| \geq 2,$$

so $\mathbf{A}$ is *tridiagonal*; i.e., only the entries on the main diagonal and the adjacent diagonals may be nonzero. In fact, the entries a_{ij} can be calculated as follows:

$$a_{ii} = \frac{1}{h_i} + \frac{1}{h_{i+1}}, \quad a_{i-1,i} = -\frac{1}{h_i}, \quad a_{i,i+1} = -\frac{1}{h_{i+1}}.$$

Also, it can be seen that $\mathbf{A}$ is *symmetric*, $a_{ij} = a_{ji}$, and *positive definite*,

$$\eta^T \mathbf{A} \eta = \sum_{i,j=1}^{M} \eta_i a_{ij} \eta_j > 0 \qquad \text{for all nonzero } \eta \in \mathbb{R}^M,$$

where η^T denotes the transpose of η. Because a positive definite matrix is nonsingular, the linear system (4.57) has a unique solution. Consequently, we have shown that (4.54) has a unique solution $p_h \in V_h$ in a different way.

The symmetry of $\mathbf{A}$ can be seen from the definition of a_{ij}. The positive definiteness can be checked as follows: with

$$\eta = \sum_{i=1}^{M} \eta_i \varphi_i \in V_h, \quad \eta^T = (\eta_1, \eta_2, \ldots, \eta_M),$$

we see that

$$\sum_{i,j=1}^{M} \eta_i a_{ij} \eta_j = \sum_{i,j=1}^{M} \eta_i \left(\frac{d\varphi_i}{dx}, \frac{d\varphi_j}{dx}\right) \eta_j$$

$$= \left(\sum_{i=1}^{M} \eta_i \frac{d\varphi_i}{dx}, \sum_{j=1}^{M} \eta_j \frac{d\varphi_j}{dx}\right) = \left(\frac{d\eta}{dx}, \frac{d\eta}{dx}\right) \geq 0.$$

As for (4.54), the equality holds only for $\eta \equiv 0$ since a constant function η must be zero because of the boundary condition.

We remark that $\mathbf{A}$ is *sparse*; that is, only a few entries in each row of $\mathbf{A}$ are nonzero. In the present one-dimensional case, it is tridiagonal. The sparsity of $\mathbf{A}$ depends upon the fact that a basis function in V_h is different from zero only on a few intervals; that is, it has compact *support*. Thus it interferes with only a few other basis functions. That basis functions can be chosen in this manner is an important distinctive property of finite element methods.

In the case where the partition is uniform, i.e., $h = h_i$, $i = 1, 2, \ldots, M+1$, the stiffness matrix $\mathbf{A}$ takes the form

$$\mathbf{A} = \frac{1}{h}\begin{pmatrix} 2 & -1 & 0 & \ldots & 0 & 0 \\ -1 & 2 & -1 & \ldots & 0 & 0 \\ 0 & -1 & 2 & \ldots & 0 & 0 \\ \vdots & \vdots & \vdots & \ddots & \vdots & \vdots \\ 0 & 0 & 0 & \ldots & 2 & -1 \\ 0 & 0 & 0 & \ldots & -1 & 2 \end{pmatrix}.$$

With division by h in $\mathbf{A}$, (4.54) can be thought of as a variant of the *central difference scheme* where the right-hand side consists of mean values of $f\varphi_j$ over the interval (x_{j-1}, x_{j+1}) (cf. Section 4.1.5).

In general, the derivation of an *error estimate* for finite element methods is very technical. Here we briefly indicate how to obtain an estimate in one dimension. Subtract (4.54) from (4.52) to get

$$\left(\frac{dp}{dx} - \frac{dp_h}{dx}, \frac{dv}{dx}\right) = 0 \qquad \forall v \in V_h. \tag{4.58}$$

We introduce the notation

$$\|v\| = (v, v)^{1/2} = \left(\int_0^1 v^2\, dx\right)^{1/2}.$$

This is a *norm* associated with the *scalar product* $(\cdot, \cdot)$. We use the *Cauchy inequality* (cf. Exercise 4.23)

$$|(v, w)| \le \|v\|\,\|w\|. \tag{4.59}$$

Note that, using (4.58), for any $v \in V_h$ we see that

$$\begin{aligned}\left\|\frac{dp}{dx} - \frac{dp_h}{dx}\right\|^2 &= \left(\frac{dp}{dx} - \frac{dp_h}{dx}, \frac{dp}{dx} - \frac{dp_h}{dx}\right) \\ &= \left(\frac{dp}{dx} - \frac{dp_h}{dx}, \left[\frac{dp}{dx} - \frac{dv}{dx}\right] + \left[\frac{dv}{dx} - \frac{dp_h}{dx}\right]\right) \\ &= \left(\frac{dp}{dx} - \frac{dp_h}{dx}, \frac{dp}{dx} - \frac{dv}{dx}\right),\end{aligned}$$

and thus, by (4.59),

$$\left\|\frac{dp}{dx} - \frac{dp_h}{dx}\right\| \le \left\|\frac{dp}{dx} - \frac{dv}{dx}\right\| \qquad \forall v \in V_h. \tag{4.60}$$

This equation implies that p_h is the best possible approximation of p in V_h in terms of the norm in (4.60).

To obtain an error bound, we take v in (4.60) to be the *interpolant* $\tilde{p}_h \in V_h$ of p; i.e., $\tilde{p}_h$ is defined by

$$\tilde{p}_h(x_i) = p(x_i), \qquad i = 0, 1, \ldots, M+1. \tag{4.61}$$

It is an easy exercise (cf. Exercise 4.24) to see that, for $x \in [0, 1]$,

$$\begin{aligned} |(p - \tilde{p}_h)(x)| &\le \frac{h^2}{8} \max_{y \in [0,1]} \left| \frac{d^2 p(y)}{dx^2} \right|, \\ \left| \left(\frac{dp}{dx} - \frac{d\tilde{p}_h}{dx} \right)(x) \right| &\le h \max_{y \in [0,1]} \left| \frac{d^2 p(y)}{dx^2} \right|. \end{aligned} \tag{4.62}$$

With $v = \tilde{p}_h$ in (4.60) and the second equation of (4.62), we obtain

$$\left\| \frac{dp}{dx} - \frac{dp_h}{dx} \right\| \le h \max_{y \in [0,1]} \left| \frac{d^2 p(y)}{dx^2} \right|. \tag{4.63}$$

Using the fact that $p(0) - p_h(0) = 0$, we have

$$p(x) - p_h(x) = \int_0^x \left(\frac{dp}{dx} - \frac{dp_h}{dx} \right)(y)\, dy, \qquad x \in [0, 1],$$

which, together with (4.63), implies

$$|p(x) - p_h(x)| \le h \max_{y \in [0,1]} \left| \frac{d^2 p(y)}{dx^2} \right|, \qquad x \in [0, 1]. \tag{4.64}$$

Note that (4.64) is less sharp in h than the first estimate in (4.62) for the *interpolation error*. With a more delicate analysis, we can show that the first error estimate in (4.62) holds for p_h as well as $\tilde{p}_h$. In fact, it can be shown that $p_h = \tilde{p}_h$ (cf. Exercise 4.25), which is true only for one dimension.

In summary, we have obtained the quantitative estimates in (4.63) and (4.64), which show that the approximate solution of (4.54) approaches the exact solution of (4.50) as h goes to zero. This implies *convergence* of the finite element method (4.54) (cf. Section 4.1.7).

Now, we consider the equivalence between (4.51) and (4.52). Let p be a solution of (4.51). Then, for any $v \in V$ and any $\epsilon \in \mathbb{R}$, we have

$$F(p) \le F(p + \epsilon v).$$

With the definition

$$\begin{aligned} G(\epsilon) &= F(p + \epsilon v) \\ &= \frac{1}{2} \left(\frac{dp}{dx}, \frac{dp}{dx} \right) + \epsilon \left(\frac{dp}{dx}, \frac{dv}{dx} \right) + \frac{\epsilon^2}{2} \left(\frac{dv}{dx}, \frac{dv}{dx} \right) - \epsilon(f, v) - (f, p), \end{aligned}$$

we see that G has a minimum at $\epsilon = 0$, so $\frac{dG}{d\epsilon}(0) = 0$. Since

$$\frac{dG}{d\epsilon}(0) = \left(\frac{dp}{dx}, \frac{dv}{dx} \right) - (f, v),$$

p is a solution of (4.52). Conversely, suppose that p is a solution of (4.52). With any $v \in V$, set $w = v - p \in V$; we find that

$$
\begin{aligned}
F(v) = F(p+w) &= \frac{1}{2}\left(\frac{d(p+w)}{dx}, \frac{d(p+w)}{dx}\right) - (f, p+w) \\
&= \frac{1}{2}\left(\frac{dp}{dx}, \frac{dp}{dx}\right) - (f, p) + \left(\frac{dp}{dx}, \frac{dw}{dx}\right) - (f, w) + \frac{1}{2}\left(\frac{dw}{dx}, \frac{dw}{dx}\right) \\
&= \frac{1}{2}\left(\frac{dp}{dx}, \frac{dp}{dx}\right) - (f, p) + \frac{1}{2}\left(\frac{dw}{dx}, \frac{dw}{dx}\right) \geq F(p),
\end{aligned}
$$

which implies that p is a solution of (4.51). Because of the equivalence between (4.50) and (4.52), (4.51) is also equivalent to (4.50).

A two-dimensional model problem

In this subsection, we consider a stationary problem in two dimensions:

$$
\begin{aligned}
-\Delta p &= f \quad \text{in } \Omega, \\
p &= 0 \quad \text{on } \Gamma,
\end{aligned}
\tag{4.65}
$$

where Ω is a bounded domain in the plane with boundary Γ, f is a given real-valued piecewise continuous bounded function in Ω, and the *Laplacian operator* Δ is defined by

$$
\Delta p = \frac{\partial^2 p}{\partial x_1^2} + \frac{\partial^2 p}{\partial x_2^2}.
$$

We introduce the linear space

$$
V = \Big\{ v : v \text{ is a continuous function on } \Omega, \ \frac{\partial v}{\partial x_1} \text{ and } \frac{\partial v}{\partial x_2} \text{ are piecewise continuous and bounded on } \Omega, \text{ and } v = 0 \text{ on } \Gamma \Big\}.
$$

Let us recall *Green's formula*. For a vector-valued function $\mathbf{b} = (b_1, b_2)$, the *divergence theorem* reads

$$
\int_\Omega \nabla \cdot \mathbf{b} \, d\mathbf{x} = \int_\Gamma \mathbf{b} \cdot \boldsymbol{\nu} \, d\ell, \tag{4.66}
$$

where the divergence operator is given by

$$
\nabla \cdot \mathbf{b} = \frac{\partial b_1}{\partial x_1} + \frac{\partial b_2}{\partial x_2},
$$

$\boldsymbol{\nu}$ is the outward unit normal to Γ, and the dot product $\mathbf{b} \cdot \boldsymbol{\nu}$ is

$$
\mathbf{b} \cdot \boldsymbol{\nu} = b_1 \nu_1 + b_2 \nu_2.
$$

With $v, w \in V$, we take $\mathbf{b} = (\frac{\partial v}{\partial x_1} w, 0)$ and $\mathbf{b} = (0, \frac{\partial v}{\partial x_2} w)$ in (4.66), respectively, to see that

$$
\int_\Omega \frac{\partial^2 v}{\partial x_i^2} w \, d\mathbf{x} + \int_\Omega \frac{\partial v}{\partial x_i} \frac{\partial w}{\partial x_i} \, d\mathbf{x} = \int_\Gamma \frac{\partial v}{\partial x_i} w \nu_i \, d\ell, \qquad i = 1, 2. \tag{4.67}
$$

Using the definition of the *gradient operator*, i.e.,

$$\nabla v = \left(\frac{\partial v}{\partial x_1}, \frac{\partial v}{\partial x_2} \right),$$

we sum over $i = 1, 2$ in (4.67) to obtain

$$\int_\Omega \Delta v \, w \, d\mathbf{x} = \int_\Gamma \frac{\partial v}{\partial \boldsymbol{\nu}} w \, d\ell - \int_\Omega \nabla v \cdot \nabla w \, d\mathbf{x}, \tag{4.68}$$

where the *normal derivative* is

$$\frac{\partial v}{\partial \boldsymbol{\nu}} = \frac{\partial v}{\partial x_1} \nu_1 + \frac{\partial v}{\partial x_2} \nu_2.$$

Relation (4.68) is *Green's formula*, and it also holds in three dimensions (cf. Exercise 4.26).

Introduce the notation

$$a(p, v) = \int_\Omega \nabla p \cdot \nabla v \, d\mathbf{x}, \quad (f, v) = \int_\Omega f v \, d\mathbf{x}.$$

The form $a(\cdot, \cdot)$ is a *bilinear form* on $V \times V$; that is,

$$a(u, \alpha v + \beta w) = \alpha a(u, v) + \beta a(u, w),$$
$$a(\alpha u + \beta v, w) = \alpha a(u, w) + \beta a(v, w)$$

for $\alpha, \beta \in \mathbb{R}$ and $u, v, w \in V$. Also, define the functional $F : V \to \mathbb{R}$ by

$$F(v) = \frac{1}{2} a(v, v) - (f, v), \qquad v \in V.$$

As in one dimension, (4.65) can be formulated as the minimization problem

$$\text{Find } p \in V \text{ such that } F(p) \le F(v) \qquad \forall v \in V.$$

This problem is also equivalent to the variational problem (4.69) below, using the same proof as for (4.51) and (4.52).

Multiplying the first equation of (4.65) by $v \in V$ and integrating over Ω, we see that

$$-\int_\Omega \Delta p \, v \, d\mathbf{x} = \int_\Omega f v \, d\mathbf{x}.$$

Applying (4.68) to this equation and using the homogeneous boundary condition leads to

$$\int_\Omega \nabla p \cdot \nabla v \, d\mathbf{x} = \int_\Omega f v \, d\mathbf{x} \qquad \forall v \in V.$$

Thus we derive the variational form

$$\text{Find } p \in V \text{ such that } a(p, v) = (f, v) \qquad \forall v \in V. \tag{4.69}$$

We now construct finite element methods for (4.65). For simplicity, in this section, we assume that Ω is a polygonal domain. A curved domain Ω will be handled in Section 4.2.2.

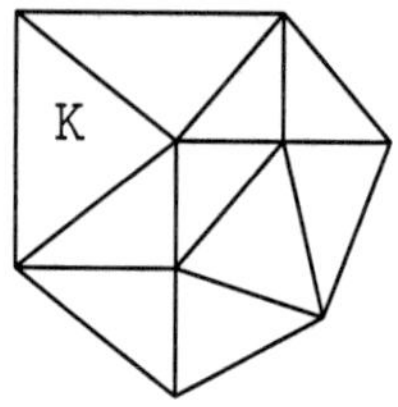

Figure 4.12. *A finite element partition in two dimensions.*

Let K_h be a partition, called a *triangulation*, of Ω into nonoverlapping (open) triangles K_i (cf. Figure 4.12):

$$\bar{\Omega} = \bar{K}_1 \cup \bar{K}_2 \cup \cdots \cup \bar{K}_{\tilde{M}},$$

such that no vertex of one triangle lies in the interior of an edge of another triangle, where $\bar{\Omega}$ represents the closure of Ω (i.e., $\bar{\Omega} = \Omega \cup \Gamma$) and a similar meaning holds for each K_i.

For (open) triangles $K \in K_h$, we define the *mesh parameters*

$$\text{diam}(K) = \text{the longest edge of } \bar{K} \quad \text{and} \quad h = \max_{K \in K_h} \text{diam}(K).$$

Now, we introduce the finite element space

$$V_h = \{v : v \text{ is a continuous function on } \Omega,\ v \text{ is linear on each triangle } K \in K_h, \text{ and } v = 0 \text{ on } \Gamma\}.$$

Note that $V_h \subset V$. The finite element method for (4.65) is formulated as

$$\text{Find } p_h \in V_h \text{ such that } a(p_h, v) = (f, v) \qquad \forall v \in V_h. \tag{4.70}$$

Existence and uniqueness of a solution to (4.70) can be checked as for (4.54). Also, in the same fashion as in the proof of the equivalence between (4.51) and (4.52), one can check that (4.70) is equivalent to a discrete minimization problem:

$$\text{Find } p_h \in V_h \text{ such that } F(p_h) \le F(v) \qquad \forall v \in V_h.$$

Denote the vertices (*nodes*) of the triangles in K_h by $\mathbf{x}_1, \mathbf{x}_2, \ldots, \mathbf{x}_{\tilde{M}}$. The basis functions φ_i in V_h, $i = 1, 2, \ldots, \tilde{M}$, are defined by

$$\varphi_i(\mathbf{x}_j) = \begin{cases} 1 & \text{if } i = j, \\ 0 & \text{if } i \neq j. \end{cases}$$

The *support* of φ_i, i.e., the set of $\mathbf{x}$ where $\varphi_i(\mathbf{x}) \neq 0$, consists of the triangles with the common node $\mathbf{x}_i$ (cf. Figure 4.13). The function φ_i is also called a hat or chapeau function.

Let M be the number of interior vertices in K_h; for convenience, let the first M vertices be the interior ones. As in the previous subsection, any function $v \in V_h$ has the unique representation

$$v(\mathbf{x}) = \sum_{i=1}^{M} v_i \varphi_i(\mathbf{x}), \qquad \mathbf{x} \in \Omega,$$

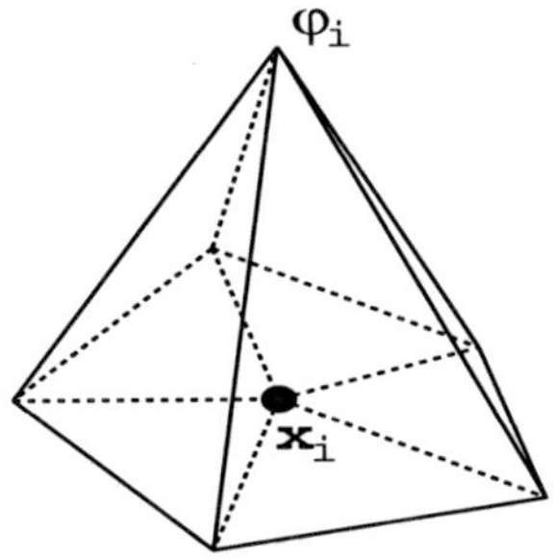

Figure 4.13. *A basis function in two dimensions.*

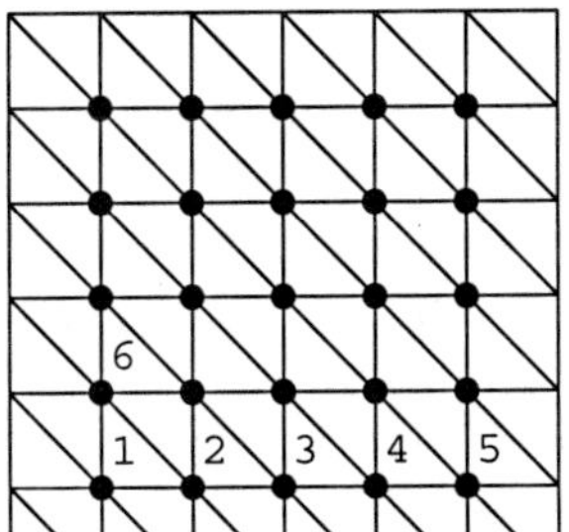

Figure 4.14. *An example of a triangulation.*

where $v_i = v(\mathbf{x}_i)$. Due to the Dirichlet boundary condition, we can exclude the vertices on the boundary of Ω.

In the same way as for (4.54), equation (4.70) can be written in matrix form (cf. Exercise 4.27)

$$\mathbf{A}\mathbf{p} = \mathbf{f}, \tag{4.71}$$

where, as before, the matrix $\mathbf{A}$ and the vectors $\mathbf{p}$ and $\mathbf{f}$ are

$$\mathbf{A} = \left(a_{ij}\right), \quad \mathbf{p} = \left(p_j\right), \quad \mathbf{f} = \left(f_j\right)$$

with

$$a_{ij} = a\left(\varphi_i, \varphi_j\right), \quad f_j = \left(f, \varphi_j\right), \quad i,\ j = 1, 2, \ldots, M.$$

As in one dimension, it can be checked that the stiffness matrix $\mathbf{A}$ is symmetric positive definite. In particular, it is nonsingular. Consequently, (4.71) and thus (4.70) have a unique solution.

As an example, we consider the case where the domain is the unit square $\Omega = (0, 1) \times (0, 1)$ and K_h is the uniform triangulation of Ω as illustrated in Figure 4.14 with the indicated *enumeration* of nodes. In this case, the matrix $\mathbf{A}$ has the form (cf. Exercise 4.28)

$$\mathbf{A} = \begin{pmatrix} 4 & -1 & 0 & 0 & \dots & 0 & -1 & 0 & \dots & 0 & 0 \\ -1 & 4 & -1 & 0 & \dots & 0 & 0 & -1 & \dots & 0 & 0 \\ 0 & -1 & 4 & -1 & \dots & 0 & 0 & 0 & \dots & -1 & 0 \\ 0 & 0 & -1 & 4 & \dots & 0 & 0 & 0 & \dots & 0 & -1 \\ \vdots & \vdots & \vdots & \vdots & \ddots & \vdots & \vdots & \vdots & \ddots & \vdots & \vdots \\ 0 & 0 & 0 & 0 & \dots & 4 & -1 & 0 & \dots & 0 & 0 \\ -1 & 0 & 0 & 0 & \dots & -1 & 4 & -1 & \dots & 0 & 0 \\ 0 & -1 & 0 & 0 & \dots & 0 & -1 & 4 & \dots & 0 & 0 \\ \vdots & \vdots & \vdots & \vdots & \ddots & \vdots & \vdots & \vdots & \ddots & \vdots & \vdots \\ 0 & 0 & -1 & 0 & \dots & 0 & 0 & 0 & \dots & 4 & -1 \\ 0 & 0 & 0 & -1 & \dots & 0 & 0 & 0 & \dots & -1 & 4 \end{pmatrix}.$$

Associated with the four corner nodes (e.g., node 1), there are only three nonzeros per row; an adjacent diagonal entry for such a node (e.g., node 5) may be zero. For other nodes adjacent to the boundary (e.g., node 2), there are solely four nonzeros per row. From this form of $\mathbf{A}$, the left-hand side of the ith equation in (4.71) is a linear combination of the values of p_h at most at the five nodes illustrated in Figure 4.15. After division by h^2, system (4.71) can be treated as a linear system generated by a *five-point difference stencil scheme* for (4.65) (cf. Section 4.1.5).

In practical computations (see the programming consideration below), the entries a_{ij} in $\mathbf{A}$ are obtained by summing the contributions from different triangles $K \in K_h$:

$$a_{ij} = a\left(\varphi_i, \varphi_j\right) = \sum_{K \in K_h} a^K\left(\varphi_i, \varphi_j\right),$$

where

$$a_{ij}^K \equiv a^K\left(\varphi_i, \varphi_j\right) = \int_K \nabla\varphi_i \cdot \nabla\varphi_j \, d\mathbf{x}. \tag{4.72}$$

Using the definition of the basis functions, we see that $a^K\left(\varphi_i, \varphi_j\right) = 0$ unless nodes $\mathbf{x}_i$ and $\mathbf{x}_j$ are both vertices of K. Thus $\mathbf{A}$ is sparse.

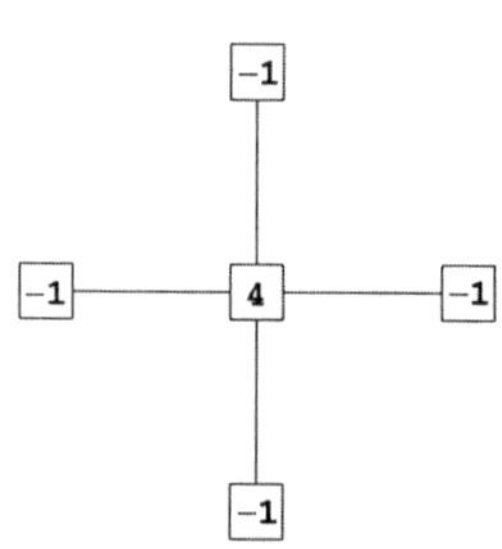

Figure 4.15. *A five-point stencil scheme.*

As noted earlier, the derivation of an estimate is very delicate. By the same argument as for (4.60), we have

$$\|\nabla p - \nabla p_h\| \leq \|\nabla p - \nabla v\| \qquad \forall v \in V_h,$$

where p and p_h are the respective solutions of (4.69) and (4.70), and we recall that $\|\cdot\|$ is the norm

$$\|\nabla p\| = \left(\int_\Omega \left(\left(\frac{\partial p}{\partial x_1}\right)^2 + \left(\frac{\partial p}{\partial x_2}\right)^2\right) d\mathbf{x}\right)^{1/2}.$$

This implies that p_h is the best possible approximation of p in V_h in terms of the norm deduced from the bilinear form $a(\cdot,\cdot)$. Applying an approximation theorem (Chen, 2005), we have

$$\|p - p_h\| + h\,\|\nabla p - \nabla p_h\| \leq Ch^2, \tag{4.73}$$

where the constant C depends on the second partial derivatives of p and the smallest angle of the triangles $K \in K_h$, but does not depend on h (Ciarlet, 1978; Chen, 2005). Error estimate (4.73) indicates that if the solution is sufficiently smooth, p_h tends to p in the norm $\|\cdot\|$ as h approaches zero.

An extension to general boundary conditions

We now extend the finite element methods to the stationary problem with the boundary condition of the third kind

$$\begin{aligned} -\Delta p &= f \qquad \text{in } \Omega, \\ bp + \frac{\partial p}{\partial \boldsymbol{\nu}} &= g \qquad \text{on } \Gamma, \end{aligned} \tag{4.74}$$

where b and g are given functions and $\partial p/\partial \boldsymbol{\nu}$ is the outward normal derivative. When $b = 0$, the boundary condition is the *second kind* or *Neumann condition*. When b is infinite, the boundary condition reduces to the *first kind* or *Dirichlet condition*, which was considered in the previous subsection. A *fourth kind* of boundary condition (i.e., a *periodic* boundary condition) will be considered in Section 4.6. In this subsection, we consider the case where b is bounded.

Note that if $b = 0$ on Γ, Green's formula (4.68) with (4.74) implies (cf. Exercise 4.30)

$$\int_\Omega f\, d\mathbf{x} + \int_\Gamma g\, d\ell = 0. \tag{4.75}$$

For (4.74) to have a solution, the *compatibility condition* (4.75) must be satisfied. In this case, p is unique only up to an additive constant.

Introducing the linear space

$$V = \Big\{ v : v \text{ is a continuous function on } \Omega, \text{ and } \frac{\partial v}{\partial x_1} \text{ and } \frac{\partial v}{\partial x_2}$$
$$\text{are piecewise continuous and bounded on } \Omega \Big\},$$

and the notation

$$a(v, w) = \int_\Omega \nabla v \cdot \nabla w \, d\mathbf{x} + \int_\Gamma bvw \, d\ell, \qquad v, w \in V,$$

$$(f, v) = \int_\Omega fv \, d\mathbf{x}, \quad (g, v)_\Gamma = \int_\Gamma gv \, d\ell, \qquad v \in V,$$

on the same lines as in the previous subsection, problem (4.74) can be written (cf. Exercise 4.31) as

$$\text{Find } p \in V \text{ such that } a(p, v) = (f, v) + (g, v)_\Gamma \qquad \forall v \in V. \tag{4.76}$$

Note that the boundary condition in (4.74) is not imposed in the definition of V. It appears implicitly in (4.76). A boundary condition that need not be imposed is called a *natural condition*. The pure Neumann boundary condition is natural. The Dirichlet boundary condition has been imposed explicitly in V earlier, and is termed an *essential condition*.

If $b \equiv 0$, the definition of V needs to be modified to take into account the up-to-a-constant uniqueness of solution to (4.74). That is, the space V can be modified to, say,

$$V = \Big\{ v : \ v \text{ is a continuous function on } \Omega, \ \frac{\partial v}{\partial x_1} \text{ and } \frac{\partial v}{\partial x_2}$$
$$\text{are piecewise continuous and bounded on } \Omega, \text{ and } \int_\Omega v \, d\mathbf{x} = 0 \Big\}.$$

To construct finite element methods for (4.74), let K_h be a triangulation of Ω as in the previous subsection. The finite element space V_h is

$$V_h = \{ v : v \text{ is a continuous function on } \Omega \text{ and}$$
$$\text{is linear on each triangle } K \in K_h \}.$$

Note that the functions in V_h are not required to satisfy any boundary condition. Now, the finite element solution satisfies

$$\text{Find } p_h \in V_h \text{ such that } a(p_h, v) = (f, v) + (g, v)_\Gamma \qquad \forall v \in V_h. \tag{4.77}$$

Again, for the pure Neumann boundary condition, V_h must be modified to

$$V_h = \Big\{ v : v \text{ is a continuous function on } \Omega \text{ and is linear}$$
$$\text{on each triangle } K \in K_h, \text{ and } \int_\Omega v \, d\mathbf{x} = 0 \Big\}.$$

As in the last two subsections, (4.77) can be formulated in matrix form, and an error estimate can be similarly stated under an appropriate smoothness assumption on the solution p that involves its second partial derivatives.

The *Poisson equation* has been considered in (4.65) and (4.74). More general partial differential equations will be treated in subsequent sections and chapters.

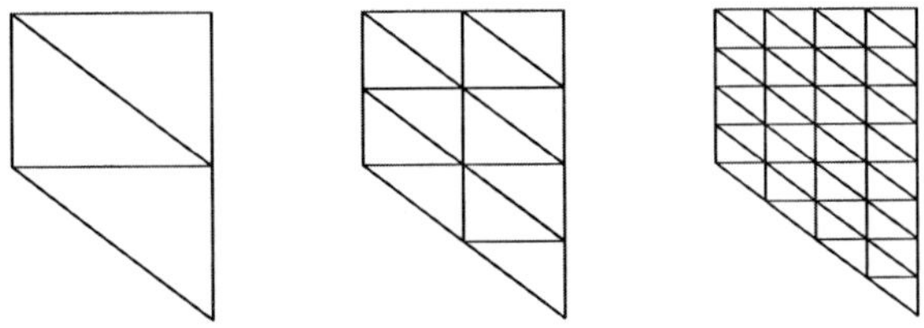

Figure 4.16. *Uniform refinement.*

Programming considerations

The essential features of a typical computer program implementing the finite element method are the following:

- Input of data such as the domain Ω, the right-hand side function f, the boundary data b and g (cf. (4.74)), and the coefficients that may appear in a differential problem.
- Construction of the triangulation K_h.
- Computation and assembly of the stiffness matrix $\mathbf{A}$ and the right-hand side vector $\mathbf{f}$.
- Solution of the linear system of algebraic equations $\mathbf{Ap} = \mathbf{f}$.
- Output of the computational results.

The data input can be easily implemented in a small subroutine, and the result output depends on the computer system and software used. Here we briefly discuss the other three parts. As an illustration, we focus on two dimensions.

(i) Construction of the triangulation K_h

The triangulation K_h can be constructed from a successive refinement of an initial coarse partition of Ω; fine triangles can be obtained by connecting the midpoints of edges of coarse triangles, for example. A sequence of uniform refinements leads to *quasi-uniform* grids where the triangles in K_h essentially have the same size in all regions of Ω (cf. Figure 4.16). If the boundary Γ of Ω is a curve, special care needs to be taken near Γ (cf. Section 4.2.2).

In practical applications, it is often necessary to use triangles in K_h that vary considerably in size in different regions of Ω. For example, one utilizes smaller triangles in regions where the exact solution has a fast variation or where its certain derivatives are large (cf. Figure 4.17, where a *local refinement* strategy is carried out). In this strategy, proper care is taken in the transition zone between regions with triangles of different sizes so that a *regular* local refinement results (i.e, no vertex of one triangle lies in the interior of an edge of another triangle; see Section 4.7). Methods that automatically refine grids where needed are called *adaptive methods*, and will be studied in detail in Section 4.7.

Let a triangulation K_h have M nodes and $\mathcal{M}$ triangles. The triangulation can be represented by two arrays $\mathbf{Z}(2, M)$ and $\mathcal{Z}(3, \mathcal{M})$, where $\mathbf{Z}(i, j)$ $(i = 1, 2)$ indicates the coordinates of the jth node, $j = 1, 2, \ldots, M$, and $\mathcal{Z}(i, k)$ $(i = 1, 2, 3)$ enumerates the nodes of the kth triangle, $k = 1, 2, \ldots, \mathcal{M}$. An example is given in Figure 4.18, where

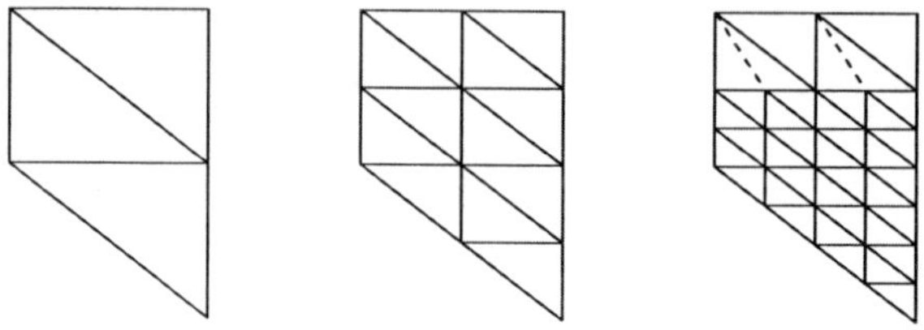

Figure 4.17. *Nonuniform refinement.*

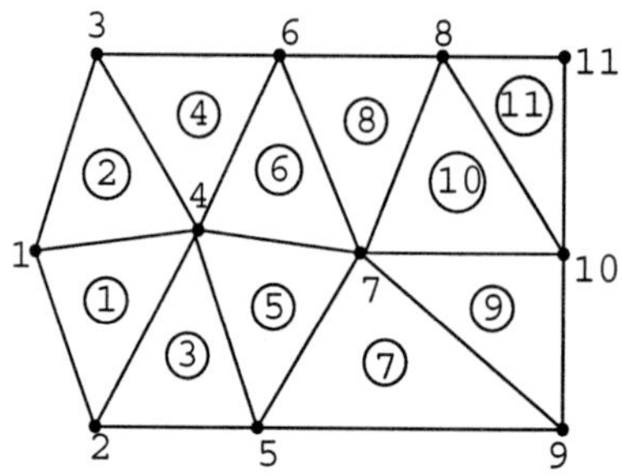

Figure 4.18. *Node and triangle enumeration.*

the triangle numbers are in circles. For this example, the array $\mathcal{Z}(3, \mathcal{M})$ has the form, where $M = \mathcal{M} = 11$,

$$\mathcal{Z} = \begin{pmatrix} 1 & 1 & 2 & 3 & 4 & 4 & 5 & 6 & 7 & 7 & 8 \\ 2 & 4 & 5 & 4 & 5 & 7 & 9 & 7 & 9 & 10 & 10 \\ 4 & 3 & 4 & 6 & 7 & 6 & 7 & 8 & 10 & 8 & 11 \end{pmatrix}.$$

If a direct method (Gaussian elimination) is employed to solve the linear system $\mathbf{Ap} = \mathbf{f}$, the nodes should be enumerated in such a way that the *bandwidth* of each row in $\mathbf{A}$ is as small as possible. This matter will be studied in the next chapter, in connection with the discussion of solution algorithms for linear systems.

In general, when local refinement is involved in a triangulation K_h, it is very difficult to enumerate the nodes and triangles efficiently; some strategies will be given in Section 4.7. For a simple domain Ω (e.g., a convex polygonal Ω), it is rather easy to construct and represent a triangulation that utilizes uniform refinement in the whole domain.

(ii) Assembly of the stiffness matrix

After the triangulation K_h is constructed, one computes the *element stiffness matrices* with entries a_{ij}^K given by (4.72). We recall that $a_{ij}^K = 0$ unless nodes $\mathbf{x}_i$ and $\mathbf{x}_j$ are both vertices of $K \in K_h$.

For a kth triangle K_k, $\mathcal{Z}(m, k)$ $(m = 1, 2, 3)$ are the numbers of the vertices of K_k, and the element stiffness matrix $\mathbf{A}^{(k)} = \left(a_{mn}^k\right)_{m,n=1}^3$ is calculated as

$$a_{mn}^k = \int_{K_k} \nabla\varphi_m \cdot \nabla\varphi_n \, d\mathbf{x}, \qquad m,\ n = 1, 2, 3,$$

where the (linear) basis function φ_m over K_k satisfies

$$\varphi_m(\mathbf{x}_{\mathcal{Z}(n,k)}) = \begin{cases} 1 & \text{if } m = n, \\ 0 & \text{if } m \neq n. \end{cases}$$

The right-hand side $\mathbf{f}$ over K_k is computed by

$$f_m^k = \int_{K_k} f \varphi_m \, d\mathbf{x}, \qquad m = 1, 2, 3.$$

Note that m and n are the local numbers of the three vertices of K_k, while i and j used in (4.72) are the global numbers of vertices in K_h.

To assemble the global matrix $\mathbf{A} = (a_{ij})$ and the right-hand-side vector $\mathbf{f} = (f_j)$, one loops over all triangles K_k and successively adds the contributions from different K_k's:

$$\begin{aligned} &\text{For } k = 1, 2, \dots, \mathcal{M}, \text{ compute} \\ &a_{\mathcal{Z}(m,k),\mathcal{Z}(n,k)} = a_{\mathcal{Z}(m,k),\mathcal{Z}(n,k)} + a_{mn}^k, \\ &f_{\mathcal{Z}(m,k)} = f_{\mathcal{Z}(m,k)} + f_m^k, \qquad m, \; n = 1, 2, 3. \end{aligned}$$

The approach used is *element-oriented*; that is, we loop over elements (i.e., triangles). Experience shows that this approach is more efficient than the *node-oriented* approach (i.e., looping over all nodes); the latter approach wastes much time in repeated computations of $\mathbf{A}$ and $\mathbf{f}$.

(iii) Solution of a linear system

The solution of the linear system $\mathbf{Ap} = \mathbf{f}$ can be performed via a direct algorithm (Gaussian elimination) or an iterative algorithm (e.g., the conjugate gradient algorithm), which will be discussed in the next chapter. Here we just mention that in using these two algorithms, it is not necessary to exploit an array $\mathbf{A}(M, M)$ to store the stiffness matrix $\mathbf{A}$. Instead, since $\mathbf{A}$ is sparse and usually a banded matrix, only the nonzero entries of $\mathbf{A}$ need to be stored, say, in a one-dimensional array.

Finite element spaces

In the previous subsections, we have considered the finite element space of piecewise linear functions. Here we describe more general finite element spaces.

(i) Triangles

We first treat the case where $\Omega \subset \mathbb{R}^2$ is a polygonal domain in the plane. Let K_h be a triangulation of Ω into triangles K as earlier. We introduce the notation

$$P_r(K) = \{v : \; v \text{ is a polynomial of degree at most } r \text{ on } K\},$$

where $r = 0, 1, 2, \dots$. For $r = 1$, $P_1(K)$ is the space of *linear functions*, used previously, of the form

$$v(\mathbf{x}) = v_{00} + v_{10}x_1 + v_{01}x_2, \qquad \mathbf{x} = (x_1, x_2) \in K, \; v \in P_1(K),$$

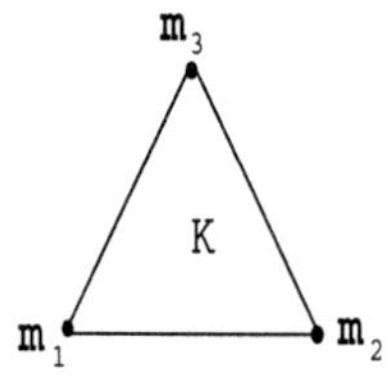

Figure 4.19. *The element degrees of freedom for $P_1(K)$.*

where $v_{ij} \in \mathbb{R}$, $i, j = 0, 1$. Note that $\dim(P_1(K)) = 3$; i.e., its *dimension* is three.

For $r = 2$, $P_2(K)$ is the space of *quadratic* functions on K:

$$v(\mathbf{x}) = v_{00} + v_{10}x_1 + v_{01}x_2 + v_{20}x_1^2 + v_{11}x_1x_2 + v_{02}x_2^2, \qquad v \in P_2(K),$$

where $v_{ij} \in \mathbb{R}$, $i, j = 0, 1, 2$. We see that $\dim(P_2(K)) = 6$.

In general, we have

$$P_r(K) = \left\{ v : v(\mathbf{x}) = \sum_{0 \le i+j \le r} v_{ij} x_1^i x_2^j, \ \mathbf{x} \in K, \ v_{ij} \in \mathbb{R} \right\}, \qquad r \ge 0,$$

so

$$\dim(P_r(K)) = \frac{(r+1)(r+2)}{2}.$$

Example 4.1. Define

$$V_h = \{v : v \text{ is continuous on } \Omega \text{ and } v|_K \in P_1(K), \ K \in K_h\},$$

where $v|_K$ represents the restriction of v to K. As parameters, or *global degrees of freedom*, to describe the functions in V_h, we use the values at the vertices (nodes) of K_h. To see that this is a legitimate choice, for each triangle $K \in K_h$, let its vertices be indicated by $\mathbf{m}_1$, $\mathbf{m}_2$, and $\mathbf{m}_3$ (cf. Figure 4.19). Also, let the (local) basis functions of $P_1(K)$ be λ_i, $i = 1, 2, 3$, which are defined by

$$\lambda_i(\mathbf{m}_j) = \begin{cases} 1 & \text{if } i = j, \\ 0 & \text{if } i \ne j, \end{cases} \qquad i, j = 1, 2, 3.$$

These basis functions can be determined in the following approach: Let an equation of the straight line through the vertices $\mathbf{m}_2$ and $\mathbf{m}_3$ be given by

$$c_0 + c_1x_1 + c_2x_2 = 0,$$

and then define

$$\lambda_1(\mathbf{x}) = \gamma(c_0 + c_1x_1 + c_2x_2), \qquad \mathbf{x} = (x_1, x_2),$$

where the constant γ is chosen such that $\lambda_1(\mathbf{m}_1) = 1$. The functions λ_2 and λ_3 can be determined in the same approach. These functions λ_1, λ_2, and λ_3 are sometimes called the *barycentric coordinates* of a triangle. If K is the reference triangle with vertices $(1, 0)$,

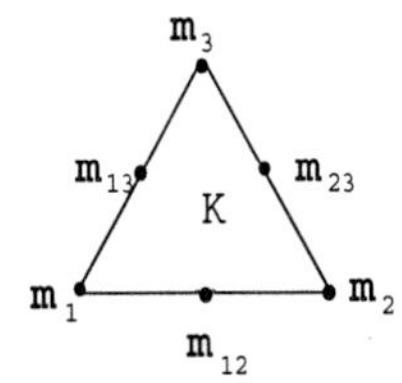

Figure 4.20. *The element degrees of freedom for $P_2(K)$.*

$(0, 1)$, and $(0, 0)$, then λ_1, λ_2, and λ_3 are, respectively, x_1, x_2, and $1 - x_1 - x_2$. Now, any function $v \in P_1(K)$ has the unique representation

$$v(\mathbf{x}) = \sum_{i=1}^{3} v(\mathbf{m}_i)\lambda_i(\mathbf{x}), \qquad \mathbf{x} \in K.$$

Thus $v \in P_1(K)$ is uniquely determined by its values at the three vertices. Therefore, on each triangle $K \in K_h$, the degrees of freedom, *element degrees of freedom*, can be these (nodal) values. These degrees of freedom are the global degrees of freedom and were used to construct the basis functions in V_h before.

For v such that $v|_K \in P_1(K)$, $K \in K_h$, if it is continuous at internal vertices, then $v \in C^0(\bar{\Omega})$ (Chen, 2005), where $C^0(\bar{\Omega})$ is the set of continuous functions on $\bar{\Omega}$.

Example 4.2. Let

$$V_h = \{v : \ v \text{ is continuous on } \Omega \text{ and } v|_K \in P_2(K), \ K \in K_h\}.$$

Namely, V_h is the space of continuous piecewise quadratic functions. The global degrees of freedom of a function $v \in V_h$ are chosen by the values of v at the vertices and the midpoints of edges in K_h. It can be shown that v is uniquely defined by these degrees of freedom (Chen, 2005). For each $K \in K_h$, the element degrees of freedom are shown in Figure 4.20, where the midpoints of edges of K are denoted by $\mathbf{m}_{ij}$, $i < j, i, j = 1, 2, 3$.

It can be seen (cf. Exercise 4.32) that a function $v \in P_2(K)$ has the representation

$$\begin{aligned} v(\mathbf{x}) &= \sum_{i=1}^{3} v(\mathbf{m}_i)\lambda_i(\mathbf{x})\big(2\lambda_i(\mathbf{x}) - 1\big) \\ &\quad + \sum_{i,j=1;\ i<j}^{3} 4v(\mathbf{m}_{ij})\lambda_i(\mathbf{x})\lambda_j(\mathbf{x}), \qquad \mathbf{x} \in K. \end{aligned} \tag{4.78}$$

Also, as in Example 4.1, one can prove that if v is continuous at the internal vertices and midpoints of edges and $v \in P_2(K)$, $K \in K_h$, then $v \in C^0(\bar{\Omega})$ (Chen, 2005).

Example 4.3. Set

$$V_h = \{v : \ v \text{ is continuous on } \Omega \text{ and } v|_K \in P_3(K), \ K \in K_h\}.$$

That is, V_h is the space of continuous piecewise cubic functions. Let $K \in K_h$ have vertices $\mathbf{m}_i$, $i = 1, 2, 3$. Define, for $i, \ j = 1, 2, 3, \ i \neq j$,

$$\mathbf{m}_0 = \frac{1}{3}(\mathbf{m}_1 + \mathbf{m}_2 + \mathbf{m}_3), \quad \mathbf{m}_{i,j} = \frac{1}{3}\big(2\mathbf{m}_i + \mathbf{m}_j\big),$$

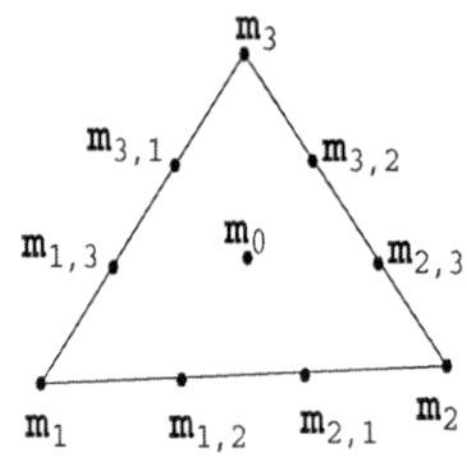

Figure 4.21. *The element degrees of freedom for $P_3(K)$.*

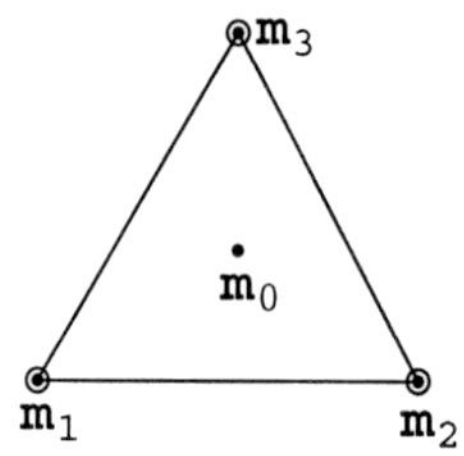

Figure 4.22. *The second set of degrees of freedom for $P_3(K)$.*

where $\mathbf{m}_0$ is the *center of gravity* of K (centroid); see Figure 4.21. It can be proven that a function $v \in P_3(K)$ is uniquely determined by the values (Chen, 2005)

$$v(\mathbf{m}_i), \quad v(\mathbf{m}_0), \quad v\left(\mathbf{m}_{i,j}\right), \quad i,\ j = 1, 2, 3,\ i \neq j.$$

These values can be used as the degrees of freedom.

Example 4.4. The degrees of freedom for $P_3(K)$ (and thus for V_h) can be chosen in a different way. A function $v \in P_3(K)$ is also uniquely defined by (cf. Figure 4.22)

$$v(\mathbf{m}_i), \quad v(\mathbf{m}_0), \quad \frac{\partial v}{\partial x_j}(\mathbf{m}_i), \quad i = 1, 2, 3,\ j = 1, 2.$$

The corresponding finite element space $V_h \subset C^0(\bar{\Omega})$ is defined by

$$V_h = \left\{ v : v \text{ and } \frac{\partial v}{\partial x_i}\ (i = 1, 2) \text{ are continuous at} \right.$$
$$\left. \text{vertices of } K_h;\ v|_K \in P_3(K),\ K \in K_h \right\}.$$

We have considered the cases $r \leq 3$. In general, for any $r \geq 1$, we define

$$V_h = \{v :\ v \text{ is continuous on } \Omega \text{ and } v|_K \in P_r(K),\ K \in K_h\}.$$

A function $v \in P_r(K)$ can be uniquely determined by its values at the three vertices, $3(r-1)$ distinct points on the edges, and $(r-1)(r-2)/2$ interior points in K. The values at these points can be employed as the degrees of freedom in V_h.

Figure 4.23. *The element degrees of freedom for* $Q_1(K)$.

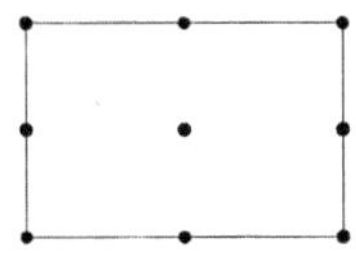

Figure 4.24. *The element degrees of freedom for* $Q_2(K)$.

(ii) Rectangles

We now consider the case where Ω is a rectangular domain and K_h is a partition of Ω into nonoverlapping rectangles such that the horizontal and vertical edges of rectangles are parallel to the x_1- and x_2-coordinate axes, respectively. We also require that no vertex of any rectangle lie in the interior of an edge of another rectangle. We introduce the notation

$$Q_r(K) = \left\{ v : v(\mathbf{x}) = \sum_{i,j=0}^{r} v_{ij} x_1^i x_2^j,\ \mathbf{x} \in K,\ v_{ij} \in \mathbb{R} \right\}, \qquad r \geq 0.$$

Note that $\dim(Q_r(K)) = (r+1)^2$.

For $r = 1$, we define

$$V_h = \{ v : v \text{ is continuous on } \Omega \text{ and } v|_K \in Q_1(K),\ K \in K_h \}.$$

A function $v \in Q_1(K)$ is *bilinear* and of the form

$$v(\mathbf{x}) = v_{00} + v_{10}x_1 + v_{01}x_2 + v_{11}x_1x_2, \qquad \mathbf{x} = (x_1, x_2) \in K,\ v_{ij} \in \mathbb{R}.$$

As in the triangular case, it can be checked that v is uniquely determined by its values at the four vertices of K, which can be chosen as the degrees of freedom for V_h (cf. Figure 4.23).

For $r = 2$, define

$$V_h = \{ v : v \text{ is continuous on } \Omega \text{ and } v|_K \in Q_2(K),\ K \in K_h \},$$

where $Q_2(K)$ is the set of *biquadratic* functions on K. The degrees of freedom can be chosen by the values of functions at the vertices, midpoints of edges, and center of each rectangle (cf. Figure 4.24). Other cases $r \geq 3$ can be analogously discussed.

The use of rectangles requires that the geometry of Ω be special. Thus it is of interest to utilize more general *quadrilaterals*, which will be considered in Section 4.2.2, in connection with *isoparametric finite elements* (cf. Exercise 4.33).

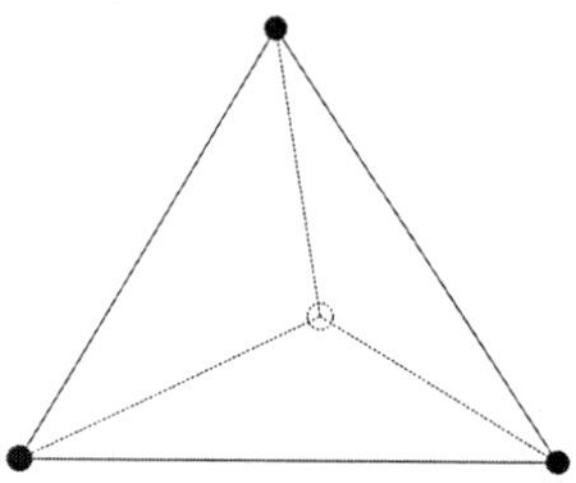

Figure 4.25. *The element degrees of freedom for $P_1(K)$ on a tetrahedron.*

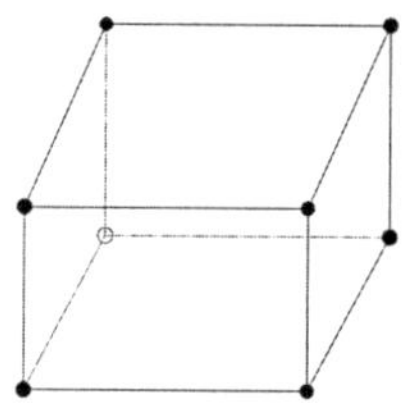

Figure 4.26. *The element degrees of freedom for $Q_1(K)$ on a parallelepiped.*

(iii) Three dimensions

Example 4.5. In three dimensions, for a polygonal domain $\Omega \subset \mathbb{R}^3$, let K_h be a partition of Ω into nonoverlapping *tetrahedra* such that no vertex of any tetrahedron lies in the interior of an edge or face of another tetrahedron. For each $K \in K_h$ and $r \geq 0$, set

$$P_r(K) = \left\{ v : v(\mathbf{x}) = \sum_{0 \leq i+j+k \leq r} v_{ijk} x_1^i x_2^j x_3^k, \ \mathbf{x} \in K, \ v_{ijk} \in \mathbb{R} \right\},$$

where $\mathbf{x} = (x_1, x_2, x_3)$ and

$$\dim(P_r(K)) = \frac{(r+1)(r+2)(r+3)}{6}.$$

For $r = 1$, the function values of $v \in P_1(K)$ at the four vertices of K can be utilized as the degrees of freedom (cf. Figure 4.25). Other cases $r \geq 2$ can be also handled.

Example 4.6. Let Ω be a rectangular domain in $\mathbb{R}^3$, and K_h be a partition of Ω into nonoverlapping *rectangular parallelepipeds* such that the faces are parallel to the x_1-, x_2-, and x_3-coordinate axes, respectively. For each $K \in K_h$, the polynomials we use for V_h are of the type

$$Q_r(K) = \left\{ v : v(\mathbf{x}) = \sum_{i,j,k=0}^{r} v_{ijk} x_1^i x_2^j x_3^k, \ \mathbf{x} \in K, \ v_{ijk} \in \mathbb{R} \right\}, \qquad r \geq 0.$$

Note that $\dim(Q_r(K)) = (r+1)^3$. For $r = 1$, the function values of $v \in Q_1(K)$ at the eight vertices of K can be utilized as the degrees of freedom (cf. Figure 4.26).

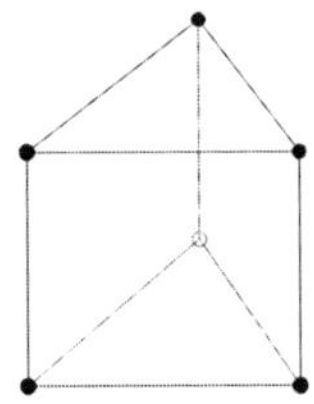

Figure 4.27. *The element degrees of freedom for $P_{1,1}(K)$ on a prism.*

Example 4.7. Let $\Omega \subset \mathbb{R}^3$ be a domain of the form $\Omega = G \times [l_1, l_2]$, where $G \subset \mathbb{R}^2$ and l_1 and l_2 are real numbers. Let K_h be a partition of Ω into *prisms* such that their base is a triangle in the (x_1, x_2)-plane with three vertical edges parallel to the x_3-axis. Define $P_{l,r}$ to be the space of polynomials of degree l in the two variables x_1 and x_2 and of degree r in the variable x_3. That is, for each $K \in K_h$ and $l,\ r \geq 0$,

$$P_{l,r}(K) = \left\{ v : \ v(\mathbf{x}) = \sum_{0 \leq i+j \leq l} \sum_{k=0}^{r} v_{ijk} x_1^i x_2^j x_3^k, \ \mathbf{x} \in K, \ v_{ijk} \in \mathbb{R} \right\}.$$

Note that $\dim(P_{l,r}(K)) = (l+1)(l+2)(r+1)/2$. For $l = 1$ and $r = 1$, the function values of $v \in P_{1,1}(K)$ at the six vertices of K can be utilized as the degrees of freedom (cf. Figure 4.27).

In summary, a *finite element* is a triple $(K, P(K), \Sigma_K)$, where K is a geometric object (i.e., element), $P(K)$ is a finite-dimensional linear space of functions on K, and Σ_K is a set of degrees of freedom such that a function $v \in P(K)$ is uniquely defined by Σ_K. For instance, in Example 4.1, K is a triangle, $P(K) = P_1(K)$, and Σ_K is the set of the values at the vertices of K. When Σ_K includes the values of partial derivatives of functions, the finite element is said to be of *Hermite type*, as in Examples 4.4. When all degrees of freedom are given by function values, the finite element is called a *Lagrange* element.

Error estimates

For a bounded domain Ω in the plane, we define the space of *square integrable functions* on Ω

$$L^2(\Omega) = \left\{ v : \ v \text{ is defined on } \Omega \text{ and } \int_\Omega v^2 \, d\mathbf{x} < \infty \right\}.$$

This space is equipped with the *scalar product*

$$(v, w) = \int_\Omega vw \, d\mathbf{x}$$

and the corresponding *norm* (the $L^2(\Omega)$-norm)

$$\|v\| \equiv \|v\|_{L^2(\Omega)} = (v, v)^{1/2}.$$

To introduce other spaces used in this section, we define

$$D^\alpha v = \frac{\partial^{|\alpha|} v}{\partial x_1^{\alpha_1} \partial x_2^{\alpha_2}},$$

where $\alpha = (\alpha_1, \alpha_2)$ is an index, with α_1 and α_2 nonnegative integers, and $|\alpha| = \alpha_1 + \alpha_2$. This notation indicates a partial derivative of v. For example, a second partial derivative can be written as $D^\alpha v$ with $\alpha = (2, 0)$, $\alpha = (1, 1)$, or $\alpha = (0, 2)$. For $r = 1, 2, \ldots$, we define

$$H^r(\Omega) = \left\{ v \in L^2(\Omega) : \ D^\alpha v \in L^2(\Omega) \ \forall |\alpha| \le r \right\}$$

with the norm

$$\|v\|_{H^r(\Omega)} = \left(\sum_{|\alpha| \le r} \int_\Omega |D^\alpha v|^2 \, d\mathbf{x} \right)^{1/2}.$$

A seminorm is defined by

$$|v|_{H^r(\Omega)} = \left(\sum_{|\alpha| = r} \int_\Omega |D^\alpha v|^2 \, d\mathbf{x} \right)^{1/2}.$$

Namely, the functions in $H^r(\Omega)$, together with their derivatives $D^\alpha v$ of order $|\alpha|$ at most r, are square integrable. These are examples of *Sobolev spaces* (Adams, 1975). They can be defined in three dimensions in the same fashion.

For $h > 0$, let K_h be a triangulation of Ω into triangles. For $K \in K_h$, as previously we define the mesh parameters

$$h_K = \text{diam}(K) = \text{the longest edge of } \bar{K}, \qquad h = \max_{K \in K_h} \text{diam}(K).$$

We also need the quantity

$$\rho_K = \text{the diameter of the circle inscribed in } K.$$

We say that a triangulation is *regular* if there is a constant β_1, independent of h, such that

$$\frac{h_K}{\rho_K} \le \beta_1 \qquad \forall K \in K_h. \tag{4.79}$$

This condition says that the triangles in K_h are not arbitrarily thin, or equivalently, the angles of the triangles are not arbitrarily small. The constant β_1 is a measure of the smallest angle over all $K \in K_h$.

As an example of an error estimate, let us consider problem (4.65) and its discrete version (4.70), where the finite element space V_h is

$$V_h = \left\{ v \in H^1(\Omega) : \ v|_K \in P_r(K), \ K \in K_h, \ \text{and } v\big|_\Gamma = 0 \right\}$$

for $r \ge 1$. Then a typical *error estimate* is (Ciarlet, 1978; Chen, 2005)

$$\|p - p_h\|_{H^1(\Omega)} \le C h^r |p|_{H^{r+1}(\Omega)}, \tag{4.80}$$

where the constant C depends only on r and β_1 in (4.79). To state an estimate in the $L^2(\Omega)$-norm, we require that the polygonal domain Ω be convex; if Ω has a smooth boundary, convexity is not required. In the convex case, we have

$$\|p - p_h\|_{L^2(\Omega)} \le C h^{r+1} |p|_{H^{r+1}(\Omega)}. \tag{4.81}$$

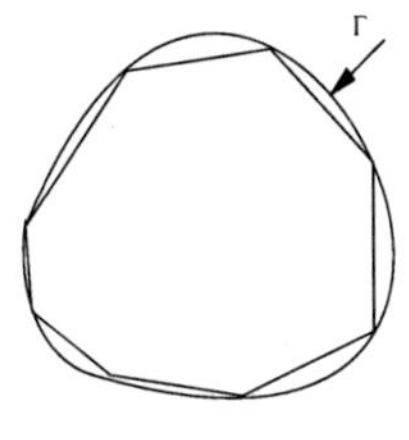

Figure 4.28. *A polygonal line approximation of* Γ.

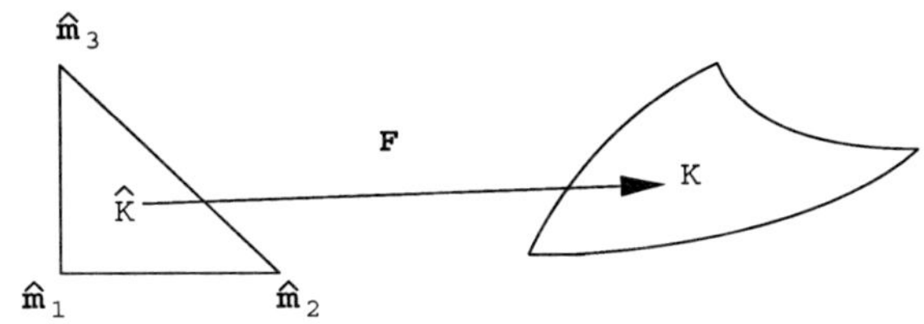

Figure 4.29. *The mapping* **F**.

These estimates are valid for other finite element spaces considered in this section, and are *optimal* (i.e., the estimates with the largest power of h one can get between the exact solution and its approximate solution). A proof of estimates (4.80) and (4.81) can be found in Chen (2005).

4.2.2 General domains

In the construction of finite element spaces so far, we have assumed that the domain Ω is polygonal. In this section, we consider the case where Ω is curved. For simplicity, we focus on two space dimensions.

For a two-dimensional domain Ω, the simplest approximation Γ_h for its curved boundary Γ is a polygonal line (cf. Figure 4.28). The resulting error (the maximal distance from Γ to Γ_h) due to this approximation is of order $\mathcal{O}(h^2)$, where h is the mesh size as usual (cf. Exercise 4.34). To obtain a more accurate approximation, we can approximate Γ with piecewise polynomials of degree $r \geq 2$. The error in this approximation becomes $\mathcal{O}(h^{r+1})$. In the partition of such an approximated domain, the elements closest to Γ then have at least one curved edge.

As an example, let $(\hat{K}, P(\hat{K}), \Sigma_{\hat{K}})$ be a finite element, where $\hat{K}$ is the *reference triangle* with vertices $\hat{\mathbf{m}}_1 = (0,0)$, $\hat{\mathbf{m}}_2 = (1,0)$, and $\hat{\mathbf{m}}_3 = (0,1)$ in the $\hat{\mathbf{x}}$-plane. Furthermore, assume that this element is of the Lagrange type; that is, all degrees of freedom are defined by the function values at certain points $\hat{\mathbf{m}}_i$, $i = 1, 2, \ldots, l$ (cf. Section 4.2.1). Suppose that $\mathbf{F}$ is a one-to-one mapping of $\hat{K}$ onto a curved triangle K in the $\mathbf{x}$-plane with inverse $\mathbf{F}^{-1}$; i.e., $K = \mathbf{F}(\hat{K})$ (cf. Figure 4.29). Then we define

$$P(K) = \left\{ v : v(\mathbf{x}) = \hat{v}\left(\mathbf{F}^{-1}(\mathbf{x})\right),\ \mathbf{x} \in K,\ \hat{v} \in P(\hat{K}) \right\},$$
$$\Sigma_K \text{ consists of function values at } \mathbf{m}_i = \mathbf{F}(\hat{\mathbf{m}}_i),\ i = 1, 2, \ldots, l.$$

If $\mathbf{F} = (F_1, F_2)$ is of the same type as the functions in $P(K)$, i.e., $F_1,\ F_2 \in P(K)$, then we say that the element $(K, P(K), \Sigma_K)$ is an *isoparametric element.* In general, $\mathbf{F}^{-1}$ is not a polynomial, and thus the functions $v \in P(K)$ for a curved element are not polynomials either.

Let $K_h = \{K\}$ be a triangulation of Ω into "triangles," where some of them may have one or more curved edges, and let Ω_h be the union of these triangles in K_h. Note that Ω_h is an approximation of Ω with a piecewise smooth boundary. Now, the finite element space V_h is

$$V_h = \left\{v \in H^1(\Omega_h) : v|_K \in P(K),\ K \in K_h\right\}.$$

With this space, the finite element method can be defined as in (4.70) for the Poisson equation (4.65), for example. Moreover, error estimates analogous to (4.80) and (4.81) hold.

We now consider the computation of a stiffness matrix. Let $\{\hat{\varphi}_i\}_{i=1}^l$ be a basis of $P(\hat{K})$. We define

$$\varphi_i(\mathbf{x}) = \hat{\varphi}_i\left(\mathbf{F}^{-1}(\mathbf{x})\right), \qquad \mathbf{x} \in K,\ i = 1, 2, \ldots, l.$$

For (4.65), we need to compute (cf. Section 4.2.1)

$$a^K(\varphi_i, \varphi_j) = \int_K \nabla\varphi_i \cdot \nabla\varphi_j \, d\mathbf{x}, \qquad i,\ j = 1, 2, \ldots, l. \tag{4.82}$$

It follows from the chain rule that

$$\frac{\partial \varphi_i}{\partial x_k} = \frac{\partial}{\partial x_k}\left(\hat{\varphi}_i\left(\mathbf{F}^{-1}(\mathbf{x})\right)\right) = \frac{\partial \hat{\varphi}_i}{\partial \hat{x}_1}\frac{\partial \hat{x}_1}{\partial x_k} + \frac{\partial \hat{\varphi}_i}{\partial \hat{x}_2}\frac{\partial \hat{x}_2}{\partial x_k}$$

for $k = 1, 2$. Consequently, we see that

$$\nabla\varphi_i = \mathbf{G}^{-T}\nabla\hat{\varphi}_i,$$

where $\mathbf{G}^{-T}$ is the transpose of the Jacobian of $\mathbf{F}^{-1}$:

$$\mathbf{G}^{-T} = \begin{pmatrix} \dfrac{\partial \hat{x}_1}{\partial x_1} & \dfrac{\partial \hat{x}_2}{\partial x_1} \\ \dfrac{\partial \hat{x}_1}{\partial x_2} & \dfrac{\partial \hat{x}_2}{\partial x_2} \end{pmatrix}.$$

When we apply the change of variable $\mathbf{F} : \hat{K} \to K$ to (4.82), we have

$$a^K(\varphi_i, \varphi_j) = \int_{\hat{K}} \left(\mathbf{G}^{-T}\nabla\hat{\varphi}_i\right) \cdot \left(\mathbf{G}^{-T}\nabla\hat{\varphi}_j\right) |\det \mathbf{G}| \, d\hat{\mathbf{x}} \tag{4.83}$$

for $i, j = 1, 2, \ldots, l$, where $|\det \mathbf{G}|$ is the absolute value of the determinant of the Jacobian $\mathbf{G}$:

$$\mathbf{G} = \begin{pmatrix} \dfrac{\partial x_1}{\partial \hat{x}_1} & \dfrac{\partial x_1}{\partial \hat{x}_2} \\ \dfrac{\partial x_2}{\partial \hat{x}_1} & \dfrac{\partial x_2}{\partial \hat{x}_2} \end{pmatrix}.$$

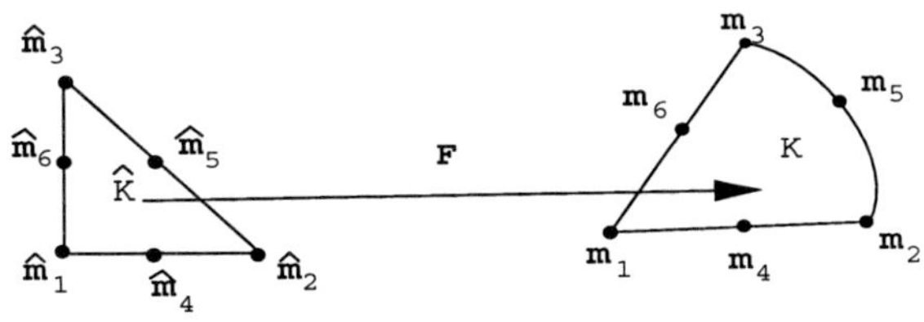

Figure 4.30. *An example of the mapping* **F**.

Applying an algebraic computation, we see that

$$\mathbf{G}^{-T} = \left(\mathbf{G}^{-1}\right)^T = \frac{1}{\det \mathbf{G}} \mathbf{G}',$$

where

$$\mathbf{G}' = \begin{pmatrix} \dfrac{\partial x_2}{\partial \hat{x}_2} & -\dfrac{\partial x_2}{\partial \hat{x}_1} \\ -\dfrac{\partial x_1}{\partial \hat{x}_2} & \dfrac{\partial x_1}{\partial \hat{x}_1} \end{pmatrix}.$$

Hence (4.83) becomes

$$a^K(\varphi_i, \varphi_j) = \int_{\hat{K}} \left(\mathbf{G}'\nabla\hat{\varphi}_i\right) \cdot \left(\mathbf{G}'\nabla\hat{\varphi}_j\right) \frac{1}{|\det \mathbf{G}|} \, d\hat{\mathbf{x}} \tag{4.84}$$

for $i, j = 1, 2, \ldots, l$. Therefore, the matrix entry a_{ij} on K can be calculated by either (4.83) or (4.84). In general, it is difficult to evaluate these two integrals analytically. However, they can be relatively easily evaluated using a numerical integration formula (or a *quadrature rule*); see the next section for more details.

Now, we give an example of constructing the mapping $\mathbf{F} : \hat{K} \to K$. Let the reference triangle $\hat{K}$ have vertices $\hat{\mathbf{m}}_i$, $i = 1, 2, 3$, and midpoints $\hat{\mathbf{m}}_i$ of the edges, $i = 4, 5, 6$. Furthermore, let $P(\hat{K}) = P_2(\hat{K})$ and let $\Sigma_{\hat{K}}$ be composed of the function values at $\hat{\mathbf{m}}_i$, $i = 1, 2, \ldots, 6$. Define the basis functions $\hat{\varphi}_i \in P_2(\hat{K})$ by

$$\hat{\varphi}_i(\hat{\mathbf{m}}_j) = \delta_{ij}, \qquad i, j = 1, 2, \ldots, 6.$$

Also, let the points $\mathbf{m}_i$, $i = 1, 2, \ldots, 6$, in the $\mathbf{x}$-plane satisfy that $\mathbf{m}_4$ and $\mathbf{m}_6$ are the midpoints of the line segments $\mathbf{m}_1\mathbf{m}_2$ and $\mathbf{m}_1\mathbf{m}_3$, respectively, and $\mathbf{m}_5$ is slightly displaced from the line segment $\mathbf{m}_2\mathbf{m}_3$ (cf. Figure 4.30). We now define $\mathbf{F}$ by

$$\mathbf{F}(\hat{\mathbf{x}}) = \sum_{i=1}^{6} \mathbf{m}_i \hat{\varphi}_i(\hat{\mathbf{x}}), \qquad \hat{\mathbf{x}} \in \hat{K}.$$

Clearly, $\mathbf{m}_i = \mathbf{F}(\hat{\mathbf{m}}_i)$, $i = 1, 2, \ldots, 6$. Moreover, it can be shown that $\mathbf{F}$ is one-to-one for sufficiently small h_K (Johnson, 1994), i.e., for sufficiently fine triangulations near Γ.

4.2.3 Quadrature rules

As mentioned previously, some integrals such as (4.83) and (4.84) can be evaluated only approximately. We can use a *quadrature rule* of the type

$$\int_K g(\mathbf{x})\,d\mathbf{x} \approx \sum_{i=1}^{m} w_i g(\mathbf{x}_i), \tag{4.85}$$

where $w_i > 0$ are certain weights and the points $\mathbf{x}_i$ are in the element K. If the quadrature rule (4.85) is exact for polynomials of degree r,

$$\int_K g(\mathbf{x})\,d\mathbf{x} = \sum_{i=1}^{m} w_i g(\mathbf{x}_i), \qquad g \in P_r(K), \tag{4.86}$$

then the error in using (4.85) can be bounded by (Ciarlet and Raviart, 1972)

$$\left| \int_K g(\mathbf{x})\,d\mathbf{x} - \sum_{i=1}^{m} w_i g(\mathbf{x}_i) \right| \leq C h_K^{r+1} \sum_{|\alpha|=r+1} \int_K |D^\alpha g(\mathbf{x})|\,d\mathbf{x},$$

where $r > 0$; refer to Section 4.2.1 for the definition of $D^\alpha g$. Several examples are presented below, where r indicates the maximum degree of polynomials for which (4.86) holds.

Example 4.8. Let K be a triangle with vertices $\mathbf{m}_i$, midpoints $\mathbf{m}_{ij}$, $i, j = 1, 2, 3$, $i < j$, and the center of gravity $\mathbf{m}_0$. Also, let $|K|$ indicate the area of K. Then we have

$$\int_K g(\mathbf{x})\,d\mathbf{x} \approx |K| g(\mathbf{m}_0), \qquad \text{where } r = 1,$$

$$\int_K g(\mathbf{x})\,d\mathbf{x} \approx \frac{|K|}{3} \left(g(\mathbf{m}_{12}) + g(\mathbf{m}_{23}) + g(\mathbf{m}_{13})\right), \quad \text{where } r = 2,$$

$$\begin{aligned} \int_K g(\mathbf{x})\,d\mathbf{x} \approx |K| \Bigg\{ &\sum_{i=1}^{3} \frac{g(\mathbf{m}_i)}{20} + \frac{9g(\mathbf{m}_0)}{20} \\ &+ \frac{2}{15}\left(g(\mathbf{m}_{12}) + g(\mathbf{m}_{23}) + g(\mathbf{m}_{13})\right) \Bigg\}, \quad \text{where } r = 3. \end{aligned}$$

Example 4.9. Let K be a rectangle centered at the origin and with edges parallel to the x_1- and x_2-coordinate axes of lengths $2h_1$ and $2h_2$, respectively. Then

$$\int_K g(\mathbf{x})\,d\mathbf{x} \approx |K| g(0), \qquad \text{where } r = 1,$$

$$\begin{aligned} \int_K g(\mathbf{x})\,d\mathbf{x} \approx \frac{|K|}{4} \Bigg\{ &g\left(\frac{h_1}{\sqrt{3}}, \frac{h_2}{\sqrt{3}}\right) + g\left(\frac{h_1}{\sqrt{3}}, -\frac{h_2}{\sqrt{3}}\right) \\ &+ g\left(-\frac{h_1}{\sqrt{3}}, \frac{h_2}{\sqrt{3}}\right) + g\left(-\frac{h_1}{\sqrt{3}}, -\frac{h_2}{\sqrt{3}}\right) \Bigg\}, \\ &\text{where } r = 3. \end{aligned}$$

4.2.4 Finite element methods for transient problems

In this section, we briefly study the finite element method for a *transient* (parabolic) problem in a bounded domain $\Omega \subset \mathbb{R}^d$, $d \geq 1$:

$$\begin{aligned} \phi\frac{\partial p}{\partial t} - \nabla \cdot (\mathbf{a}\nabla p) &= f \quad &&\text{in } \Omega \times J, \\ p &= 0 &&\text{on } \Gamma \times J, \\ p(\cdot, 0) &= p_0 &&\text{in } \Omega, \end{aligned} \tag{4.87}$$

where $J = (0, T]$ $(T > 0)$ is the time interval of interest and ϕ, f, $\mathbf{a}$, and p_0 are given functions. The function ϕ is assumed to be nonnegative on Ω, and the tensor function $\mathbf{a}$ is assumed to satisfy

$$0 < a_* \leq |\eta|^2 \sum_{i,j=1}^{d} a_{ij}(\mathbf{x})\eta_i\eta_j \leq a^* < \infty, \qquad \mathbf{x} \in \Omega,\ \boldsymbol{\eta} \neq \mathbf{0} \in \mathbb{R}^d. \tag{4.88}$$

We first present a *semidiscrete* approximation scheme where (4.87) is discretized only in space using the finite element method. Then we consider *fully discrete* approximation schemes where the time discretization is based on the *backward Euler method*, the *forward Euler method*, and the *Crank–Nicholson method*, respectively. For more details on the finite element method for transient problems, refer to Thomée (1984).

A one-dimensional model problem

To understand some of the major properties of the solution to problem (4.87), we consider the one-dimensional version

$$\begin{aligned} \frac{\partial p}{\partial t} - \frac{\partial^2 p}{\partial x^2} &= 0, \qquad && 0 < x < \pi,\ t \in J, \\ p(0,t) = p(\pi, t) &= 0, && t \in J, \\ p(x, 0) &= p_0(x), && 0 < x < \pi. \end{aligned} \tag{4.89}$$

Application of separation of variables yields

$$p(x,t) = \sum_{j=1}^{\infty} p_0^j e^{-j^2 t} \sin(jx), \tag{4.90}$$

where the *Fourier coefficients* p_0^j of the initial datum p_0 are given by

$$p_0^j = \sqrt{\frac{2}{\pi}} \int_0^\pi p_0(x)\sin(jx)\,dx, \qquad j = 1, 2, \ldots.$$

Note that $\{\sqrt{\frac{2}{\pi}}\sin(jx)\}_{j=1}^{\infty}$ forms an *orthonormal system* in the sense that

$$\frac{2}{\pi}\int_0^\pi \sin(jx)\sin(kx)\,dx = \begin{cases} 1 & \text{if } j = k, \\ 0 & \text{if } j \neq k. \end{cases} \tag{4.91}$$

It follows from (4.90) that the solution p is a linear combination of sine waves $\sin(jx)$ with amplitudes $p_0^j e^{-j^2 t}$ and frequencies j. Because $e^{-j^2 t}$ is very small for $j^2 t$ moderately large, each component $\sin(jx)$ lives on a time scale of order $\mathcal{O}(j^{-2})$. Consequently, high-frequency components are quickly damped, and the solution p becomes smoother as t increases. This property can be also understood from the following stability estimates:

$$
\begin{aligned}
\|p(t)\|_{L^2(\Omega)} &\leq \|p_0\|_{L^2(\Omega)}, && t \in J,\\
\left\|\frac{\partial p}{\partial t}(t)\right\|_{L^2(\Omega)} &\leq \frac{C}{t}\|p_0\|_{L^2(\Omega)}, && t \in J.
\end{aligned}
\tag{4.92}
$$

We prove these two estimates formally (a proof that is not concerned with any of the convergence questions). From (4.90) and (4.91) it follows that

$$
\begin{aligned}
\|p(t)\|^2_{L^2(\Omega)} &= \int_0^\pi (p(x,t))^2\, dx = \frac{\pi}{2}\sum_{j=1}^{\infty}\left(p_0^j\right)^2 e^{-2j^2 t}\\
&\leq \frac{\pi}{2}\sum_{j=1}^{\infty}\left(p_0^j\right)^2 = \|p_0\|^2_{L^2(\Omega)}.
\end{aligned}
$$

Also, note that

$$
\frac{\partial p}{\partial t} = \sum_{j=1}^{\infty} p_0^j(-j^2)e^{-j^2 t}\sin(jx),
$$

so that

$$
\left\|\frac{\partial p}{\partial t}(t)\right\|^2_{L^2(\Omega)} = \frac{\pi}{2}\sum_{j=1}^{\infty}\left(p_0^j\right)^2(-j^2)^2 e^{-2j^2 t}.
$$

Using the fact that there is a constant C such that $0 \leq \gamma^2 e^{-\gamma} \leq C$ for any $\gamma \geq 0$, we see that

$$
\left\|\frac{\partial p}{\partial t}(t)\right\|^2_{L^2(\Omega)} \leq \frac{C}{t^2}\|p_0\|^2_{L^2(\Omega)}.
$$

It follows from the second estimate in (4.92) that if $\|p_0\|_{L^2(\Omega)} < \infty$, then

$$
\left\|\frac{\partial p}{\partial t}(t)\right\|_{L^2(\Omega)} = \mathcal{O}(t^{-1})
$$

as $t \to 0$. An initial phase (for t small) where certain derivatives of p are large is referred to as an *initial transient*. In general, the solution p of a parabolic problem has an initial transient. It will become smoother as t increases. This observation is very important when the parabolic problem is numerically solved. It is desirable to vary the grid size (in space and time) according to the smoothness of p. For a region where p is nonsmooth, a fine grid is used; for a region where p becomes smoother, the grid size is increased. That is, an *adaptive finite element method* should be employed; see Section 4.7. Transients may also occur at times $t > 0$ if the boundary data or the source term f changes abruptly in time.

A semidiscrete scheme in space

We now return to problem (4.87). For simplicity, we study a special case of this problem where $\phi = 1$. Set

$$V = H_0^1(\Omega) = \left\{v \in H^1(\Omega) : \; v\big|_\Gamma = 0\right\}.$$

As in Section 4.2.1, we exploit the notation

$$a(p, v) = \int_\Omega \mathbf{a}\nabla p \cdot \nabla v \, d\mathbf{x}, \quad (f, v) = \int_\Omega f v \, d\mathbf{x}.$$

Then (4.87) is written in the variational form: Find $p : J \to V$ such that

$$\begin{aligned} \left(\frac{\partial p}{\partial t}, v\right) + a(p, v) &= (f, v) \qquad && \forall v \in V, \; t \in J, \\ p(\mathbf{x}, 0) &= p_0(\mathbf{x}) && \forall \mathbf{x} \in \Omega. \end{aligned} \tag{4.93}$$

Let V_h be a finite element subspace of V. Replacing V in (4.93) by V_h, we have the finite element method: Find $p_h : J \to V_h$ such that

$$\begin{aligned} \left(\frac{\partial p_h}{\partial t}, v\right) + a(p_h, v) &= (f, v) \qquad && \forall v \in V_h, \; t \in J, \\ (p_h(\cdot, 0), v) &= (p_0, v) && \forall v \in V_h. \end{aligned} \tag{4.94}$$

This system is discretized in space but continuous in time. For this reason, it is called a *semidiscrete scheme*. Let the basis functions in V_h be denoted by φ_i, $i = 1, 2, \ldots, M$, and express p_h as

$$p_h(\mathbf{x}, t) = \sum_{i=1}^{M} p_i(t)\varphi_i(\mathbf{x}), \qquad (\mathbf{x}, t) \in \Omega \times J. \tag{4.95}$$

For $j = 1, 2, \ldots, M$, we take $v = \varphi_j$ in (4.94) and utilize (4.95) to see that, for $t \in J$,

$$\begin{aligned} &\sum_{i=1}^{M} \left(\varphi_i, \varphi_j\right) \frac{dp_i}{dt} + \sum_{i=1}^{M} a\left(\varphi_i, \varphi_j\right) p_i = \left(f, \varphi_j\right), \qquad && j = 1, 2, \ldots, M, \\ &\sum_{i=1}^{M} \left(\varphi_i, \varphi_j\right) p_i(0) = \left(p_0, \varphi_j\right), && j = 1, 2, \ldots, M, \end{aligned}$$

which, in matrix form, is

$$\begin{aligned} &\mathbf{B}\frac{d\mathbf{p}(t)}{dt} + \mathbf{A}\mathbf{p}(t) = \mathbf{f}(t), \qquad t \in J, \\ &\mathbf{B}\mathbf{p}(0) = \mathbf{p}_0, \end{aligned} \tag{4.96}$$

where the $M \times M$ matrices $\mathbf{A}$ and $\mathbf{B}$ and the vectors $\mathbf{p}$, $\mathbf{f}$, and $\mathbf{p}_0$ are

$$\begin{aligned} \mathbf{A} &= \left(a_{ij}\right), \qquad & a_{ij} &= a\left(\varphi_i, \varphi_j\right), \\ \mathbf{B} &= \left(b_{ij}\right), & b_{ij} &= \left(\varphi_i, \varphi_j\right), \\ \mathbf{p} &= \left(p_j\right), & \mathbf{f} &= \left(f_j\right), \quad f_j = \left(f, \varphi_j\right), \\ \mathbf{p}_0 &= \left((p_0)_j\right), & (p_0)_j &= \left(p_0, \varphi_j\right). \end{aligned}$$

Both **A** and **B** are symmetric and positive definite, as was shown in the stationary case. Their *condition numbers* are of the order $\mathcal{O}(h^{-2})$ and $\mathcal{O}(1)$ as $h \to 0$ (Chen, 2005), respectively, where we recall that for a symmetric matrix its condition number is defined as the ratio of its largest eigenvalue to its smallest eigenvalue. For this reason, the matrices **A** and **B** are referred to as the *stiffness* and *mass* matrices, respectively. Thus (4.96) is a *stiff system* of ordinary differential equations (ODEs). To solve the ODE system we discretize the time derivative. One approach is to exploit the numerical methods developed already for ODEs. Because of the large number of simultaneous equations, however, simple numerical methods for transient partial differential problems have been developed independent of the methods for ODEs, as discussed in the next subsection.

We mention that the terms "stiffness" and "mass" really come by analogy to modeling a *mass-spring* system. Matrix **B** would model the mass, while matrix **A** would model the spring, which has a poor condition number when it is "stiff."

We show a stability result for the semidiscrete scheme (4.94) with $f = 0$. We choose $v = p_h(t)$ in the first equation of (4.94) to obtain

$$\left(\frac{\partial p_h}{\partial t}, p_h\right) + a(p_h, p_h) = 0,$$

which gives

$$\frac{1}{2}\frac{d}{dt}\|p_h(t)\|^2_{L^2(\Omega)} + a(p_h, p_h) = 0.$$

Also, take $v = p_h(0)$ in the second equation of (4.94) and use Cauchy's inequality (4.59) to see that

$$\|p_h(0)\|_{L^2(\Omega)} \leq \|p_0\|_{L^2(\Omega)}.$$

Then it follows that

$$\|p_h(t)\|^2_{L^2(\Omega)} + 2\int_0^t a(p_h(\ell), p_h(\ell))\, d\ell = \|p_h(0)\|^2_{L^2(\Omega)} \leq \|p_0\|^2_{L^2(\Omega)}.$$

Consequently, we obtain

$$\|p_h(t)\|_{L^2(\Omega)} \leq \|p_0\|_{L^2(\Omega)}, \qquad t \in J. \tag{4.97}$$

This inequality is similar to the first inequality in (4.92). In fact, the latter inequality can be shown in the same manner. The derivation of an error estimate for (4.94) is much more elaborate than that for a stationary problem. We just state an estimate for the case where V_h is the space of piecewise linear functions on a *quasi-uniform* triangulation of Ω in the sense that there is a positive constant β_2, independent of h, such that

$$h_K \geq \beta_2 h \qquad \forall K \in K_h, \tag{4.98}$$

where we recall that $h_K = \text{diam}(K)$, $K \in K_h$, and $h = \max\{h_K : K \in K_h\}$. Condition (4.98) requires that all elements $K \in K_h$ be of roughly the same size. The error estimate is (Thomée, 1984; Johnson, 1994)

$$\max_{t\in J} \|(p - p_h)(t)\|_{L^2(\Omega)} \leq C\left(1 + \left|\ln \frac{T}{h^2}\right|\right) \max_{t\in J} h^2 \|p(t)\|_{H^2(\Omega)}. \tag{4.99}$$

Due to the presence of the factor $\ln h^{-2}$, this estimate is only *almost optimal.*

Fully discrete schemes

We consider three fully discrete schemes: the backward and forward Euler methods and the Crank–Nicholson method.

(i) The backward Euler method

Let $0 = t^0 < t^1 < \cdots < t^N = T$ be a partition of J into subintervals $J^n = (t^{n-1}, t^n)$ with length $\Delta t^n = t^n - t^{n-1}$. For a generic function v of time, set $v^n = v(t^n)$. The *backward Euler method* for the semidiscrete version (4.94) is: Find $p_h^n \in V_h$, $n = 1, 2, \ldots, N$, such that

$$\begin{aligned} &\left(\frac{p_h^n - p_h^{n-1}}{\Delta t^n}, v\right) + a\left(p_h^n, v\right) = (f^n, v) && \forall v \in V_h, \\ &\left(p_h^0, v\right) = (p_0, v) && \forall v \in V_h. \end{aligned} \tag{4.100}$$

Note that (4.100) comes from replacing the time derivative in (4.94) by the difference quotient $(p_h^n - p_h^{n-1})/\Delta t^n$. This replacement results in a discretization error of order $\mathcal{O}(\Delta t^n)$ (cf. Section 4.1.1). As in (4.96), equation (4.100) can be expressed in matrix form as

$$\begin{aligned} &(\mathbf{B} + \mathbf{A}\Delta t^n)\,\mathbf{p}^n = \mathbf{B}\mathbf{p}^{n-1} + \mathbf{f}^n \Delta t^n, \\ &\mathbf{B}\mathbf{p}(0) = \mathbf{p}_0, \end{aligned} \tag{4.101}$$

where

$$p_h^n = \sum_{i=1}^{M} p_i^n \varphi_i, \qquad n = 0, 1, \ldots, N,$$

and

$$\mathbf{p}^n = \left(p_1^n, p_2^n, \ldots, p_M^n\right)^T.$$

Clearly, (4.101) is an *implicit* scheme; that is, we need to solve a system of linear equations at each time step.

Let us state a basic *stability* estimate for (4.100) in the case $f = 0$. Choosing $v = p_h^n$ in (4.100), we see that

$$\|p_h^n\|^2 - \left(p_h^{n-1}, p_h^n\right) + a\left(p_h^n, p_h^n\right)\Delta t^n = 0.$$

It follows from Cauchy's inequality (4.59) that

$$\left(p_h^{n-1}, p_h^n\right) \le \|p_h^{n-1}\|\,\|p_h^n\| \le \frac{1}{2}\|p_h^{n-1}\|^2 + \frac{1}{2}\|p_h^n\|^2.$$

Consequently, we get

$$\frac{1}{2}\|p_h^n\|^2 - \frac{1}{2}\|p_h^{n-1}\|^2 + a\left(p_h^n, p_h^n\right)\Delta t^n \le 0.$$

We sum over n and use the second equation in (4.100) to give

$$\|p_h^j\|^2 + 2\sum_{n=1}^{j} a\left(p_h^n, p_h^n\right)\Delta t^n \le \|p_h^0\|^2 \le \|p^0\|^2.$$

Because $a\left(p_h^n, p_h^n\right) \geq 0$, we obtain the stability result

$$\|p_h^j\| \leq \|p^0\|, \qquad j = 0, 1, \ldots, N. \tag{4.102}$$

Note that (4.102) holds regardless of the size of the time steps Δt^j. In other words, the backward Euler method (4.100) is *unconditionally stable*. This is a very desirable feature of a time discretization scheme for a parabolic problem (cf. Section 4.1.6).

We remark that an estimate for the error $p - p_h$ can be derived. The error stems from a combination of the space and time discretizations. When V_h is the finite element space of piecewise linear functions, for example, the error $p^n - p_h^n$ $(0 \leq n \leq N)$ in the $L^2(\Omega)$-norm is of order $\mathcal{O}\left(\Delta t + h^2\right)$ (Thomée, 1984) under appropriate smoothness assumptions on p, where $\Delta t = \max\{\Delta t^j, 1 \leq j \leq N\}$.

(ii) The Crank–Nicholson method

The *Crank–Nicholson method* for (4.94) is defined as follows: Find $p_h^n \in V_h$, $n = 1, 2, \ldots, N$, such that

$$\begin{aligned} &\left(\frac{p_h^n - p_h^{n-1}}{\Delta t^n}, v\right) + a\left(\frac{p_h^n + p_h^{n-1}}{2}, v\right) = \left(\frac{f^n + f^{n-1}}{2}, v\right) \\ &\qquad\qquad\qquad\qquad\qquad\qquad\qquad \forall v \in V_h, \\ &\left(p_h^0, v\right) = (p_0, v) \qquad \forall v \in V_h. \end{aligned} \tag{4.103}$$

In the present case, the difference quotient $(p_h^n - p_h^{n-1})/\Delta t^n$ now replaces the average $\left(\partial p(t^n)/\partial t + \partial p(t^{n-1})/\partial t\right)/2$. The resulting discretization error is $\mathcal{O}\left((\Delta t^n)^2\right)$ (cf. Section 4.1.1). Similarly to (4.101), the linear system from (4.103) is

$$\begin{aligned} &\left(\mathbf{B} + \frac{\Delta t^n}{2}\mathbf{A}\right)\mathbf{p}^n = \left(\mathbf{B} - \frac{\Delta t^n}{2}\mathbf{A}\right)\mathbf{p}^{n-1} + \frac{\mathbf{f}^n + \mathbf{f}^{n-1}}{2}\Delta t^n, \\ &\mathbf{B}\mathbf{p}(0) = \mathbf{p}_0 \end{aligned} \tag{4.104}$$

for $n = 1, 2, \ldots, N$. Again, this is an implicit method. When $f = 0$, by taking $v = (p_h^n + p_h^{n-1})/2$ in (4.103) one can show that the stability result (4.102) unconditionally holds for the Crank–Nicholson method as well (cf. Exercise 4.35). For the piecewise linear finite element space V_h, for each n the error $p^n - p_h^n$ in the $L^2(\Omega)$-norm is $\mathcal{O}\left((\Delta t)^2 + h^2\right)$ this time. Note that the Crank–Nicholson method is more accurate in time than the backward Euler method and is slightly more expensive from the computational point of view.

(iii) The forward Euler method

We conclude with the *forward Euler method*. This method takes the form: Find $p_h^n \in V_h$, $n = 1, 2, \ldots, N$, such that

$$\begin{aligned} &\left(\frac{p_h^n - p_h^{n-1}}{\Delta t^n}, v\right) + a\left(p_h^{n-1}, v\right) = \left(f^{n-1}, v\right) \qquad \forall v \in V_h, \\ &\left(p_h^0, v\right) = (p_0, v) \qquad\qquad\qquad\qquad\qquad\quad\; \forall v \in V_h, \end{aligned} \tag{4.105}$$

and the corresponding matrix form is

$$\begin{aligned}\mathbf{B}\mathbf{p}^n &= \left(\mathbf{B} - \mathbf{A}\Delta t^n\right)\mathbf{p}^{n-1} + \mathbf{f}^{n-1}\Delta t^n,\\ \mathbf{B}\mathbf{p}(0) &= \mathbf{p}_0.\end{aligned} \tag{4.106}$$

Introducing the Cholesky decomposition $\mathbf{B} = \mathbf{D}\mathbf{D}^T$ (see the next chapter) and using the new variable $\mathbf{q} = \mathbf{D}^T\mathbf{p}$, where $\mathbf{D}^T$ is the transpose of $\mathbf{D}$, problem (4.106) is of the simpler form

$$\begin{aligned}\mathbf{q}^n &= \left(\mathbf{I} - \tilde{\mathbf{A}}\Delta t^n\right)\mathbf{q}^{n-1} + \mathbf{D}^{-1}\mathbf{f}^{n-1}\Delta t^n,\\ \mathbf{q}(0) &= \mathbf{D}^{-1}\mathbf{p}_0,\end{aligned} \tag{4.107}$$

where $\tilde{\mathbf{A}} = \mathbf{D}^{-1}\mathbf{A}\mathbf{D}^{-T}$. Clearly, (4.107) is an *explicit scheme* in $\mathbf{q}$. A stability result similar to (4.102) can be proven only under the *stability condition*

$$\Delta t^n \leq Ch^2, \qquad n = 1, 2, \ldots, N, \tag{4.108}$$

where C is a constant independent of Δt and h. This can be seen as follows: with $f = 0$, the first equation of (4.107) becomes

$$\mathbf{q}^n = \left(\mathbf{I} - \tilde{\mathbf{A}}\Delta t^n\right)\mathbf{q}^{n-1}. \tag{4.109}$$

Define the *matrix norm*

$$\|\tilde{\mathbf{A}}\| = \max_{\boldsymbol{\eta}\in\mathbb{R}^M, \boldsymbol{\eta}\neq\mathbf{0}} \frac{\|\tilde{\mathbf{A}}\boldsymbol{\eta}\|}{\|\boldsymbol{\eta}\|},$$

where $\|\boldsymbol{\eta}\|$ is the Euclidean norm of $\boldsymbol{\eta} = (\eta_1, \eta_2, \ldots, \eta_M)$: $\|\boldsymbol{\eta}\|^2 = \eta_1^2 + \eta_2^2 + \cdots + \eta_M^2$. Assume that the symmetric, positive definite matrix $\tilde{\mathbf{A}}$ has eigenvalues $\mu_i > 0$, $i = 1, 2, \ldots, M$. Then we see that (Axelsson, 1994)

$$\|\tilde{\mathbf{A}}\| = \max_{i=1,2,\ldots,M} \mu_i.$$

Thus it follows that

$$\|\mathbf{I} - \tilde{\mathbf{A}}\Delta t^n\| = \max_{i=1,2,\ldots,M} |1 - \mu_i\Delta t^n|.$$

Let the maximum occur as $i = M$, for example. Then

$$\|\mathbf{I} - \tilde{\mathbf{A}}\Delta t^n\| \leq 1$$

only if $\mu_M\Delta t^n \leq 2$. Since $\mu_M = \mathcal{O}(h^{-2})$ (Chen, 2005), $\Delta t^n \leq 2/\mu_M = \mathcal{O}(h^2)$, which is (4.108).

The stability condition (4.108) requires that the time step be sufficiently small (cf. condition (4.23)). In other words, the forward Euler method (4.105) is *conditionally stable*. This condition is very restrictive, particularly for long-time integration. In contrast, the backward Euler and Crank–Nicholson methods are unconditionally stable but require more work per time step. These two methods are more efficient for parabolic problems since the extra cost involved at each step for an implicit method is more than compensated by the fact that larger time steps can be utilized.

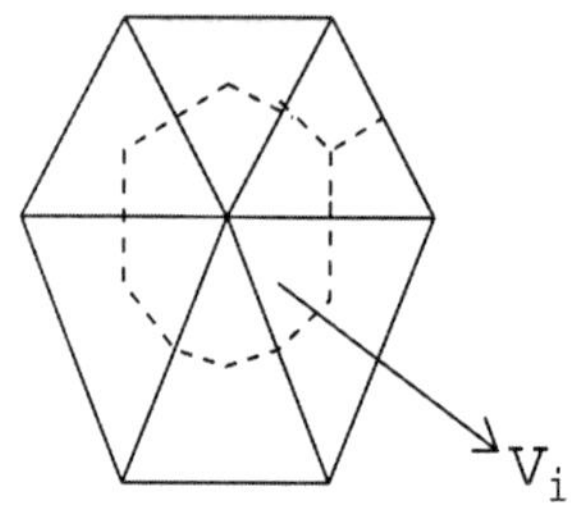

Figure 4.31. *A control volume.*

4.3 Control Volume Finite Element Methods

The finite difference methods presented in Section 4.1 are locally conservative but are not flexible in the treatment of complex reservoirs. On the other hand, the standard finite element methods described in Section 4.2 are more flexible but not conservative on local elements (e.g., on triangles). They are globally conservative. In this section, we introduce a variation of finite element methods so that they are locally conservative on each *control volume*. Control volumes can be formed around grid nodes by joining the midpoints of the edges of a triangle with a point inside the triangle, for example (cf. Figure 4.31). Different locations of the point give rise to different forms of the flow term between grid nodes. When it is the barycenter of the triangle, the resulting grid is of CVFE (*control volume finite element*) type, and the resulting finite element methods are the CVFE methods. These methods were first introduced by Lemonnier (1979) for reservoir simulation. The CVFE grids are different from the PEBI (*perpendicular bisection*) grids (also called *Voronoi* grids (Heinrich, 1987)) in that the latter are locally orthogonal. The CVFE grids are more flexible.

4.3.1 The basic CVFE

To see the CVFE idea, we focus on linear triangular elements in two dimensions. A conceptual extension to three dimensions is straightforward. We consider the stationary problem

$$-\nabla \cdot (\mathbf{a}\nabla p) = f(x_1, x_2) \quad \text{in } \Omega, \tag{4.110}$$

where Ω is a bounded domain in the plane and p is pressure, for example.

Let V_i be a control volume. Replacing p by $p_h \in V_h$ (the space of continuous piecewise linear functions on $\bar{\Omega}$; cf. Section 4.2.1) in (4.110) and integrating over V_i, we see that

$$-\int_{V_i} \nabla \cdot (\mathbf{a}\nabla p_h)\, d\mathbf{x} = \int_{V_i} f\, d\mathbf{x}.$$

The divergence theorem implies

$$-\int_{\partial V_i} \mathbf{a}\nabla p_h \cdot \boldsymbol{\nu}\, d\ell = \int_{V_i} f\, d\mathbf{x}. \tag{4.111}$$

Note that $\nabla p_h \cdot \boldsymbol{\nu}$ is continuous across each segment of ∂V_i (that lies inside a triangle). Thus, if $\mathbf{a}$ is continuous across that segment, so is the flux $\mathbf{a}\nabla p_h \cdot \boldsymbol{\nu}$. Therefore, the flux is

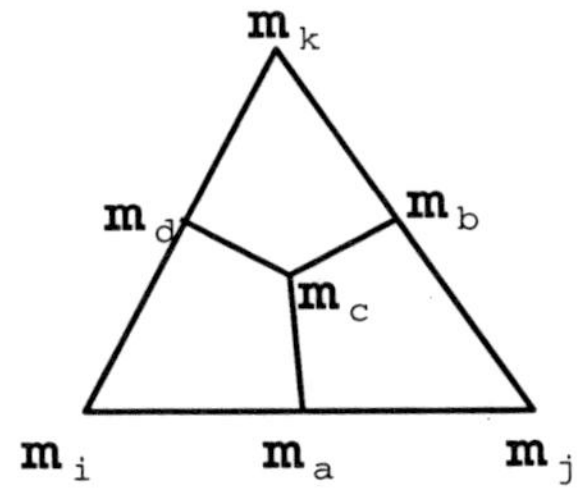

Figure 4.32. *A base triangle.*

continuous across the edges of the control volume V_i. Furthermore, (4.111) indicates that the CVFE method is locally (i.e., on each control volume) conservative.

Given a triangle K with vertices $\mathbf{m}_i$, $\mathbf{m}_j$, and $\mathbf{m}_k$, edge midpoints $\mathbf{m}_a$, $\mathbf{m}_b$, and $\mathbf{m}_d$, and center $\mathbf{m}_c$ (cf. Figure 4.32), it follows from Example 4.1 that the approximation p_h to p on K is given by

$$p_h = p_i\lambda_i + p_j\lambda_j + p_k\lambda_k, \tag{4.112}$$

where we recall that the local basis functions λ_i are defined by

$$\lambda_i(\mathbf{m}_j) = \begin{cases} 1 & \text{if } i = j, \\ 0 & \text{if } i \neq j \end{cases}$$

with

$$\lambda_i + \lambda_j + \lambda_k = 1. \tag{4.113}$$

These basis functions are the *barycentric coordinates* of the triangle K. Define

$$a_i = m_{j,2} - m_{k,2}, \quad b_i = -(m_{j,1} - m_{k,1}),$$
$$c_i = m_{j,1}m_{k,2} - m_{j,2}m_{k,1},$$

where $\mathbf{m}_i = (m_{i,1}, m_{i,2})$ and $\{i, j, k\}$ is cyclically permuted. Then the local basis functions λ_i, λ_j, and λ_k are given by (cf. Exercise 4.36)

$$\begin{pmatrix} \lambda_i \\ \lambda_j \\ \lambda_k \end{pmatrix} = \frac{1}{2|K|} \begin{pmatrix} c_i & a_i & b_i \\ c_j & a_j & b_j \\ c_k & a_k & b_k \end{pmatrix} \begin{pmatrix} 1 \\ x_1 \\ x_2 \end{pmatrix}, \tag{4.114}$$

where $|K|$ is the area of the triangle K. Consequently,

$$\frac{\partial \lambda_l}{\partial x_1} = \frac{a_l}{2|K|}, \quad \frac{\partial \lambda_l}{\partial x_2} = \frac{b_l}{2|K|}, \quad l = i, j, k. \tag{4.115}$$

We consider the computation of the left-hand side of (4.111) on $\mathbf{m}_a\mathbf{m}_c\mathbf{m}_d$ (cf. Figure 4.32):

$$f_i \equiv -\int_{\mathbf{m}_a\mathbf{m}_c+\mathbf{m}_c\mathbf{m}_d} \mathbf{a}\nabla p_h \cdot \boldsymbol{\nu}\, d\ell. \tag{4.116}$$

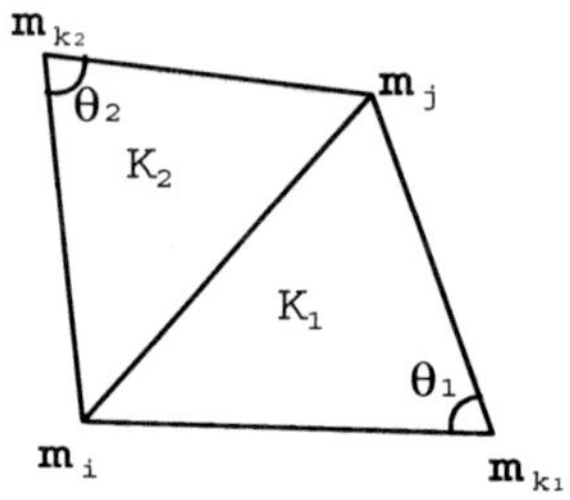

Figure 4.33. *Two adjacent triangles.*

On $\mathbf{m}_a\mathbf{m}_c$,

$$\nu = \frac{(m_{c,2} - m_{a,2}, m_{a,1} - m_{c,1})}{|\mathbf{m}_a\mathbf{m}_c|},$$

and, on $\mathbf{m}_c\mathbf{m}_d$,

$$\nu = \frac{(m_{d,2} - m_{c,2}, m_{c,1} - m_{d,1})}{|\mathbf{m}_c\mathbf{m}_d|},$$

where $|\mathbf{m}_a\mathbf{m}_c|$ denotes the length of edge $\mathbf{m}_a\mathbf{m}_c$. Consequently, if $\mathbf{a}$ is a constant tensor on the triangle K, it follows from (4.112), (4.115), (4.116), the definition of a_i and b_i, and simple algebraic calculations (cf. Exercise 4.37) that

$$f_i = |K| \sum_{l=i}^{k} \mathbf{a}\nabla\lambda_l \cdot \nabla\lambda_i \; p_l, \tag{4.117}$$

which shows that the CVFE and standard finite element methods using piecewise linear functions produce the same stiffness matrix (cf. Section 4.2.1).

Using (4.113), equation (4.117) can be recast in the finite difference form

$$f_i = -T_{ij}(p_j - p_i) - T_{ik}(p_k - p_i), \tag{4.118}$$

where the *transmissibility coefficients* T_{ij} and T_{ik} are

$$T_{ij} = -|K|\mathbf{a}\nabla\lambda_j \cdot \nabla\lambda_i, \quad T_{ik} = -|K|\mathbf{a}\nabla\lambda_k \cdot \nabla\lambda_i.$$

We now consider the assembly of the *global transmissibility matrix*. Each connection between any two adjacent nodes $\mathbf{m}_i$ and $\mathbf{m}_j$ includes the contributions from two triangles K_1 and K_2 that share the common edge with endpoints $\mathbf{m}_i$ and $\mathbf{m}_j$ (cf. Figure 4.33). The transmissibility between $\mathbf{m}_i$ and $\mathbf{m}_j$, where at least one of them is not on the external boundary, is

$$T_{ij} = -\sum_{l=1}^{2} \left(|K|\mathbf{a}\nabla\lambda_j \cdot \nabla\lambda_i \right)\Big|_{K_l}. \tag{4.119}$$

Applying (4.111) and (4.118), we obtain the linear system on the control volume V_i in terms of pressure values at the vertices of triangles

$$-\sum_{j\in\Omega_i} T_{ij}\left(p_j - p_i\right) = F_i, \tag{4.120}$$

where Ω_i is the set of all neighboring nodes of $\mathbf{m}_i$ and $F_i = \int_{V_i} f \, d\mathbf{x}$.

If ∂V_i contains part of the Neumann boundary, then the flux on that part is given; if it contains part of the Dirichlet boundary, the pressure on the corresponding part is given. The third boundary condition can be also incorporated as in Section 4.2.1. Since linear elements are used, an *error estimate* as in (4.73) holds for the CVFE method considered. Finally, the CVFE method can be extended to transient problems as in Section 4.2.4.

4.3.2 Positive transmissibilities

The *transmissibility coefficient* T_{ij} defined in (4.119) must be positive. Positive transmissibilities or *positive flux linkages* always yield a direction of the discrete flux in the physical direction. Negative transmissibilities are not physically meaningful and generate unsatisfactory solutions.

For simplicity, consider a *homogeneous anisotropic* medium (cf. Section 2.2.1): $\mathbf{a} = \operatorname{diag}(a_{11}, a_{22})$ (i.e., a_{11} and a_{22} are positive constants). In this case, using (4.115) and (4.119), T_{ij} restricted to each triangle K (cf. Figure 4.33) is

$$T_{ij} = -\frac{a_{11}a_j a_i + a_{22}b_j b_i}{4|K|}.$$

Introduce a coordinate transform:

$$x_1' = \frac{x_1}{\sqrt{a_{11}}}, \quad x_2' = \frac{x_2}{\sqrt{a_{22}}}.$$

Under this transform, the area of the transformed triangle K' is

$$|K'| = \frac{|K|}{\sqrt{a_{11}a_{22}}}.$$

Consequently, T_{ij} becomes

$$T_{ij} = \sqrt{a_{11}a_{22}}\frac{\left|\mathbf{m}_{k'}\mathbf{m}_{j'}\right| \left|\mathbf{m}_{k'}\mathbf{m}_{i'}\right| \cos\theta_{k'}}{4|K'|} = \sqrt{a_{11}a_{22}}\frac{\cot\theta_{k'}}{2},$$

where $\theta_{k'}$ is the angle of the triangle at node $\mathbf{m}_{k'}$ in the transformed plane. Because each global transmissibility consists of the contributions from two adjacent triangles, the global T_{ij} between nodes $\mathbf{m}_i$ and $\mathbf{m}_j$ (cf. Figure 4.33) is

$$T_{ij} = \sqrt{a_{11}a_{22}}\left(\frac{\cot\theta_{k_1'} + \cot\theta_{k_2'}}{2}\right), \tag{4.121}$$

where $\theta_{k_1'}$ and $\theta_{k_2'}$ are the opposite angles of the two triangles. Thus the requirement $T_{ij} > 0$ is equivalent to

$$\theta_{k_1'} + \theta_{k_2'} < \pi. \tag{4.122}$$

For an edge on the external boundary, the requirement for the angle opposite this edge is

$$\theta_{k'} < \frac{\pi}{2}. \tag{4.123}$$

Note that all these angles are measured in the (x_1', x_2')-coordinate plane.

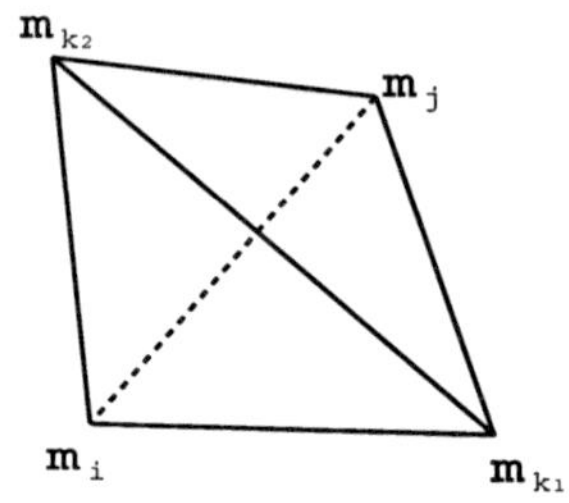

Figure 4.34. *An edge swap.*

4.3.3 The CVFE grid construction

It is interesting to note that condition (4.122) is related to a *Delaunay triangulation*. A Delaunay triangulation satisfies the *empty circle criterion*: The circumcircle of each triangle must not contain any other nodes in its interior. Given a shape-regular triangulation K_h (cf. Section 4.2.1) of a convex domain, K_h can be converted to a Delaunay triangulation in a sequence of local edge swaps as follows (Joe, 1986; D'Azevedo and Simpson, 1989): Each internal edge in K_h is examined. If it is a part of a convex quadrilateral (cf. Figure 4.33), then the circumcircles of the two triangles are checked. If one of the circumcircles contains the fourth vertex of the quadrilateral, then the diagonal of this quadrilateral is swapped (cf. Figure 4.34). The resulting local triangulation then satisfies the empty circle criterion, i.e., the *local optimality condition* (Joe, 1986; D'Azevedo and Simpson, 1989). A sequence of local edge swaps eventually converges, so that every internal edge is locally optimal. All internal edges of a triangulation are locally optimal if and only if it is a Delaunay triangulation (Joe, 1986).

On the other hand, the local optimal condition is equivalent to condition (4.122) (D'Azevedo and Simpson, 1989). Hence the edge swapping procedure can be given geometrically: Given a regular triangulation K_h, if the sum of the two angles opposite edge $\mathbf{m}_i\mathbf{m}_j$ (cf. Figure 4.33) is larger than π, then this edge is replaced by edge $\mathbf{m}_{k_1}\mathbf{m}_{k_2}$. This edge swap can be carried out only if the quadrilateral is convex. If it is not, then condition (4.122) must necessarily be true. For a convex domain, no addition or movement of nodes is required to convert K_h to a Delaunay triangulation.

For problem (4.110), the edge swapping procedure can be generalized (Forsyth, 1991): Each edge $\mathbf{m}_i\mathbf{m}_j$ is examined, and the transmissibility T_{ij} is computed using (4.119). If T_{ij} is negative, then this edge is replaced by $\mathbf{m}_{k_1}\mathbf{m}_{k_2}$. If the solution domain is convex and $\mathbf{a}$ is constant, this procedure is equivalent to establishing a Delaunay triangulation in the (x_1', x_2')-plane where $\mathbf{a}'$ is the identity tensor. The equivalence of positive transmissibilities with a Delaunay triangulation is true only for internal edges in the transformed plane when $\mathbf{a}$ is constant. In general, a Delaunay triangulation of the physical plane cannot ensure positive transmissibilities, even for internal edges. However, because most domains that arise in practical applications can be treated as unions of convex regions with a constant permeability tensor $\mathbf{a}$, the local edge swap procedure should tend to minimize the number of internal edges having negative transmissibilities.

In general, edges on the external boundary of a domain can have negative transmis-

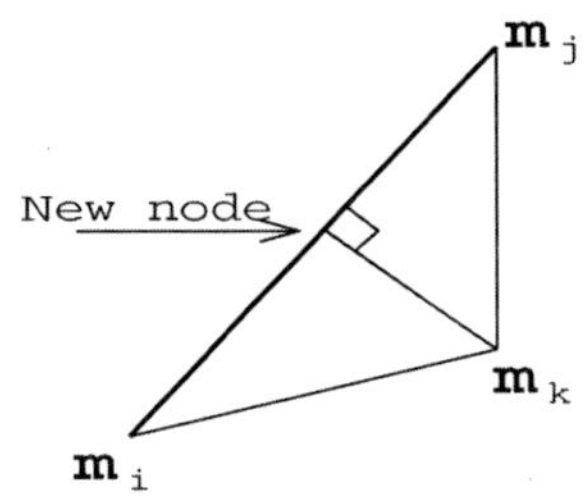

Figure 4.35. *An addition of a new boundary node.*

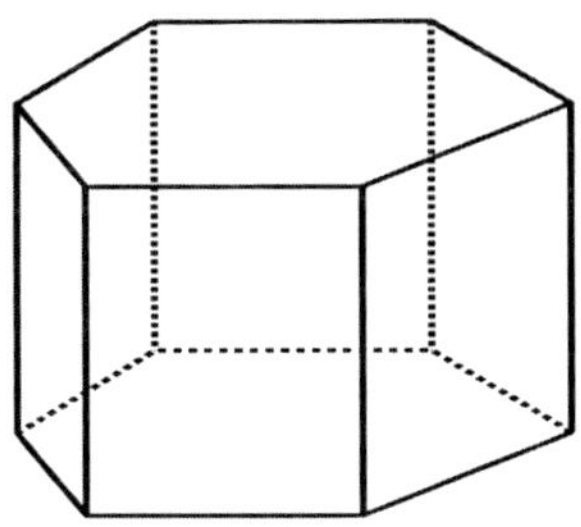

Figure 4.36. *A hexagonal prism.*

sibilities. This problem can be overcome by adding a boundary node, as in Figure 4.35. Suppose that $T_{ij} < 0$ on edge $\mathbf{m}_i\mathbf{m}_j$; i.e., in the (x_1', x_2')-plane, the angle opposite this edge is larger than $\pi/2$. A new node is added at the intersection of $\mathbf{m}_i\mathbf{m}_j$ with the orthogonal line segment to $\mathbf{m}_i\mathbf{m}_j$ drawn from $\mathbf{m}_k$. Note that there is no edge swap for a boundary edge.

In three dimensions, the Delaunay empty sphere criterion is not equivalent to positive transmissibilities (Letniowski, 1992). Because a reservoir domain generally has a layer structure, the grid in the x_3-direction is often obtained by a vertical projection of two-dimensional grids (cf. Figure 4.36).

4.3.4 The upstream weighted CVFE

The basic idea of *upstream weighting* is to choose the value of a property coefficient according to the upstream direction of a flux. The same idea has been used in the upwind finite difference methods (cf. Section 4.1.8). In this section, we consider two upstream weighting strategies for (4.120): potential- and flux-based strategies.

The potential-based upstream weighting scheme

Suppose that (4.110) is of the form

$$-\nabla \cdot (\lambda \mathbf{a} \nabla p) = f(x_1, x_2) \quad \text{in } \Omega, \tag{4.124}$$

where $\mathbf{a}$ and λ can be a permeability tensor and a mobility coefficient, respectively, for example. For this problem, a CVFE method analogous to (4.120) can be derived. If $\mathbf{a}$ is a scalar a and is different on the two sides of an edge of V_i, across that edge it should be approximated by the *harmonic average*

$$a_{har}(\mathbf{x}) = \frac{2a^+(\mathbf{x})a^-(\mathbf{x})}{a^+(\mathbf{x}) + a^-(\mathbf{x})},$$

where a^+ and a^- indicate the respective values from the two sides. The reason for using a harmonic average is that for an inactive node (i.e., the node where $a = 0$), this average gives the correct value (i.e., $a = 0$), in contrast with the arithmetic average. If $\mathbf{a}$ is a tensor, this harmonic average is used for each component of $\mathbf{a}$, and the result is denoted by $\mathbf{a}_{har}$. For the mobility coefficient λ, in practice, upstream weighting must be used to maintain stability for the CVFE methods. As a result of these two observations, the transmissibility between nodes $\mathbf{m}_i$ and $\mathbf{m}_j$ restricted to each triangle K becomes

$$T_{ij} = -|K|\, \lambda_{ij}^{up} \mathbf{a}_{har} \nabla\lambda_j \cdot \nabla\lambda_i, \tag{4.125}$$

where the *potential-based upstream weighting* scheme is defined by

$$\lambda_{ij}^{up} = \begin{cases} \lambda(\mathbf{m}_i) & \text{if } p_i > p_j, \\ \lambda(\mathbf{m}_j) & \text{if } p_i < p_j. \end{cases} \tag{4.126}$$

In fact, it is a pressure-based approach in the current context. The name *potential-based* is due to the fact that potentials are usually used in place of p in reservoir simulation (cf. Section 2.2.2).

This potential-based upstream weighting scheme is easy to implement. However, it violates the important flux continuity property across the interfaces between control volumes. To see this, consider the case $\mathbf{a} = \operatorname{diag}(a_{11}, a_{22})$, where $\mathbf{a}$ is a constant diagonal tensor on the triangle K (cf. Figure 4.32). Applying (4.115) and (4.125), the flux on edge $\mathbf{m}_a\mathbf{m}_c$ is

$$\begin{aligned} f_{i,\mathbf{m}_a\mathbf{m}_c} = &-\lambda_{ij}^{up}\left(a_{11}(m_{c,2} - m_{a,2})\frac{\partial\lambda_j}{\partial x_1} + a_{22}(m_{a,1} - m_{c,1})\frac{\partial\lambda_j}{\partial x_2}\right)(p_j - p_i) \\ &-\lambda_{ik}^{up}\left(a_{11}(m_{c,2} - m_{a,2})\frac{\partial\lambda_k}{\partial x_1} + a_{22}(m_{a,1} - m_{c,1})\frac{\partial\lambda_k}{\partial x_2}\right)(p_k - p_i), \end{aligned}$$

and on edge $\mathbf{m}_c\mathbf{m}_d$,

$$\begin{aligned} f_{i,\mathbf{m}_c\mathbf{m}_d} = &-\lambda_{ij}^{up}\left(a_{11}(m_{d,2} - m_{c,2})\frac{\partial\lambda_j}{\partial x_1} + a_{22}(m_{c,1} - m_{d,1})\frac{\partial\lambda_j}{\partial x_2}\right)(p_j - p_i) \\ &-\lambda_{ik}^{up}\left(a_{11}(m_{d,2} - m_{c,2})\frac{\partial\lambda_k}{\partial x_1} + a_{22}(m_{c,1} - m_{d,1})\frac{\partial\lambda_k}{\partial x_2}\right)(p_k - p_i). \end{aligned}$$

Similarly, the fluxes on edges $\mathbf{m}_b\mathbf{m}_c$ and $\mathbf{m}_c\mathbf{m}_a$ at node $\mathbf{m}_j$ are, respectively,

$$\begin{aligned} f_{j,\mathbf{m}_b\mathbf{m}_c} = &-\lambda_{jk}^{up}\left(a_{11}(m_{c,2} - m_{b,2})\frac{\partial\lambda_k}{\partial x_1} + a_{22}(m_{b,1} - m_{c,1})\frac{\partial\lambda_k}{\partial x_2}\right)(p_k - p_j) \\ &-\lambda_{ji}^{up}\left(a_{11}(m_{c,2} - m_{b,2})\frac{\partial\lambda_i}{\partial x_1} + a_{22}(m_{b,1} - m_{c,1})\frac{\partial\lambda_i}{\partial x_2}\right)(p_i - p_j) \end{aligned}$$

and

$$
\begin{aligned}
f_{j,\mathbf{m}_c\mathbf{m}_a} = &-\lambda_{jk}^{up}\left(a_{11}(m_{a,2}-m_{c,2})\frac{\partial \lambda_k}{\partial x_1}+a_{22}(m_{c,1}-m_{a,1})\frac{\partial \lambda_k}{\partial x_2}\right)(p_k-p_j)\\
&-\lambda_{ji}^{up}\left(a_{11}(m_{a,2}-m_{c,2})\frac{\partial \lambda_i}{\partial x_1}+a_{22}(m_{c,1}-m_{a,1})\frac{\partial \lambda_i}{\partial x_2}\right)(p_i-p_j),
\end{aligned}
$$

and the fluxes on edges $\mathbf{m}_d\mathbf{m}_c$ and $\mathbf{m}_c\mathbf{m}_b$ at node $\mathbf{m}_k$ are, respectively,

$$
\begin{aligned}
f_{k,\mathbf{m}_d\mathbf{m}_c} = &-\lambda_{ki}^{up}\left(a_{11}(m_{c,2}-m_{d,2})\frac{\partial \lambda_i}{\partial x_1}+a_{22}(m_{d,1}-m_{c,1})\frac{\partial \lambda_i}{\partial x_2}\right)(p_i-p_k)\\
&-\lambda_{kj}^{up}\left(a_{11}(m_{c,2}-m_{d,2})\frac{\partial \lambda_j}{\partial x_1}+a_{22}(m_{d,1}-m_{c,1})\frac{\partial \lambda_j}{\partial x_2}\right)(p_j-p_k)
\end{aligned}
$$

and

$$
\begin{aligned}
f_{k,\mathbf{m}_c\mathbf{m}_b} = &-\lambda_{ki}^{up}\left(a_{11}(m_{b,2}-m_{c,2})\frac{\partial \lambda_i}{\partial x_1}+a_{22}(m_{c,1}-m_{b,1})\frac{\partial \lambda_i}{\partial x_2}\right)(p_i-p_k)\\
&-\lambda_{kj}^{up}\left(a_{11}(m_{b,2}-m_{c,2})\frac{\partial \lambda_j}{\partial x_1}+a_{22}(m_{c,1}-m_{b,1})\frac{\partial \lambda_j}{\partial x_2}\right)(p_j-p_k).
\end{aligned}
$$

For the flux to be continuous across edge $\mathbf{m}_a\mathbf{m}_c$, it is required that $f_{i,\mathbf{m}_a\mathbf{m}_c}+f_{j,\mathbf{m}_c\mathbf{m}_a}=0$; i.e.,

$$
\begin{aligned}
&-\lambda_{ij}^{up}\left(a_{11}(m_{c,2}-m_{a,2})\frac{\partial \lambda_j}{\partial x_1}+a_{22}(m_{a,1}-m_{c,1})\frac{\partial \lambda_j}{\partial x_2}\right)(p_j-p_i)\\
&-\lambda_{ik}^{up}\left(a_{11}(m_{c,2}-m_{a,2})\frac{\partial \lambda_k}{\partial x_1}+a_{22}(m_{a,1}-m_{c,1})\frac{\partial \lambda_k}{\partial x_2}\right)(p_k-p_i)\\
&-\lambda_{jk}^{up}\left(a_{11}(m_{a,2}-m_{c,2})\frac{\partial \lambda_k}{\partial x_1}+a_{22}(m_{c,1}-m_{a,1})\frac{\partial \lambda_k}{\partial x_2}\right)(p_k-p_j)\\
&-\lambda_{ji}^{up}\left(a_{11}(m_{a,2}-m_{c,2})\frac{\partial \lambda_i}{\partial x_1}+a_{22}(m_{c,1}-m_{a,1})\frac{\partial \lambda_i}{\partial x_2}\right)(p_i-p_j)=0.
\end{aligned}
$$

Because it must be satisfied for all choices of $\mathbf{a}$, this equation reduces to

$$
\begin{aligned}
a_{11}(m_{a,2}-m_{c,2})\bigg[&\lambda_{ij}^{up}\frac{\partial \lambda_j}{\partial x_1}(p_j-p_i)+\lambda_{ji}^{up}\frac{\partial \lambda_i}{\partial x_1}(p_j-p_i)\\
&+\lambda_{ik}^{up}\frac{\partial \lambda_k}{\partial x_1}(p_k-p_i)+\lambda_{jk}^{up}\frac{\partial \lambda_k}{\partial x_1}(p_j-p_k)\bigg]=0
\end{aligned}
$$

and

$$
\begin{aligned}
a_{22}(m_{c,1}-m_{a,1})\bigg[&\lambda_{ij}^{up}\frac{\partial \lambda_j}{\partial x_2}(p_j-p_i)+\lambda_{ji}^{up}\frac{\partial \lambda_i}{\partial x_2}(p_j-p_i)\\
&+\lambda_{ik}^{up}\frac{\partial \lambda_k}{\partial x_2}(p_k-p_i)+\lambda_{jk}^{up}\frac{\partial \lambda_k}{\partial x_2}(p_j-p_k)\bigg]=0.
\end{aligned}
$$

For these two equations to hold simultaneously for any type of triangle, the only possibility is

$$
p_k \geq p_i = p_j.
$$

In the same manner, we can prove

$$p_i \geq p_j = p_k \quad \text{and} \quad p_j \geq p_i = p_k.$$

Hence, for the flux to be continuous across the edges of control volumes, $p_i = p_j = p_k$. That is, the flux is continuous across the edges of all control volumes if and only if the approximate solution p_h has the same value at all vertices, which is generally not true. Therefore, in general, the potential-based upstream weighted CVFE method generates a discontinuous flux across the edges of control volumes. On the other hand, the above argument leads to another upstream weighting strategy: flux-based.

The flux-based upstream weighting scheme

For the flux-based approach, the upstream direction is determined by the sign of a flux. It follows from (4.116) and (4.125) that the flux on edge $\mathbf{m}_a\mathbf{m}_c$ at node $\mathbf{m}_i$ (cf. Figure 4.32) is

$$f_{i,\mathbf{m}_a\mathbf{m}_c} = -\sum_{l=i}^{k} \lambda^{up}\mathbf{a}_{har}\nabla\lambda_l \cdot \left(m_{c,2} - m_{a,2}, m_{a,1} - m_{c,1}\right) p_l,$$

and, at node $\mathbf{m}_j$,

$$f_{j,\mathbf{m}_c\mathbf{m}_a} = -\sum_{l=i}^{k} \lambda^{up}\mathbf{a}_{har}\nabla\lambda_l \cdot \left(m_{a,2} - m_{c,2}, m_{c,1} - m_{a,1}\right) p_l,$$

where the upstream weighting is now defined by

$$\lambda^{up} = \begin{cases} \lambda(\mathbf{m}_i) & \text{if } f_{i,\mathbf{m}_a\mathbf{m}_c} > 0, \\ \lambda(\mathbf{m}_j) & \text{if } f_{i,\mathbf{m}_a\mathbf{m}_c} < 0. \end{cases} \tag{4.127}$$

From this definition it follows that

$$f_{i,\mathbf{m}_a\mathbf{m}_c} + f_{j,\mathbf{m}_c\mathbf{m}_a} = 0. \tag{4.128}$$

The fluxes on other edges can be defined in the same fashion. It is evident from (4.128) that the flux-based upstream weighted CVFE method has a continuous flux across the edges of control volumes.

4.3.5 Control volume function approximation methods

The CVFE methods can be generalized in a variety of ways. The simplest generalization is to finite elements of higher order as in Section 4.2.1, i.e., to piecewise polynomials of higher degree. Here we consider their generalization to nonpolynomial functions, such as *spline functions*. The resulting control volume methods are called *control volume function approximation (CVFA) methods* (Li et al., 2003A). Compared with the CVFE, these methods can be more easily applied to arbitrarily shaped control volumes. They are particularly suitable for hybrid grid reservoir simulation.

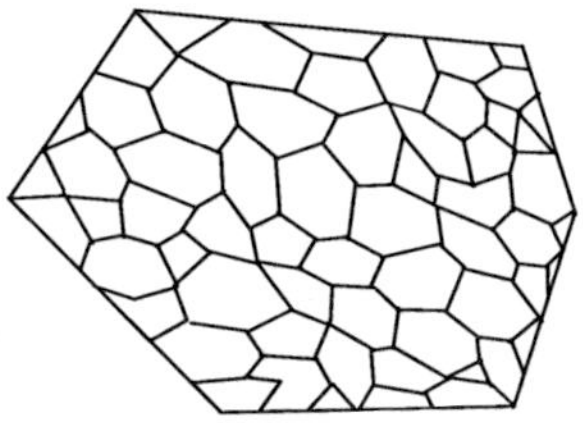

Figure 4.37. *A partition of Ω into control volumes.*

Assume that a partition K_h of Ω consists of a set of (open) control volumes V_i:

$$\bar{\Omega} = \bigcup_{i=1}^{N} \bar{V}_i, \quad V_i \cap V_j = \emptyset, \ i \neq j,$$

where N is the total number of control volumes. Different control volumes can have different shapes (cf. Figure 4.37). They can be generated from basic triangular, quadrilateral, and/or elliptic elements; they can also stand alone as the elements of the partition K_h of Ω. We define the boundary of each V_i by

$$\partial V_i = \bigcup_{k=1}^{N_i} e_{ik}, \tag{4.129}$$

where N_i is the number of edges e_{ik} on ∂V_i. For each V_i, the integral equation of problem (4.110) is given as in (4.111). On $e_{ik} \subset \partial V_i$, an interpolant p_h is used to approximate p:

$$p_h(\mathbf{x}) = \sum_{j=1}^{R_{ik}} p_{ik}^j \varphi_{ik}^j(\mathbf{x}), \qquad \mathbf{x} \in e_{ik}, \ i = 1, 2, \ldots, N, \tag{4.130}$$

where R_{ik} is the number of interpolation nodes $\mathbf{x}_{ik}^j$ for e_{ik} and these nodes can be located on or surrounding V_i (cf. Figure 4.38). The basis functions φ_{ik}^j are assumed to satisfy

$$\varphi_{ik}^j(\mathbf{x}) = \begin{cases} 1 & \text{at node } \mathbf{x}_{ik}^j \\ 0 & \text{at other nodes,} \end{cases}$$

and

$$\sum_{j=1}^{R_{ik}} \varphi_{ik}^j(\mathbf{x}) = 1, \qquad \mathbf{x} \in e_{ik}, \ k = 1, 2, \ldots, N_i, \ i = 1, 2, \ldots, N. \tag{4.131}$$

As a result, we see that p_{ik}^j represents the pressure at the jth interpolation node $\mathbf{x}_{ik}^j$ for e_{ik}, and a constant pressure is also represented by (4.111). The latter property is important in local mass conservation of the CVFA methods.

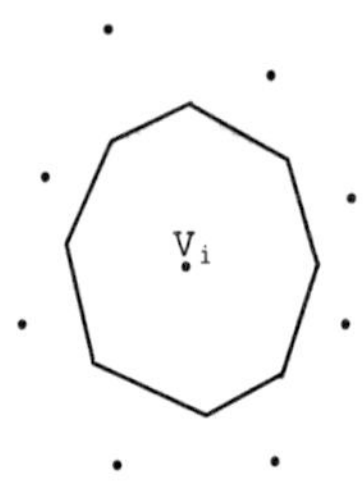

Figure 4.38. *A control volume with interpolation nodes.*

Application of (4.129) to (4.111) yields

$$-\sum_{k=1}^{N_i} \int_{e_{ik}} \mathbf{a}\nabla p \cdot \boldsymbol{\nu} \, d\ell = \int_{V_i} f \, d\mathbf{x}, \qquad i = 1, 2, \ldots, N. \tag{4.132}$$

Substituting (4.130) into (4.132) gives

$$-\sum_{k=1}^{N_i} \sum_{j=1}^{R_{ik}} \int_{e_{ik}} a(\mathbf{x}) p_{ik}^j \nabla \varphi_{ik}^j(\mathbf{x}) \cdot \boldsymbol{\nu} \, d\ell = \int_{V_i} f \, d\mathbf{x}, \qquad i = 1, 2, \ldots, N. \tag{4.133}$$

Set

$$T_{ik}^j = -\int_{e_{ik}} a(\mathbf{x}) \nabla \varphi_{ik}^j(\mathbf{x}) \cdot \boldsymbol{\nu} \, d\ell$$

for $j = 1, 2, \ldots, R_{ik}$, $k = 1, 2, \ldots, N_i$, and $i = 1, 2, \ldots, N$. Then (4.133) becomes

$$\sum_{k=1}^{N_i} \sum_{j=1}^{R_{ik}} T_{ik}^j p_{ik}^j = F_i, \qquad i = 1, 2, \ldots, N. \tag{4.134}$$

This is a linear system in terms of p_{ik}^j. The upstream weighting versions of the CVFA methods can be defined and analyzed as for the CVFE methods in Section 4.3.4.

It remains to construct the basis functions φ_{ik}^j. As an example, we describe spline basis functions. These functions have very good smoothness properties (Schumaker, 1981). Other nonpolynomial functions, such as distance weighted functions (Li et al., 2003A), can be also applied.

First, we define

$$\omega_{ik}^j(\mathbf{x}) = a_{ik}^j + b_{ik}^j x_1 + c_{ik}^j x_2 + \sum_{l=1}^{R_{ik}} f_{ik,l}^j h_{ik}^l(\mathbf{x}), \qquad \mathbf{x} = (x_1, x_2) \in e_{ik},$$

where $a_{ik}^j, b_{ik}^j, c_{ik}^j, f_{ik,l}^j \in \mathbb{R}$, and

$$h_{ik}^l(\mathbf{x}) = 2(r_{ik}^l)^2 \ln r_{ik}^l,$$
$$r_{ik}^l(x_1, x_2) = \left((x_1 - x_{ik,1}^l)^2 + (x_2 - x_{ik,2}^l)^2\right)^{1/2}$$

with $\mathbf{x}_{ik}^l = (x_{ik,1}^l, x_{ik,2}^l)$ the coordinates of nodes, $j, l = 1, 2, \ldots, R_{ik}$, $k = 1, 2, \ldots, N_i$, $i = 1, \ldots, N$. These spline functions are required to satisfy the following properties:

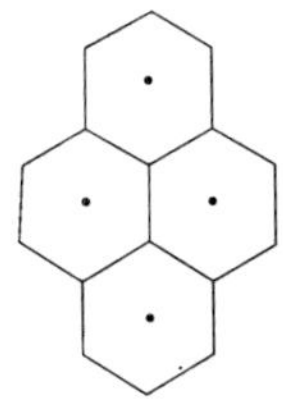

Figure 4.39. *The neighboring nodes of edge e_{ik} (the central vertical edge).*

- nodal values:

$$\omega_{ik}^{j}(\mathbf{x}) = \begin{cases} 1 & \text{at node } \mathbf{x}_{ik}^{j}, \\ 0 & \text{at other nodes}, \end{cases}$$

- zero total force:

$$\sum_{l=1}^{R_{ik}} f_{ik,l}^{j} = 0,$$

- zero total force moment:

$$\sum_{l=1}^{R_{ik}} f_{ik,l}^{j} \mathbf{x}_{ik}^{l} = \mathbf{0}.$$

It can be checked that these three constraints determine the coefficients a_{ik}^{j}, b_{ik}^{j}, c_{ik}^{j}, and $f_{ik,l}^{j}$ with an appropriate choice of the interpolation nodes $\mathbf{x}_{ik}^{j}$. The simplest choice is to use four neighboring centers of control volumes for each edge e_{ik} (cf. Figure 4.39).

Now, the basis functions φ_{ik}^{j} are defined by

$$\varphi_{ik}^{j}(\mathbf{x}) = \frac{\omega_{ik}^{j}(\mathbf{x})}{\sum_{l=1}^{R_{ik}} \omega_{ik}^{l}(\mathbf{x})}, \qquad \mathbf{x} \in e_{ik}. \tag{4.135}$$

Since there is no requirement on the shape of control volumes, the CVFA methods are particularly suitable for unstructured grid reservoir simulation.

We now report a couple of examples taken from Li et al. (2003A).

Example 4.10. To compare the CVFA with the CVFE, the control volumes used are generated from triangles, as in Section 4.3.1. In (4.110), let $\Omega = (0, 1) \times (0, 1)$ be the unit square, $\mathbf{a}$ be the identity tensor, and

$$f(\mathbf{x}) = 2\pi^2 \cos(\pi x_1) \cos(\pi x_2).$$

The boundary condition is

$$\begin{aligned} &\nabla p \cdot \boldsymbol{\nu} = 0, && x_1 = 0 \text{ and } x_1 = 1, && x_2 \in (0, 1), \\ &p = \cos(\pi x_1), && x_1 \in (0, 1), && x_2 = 0, \\ &p = -\cos(\pi x_1), && x_1 \in (0, 1), && x_2 = 1. \end{aligned}$$

Then the exact solution to (4.110) is $p = \cos(\pi x_1) \cos(\pi x_2)$.

Table 4.1. *Numerical results for p in the CVFA.*

$1/h$	$\|p - p_h\|_{L^\infty(\Omega)}$	Rate	$\|p - p_h\|_{L^2(\Omega)}$	Rate
2	0.31147353	—	0.18388206	—
4	0.11490560	1.4387	4.8526985E-02	1.9219
8	3.2336764E-02	1.8292	1.2107453E-02	2.0029
16	8.3515844E-03	1.9531	3.0019570E-03	2.0199
32	2.1060989E-03	1.9875	7.4598770E-04	2.0087
64	5.2769621E-04	1.9968	1.8585293E-04	2.0050

Table 4.2. *Numerical results for* **u** *in the CVFA.*

$1/h$	$\|\mathbf{u} - \mathbf{u}_h\|_{L^\infty(\Omega)}$	Rate	$\|\mathbf{u} - \mathbf{u}_h\|_{L^2(\Omega)}$	Rate
2	1.35576698	—	1.00055733	—
4	0.79144271	0.7766	0.40574242	1.3022
8	0.41524124	0.9305	0.15380230	1.3995
16	0.21064551	0.9791	6.3754123E-02	1.2705
32	0.10577494	0.9938	2.8906865E-02	1.1411
64	5.2954964E-02	0.9982	1.3795696E-02	1.0672

Table 4.3. *Numerical results for p in the CVFE.*

$1/h$	$\|p - p_h\|_{L^\infty(\Omega)}$	Rate	$\|p - p_h\|_{L^2(\Omega)}$	Rate
2	0.35502877	—	0.18584850	—
4	0.11549486	1.6201	5.8970002E-02	1.6561
8	3.3079427E-02	1.8038	1.5744807E-02	1.9051
16	8.7789616E-03	1.9138	4.0029721E-03	1.9757
32	2.2525012E-03	1.9625	1.0049860E-03	1.9939
64	5.6991337E-04	1.9827	2.5151431E-04	1.9985

Two types of norms are used to check the convergence rates:

$$\|v\|_{L^\infty(\Omega)} = \max_{\mathbf{x}\in\Omega} |v(\mathbf{x})|, \qquad \|v\|_{L^2(\Omega)} = \left(\int_\Omega |v(\mathbf{x})|^2\, d\mathbf{x}\right)^{1/2}.$$

The interpolation nodes for the spline function approximation approach in the CVFA consist of the centers of control volumes. Numerical errors and the corresponding convergence rates for p and its gradient $\mathbf{u} = \nabla p$ are shown in Tables 4.1–4.4 for the CVFA and CVFE, where p_h and $\mathbf{u}_h$ are the approximate solutions of p and $\mathbf{u}$, respectively, h is the space step size in the x_1- and x_2-directions for the base triangulation, and the rate is the convergence rate in the corresponding norm. From these computational results, we see that the convergence rates for p and $\mathbf{u}$ are asymptotically of order $\mathcal{O}(h^2)$ and $\mathcal{O}(h)$ for both the CVFA and CVFE. However, from this and other numerical experiments (not reported here) we have observed that the approximation errors in the CVFA are smaller than those in the CVFE.

Example 4.11. We now consider an example that the CVFE method cannot easily handle:

$$\begin{aligned} -\Delta p &= \delta(\mathbf{x} - \mathbf{x}_0), & \mathbf{x} &\in \Omega, \\ p &= 0, & \mathbf{x} &\in \Gamma, \end{aligned} \tag{4.136}$$

Table 4.4. *Numerical results for* **u** *in the CVFE.*

$1/h$	$\|\mathbf{u}-\mathbf{u}_h\|_{L^\infty(\Omega)}$	Rate	$\|\mathbf{u}-\mathbf{u}_h\|_{L^2(\Omega)}$	Rate
2	1.8475225	—	1.2560773	—
4	1.3093706	0.4967	0.68846096	0.8675
8	0.70851284	0.8860	0.35305205	0.9635
16	0.36116394	0.9721	0.17767979	0.9906
32	0.18145106	0.9931	8.8985854E-02	0.9976
64	9.0834342E-02	0.9983	4.4511228E-02	0.9994

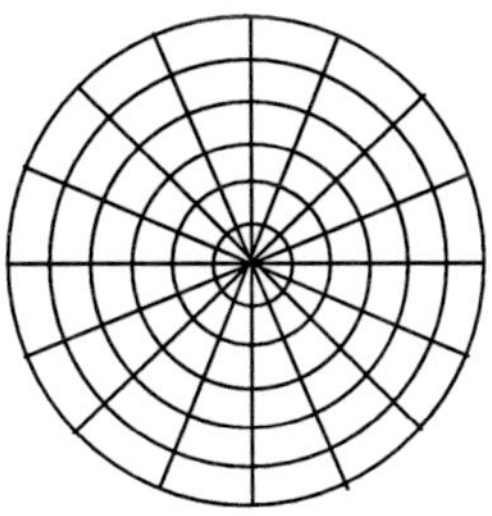

Figure 4.40. *A circular grid.*

where $\Omega = \{\mathbf{x} \in \mathbb{R}^2 : |\mathbf{x}| \le 1\}$ is the unit circle and $\delta(\mathbf{x}-\mathbf{x}_0)$ is the Dirac delta function with center $\mathbf{x}_0$. The exact solution to (4.136) is *Green's function*

$$p(\mathbf{x}) = \frac{1}{2\pi} \ln\left(\frac{|\mathbf{x}-\mathbf{x}_0|}{|\mathbf{x}_0||\mathbf{x}-\mathbf{x}_0^*|}\right), \tag{4.137}$$

where $\mathbf{x}_0^*$ is the image of $\mathbf{x}_0$ with respect to Γ:

$$\mathbf{x}_0^* = \frac{1}{|\mathbf{x}_0|^2}\mathbf{x}_0.$$

For this problem, circular grids (cf. Figure 4.40) are the most appropriate. The CVFE, however, cannot easily and accurately handle this type of grid. The flexibility of the CVFA on the shape of elements enables us to use the circular grids more easily and accurately. The numerical errors $\|p-p_h\|_{L^2(\Omega)}$ and the corresponding convergence rates for the CVFA are presented in Table 4.5, where uniform refinements in the radial and angular directions are measured by $h_r = 1/N_r$ and $h_\theta = 2\pi/N_\theta$, and $\mathbf{x}_0 = 0.5e^{\pi\bar{i}/6}$ ($\bar{i} = \sqrt{-1}$). This table shows that the convergence rate in this norm is asymptotically of order $\mathcal{O}(h)$. The reduction in the rate is due to the reduction in the regularity of the solution to (4.136) (cf. (4.137)). Because of the lack of regularity of this solution, we are not able to use the $\|\cdot\|_{L^\infty(\Omega)}$-norm.

4.3.6 Reduction of grid orientation effects

Finite difference methods were indicated to have grid orientation effects in Section 4.1.9. The example shown in Figure 4.9 is now calculated using the CVFE method and displayed in

Table 4.5. *Numerical results for the CVFA in Example* 4.11.

(N_r, N_θ)	$\|p - p_h\|_{L^2(\Omega)}$	Rate
(8, 12)	4.13287534E-03	—
(16, 24)	2.88767285E-03	0.5172
(32, 48)	1.49421476E-03	0.9595
(64, 96)	7.51098130E-04	0.9923

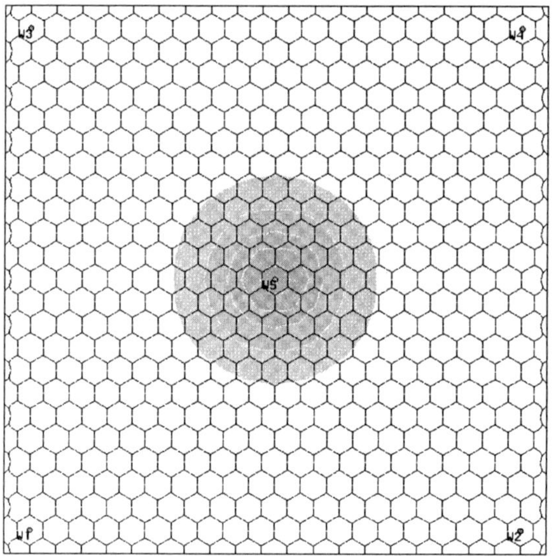

Figure 4.41. *A CVFE example.*

Figure 4.41, which indicates that the grid orientation effect disappears. The same example was also evaluated using the CVFA, and identical numerical results were obtained as for the CVFE.

4.4 Discontinuous Finite Element Methods

In the previous two sections, functions used in finite element spaces for the discretization of second-order partial differential equations were continuous across interelement boundaries. In this section, we consider the case where the functions in the finite element spaces are discontinuous across these boundaries, i.e., *discontinuous finite elements. Discontinuous Galerkin (DG) finite element methods* were originally introduced for a linear *advection (hyperbolic) problem* by Reed and Hill (1973). They have become established as an important alternative for numerically solving advection (convection) problems for which continuous finite element methods lack robustness. Important features of the DG methods are that they conserve mass locally (on each element) and are of high-order accuracy.

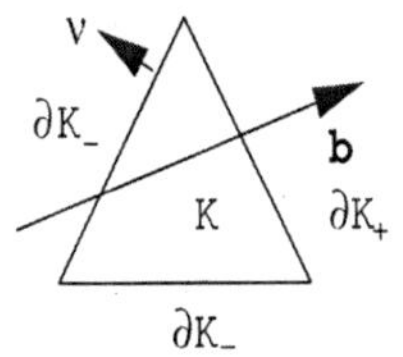

Figure 4.42. *An illustration of ∂K_- and ∂K_+.*

4.4.1 DG methods

We consider the *advection problem*:

$$\begin{aligned} \mathbf{b}\cdot\nabla p + Rp &= f, \qquad && \mathbf{x}\in\Omega, \\ p &= g, && \mathbf{x}\in\Gamma_-, \end{aligned} \tag{4.138}$$

where the functions $\mathbf{b}$, R, f, and g are given, $\Omega \subset \mathbb{R}^d$ ($d \leq 3$) is a bounded domain with boundary Γ, the *inflow boundary* Γ_- is defined by

$$\Gamma_- = \{\mathbf{x}\in\Gamma : \ (\mathbf{b}\cdot\boldsymbol{\nu})(\mathbf{x}) < 0\},$$

and $\boldsymbol{\nu}$ is the outward unit normal to Γ. The advection coefficient $\mathbf{b}$ is assumed to be smooth in $(\mathbf{x}, t)$, and the reaction coefficient R is assumed to be bounded and nonnegative. A one-dimensional version of this problem was studied in Section 4.1.8.

For $h > 0$, let K_h be a finite element partition of Ω into elements $\{K\}$. K_h is assumed to satisfy the minimum angle condition (4.79). For the DG methods, adjacent elements in K_h are not required to match; a vertex of one element can lie in the interior of the edge or face of another element, for example. Let $\mathcal{E}_h^o$ denote the set of all interior boundaries e in K_h, $\mathcal{E}_h^b$ the set of the boundaries e on Γ, and $\mathcal{E}_h = \mathcal{E}_h^o \cup \mathcal{E}_h^b$. We tacitly assume that $\mathcal{E}_h^o \neq \emptyset$.

Associated with K_h, we define the finite element space

$$V_h = \{v : v \text{ is a bounded function on } \Omega \text{ and } v|_K \in P_r(K),\ K \in K_h\},$$

where we recall that $P_r(K)$ is the space of polynomials on K of degree at most $r \geq 0$. Note that no continuity across interelement boundaries is required on functions in this space.

To introduce DG methods, we need some notation. For each $K \in K_h$, we split its boundary ∂K into the inflow and outflow parts by

$$\begin{aligned} \partial K_- &= \{\mathbf{x}\in\partial K : \ (\mathbf{b}\cdot\boldsymbol{\nu})(\mathbf{x}) < 0\}, \\ \partial K_+ &= \{\mathbf{x}\in\partial K : \ (\mathbf{b}\cdot\boldsymbol{\nu})(\mathbf{x}) \geq 0\}, \end{aligned}$$

where $\boldsymbol{\nu}$ is the outward unit normal to ∂K. A triangle K with boundary made up of ∂K_- and ∂K_+ is shown is Figure 4.42. For $e \in \mathcal{E}_h^o$, the left- and right-hand limits on e of a function $v \in V_h$ are defined by

$$v_-(\mathbf{x}) = \lim_{\epsilon\to 0^-} v(\mathbf{x}+\epsilon\mathbf{b}), \quad v_+(\mathbf{x}) = \lim_{\epsilon\to 0^+} v(\mathbf{x}+\epsilon\mathbf{b})$$

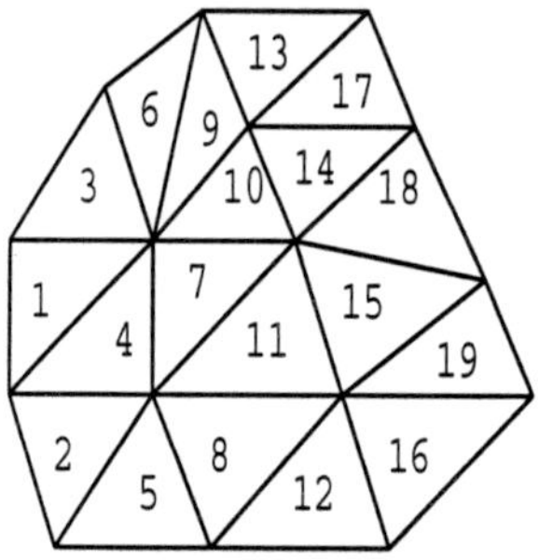

Figure 4.43. *An ordering of computation for the DG method.*

for $\mathbf{x} \in e$. The jump of v across e is given by

$$[\![v]\!] = v_+ - v_-.$$

For $e \in \mathcal{E}_h^b$, we define (from inside Ω)

$$[\![v]\!] = v.$$

Now, the DG method for (4.138) is defined: For $K \in K_h$, given $p_{h,-}$ on ∂K_-, find $p_h = p_h|_K \in P_r(K)$ such that

$$\begin{aligned}(\mathbf{b} \cdot \nabla p_h + R p_h, v)_K - \int_{\partial K_-} p_{h,+} v_+ \mathbf{b} \cdot \boldsymbol{\nu} \, d\ell \\ = (f, v)_K - \int_{\partial K_-} p_{h,-} v_+ \mathbf{b} \cdot \boldsymbol{\nu} \, d\ell \qquad \forall v \in P_r(K),\end{aligned} \tag{4.139}$$

where

$$(v, w)_K = \int_K vw \, d\mathbf{x}, \qquad p_{h,-} = g \text{ on } \Gamma_-.$$

Note that (4.139) is the standard finite element method for (4.138) on the element K, with the boundary condition being *weakly* imposed. When $p_{h,-}$ is given on ∂K_-, existence and uniqueness of a solution to (4.139) can be shown as in Section 4.2.1 (see the remarks following (4.146)). Equation (4.139) also holds for the continuous problem (4.138) (Chen, 2005). For a typical triangulation (cf. Figure 4.43), p_h can be determined first on the triangles K adjacent to Γ_-. Then this process is continued (working away from known information) until p_h is found in the whole domain Ω. Thus the computation of (4.139) is local.

If $\mathbf{b}$ is *divergence-free* (or *solenoidal*), i.e., $\nabla \cdot \mathbf{b} = 0$, we can use Green's formula (4.68) to see that (cf. Figure 4.42)

$$(\mathbf{b} \cdot \nabla p_h, 1)_K = \int_{\partial K_-} p_{h,+} \mathbf{b} \cdot \boldsymbol{\nu} \, d\ell + \int_{\partial K_+} p_{h,-} \mathbf{b} \cdot \boldsymbol{\nu} \, d\ell.$$

We substitute this into (4.139) with $v = 1$ to give

$$(R p_h, 1)_K + \int_{\partial K_+} p_{h,-} \mathbf{b} \cdot \boldsymbol{\nu} \, d\ell = (f, 1)_K - \int_{\partial K_-} p_{h,-} \mathbf{b} \cdot \boldsymbol{\nu} \, d\ell, \tag{4.140}$$

which expresses a local conservation property (i.e., the difference between inflow and outflow equals the sum of accumulation of mass).

To express (4.139) in the form used in Section 4.2, we define

$$a_K(v, w) = (\mathbf{b} \cdot \nabla v + Rv, w)_K - \int_{\partial K_-} [\![v]\!] w_+ \mathbf{b} \cdot \boldsymbol{\nu} \, d\ell, \qquad K \in K_h,$$

and

$$a(v, w) = \sum_{K \in K_h} a_K(v, w).$$

Then (4.139) is expressed as follows: Find $p_h \in V_h$ such that

$$a(p_h, v) = (f, v) \qquad \forall v \in V_h, \tag{4.141}$$

where $p_{h,-} = g$ on Γ_-. We consider a couple of examples before stating stability and convergence results for (4.141).

Example 4.12. A one-dimensional example of (4.138) is

$$\begin{aligned} &\frac{dp}{dx} + p = f, \qquad x \in (0, 1), \\ &p(0) = g. \end{aligned} \tag{4.142}$$

Let $0 = x_0 < x_1 < \cdots < x_M = 1$ be a partition of $(0, 1)$ into a set of subintervals $I_i = (x_{i-1}, x_i)$, with length $h_i = x_i - x_{i-1}$, $i = 1, 2, \ldots, M$. In this case, (4.139) becomes: For $i = 1, 2, \ldots, M$, given $(p_h(x_{i-1}))_-$, find $p_h = p_h|_{I_i} \in P_r(I_i)$ such that

$$\left(\frac{dp_h}{dx} + p_h, v\right)_{I_i} + [\![p_h(x_{i-1})]\!] (v(x_{i-1}))_+ = (f, v)_{I_i} \qquad \forall v \in P_r(I_i),$$

where $(p_h(x_0))_- = g$. In the case $r = 0$, V_h is the space of piecewise constants, and the DG method reduces to: For $i = 1, 2, \ldots, M$, find $p_i = (p_h(x_i))_-$ such that

$$\begin{aligned} &\frac{p_i - p_{i-1}}{h_i} + p_i = \frac{1}{h_i} \int_{I_i} f \, dx, \\ &p_0 = g. \end{aligned} \tag{4.143}$$

Note that (4.143) is nothing but a simple *upwind* finite difference method (cf. Section 4.1.8) with an averaged right-hand side.

Example 4.13. Set $R = f = 0$ in the advection problem (4.138). Then (4.138) simplifies to

$$\begin{aligned} \mathbf{b} \cdot \nabla p &= 0, \qquad \mathbf{x} \in \Omega, \\ p &= g, \qquad \mathbf{x} \in \Gamma_-. \end{aligned} \tag{4.144}$$

Also, let $r = 0$. Then (4.139) reads: For $K \in K_h$, given $p_{h,-}$ on ∂K_-, find $p_K = p_h|_K$ such that

$$\int_{\partial K_-} p_K \mathbf{b} \cdot \boldsymbol{\nu} \, d\ell = \int_{\partial K_-} p_{h,-} \mathbf{b} \cdot \boldsymbol{\nu} \, d\ell;$$

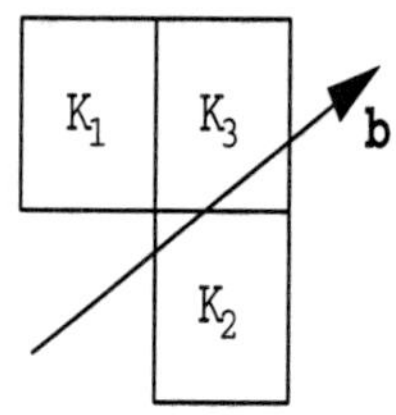

Figure 4.44. *Adjoining rectangles.*

that is,

$$p_K = \frac{\int_{\partial K_-} p_{h,-} \mathbf{b} \cdot \boldsymbol{\nu} \, d\ell}{\int_{\partial K_-} \mathbf{b} \cdot \boldsymbol{\nu} \, d\ell}. \tag{4.145}$$

Thus we see that for each $K \in K_h$ the value p_K is determined by a weighted average of the values $p_{h,-}$ on adjoining elements with edges on ∂K_-. As an example, let Ω be a rectangular domain in $\mathbb{R}^2$, K_h consist of rectangles, and $\mathbf{b} > \mathbf{0}$. In this case, for a configuration shown in Figure 4.44, we see that

$$p_3 = \frac{b_1}{b_1 + b_2} p_1 + \frac{b_2}{b_1 + b_2} p_2,$$

where $p_i = p_h|_{K_i}$, $i = 1, 2, 3$, and $\mathbf{b} = (b_1, b_2)$. Again, in this case, (4.145) corresponds to the usual upwind finite difference method for (4.144).

To state stability and convergence properties of the DG method (4.141), we define the norm

$$\|v\|_{\mathbf{b}} = \left(\|R^{1/2} v\|^2_{L^2(\Omega)} + \frac{1}{2} \sum_{K \in K_h} \int_{\partial K_-} [\![v]\!]^2 |\mathbf{b} \cdot \boldsymbol{\nu}| \, d\ell \right.$$
$$\left. + \frac{1}{2} \int_{\Gamma_+} v_-^2 \mathbf{b} \cdot \boldsymbol{\nu} \, d\ell \right)^{1/2}.$$

Then, if $\nabla \cdot \mathbf{b} = 0$, it can be shown (Chen, 2005) that

$$a(v, v) = \|v\|^2_{\mathbf{b}} - \frac{1}{2} \int_{\Gamma_-} v_-^2 |\mathbf{b} \cdot \boldsymbol{\nu}| \, d\ell, \qquad v \in V_h. \tag{4.146}$$

Using (4.146), existence and uniqueness of a solution to (4.141) can be proven in the usual way (cf. Section 4.2.1). If we assume that $R - \nabla \cdot \mathbf{b}/2 \geq 0$ (instead of $\nabla \cdot \mathbf{b} = 0$), the term $\|R^{1/2} v\|_{L^2(\Omega)}$ is replaced with the quantity $\|(R - \nabla \cdot \mathbf{b}/2)^{1/2} v\|_{L^2(\Omega)}$ in the definition of $\|v\|_{\mathbf{b}}$.

If R is strictly positive with respect to $\mathbf{x} \in \Omega$ (i.e., $R(\mathbf{x}) \geq R_0 > 0$), it can be seen from (4.141) and (4.146) that

$$\|p_h\|_{\mathbf{b}} \leq C \left(\|f\|^2_{L^2(\Omega)} + \int_{\Gamma_-} g^2 |\mathbf{b} \cdot \boldsymbol{\nu}| \, d\ell \right)^{1/2}. \tag{4.147}$$

This is a stability result for (4.141) in terms of data f and g. If the solution p to (4.138) is in $H^{r+1}(K)$ for each $K \in K_h$, an error estimate for (4.141) is given by

$$\begin{aligned}\|p - p_h\|^2_{L^2(\Omega)} + h \sum_{K \in K_h} \|\mathbf{b} \cdot \nabla(p - p_h)\|^2_{L^2(K)} \\ \leq Ch^{2r+1} \sum_{K \in K_h} \|p\|^2_{H^{r+1}(K)}\end{aligned} \tag{4.148}$$

for $r \geq 0$. Note that the $L^2(\Omega)$-estimate is half a power of h from being optimal, while the $L^2(\Omega)$-estimate of the derivative in the velocity (or *streamline*) direction is in fact optimal. For general triangulations, this $L^2(\Omega)$-estimate is sharp in the sense that the exponent of h cannot be increased (Johnson, 1994).

We end by remarking that a time-dependent advection problem can be written as a system in the same form as (4.138). To see this, consider the problem

$$\phi \frac{\partial p}{\partial t} + \mathbf{b} \cdot \nabla p + Rp = f, \qquad \mathbf{x} \in \Omega,\ t > 0,$$

and set $t = x_0$ and $b_0 = \phi$. Then we see that

$$\bar{\mathbf{b}} \cdot \nabla_{(t,\mathbf{x})} p + Rp = f,$$

where $\bar{\mathbf{b}} = (b_0, \mathbf{b})$ and $\nabla_{(t,\mathbf{x})} = (\frac{\partial}{\partial t}, \nabla_{\mathbf{x}})$ (treating time as a space-like variable). Thus the above development of the DG method for (4.138) applies.

4.4.2 Stabilized DG methods

We consider a stabilized DG (SDG) method, which modifies (4.139) as follows: For $K \in K_h$, given $p_{h,-}$ on ∂K_-, find $p_h = p_h|_K \in P_r(K)$ such that

$$\begin{aligned}(\mathbf{b} \cdot \nabla p_h + Rp_h, v + \theta \mathbf{b} \cdot \nabla v)_K - \int_{\partial K_-} p_{h,+} v_+ \mathbf{b} \cdot \boldsymbol{\nu}\, d\ell \\ = (f, v + \theta \mathbf{b} \cdot \nabla v)_K - \int_{\partial K_-} p_{h,-} v_+ \mathbf{b} \cdot \boldsymbol{\nu}\, d\ell \qquad \forall v \in P_r(K),\end{aligned} \tag{4.149}$$

where θ is a *stabilization parameter*. The difference between (4.139) and (4.149) is that a stabilized term is added in the left- and right-hand sides of (4.149). This stabilized method is also called the *streamline diffusion method* due to intuition that the added term

$$\theta\, (\mathbf{b} \cdot \nabla p_h, \mathbf{b} \cdot \nabla v)$$

corresponds to the diffusion in the direction of streamlines (or characteristics) (Johnson, 1994). The parameter θ is chosen so that $\theta = \mathcal{O}(h)$, to generate the same convergence rate as for the DG method. For $r = 0$, DG and SDG methods are the same.

Now, the bilinear forms $a_K(\cdot, \cdot)$ and $a(\cdot, \cdot)$ are defined by

$$\begin{aligned}a_K(v, w) = (\mathbf{b} \cdot \nabla v + Rv, w + \theta \mathbf{b} \cdot \nabla w)_K \\ - \int_{\partial K_-} [\![v]\!] w_+ \mathbf{b} \cdot \boldsymbol{\nu}\, d\ell, \qquad K \in K_h,\end{aligned}$$

and

$$a(v, w) = \sum_{K \in K_h} a_K(v, w).$$

Then (4.149) is expressed as follows: Find $p_h \in V_h$ such that

$$a(p_h, v) = \sum_{K \in K_h} (f, v + \theta \mathbf{b} \cdot \nabla v)_K \qquad \forall v \in V_h, \tag{4.150}$$

where $p_{h,-} = g$ on Γ_-.

If $1 - \theta R/2 \geq 0$, the norm $\| \cdot \|_{\mathbf{b}}$ is modified to

$$\|v\|_{\mathbf{b}} = \left(\left\| R^{1/2} (1 - \theta R/2)^{1/2} v \right\|^2_{L^2(\Omega)} + \frac{1}{2} \sum_{K \in K_h} \int_{\partial K_-} [\![v]\!]^2 |\mathbf{b} \cdot \boldsymbol{\nu}| \, d\ell \right.$$
$$\left. + \frac{1}{2} \sum_{K \in K_h} \|\theta^{1/2} \mathbf{b} \cdot \nabla v\|^2_{L^2(K)} + \frac{1}{2} \int_{\Gamma_+} v_-^2 \mathbf{b} \cdot \boldsymbol{\nu} \, d\ell \right)^{1/2}.$$

Then, if $\mathbf{b}$ satisfies $\nabla \cdot \mathbf{b} = 0$, it can be seen (Chen, 2005) that

$$a(v, v) \geq \|v\|^2_{\mathbf{b}} - \frac{1}{2} \int_{\Gamma_-} v_-^2 |\mathbf{b} \cdot \boldsymbol{\nu}| \, d\ell, \qquad v \in V_h. \tag{4.151}$$

Hence the stability and convergence results (4.147) and (4.148) hold also for (4.150) (Chen, 2005).

For an appropriate choice of the stabilization parameter θ, the SDG method is much more stable than the DG. For a comparison, see Chen (2005). The DG and SDG methods have been developed here only for the hyperbolic problem (4.138); they can be also used for the solution of diffusion problems (Chen, 2005).

4.5 Mixed Finite Element Methods

In this section, we study *mixed finite element methods*, which generalize the finite element methods discussed in Section 4.2. These methods were initially introduced by engineers in the 1960s (Fraeijs de Veubeke, 1965; Hellan, 1967; Hermann, 1967) for solving problems in solid continua. Since that time, they have been applied to many areas, particularly solid and fluid mechanics. Here we discuss their applications to second-order partial differential equation problems. The main reason for using mixed methods is that in some applications a vector variable (e.g., a fluid velocity) is the primary variable in which one is interested. Then the mixed methods are developed to approximate both this variable and a scalar variable (e.g., pressure) simultaneously and to give a high-order approximation of both variables. Instead of the single finite element space used in the standard finite element methods, *mixed* finite element methods employ two different spaces. These two spaces must satisfy an *inf-sup* condition for the mixed methods to be stable. Raviart and Thomas (1977) introduced the first family of mixed finite element spaces for second-order elliptic problems in the two-dimensional case. Somewhat later, Nédélec (1980) extended these spaces to three-dimensional problems. Motivated by these two papers, there are now many mixed finite element spaces available in the literature; see Brezzi et al. (1985; 1987A; 1987B) and Chen and Douglas (1989).

4.5.1 A one-dimensional model problem

As in Section 4.2, for the purpose of demonstration, we consider a stationary problem for p in one dimension:

$$
\begin{aligned}
-\frac{d^2 p}{dx^2} &= f(x), \qquad 0 < x < 1, \\
p(0) &= p(1) = 0,
\end{aligned}
\tag{4.152}
$$

where the function $f \in L^2(I)$ is given, with $I = (0, 1)$ and

$$
L^2(I) = \left\{ v : v \text{ is defined on } I \text{ and } \int_I v^2 \, dx < \infty \right\}.
$$

We recall the scalar product in $L^2(I)$:

$$
(v, w) = \int_0^1 v(x) w(x) \, dx
$$

for real-valued functions $v, w \in L^2(I)$ (cf. Section 4.2.1). We also use the linear space (cf. Section 4.2.1)

$$
H^1(I) = \left\{ v \in L^2(I) : \frac{dv}{dx} \in L^2(I) \right\}.
$$

Set

$$
V = H^1(I), \quad W = L^2(I).
$$

Observe that the functions in W are not required to be continuous on the interval I.

After introducing the variable

$$
u = -\frac{dp}{dx}, \tag{4.153}
$$

(4.152) can be recast in the form

$$
\frac{du}{dx} = f. \tag{4.154}
$$

Multiplying (4.153) by any function $v \in V$ and integrating over I, we see that

$$
(u, v) = -\left(\frac{dp}{dx}, v \right).
$$

Application of integration by parts to the right-hand side of this equation leads to

$$
(u, v) = \left(p, \frac{dv}{dx} \right),
$$

where we use the boundary conditions $p(0) = p(1) = 0$ from (4.152). Also, we multiply (4.154) by any function $w \in W$ and integrate over Ω to give

$$
\left(\frac{du}{dx}, w \right) = (f, w).
$$

Therefore, we see that the pair of functions u and p satisfies the system

$$\begin{aligned}(u, v) - \left(\frac{dv}{dx}, p\right) &= 0, \qquad v \in V,\\ \left(\frac{du}{dx}, w\right) &= (f, w), \qquad w \in W.\end{aligned} \tag{4.155}$$

This system is referred to as a *mixed variational* (or *weak*) *form* of (4.152). If the pair of functions u and p is a solution to (4.153) and (4.154), then this pair also satisfies (4.155). The converse also holds if p is sufficiently smooth (e.g., if $p \in H^2(I)$); see Exercise 4.40.

We introduce the functional $F : V \times W \to \mathbb{R}$ by

$$F(v, w) = \frac{1}{2}(v, v) - \left(\frac{dv}{dx}, w\right) + (f, w), \qquad v \in V,\ w \in W.$$

It can be shown (Chen, 2005) that problem (4.155) is equivalent to the *saddle point problem*: Find $u \in V$ and $p \in W$ such that

$$F(u, w) \leq F(u, p) \leq F(v, p) \qquad \forall v \in V,\ w \in W. \tag{4.156}$$

For this reason, problem (4.155) is also referred to as a *saddle point problem.*

To construct mixed finite element methods for solving (4.152), for a positive integer M let $0 = x_1 < x_2 < \cdots < x_M = 1$ be a partition of I into a set of subintervals $I_{i-1} = (x_{i-1}, x_i)$, with length $h_i = x_i - x_{i-1}$, $i = 2, 3, \ldots, M$. Set $h = \max\{h_i, 2 \leq i \leq M\}$. Define the *mixed finite element spaces*

$$\begin{aligned}V_h = \{v :\ & v \text{ is a continuous function on } [0, 1]\\ & \text{and is linear on each subinterval } I_i\},\\ W_h = \{w :\ & w \text{ is constant on each subinterval } I_i\}.\end{aligned}$$

Note that $V_h \subset V$ and $W_h \subset W$. Now, the *mixed finite element method* for (4.152) is defined as

Find $u_h \in V_h$ and $p_h \in W_h$ such that

$$\begin{aligned}(u_h, v) - \left(\frac{dv}{dx}, p_h\right) &= 0, \qquad v \in V_h,\\ \left(\frac{du_h}{dx}, w\right) &= (f, w), \qquad w \in W_h.\end{aligned} \tag{4.157}$$

To show that (4.157) has a unique solution, let $f = 0$; take $v = u_h$ and $w = p_h$ in (4.157), and add the resulting equations to give

$$(u_h, u_h) = 0,$$

so that $u_h = 0$. Consequently, it follows from (4.157) that

$$\left(\frac{dv}{dx}, p_h\right) = 0, \qquad v \in V_h.$$

Choose $v \in V_h$ such that $dv/dx = p_h$ (thanks to the definition of V_h and W_h) in this equation to see that $p_h = 0$. Hence the solution of (4.157) is unique. Uniqueness also yields existence since (4.157) is equivalent to a finite-dimensional linear system.

In the same fashion as for the equivalence between (4.155) and (4.156), problem (4.157) is equivalent to the saddle point problem: Find $u_h \in V_h$ and $p_h \in W_h$ such that

$$F(u_h, w) \leq F(u_h, p_h) \leq F(v, p_h) \qquad \forall v \in V_h,\ w \in W_h. \tag{4.158}$$

We introduce the *basis functions* $\varphi_i \in V_h, i = 1, 2, \ldots, M$ (cf. Figure 4.11)

$$\varphi_i(x_j) = \begin{cases} 1 & \text{if } i = j, \\ 0 & \text{if } i \neq j, \end{cases}$$

and the basis functions $\psi_i \in W_h, i = 1, 2, \ldots, M - 1$,

$$\psi_i(x) = \begin{cases} 1 & \text{if } x \in I_i, \\ 0 & \text{otherwise.} \end{cases}$$

The functions ψ_i are *characteristic* functions. Now, functions $v \in V_h$ and $w \in W_h$ have the unique representations

$$v(x) = \sum_{i=1}^{M} v_i \varphi_i(x), \quad w(x) = \sum_{i=1}^{M-1} w_i \psi_i(x), \quad 0 \leq x \leq 1,$$

where $v_i = v(x_i)$ and $w_i = w|_{I_i}$. Take v and w in (4.157) to be these basis functions to see that

$$\begin{aligned} &(u_h, \varphi_j) - \left(\frac{d\varphi_j}{dx}, p_h\right) = 0, && j = 1, 2, \ldots, M, \\ &\left(\frac{du_h}{dx}, \psi_j\right) = (f, \psi_j), && j = 1, 2, \ldots, M - 1. \end{aligned} \tag{4.159}$$

Set

$$u_h(x) = \sum_{i=1}^{M} u_i \varphi_i(x), \qquad u_i = u_h(x_i),$$

and

$$p_h(x) = \sum_{k=1}^{M-1} p_k \psi_k(x), \qquad p_k = p_h|_{I_k}.$$

Substitute these expressions into (4.159) to give

$$\begin{aligned} &\sum_{i=1}^{M} (\varphi_i, \varphi_j)\, u_i - \sum_{k=1}^{M-1} \left(\frac{d\varphi_j}{dx}, \psi_k\right) p_k = 0, && j = 1, \ldots, M, \\ &\sum_{i=1}^{M} \left(\frac{d\varphi_i}{dx}, \psi_j\right) u_i = (f, \psi_j), && j = 1, \ldots, M - 1. \end{aligned} \tag{4.160}$$

We introduce the matrices and vectors

$$\mathbf{A} = \left(a_{ij}\right)_{i,j=1,2,\ldots,M}, \quad \mathbf{B} = \left(b_{jk}\right)_{j=1,2,\ldots,M,\ k=1,2,\ldots,M-1},$$
$$\mathbf{U} = (u_i)_{i=1,2,\ldots,M}, \quad \mathbf{p} = (p_k)_{k=1,2,\ldots,M-1}, \quad \mathbf{f} = \left(f_j\right)_{j=1,2,\ldots,M-1},$$

where

$$a_{ij} = \left(\varphi_i, \varphi_j\right), \quad b_{jk} = -\left(\frac{d\varphi_j}{dx}, \psi_k\right), \quad f_j = \left(f, \psi_j\right).$$

With this notation, system (4.160) can be written in matrix form as

$$\begin{pmatrix} \mathbf{A} & \mathbf{B} \\ \mathbf{B}^T & \mathbf{0} \end{pmatrix} \begin{pmatrix} \mathbf{U} \\ \mathbf{p} \end{pmatrix} = \begin{pmatrix} \mathbf{0} \\ -\mathbf{f} \end{pmatrix}, \tag{4.161}$$

where $\mathbf{B}^T$ is the transpose of $\mathbf{B}$. Note that (4.161) is symmetric but *indefinite*. It can be shown that the matrix $\mathbf{M}$ defined by

$$\mathbf{M} = \begin{pmatrix} \mathbf{A} & \mathbf{B} \\ \mathbf{B}^T & \mathbf{0} \end{pmatrix}$$

has both positive and negative eigenvalues (cf. Exercise 4.45).

The matrix $\mathbf{A}$ is symmetric and positive definite (cf. Section 4.2.1). It is also sparse. In the one-dimensional case, it is tridiagonal. It follows from the definition of the basis functions φ_i that

$$a_{ij} = \left(\varphi_i, \varphi_j\right) = 0 \quad \text{if } |i-j| \geq 2,$$

so that

$$a_{11} = \frac{h_2}{3}, \quad a_{MM} = \frac{h_M}{3},$$

and, for $i = 2, 3, \ldots, M-1$,

$$a_{i-1,i} = \frac{h_i}{6}, \quad a_{ii} = \frac{h_i}{3} + \frac{h_{i+1}}{3}, \quad a_{i,i+1} = \frac{h_{i+1}}{6}.$$

It can be also seen that

$$b_{jj} = 1, \quad b_{j+1,j} = -1, \quad j = 1, 2, \ldots, M-1;$$

all other entries of $\mathbf{B}$ are zero. That is, the $M \times (M-1)$ matrix $\mathbf{B}$ is *bidiagonal*:

$$\mathbf{B} = \begin{pmatrix} 1 & 0 & 0 & \ldots & 0 & 0 \\ -1 & 1 & 0 & \ldots & 0 & 0 \\ 0 & -1 & 1 & \ldots & 0 & 0 \\ \vdots & \vdots & \vdots & \ddots & \vdots & \vdots \\ 0 & 0 & 0 & \ldots & 1 & 0 \\ 0 & 0 & 0 & \ldots & -1 & 1 \\ 0 & 0 & 0 & \ldots & 0 & -1 \end{pmatrix}.$$

In the case where the partition is uniform, i.e., $h = h_i$,

$$\mathbf{A} = \frac{h}{6}\begin{pmatrix} 2 & 1 & 0 & \dots & 0 & 0 \\ 1 & 4 & 1 & \dots & 0 & 0 \\ 0 & 1 & 4 & \dots & 0 & 0 \\ \vdots & \vdots & \vdots & \ddots & \vdots & \vdots \\ 0 & 0 & 0 & \dots & 4 & 1 \\ 0 & 0 & 0 & \dots & 1 & 2 \end{pmatrix}.$$

Even for the one-dimensional problem, an error analysis for the mixed finite element method (4.157) is delicate. We just point out that an error estimate of the following type can be obtained for (4.157):

$$\|p - p_h\| + \|u - u_h\| \leq Ch, \tag{4.162}$$

where u, p and u_h, p_h are the respective solutions of (4.155) and (4.157), C depends on the size of the second derivative of p, and we recall the norm (cf. Section 4.2.1)

$$\|v\| = \|v\|_{L^2(I)} = \left(\int_0^1 v^2\, dx\right)^{1/2}.$$

When u is sufficiently smooth (e.g., $u \in H^2(I)$), we can show the error estimate (Brezzi and Fortin, 1991; Chen, 2005)

$$\|u - u_h\| \leq Ch^2. \tag{4.163}$$

Error bounds (4.162) and (4.163) are optimal for p and u.

4.5.2 A two-dimensional model problem

We extend the mixed finite element method in the previous section to a stationary problem in two dimensions:

$$\begin{aligned} -\Delta p &= f \quad \text{in } \Omega, \\ p &= 0 \quad \text{on } \Gamma, \end{aligned} \tag{4.164}$$

where Ω is a bounded domain in the plane with boundary Γ and $f \in L^2(\Omega)$ is a given function. Recall that

$$L^2(\Omega) = \left\{ v : \ v \text{ is defined on } \Omega \text{ and } \int_\Omega v^2\, d\mathbf{x} < \infty \right\}$$

with the inner product

$$(v, w) = \int_\Omega v(\mathbf{x}) w(\mathbf{x})\, d\mathbf{x}, \qquad v,\ w \in L^2(\Omega).$$

We also use the space

$$\mathbf{H}(\text{div}, \Omega) = \left\{ \mathbf{v} = (v_1, v_2) \in (L^2(\Omega))^2 : \nabla \cdot \mathbf{v} \in L^2(\Omega) \right\},$$

where

$$\nabla \cdot \mathbf{v} = \frac{\partial v_1}{\partial x_1} + \frac{\partial v_2}{\partial x_2}.$$

It can be checked (cf. Exercise 4.46) that for any decomposition of Ω into subdomains such that the interiors of these subdomains are pairwise disjoint, the space $\mathbf{H}(\text{div}, \Omega)$ consists of those functions whose normal components are continuous across the interior edges in this decomposition. Define

$$\mathbf{V} = \mathbf{H}(\text{div}, \Omega), \quad W = L^2(\Omega).$$

Set

$$\mathbf{u} = -\nabla p. \tag{4.165}$$

The first equation in (4.164) becomes

$$\nabla \cdot \mathbf{u} = f. \tag{4.166}$$

Multiply (4.165) by $\mathbf{v} \in \mathbf{V}$ and integrate over Ω to see that

$$(\mathbf{u}, \mathbf{v}) = -(\mathbf{v}, \nabla p).$$

Applying Green's formula (4.68) to the right-hand side of this equation, we have

$$(\mathbf{u}, \mathbf{v}) = (\nabla \cdot \mathbf{v}, p),$$

where we use the boundary condition in (4.164). Also, multiplying (4.166) by any $w \in W$, we get

$$(\nabla \cdot \mathbf{u}, w) = (f, w).$$

Thus we have the system for $\mathbf{u}$ and p

$$\begin{aligned} (\mathbf{u}, \mathbf{v}) - (\nabla \cdot \mathbf{v}, p) &= 0, \qquad && \mathbf{v} \in \mathbf{V}, \\ (\nabla \cdot \mathbf{u}, w) &= (f, w), && w \in W. \end{aligned} \tag{4.167}$$

This is the mixed variational form of (4.164). If $\mathbf{u}$ and p satisfy (4.165) and (4.166), they also satisfy (4.167). The converse also holds if p is sufficiently smooth (e.g., if $p \in H^2(\Omega)$); see Exercise 4.47. In a similar fashion as for (4.155) and (4.156), (4.167) can be written as a saddle point problem.

For a polygonal domain Ω, let K_h be a partition of Ω into nonoverlapping (open) triangles such that no vertex of one triangle lies in the interior of an edge of another triangle. Define the mixed finite element spaces

$$\begin{aligned} \mathbf{V}_h &= \{\mathbf{v} \in \mathbf{V} : \ \mathbf{v}|_K = (b_K x_1 + a_K, b_K x_2 + c_K), \\ &\qquad a_K, b_K, c_K \in \mathbb{R}, \ K \in K_h\}, \\ W_h &= \{w : \ w \text{ is constant on each triangle in } K_h\}. \end{aligned}$$

As noted, $\mathbf{V}_h$ can be also described as follows:

$$\begin{aligned} \mathbf{V}_h = \{\mathbf{v} : \mathbf{v}|_K &= (b_K x_1 + a_K, b_K x_2 + c_K), \ K \in K_h, \\ &a_K, b_K, c_K \in \mathbb{R}, \text{ and the normal components of } \mathbf{v} \\ &\text{are continuous across the interior edges in } K_h\}. \end{aligned}$$

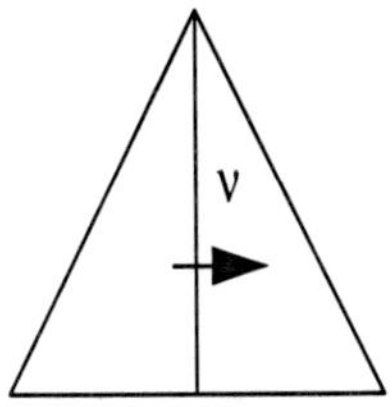

Figure 4.45. *An illustration of the unit normal* $\boldsymbol{\nu}$.

Note that $\mathbf{V}_h \subset \mathbf{V}$ and $W_h \subset W$. The mixed finite element method for (4.164) is defined as

$$
\begin{aligned}
&\text{Find } \mathbf{u}_h \in \mathbf{V}_h \text{ and } p_h \in W_h \text{ such that} \\
&(\mathbf{u}_h, \mathbf{v}) - (\nabla \cdot \mathbf{v}, p_h) = 0, \qquad \mathbf{v} \in \mathbf{V}_h, \\
&(\nabla \cdot \mathbf{u}_h, w) = (f, w), \qquad w \in W_h.
\end{aligned}
\tag{4.168}
$$

As for (4.157), it can be proven that (4.168) has a unique solution.

Let $\{\mathbf{x}_i\}$ be the set of the midpoints of edges in K_h, $i = 1, 2, \ldots, M$. With each point $\mathbf{x}_i$, we associate a unit normal vector $\boldsymbol{\nu}_i$. For $\mathbf{x}_i \in \Gamma$, $\boldsymbol{\nu}_i$ is just the outward unit normal to Γ; for $\mathbf{x}_i \in e = \bar{K}_1 \cap \bar{K}_2$, $K_1, K_2 \in K_h$, let $\boldsymbol{\nu}_i$ be any unit vector orthogonal to e (cf. Figure 4.45). We now define the basis functions of $\mathbf{V}_h$, $i = 1, 2, \ldots, M$, by

$$
\left(\boldsymbol{\varphi}_i \cdot \boldsymbol{\nu}_i\right)(\mathbf{x}_j) = \begin{cases} 1 & \text{if } i = j, \\ 0 & \text{if } i \neq j. \end{cases}
$$

Any function $\mathbf{v} \in \mathbf{V}_h$ has the unique representation

$$
\mathbf{v}(\mathbf{x}) = \sum_{i=1}^{M} v_i \boldsymbol{\varphi}_i(\mathbf{x}), \qquad \mathbf{x} \in \Omega,
$$

where $v_i = (\mathbf{v} \cdot \boldsymbol{\nu}_i)(\mathbf{x}_i)$. Also, the basis functions $\psi_i \in W_h$, $i = 1, 2, \ldots, N$, can be defined as in the previous section; i.e.,

$$
\psi_i(\mathbf{x}) = \begin{cases} 1 & \text{if } \mathbf{x} \in K_i, \\ 0 & \text{otherwise}, \end{cases}
$$

where $\bar{\Omega} = \bigcup_{i=1}^{N} \bar{K}_i$ and N is the number of triangles in K_h. Any function $w \in W_h$ has the representation

$$
w(\mathbf{x}) = \sum_{i=1}^{N} w_i \psi_i(\mathbf{x}), \qquad \mathbf{x} \in \Omega, \ w_i = w|_{K_i}.
$$

In the same manner as in the previous section, system (4.168) can be recast in matrix form (cf. Exercise 4.48):

$$
\begin{pmatrix} \mathbf{A} & \mathbf{B} \\ \mathbf{B}^T & \mathbf{0} \end{pmatrix} \begin{pmatrix} \mathbf{U} \\ \mathbf{p} \end{pmatrix} = \begin{pmatrix} \mathbf{0} \\ -\mathbf{f} \end{pmatrix},
\tag{4.169}
$$

where

$$\mathbf{A} = \left(a_{ij}\right)_{i,j=1,2,\ldots,M}, \quad \mathbf{B} = \left(b_{jk}\right)_{j=1,2,\ldots,M,\ k=1,2,\ldots,N},$$
$$\mathbf{U} = (u_i)_{i=1,2,\ldots,M}, \quad \mathbf{p} = (p_k)_{k=1,2,\ldots,N}, \quad \mathbf{f} = \left(f_j\right)_{j=1,2,\ldots,N},$$

with

$$a_{ij} = \left(\boldsymbol{\varphi}_i, \boldsymbol{\varphi}_j\right), \quad b_{jk} = -\left(\nabla \cdot \boldsymbol{\varphi}_j, \psi_k\right), \quad f_j = \left(f, \psi_j\right).$$

Again, the matrix $\mathbf{M}$ defined by

$$\mathbf{M} = \begin{pmatrix} \mathbf{A} & \mathbf{B} \\ \mathbf{B}^T & \mathbf{0} \end{pmatrix}$$

has both positive and negative eigenvalues. The matrix $\mathbf{A}$ is symmetric, positive definite, and sparse. In fact, it has at most five nonzero entries in each row in the present case (cf. Exercise 4.48). The matrix $\mathbf{B}$ is also sparse, with two nonzero entries in each row.

Let $\mathbf{u}$, p and $\mathbf{u}_h$, p_h be the respective solutions of (4.167) and (4.168). Then the following error estimate holds (Brezzi and Fortin, 1991; Chen, 2005):

$$\|p - p_h\| + \|\mathbf{u} - \mathbf{u}_h\| \leq Ch, \tag{4.170}$$

where C depends on the size of the second partial derivatives of p. The estimate is optimal for this pair of mixed finite element spaces.

4.5.3 Extension to boundary conditions of other kinds

A Neumann boundary condition

In the previous section, we considered the Dirichlet boundary condition in (4.164). We now extend the mixed finite element method to the stationary problem with the *homogeneous Neumann boundary condition*:

$$\begin{aligned} -\Delta p &= f && \text{in } \Omega, \\ \frac{\partial p}{\partial \boldsymbol{\nu}} &= 0 && \text{on } \Gamma, \end{aligned} \tag{4.171}$$

where $\partial p / \partial \boldsymbol{\nu}$ is the derivative of p normal to boundary Γ.

Application of Green's formula (4.68) to (4.171) yields

$$\int_\Omega f \, d\mathbf{x} = 0.$$

This is a *compatibility condition*. In this case, p is unique up to an additive constant.

We define the spaces

$$\mathbf{V} = \{\mathbf{v} = (v_1, v_2) \in \mathbf{H}(\text{div}, \Omega) : \ \mathbf{v} \cdot \boldsymbol{\nu} = 0 \text{ on } \Gamma\},$$
$$W = \left\{ w \in L^2(\Omega) : \ \int_\Omega w \, d\mathbf{x} = 0 \right\}.$$

With the choice of these two spaces, the mixed variational form of (4.171) is

$$\begin{aligned}
&\text{Find } \mathbf{u} \in \mathbf{V} \text{ and } p \in W \text{ such that} \\
&(\mathbf{u}, \mathbf{v}) - (\nabla \cdot \mathbf{v}, p) = 0, \qquad \mathbf{v} \in \mathbf{V}, \\
&(\nabla \cdot \mathbf{u}, w) = (f, w), \qquad w \in W.
\end{aligned} \tag{4.172}$$

Note that the Neumann boundary condition becomes the *essential* condition that must be incorporated into the definition of the space $\mathbf{V}$. In contrast, the Dirichlet boundary condition is the essential condition in finite element methods (cf. Section 4.2.1).

Let K_h be a partition of Ω into nonoverlapping triangles, as defined in the previous section. We define the mixed finite element spaces

$$\begin{aligned}
\mathbf{V}_h = \{&\mathbf{v} \in \mathbf{H}(\mathrm{div}, \Omega) : \mathbf{v}|_K = (b_K x_1 + a_K, b_K x_2 + c_K), \\
&a_K, b_K, c_K \in \mathbb{R}, \ K \in K_h, \text{ and } \mathbf{v} \cdot \boldsymbol{\nu} = 0 \text{ on } \Gamma\}, \\
W_h = &\left\{ w : w|_K \text{ is constant on each } K \in K_h \text{ and } \int_\Omega w \, d\mathbf{x} = 0 \right\}.
\end{aligned}$$

Again, $\mathbf{V}_h \subset \mathbf{V}$ and $W_h \subset W$. The mixed finite element method for (4.171) reads as follows:

$$\begin{aligned}
&\text{Find } \mathbf{u}_h \in \mathbf{V}_h \text{ and } p_h \in W_h \text{ such that} \\
&(\mathbf{u}_h, \mathbf{v}) - (\nabla \cdot \mathbf{v}, p_h) = 0, \qquad \mathbf{v} \in \mathbf{V}_h, \\
&(\nabla \cdot \mathbf{u}_h, w) = (f, w), \qquad w \in W_h.
\end{aligned} \tag{4.173}$$

This system can be rewritten in matrix form as in (4.169), and the error estimate (4.170) also holds.

A boundary condition of the third kind

We now consider a boundary condition of the *third kind*:

$$\begin{aligned}
-\Delta p &= f \qquad \text{in } \Omega, \\
bp + \frac{\partial p}{\partial \boldsymbol{\nu}} &= g \qquad \text{on } \Gamma,
\end{aligned} \tag{4.174}$$

where b is a strictly positive function on Γ and g is a given function.

With the linear spaces $\mathbf{V}$ and W defined as in Section 4.5.2, the mixed variational form of (4.174) is

$$\begin{aligned}
&\text{Find } \mathbf{u} \in \mathbf{V} \text{ and } p \in W \text{ such that} \\
&(\mathbf{u}, \mathbf{v}) + \int_\Gamma b^{-1} \mathbf{u} \cdot \boldsymbol{\nu} \, \mathbf{v} \cdot \boldsymbol{\nu} \, d\ell - (\nabla \cdot \mathbf{v}, p) \\
&\qquad = -\int_\Gamma b^{-1} g \mathbf{v} \cdot \boldsymbol{\nu} \, d\ell, \qquad \mathbf{v} \in \mathbf{V}, \\
&(\nabla \cdot \mathbf{u}, w) = (f, w), \qquad w \in W.
\end{aligned} \tag{4.175}$$

Similarly, with the mixed finite element spaces in Section 4.5.2, the mixed finite element method for (4.174) is

$$\text{Find } \mathbf{u}_h \in \mathbf{V}_h \text{ and } p_h \in W_h \text{ such that}$$

$$\begin{aligned} (\mathbf{u}_h, \mathbf{v}) + \int_\Gamma b^{-1} \mathbf{u}_h \cdot \boldsymbol{\nu}\, \mathbf{v} \cdot \boldsymbol{\nu}\, d\ell - (\nabla \cdot \mathbf{v}, p_h) & \\ = -\int_\Gamma b^{-1} g \mathbf{v} \cdot \boldsymbol{\nu}\, d\ell, & \quad \mathbf{v} \in \mathbf{V}_h, \\ (\nabla \cdot \mathbf{u}_h, w) = (f, w), & \quad w \in W_h. \end{aligned} \tag{4.176}$$

The matrix form and error estimate of (4.176) can be obtained in the same fashion as in Section 4.5.2 (cf. Exercise 4.51).

4.5.4 Mixed finite element spaces

We consider the model problem for p:

$$\begin{aligned} -\nabla \cdot (\mathbf{a} \nabla p) = f & \quad \text{in } \Omega, \\ p = g & \quad \text{on } \Gamma, \end{aligned} \tag{4.177}$$

where $\Omega \subset \mathbb{R}^d$ ($d = 2$ or 3) is a bounded two- or three-dimensional domain with boundary Γ, the diffusion tensor $\mathbf{a}$ is assumed to satisfy condition (4.88), and f and g are given real-valued piecewise continuous bounded functions in Ω and Γ, respectively. This problem was considered in the previous sections. To write (4.177) in a mixed variational form, the *Sobolev spaces* introduced in Section 4.5.2 are exploited. The norms of these two spaces $W = L^2(\Omega)$ and $\mathbf{V} = \mathbf{H}(\text{div}, \Omega)$ are, respectively, defined by

$$\|w\| \equiv \|w\|_{L^2(\Omega)} = \left(\int_\Omega w^2 \, d\mathbf{x} \right)^{1/2}, \qquad w \in W,$$

and

$$\|\mathbf{v}\|_\mathbf{V} \equiv \|\mathbf{v}\|_{\mathbf{H}(\text{div},\Omega)} = \left\{ \|\mathbf{v}\|^2 + \|\nabla \cdot \mathbf{v}\|^2 \right\}^{1/2}, \qquad \mathbf{v} \in \mathbf{V}.$$

The definition of $\mathbf{H}(\text{div}, \Omega)$ for $\Omega \subset \mathbb{R}^3$ is similar to that in Section 4.5.2; in this case, recall that

$$\nabla \cdot \mathbf{v} = \frac{\partial v_1}{\partial x_1} + \frac{\partial v_2}{\partial x_2} + \frac{\partial v_3}{\partial x_3}, \qquad \mathbf{v} = (v_1, v_2, v_3).$$

Let

$$\mathbf{u} = -\mathbf{a} \nabla p. \tag{4.178}$$

In the same way as in the derivation of (4.167), problem (4.177) is written in the mixed variational form:

$$\text{Find } \mathbf{u} \in \mathbf{V} \text{ and } p \in W \text{ such that}$$

$$\begin{aligned} (\mathbf{a}^{-1}\mathbf{u}, \mathbf{v}) - (\nabla \cdot \mathbf{v}, p) = -\int_\Gamma g \mathbf{v} \cdot \boldsymbol{\nu}\, d\ell, & \quad \mathbf{v} \in \mathbf{V}, \\ (\nabla \cdot \mathbf{u}, w) = (f, w), & \quad w \in W. \end{aligned} \tag{4.179}$$

There is a constant $C_1 > 0$ such that the *inf-sup* condition between the spaces $\mathbf{V}$ and W holds (Chen, 2005):

$$\sup_{0 \neq \mathbf{v} \in \mathbf{V}} \frac{|(\nabla \cdot \mathbf{v}, w)|}{\|\mathbf{v}\|_{\mathbf{V}}} \geq C_1 \|w\| \qquad \forall w \in W. \tag{4.180}$$

Because of (4.88) and (4.180), problem (4.179) has a unique solution $\mathbf{u} \in \mathbf{V}$ and $p \in W$ (Brezzi and Fortin, 1991), with $\mathbf{u}$ given by (4.178).

Let $\mathbf{V}_h \subset \mathbf{V}$ and $W_h \subset W$ be certain finite-dimensional subspaces. The discrete version of (4.179) is

$$\begin{aligned}
&\text{Find } \mathbf{u}_h \in \mathbf{V}_h \text{ and } p_h \in W_h \text{ such that} \\
&(\mathbf{a}^{-1}\mathbf{u}_h, \mathbf{v}) - (\nabla \cdot \mathbf{v}, p_h) = -\int_\Gamma g\mathbf{v} \cdot \boldsymbol{\nu}\, d\ell, \qquad \mathbf{v} \in \mathbf{V}_h, \\
&(\nabla \cdot \mathbf{u}_h, w) = (f, w), \qquad w \in W_h.
\end{aligned} \tag{4.181}$$

For this problem to have a unique solution, it is natural to impose a *discrete inf-sup* condition between $\mathbf{V}_h$ and W_h similar to (4.180):

$$\sup_{0 \neq \mathbf{v} \in \mathbf{V}_h} \frac{|(\nabla \cdot \mathbf{v}, w)|}{\|\mathbf{v}\|_{\mathbf{V}}} \geq C_2 \|w\| \qquad \forall w \in W_h, \tag{4.182}$$

where $C_2 > 0$ is a constant independent of h.

In the previous two sections, we considered the mixed finite element spaces $\mathbf{V}_h$ and W_h over triangles. These spaces are the lowest-order triangular spaces introduced by Raviart and Thomas (1977), and they satisfy condition (4.182). In this section, we describe other mixed finite element spaces that satisfy this *stability condition*. These spaces are RTN (Raviart and Thomas, 1977; Nédélec, 1980), BDM (Brezzi et al., 1985), BDDF (Brezzi et al., 1987A), BDFM (Brezzi et al., 1987B), and CD (Chen and Douglas, 1989) spaces.

Condition (4.182) is also called the *Babuška–Brezzi condition* or sometimes the *Ladyshenskaja–Babuška–Brezzi condition*.

For simplicity, let Ω be a polygonal domain in this section. For a curved domain, the definition of the mixed finite element spaces under consideration is the same, but the degrees of freedom for $\mathbf{V}_h$ need to be modified (Brezzi and Fortin, 1991).

Mixed finite element spaces on triangles

For $\Omega \subset \mathbb{R}^2$, let K_h be a partition of Ω into triangles such that adjacent elements completely share their common edge. For a triangle $K \in K_h$, let

$$P_r(K) = \{v : \ v \text{ is a polynomial of degree at most } r \text{ on } K\},$$

where $r \geq 0$ is an integer. Mixed finite element spaces $\mathbf{V}_h \times W_h$ are defined locally on each element $K \in K_h$, so let $\mathbf{V}_h(K) = \mathbf{V}_h|_K$ (the restriction of $\mathbf{V}_h$ to K) and $W_h(K) = W_h|_K$.

(i) The RT spaces on triangles

As noted, these spaces are the first mixed finite element spaces introduced by Raviart and Thomas (1977). They are defined for each $r \geq 0$ by

$$\mathbf{V}_h(K) = \left(P_r(K)\right)^2 \oplus \left((x_1, x_2)P_r(K)\right), \quad W_h(K) = P_r(K),$$

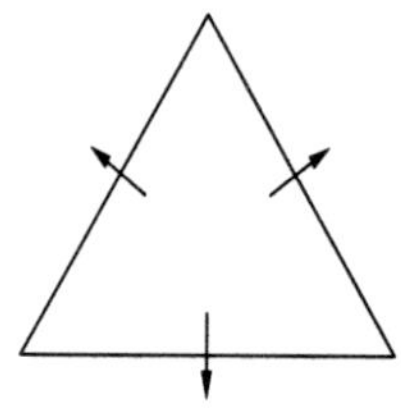

Figure 4.46. *The triangular RT.*

where the notation $\oplus$ indicates a direct sum and $(x_1, x_2)P_r(K) = (x_1 P_r(K), x_2 P_r(K))$. The case $r = 0$ was used in the previous sections. In this case, we observe that $\mathbf{V}_h(K)$ has the form

$$\mathbf{V}_h(K) = \{v : v = (a_K + b_K x_1, c_K + b_K x_2),\ a_K,\ b_K,\ c_K \in \mathbb{R}\},$$

and its dimension is three. As discussed in Section 4.5.2, as parameters, or the *degrees of freedom*, to describe the functions in $\mathbf{V}_h$ we use the values of normal components of the functions at the midpoints of edges in K_h (cf. Figure 4.46). Also, in the case $r = 0$, the degrees of freedom for W_h can be the averages of functions over K, as in Section 4.5.2.

In general, for $r \geq 0$ the dimensions of $\mathbf{V}_h(K)$ and $W_h(K)$ are

$$\dim\big(\mathbf{V}_h(K)\big) = (r+1)(r+3), \quad \dim\big(W_h(K)\big) = \frac{(r+1)(r+2)}{2}.$$

The degrees of freedom for the space $\mathbf{V}_h(K)$, with $r \geq 0$, are given by (Raviart and Thomas, 1977)

$$\begin{aligned} &(\mathbf{v}\cdot\boldsymbol{\nu}, w)_e && \forall w \in P_r(e),\ e \in \partial K, \\ &(\mathbf{v}, \mathbf{w})_K && \forall \mathbf{w} \in (P_{r-1}(K))^2, \end{aligned}$$

where $\boldsymbol{\nu}$ is the outward unit normal to $e \in \partial K$. This is a legitimate choice; i.e., a function in $\mathbf{V}_h$ is uniquely determined by these degrees of freedom.

(ii) The BDM spaces on triangles

The BDM spaces on triangles (Brezzi et al., 1985) lie between corresponding RT spaces, are of smaller dimension than the RT space of the same index, and provide asymptotic error estimates for the vector variable of the same order as the corresponding RT space. They are defined for each $r \geq 1$ by

$$\mathbf{V}_h(K) = \big(P_r(K)\big)^2, \quad W_h(K) = P_{r-1}(K).$$

The simplest BDM spaces on triangles are those with $r = 1$. In this case, $\mathbf{V}_h(K)$ is

$$\begin{aligned} \mathbf{V}_h(K) = \{v : v = (a_K^1 + a_K^2 x_1 + a_K^3 x_2, a_K^4 + a_K^5 x_1 + a_K^6 x_2),\\ a_K^i \in \mathbb{R},\ i = 1, 2, \ldots, 6\}, \end{aligned}$$

so its dimension is six. The degrees of freedom for $\mathbf{V}_h$ are the values of normal components of functions at the two quadratic Gauss points on each edge in K_h (cf. Figure 4.47). The space $W_h(K)$ with $r = 1$ consists of constants.

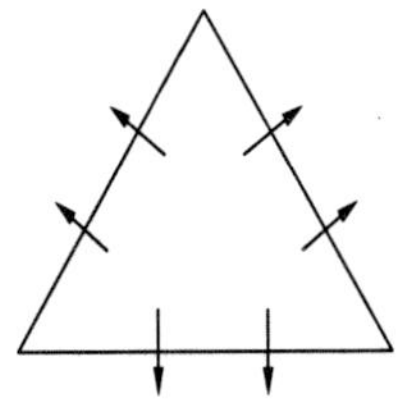

Figure 4.47. *The triangular BDM.*

In general, for $r \geq 1$ the dimensions of $\mathbf{V}_h(K)$ and $W_h(K)$ are

$$\dim\big(\mathbf{V}_h(K)\big) = (r+1)(r+2), \quad \dim\big(W_h(K)\big) = \frac{r(r+1)}{2}.$$

Set

$$B_{r+1}(K) = \{v \in P_{r+1}(K) : v|_{\partial K} = 0\} = \lambda_1\lambda_2\lambda_3 P_{r-2}(K),$$

where λ_1, λ_2, and λ_3 are the barycentric coordinates of the triangle K (cf. Section 4.2.1). The degrees of freedom for $\mathbf{V}_h(K)$ are (Brezzi et al., 1985)

$$\begin{aligned} &(\mathbf{v}\cdot\boldsymbol{\nu}, w)_e && \forall w \in P_r(e),\ e \in \partial K, \\ &(\mathbf{v}, \nabla w)_K && \forall w \in P_{r-1}(K), \\ &(\mathbf{v},\ \mathbf{curl}\ w)_K && \forall w \in B_{r+1}(K), \end{aligned}$$

where $\mathbf{curl}\, w = (-\partial w/\partial x_2, \partial w/\partial x_1)$.

Mixed finite element spaces on rectangles

We now consider the case where Ω is a rectangular domain and K_h is a partition of Ω into rectangles such that the horizontal and vertical edges of rectangles are parallel to the x_1- and x_2-coordinate axes, respectively, and adjacent elements completely share their common edge. Define

$$Q_{l,r}(K) = \left\{v :\ v(\mathbf{x}) = \sum_{i=0}^{l}\sum_{j=0}^{r} v_{ij}x_1^i x_2^j,\ \mathbf{x} = (x_1, x_2) \in K,\ v_{ij} \in \mathbb{R}\right\};$$

i.e., $Q_{l,r}(K)$ is the space of polynomials of degree at most l in x_1 and r in x_2, $l, r \geq 0$.

(i) The RT spaces on rectangles

These spaces are an extension of the RT spaces on triangles to rectangles (Raviart and Thomas, 1977), and for each $r \geq 0$ they are defined by

$$\mathbf{V}_h(K) = Q_{r+1,r}(K) \times Q_{r,r+1}(K), \quad W_h(K) = Q_{r,r}(K).$$

In the case $r = 0$, $\mathbf{V}_h(K)$ takes the form

$$\mathbf{V}_h(K) = \left\{v : v = \left(a_K^1 + a_K^2 x_1, a_K^3 + a_K^4 x_2\right),\ a_K^i \in \mathbb{R},\ i = 1, 2, 3, 4\right\},$$

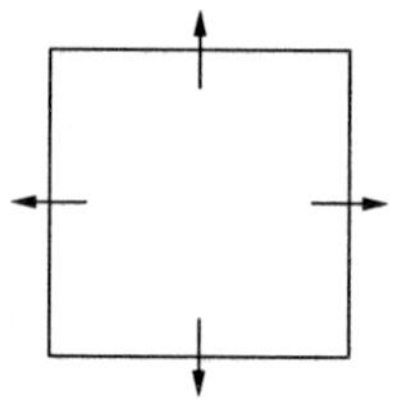

Figure 4.48. *The rectangular RT.*

and its dimension is four. The degrees of freedom for $\mathbf{V}_h$ are the values of normal components of functions at the midpoint on each edge in K_h (cf. Figure 4.48). In this case, $Q_{0,0}(K) = P_0(K)$.

For a general $r \geq 0$, the dimensions of $\mathbf{V}_h(K)$ and $W_h(K)$ are

$$\dim\big(\mathbf{V}_h(K)\big) = 2(r+1)(r+2), \quad \dim\big(W_h(K)\big) = (r+1)^2.$$

The degrees of freedom for $\mathbf{V}_h(K)$ are

$$\begin{aligned} &(\mathbf{v}\cdot\boldsymbol{\nu}, w)_e && \forall w \in P_r(e),\ e \in \partial K,\\ &(\mathbf{v}, \mathbf{w})_K && \forall \mathbf{w} = (w_1, w_2),\ w_1 \in Q_{r-1,r}(K),\ w_2 \in Q_{r,r-1}(K). \end{aligned}$$

(ii) The BDM spaces on rectangles

The BDM spaces (Brezzi et al., 1985) on rectangles differ considerably from the RT spaces on rectangles in that the vector elements are based on augmenting the space of vector polynomials of total degree r by exactly two additional vectors in place of augmenting the space of vector tensor products of polynomials of degree r by $2r+2$ polynomials of higher degree. A lower-dimensional space for the scalar variable is also used. These spaces, for any $r \geq 1$, are given by

$$\begin{aligned} \mathbf{V}_h(K) &= \big(P_r(K)\big)^2 \oplus \operatorname{span}\left\{\mathbf{curl}\,\left(x_1^{r+1}x_2\right), \mathbf{curl}\,\left(x_1x_2^{r+1}\right)\right\},\\ W_h(K) &= P_{r-1}(K). \end{aligned}$$

In the case $r = 1$, $\mathbf{V}_h(K)$ is

$$\begin{aligned} \mathbf{V}_h(K) = \big\{v : v = \big(&a_K^1 + a_K^2x_1 + a_K^3x_2 - a_K^4x_1^2 - 2a_K^5x_1x_2,\\ &a_K^6 + a_K^7x_1 + a_K^8x_2 + 2a_K^4x_1x_2 + a_K^5x_2^2\big),\\ &a_K^i \in \mathbb{R},\ i = 1, 2, \ldots, 8\big\}, \end{aligned}$$

and its dimension is eight. The degrees of freedom for $\mathbf{V}_h$ are the values of normal components of functions at the two quadratic Gauss points on each edge in K_h (cf. Figure 4.49).

For any $r \geq 1$, the dimensions of $\mathbf{V}_h(K)$ and $W_h(K)$ are

$$\dim\big(\mathbf{V}_h(K)\big) = (r+1)(r+2) + 2, \quad \dim\big(W_h(K)\big) = \frac{r(r+1)}{2}.$$

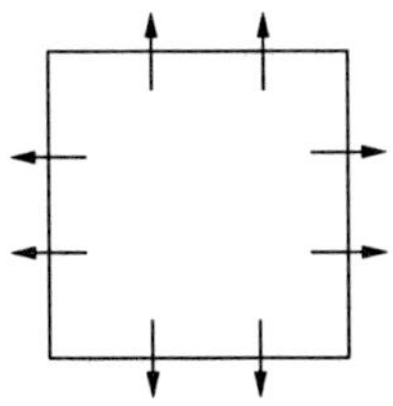

Figure 4.49. *The rectangular BDM.*

The degrees of freedom for $\mathbf{V}_h(K)$ are

$$(\mathbf{v} \cdot \boldsymbol{\nu}, w)_e \qquad \forall w \in P_r(e),\ e \in \partial K,$$
$$(\mathbf{v}, \mathbf{w})_K \qquad \forall \mathbf{w} \in (P_{r-2}(K))^2 .$$

(iii) The BDFM spaces on rectangles

These spaces (Brezzi et al., 1987B) are related to the BDM spaces on rectangles and are also called reduced BDM spaces. They give the same rates of convergence as the corresponding RT spaces with fewer parameters per rectangle except for the lowest degree space. For each $r \geq 0$, they are defined by

$$\begin{aligned} \mathbf{V}_h(K) &= \{w \in P_{r+1}(K) : \text{ the coefficient of } x_2^{r+1} \text{ vanishes}\} \\ &\quad \times \{w \in P_{r+1}(K) : \text{ the coefficient of } x_1^{r+1} \text{ vanishes}\}, \\ W_h(K) &= P_r(K). \end{aligned}$$

In the case $r = 0$, the BDFM spaces are just the RT spaces on rectangles. For a general $r \geq 0$, the dimensions of $\mathbf{V}_h(K)$ and $W_h(K)$ are

$$\dim\big(\mathbf{V}_h(K)\big) = (r+2)(r+3) - 2, \quad \dim\big(W_h(K)\big) = \frac{(r+1)(r+2)}{2}.$$

The degrees of freedom for $\mathbf{V}_h(K)$ are

$$(\mathbf{v} \cdot \boldsymbol{\nu}, w)_e \qquad \forall w \in P_r(e),\ e \in \partial K,$$
$$(\mathbf{v}, \mathbf{w})_K \qquad \forall \mathbf{w} \in (P_{r-1}(K))^2 .$$

While rectangular elements have been presented, an extension to general quadrilaterals can be made through change of variables from a reference rectangular element to quadrilaterals (Wang and Mathew, 1994; Arnold et al., 2005); refer to Section 4.2.2.

Mixed finite element spaces on tetrahedra

Let K_h be a partition of $\Omega \subset \mathbb{R}^3$ into tetrahedra such that adjacent elements completely share their common face. In three dimensions, P_r is now the space of polynomials of degree r in three variables x_1, x_2, and x_3.

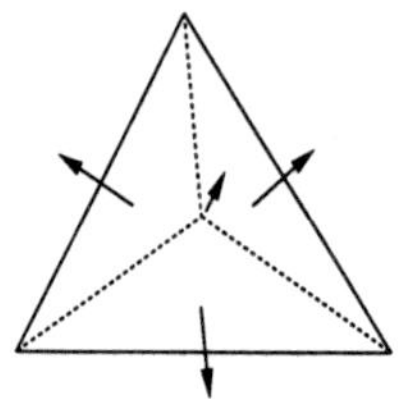

Figure 4.50. *The RTN on a tetrahedron.*

(i) The RTN spaces on tetrahedra

These spaces (Nédélec, 1980) are the three-dimensional analogues of the RT spaces on triangles, and they are defined for each $r \geq 0$ by

$$\mathbf{V}_h(K) = \left(P_r(K)\right)^3 \oplus \left((x_1, x_2, x_3)P_r(K)\right), \quad W_h(K) = P_r(K),$$

where $(x_1, x_2, x_3)P_r(K) = (x_1 P_r(K), x_2 P_r(K), x_3 P_r(K))$. As in two dimensions, for $r = 0$, $\mathbf{V}_h$ is

$$\mathbf{V}_h(K) = \{v : v = (a_K + b_K x_1, c_K + b_K x_2, d_K + b_K x_3), \\ a_K,\ b_K,\ c_K \in \mathbb{R}\},$$

and its dimension is four. The degrees of freedom are the values of normal components of functions at the centroid of each face in K (cf. Figure 4.50).

In general, for $r \geq 0$ the dimensions of $\mathbf{V}_h(K)$ and $W_h(K)$ are

$$\dim\left(\mathbf{V}_h(K)\right) = \frac{(r+1)(r+2)(r+4)}{2},$$

$$\dim\left(W_h(K)\right) = \frac{(r+1)(r+2)(r+3)}{6}.$$

The degrees of freedom for $\mathbf{V}_h(K)$ are

$$(\mathbf{v} \cdot \boldsymbol{\nu}, w)_e \qquad \forall w \in P_r(e),\ e \in \partial K,$$
$$(\mathbf{v}, \mathbf{w})_K \qquad \forall \mathbf{w} \in (P_{r-1}(K))^3.$$

(ii) The BDDF spaces on tetrahedra

The BDDF spaces (Brezzi et al., 1987A) are an extension of the BDM spaces on triangles to tetrahedra, and they are given for each $r \geq 1$ by

$$\mathbf{V}_h(K) = \left(P_r(K)\right)^3, \quad W_h(K) = P_{r-1}(K).$$

The dimensions of $\mathbf{V}_h(K)$ and $W_h(K)$ are

$$\dim\left(\mathbf{V}_h(K)\right) = \frac{(r+1)(r+2)(r+3)}{2},$$

$$\dim\left(W_h(K)\right) = \frac{r(r+1)(r+2)}{6}.$$

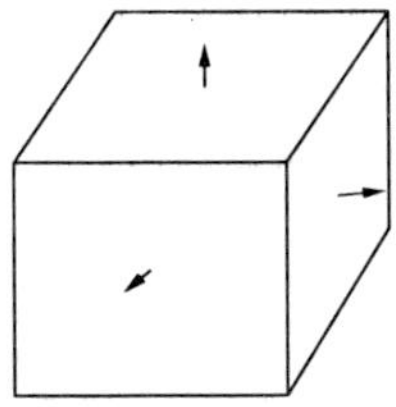

Figure 4.51. *The RTN on a rectangular parallelepiped.*

The degrees of freedom for $\mathbf{V}_h(K)$ are

$$\begin{aligned}
&(\mathbf{v}\cdot\boldsymbol{\nu}, w)_e && \forall w \in P_r(e),\ e \in \partial K,\\
&(\mathbf{v}, \nabla w)_K && \forall w \in P_{r-1}(K),\\
&(\mathbf{v}, \mathbf{w})_K && \forall \mathbf{w} \in \{\mathbf{z} \in (P_r(K))^3 : \mathbf{z}\cdot\boldsymbol{\nu} = 0 \text{ on } \partial K\\
& && \qquad \text{and } (\mathbf{z}, \nabla w)_K = 0,\ w \in P_{r-1}(K)\}.
\end{aligned}$$

Mixed finite element spaces on parallelepipeds

Let $\Omega \subset \mathbb{R}^3$ be a rectangular domain and K_h be a partition of Ω into rectangular parallelepipeds such that their faces are parallel to the coordinate axes and adjacent elements completely share their common face. Define, with $\mathbf{x} = (x_1, x_2, x_3)$,

$$Q_{l,m,r}(K) = \left\{ v : v(\mathbf{x}) = \sum_{i=0}^{l}\sum_{j=0}^{m}\sum_{k=0}^{r} v_{ijk}x_1^i x_2^j x_3^k,\ \mathbf{x} \in K,\ v_{ijk} \in \mathbb{R} \right\};$$

i.e., $Q_{l,m,r}(K)$ is the space of polynomials of degree at most l in x_1, m in x_2, and r in x_3 on K, respectively, $l, m, r \geq 0$.

(i) The RTN spaces on rectangular parallelepipeds

These spaces (Nédélec, 1980) are the three-dimensional analogues of the RT spaces on rectangles, and for each $r \geq 0$ they are defined by

$$\begin{aligned}
\mathbf{V}_h(K) &= Q_{r+1,r,r}(K) \times Q_{r,r+1,r}(K) \times Q_{r,r,r+1}(K),\\
W_h(K) &= Q_{r,r,r}(K).
\end{aligned}$$

For $r = 0$, $\mathbf{V}_h$ is

$$\mathbf{V}_h(K) = \big\{ v : v = \big(a_K^1 + a_K^2 x_1, a_K^3 + a_K^4 x_2, a_K^5 + a_K^6 x_3\big),\quad a_K^i \in \mathbb{R},\ i = 1, 2, \ldots, 6 \big\},$$

and its dimension is six. The degrees of freedom are the values of normal components of functions at the centroid of each face in K (cf. Figure 4.51).

For $r \geq 0$, the dimensions of $\mathbf{V}_h(K)$ and $W_h(K)$ are

$$\dim\big(\mathbf{V}_h(K)\big) = 3(r+1)^2(r+2), \quad \dim\big(W_h(K)\big) = (r+1)^3,$$

and the degrees of freedom for $\mathbf{V}_h(K)$ are

$$\begin{aligned}
&(\mathbf{v}\cdot\boldsymbol{\nu}, w)_e && \forall w \in Q_{r,r}(e),\ e \in \partial K,\\
&(\mathbf{v}, \mathbf{w})_K && \forall \mathbf{w} = (w_1, w_2, w_3),\ w_1 \in Q_{r-1,r,r}(K),\\
& && \quad w_2 \in Q_{r,r-1,r}(K),\ w_3 \in Q_{r,r,r-1}(K).
\end{aligned}$$

(ii) The BDDF spaces on rectangular parallelepipeds

These spaces (Brezzi et al., 1987A) are the three-dimensional analogues of the BDM spaces on rectangles. They are defined for $r \geq 1$ by

$$\begin{aligned}
\mathbf{V}_h(K) = \left(P_r(K)\right)^3 \oplus \mathrm{span}\big\{ &\mathbf{curl}(0, 0, x_1^{r+1}x_2),\ \mathbf{curl}(0, x_1 x_3^{r+1}, 0),\\
&\mathbf{curl}(x_2^{r+1}x_3, 0, 0),\ \mathbf{curl}(0, 0, x_1 x_2^{i+1}x_3^{r-i}),\\
&\mathbf{curl}(0, x_1^{i+1}x_2^{r-i}x_3, 0),\ \mathbf{curl}(x_1^{r-i}x_2 x_3^{i+1}, 0, 0)\big\},\\
W_h(K) = P_{r-1}(K),&
\end{aligned}$$

where $i = 1, 2, \ldots, r$ and, with $\mathbf{v} = (v_1, v_2, v_3)$,

$$\mathbf{curl}\ \mathbf{v} = \left(\frac{\partial v_3}{\partial x_2} - \frac{\partial v_2}{\partial x_3}, \frac{\partial v_1}{\partial x_3} - \frac{\partial v_3}{\partial x_1}, \frac{\partial v_2}{\partial x_1} - \frac{\partial v_1}{\partial x_2}\right).$$

The dimensions of $\mathbf{V}_h(K)$ and $W_h(K)$ are

$$\begin{aligned}
\dim\big(\mathbf{V}_h(K)\big) &= \frac{(r+1)(r+2)(r+3)}{2} + 3(r+1),\\
\dim\big(W_h(K)\big) &= \frac{r(r+1)(r+2)}{6}.
\end{aligned}$$

The degrees of freedom for $\mathbf{V}_h(K)$ are

$$\begin{aligned}
&(\mathbf{v}\cdot\boldsymbol{\nu}, w)_e && \forall w \in P_r(e),\ e \in \partial K,\\
&(\mathbf{v}, \mathbf{w})_K && \forall \mathbf{w} \in (P_{r-2}(K))^3.
\end{aligned}$$

(iii) The BDFM spaces on rectangular parallelepipeds

These spaces (Brezzi et al., 1987B) are related to the BDDF spaces on rectangular parallelepipeds and are also called the reduced BDDF spaces. They are defined for each $r \geq 0$ as

$$\begin{aligned}
\mathbf{V}_h(K) = &\left\{w \in P_{r+1}(K) :\ \text{the coefficient of } \sum_{i=0}^{r+1} x_2^{r+1-i}x_3^i \text{ vanishes}\right\}\\
&\times\left\{w \in P_{r+1}(K) :\ \text{the coefficient of } \sum_{i=0}^{r+1} x_3^{r+1-i}x_1^i \text{ vanishes}\right\}\\
&\times\left\{w \in P_{r+1}(K) :\ \text{the coefficient of } \sum_{i=0}^{r+1} x_1^{r+1-i}x_2^i \text{ vanishes}\right\},\\
W_h(K) = &\ P_r(K).
\end{aligned}$$

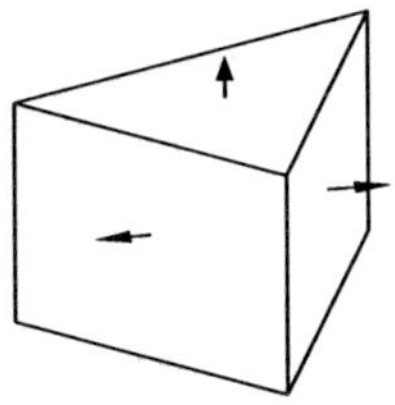

Figure 4.52. *The RTN on a prism.*

The dimensions of $\mathbf{V}_h(K)$ and $W_h(K)$ are

$$\dim\big(\mathbf{V}_h(K)\big) = \frac{(r+2)(r+3)(r+4)}{2} - 3(r+2),$$

$$\dim\big(W_h(K)\big) = \frac{(r+1)(r+2)(r+3)}{6}.$$

The degrees of freedom for $\mathbf{V}_h(K)$ are

$$(\mathbf{v}\cdot\boldsymbol{\nu}, w)_e \qquad \forall w \in P_r(e),\ e \in \partial K,$$
$$(\mathbf{v}, \mathbf{w})_K \qquad \forall \mathbf{w} \in (P_{r-1}(K))^3\,.$$

Mixed finite element spaces on prisms

Let $\Omega \subset \mathbb{R}^3$ be a domain of the form $\Omega = G \times (l_1, l_2)$, where $G \subset \mathbb{R}^2$ and l_1 and l_2 are real numbers. Let K_h be a partition of Ω into prisms such that their bases are triangles in the (x_1, x_2)-plane with three vertical edges parallel to the x_3-axis and adjacent prisms completely share their common face. $P_{l,r}$ denotes the space of polynomials of degree l in the two variables x_1 and x_2 and of degree r in the variable x_3.

(i) The RTN spaces on prisms

These spaces (Nédélec, 1986) are an extension of the RTN spaces on rectangular parallelepipeds to prisms, and they are defined for each $r \ge 0$ by

$$\mathbf{V}_h(K) = \big\{\mathbf{v} = (v_1, v_2, v_3) : \ v_3 \in P_{r,r+1}(K)\big\}, \quad W_h(K) = P_{r,r}(K),$$

where (v_1, v_2) satisfies that, for x_3 fixed,

$$(v_1, v_2) \in \big(P_r(K)\big)^2 \oplus \big((x_1, x_2)P_r(K)\big),$$

and v_1 and v_2 are of degree r in x_3. For $r = 0$, $\mathbf{V}_h$ has the form

$$\mathbf{V}_h(K) = \big\{v : \ v = \big(a_K^1 + a_K^2 x_1, a_K^3 + a_K^2 x_2, a_K^4 + a_K^5 x_3\big),$$
$$a_K^i \in \mathbb{R},\ i = 1, 2, \ldots, 5\big\},$$

and its dimension is five. The degrees of freedom are the values of normal components of functions at the centroid of each face in K (cf. Figure 4.52).

For $r \geq 0$, the dimensions of $\mathbf{V}_h(K)$ and $W_h(K)$ are

$$\dim\big(\mathbf{V}_h(K)\big) = (r+1)^2(r+3) + \frac{(r+1)(r+2)^2}{2},$$
$$\dim\big(W_h(K)\big) = \frac{(r+1)^2(r+2)}{2}.$$

The degrees of freedom for $\mathbf{V}_h(K)$ are

$$\begin{aligned}
&(\mathbf{v}\cdot\boldsymbol{\nu}, w)_e && \forall w \in P_r(e) \text{ for the two horizontal faces},\\
&(\mathbf{v}\cdot\boldsymbol{\nu}, w)_e && \forall w \in Q_{r,r}(e) \text{ for the three vertical faces},\\
&\big((v_1, v_2), (w_1, w_2)\big)_K && \forall (w_1, w_2) \in \big(P_{r-1,r}(K)\big)^2,\\
&(v_3, w_3)_K && \forall w_3 \in P_{r,r-1}(K).
\end{aligned}$$

(ii) The first CD spaces on prisms

The first CD spaces (Chen and Douglas, 1989) are an analogue of the RTN spaces on prisms, but different degrees of freedom are used, and the number of these degrees is less than required by the RNT spaces. They are defined for each $r \geq 0$ by

$$\begin{aligned}
\mathbf{V}_h(K) = \Big\{\mathbf{v} = (v_1, v_2, v_3) : \ (v_1, v_2) \in \big(P_{r+1,r}(K)\big)^2,&\\
v_3 \in P_{r,r+1}(K)\Big\},&\\
W_h(K) = P_{r,r}(K),&
\end{aligned}$$

where the dimensions of $\mathbf{V}_h(K)$ and $W_h(K)$ are

$$\dim\big(\mathbf{V}_h(K)\big) = (r+1)(r+2)(r+3) + \frac{(r+1)(r+2)^2}{2},$$
$$\dim\big(W_h(K)\big) = \frac{(r+1)^2(r+2)}{2}.$$

Let

$$B_{r+2,r}(K) = \{v \in P_{r+2,r}(K) : v|_e = 0 \text{ on the three vertical faces}\}.$$

The degrees of freedom for $\mathbf{V}_h(K)$ are

$$\begin{aligned}
&(\mathbf{v}\cdot\boldsymbol{\nu}, w)_e && \forall w \in P_r(e) \text{ for the two horizontal faces},\\
&(\mathbf{v}\cdot\boldsymbol{\nu}, w)_e && \forall w \in Q_{r+1,r}(e) \text{ for the three vertical faces},\\
&\big((v_1, v_2), \nabla_{(x_1,x_2)} w\big)_K && \forall w \in P_{r,r}(K),\\
&\big((v_1, v_2), \mathbf{curl}_{(x_1,x_2)} w\big)_K && \forall w \in B_{r+2,r}(K),\\
&(v_3, w_3)_K && \forall w_3 \in P_{r,r-1}(K),
\end{aligned}$$

where $\nabla_{(x_1,x_2)}$ and $\mathbf{curl}_{(x_1,x_2)}$ indicate the corresponding operators with respect to x_1 and x_2.

(iii) The second CD spaces on prisms

The second CD spaces (Chen and Douglas, 1989) are based on the BDDF spaces on rectangular parallelepipeds and use a much smaller number of degrees of freedom than the RTN and first CD spaces on prisms. They are defined for each $r \geq 1$ by

$$\begin{aligned} \mathbf{V}_h(K) &= \big(P_r(K)\big)^3 \oplus \text{span}\big\{\mathbf{curl}(x_2^{r+1}x_3, 0, 0), \\ &\qquad \mathbf{curl}(x_2x_3^{r+1}, -x_1x_3^{r+1}, 0), \\ &\qquad \mathbf{curl}(0, x_1^{i+1}x_2^{r-i}x_3, 0),\ i = 1, 2, \dots, r\big\}, \\ W_h(K) &= P_{r-1}(K). \end{aligned}$$

The dimensions of $\mathbf{V}_h(K)$ and $W_h(K)$ are

$$\begin{aligned} \dim\big(\mathbf{V}_h(K)\big) &= \frac{(r+1)(r+2)(r+3)}{2} + r + 2, \\ \dim\big(W_h(K)\big) &= \frac{r(r+1)(r+2)}{6}. \end{aligned}$$

Let

$$B_{r+1}(K) = \{v \in P_{r+1}(K) : v|_e = 0 \text{ on the three vertical faces of } K\}.$$

The degrees of freedom for $\mathbf{V}_h(K)$ are

$$\begin{aligned} &(\mathbf{v} \cdot \boldsymbol{\nu}, w)_e && \forall w \in P_r(e),\ e \in \partial K, \\ &\big((v_1, v_2), \nabla_{(x_1,x_2)} w\big)_K && \forall w \in P_{r-1}(K), \\ &\big((v_1, v_2), \mathbf{curl}_{(x_1,x_2)} w\big)_K && \forall w \in B_{r+1}(K), \\ &(v_3, w_3)_K && \forall w_3 \in P_{r-2}(K). \end{aligned}$$

(iv) The third CD spaces on prisms

The third CD spaces (Chen and Douglas, 1989) are based on the BDFM spaces on rectangular parallelepipeds and also use a much smaller number of degrees of freedom than the RTN and first CD spaces on prisms. They are defined for each $r \geq 0$ by

$$\begin{aligned} \mathbf{V}_h(K) &= \big\{w \in P_{r+1}(K) : \text{ the coefficient of } x_3^{r+1} \text{ vanishes}\big\} \\ &\quad \times \big\{w \in P_{r+1}(K) : \text{ the coefficient of } x_3^{r+1} \text{ vanishes}\big\} \\ &\quad \times \left\{w \in P_{r+1}(K) : \text{ the coefficient of } \sum_{i=0}^{r+1} x_1^{r+1-i} x_2^i \text{ vanishes}\right\}, \\ W_h(K) &= P_r(K). \end{aligned}$$

The dimensions of $\mathbf{V}_h(K)$ and $W_h(K)$ are

$$\begin{aligned} \dim\big(\mathbf{V}_h(K)\big) &= \frac{(r+2)(r+3)(r+4)}{2} - r - 4, \\ \dim\big(W_h(K)\big) &= \frac{(r+1)(r+2)(r+3)}{6}. \end{aligned}$$

The degrees of freedom for $\mathbf{V}_h(K)$ are

$$
\begin{aligned}
&(\mathbf{v}\cdot\nu, w)_e && \forall w \in P_r(e) \text{ for the two horizontal faces},\\
&(\mathbf{v}\cdot\nu, w)_e && \forall w \in P_{r+1} \setminus \{x_3^{r+1}\}|_e \text{ for the three vertical faces},\\
&\left((v_1, v_2), \nabla_{(x_1,x_2)} w\right)_K && \forall w \in P_{r-1}(K),\\
&\left((v_1, v_2), \mathbf{curl}_{(x_1,x_2)} w\right)_K && \forall w \in B_{r+2}(K),\\
&(v_3, w_3)_K && \forall w_3 \in P_{r-1}(K).
\end{aligned}
$$

In summary, the mixed finite element spaces on various geometrical elements in both two and three dimensions have been presented in this section. These spaces satisfy the *inf-sup* condition (4.182) (Brezzi and Fortin, 1991; Chen, 2005) and lead to optimal approximation properties (see the next section). We have considered only a polygonal domain Ω. For a more general domain, the partition T_h can have curved edges or faces on the boundary Γ, and the mixed spaces are constructed in a similar fashion (Raviart and Thomas, 1977; Nédélec, 1980; Brezzi et al., 1985; 1987A; 1987B; Chen and Douglas, 1989).

4.5.5 Approximation properties

The RTN, BDM, BDFM, BDDF, and CD mixed finite element spaces have the approximation properties

$$
\begin{aligned}
&\inf_{\mathbf{v}_h\in\mathbf{V}_h} \|\mathbf{v}-\mathbf{v}_h\| \le Ch^l\|\mathbf{v}\|_{\mathbf{H}^l(\Omega)}, && 1 \le l \le r+1,\\
&\inf_{\mathbf{v}_h\in\mathbf{V}_h} \|\nabla\cdot(\mathbf{v}-\mathbf{v}_h)\| \le Ch^l\|\nabla\cdot\mathbf{v}\|_{H^l(\Omega)}, && 0 \le l \le r^*,\\
&\inf_{w_h\in W_h} \|w-w_h\| \le Ch^l\|w\|_{H^l(\Omega)}, && 0 \le l \le r^*,
\end{aligned}
\tag{4.183}
$$

where $r^* = r+1$ for the RTN, BDFM, and first and third CD spaces and $r^* = r$ for the BDM, BDDF, and second CD spaces. Using (4.183), we can establish the corresponding error estimates for the mixed finite element method (4.181) when $\mathbf{V}_h$ and W_h are these mixed spaces (Chen, 2005).

We have presented the mixed finite element methods only for stationary problems. These methods can be extended to transient problems as in Section 4.2.4; i.e., the discretization in time can be carried out using either the backward Euler method or the Crank–Nicolson method and in space using the mixed methods. The linear systems of algebraic equations arising from the mixed methods are of *saddle type*; i.e., the system matrices have both positive and negative eigenvalues. Thus the solution of these systems needs special care. For a collection of iterative algorithms suitable for saddle linear systems, the reader should refer to Chen (2005). When $\mathbf{V}_h \times W_h$ are the lowest-order RTN spaces over rectangular parallelepipeds, the linear system arising from the mixed method can be written as a system generated by a *cell-centered* (or *block-centered*) finite difference scheme using certain quadrature rules (Russell and Wheeler, 1983).

4.6 Characteristic Finite Element Methods

In this section, we consider an application of finite element methods to the *reaction-diffusion-advection problem*:

$$\frac{\partial(\phi p)}{\partial t} + \nabla \cdot \left(\mathbf{b}p - \mathbf{a}\nabla p\right) + Rp = f \tag{4.184}$$

for the unknown solution p, where ϕ, $\mathbf{b}$ (vector), $\mathbf{a}$ (tensor), R, and f are given functions. Note that (4.184) involves advection ($\mathbf{b}$), diffusion ($\mathbf{a}$), and reaction (R). Many equations arise in this form, e.g., saturation and concentration equations for multiphase, multicomponent flows in porous media (cf. Chapter 2).

When diffusion dominates advection, the finite element methods developed in Section 4.2 perform well for (4.184). When advection dominates diffusion, however, they do not perform well. In particular, they exhibit excessive nonphysical oscillations when the solution to (4.184) is not smooth. Standard upstream weighting approaches have been applied to the finite element methods with the purpose of eliminating the nonphysical oscillations (cf. Section 4.3), but these approaches smear sharp fronts in the solution. Although extremely fine mesh refinement is possible to overcome this difficulty, it is not feasible due to the excessive computational effort involved.

Many numerical methods have been developed for solving (4.184) where advection dominates, such as the *optimal spatial method*. This method employs an *Eulerian approach* that is based on the minimization of the error in the approximation of spatial derivatives and the use of optimal test functions satisfying a local adjoint problem (Brooks and Hughes, 1982; Barrett and Morton, 1984). It yields an upstream bias in the resulting approximation and has the following features: (i) time truncation errors dominate the solution; (ii) the solution has significant numerical diffusion and phase errors; (iii) the *Courant number* (i.e., $|\mathbf{b}|\Delta t/(\phi h)$) is generally restricted to being less than one (cf. (4.40) for the definition of this number).

Other Eulerian methods, such as the *Petrov–Galerkin finite element method*, have been developed to use nonzero spatial truncation errors to cancel temporal errors and thereby reduce the overall truncation errors (Christie et al., 1976; Westerink and Shea, 1989). While these methods improve accuracy in the approximation of the solution, they still suffer from a strict Courant number limitation.

Another class of numerical methods for the solution of (4.184) is the *Eulerian–Lagrangian methods*. Because of the Lagrangian nature of advection, these methods treat the advection by a characteristic tracking approach. They have shown great potential. This class is rich and bears a variety of names, the *method of characteristics* (Garder et al., 1964), the *modified method of characteristics* (Douglas and Russell, 1982), the *transport diffusion method* (Pironneau, 1982), the *Eulerian–Lagrangian method* (Neuman, 1981), the *operator splitting method* (Espedal and Ewing, 1987), the *Eulerian–Lagrangian localized adjoint method* (Celia et al., 1990; Russell, 1990), the *characteristic mixed finite element method* (Yang, 1992; Arbogast and Wheeler, 1995), and the *Eulerian–Lagrangian mixed discontinuous method* (Chen, 2002B). The common features of this class are (i) the Courant number restriction of the purely Eulerian methods is alleviated because of the Lagrangian nature of the advection step; (ii) since the spatial and temporal dimensions are coupled through the characteristic tracking, the effect of time truncation errors present in the optimal spatial method is greatly reduced; (iii) they produce nonoscillatory solutions without numerical diffusion, using reasonably large time steps on grids no finer than necessary to resolve the solution on the moving fronts. In this section, we describe the Eulerian–Lagrangian methods.

4.6.1 The modified method of characteristics

The modified method of characteristics (MMOC) was independently developed by Douglas and Russell (1982) and Pironneau (1982) and is based on a nondivergence form of (4.184). It was called the *transport-diffusion method* by Pironneau. In the engineering literature the name *Eulerian–Lagrangian method* is often used (Neuman, 1981).

A one-dimensional model problem

For the purpose of introduction, we consider a one-dimensional model problem on the whole real line:

$$\begin{aligned}&\phi(x)\frac{\partial p}{\partial t} + b(x)\frac{\partial p}{\partial x} - \frac{\partial}{\partial x}\left(a(x,t)\frac{\partial p}{\partial x}\right) + R(x,t)p = f(x,t),\\ &\qquad\qquad\qquad\qquad x \in \mathbb{R},\ t > 0,\\ &p(x,0) = p_0(x), \qquad x \in \mathbb{R}.\end{aligned} \tag{4.185}$$

Set

$$\psi(x) = \left(\phi^2(x) + b^2(x)\right)^{1/2}.$$

Assume that

$$\phi(x) > 0, \qquad x \in \mathbb{R},$$

so that $\psi(x) > 0$, $x \in \mathbb{R}$. Let the characteristic direction associated with the hyperbolic part of (4.185), $\phi\partial p/\partial t + b\partial p/\partial x$, be denoted by $\tau(x)$, so that

$$\frac{\partial}{\partial \tau(x)} = \frac{\phi(x)}{\psi(x)}\frac{\partial}{\partial t} + \frac{b(x)}{\psi(x)}\frac{\partial}{\partial x}.$$

Then (4.185) can be rewritten as

$$\begin{aligned}&\psi(x)\frac{\partial p}{\partial \tau} - \frac{\partial}{\partial x}\left(a(x,t)\frac{\partial p}{\partial x}\right) + R(x,t)p = f(x,t),\\ &\qquad\qquad\qquad\qquad x \in \mathbb{R},\ t > 0,\\ &p(x,0) = p_0(x), \qquad x \in \mathbb{R}.\end{aligned} \tag{4.186}$$

We assume that the coefficients a, b, R, and ϕ are bounded and satisfy

$$\left|\frac{b(x)}{\phi(x)}\right| + \left|\frac{d}{dx}\left(\frac{b(x)}{\phi(x)}\right)\right| \leq C, \qquad x \in \mathbb{R},$$

where C is a positive constant. We introduce the linear space (cf. Section 4.2.1)

$$V = W^{1,2}(\mathbb{R}).$$

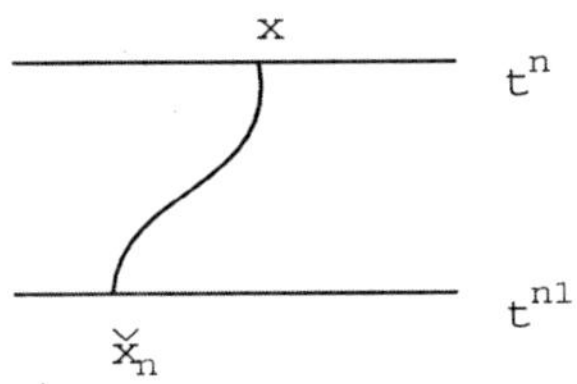

Figure 4.53. *An illustration of the definition* $\check{x}_n$.

The reader can refer to Adams (1975) for the definition of the Sobolev space $W^{1,2}(\mathbb{R})$ (alternatively, as in Section 4.2.1, V can be taken to be the space of continuous functions on $\mathbb{R}$ that have piecewise continuous and bounded first derivatives in $\mathbb{R}$ and approach zero at $\pm\infty$). We recall the scalar product in $L^2(\mathbb{R})$

$$(v, w) = \int_{\mathbb{R}} v(x)w(x)\,dx.$$

Now, multiplying the first equation of (4.186) by any $v \in V$ and applying integration by parts in space, problem (4.186) can be written in the equivalent variational form

$$\begin{aligned} &\left(\psi\frac{\partial p}{\partial \tau}, v\right) + \left(a\frac{\partial p}{\partial x}, \frac{dv}{dx}\right) + (Rp, v) = (f, v), \quad && v \in V,\ t > 0, \\ &p(x, 0) = p_0(x), && x \in \mathbb{R}. \end{aligned} \tag{4.187}$$

Let $0 = t^0 < t^1 < \cdots < t^n < \cdots$ be a partition in time, with $\Delta t^n = t^n - t^{n-1}$. For a generic function v of time, set $v^n = v(t^n)$. The characteristic derivative is approximated in the following way: let

$$\check{x}_n = x - \frac{\Delta t^n}{\phi(x)} b(x), \tag{4.188}$$

and note that, at $t = t^n$,

$$\begin{aligned} \psi\frac{\partial p}{\partial \tau} &\approx \psi(x)\frac{p(x, t^n) - p(\check{x}_n, t^{n-1})}{\left((x - \check{x}_n)^2 + (\Delta t^n)^2\right)^{1/2}} \\ &= \phi(x)\frac{p(x, t^n) - p(\check{x}_n, t^{n-1})}{\Delta t^n}. \end{aligned} \tag{4.189}$$

That is, a backtracking algorithm is used to approximate the characteristic derivative; $\check{x}_n$ is the foot (at level t^{n-1}) of the characteristic corresponding to x at the head (at level t^n) (cf. Figure 4.53).

Let V_h be a finite element subspace of $V \cap W^{1,\infty}(\mathbb{R})$ (cf. Section 4.2.1). Because we are considering the whole line, V_h is necessarily infinite-dimensional. In practice, we can assume that the support of p_0 is compact, the portion of the line on which we need to know p is bounded, and p is very small outside that set. Then V_h can be taken to be finite-dimensional.

The MMOC for (4.185) is defined: For $n = 1, 2, \ldots$, find $p_h^n \in V_h$ such that

$$\left(\phi \frac{p_h^n - \check{p}_h^{n-1}}{\Delta t^n}, v\right) + \left(a^n \frac{dp_h^n}{dx}, \frac{dv}{dx}\right) + (R^n p_h^n, v) = (f^n, v) \qquad \forall v \in V_h, \tag{4.190}$$

where

$$\check{p}_h^{n-1} = p_h\left(\check{x}_n, t^{n-1}\right) = p_h\left(x - \frac{\Delta t^n}{\phi(x)} b(x), t^{n-1}\right). \tag{4.191}$$

The initial approximation p_h^0 can be defined as the interpolant of p_0 in V_h, for example.

Note that (4.190) determines $\{p_h^n\}$ uniquely in terms of the data p_0 and f (at least, for reasonable a and R such that a is uniformly positive with respect to x and t and R is nonnegative). This can be seen as follows: Since (4.190) is a finite-dimensional system, it suffices to show uniqueness of the solution. Let $f = p_0 = 0$, and take $v = p_h^n$ in (4.190) to see that

$$\left(\phi \frac{p_h^n - \check{p}_h^{n-1}}{\Delta t^n}, p_h^n\right) + \left(a^n \frac{dp_h^n}{dx}, \frac{dp_h^n}{dx}\right) + (R^n p_h^n, p_h^n) = 0;$$

with an induction assumption that $p_h^{n-1} = 0$, this equation implies $p_h^n = 0$.

It is obvious that the linear system arising from (4.190) is symmetric positive definite (cf. Section 4.2.1), even in the presence of the advection term. This system has an improved (over that arising from a direct application to (4.184) of the finite element method described in Section 4.2.4) condition number of order (cf. Exercise 4.52)

$$\mathcal{O}\left(1 + \max_{x \in \mathbb{R},\, t \geq 0} |a(x,t)| h^{-2} \Delta t\right), \qquad \Delta t = \max_{n=1,2,\ldots} \Delta t^n.$$

Thus the system arising from (4.190) is well suited for the iterative linear solution algorithms discussed in the next chapter.

We end with a remark on a convergence result for (4.190). Let $V_h \subset V$ be a finite element space (cf. Section 4.2.1) with the following approximation property:

$$\inf_{v_h \in V_h} \left(\|v - v_h\|_{L^2(\mathbb{R})} + h\|v - v_h\|_{W^{1,2}(\mathbb{R})}\right) \leq C h^{r+1} |v|_{W^{r+1,2}(\mathbb{R})}, \tag{4.192}$$

where the constant $C > 0$ is independent of h and $r > 0$ is an integer; refer to Section 4.2.1 for the definition of spaces and their norms. Then, under appropriate assumptions on the smoothness of the solution p and a suitable choice of p_h^0 it can be shown (Douglas and Russell 1982) that

$$\max_{1 \leq n \leq N} \left(\|p^n - p_h^n\|_{L^2(\mathbb{R})} + h\|p^n - p_h^n\|_{W^{1,2}(\mathbb{R})}\right) \leq C(p)\left(h^{r+1} + \Delta t\right), \tag{4.193}$$

where N is an integer such that $t^N = T < \infty$ and $J = (0, T]$ is the time interval of interest.

This result, by itself, is not different from what we have obtained with the standard finite element methods in Section 4.2. However, the constant C is greatly improved when the MMOC is applied to (4.185). In time, C depends on a norm of $\frac{\partial^2 p}{\partial \tau^2}$ with the standard methods, but on a norm of $\frac{\partial^2 p}{\partial \tau^2}$ with the MMOC (Chen, 2005). The latter norm is much smaller, and thus long time steps with large Courant numbers are possible.

Some matters are raised by (4.190) and its analogues for more complicated differential problems. The first concern is the backtracking scheme that determines $\check{x}_n$ and a numerical quadrature rule that computes the associated integral. For the problem considered in this subsection, this matter can be resolved; the required computations can be performed exactly. For more complicated problems, there are discussions by Russell and Trujillo (1990). The second matter is the treatment of boundary conditions. In this section, we work on the whole line or on periodic boundary conditions (see the next subsection). For a bounded domain, if a backtracked characteristic crosses a boundary of the domain, it is not obvious what the meaning of $\check{x}_n$ or of $p_h(\check{x}_n)$ will be. The last matter, and perhaps the greatest drawback of the MMOC, is its failure to conserve mass. This issue will be discussed in detail in Section 4.6.1.

Periodic boundary conditions

In the previous subsection, (4.185) was considered on the whole line. For a bounded interval, say (0, 1), the MMOC has a difficulty handling general boundary conditions. In this case, it is normally developed for *periodic boundary conditions* (cf. Exercise 4.53):

$$p(0,t) = p(1,t), \quad \frac{\partial p}{\partial x}(0,t) = \frac{\partial p}{\partial x}(1,t). \tag{4.194}$$

These conditions are also called *cyclic boundary conditions*. In the periodic case, assume that all functions in (4.185) are spatially (0, 1)-periodic. Accordingly, the linear space V is modified to

$$V = \{v \in H^1(I) : v \text{ is } I\text{-periodic}\}, \quad I = (0,1).$$

With this modification, the developments in (4.187) and (4.190) remain unchanged.

Extension to multidimensional problems

We now extend the MMOC to (4.184) defined on a multidimensional domain. Let $\Omega \subset \mathbb{R}^d$ $(d \le 3)$ be a rectangle (respectively, a rectangular parallelepiped), and assume that (4.184) is Ω-periodic; i.e., all functions in (4.184) are spatially Ω-periodic. We write (4.184) in nondivergence form:

$$\begin{aligned} \phi(\mathbf{x})\frac{\partial p}{\partial t} &+ \mathbf{b}(\mathbf{x},t)\cdot\nabla p - \nabla\cdot\big(\mathbf{a}(\mathbf{x},t)\nabla p\big) \\ &+ R(\mathbf{x},t)p = f(\mathbf{x},t), \qquad \mathbf{x}\in\Omega,\ t>0, \\ p(\mathbf{x},0) &= p_0(\mathbf{x}), \qquad \mathbf{x}\in\Omega. \end{aligned} \tag{4.195}$$

Set

$$\psi(\mathbf{x},t) = \big(\phi^2(\mathbf{x}) + |\mathbf{b}(\mathbf{x},t)|^2\big)^{1/2},$$

where $|\mathbf{b}|^2 = b_1^2 + b_2^2 + \cdots + b_d^2$, with $\mathbf{b} = (b_1, b_2, \ldots, b_d)$. Assume that

$$\phi(\mathbf{x}) > 0, \qquad \mathbf{x}\in\Omega.$$

Now, the characteristic direction corresponding to the hyperbolic part of (4.195), $\phi\partial p/\partial t + \mathbf{b}\cdot\nabla p$, is $\boldsymbol{\tau}$, so

$$\frac{\partial}{\partial \boldsymbol{\tau}} = \frac{\phi(\mathbf{x})}{\psi(\mathbf{x},t)}\frac{\partial}{\partial t} + \frac{1}{\psi(\mathbf{x},t)}\mathbf{b}(\mathbf{x},t)\cdot\nabla.$$

With this definition, (4.195) becomes

$$\begin{aligned} &\psi(\mathbf{x},t)\frac{\partial p}{\partial \boldsymbol{\tau}} - \nabla\cdot\big(\mathbf{a}(\mathbf{x},t)\nabla p\big) + R(\mathbf{x},t)p = f(\mathbf{x},t), \\ &\qquad\qquad\qquad\qquad\qquad\qquad \mathbf{x}\in\Omega,\ t>0, \\ &p(\mathbf{x},0) = p_0(\mathbf{x}), \qquad \mathbf{x}\in\Omega. \end{aligned} \tag{4.196}$$

We define the linear space

$$V = \{v \in H^1(\Omega) : v \text{ is } \Omega\text{-periodic}\}.$$

Recall the notation

$$(v, w)_S = \int_S v(\mathbf{x})w(\mathbf{x})\,d\mathbf{x}.$$

If $S = \Omega$, we omit it in this notation. Now, applying Green's formula (4.68) in space and the periodic boundary conditions, (4.196) can be written in the equivalent variational form

$$\begin{aligned} &\left(\psi\frac{\partial p}{\partial \boldsymbol{\tau}}, v\right) + (\mathbf{a}\nabla p, \nabla v) + (Rp, v) = (f, v), \qquad v\in V,\ t>0, \\ &p(\mathbf{x},0) = p_0(\mathbf{x}), \qquad\qquad\qquad\qquad\qquad\qquad \mathbf{x}\in\Omega. \end{aligned} \tag{4.197}$$

The characteristic is approximated by

$$\check{\mathbf{x}}_n = \mathbf{x} - \frac{\Delta t^n}{\phi(\mathbf{x})}\mathbf{b}(\mathbf{x}, t^n). \tag{4.198}$$

Furthermore, we see that, at $t = t^n$,

$$\begin{aligned} \psi\frac{\partial p}{\partial \boldsymbol{\tau}} &\approx \psi(\mathbf{x},t^n)\frac{p(\mathbf{x},t^n) - p(\check{\mathbf{x}}_n, t^{n-1})}{\big(|\mathbf{x}-\check{\mathbf{x}}_n|^2 + (\Delta t^n)^2\big)^{1/2}} \\ &= \phi(\mathbf{x})\frac{p(\mathbf{x},t^n) - p(\check{\mathbf{x}}_n, t^{n-1})}{\Delta t^n}. \end{aligned} \tag{4.199}$$

A backtracking algorithm similar to that employed in one dimension is used to approximate the characteristic derivative (cf. Figure 4.54).

Let $V_h \subset V$ be a finite element space associated with a regular partition K_h of Ω (cf. Section 4.2.1). The MMOC for (4.195) is given: For $n = 1, 2, \ldots$, find $p_h^n \in V_h$ such that

$$\begin{aligned} &\left(\phi\frac{p_h^n - \check{p}_h^{n-1}}{\Delta t^n}, v\right) + \big(\mathbf{a}^n\nabla p_h^n, \nabla v\big) \\ &\qquad + (R^n p_h^n, v) = (f^n, v) \qquad \forall v \in V_h, \end{aligned} \tag{4.200}$$

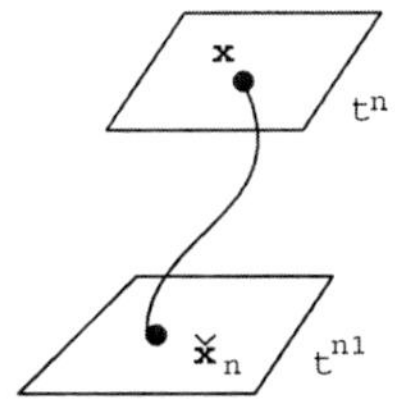

Figure 4.54. *An illustration of the definition* $\check{\mathbf{x}}_n$.

where

$$\check{p}_h^{n-1} = p_h\left(\check{\mathbf{x}}_n, t^{n-1}\right) = p_h\left(\mathbf{x} - \frac{\Delta t^n}{\phi(\mathbf{x})}\mathbf{b}(\mathbf{x}, t^n), t^{n-1}\right). \tag{4.201}$$

The remarks made at the end of Section 4.6.1 for (4.190) also apply to (4.200). In particular, existence and uniqueness of a solution for reasonable choices of **a** and R can be shown in the same way (cf. Exercise 4.54), and the error estimate (4.193) under appropriate assumptions on p also holds for (4.200) (Chen, 2005):

$$\max_{1\le n\le N}\left(\|p^n - p_h^n\|_{L^2(\Omega)} + h\|p^n - p_h^n\|_{H^1(\Omega)}\right) \le C(p)\left(h^{r+1} + \Delta t\right),$$

provided that an approximation property similar to (4.192) holds for V_h in the multiple dimensions.

Discussion of a conservation relation

We discuss the MMOC in the simple case where

$$R = f = 0, \quad \nabla\cdot\mathbf{b} = 0 \quad \text{in } \Omega. \tag{4.202}$$

That is, **b** is *divergence-free* (or *solenoidal*). Application of condition (4.202), the periodicity assumption, and the divergence theorem (4.66) to (4.195) yields the *conservation relation*

$$\int_\Omega \phi(\mathbf{x})p(\mathbf{x}, t)\, d\mathbf{x} = \int_\Omega \phi(\mathbf{x})p_0(\mathbf{x})\, d\mathbf{x}, \qquad t > 0. \tag{4.203}$$

In applications, it is desirable to conserve at least a discrete form of this relation in any numerical approximation of (4.195). However, in general, the MMOC does not conserve it. To see this, we take $v = 1$ in (4.200) and apply (4.202) to give

$$\begin{aligned}\int_\Omega \phi(\mathbf{x})p(\mathbf{x}, t^n)\, d\mathbf{x} &= \int_\Omega \phi(\mathbf{x})p(\check{\mathbf{x}}_n, t^{n-1})\, d\mathbf{x}\\ &\neq \int_\Omega \phi(\mathbf{x})p(\mathbf{x}, t^{n-1})\, d\mathbf{x}.\end{aligned} \tag{4.204}$$

For each n, define the transformation

$$\mathbf{G}(\mathbf{x}) \equiv \mathbf{G}(\mathbf{x}, t^n) = \mathbf{x} - \frac{\Delta t^n}{\phi(\mathbf{x})}\mathbf{b}(\mathbf{x}, t^n). \tag{4.205}$$

We assume that $\mathbf{b}/\phi$ has bounded first partial derivatives in space. Then, for $d = 3$, the *Jacobian of this transformation*, $\mathbf{J}(\mathbf{G})$, is

$$\begin{pmatrix} 1 - \frac{\partial}{\partial x_1}\left(\frac{b_1^n}{\phi}\right)\Delta t^n & -\frac{\partial}{\partial x_2}\left(\frac{b_1^n}{\phi}\right)\Delta t^n & -\frac{\partial}{\partial x_3}\left(\frac{b_1^n}{\phi}\right)\Delta t^n \\ -\frac{\partial}{\partial x_1}\left(\frac{b_2^n}{\phi}\right)\Delta t^n & 1 - \frac{\partial}{\partial x_2}\left(\frac{b_2^n}{\phi}\right)\Delta t^n & -\frac{\partial}{\partial x_3}\left(\frac{b_2^n}{\phi}\right)\Delta t^n \\ -\frac{\partial}{\partial x_1}\left(\frac{b_3^n}{\phi}\right)\Delta t^n & -\frac{\partial}{\partial x_2}\left(\frac{b_3^n}{\phi}\right)\Delta t^n & 1 - \frac{\partial}{\partial x_1}\left(\frac{b_3^n}{\phi}\right)\Delta t^n \end{pmatrix},$$

and its determinant equals (cf. Exercise 4.55)

$$|\mathbf{J}(\mathbf{G})| = 1 - \nabla \cdot \left(\frac{\mathbf{b}^n}{\phi}\right)\Delta t^n + \mathcal{O}\left((\Delta t^n)^2\right). \tag{4.206}$$

Thus, even in the case where ϕ is constant, for the second equality of (4.204) to hold requires that the Jacobian of the transformation (4.205) be identically one. While this is true for constant ϕ and $\mathbf{b}$, it cannot be expected to be true for variable coefficients. In the case where ϕ is constant and $\nabla \cdot \mathbf{b} = 0$, it follows from (4.206) that the determinant of this transformation is $1 + \mathcal{O}((\Delta t^n)^2)$, so a systematic error of size $\mathcal{O}((\Delta t^n)^2)$ should be expected. On the other hand, if $\nabla \cdot (\mathbf{b}/\phi) \neq 0$, the determinant is $1 + \mathcal{O}(\Delta t^n)$, and a systematic error of size $\mathcal{O}(\Delta t^n)$ can occur. In particular, in using the MMOC in the solution of a two-phase immiscible flow problem (cf. Chapter 7), Douglas et al. (1997) found that conservation of mass failed by as much as 10% in simulations with stochastic rock properties and about half that much with uniform rock properties. Errors of this magnitude obscure the relevance of numerical approximations to an unacceptable level and motivated the search for a modification of the MMOC that both conserves (4.203) and is at most very little more computationally expensive than the MMOC. Another method, the *modified method of characteristics with adjusted advection*, was defined by Douglas et al. (1997) and satisfies these criteria. This method is derived from the MMOC by perturbing the foot of characteristics in an ad hoc fashion. We do not introduce this method in this chapter. Instead, we describe the *Eulerian–Lagrangian localized adjoint method* (ELLAM) (Celia et al., 1990; Russell, 1990).

4.6.2 The Eulerian–Lagrangian localized adjoint method

We consider the ELLAM for problem (4.184) in divergence form:

$$\begin{aligned} &\frac{\partial(\phi p)}{\partial t} + \nabla \cdot \left(\mathbf{b}p - \mathbf{a}\nabla p\right) + Rp = f, && \mathbf{x} \in \Omega,\ t > 0, \\ &(\mathbf{b}p - \mathbf{a}\nabla p) \cdot \nu = g, && \mathbf{x} \in \Gamma,\ t > 0, \\ &p(\mathbf{x}, 0) = p_0(\mathbf{x}), && \mathbf{x} \in \Omega, \end{aligned} \tag{4.207}$$

where $\Omega \subset \mathbb{R}^d$ ($d \leq 3$) is a bounded domain and $\phi = \phi(\mathbf{x}, t)$ and $\mathbf{b} = \mathbf{b}(\mathbf{x}, t)$ are now variable. We consider a flux boundary condition in (4.207); an extension to Dirichlet conditions is possible (Chen, 2005).

For any $\mathbf{x} \in \Omega$ and two times $0 \leq t^{n-1} < t^n$, the hyperbolic part of problem (4.207), $\phi \partial p/\partial t + \mathbf{b} \cdot \nabla p$, defines the characteristic $\check{\mathbf{x}}_n(\mathbf{x}, t)$ along the interstitial velocity $\varphi = \mathbf{b}/\phi$ (cf. Figure 4.54):

$$\begin{aligned} &\frac{\partial}{\partial t}\check{\mathbf{x}}_n = \varphi(\check{\mathbf{x}}_n, t), \qquad t \in J^n, \\ &\check{\mathbf{x}}_n(\mathbf{x}, t^n) = \mathbf{x}. \end{aligned} \tag{4.208}$$

In general, the characteristics in (4.208) can be determined only approximately. There are many methods to solve this first-order ODE for the approximate characteristics. We consider only the Euler method.

The Euler method for solving (4.208) for the approximate characteristics is: For any $\mathbf{x} \in \Omega$,

$$\check{\mathbf{x}}_n(\mathbf{x}, t) = \mathbf{x} - \varphi(\mathbf{x}, t^n)(t^n - t), \qquad t \in [\check{t}(\mathbf{x}), t^n], \tag{4.209}$$

where $\check{t}(\mathbf{x}) = t^{n-1}$ if $\check{\mathbf{x}}_n(\mathbf{x}, t)$ does not backtrack to the boundary Γ for $t \in [t^{n-1}, t^n]$; $\check{t}(\mathbf{x}) \in J^n = (t^{n-1}, t^n]$ is the time instant when $\check{\mathbf{x}}_n(\mathbf{x}, t)$ intersects Γ, i.e., $\check{\mathbf{x}}_n(\mathbf{x}, \check{t}(\mathbf{x})) \in \Gamma$, otherwise. Let

$$\Gamma_+ = \{\mathbf{x} \in \Gamma : \ (\mathbf{b} \cdot \boldsymbol{\nu})(\mathbf{x}) \geq 0\}.$$

For $(\mathbf{x}, t) \in \Gamma_+ \times J^n$, the approximate characteristic emanating backward from $(\mathbf{x}, t)$ is

$$\check{\mathbf{x}}_n(\mathbf{x}, \theta) = \mathbf{x} - \varphi(\mathbf{x}, t)(t - \theta), \qquad \theta \in [\check{t}(\mathbf{x}, t), t], \tag{4.210}$$

where $\check{t}(\mathbf{x}, t) = t^{n-1}$ if $\check{\mathbf{x}}_n(\mathbf{x}, \theta)$ does not backtrack to the boundary Γ for $\theta \in [t^{n-1}, t]$; $\check{t}(\mathbf{x}, t) \in (t^{n-1}, t]$ is the time instant when $\check{\mathbf{x}}_n(\mathbf{x}, \theta)$ intersects Γ otherwise.

If Δt^n is sufficiently small (depending upon the smoothness of φ), the approximate characteristics do not cross each other, which is assumed. Then $\check{\mathbf{x}}_n(\cdot, t)$ is a one-to-one mapping of $\mathbb{R}^d$ to $\mathbb{R}^d$ ($d \leq 3$); we indicate its inverse by $\hat{\mathbf{x}}_n(\cdot, t)$.

For any $t \in J^n$, we define

$$\tilde{\varphi}(\mathbf{x}, t) = \varphi(\hat{\mathbf{x}}_n(\mathbf{x}, t), t^n), \quad \tilde{\mathbf{b}} = \tilde{\varphi}\phi. \tag{4.211}$$

We assume that $\tilde{\mathbf{b}} \cdot \boldsymbol{\nu} \geq 0$ on Γ_+.

Let K_h be a partition of Ω into elements $\{K\}$. For each $K \in K_h$, let $\check{K}(t)$ represent the trace back of K to time t, $t \in J^n$,

$$\check{K}(t) = \{\mathbf{x} \in \Omega : \mathbf{x} = \check{\mathbf{x}}_n(\mathbf{y}, t) \text{ for some } \mathbf{y} \in K\},$$

and $\mathcal{K}^n$ be the space-time region that follows the characteristics (cf. Figure 4.55),

$$\mathcal{K}^n = \{(\mathbf{x}, t) \in \Omega \times J : t \in J^n \text{ and } \mathbf{x} \in \check{K}(t)\}.$$

Also, define $\mathcal{B}^n = \{(\mathbf{x}, t) \in \partial\mathcal{K}^n : \mathbf{x} \in \partial\Omega\}$.

We write the hyperbolic part of (4.207) as

$$\frac{\partial(\phi p)}{\partial t} + \nabla \cdot \big(\mathbf{b}p\big) = \frac{\partial(\phi p)}{\partial t} + \nabla \cdot \big(\tilde{\mathbf{b}}p\big) + \nabla \cdot \big([\mathbf{b} - \tilde{\mathbf{b}}]p\big). \tag{4.212}$$

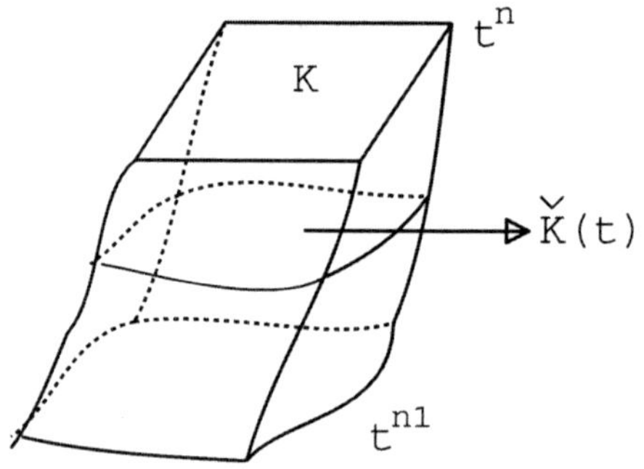

Figure 4.55. *An illustration of $\mathcal{K}^n$.*

With $\boldsymbol{\tau}(x,t) = (\tilde{\mathbf{b}}, \phi)$ and a smooth test function $v(\mathbf{x},t)$, application of Green's formula in space and time gives (cf. Exercise 4.56)

$$\begin{aligned}\int_{\mathcal{K}^n} \left(\frac{\partial(\phi p)}{\partial t} + \nabla \cdot (\tilde{\mathbf{b}} p) \right) v \, d\mathbf{x}\, dt \\ = \int_K \phi^n p^n v^n \, d\mathbf{x} - \int_{\check{K}(t^{n-1})} \phi^{n-1} p^{n-1} v^{n-1,+} \, d\mathbf{x} \\ + \int_{\mathcal{B}^n} p \tilde{\mathbf{b}} \cdot \boldsymbol{\nu} v \, d\ell - \int_{\mathcal{K}^n} p \boldsymbol{\tau} \cdot \left(\nabla v, \frac{\partial v}{\partial t} \right) d\mathbf{x}\, dt,\end{aligned} \tag{4.213}$$

where we used the fact that $\boldsymbol{\tau} \cdot \boldsymbol{\nu}_{\mathcal{K}^n} = 0$ on the space-time edges $(\partial \mathcal{K}^n \cap (\check{K} \times J^n)) \setminus \mathcal{B}^n$ and $v^{n-1,+} = v(x, t^{n-1,+}) = \lim_{\epsilon \to 0^+} v(x, t^{n-1} + \epsilon)$ to account for the fact that $v(x,t)$ can be discontinuous at the time levels.

Similarly, the diffusion part of (4.207) gives

$$\begin{aligned}\int_{\mathcal{K}^n} \nabla \cdot (\mathbf{a} \nabla p) v \, d\mathbf{x}\, dt \\ = \int_{J^n} \left\{ \int_{\partial \check{K}(t)} \mathbf{a} \nabla p \cdot \boldsymbol{\nu}_{\check{K}(t)} v \, d\ell - \int_{\check{K}(t)} (\mathbf{a} \nabla p) \cdot \nabla v \, d\mathbf{x} \right\} dt.\end{aligned} \tag{4.214}$$

We assume that the test function $v(\mathbf{x},t)$ is constant along the approximate characteristics. Then combining (4.212)–(4.214) yields the space-time variational form of (4.207):

$$\begin{aligned}(\phi^n p^n, v^n) - \left(\phi^{n-1} p^{n-1}, v^{n-1,+}\right) \\ + \int_{J^n} \{(\mathbf{a} \nabla p, \nabla v) + (Rp, v)\} \, dt = \int_{J^n} \left\{ (f, v) - (g, v)_\Gamma \right\} dt \\ + \int_{J^n} \left\{ \left(\nabla \cdot \left[(\tilde{\mathbf{b}} - \mathbf{b}) p \right], \hat{v} \right) - \left(p \left[\tilde{\mathbf{b}} - \mathbf{b} \right] \cdot \boldsymbol{\nu}, v \right)_\Gamma \right\} dt,\end{aligned} \tag{4.215}$$

where the inner product notation in space is used. If we apply backward Euler time integration along characteristics to the diffusion, reaction, and source term in (4.215), we see that

$$
\begin{aligned}
&(\phi^n p^n, v^n) + (\Delta t^n \mathbf{a}^n \nabla p^n, \nabla v^n) + (\Delta t^n R^n p^n, v^n) \\
&\quad = \left(\phi^{n-1} p^{n-1}, v^{n-1,+}\right) + (\Delta t^n f^n, v^n) - \int_{J^n} (g, v)_\Gamma \, dt \\
&\qquad + \int_{J^n} \left\{ \left(\nabla \cdot \left[(\tilde{\mathbf{b}} - \mathbf{b}) p\right], \hat{v}\right) - \left(p\left[\tilde{\mathbf{b}} - \mathbf{b}\right] \cdot \boldsymbol{\nu}, v\right)_\Gamma \right\} dt,
\end{aligned}
\tag{4.216}
$$

where $\Delta t^n(\mathbf{x}) = t^n - \check{t}(\mathbf{x})$. The $\mathbf{x}$-dependent Δt^n seems quite appropriate, since the diffusion at each point is weighted by the length of time over which it acts.

Let $V_h \subset H^1(\Omega)$ be a finite element space (cf. Section 4.2.1). For any $w \in V_h$, we define a test function $v(\mathbf{x}, t)$ to be a constant extension of $w(\mathbf{x})$ into the space-time region $\Omega \times J^n$ along the approximate characteristics (cf. (4.209) and (4.210))

$$
\begin{aligned}
v(\check{\mathbf{x}}_n(\mathbf{x}, t), t) &= w(\mathbf{x}), \qquad t \in [\check{t}(\mathbf{x}), t^n], \ \mathbf{x} \in \Omega, \\
v(\check{\mathbf{x}}_n(\mathbf{x}, \theta), \theta) &= w(\mathbf{x}), \qquad \theta \in [\check{t}(\mathbf{x}, t), t], \ (\mathbf{x}, t) \in \Gamma_+ \times J^n.
\end{aligned}
\tag{4.217}
$$

Now, based on (4.216), an ELLAM procedure is defined: For $n = 1, 2, \ldots,$ find $p_h^n \in V_h$ such that

$$
\begin{aligned}
&\left(\phi^n p_h^n, v^n\right) + \left(\Delta t^n \mathbf{a}^n \nabla p_h^n, \nabla v^n\right) + \left(\Delta t^n R^n p_h^n, v^n\right) \\
&\quad = \left(\phi^{n-1} p_h^{n-1}, v^{n-1,+}\right) + (\Delta t^n f^n, v^n) - \int_{J^n} (g, v)_\Gamma \, dt.
\end{aligned}
\tag{4.218}
$$

Taking $v = 1$ in (4.218) yields the statement of global mass conservation. The remarks made on accuracy and efficiency of the MMOC also apply to (4.218) (cf. Exercise 4.57). In particular, when V_h is the space of piecewise linear functions defined on a regular triangulation K_h, the following convergence result holds (Wang, 2000).

Assume that Ω is a convex polygonal domain or has a smooth boundary Γ, and the coefficients $\mathbf{a}$, $\mathbf{b}$, ϕ, f, and R satisfy

$$
\begin{aligned}
&\mathbf{a} \in \left(W^{1,\infty}(\Omega \times J)\right)^{d \times d}, \quad \mathbf{b} \in \left(W^{1,\infty}(\Omega \times J)\right)^{d}, \\
&\phi, \ f \in W^{1,\infty}(\Omega \times J), \qquad R \in L^\infty(J; W^{1,\infty}(\Omega)).
\end{aligned}
$$

If the solution p to (4.207) satisfies $p \in L^\infty(J; W^{2,\infty}(\Omega))$ and $\partial p/\partial t \in L^2(J; H^2(\Omega))$, the initialization error satisfies

$$
\|p_0 - p_h^0\|_{L^2(\Omega)} \le Ch^2 \|p_0\|_{H^2(\Omega)},
$$

and Δt is sufficiently small, then

$$
\begin{aligned}
\max_{1\le n\le N} &\|p^n - p_h^n\|_{L^2(\Omega)} \\
&\le C\left\{\Delta t\left(\left\|\frac{dp}{d\tau}\right\|_{L^2(J;H^1(\Omega))} + \|p\|_{L^\infty(J;W^{2,\infty}(\Omega))} + \left\|\frac{df}{d\tau}\right\|_{L^2(\Omega\times J)} + \|f\|_{L^2(\Omega\times J)}\right)\right. \\
&\quad \left. + h^2\left(\|p\|_{L^\infty(J;W^{2,\infty}(\Omega))} + \left\|\frac{\partial p}{\partial t}\right\|_{L^2(J;H^2(\Omega))} + \|p_0\|_{H^2(\Omega)}\right)\right\},
\end{aligned}
$$

where p_h is the solution of (4.218), and for real numbers $q, r \ge 0$,

$$
\begin{aligned}
\|v\|_{L^2(J;W^{q,r}(\Omega))} &= \left\| \|v(\cdot,t)\|_{W^{q,r}(\Omega)} \right\|_{L^2(J)}, \\
\|v\|_{L^\infty(J;W^{q,r}(\Omega))} &= \max_{t\in J} \|v(\cdot,t)\|_{W^{q,r}(\Omega)}.
\end{aligned}
$$

With advection on the right-hand side of (4.218) only, the linear system arising from (4.218) is well suited for iterative linear solution algorithms in multiple space dimensions (see the next chapter). The characteristic idea can be combined with other finite element methods presented in Sections 4.3–4.5; see Yang (1992) and Arbogast and Wheeler (1995) for characteristic mixed methods and Chen (2002B) for Eulerian–Lagrangian discontinuous methods, for example.

4.7 Adaptive Finite Element Methods

In reservoir simulation, many important physical and chemical phenomena are sufficiently localized and transient that *adaptive numerical methods* are necessary to resolve them. Adaptive numerical methods have become increasingly important because researchers have realized the great potential of the concepts underlying these methods. They are numerical schemes that automatically adjust themselves to improve approximate solutions. These methods are not exactly new in the computational area, even in the finite element literature. The adaptive adjustment of time steps in the numerical solution of ODEs has been the subject of research for many decades. Furthermore, the search for optimal finite element grids dates back to the early 1970s (Oliveira, 1971). But modern interest in this subject began in the late 1970s, mainly thanks to important contributions by Babuška and Rheinboldt (1978A; 1978B) and many others.

The overall accuracy of numerical approximations often deteriorates due to local singularities like those arising from reentrant corners of domains, interior or boundary layers, and sharp moving fronts. An obvious strategy is to refine the grids near these critical regions, i.e., to insert more grid points near where the singularities occur. The question is then how we identify those regions, refine them, and obtain a good balance between the refined and unrefined regions such that the overall accuracy is optimal. To answer this question, we need to utilize *adaptivity*. That is, we need somehow to *restructure a numerical scheme* to improve the quality of its approximate solutions. This puts a great demand on the choice of numerical methods. Restructuring a numerical scheme includes changing the number of elements, refining local grids, increasing the local order of approximation, moving nodal points, and modifying algorithm structures.

Another closely related question is how to obtain reliable estimates of the accuracy of computed approximate solutions. A priori error estimates, as obtained in the previous five sections, are often insufficient because they produce information only on the asymptotic behavior of errors and they require a solution regularity that is not satisfied in the presence of the above-mentioned singularities. To answer this question, we need to assess the quality of approximate solutions a posteriori, i.e., after an initial approximation is obtained. This requires that we compute a posteriori error estimates. Of course, the computation of the a posteriori estimates should be far less expensive than that of the approximate solutions. Moreover, it must be possible to compute *dynamically* local error indicators that lead to some estimate of the local quality of the solution.

The aim of this section is to present a brief introduction of some of basic topics on the two components of the adaptive finite element methods: the *adaptive strategy* and *a posteriori error estimation*. We focus on these two components for the standard finite element methods considered in Section 4.2.

4.7.1 Local grid refinement in space

There are three basic types of adaptive strategies: (1) local refinement of a fixed grid, (2) addition of more degrees of freedom locally by utilizing higher-order basis functions in certain elements, and (3) adaptively moving a computational grid to achieve better local resolution.

Local grid refinement of a fixed grid is called an *h-scheme*. In this scheme, the mesh is automatically refined or unrefined depending upon a local error indicator. Such a scheme leads to a very complex data management problem because it involves the dynamic regeneration of a grid, renumbering of nodal points and elements, and element connectivity. However, the h-scheme can be very effective in generating near-optimal grids for a given error tolerance. Efficient h-schemes with fast data management procedures have been developed for complex problems (Diaz et al., 1984; Ewing, 1986; Bank, 1990). Moreover, the h-scheme can be also employed to *unrefine* a grid (or *coarsen* a grid) when a local error indicator becomes smaller than a preassigned tolerance.

Addition of more degrees of freedom locally by utilizing higher-order basis functions in certain elements is referred to as a *p-scheme* (Babuška et al., 1983; Szabo, 1986). As discussed in Section 4.2, finite element methods for a given problem attempt to approximate a solution by functions in a finite-dimensional space of polynomials. The p-scheme generally utilizes a fixed grid and a fixed number of grid elements. If the error indicator in any element exceeds a given tolerance, the local order of the polynomial degree is increased to reduce the error. This scheme can be very effective in modeling thin boundary layers around bodies moving in a flow field, where the use of very fine grids is impractical and costly. However, the data management problem associated with the p-scheme, especially for regions of complex geometry, can be very difficult.

Adaptively moving a computational grid to get better local resolution is usually termed an *r-scheme* (Miller and Miller, 1981). It employs a fixed number of grid points and attempts to move them dynamically to areas where the error indicator exceeds a preassigned tolerance. The r-scheme can be easily implemented, and does not have the difficult data management problem associated with the h- and p-schemes. On the other hand, it suffers from several deficiencies. Without special care in its implementation, it can be unstable and result in grid tangling and local degradation of approximate solutions. It can never reduce the error below a fixed limit since it is not capable of handling the migration of regions where the solution is singular. However, by an appropriate combination with other adaptive strategies, the r-scheme can lead to a useful scheme for controlling solution errors.

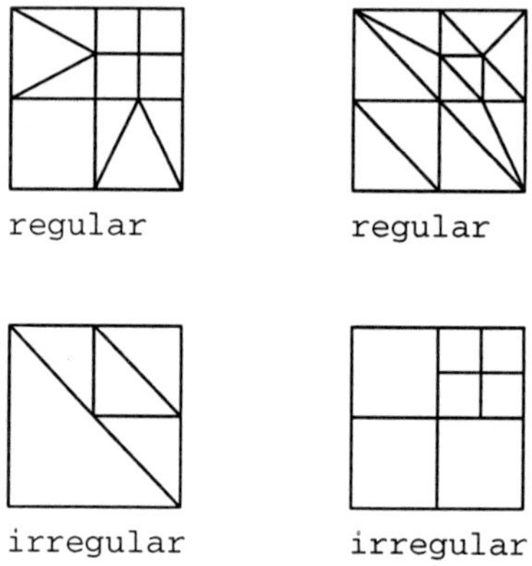

Figure 4.56. *Examples of regular and irregular vertices.*

Combinations of these three basic strategies such as the *hr-, hp-*, and *hpr-schemes* are also possible (Babuška and Dorr, 1981; Oden et al., 1989). In this chapter, as an example, we study the widely applied h-scheme.

Regular h-schemes

We focus on a two-dimensional domain. An extension of the concept in this section to three dimensions is simple to visualize. However, the modification of the supporting algorithms in the next section is not straightforward.

In the two-dimensional case, a grid can be triangular, quadrilateral, or of mixed type (i.e., consisting of both triangles and quadrilaterals); see Section 4.2. A vertex is *regular* if it is a vertex of each of its neighboring elements, and a grid is *regular* if its every vertex is regular. All other vertices are said to be *irregular* (*slave nodes* or *hanging nodes*); see Figure 4.56. The *irregularity index* of a grid is the maximum number of irregular vertices belonging to the same edge of an element.

If all elements in a grid are subdivided into an equal number (usually four) of smaller elements simultaneously, the refinement is referred to as *global*. For example, a refinement is global by connecting the opposite midpoints of the edges of each triangle or quadrilateral in the grid. Global refinement does not introduce irregular vertices. In the previous five sections, all the refinements were global and regular. In contrast, in the case of a *local refinement* where only some of the elements in a grid are subdivided into smaller elements, irregular vertices may appear; refer to Figure 4.56.

In this subsection, we study a regular local refinement. The following *refinement rule* can be used to convert irregular vertices to regular ones (Bank, 1990; Braess, 1997). This rule is designed for a triangular grid and guarantees that each of the angles in the original grid is bisected at most once. We may think of starting with a triangulation as in Figure 4.57. It contains six irregular vertices, which need to be converted to regular vertices.

A refinement rule for a triangulation is defined as follows:

1. If an edge of a triangle contains two or more vertices of other triangles (not counting its own vertices), then this triangle is subdivided into four equal smaller triangles. This procedure is repeated until such triangles no longer exist.

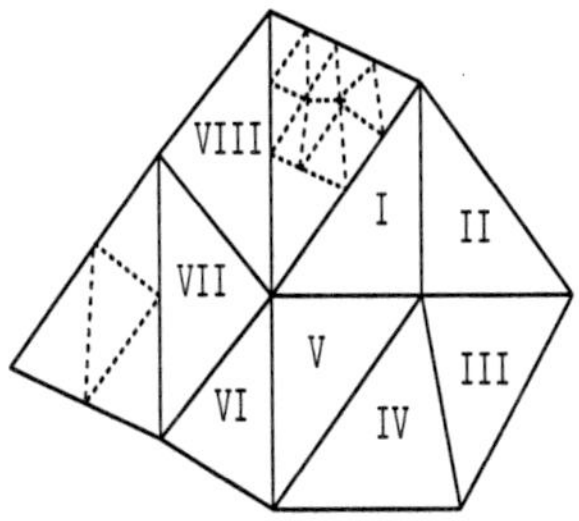

Figure 4.57. *A coarse grid (solid lines) and a refinement (dotted lines).*

2. If the midpoint of an edge of a triangle contains a vertex of another triangle, this triangle is subdivided into two parts. The new edge is called a *green edge*.

3. If a further refinement is needed, the green edges are first eliminated before the next iteration.

For the triangulation in Figure 4.57, we apply the first step to triangles I and VIII. This requires the use of the refinement rule twice on triangle VII. Next, we construct green edges on triangles II, V, and VI and on three subtriangles (cf. Exercise 4.58).

Despite its recursive nature, this procedure stops after a finite number of iterations. Let k be the maximum number of levels in the underlying refinement, where the maximum is taken over all elements ($k = 2$ in Figure 4.57). Then every element is subdivided at most k times, which presents an upper bound on the number of times step 1 is used. We emphasize that this procedure is purely two dimensional. A generalization to three dimensions is not straightforward. For a triangulation of Ω into tetrahedra, see a technique due to Rivara (1984A).

Irregular h-schemes

Irregular grids leave more freedom for local refinement. In the general case of arbitrary irregular grids, an element may be refined locally without any interference with its neighbors. As for regular local refinements, some desirable properties should be preserved for irregular refinements.

First, in the process of consecutive refinements no distorted elements should be generated. That is, the minimal angle of every element should be bounded away from zero by a common bound that probably depends only on the initial grid (cf. (4.79)).

Second, a new grid resulting from a local refinement should contain all the nodes of the old grid. In particular, if continuous finite element spaces $\{V_{h_k}\}$ are exploited for a second-order partial differential problem in all levels, consecutive refinements should lead to a *nested* sequence of these spaces:

$$V_{h_1} \subset V_{h_2} \subset \cdots \subset V_{h_k} \subset V_{h_{k+1}} \subset \cdots ,$$

where $h_{k+1} < h_k$ and h_k is the mesh size at the kth grid level. In the case of irregular local refinements, to preserve continuity of functions in these spaces the function values at the irregular nodes of a new grid are obtained by *polynomial interpolation* of the values at the old grid nodes.

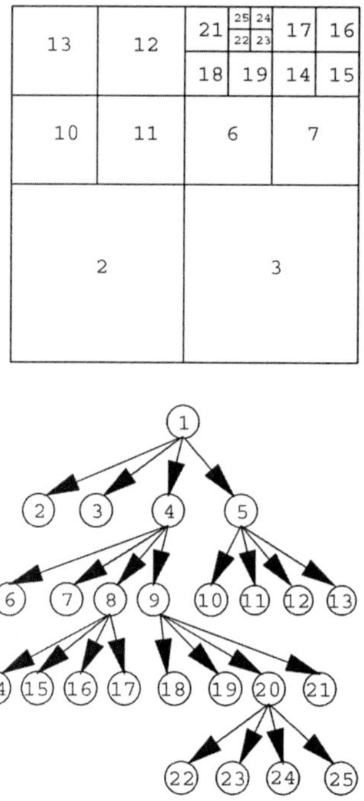

Figure 4.58. *A local refinement and the corresponding tree structure.*

Third, as defined before, the *irregularity index* of a grid is the maximum number of irregular vertices belonging to an edge of an element. There are reasons to restrict ourselves to 1-irregular grids. In practice, it seems to be very unlikely that grids with a higher irregularity index can be useful for a local h-scheme. Also, in general, the stiffness matrix arising from the finite element discretization of a problem should be sparse. It turns out that the sparsity cannot be guaranteed for a general irregular grid (Bank et al., 1983). To produce 1-irregular grids, we can employ the 1*-irregular rule*: Refine any unrefined element for which any of the edges contains more than one irregular node.

Unrefinements

As noted, an h-scheme can be also employed to *unrefine* a grid. There are two factors that decide if an element needs to be unrefined: (1) a local error indicator and (2) a structural condition imposed on the grid resulting from the regularity or 1-irregularity requirement. Both these factors must be examined before an element is unrefined.

When an element is refined, it produces a number of new smaller elements; the old element is called a *father* and the smaller ones are termed its *sons*. A *tree structure* (or *family structure*) consists of remembering for each element its father (if there is one) and its sons. Figure 4.58 shows a typical tree structure, together with a corresponding current grid generated by consecutive refinements of a single square. The *root* of the tree originates at the initial element and the *leaves* are those elements being not refined.

The tree structure provides for easy and fast unrefinements. When the tree information is stored, a local unrefinement can be done by simply "cutting the corresponding branch" of the tree, i.e., unrefining previously refined elements and restoring locally the previous grid.

4.7.2 Data structures

In the finite element methods developed in Section 4.2, all elements and nodes are usually numbered in a consecutive fashion so that a minimal band in the stiffness matrix of a finite element system can be produced. When a computational code identifies an element to evaluate its contribution to this matrix, the minimal information required is the set of node numbers corresponding to this element (cf. Section 4.2.1).

Adaptive local refinements and unrefinements require much more complex data structures than the classical global ones in Section 4.2. Because elements and nodes are added and deleted adaptively, it is often impossible to number them in a consecutive fashion. Hence we need to establish some kind of *natural ordering of elements*. In particular, all elements must be placed in an order, and a code must recognize, for a given element, the next element (or the previous element if necessary) in the sequence. Therefore, for an element, the following information should be stored:

- nodes,
- neighbors,
- father,
- sons,
- level of refinement.

For a given node, its coordinates are also needed. The logic of a data structure corresponding to a particular local refinement may need additional information. However, the above-listed information seems to be the minimal requirement for all existing data structures. Several data structures are available for adaptive local grid refinements and unrefinements (Rheinboldt and Mesztenyi, 1980; Bank et al., 1983; Rivara, 1984B).

4.7.3 A posteriori error estimates

We now study the second component of the adaptive finite element method: a posteriori error estimation. *A posteriori error estimators* and *indicators* can be utilized to give a specific assessment of errors and to form a solid basis for local refinements and unrefinements. A posteriori error estimators can be roughly classified as follows (Verfürth, 1996).

1. *Residual estimators.* These estimators bound the error of the computed approximate solution by a suitable norm of its residual with respect to the strong form of a differential equation (Babuška and Rheinboldt, 1978A).

2. *Local problem-based estimators.* This approach solves locally discrete problems, which are similar to, but simpler than, the original problem, and uses appropriate norms of the local solutions for error estimation (Babuška and Rheinboldt, 1978B; Bank and Weiser, 1985).

3. *Averaging-based estimators.* This approach utilizes some local extrapolation or averaging technique to define error estimation (Zienkiewicz and Zhu, 1987).

4. *Hierarchical basis estimators.* This approach calculates the residual of the computed approximate solution with respect to another finite element space of higher-order polynomials or with respect to a refined grid (Deuflhard et al., 1989).

As an example, we briefly study the residual estimators for the model problem in two dimensions:

$$\begin{aligned} -\Delta p &= f && \text{in } \Omega, \\ p &= 0 && \text{on } \Gamma_D, \\ \frac{\partial p}{\partial \boldsymbol{\nu}} &= g && \text{on } \Gamma_N, \end{aligned} \tag{4.219}$$

where Ω is a bounded domain in the plane with boundary $\bar{\Gamma} = \bar{\Gamma}_D \cup \bar{\Gamma}_N$, $\Gamma_D \cap \Gamma_N = \emptyset$, $f \in L^2(\Omega)$ and $g \in L^2(\Gamma_N)$ are given functions, and the *Laplacian operator* Δ is defined as in Section 4.2.1. We only study this simple problem; for generalizations to more general problems, refer to Section 4.2 or the references cited in this chapter.

Assume that Γ_D is closed relative to Γ and has a positive length. Define (cf. Section 4.2.1)

$$V = \{v \in H^1(\Omega) : v = 0 \text{ on } \Gamma_D\}.$$

Also, introduce the notation

$$a(p, v) = \int_\Omega \nabla p \cdot \nabla v \, d\mathbf{x}, \quad L(v) = \int_\Omega f v \, d\mathbf{x} + \int_{\Gamma_N} g v \, d\ell, \qquad v \in V.$$

As in (4.69), problem (4.219) can be recast in the variational form:

$$\text{Find } p \in V \text{ such that } a(p, v) = L(v) \qquad \forall v \in V. \tag{4.220}$$

Let Ω be a convex polygonal domain (or its boundary Γ is smooth), and let K_h be a triangulation of Ω into triangles K of diameter h_K, as in Section 4.2.1. With the triangulation K_h, associate a grid function $h(\mathbf{x})$ such that, for some positive constant C_1,

$$C_1 h_K \le h(\mathbf{x}) \le h_K \qquad \forall \mathbf{x} \in K, \quad K \in K_h. \tag{4.221}$$

Moreover, assume that there exists a positive constant C_2 such that

$$C_2 h_K^2 \le |K| \qquad \forall K \in K_h, \tag{4.222}$$

where $|K|$ is the area of K. Recall that (4.222) is the *minimum angle* condition stating that the angles of triangles in K_h are bounded below by C_2 (cf. (4.79)).

To keep the notation to a minimum, let $V_h \subset V$ be defined by

$$V_h = \{v \in V : v|_K \in P_1(K), \ K \in K_h\}.$$

An extension to finite element spaces of higher-order polynomials will be noted at the end of this subsection. The finite element method for (4.219) is formulated:

$$\text{Find } p_h \in V_h \text{ such that } a(p_h, v) = L(v) \qquad \forall v \in V_h. \tag{4.223}$$

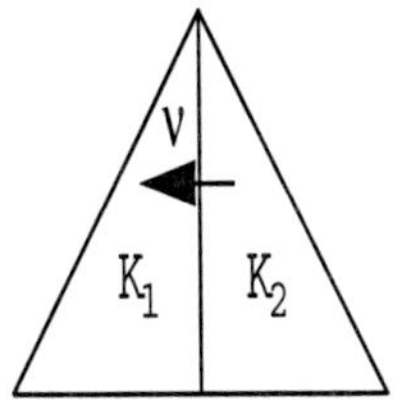

Figure 4.59. *An illustration of* $\boldsymbol{\nu}$.

It follows from (4.220) and (4.223) that

$$a(p - p_h, v) = L(v) - a(p_h, v) \qquad \forall v \in V. \tag{4.224}$$

The right-hand side of (4.224) implicitly defines the *residual* of p_h as an element in the dual space of V. Because Γ_D has a positive length, *Poincaré's inequality* (Chen, 2005) holds:

$$\|v\|_{L^2(\Omega)} \leq C(\Omega)\, \|\nabla v\|_{\mathbf{L}^2(\Omega)} \qquad \forall v \in V, \tag{4.225}$$

where C depends on Ω and the length of Γ_D. Using (4.225) and Cauchy's inequality (4.59), we have

$$\begin{aligned} \frac{1}{1+C^2(\Omega)}\|v\|_{H^1(\Omega)} &\leq \sup\{a(v, w) : w \in V,\ \|w\|_{H^1(\Omega)} = 1\} \\ &\leq \|v\|_{H^1(\Omega)}. \end{aligned} \tag{4.226}$$

Consequently, it follows from (4.224) and (4.226) that

$$\begin{aligned} &\sup\left\{L(v) - a(p_h, v) : v \in V,\ \|v\|_{H^1(\Omega)} = 1\right\} \\ &\leq \|p - p_h\|_{H^1(\Omega)} \\ &\leq \left(1 + C^2(\Omega)\right) \sup\left\{L(v) - a(p_h, v) : v \in V,\ \|v\|_{H^1(\Omega)} = 1\right\}. \end{aligned} \tag{4.227}$$

Since the supremum term in (4.227) is equivalent to the norm of the residual in the dual space of V, this inequality implies that the norm in V of the error is, up to multiplicative constants, bounded from above and below by the norm of the residual in the dual space of V. Most a posteriori error estimators attempt to bound this dual norm of the residual by quantities that can be more easily evaluated from f, g, and p_h.

Let $\mathcal{E}_h^o$ denote the set of all interior edges e in K_h, $\mathcal{E}_h^b$ the set of the edges e on Γ, and $\mathcal{E}_h = \mathcal{E}_h^o \cup \mathcal{E}_h^b$. Furthermore, let $\mathcal{E}_h^D$ and $\mathcal{E}_h^N$ be the sets of edges e on Γ_D and Γ_N, respectively.

With each $e \in \mathcal{E}_h$, associate a unit normal vector $\boldsymbol{\nu}$. For $e \in \mathcal{E}_h^b$, $\boldsymbol{\nu}$ is just the outward unit normal to Γ. For $e \in \mathcal{E}_h^o$, with $e = \bar{K}_1 \cap \bar{K}_2$, $K_1, K_2 \in K_h$, the direction of $\boldsymbol{\nu}$ is associated with the definition of jumps across e; if the jump of function v across e is defined by

$$[\![v]\!] = (v|_{K_2})|_e - (v|_{K_1})|_e, \tag{4.228}$$

then $\boldsymbol{\nu}$ is defined as the unit normal exterior to K_2 (cf. Figure 4.59).

We recall the scalar product notation

$$(v, w)_{\mathcal{S}} = \int_{\mathcal{S}} v(\mathbf{x}) w(\mathbf{x})\, d\mathbf{x}, \qquad v,\, w \in L^2(\mathcal{S}).$$

If $\mathcal{S} = \Omega$, we omit it in this notation. Note that, by Green's formula (4.68), the definition of $L(\cdot)$, and the fact that $\Delta p_h = 0$ on all $K \in K_h$,

$$\begin{aligned} L(v) - a(p_h, v) &= L(v) - \sum_{K \in K_h} (\nabla p_h, \nabla v)_K \\ &= L(v) - \sum_{K \in K_h} \left[(\nabla p_h \cdot \boldsymbol{\nu}_K, v)_{\partial K} - (\Delta p_h, v)_K \right] \\ &= (f, v) + \sum_{e \in \mathcal{E}_h^N} (g - \nabla p_h \cdot \boldsymbol{\nu}, v)_e - \sum_{e \in \mathcal{E}_h^o} ([\![\nabla p_h \cdot \boldsymbol{\nu}]\!], v)_e . \end{aligned} \tag{4.229}$$

Applying (4.227) and (4.229), one can show that (cf. Exercise 4.60)

$$\begin{aligned} \|p - p_h\|_{H^1(\Omega)} \le C \Bigg\{ &\sum_{K \in K_h} h_K^2 \|f\|_{L^2(K)}^2 \\ &+ \sum_{e \in \mathcal{E}_h^N} h_e \|g - \nabla p_h \cdot \boldsymbol{\nu}\|_{L^2(e)}^2 + \sum_{e \in \mathcal{E}_h^o} h_e \|[\![\nabla p_h \cdot \boldsymbol{\nu}]\!]\|_{L^2(e)}^2 \Bigg\}^{1/2}, \end{aligned} \tag{4.230}$$

where C depends on C_2 in (4.220) and $C(\Omega)$ in (4.225), and h_K and h_e represent the diameter and length, respectively, of K and e.

The right-hand side in (4.230) can be utilized as an a posteriori error estimator because it involves only the known data f and g, the approximate solution p_h, and the geometrical data of the triangulation K_h. For general functions f and g, the exact computation of the integrals in the first and second terms of the right-hand side of (4.230) is often impossible. These integrals must be approximated by appropriate quadrature formulas (cf. Section 4.2.3). On the other hand, it is also possible to approximate f and g by polynomials in suitable finite element spaces. Both approaches, numerical quadrature and approximation by simpler functions combined with exact integration of the latter functions, are often equivalent and generate analogous a posteriori estimators. We restrict ourselves to the simpler function approximation approach. In particular, let f_h and g_h be the L^2-projections of f and g into the spaces of piecewise constants with respect to K_h and $\mathcal{E}_h^N$, respectively; i.e., on each $K \in K_h$ and $e \in \mathcal{E}_h^N$, $f_K = f_h|_K$ and $g_e = g_h|_e$ are given by the local mean values

$$f_K = \frac{1}{|K|} \int_K f\, d\mathbf{x}, \quad g_e = \frac{1}{h_e} \int_e g\, d\ell. \tag{4.231}$$

Then we define a *residual a posteriori error estimator*:

$$\begin{aligned} \mathcal{R}_K = \Bigg\{ &h_K^2 \|f_K\|_{L^2(K)}^2 + \sum_{e \in \partial K \cap \mathcal{E}_h^N} h_e \|g_e - \nabla p_h \cdot \boldsymbol{\nu}\|_{L^2(e)}^2 \\ &+ \frac{1}{2} \sum_{e \in \partial K \cap \mathcal{E}_h^o} h_e \|[\![\nabla p_h \cdot \boldsymbol{\nu}]\!]\|_{L^2(e)}^2 \Bigg\}^{1/2}. \end{aligned} \tag{4.232}$$

The first term in $\mathcal{R}_K$ is related to the residual of p_h with respect to the strong form of the differential equation. The second and third terms reflect the facts that p_h does not exactly satisfy the Neumann boundary condition and that $p_h \notin H^2(\Omega)$. Since interior edges are counted twice, combining (4.230), (4.232), and the triangle inequality, we obtain (cf. Exercise 4.61)

$$\|p - p_h\|_{H^1(\Omega)} \leq C\left\{\sum_{K\in K_h}\left(\mathcal{R}_K^2 + h_K^2\|f - f_K\|_{L^2(K)}^2\right) + \sum_{e\in\mathcal{E}_h^N} h_e\|g - g_e\|_{L^2(e)}^2\right\}^{1/2}. \tag{4.233}$$

Based on (4.233), with a given tolerance $\epsilon > 0$, the following *adaptive algorithm* can be defined (below RHS denotes the right-hand side of (4.233)).

Algorithm I.

- Choose an initial grid K_{h_0} with grid size h_0, and find a finite element solution p_{h_0} using (4.223) with $V_h = V_{h_0}$;
- Given a solution p_{h_k} in V_{h_k} with grid size h_k, stop if the following stopping criterion is satisfied:

$$\text{RHS} \leq \epsilon; \tag{4.234}$$

- If (4.234) is violated, find a new grid K_{h_k} with grid size h_k such that the following equation is satisfied:

$$\text{RHS} = \epsilon, \tag{4.235}$$

and continue.

Inequality (4.234) is the stopping criterion, and (4.235) defines the adaptive strategy. It follows from (4.233) that the estimate $\|p - p_h\|_{H^1(\Omega)}$ is bounded by ϵ if (4.234) is reached with $p_h = p_{h_k}$. Equation (4.235) determines a new grid size h_k by maximality. Namely, we seek a grid size h_k as large as possible (to maintain efficiency) such that (4.235) is satisfied. The maximality is generally determined by *equidistribution* of an error such that the error contributions from the individual elements K are approximately equal. Let M_{h_k} be the number of elements in K_{h_k}; equidistribution means that

$$(\text{RHS}|_K)^2 = \frac{\epsilon^2}{M_{h_k}}, \qquad K \in K_{h_k}.$$

Since the solution p_{h_k} depends on K_{h_k}, this is a nonlinear problem. The nonlinearity can be simplified by replacing M_{h_k} by $M_{h_{k-1}}$ (the number at the previous level), for example.

The following inequality implies, in a sense, that the converse of (4.233) also holds (Verfürth, 1996; Chen, 2005): for $K \in K_h$,

$$\mathcal{R}_K \leq C\left\{\sum_{K'\in\Omega_K}\left(\|p - p_h\|_{H^1(K')}^2 + h_{K'}^2\|f - f_{K'}\|_{L^2(K')}^2\right) + \sum_{e\in\partial K\cap\mathcal{E}_h^N} h_e\|g - g_e\|_{L^2(e)}^2\right\}^{1/2}, \tag{4.236}$$

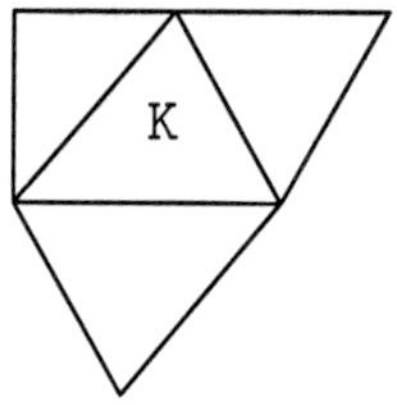

Figure 4.60. *An illustration of* Ω_K.

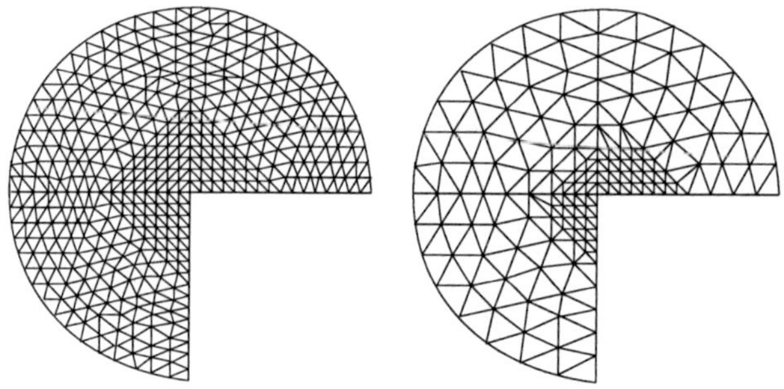

Figure 4.61. *Uniform (left) and adaptive (right) triangulations.*

where (cf. Figure 4.60)

$$\Omega_K = \bigcup \{K' \in K_h \ : \ \partial K' \cap \partial K \neq \emptyset\}.$$

Estimate (4.236) indicates that Algorithm I is *efficient* in the sense that the computational grid produced by this algorithm is not overly refined for a given accuracy, while (4.234) implies that this algorithm is *reliable* in the sense that the H^1-error is guaranteed to be within a given tolerance.

We end this section with three remarks. First, it is possible to control the error in norms other than the H^1-norm; we can control the gradient error in the maximum norm (the $L^\infty(\Omega)$-norm; cf. Johnson, 1994), for example. Second, the results in this section carry over to finite element spaces of polynomials of degree $r \geq 2$. In this case, f_h and g_h are the L^2-projections of f and g into the spaces of piecewise polynomials of degree $r-1$ with respect to K_h and $\mathcal{E}_h^N$, respectively, and f_K in the first term of $\mathcal{R}_K$ is replaced by $\Delta p_h|_K + f_K$ (cf. Exercise 4.62). Finally, the adaptive finite element methods presented in this section can be extended to transient problems (Chen, 2005).

Example 4.14. This example follows Verfürth (1996). Consider problem (4.219) on a circular segment centered at the origin, with radius one and angle $3\pi/2$ (cf. Figure 4.61). The function f is zero, and the solution p vanishes on the straight parts of the boundary Γ and has a normal derivative $\frac{2}{3}\cos(\frac{2}{3}\theta)$ on the curved part of Γ. In terms of polar coordinates, the exact solution p to (4.219) is $p = r^{2/3}\sin(\frac{2}{3}\theta)$. We calculate the finite element solution p_h using (4.223) with the space of piecewise linear functions V_h associated with the two triangulations shown in Figure 4.61. The left triangulation is constructed by five uniform refinements of

Table 4.6. *A comparison of uniform and adaptive refinements.*

Refinement	NT	NN	e_r	m_q
Uniform	3072	1552	3.8%	0.7
Adaptive	298	143	2.8%	0.6

an initial triangulation K_{h_0}, which is composed of three right-angled isosceles triangles with short edges of unit length. In each refinement step, every triangle is divided into four smaller triangles by connecting the midpoints of its edges. The midpoint of an edge having its two endpoints on Γ is projected onto Γ. The right triangulation in Figure 4.61 is obtained from K_{h_0} by using Algorithm I based on the error estimator in (4.232). A triangle $K \in K_h$ is divided into four smaller triangles if $\mathcal{R}_K \geq 0.5 \max_{K' \in K_h} \mathcal{R}_{K'}$. Again, the midpoint of an edge having its two endpoints on Γ is projected onto Γ. For both triangulations, Table 4.6 lists the number of triangles (NT), the number of unknowns (NN), the relative error $e_r = \|p - p_h\|_{H^1(\Omega)} / \|p\|_{H^1(\Omega)}$, and the measurement $m_q = (\sum_{K \in K_h} \mathcal{R}_K^2)^{1/2} / \|p - p_h\|_{H^1(\Omega)}$ of the quality of the error estimator. From this table we clearly see the advantage of the adaptive method and the reliability of the error estimator.

4.7.4 The eighth SPE project: Gridding techniques

Spatial grids of many different types have been presented in this chapter: rectangles (rectangular parallelepipeds), triangles (tetrahedra), CVFE grids, prisms, and their various flexible variations. To see a grid number reduction that can be obtained using flexible grids in an application, the eighth SPE comparative solution project (CSP) is used (Quandalle, 1993). The objectives of this project are

- to compare numerical predictions using flexible grids vs. regular grids,
- to compare numerical predictions using different flexible grids,
- to evaluate the grid number reduction that can be obtained using flexible grids.

The problem is a three-dimensional simulation of oil production associated with gas injection in a four-layer reservoir, as described in Figures 4.62 and 4.63 and Tables 4.7–4.11, where B_o stands for the formation volume factor. Fluid and rock property data are those of the first CSP (Odeh (1981)) except that there is no water in the present project. This problem is run twice with the same simulator:

- a first time with the $10 \times 10 \times 4$ regular discretization grid shown in Figure 4.62,
- a second time with a four-layer grid flexible in the horizontal direction. The aim of the second run is to reduce the number of gridblocks as much as possible using the grid flexibility while the following constraints on both producers are respected:
  - the gas breakthrough time predicted with the flexible grid (fixed as the time corresponding to a gas-oil ratio (GOR) of 2,000 SCF/STB) must match within 10% the breakthrough time of the $10 \times 10 \times 4$ grid;
  - at the time when the regular grid model reaches a GOR of 10,000 SCF/STB, the flexible grid model must predict a GOR within 10% of the same GOR value.

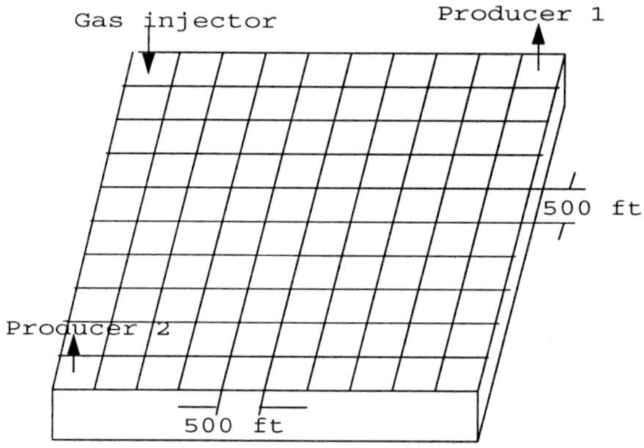

Figure 4.62. *Reservoir and grid system.*

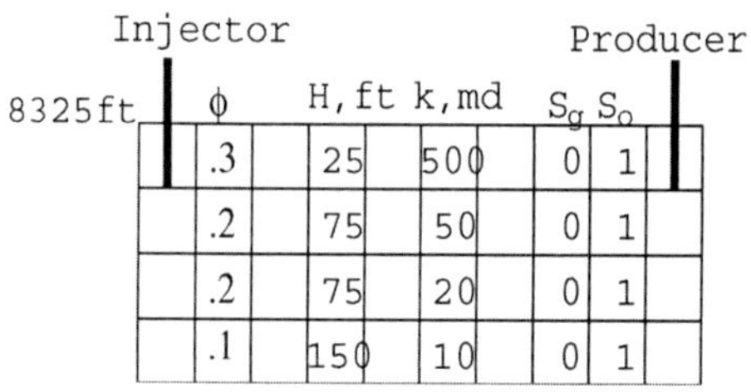

Figure 4.63. *Vertical cross section.*

The well boundary conditions are defined in terms of an oil rate at surface conditions for both producers and a gas rate at surface conditions for the gas injector. The producer bottom hole pressure limit is set to be so low that the simulation time never reaches this value.

A black oil simulator (cf. Section 2.6 and Chapter 8) is used for the present simulation with two types of flexible grids for the second run: a local grid refinement based on rectangles (cf. Figure 4.64) and a CVFE grid (cf. Figure 4.65, in fact, a hybrid grid). The number of gridblocks in these two cases is 96 and 68, respectively, which correspond to the cases used by INTERA Information Technologies in the eighth CSP (Quandalle, 1993). The five-point finite difference stencil scheme is used for the rectangular grids, while the CVFE method is applied for the CVFE grid. Tables 4.12 and 4.13 show the gas breakthrough time for the basic $10 \times 10 \times 4$ grid and two flexible grids, TGR (the time at which the basic grid reaches a GOR of 10,000 SCF/STB), and the production GOR at TGR for both flexible grids, respectively, for producers 1 and 2. Figures 4.66 and 4.67 indicate a comparison of the production GOR values and flowing bottom hole pressures among the three grids for producer 1. From Tables 4.12 and 4.13 and these figures, we see that while the number of

Table 4.7. *Reservoir data and constraints.*

Initial reservoir pressure at 8,400 ft: 4,800 psia
The gas injector perforated in the upper layer only, at a distance of 250 ft in both x_1 and x_2 directions.
Producer 1 perforated in the upper layer only, at a distance of 4,750 ft in the x_1 direction and 250 ft in x_2.
Producer 2 perforated in the upper layer only, at a distance of 250 ft in the x_1 direction and 4,750 ft in x_2.
Gas injection rate: 12.5 MM SCF/D
Maximum oil production rate for each producer: 1,875 STB/D
Minimum oil production rate for each producer: 1,000 STB/D
Minimum bottom hole pressure for each producer: 1,000 psi
Rock compressibility: 3×10^{-6} 1/psi
Porosity measured at 14.7 psi: 0.3
Wellbore radius: 0.25 ft
Capillary pressure: 0 psi
Reservoir temperature: 200° F
Gas specific gravity: 0.792
Runs terminated either at the end of 10 years or when both producers have reached a GOR of 30,000 SCF/STB.

Table 4.8. *Saturated oil PVT data.*

Reservoir pressure (psia)	B_o (RB/STB)	μ_o (cp)	ρ_o (lbm/ft^3)	Solution GOR (SCF/STB)
14.7	1.062	1.040	46.244	1.0
264.7	1.150	0.975	43.544	90.5
514.7	1.207	0.910	42.287	180.0
1014.7	1.295	0.830	41.004	371.0
2014.7	1.435	0.695	38.995	636.0
2514.7	1.500	0.641	38.304	775.0
3014.7	1.565	0.594	37.781	930.0
4014.7	1.695	0.510	37.046	1270.0
5014.7	1.827	0.449	36.424	1618.0
9014.7	2.357	0.203	34.482	2984.0

Table 4.9. *Undersaturated oil PVT data.*

Pressure (psia)	B_o (RB/STB)	μ_o (cp)	ρ_o (lbm/ft^3)
4014.7	1.695	0.510	37.046
9014.7	1.579	0.740	39.768

gridblocks is reduced by a factor of four with the local rectangular grid refinement or six with the CVFE grid, the production GOR values and pressures are close to those obtained by the $10 \times 10 \times 4$ grid. This result demonstrates the potential of using flexible grids in reservoir simulation.

Table 4.10. *Gas PVT data.*

Pressure (psia)	B_g (RB/STB)	μ_g (cp)	ρ_g (lbm/ft^3)
14.7	0.935829	0.0080	0.0647
264.7	0.067902	0.0096	0.8916
514.7	0.035228	0.0112	1.7185
1014.7	0.017951	0.0140	3.3727
2014.7	0.009063	0.0189	6.6806
2514.7	0.007266	0.0208	8.3326
3014.7	0.006064	0.0228	9.9837
4014.7	0.004554	0.0268	13.2952
5014.7	0.003644	0.0309	16.6139
9014.7	0.002167	0.0470	27.9483

Table 4.11. *Relative permeability data.*

S_g	k_{rg}	k_{ro}
0.0	0.0	1.0
0.001	0.0	1.0
0.02	0.0	0.997
0.05	0.005	0.980
0.12	0.025	0.700
0.2	0.075	0.350
0.25	0.125	0.200
0.30	0.190	0.090
0.40	0.410	0.021
0.45	0.60	0.010
0.50	0.72	0.001
0.60	0.87	0.0001
0.70	0.94	0.000
0.85	0.98	0.000
1.0	1.0	0.000

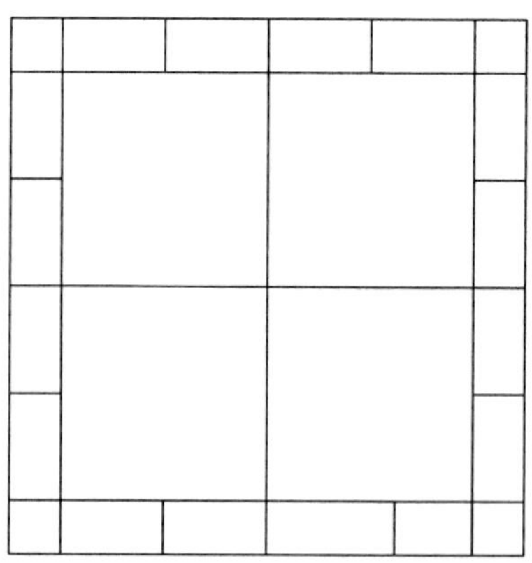

Figure 4.64. *Local rectangular grid refinement.*

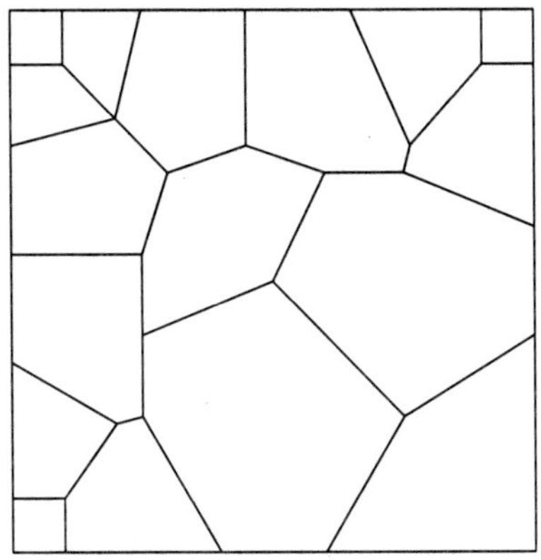

Figure 4.65. *CVFE grid.*

Table 4.12. *Gas breakthrough time for producer* 1.

Grid	Breakthrough time (days)	TGR (days)	GOR (SCF/STB)
10 × 10 × 4	807	2,256	10,000
Local refinement	774		10,403
CVFE grid	857		9,552

Table 4.13. *Gas breakthrough time for producer* 2.

Grid	Breakthrough time (days)	TGR (days)	GOR (SCF/STB)
10 × 10 × 4	760	2,196	10,000
Local refinement	726		10,055
CVFE grid	823		9,560

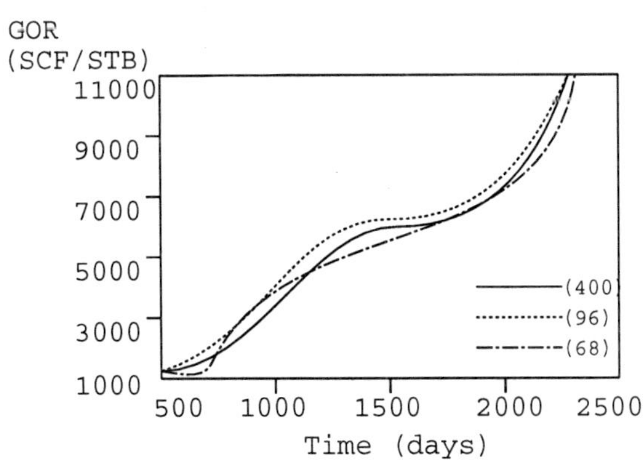

Figure 4.66. *Gas-oil ratio for producer* 1.

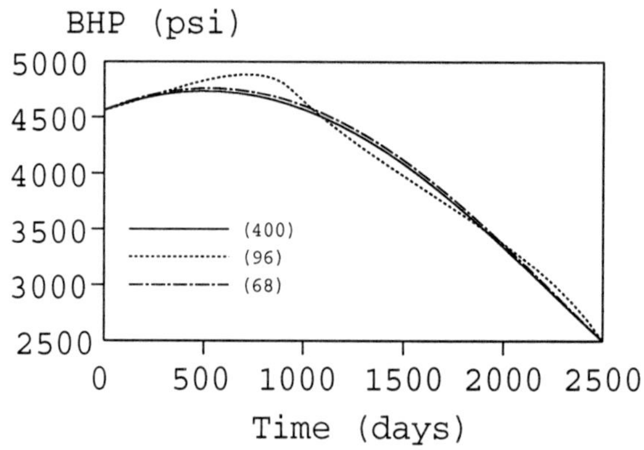

Figure 4.67. *Bottom hole pressure for producer* 1.

4.8 Bibliographical Remarks

The finite difference methods presented in Section 4.1 are locally conservative, but are not flexible in the treatment of complex reservoirs. On the other hand, the standard finite element methods described in Section 4.2 are more flexible, but not conservative on local elements (e.g., on triangles). They are globally conservative. The CVFE methods developed in Section 4.3 conserve mass locally on each control volume. The discontinuous finite element methods given in Section 4.4 possess this local property. These discontinuous methods are particularly suitable for numerical solution of advection problems and can be easily used in the adaptive methods introduced in Section 4.7. The mixed finite element methods discussed in Section 4.5 are designed to give a high-order approximation for a velocity vector. Finally, the characteristic finite element methods studied in Section 4.6 are suitable for advection-dominated flow and transport equation problems.

The literature on finite difference methods is rich (e.g., Richtmyer and Morton, 1967; Thomas, 1995). For applications of these methods to reservoir simulation, the reader should refer to Peaceman (1977) and Aziz and Settari (1979). There are numerous books on finite element methods (e.g., Strang and Fix, 1973; Ciarlet, 1978; Thomée, 1984; Brezzi and Fortin, 1991; Brenner and Scott, 1994; Johnson, 1994; Braess, 1997; Quarteroni and Valli, 1997). In Section 4.2.4, we briefly treated transient problems. The book by Thomée (1984) exclusively treats time-dependent problems. The content of Sections 4.2 and 4.4–4.7 is taken from Chen (2005). Finally, for more information on the eighth CSP, refer to Quandalle (1993).

Exercises

4.1. Consider problem (4.18) with $a = 1$ and $\Omega = (0, 1) \times (0, 1)$ (the unit square):

$$-\frac{\partial^2 p}{\partial x_1^2} - \frac{\partial^2 p}{\partial x_2^2} = q(x_1, x_2), \qquad (x_1, x_2) \in \Omega, \tag{4.237}$$

where q indicates an injector located at $(0.1667, 0.1667)$ or a producer located at $(0.8333, 0.8333)$. A homogeneous Neumann boundary condition (*no-flow boundary condition*) is

$$\frac{\partial p}{\partial \boldsymbol{\nu}} = 0,$$

where $\partial p/\partial \boldsymbol{\nu}$ is the normal derivative and $\boldsymbol{\nu}$ is the outward unit normal to $\Gamma = \partial\Omega$ (the boundary of Ω). (i) Formulate a finite difference scheme for (4.237) similar to scheme (4.20) using a block-centered grid with three equal subintervals in each of the x_1- and x_2-directions. (ii) Discretize the Neumann boundary condition using a first-order scheme analogous to (4.14) with $g = 0$. (iii) The well term q is evaluated:

$$q_{i,j} = \frac{2\pi}{\ln(r_e/r_w)}(p_{bh} - p_{i,j}) \quad \text{with } (i, j) = (1, 1) \text{ or } (3, 3),$$

where the *wellbore radius* r_w equals 0.001, the *drainage radius* r_e of both wells is given by $r_e = 0.2h$ with h the step size in the x_1- and x_2-directions, and the *wellbore pressure* p_{bh} equals 1.0 at the injector and -1.0 at the producer. Write the finite difference scheme derived in (i) in matrix form $\mathbf{Ap} = \mathbf{q}$ (with $\mathbf{q}$ denoting the well vector) and find the matrix $\mathbf{A}$ and vector $\mathbf{q}$.

4.2. Extend the definition of *consistency* given in Section 4.1.7 to the initial parabolic problem (4.21) in two dimensions.

4.3. Using the definition in Exercise 4.2, show that the forward, backward, and Crank–Nicholson difference schemes introduced in Section 4.1.6 are consistent with problem (4.21).

4.4. For problem (4.27), show that the one-dimensional counterpart of the Crank–Nicholson difference scheme defined in Section 4.1.6 is unconditionally stable.

4.5. Prove that the one-dimensional counterpart of the backward difference scheme defined in Section 4.1.6 is convergent for problem (4.27).

4.6. Prove that the one-dimensional counterpart of the Crank–Nicholson difference scheme defined in Section 4.1.6 is convergent for problem (4.27).

4.7. Show that the explicit scheme (4.39) is consistent with problem (4.38).

4.8. Show that the amplification factor γ for scheme (4.39) is

$$\gamma = 1 + \frac{b\Delta t}{h}(1 - \cos(kh)) - \bar{\imath}\frac{b\Delta t}{h}\sin(kh). \tag{4.238}$$

4.9. Prove that in the case $b > 0$, the factor γ in equation (4.238) satisfies $|\gamma| > 1$.

4.10. Prove that in the case $b < 0$, the factor γ in equation (4.238) satisfies $|\gamma| \le 1$, provided that the CFL condition (4.40) holds.

4.11. In the case $b > 0$, show that the explicit scheme (4.41) is stable under condition (4.40).

4.12. Show that the amplification factor γ of scheme (4.42) is

$$\gamma = 1 - \bar{\imath}\frac{b\Delta t}{h}\sin(kh).$$

4.13. Prove that the amplification factor γ of scheme (4.43) is

$$\gamma = \left(1 - \frac{b\Delta t}{h}(1 - \cos(kh)) + \bar{i}\frac{b\Delta t}{h}\sin(kh)\right)^{-1}.$$

4.14. Prove that scheme (4.45) has the amplification factor γ

$$\gamma = \left(1 + \bar{i}\frac{b\Delta t}{h}\sin(kh)\right)^{-1}.$$

4.15. Define a Crank–Nicholson analog to scheme (4.39) for problem (4.38) with $b < 0$, and study its stability.

4.16. Define a Crank–Nicholson analog to scheme (4.41) for problem (4.38) with $b > 0$, and study its stability.

4.17. Define a Crank–Nicholson analog to scheme (4.42) for problem (4.38), and study its stability in both cases $b < 0$ and $b > 0$.

4.18. Derive the local truncation error associated with the difference scheme (4.39) for problem (4.38) with $b < 0$ (cf. (4.46)).

4.19. Express numerical dispersion a_{num} for the Crank–Nicholson scheme defined in Exercise 4.16 in terms of b, h, and Δt.

4.20. Consider the diffusion-convection problem

$$\begin{aligned} &\frac{\partial p}{\partial t} + b\frac{\partial p}{\partial x} - a\frac{\partial^2 p}{\partial x^2} = 0, \qquad 0 < x < \infty,\ t > 0, \\ &p(x,0) = 0, \qquad 0 < x < \infty, \\ &p(0,t) = 1, \qquad p(\infty,t) = 0, \quad t > 0, \end{aligned} \tag{4.239}$$

where $a > 0$ and b are constants. This problem has the exact solution

$$p = \frac{1}{2}\left\{\operatorname{erfc}\left(\frac{x - bt}{2(at)^{1/2}}\right) + \exp\left(\frac{bx}{a}\right)\operatorname{erfc}\left(\frac{x + bt}{2(at)^{1/2}}\right)\right\},$$

where the *complementary error function* erfc is

$$\operatorname{erfc}(x) = 1 - \frac{2}{\pi^{1/2}}\int_0^x \exp\left(-\ell^2\right)\, d\ell.$$

For problem (4.239), consider the difference scheme

$$\frac{p_i^{n+1} - p_i^n}{\Delta t} + b\frac{p_i^n - p_{i-1}^n}{h} - a\frac{p_{i+1}^{n+1} - 2p_i^{n+1} + p_{i-1}^{n+1}}{h^2} = 0 \tag{4.240}$$

with the initial and boundary conditions

$$\begin{aligned} &p_i^0 = 0, \qquad i \geq 1, \\ &p_0^n = 1, \qquad p_I^n = 0, \quad n \geq 1, \end{aligned}$$

where the last equation is an adequate representation of the boundary condition at $x = \infty$ if I is large enough. In computations, we choose $I = 5/h$.

The exact solution is undefined at $x = 0$ and $t = 0$. The difference scheme, however, requires a value for p_0^0. In the computations, an arbitrary choice $p_0^0 = 0.5$ is made. Further data are given by

$$h = 0.1, \quad b = 1.0, \quad a = 0.01.$$

Use (4.240) to find the numerical solutions of (4.239) in the two cases $\Delta t = 0.05$ and $\Delta t = 0.1$, and compare the corresponding numerical dispersions a_{num} (refer to (4.48)).

4.21. Show that if $p \in V$ satisfies (4.52) and if p is twice continuously differentiable, where the space V is defined in Section 4.2.1, then p satisfies (4.50).

4.22. Write a code to solve the one-dimensional problem (4.50) approximately using the finite element method developed in Section 4.2.1. Use the function $f(x) = 4\pi^2 \sin(2\pi x)$ and a uniform partition of (0, 1) with $h = 0.1$. Also, compute the errors

$$\left\| \frac{dp}{dx} - \frac{dp_h}{dx} \right\| = \left(\int_0^1 \left(\frac{dp}{dx} - \frac{dp_h}{dx} \right)^2 dx \right)^{1/2},$$

with $h = 0.1, 0.01$, and 0.001, and compare them. Here p and p_h are the exact and approximate solutions, respectively (cf. Section 4.2.1).

4.23. Show Cauchy's inequality (4.59).

4.24. Prove the estimates (4.62).

4.25. Referring to Section 4.2.1, show that the interpolant $\tilde{p}_h \in V_h$ of p defined in (4.61) equals the finite element solution p_h obtained by (4.54).

4.26. Prove Green's formula (4.68) in three space dimensions.

4.27. Carry out the derivation of system (4.71).

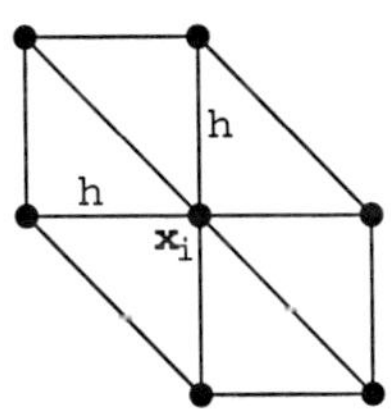

Figure 4.68. *The support of a basis function at node* $\mathbf{x}_i$.

4.28. For the figure given in Figure 4.68, construct the linear basis function at node $\mathbf{x}_i$ according to the definition in Section 4.2.1. Then use this result to show that the stiffness matrix $\mathbf{A}$ in (4.71) for the uniform partition of the unit square $(0, 1) \times (0, 1)$ given in Figure 4.14 is determined as in Section 4.2.1.

4.29. Write a code to solve the Poisson equation (4.65) approximately using the finite element method developed in Section 4.2.1. Use $f(x_1, x_2) = 8\pi^2 \sin(2\pi x_1) \sin(2\pi x_2)$ and a uniform partition of $\Omega = (0, 1) \times (0, 1)$, as given in Figure 4.14. Also, compute the errors

$$\|\nabla p - \nabla p_h\| = \left(\int_\Omega |\nabla p - \nabla p_h|^2 \, d\mathbf{x} \right)^{1/2},$$

with $h = 0.1$, 0.01, and 0.001, and compare them. Here p and p_h are the exact and approximate solutions, respectively, and h is the mesh size in the x_1- and x_2-directions.

4.30. Prove equation (4.75) for equation (4.74).

4.31. Derive equation (4.76) from equation (4.74) in detail.

4.32. Prove equation (4.78).

4.33. Let $\hat{K} = (0, 1) \times (0, 1)$ be the unit square with vertices $\hat{\mathbf{m}}_i$, $i = 1, 2, 3, 4$, $P(\hat{K}) = Q_1(\hat{K})$, and $\Sigma_{\hat{K}}$ be the degrees of freedom corresponding to the values at $\hat{\mathbf{m}}_i$. If K is a convex quadrilateral, define an appropriate mapping $\mathbf{F} : \hat{K} \to K$ so that an isoparametric finite element $(K, P(K), \Sigma_K)$ can be defined in the form

$$P(K) = \left\{ v : v(\mathbf{x}) = \hat{v}\left(\mathbf{F}^{-1}(\mathbf{x})\right), \ \mathbf{x} \in K, \ \hat{v} \in P(\hat{K}) \right\},$$

$$\Sigma_K \text{ consists of function values at } \mathbf{m}_i = \mathbf{F}(\hat{\mathbf{m}}_i), \ i = 1, 2, 3, 4.$$

4.34. Suppose that Γ is a circle with diameter L and that Γ_h is a polygonal approximation of Γ with vertices on Γ and maximal edge length equal to h. Show that the maximal distance from Γ to Γ_h is $\mathcal{O}(h^2/4L)$ (cf. Section 4.2.2).

4.35. Show the stability result (4.102) for Crank–Nicholson's method (4.103) with $f = 0$. What can be shown if $f \neq 0$?

4.36. Prove that the barycentric coordinates λ_i, λ_j, and λ_k of triangle K satisfy equations (4.114).

4.37. Derive equation (4.117) in detail.

4.38. As pointed out in Section 4.3.2, positive transmissibilities (or positive flux linkages) are very important in numerical reservoir simulation. This is particularly so when dealing with gravity-dominated flows involving fluids having different densities. Suppose that node $\mathbf{m}_i$ is physically located above node $\mathbf{m}_j$ in the vertical direction (depth increases as one moves from $\mathbf{m}_i$ to $\mathbf{m}_j$); initially, both nodes have equal (mobile) saturations of a dense fluid (called fluid A) and a light fluid (fluid B). Physically, fluid A must sink, and fluid B must rise. Explain the meanings of a positive discrete transmissibility between $\mathbf{m}_i$ and $\mathbf{m}_j$ and a negative discrete transmissibility between these two nodes. Which one corresponds to the physically correct motion?

4.39. The concept of irreducible saturation S_{ir} was introduced in Chapter 3. A fluid phase is mobile only when its saturation value is larger than its S_{ir}, which is reflected in its mobility λ (i.e., its relative permeability):

$$\lambda(S) \begin{cases} > 0 & \text{if } S > S_{ir}, \\ = 0 & \text{if } S \leq S_{ir}. \end{cases}$$

Consider problem (4.124), where the permeability tensor $\mathbf{a}$ is identity and Ω is a single triangle given in Figure 4.32. Suppose that the pressure values at the three vertices satisfy $p_k > p_i = p_j$ and the saturation values satisfy $S_k > S_i = S_j = S_{ir}$. This implies that the flux direction is in the negative x_2-direction, and the flux flowing out of the quadrilateral $\mathbf{m}_i\mathbf{m}_a\mathbf{m}_c\mathbf{m}_d$ through edge $\mathbf{m}_a\mathbf{m}_c$ in the x_2-direction is zero since $S_j = S_{ir}$. Find the flux on edge $\mathbf{m}_a\mathbf{m}_c$ in the x_2-direction using the potential-based upstream weighting CVFE (cf. Section 4.3.4) and the same flux using the flux-based upstream weighting CVFE (cf. Section 4.3.4). What do these two results tell us?

4.40. Show that if $u \in V = H^1(I)$ and $p \in W = L^2(I)$ satisfy (4.155) and if p is twice continuously differentiable, then p satisfies (4.152).

4.41. Write a code to solve problem (4.152) approximately using the mixed finite element method introduced in Section 4.5.1. Use $f(x) = 4\pi^2 \sin(2\pi x)$ and a uniform partition of (0, 1) with $h = 0.1$. Also, compute the errors

$$\|p - p_h\| = \left(\int_0^1 (p - p_h)^2 \, dx \right)^{1/2},$$

$$\|u - u_h\| = \left(\int_0^1 (u - u_h)^2 \, dx \right)^{1/2},$$

with $h = 0.1$, 0.01, and 0.001, and compare them. Here p, u and p_h, u_h are the solutions to (4.155) and (4.157), respectively (cf. Section 4.5.1). (If necessary, refer to Chen (2005) for a linear solver.)

4.42. Consider the following problem with an inhomogeneous boundary condition:

$$-\frac{d^2 p}{dx^2} = f(x), \qquad 0 < x < 1,$$

$$p(0) = p_{D0}, \quad p(1) = p_{D1},$$

where f is a given real-valued piecewise continuous bounded function in (0, 1), and p_{D0} and p_{D1} are real numbers. Write this problem in a mixed variational formulation, and construct a mixed finite element method using the finite element spaces described in Section 4.5.1. Determine the corresponding linear system of algebraic equations for a uniform partition.

4.43. Consider the following problem with a Neumann boundary condition at $x = 1$:

$$-\frac{d^2 p}{dx^2} = f(x), \qquad 0 < x < 1,$$

$$p(0) = \frac{dp}{dx}(1) = 0.$$

Express this problem in a mixed variational formulation, formulate a mixed finite element method using the finite element spaces considered in Section 4.5.1, and determine the corresponding linear system of algebraic equations for a uniform partition.

4.44. Construct finite element subspaces $V_h \times W_h$ of $H^1(I) \times L^2(I)$ that consist, respectively, of piecewise quadratic and linear functions on a partition of $I = (0, 1)$. How can the parameters (degrees of freedom) be chosen to describe such functions in V_h and W_h? Find the corresponding basis functions. Then define a mixed finite element method for equation (4.152) using these spaces $V_h \times W_h$ and express the corresponding linear system of algebraic equations for a uniform partition of I.

4.45. Show that the matrix **M** defined in Section 4.5.1 has both positive and negative eigenvalues.

4.46. Define the space

$$\mathbf{H}(\mathrm{div}, \Omega) = \left\{ \mathbf{v} = (v_1, v_2) \in (L^2(\Omega))^2 : \nabla \cdot \mathbf{v} \in L^2(\Omega) \right\}.$$

Show that for any decomposition of $\Omega \subset \mathbb{R}^2$ into subdomains such that the interiors of these subdomains are pairwise disjoint, $\mathbf{v} \in \mathbf{H}(\mathrm{div}, \Omega)$ if and only if its normal components are continuous across the interior edges in this decomposition.

4.47. Prove that if $\mathbf{u} \in \mathbf{V} = \mathbf{H}(\mathrm{div}, \Omega)$ and $p \in W = L^2(\Omega)$ satisfy (4.167) and if $p \in H^2(\Omega)$, then p satisfies (4.164).

4.48. Let the basis functions $\{\varphi_i\}$ and $\{\psi_i\}$ of $\mathbf{V}_h$ and W_h be defined as in Section 4.5.2. For a uniform partition of $\Omega = (0, 1) \times (0, 1)$ given as in Figure 4.14, determine the matrices $\mathbf{A}$ and $\mathbf{B}$ in system (4.169).

4.49. Consider problem (4.164) with an inhomogeneous boundary condition, i.e.,

$$\begin{aligned} -\Delta p &= f && \text{in } \Omega, \\ p &= g && \text{on } \Gamma, \end{aligned}$$

where Ω is a bounded domain in the plane with boundary Γ, and f and g are given. Express this problem in a mixed variational formulation, formulate a mixed finite element method using the finite element spaces given in Section 4.5.2, and determine the corresponding linear system of algebraic equations for a uniform partition of $\Omega = (0, 1) \times (0, 1)$ as displayed in Figure 4.14.

4.50. Consider the problem

$$\begin{aligned} -\Delta p &= f && \text{in } \Omega, \\ p &= g_D && \text{on } \Gamma_D, \\ \frac{\partial p}{\partial \boldsymbol{\nu}} &= g_N && \text{on } \Gamma_N, \end{aligned}$$

where Ω is a bounded domain in the plane with boundary Γ, $\bar{\Gamma} = \bar{\Gamma}_D \cup \bar{\Gamma}_N$, $\Gamma_D \cap \Gamma_N = \emptyset$, and f, g_D, and g_N are given functions. Write down a mixed variational formulation for this problem and formulate a mixed finite element method using the finite element spaces given in Section 4.5.2.

4.51. Let $\{\varphi_i\}$ and $\{\psi_i\}$ be the basis functions of $\mathbf{V}_h$ and W_h respectively, in system (4.176). Write (4.176) in matrix form.

4.52. Show that after multiplying both sides of (4.190) by Δt^n, the condition number of the stiffness matrix corresponding to the left-hand side of (4.190) is of order

$$\mathcal{O}\left(1 + \max_{x \in \mathbb{R},\, t \geq 0} |a(x, t)| h^{-2} \Delta t\right), \qquad \Delta t = \max_{n=1,2,\ldots} \Delta t^n.$$

4.53. Let $v \in C^1(\mathbb{R})$ (the set of continuously differentiable functions) be a $(0, 1)$-periodic function. Show that the condition $v(0) = v(1)$ implies

$$\frac{\partial v(0)}{\partial x} = \frac{\partial v(1)}{\partial x}.$$

4.54. Let $\mathbf{a}$ be positive semidefinite, ϕ be uniformly positive with respect to x and t, and R be nonnegative. Show that (4.200) has a unique solution $p_h^n \in V_h$ for each n.

4.55. Prove relation (4.206).

4.56. Derive equation (4.213) in detail.

4.57. Let $\mathbf{a}$ be positive semidefinite, ϕ be uniformly positive with respect to x and t, and R be nonnegative. Show that (4.218) has a unique solution $p_h^n \in V_h$ for each n.

4.58. For the example in Figure 4.57, use the refinement rule defined in Section 4.7.1 to convert irregular vertices to regular vertices.

4.59. For the problem

$$
\begin{aligned}
-\nabla \cdot (\mathbf{a}\nabla p) &= f && \text{in } \Omega, \\
p &= 0 && \text{on } \Gamma_D, \\
\mathbf{a}\nabla p \cdot \boldsymbol{\nu} &= g_N && \text{on } \Gamma_N,
\end{aligned}
$$

derive an inequality similar to (4.230).

4.60. Show inequality (4.230) using (4.227) and (4.229).

4.61. Apply (4.231) and (4.232) to derive (4.233) from (4.230).

4.62. For the problem

$$
\begin{aligned}
-\nabla \cdot (\mathbf{a}\nabla p) &= f && \text{in } \Omega, \\
p &= 0 && \text{on } \Gamma_D, \\
\mathbf{a}\nabla p \cdot \boldsymbol{\nu} &= g_N && \text{on } \Gamma_N,
\end{aligned}
$$

define an error estimator similar to (4.232).

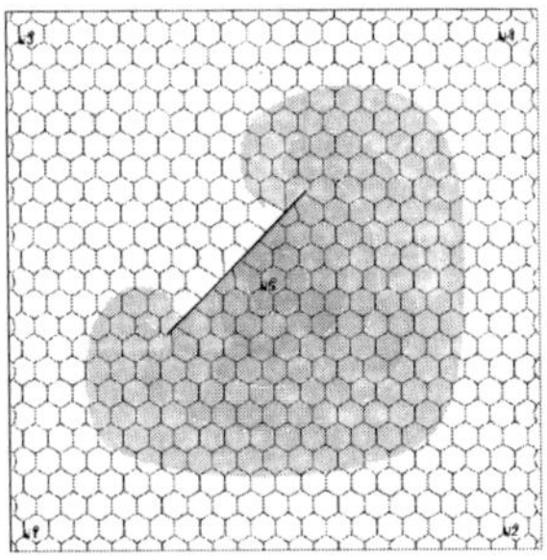

Chapter 5

Solution of Linear Systems

We have seen that an application of finite difference or finite element methods to a stationary problem or to an implicit scheme for a transient problem produces a linear system of equations of the form

$$\mathbf{Ap} = \mathbf{f}, \tag{5.1}$$

where $\mathbf{A}$ is an $M \times M$ matrix. In general, the matrix $\mathbf{A}$ arising in numerical reservoir simulation is sparse, highly nonsymmetric, and ill-conditioned. Its dimension M often ranges from hundreds to millions. For the solution of systems of the latter size, Krylov subspace algorithms are the sole option. In this chapter, we consider these iterative algorithms for solving (5.1) for various types of matrix $\mathbf{A}$. For completeness, in the first two sections (Sections 5.1 and 5.2), we discuss direct algorithms (Gaussian elimination or Cholesky's method). These algorithms are first studied for a tridiagonal matrix, and then extended to a general sparse matrix. Because the form of matrix $\mathbf{A}$ depends on the ordering of nodes, Section 5.3 briefly touches on this topic; several common ordering techniques used in reservoir simulation are reviewed. The CG (conjugate gradient), GMRES (generalized minimum residual), ORTHOMIN (orthogonal minimum residual), and BiCGSTAB (biconjugate gradient stabilized) iterative algorithms are discussed, respectively, in Sections 5.4–5.7. The discussion of these algorithms is for algorithms of general applicability. Some guidelines are also provided about the choice of a suitable algorithm for a given problem. The Krylov subspace algorithms are often useless without *preconditioning*. Therefore, the preconditioned versions of these algorithms and the choice of preconditioners are studied in Sections 5.8 and 5.9. Practical considerations for the choice of preconditioners in reservoir simulation are given in Section 5.10. Finally, comparisons between direct and iterative algorithms and bibliographical information are presented in Sections 5.11 and 5.12, respectively.

Generally speaking, uppercase letters of bold type will indicate matrices, while lowercase letters of bold type will represent vectors.

5.1 Tridiagonal Systems

In some cases, particularly for one-dimensional, single phase flow problems, the matrix $\mathbf{A}$ is *tridiagonal*:

$$\mathbf{A} = \begin{pmatrix} a_1 & b_1 & 0 & \dots & 0 & 0 \\ c_2 & a_2 & b_2 & \dots & 0 & 0 \\ 0 & c_3 & a_3 & \dots & 0 & 0 \\ \vdots & \vdots & \vdots & \ddots & \vdots & \vdots \\ 0 & 0 & 0 & \dots & a_{M-1} & b_{M-1} \\ 0 & 0 & 0 & \dots & c_M & a_M \end{pmatrix}.$$

System (5.1) with such a tridiagonal matrix can be solved either by a direct elimination algorithm or by an iterative algorithm. For this type of system, no known iterative algorithm can compete with direct elimination.

A positive definite matrix **A** has a unique *LU factorization* (Golub and van Loan, 1996)

$$\mathbf{A} = \mathbf{LU}, \tag{5.2}$$

where $\mathbf{L} = (l_{ij})$ is a *lower triangular* $M \times M$ matrix, i.e., $l_{ij} = 0$ if $j > i$, and $\mathbf{U} = (u_{ij})$ is an *upper triangular* $M \times M$ matrix, i.e., $u_{ij} = 0$ if $j < i$. For the special tridiagonal matrix under consideration, the matrices **L** and **U** are sought to have the forms

$$\mathbf{L} = \begin{pmatrix} l_1 & 0 & 0 & \dots & 0 & 0 \\ c_2 & l_2 & 0 & \dots & 0 & 0 \\ 0 & c_3 & l_3 & \dots & 0 & 0 \\ \vdots & \vdots & \vdots & \ddots & \vdots & \vdots \\ 0 & 0 & 0 & \dots & l_{M-1} & 0 \\ 0 & 0 & 0 & \dots & c_M & l_M \end{pmatrix}$$

and

$$\mathbf{U} = \begin{pmatrix} 1 & u_1 & 0 & \dots & 0 & 0 \\ 0 & 1 & u_2 & \dots & 0 & 0 \\ 0 & 0 & 1 & \dots & 0 & 0 \\ \vdots & \vdots & \vdots & \ddots & \vdots & \vdots \\ 0 & 0 & 0 & \dots & 1 & u_{M-1} \\ 0 & 0 & 0 & \dots & 0 & 1 \end{pmatrix}.$$

Note that the lower diagonal of **L** is the same as that of **A**, and the main diagonal of **U** is all ones. The identity (5.2) gives $2M - 1$ equations for the unknowns $l_1, l_2, \dots, l_M$ and $u_1, u_2, \dots, u_{M-1}$. The solution is

$$\begin{aligned} l_1 &= a_1, & \\ u_{i-1} &= b_{i-1}/l_{i-1}, & i &= 2, 3, \dots, M, \\ l_i &= a_i - c_i u_{i-1}, & i &= 2, 3, \dots, M. \end{aligned}$$

This algorithm is *Thomas' algorithm*.

With the factorization (5.2), system (5.1) can be easily solved using *forward elimination* and *backward substitution*:

$$\begin{aligned} \mathbf{Lv} &= \mathbf{f}, \\ \mathbf{Up} &= \mathbf{v}. \end{aligned} \tag{5.3}$$

Namely, since $\mathbf{L}$ is lower triangular, the first equation in (5.3) can be solved by forward elimination:

$$v_1 = \frac{f_1}{l_1}, \quad v_i = \frac{f_i - c_i v_{i-1}}{l_i}, \quad i = 2, 3, \ldots, M.$$

Next, since $\mathbf{U}$ is upper triangular, the second equation in (5.3) can be solved by backward substitution:

$$p_M = v_M, \quad p_i = v_i - u_i p_{i+1}, \quad i = M-1, M-2, \ldots, 1.$$

As discussed in the preceding chapter, for many practical problems, the matrix $\mathbf{A}$ is symmetric:

$$\mathbf{A} = \begin{pmatrix} a_1 & b_1 & 0 & \ldots & 0 & 0 \\ b_1 & a_2 & b_2 & \ldots & 0 & 0 \\ 0 & b_2 & a_3 & \ldots & 0 & 0 \\ \vdots & \vdots & \vdots & \ddots & \vdots & \vdots \\ 0 & 0 & 0 & \ldots & a_{M-1} & b_{M-1} \\ 0 & 0 & 0 & \ldots & b_{M-1} & a_M \end{pmatrix}.$$

In the symmetric case, $\mathbf{A}$ can be factorized:

$$\mathbf{A} = \mathbf{L}\mathbf{L}^T,$$

where $\mathbf{L}^T$ is the transpose of $\mathbf{L}$, and $\mathbf{L}$ now takes the form

$$\mathbf{L} = \begin{pmatrix} l_1 & 0 & 0 & \ldots & 0 & 0 \\ u_1 & l_2 & 0 & \ldots & 0 & 0 \\ 0 & u_2 & l_3 & \ldots & 0 & 0 \\ \vdots & \vdots & \vdots & \ddots & \vdots & \vdots \\ 0 & 0 & 0 & \ldots & l_{M-1} & 0 \\ 0 & 0 & 0 & \ldots & u_{M-1} & l_M \end{pmatrix}.$$

With this factorization, the entries are computed as follows:

$$\begin{aligned} l_1 &= \sqrt{a_1}, & & \\ u_i &= b_i / l_i, & & i = 1, 2, \ldots, M-1, \\ l_{i+1} &= \sqrt{a_{i+1} - u_i^2}, & & i = 1, 2, \ldots, M-1. \end{aligned}$$

Now, system (5.1) can be solved similarly to (5.3) using forward elimination and backward substitution.

In using the LU factorization algorithm we must assure that

$$l_i \neq 0, \qquad i = 1, 2, \ldots, M.$$

It can be shown that if $\mathbf{A}$ is symmetric positive definite, $l_i > 0$, $i = 1, 2, \ldots, M$ (Axelsson, 1994; Golub and van Loan, 1996). The quantities l_i are referred to as the *pivots*.

Thomas' algorithm can be extended to the solution of *block tridiagonal systems* (cf. Exercise 5.1). These systems may arise from the discretization of one-dimensional, two- or three-phase flow problems. For example, the simultaneous solution approach for two-phase flow generates two unknowns per grid point (node) and for three-phase flow three unknowns per grid point.

The most general block tridiagonal matrix for three-phase flow is

$$\mathbf{A} = \begin{pmatrix} \mathbf{a}_1 & \mathbf{b}_1 & \mathbf{0} & \ldots & \mathbf{0} & \mathbf{0} \\ \mathbf{c}_2 & \mathbf{a}_2 & \mathbf{b}_2 & \ldots & \mathbf{0} & \mathbf{0} \\ \mathbf{0} & \mathbf{c}_3 & \mathbf{a}_3 & \ldots & \mathbf{0} & \mathbf{0} \\ \vdots & \vdots & \vdots & \ddots & \vdots & \vdots \\ \mathbf{0} & \mathbf{0} & \mathbf{0} & \ldots & \mathbf{a}_{M-1} & \mathbf{b}_{M-1} \\ \mathbf{0} & \mathbf{0} & \mathbf{0} & \ldots & \mathbf{c}_M & \mathbf{a}_M \end{pmatrix}, \tag{5.4}$$

where $\mathbf{a}_i$, $\mathbf{b}_i$, and $\mathbf{c}_i$ are now 3×3 matrices. The unknown and right-hand side vectors $\mathbf{p}$ and $\mathbf{f}$ are

$$\mathbf{p} = \begin{pmatrix} \mathbf{p}_1 \\ \mathbf{p}_2 \\ \vdots \\ \mathbf{p}_M \end{pmatrix}, \qquad \mathbf{f} = \begin{pmatrix} \mathbf{f}_1 \\ \mathbf{f}_2 \\ \vdots \\ \mathbf{f}_M \end{pmatrix},$$

where

$$\mathbf{p}_i = \begin{pmatrix} p_i^1 \\ p_i^2 \\ p_i^3 \end{pmatrix}, \qquad \mathbf{f}_i = \begin{pmatrix} f_i^1 \\ f_i^2 \\ f_i^3 \end{pmatrix}, \qquad i = 1, 2, \ldots, M.$$

5.2 Gaussian Elimination

Gaussian elimination transforms a general linear system into an upper triangular system through elementary row (or column) operations. To see the idea, we begin with the solution of a 3×3 system:

$$\begin{aligned} a_{11}p_1 + a_{12}p_2 + a_{13}p_3 &= f_1, \\ a_{21}p_1 + a_{22}p_2 + a_{23}p_3 &= f_2, \\ a_{31}p_1 + a_{32}p_2 + a_{33}p_3 &= f_3. \end{aligned} \tag{5.5}$$

Assume that $a_{11} \neq 0$. The first step is to eliminate p_1 in the last two equations of (5.5). For this, set

$$m_{21} = \frac{a_{21}}{a_{11}}, \qquad m_{31} = \frac{a_{31}}{a_{11}}.$$

Multiplying the first equation of (5.5) by m_{21} and subtracting the resulting equation from the second equation of (5.5) yields

$$a_{22}^{(2)} p_2 + a_{23}^{(2)} p_3 = f_2^{(2)}, \tag{5.6}$$

where

$$a_{22}^{(2)} = a_{22} - m_{21}a_{12}, \quad a_{23}^{(2)} = a_{23} - m_{21}a_{13}, \quad f_2^{(2)} = f_2 - m_{21} f_1.$$

The same argument applied to the third equation of (5.5) implies

$$a_{32}^{(2)} p_2 + a_{33}^{(2)} p_3 = f_3^{(2)}, \tag{5.7}$$

where

$$a_{32}^{(2)} = a_{32} - m_{31}a_{12}, \quad a_{33}^{(2)} = a_{33} - m_{31}a_{13}, \quad f_3^{(2)} = f_3 - m_{31} f_1.$$

After the first step, system (5.5) becomes

$$\begin{aligned} a_{11}p_1 + a_{12}p_2 + a_{13}p_3 &= f_1, \\ a_{22}^{(2)} p_2 + a_{23}^{(2)} p_3 &= f_2^{(2)}, \\ a_{32}^{(2)} p_2 + a_{33}^{(2)} p_3 &= f_3^{(2)}. \end{aligned} \tag{5.8}$$

The second step is to eliminate p_2 in the third equation of (5.8). Assume that $a_{22}^{(2)} \neq 0$, and set

$$m_{32} = a_{32}^{(2)} / a_{22}^{(2)}.$$

Multiplying the second equation of (5.8) by m_{32} and subtracting the resulting equation from the third equation of (5.8) gives

$$a_{33}^{(3)} p_3 = f_3^{(3)}, \tag{5.9}$$

where

$$a_{33}^{(3)} = a_{33}^{(2)} - m_{32}a_{23}^{(2)}, \quad f_3^{(3)} = f_3^{(2)} - m_{32} f_2^{(2)}.$$

As a result, forward elimination reduces system (5.5) to the upper triangular system

$$\begin{aligned} a_{11}p_1 + a_{12}p_2 + a_{13}p_3 &= f_1, \\ a_{22}^{(2)} p_2 + a_{23}^{(2)} p_3 &= f_2^{(2)}, \\ a_{33}^{(3)} p_3 &= f_3^{(3)}. \end{aligned} \tag{5.10}$$

Now, backward substitution can easily solve for p_3, p_2, and p_1. Gaussian elimination works in the same way for a general $M \times M$ system.

For a general system, Gaussian elimination can be described more easily in terms of an LU factorization of matrix $\mathbf{A}$ as in the previous section. As noted, for a general positive definite matrix $\mathbf{A}$, it has the factorization (5.2), where $\mathbf{L} = (l_{ij})$ is a unit lower triangular matrix, i.e., $l_{ii} = 1$ and $l_{ij} = 0$ if $j > i$, and $\mathbf{U} = (u_{ij})$ is an upper triangular, i.e., $u_{ij} = 0$ if $j < i$. We compute $\mathbf{L}$ and $\mathbf{U} = \mathbf{A}^{(M)}$, where the matrices $\mathbf{A}^{(k)}$, $k = 1, 2, \ldots, M$, are successively calculated as follows:

Set $\mathbf{A}^{(1)} = \mathbf{A}$;

Given $\mathbf{A}^{(k)}$ of the form

$$\mathbf{A}^{(k)} = \begin{pmatrix} a_{11}^{(k)} & a_{12}^{(k)} & \dots & a_{1k}^{(k)} & \dots & a_{1M}^{(k)} \\ 0 & a_{22}^{(k)} & \dots & a_{2k}^{(k)} & \dots & a_{2M}^{(k)} \\ \vdots & \vdots & \ddots & \vdots & \ddots & \vdots \\ 0 & 0 & \dots & a_{kk}^{(k)} & \dots & a_{kM}^{(k)} \\ \vdots & \vdots & \ddots & \vdots & \ddots & \vdots \\ 0 & 0 & \dots & a_{Mk}^{(k)} & \dots & a_{MM}^{(k)} \end{pmatrix},$$

set $l_{ik} = -a_{ik}^{(k)}/a_{kk}^{(k)}, \quad i = k+1, k+2, \dots, M,$

calculate $\mathbf{A}^{(k+1)} = \left(a_{ij}^{(k+1)}\right)$ by

$$a_{ij}^{(k+1)} = a_{ij}^{(k)}, \quad i = 1, 2, \dots, k \text{ or } j = 1, 2, \dots, k-1,$$

$$a_{ij}^{(k+1)} = a_{ij}^{(k)} + l_{ik}a_{kj}^{(k)}, \quad i = k+1, \dots, M, \ j = k, \dots, M.$$

Obviously, Gaussian elimination requires that each diagonal entry $a_{kk}^{(k)}$ be nonzero. For the symmetric positive definite matrix $\mathbf{A}$, $a_{kk}^{(k)} > 0$, $k = 1, 2, \dots, M$. To minimize *round-off* errors, this entry should be chosen as large as possible. Partial *pivoting* means that at every stage of elimination one searches for the largest entry in magnitude among $a_{kk}^{(k)}, a_{k+1,k}^{(k)}, \dots, a_{Mk}^{(k)}$ and then interchanges the row with the largest entry with the kth row to maximize the diagonal entry. While pivoting may be required for ill-conditioned matrices, it is usually not necessary for matrices arising in reservoir simulation. For a theory on round-off errors of Gaussian elimination, the reader may refer to Higham (1996).

When $\mathbf{A}$ is symmetric, this matrix can be alternatively factorized as

$$\mathbf{A} = \mathbf{L}\mathbf{L}^T; \tag{5.11}$$

i.e.,

$$\sum_{k=1}^{j} l_{ik}l_{jk} = a_{ij}, \qquad j = 1, 2, \dots, i, \ i = 1, 2, \dots, M.$$

In this case, the entries l_{ij} of $\mathbf{L}$ in (5.11) can be computed directly using *Cholesky's approach*, $i = 1, 2, \dots, M$,

$$l_{ii} = \sqrt{a_{ii} - \sum_{k=1}^{i-1} l_{ik}^2},$$

$$l_{ij} = \left(a_{ij} - \sum_{k=1}^{j-1} l_{ik}l_{jk}\right) / l_{jj}, \qquad j = 1, 2, \dots, i-1.$$

Note that in the above computation of $\mathbf{L}$, M square root operations are required. To avoid this, we can write $\mathbf{L}$ as

$$\mathbf{L} = \tilde{\mathbf{L}}\mathbf{D}, \tag{5.12}$$

where $\tilde{\mathbf{L}}$ is a unit lower triangular matrix (i.e., $\tilde{l}_{ii} = 1, i = 1, 2, \ldots, M$) and $\mathbf{D}$ is a diagonal matrix:

$$\mathbf{D} = \operatorname{diag}\left(\sqrt{d_1}, \sqrt{d_2}, \ldots, \sqrt{d_M}\right).$$

In this factorization we see that

$$\sum_{k=1}^{j} \tilde{l}_{ik} d_k \tilde{l}_{jk} = a_{ij}, \qquad j = 1, 2, \ldots, i,\ i = 1, 2, \ldots, M,$$

which implies, for $i = 1, \ldots, M$,

$$\begin{aligned} d_i &= a_{ii} - \sum_{k=1}^{i-1} \tilde{l}_{ik}^2 d_k, \\ \tilde{l}_{ij} &= \left(a_{ij} - \sum_{k=1}^{j-1} \tilde{l}_{ik} d_k \tilde{l}_{jk}\right) / d_j, \qquad j = 1, 2, \ldots, i-1. \end{aligned} \tag{5.13}$$

The number of arithmetic operations in (5.13) is asymptotically of order $M^3/6$ (cf. Exercise 5.2). If the matrix $\mathbf{A}$ is sparse, one can greatly reduce the number of operations by exploiting the sparsity. This is the case when $\mathbf{A}$ is a *banded matrix*. In this case, for its ith row, there is an integer m_i such that

$$a_{ij} = 0 \quad \text{if } j < m_i, \quad i = 1, 2, \ldots, M.$$

Note that m_i is the column number of the first nonzero entry in the ith row. Then the *bandwidth* L_i of the ith row satisfies

$$L_i = i - m_i, \qquad i = 1, 2, \ldots, M.$$

We warn the reader that $2L_i + 1$ is sometimes called the bandwidth. It can be checked from (5.13) that $\mathbf{A}$ and $\tilde{\mathbf{L}}$ have the same value m_i. Thus, in the banded case, (5.13) can be modified to ($i = 1, 2, \ldots, M$)

$$\begin{aligned} d_i &= a_{ii} - \sum_{k=m_i}^{i-1} \tilde{l}_{ik}^2 d_k, \\ \tilde{l}_{ij} &= \left(a_{ij} - \sum_{k=\max(m_i, m_j)}^{j-1} \tilde{l}_{ik} d_k \tilde{l}_{jk}\right) \Big/ d_j, \\ &\qquad j = m_i, m_i - 1, \ldots, i-1. \end{aligned} \tag{5.14}$$

We remark that the number of arithmetic operations to factor a banded matrix is asymptotically of the order $ML^2/2$, where $L = \max_{1 \le i \le M} L_i$ (cf. Exercise 5.3). This

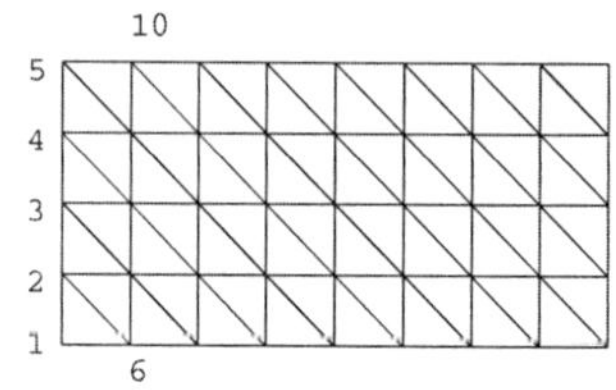

Figure 5.1. *An example of enumeration.*

number is much smaller than $M^3/6$ if L is smaller than M. For the finite element methods presented in Section 4.2, we have

$$a_{ij} = a(\varphi_i, \varphi_j), \qquad i,\ j = 1, 2, \ldots, M,$$

where $\{\varphi_i\}_{i=1}^M$ is a basis of V_h. Then we see that

$$L = \max\{|i - j| : \varphi_i \text{ and } \varphi_j \text{ correspond to degrees of freedom belonging to the same element}\}.$$

Consequently, the bandwidth depends on the enumeration of nodes. If direct elimination is used, the nodes should be enumerated in such a way that the bandwidth is as small as possible. For example, with a vertical enumeration of nodes in Figure 5.1, L is 5 (assuming that one degree of freedom is associated with each node). With a horizontal enumeration, L would be 10. The *standard* or *natural ordering* of unknowns is obtained if the unknowns are ordered by lines (vertically or horizontally); see Figure 5.1. There are other ordering methods that can save computational time and computer storage; see the next section.

Now, we return to (5.1) with the factorization (5.11) of $\mathbf{A}$, where $\mathbf{L}$ is given by (5.12). With this factorization, system (5.1) becomes

$$\begin{aligned} \tilde{\mathbf{L}}\mathbf{D}^2\mathbf{v} &= \mathbf{f}, \\ \tilde{\mathbf{L}}^T\mathbf{p} &= \mathbf{v}. \end{aligned} \tag{5.15}$$

We emphasize that these systems are triangular. The first system is

$$\sum_{k=1}^{i} \tilde{l}_{ik} d_k v_k = f_i, \qquad i = 1, 2, \ldots, M.$$

Thus forward elimination implies

$$v_1 = \frac{f_1}{d_1}, \quad v_i = \frac{f_i - \sum_{k=1}^{i-1} \tilde{l}_{ik} d_k v_k}{d_i}, \quad i = 2, 3, \ldots, M. \tag{5.16}$$

Similarly, the second system is solved by backward substitution:

$$\begin{aligned} p_M = v_M, \quad p_i = v_i - \sum_{k=i+1}^{M} \tilde{l}_{ki}\, p_k, \\ i = M-1, M-2, \ldots, 1. \end{aligned} \tag{5.17}$$

If **A** is banded, we apply (5.14) to (5.16) to give

$$v_1 = \frac{f_1}{d_1}, \quad v_i = \frac{f_i - \sum_{k=m_i}^{i-1} \tilde{l}_{ik} d_k v_k}{d_i}, \quad i = 2, 3, \ldots, M.$$

Also, it follows from (5.17) that

$$\begin{aligned} p_M &= v_M, \\ p_{M-1} &= v_{M-1} - \tilde{l}_{M,M-1}\, p_M, \\ p_{M-2} &= v_{M-2} - \tilde{l}_{M-1,M-2}\, p_{M-1} - \tilde{l}_{M,M-2}\, p_M, \\ &\vdots \\ p_1 &= v_1 - \tilde{l}_{2,1}\, p_2 - \tilde{l}_{3,1}\, p_3 - \cdots - \tilde{l}_{M,1}\, p_M. \end{aligned}$$

Note that one subtracts $\tilde{l}_{M,k}\, p_M$ from v_k, $k = M-1, M-2, \ldots, 1$. Due to the banded structure of **A**, i.e.,

$$\tilde{l}_{M,k} = 0 \quad \text{if } k < m_M,$$

$\tilde{l}_{M,k}\, p_M$ is subtracted from v_k only when $k \geq m_M$. As a result, one can first find v_k successively by

$$v_k := v_k - \tilde{l}_{ik}\, p_i, \quad k = m_i, m_i + 1, \ldots, i-1, \; i = M, M-1, \ldots, 1,$$

and then

$$p_i = v_i, \qquad i = M, M-1, \ldots, 1.$$

5.3 Ordering of the Nodes

As noted in the previous section, the form of the stiffness matrix **A** depends on the ordering of the nodes. Different orderings of nodes have been in use for a long time in connection with finite differences. Classical orderings include lexicographical, rotated lexicographical, red-black (chequerboard), zebra-line, and four-color orderings (Hackbusch, 1985). In this section, we very briefly touch on a few common ordering techniques used in finite difference reservoir simulation (Price and Coats, 1974). These techniques can be extended to the finite element setting. For simplicity, we consider a triangulation of a reservoir domain Ω into triangles, and assume that one degree of freedom is associated with each node (cf. Figure 5.1).

For a two-dimensional problem the work requirement for standard Gaussian elimination can be written in terms of the total number of nodes in the x_1-direction (I) and the total number of nodes in the x_2-direction (J). If $J < I$ (cf. Figure 5.2), then the work W for Gaussian elimination in the standard ordering (Price and Coats, 1974) is

$$W = \mathcal{O}\big((IJ)J^2\big),$$

and the corresponding storage requirement is

$$S = \mathcal{O}\big((IJ)J\big).$$

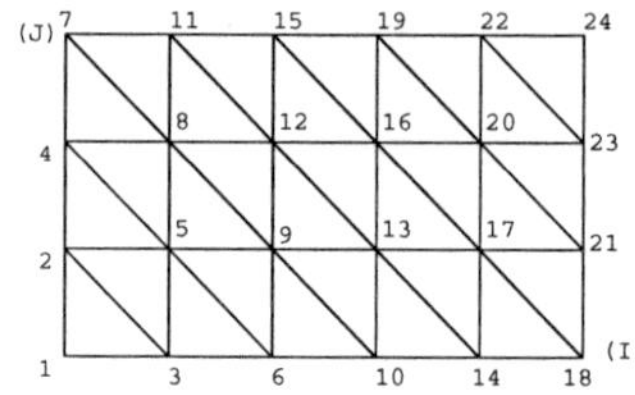

Figure 5.2. *A D_2 ordering.*

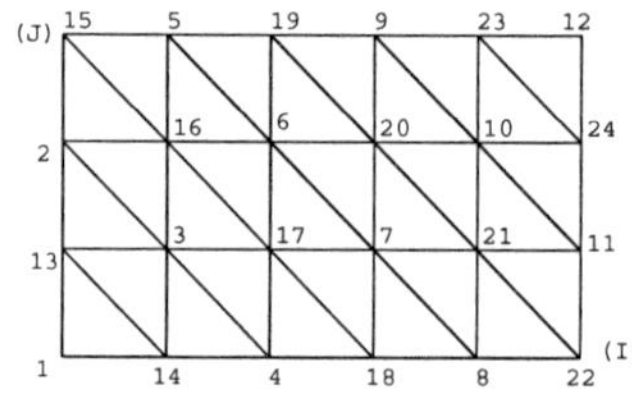

Figure 5.3. *A D_4 ordering.*

For the diagonal (called D_2) ordering shown in Figure 5.2 and $J < I$, the work W and storage S are (Price and Coats, 1974)

$$W = \mathcal{O}\left(IJ^3 - \frac{J^4}{2}\right), \quad S = \mathcal{O}\left(IJ^2 - \frac{J^3}{3}\right).$$

In the case $I = J$, this ordering method roughly requires one-half the work and two-thirds the storage of the standard ordering.

For the alternating diagonal (called D_4) ordering shown in Figure 5.3 and $J < I$, the estimates for W and S are (Price and Coats, 1974)

$$W = \mathcal{O}\left(\frac{IJ^3}{2} - \frac{J^4}{4}\right), \quad S = \mathcal{O}\left(\frac{IJ^2}{2} - \frac{J^3}{6}\right).$$

Then we see that in the case $I = J$, the D_4 ordering roughly needs one-quarter the work and one-third storage of the standard ordering. Therefore, among the three ordering techniques, D_4 is the most superior in terms of computational time and computer storage.

Having observed the advantage of the D_4 ordering, we now consider its implementation. The matrix $\mathbf{A}$ for this ordering is of the type shown in Figure 5.4, which can be written in the block form

$$\mathbf{Ap} = \begin{pmatrix} \mathbf{A}_{11} & \mathbf{A}_{12} \\ \mathbf{A}_{21} & \mathbf{A}_{22} \end{pmatrix} \begin{pmatrix} \mathbf{p}_1 \\ \mathbf{p}_2 \end{pmatrix} = \begin{pmatrix} \mathbf{f}_1 \\ \mathbf{f}_2 \end{pmatrix},$$

where $\mathbf{A}_{11}$ and $\mathbf{A}_{22}$ are diagonal matrices and $\mathbf{A}_{12}$ and $\mathbf{A}_{21}$ are sparse matrices. Because $\mathbf{A}_{11}$ is diagonal, performing forward elimination on the lower half of $\mathbf{A}$ gives

$$\mathbf{Ap} = \begin{pmatrix} \mathbf{A}_{11} & \mathbf{A}_{12} \\ \mathbf{0} & \bar{\mathbf{A}}_{22} \end{pmatrix} \begin{pmatrix} \mathbf{p}_1 \\ \mathbf{p}_2 \end{pmatrix} = \begin{pmatrix} \mathbf{f}_1 \\ \bar{\mathbf{f}}_2 \end{pmatrix}, \tag{5.18}$$

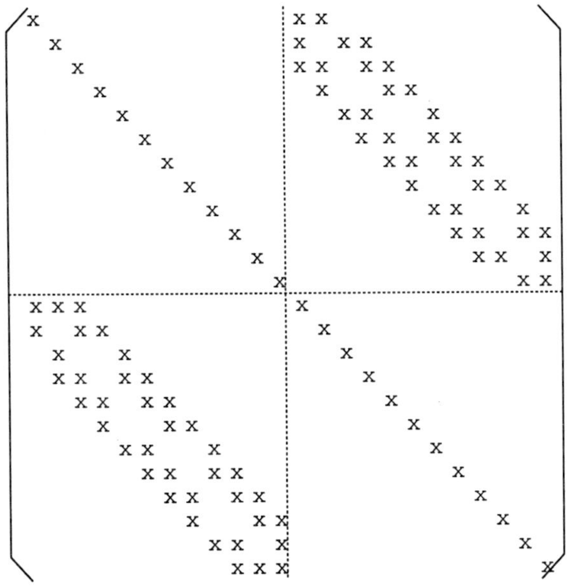

Figure 5.4. *Matrix* **A** *in the* D_4 *ordering.*

where $\bar{\mathbf{A}}_{22} = \mathbf{A}_{22} - \mathbf{A}_{11}^{-1}\mathbf{A}_{12}$ and $\bar{\mathbf{f}}_2 = \mathbf{f}_2 - \mathbf{A}_{11}^{-1}\mathbf{f}_1$. We now solve the equations for the lower half

$$\bar{\mathbf{A}}_{22}\mathbf{p}_2 = \bar{\mathbf{f}}_2. \tag{5.19}$$

After $\mathbf{p}_2$ is computed, $\mathbf{p}_1$ can be recovered by back substitution

$$\mathbf{p}_1 = \mathbf{A}_{11}^{-1}\left(\mathbf{f}_1 - \mathbf{A}_{12}\mathbf{p}_2\right). \tag{5.20}$$

Compared with the original problem (5.1), the size of system (5.19) is reduced by half. Hence the work for half the unknowns will be reduced by a factor of two for a constant bandwidth matrix.

5.4 CG

We recall that the *condition number* of matrix **A** is defined by

$$\text{cond}(\mathbf{A}) = \|\mathbf{A}\|\,\|\mathbf{A}^{-1}\|,$$

where $\|\mathbf{A}\|$ is the matrix norm of **A** induced by a norm $\|\cdot\|$ on $\mathbb{R}^M$ (e.g., the l^2-norm $\|\cdot\|_2$ on $\mathbb{R}^M$: $\|\mathbf{v}\|_2 = (\sum_{i=1}^{M}|v_i|^2)^{1/2}$, $\mathbf{v} = (v_1, v_2, \ldots, v_M)$). Here cond(**A**) is understood to be infinite if **A** is singular. The matrix **A** in system (5.1) arising from the standard finite element discretization of a second-order elliptic problem, for example, has a condition number proportional to h^{-2} (Johnson, 1994; Chen, 2005) as $h \to 0$, where h is the spatial mesh size. For the application of the finite element method to a large-scale problem, it would be very expensive to solve the resulting system of equations via Gaussian elimination discussed in the previous sections. The usual practice for obtaining the solution of a large-scale system is to use an iterative algorithm.

It is beyond the scope of this book to provide even a brief introduction to all available iterative algorithms for the solution of system (5.1). Some simple iterative algorithms such as *stationary* point and block Jacobi, Gauss–Seidel, and successive over relaxation (SOR) algorithms as applied to reservoir simulation were discussed by Peaceman (1977) and Aziz and Settari (1979) in the finite difference setting. In this chapter, we study Krylov subspace algorithms for linear systems. The two such algorithms we study in depth are the CG and GMRES algorithms. Because the ORTHOMIN algorithm has been widely employed in reservoir simulation, we also briefly discuss this algorithm.

CG was introduced by Hestenes and Stiefel in 1952 as a direct algorithm. It has been in wide use as an iterative algorithm, and has generally superseded the Jacobi, Gauss–Seidel, and SOR iterative algorithms.

Unlike the stationary iterative algorithms, the Krylov subspace algorithms do not have an iteration matrix. They minimize, at the kth iteration, some measure of errors over the affine space

$$\mathbf{p}^0 + \mathcal{K}^k,$$

where $\mathbf{p}^0$ is an initial guess to (5.1) and the kth *Krylov space* $\mathcal{K}^k$ is defined by

$$\mathcal{K}^k = \text{span}\left(\mathbf{r}^0, \mathbf{A}\mathbf{r}^0, \ldots, \mathbf{A}^{k-1}\mathbf{r}^0\right), \qquad k \geq 1.$$

The residual $\mathbf{r}^k$ for the kth iterate $\mathbf{p}^k$ is

$$\mathbf{r}^k = \mathbf{f} - \mathbf{A}\mathbf{p}^k, \qquad k \geq 0.$$

If $\mathbf{A}$ is symmetric positive definite, it deduces a scalar product $\langle \cdot, \cdot \rangle$ on $\mathbb{R}^M$:

$$\langle \mathbf{v}, \mathbf{w} \rangle = \mathbf{v}^T \mathbf{A} \mathbf{w} = \sum_{i,j=1}^{M} v_i a_{ij} w_j, \qquad \mathbf{v}, \mathbf{w} \in \mathbb{R}^M.$$

The norm $\|\cdot\|_{\mathbf{A}}$ corresponding to $\langle \cdot, \cdot \rangle$ is the *energy norm*

$$\|\mathbf{v}\|_{\mathbf{A}} = \langle \mathbf{v}, \mathbf{v} \rangle^{1/2}, \qquad \mathbf{v} \in \mathbb{R}^M.$$

The kth iterate $\mathbf{p}^k$ of CG minimizes the functional

$$F(\mathbf{p}) = \frac{1}{2} \langle \mathbf{p}, \mathbf{p} \rangle - \mathbf{p}^T \mathbf{f}$$

over $\mathbf{p}^0 + \mathcal{K}^k$. Note that if $F(\mathbf{p}^*)$ is the minimal value in $\mathbb{R}^M$, then

$$\nabla F(\mathbf{p}^*) = \mathbf{A}\mathbf{p}^* - \mathbf{f} = \mathbf{0};$$

i.e., $\mathbf{p}^*$ is the solution.

Given $\mathbf{p}^0$, CG seeks successive approximations $\mathbf{p}^k$ of the form

$$\mathbf{p}^k = \mathbf{p}^{k-1} + \alpha_{k-1} \mathbf{d}^{k-1}, \qquad k = 1, 2, \ldots, \tag{5.21}$$

CG Algorithm

Given $\mathbf{p}^0 \in \mathbb{R}^M$, set $\mathbf{r}^0 = \mathbf{f} - \mathbf{A}\mathbf{p}^0$ and $\mathbf{d}^0 = \mathbf{r}^0$.
For $k = 1, 2, \ldots,$ determine $\mathbf{p}^k$ and $\mathbf{d}^k$ by

$$\alpha_{k-1} = \frac{(\mathbf{r}^{k-1})^T \mathbf{r}^{k-1}}{\langle \mathbf{d}^{k-1}, \mathbf{d}^{k-1} \rangle};$$

$$\mathbf{p}^k = \mathbf{p}^{k-1} + \alpha_{k-1}\mathbf{d}^{k-1};$$

$$\mathbf{r}^k = \mathbf{r}^{k-1} - \alpha_{k-1}\mathbf{A}\mathbf{d}^{k-1};$$

$$\beta_{k-1} = \frac{(\mathbf{r}^k)^T \mathbf{r}^k}{(\mathbf{r}^{k-1})^T \mathbf{r}^{k-1}};$$

$$\mathbf{d}^k = \mathbf{r}^k + \beta_{k-1}\mathbf{d}^{k-1}.$$

Figure 5.5. *The algorithm CG.*

where $\mathbf{d}^{k-1}$ is a *search direction* and α_{k-1} is a *step length*. Once $\mathbf{d}^{k-1}$ is found, α_{k-1} is easy to compute from the minimization property of the iteration:

$$\left. \frac{dF(\mathbf{p}^{k-1} + \alpha \mathbf{d}^{k-1})}{d\alpha} \right|_{\alpha = \alpha_{k-1}} = 0.$$

The search directions $\mathbf{d}^{k-1}$ are supposed to satisfy the $\mathbf{A}$-*conjugacy* condition

$$\left(\mathbf{d}^{k_1}\right)^T \mathbf{A}\mathbf{d}^{k_2} = 0 \qquad \text{if } k_1 \neq k_2.$$

The usual CG implementation reflects the minimization property and the $\mathbf{A}$-conjugacy condition. The input for the CG algorithm is the initial iterate $\mathbf{p}^0$, which can be overwritten by the solution, the right hand side $\mathbf{f}$, and the coefficient matrix $\mathbf{A}$ (or a routine that computes the action of $\mathbf{A}$ on a vector). Then this algorithm for the solution of (5.1) can be defined as in Figure 5.5.

The matrix $\mathbf{A}$ itself need not be formed or stored; only a routine for matrix-vector products is required. For this reason, the Krylov space algorithms are usually called the *matrix-free* algorithms.

It can be shown that the CG algorithm gives, in the absence of round-off errors, the exact solution after at most M steps; i.e.,

$$\mathbf{A}\mathbf{p}^k = \mathbf{f} \qquad \text{for some } k \leq M.$$

In practice, the required number of iterations is sometimes smaller than M. In fact, for a given *tolerance* $\epsilon > 0$, to satisfy

$$\|\mathbf{p} - \mathbf{p}^k\|_{\mathbf{A}} \leq \epsilon \|\mathbf{p} - \mathbf{p}^0\|_{\mathbf{A}}$$

it suffices to choose k such that (Axelsson, 1994)

$$k \geq \frac{1}{2}\sqrt{\text{cond}(\mathbf{A})}\,\ln\frac{2}{\epsilon}.$$

Hence the required number of iterations for the CG algorithm is proportional to $\sqrt{\text{cond}(\mathbf{A})}$. As shown above, in a typical finite element application to a second-order elliptic problem, $\text{cond}(\mathbf{A}) = \mathcal{O}(h^{-2})$, and so the required number of iterations is of order $\mathcal{O}(h^{-1})$.

5.5 GMRES

Systems of algebraic equations arising from the discretization of the governing equations in reservoir simulation have special properties. The coefficient (stiffness) matrices of these systems are sparse but nonsymmetric and indefinite. While sparse, their natural banded structure is usually spoiled by wells that perforate into many gridblocks and/or by irregular gridblock structure. For such systems, the CG algorithm can suffer severe deterioration in performance.

Over a dozen parameter-free Krylov subspace algorithms have been proposed for solving nonsymmetric systems of linear equations. Three leading iterative algorithms are the CGN (the CG iteration applied to the normal equations (cf. Hestenes and Stiefel, 1952)), GMRES (residual minimization in a Krylov space (cf. Kuznetsov, 1969; Saad and Schultz, 1986)), and BiCGSTAB (a biorthogonalization method adapted from the biconjugate gradient iteration (cf. van der Vorst, 1992)). These three algorithms differ fundamentally in their capabilities. As shown by Nachtigal et al. (1992), examples of matrices can be constructed to show that each type of iteration can outperform the others by a factor on the order of $\sqrt{M}$ or M (or even more). As examples, in this chapter we study GMRES and BiCGSTAB.

The GMRES algorithm is known to be a very efficient algorithm for solving general sparse, nonsymmetric systems (Kuznetsov, 1969; Saad and Schultz, 1986). The kth iterate of GMRES is the solution to the least squares problem

$$\min_{\mathbf{p}\in\mathbf{p}^0+\mathcal{K}^k} \|\mathbf{f} - \mathbf{A}\mathbf{p}\|_2. \tag{5.22}$$

Suppose that one has an orthogonal projector $\mathbf{V}^k$ onto $\mathcal{K}^k$. Then any $\mathbf{z} \in \mathcal{K}^k$ can be represented:

$$\mathbf{z} = \sum_{i=1}^{k} q_i \mathbf{v}^i$$

for some $\mathbf{q} = (q_1, q_2, \ldots, q_k)^T \in \mathbb{R}^k$, where $\mathbf{v}^i$ is the ith column of $\mathbf{V}^k$. Set

$$\mathbf{p} - \mathbf{p}^0 = \mathbf{V}^k\mathbf{q}$$

for some $\mathbf{q} \in \mathbb{R}^k$. Since

$$\mathbf{f} - \mathbf{A}\mathbf{p} = \mathbf{f} - \mathbf{A}\mathbf{p}^0 - \mathbf{A}\mathbf{V}^k\mathbf{q} = \mathbf{r}^0 - \mathbf{A}\mathbf{V}^k\mathbf{q},$$

problem (5.22) can be converted to the least squares problem

$$\min_{\mathbf{q}\in\mathbb{R}^k} \|\mathbf{r}^0 - \mathbf{A}\mathbf{V}^k\mathbf{q}\|_2. \tag{5.23}$$

Arnoldi's Algorithm

Given $\mathbf{p}^0$, set $\mathbf{r}^0 = \mathbf{f} - \mathbf{A}\mathbf{p}^0$ and $\mathbf{v}^1 = \mathbf{r}^0/\|\mathbf{r}^0\|_2$.
For $j = 1, 2, \ldots, k$, compute

$$h_{ij} = (\mathbf{v}^i)^T \mathbf{A}\mathbf{v}^j \text{ for } i = 1, 2, \ldots, j;$$

$$\mathbf{w}^j = \mathbf{A}\mathbf{v}^j - \sum_{i=1}^{j} h_{ij}\mathbf{v}^i;$$

$$h_{j+1,j} = \|\mathbf{w}^j\|_2;$$

If $h_{j+1,j} = 0$, then stop;

$$\mathbf{v}^{j+1} = \mathbf{w}^j / h_{j+1,j}.$$

Figure 5.6. *The Arnoldi algorithm.*

This is a standard least squares problem that can be solved by QR factorization, for example. The problem with such a direct method is that the matrix vector product of $\mathbf{A}$ with $\mathbf{V}^k$ must be performed at each iteration.

If the Gram–Schmidt orthogonalization technique is applied to (5.23), the resulting least squares problem does not require any extra product of $\mathbf{A}$ with vectors. The technique for constructing an orthonormal basis for $\mathcal{K}^k$ is referred to as the *Arnoldi algorithm* (Arnoldi, 1951); cf. Figure 5.6. The input data for this algorithm are $\mathbf{p}^0$, $\mathbf{f}$, $\mathbf{A}$, and a dimension k.

If the Arnoldi algorithm does not stop before the kth step, the vectors $\mathbf{v}^1, \mathbf{v}^2, \ldots, \mathbf{v}^k$ form an orthonormal basis for $\mathcal{K}^k$ (cf. Exercise 5.4). Denote by $\mathbf{V}^k$ the $M \times k$ matrix with these column vectors, and by $\mathbf{H}^k$ the $(k+1) \times k$ *upper Hessenberg matrix* whose nonzero entries h_{ij} are computed by the Arnoldi algorithm. This algorithm (unless it terminates prematurely with a solution) generates the relation (cf. Exercise 5.5)

$$\mathbf{A}\mathbf{V}^k = \mathbf{V}^{k+1}\mathbf{H}^k. \tag{5.24}$$

Let $\mathbf{e}_1 = (1, 0, \ldots, 0)^T \in \mathbb{R}^{k+1}$ and $\beta = \|\mathbf{r}^0\|_2$. For the kth iterate $\mathbf{p}^k$ of GMRES, define

$$\mathbf{p}^k = \mathbf{p}^0 + \mathbf{V}^k\mathbf{q}^k \tag{5.25}$$

for some $\mathbf{q}^k \in \mathbb{R}^k$. Then it follows from (5.24) and (5.25) that

$$\mathbf{r}^k = \mathbf{f} - \mathbf{A}\mathbf{p}^k = \mathbf{r}^0 - \mathbf{A}(\mathbf{p}^k - \mathbf{p}^0) = \mathbf{V}^{k+1}\left(\beta\mathbf{e}_1 - \mathbf{H}^k\mathbf{q}^k\right).$$

Using the orthogonality of $\mathbf{V}^{k+1}$,

$$\|\mathbf{r}^k\|_2 = \left\|\mathbf{V}^{k+1}\left(\beta\mathbf{e}_1 - \mathbf{H}^k\mathbf{q}^k\right)\right\|_2 = \|\beta\mathbf{e}_1 - \mathbf{H}^k\mathbf{q}^k\|_2.$$

That is, $\mathbf{q}^k$ minimizes $\|\beta\mathbf{e}_1 - \mathbf{H}^k\mathbf{q}^k\|_2$. The minimizer $\mathbf{q}^k$ is inexpensive to obtain because it requires the solution of a $(k+1) \times k$ least squares problem when k is small.

GMRES Algorithm

Given $\mathbf{p}^0 \in \mathbb{R}^M$, set $\mathbf{r}^0 = \mathbf{f} - \mathbf{A}\mathbf{p}^0$, $\beta = \|\mathbf{r}^0\|_2$, and $\mathbf{v}^1 = \mathbf{r}^0/\beta$.
For the $(k+1) \times k$ matrix $\mathbf{H}^k = (h_{ij})$, set $\mathbf{H}^k = \mathbf{0}$.
For $j = 1, 2, \dots, k$, compute

$\mathbf{w}^j = \mathbf{A}\mathbf{v}^j$;

$h_{ij} = (\mathbf{v}^i)^T \mathbf{w}^j$ for $i = 1, 2, \dots, j$;

$$\mathbf{w}^j = \mathbf{w}^j - \sum_{i=1}^{j} h_{ij}\mathbf{v}^i;$$

$h_{j+1,j} = \|\mathbf{w}^j\|_2$;

If $h_{j+1,j} = 0$, set $k = j$ and skip the next step;

$\mathbf{v}^{j+1} = \mathbf{w}^j / h_{j+1,j}$.

Determine the minimizer $\mathbf{q}^k$ of $\|\beta \mathbf{e}_1 - \mathbf{H}^k \mathbf{q}^k\|_2$.
Set $\mathbf{p}^k = \mathbf{p}^0 + \mathbf{V}^k \mathbf{q}^k$.

Figure 5.7. *The GMRES algorithm.*

The input data for GMRES are $\mathbf{p}^0$, $\mathbf{f}$, and $\mathbf{A}$ (or a routine that computes the action of $\mathbf{A}$ on a vector); cf. Figure 5.7.

As for CG, if $\mathbf{A}$ is nonsingular, the GMRES algorithm will find, in the absence of round-off errors, the solution within M iterations. To obtain more precise information on convergence rates, we consider the case where $\mathbf{A}$ is diagonalizable. Recall that $\mathbf{A}$ is *diagonalizable* if there is a nonsingular matrix $\mathbf{E}$ such that

$$\mathbf{A} = \mathbf{E}\mathbf{\Lambda}\mathbf{E}^{-1},$$

where $\mathbf{\Lambda}$ is a diagonal matrix with the eigenvalues of $\mathbf{A}$ on its diagonal. In this case, the kth GMRES iterate $\mathbf{p}^k$ satisfies (Saad, 2004)

$$\frac{\|\mathbf{r}^k\|_2}{\|\mathbf{r}^0\|_2} \le \text{cond}(\mathbf{E}) \inf_{p_k \in P_k, p_k(0)=1} \left\{ \max_{z \in \sigma(\mathbf{A})} |p_k(z)| \right\}, \tag{5.26}$$

where $\text{cond}(\mathbf{E})$ is the condition number of $\mathbf{E}$, P_k is the set of polynomials of degree at most k, and $\sigma(\mathbf{A})$ is the set of eigenvalues of $\mathbf{A}$ (the *spectrum* of $\mathbf{A}$). It is unclear how to estimate $\text{cond}(\mathbf{E})$. If $\mathbf{A}$ is normal, of course, $\text{cond}(\mathbf{E}) = 1$.

In the GMRES algorithm, $\mathbf{p}^k$ is evaluated only upon termination and is not required within the iteration. It is important that the basis for the Krylov space must be stored as the iteration progresses. This implies that to perform k GMRES iterations, k vectors of length M must be stored and that GMRES becomes impractical when k is large because of computer memory requirements. There are two remedies. The first is to "truncate" the orthogonalization in the Arnoldi algorithm; i.e., an integer k is selected and fixed, and an

GCR Algorithm

Given $\mathbf{p}^0 \in \mathbb{R}^M$, set $\mathbf{r}^0 = \mathbf{f} - \mathbf{A}\mathbf{p}^0$ and $\mathbf{d}^0 = \mathbf{r}^0$.

For $k = 1, 2, \ldots,$ compute $\mathbf{p}^k$ and $\mathbf{d}^k$ by

$$\alpha_{k-1} = \frac{(\mathbf{r}^{k-1})^T \left(\mathbf{A}\mathbf{d}^{k-1}\right)}{\left(\mathbf{A}\mathbf{d}^{k-1}\right)^T \left(\mathbf{A}\mathbf{d}^{k-1}\right)};$$

$$\mathbf{p}^k = \mathbf{p}^{k-1} + \alpha_{k-1}\mathbf{d}^{k-1};$$

$$\mathbf{r}^k = \mathbf{r}^{k-1} - \alpha_{k-1}\mathbf{A}\mathbf{d}^{k-1};$$

$$\beta_{i,k-1} = -\frac{\left(\mathbf{A}\mathbf{r}^k\right)^T \left(\mathbf{A}\mathbf{d}^i\right)}{\left(\mathbf{A}\mathbf{d}^i\right)^T \left(\mathbf{A}\mathbf{d}^i\right)} \qquad \text{for } i = 1, 2, \ldots, k-1;$$

$$\mathbf{d}^k = \mathbf{r}^k + \sum_{i=1}^{k-1} \beta_{i,k-1}\mathbf{d}^i.$$

Figure 5.8. *The GCR algorithm.*

"incomplete" orthogonalization is performed, which will be described in the next section, in connection with ORTHOMIN. The second remedy is to *restart* the iteration after every k steps for some integer k (e.g., 5, 10, or 20), with $\mathbf{p}^k$ used as the initial guess in the next iteration. This restarted version of the algorithm is termed GMRES(k) (Saad and Schultz, 1986). There is no general convergence theory for restarted GMRES; for a positive definite matrix $\mathbf{A}$, however, GMRES(k) converges for any $k \geq 1$. Restarting will slow convergence; when it works, however, it will significantly reduce storage.

5.6 ORTHOMIN

The ORTHOMIN algorithm (Vinsome, 1976) has been applied to reservoir simulation and is still widely used in this area due to its ability to solve efficiently nonsymmetric, sparse systems of algebraic equations. In this section, we briefly discuss this algorithm; comparison with GMRES will be presented in Section 5.11. ORTHOMIN is a truncated version of the GCR (generalized conjugate residual) algorithm. Hence, to introduce ORTHOMIN, we first describe GCR.

The two algorithms, CG and GMRES, are based on the choice of a basis of the Krylov subspace $\mathcal{K}^k$. In CG, the search directions $\mathbf{d}^k$ are $\mathbf{A}$-orthogonal, i.e., conjugate. GMRES utilizes an orthogonal basis of $\mathcal{K}^k$. In fact, all Krylov subspace algorithms are strongly related to the choice of a basis of this Krylov subspace. In GCR, for example, the $\mathbf{d}^k$'s are sought to be $\mathbf{A}^T\mathbf{A}$-orthogonal, and the algorithm can be defined as in Figure 5.8.

Compared with the CG algorithm in Section 5.4, the $\mathbf{d}^k$'s are now $\mathbf{A}^T\mathbf{A}$-orthogonal, as noted. Also, to compute the scalars $\beta_{i,k-1}$ in GCR, the vector $\mathbf{A}\mathbf{r}^k$ and the previous $\mathbf{A}\mathbf{d}^i$'s are required. To limit the number of matrix-vector products per step to one, we can proceed

ORTHOMIN (m) Algorithm

Given $\mathbf{p}^0 \in \mathbb{R}^M$ and m, set $\mathbf{r}^0 = \mathbf{f} - \mathbf{A}\mathbf{p}^0$ and $\mathbf{d}^0 = \mathbf{r}^0$.
For $k = 1, 2, \ldots,$ compute $\mathbf{p}^k$ and $\mathbf{d}^k$ by

$$\alpha_{k-1} = \frac{(\mathbf{r}^{k-1})^T \left(\mathbf{A}\mathbf{d}^{k-1}\right)}{\left(\mathbf{A}\mathbf{d}^{k-1}\right)^T \left(\mathbf{A}\mathbf{d}^{k-1}\right)};$$

$$\mathbf{p}^k = \mathbf{p}^{k-1} + \alpha_{k-1}\mathbf{d}^{k-1};$$

$$\mathbf{r}^k = \mathbf{r}^{k-1} - \alpha_{k-1}\mathbf{A}\mathbf{d}^{k-1};$$

$$\beta_{i,k-1} = -\frac{\left(\mathbf{A}\mathbf{r}^k\right)^T \left(\mathbf{A}\mathbf{d}^i\right)}{\left(\mathbf{A}\mathbf{d}^i\right)^T \left(\mathbf{A}\mathbf{d}^i\right)} \qquad \text{for } i = k - m, 2, \ldots, k - 1;$$

$$\mathbf{d}^k = \mathbf{r}^k + \sum_{i=k-m}^{k-1} \beta_{i,k-1}\mathbf{d}^i.$$

Figure 5.9. *The algorithm ORTHOMIN(m).*

as follows: Follow the computation of $\mathbf{r}^k$ by a calculation of $\mathbf{A}\mathbf{r}^k$ and then calculate $\mathbf{A}\mathbf{d}^k$ after the last line of the GCR algorithm from the equation

$$\mathbf{A}\mathbf{d}^k = \mathbf{A}\mathbf{r}^k + \sum_{i=1}^{k-1} \beta_{i,k-1}\mathbf{A}\mathbf{d}^i.$$

Both the set of the $\mathbf{d}^i$'s and that of the $\mathbf{A}\mathbf{d}^i$'s need to be stored. This doubles the storage requirement compared with CG (and GMRES). The number of arithmetic operations per iteration is also roughly 50% higher than for GMRES.

GCR suffers from the same practical limitations as GMRES. A restarted version GCR(k) can be defined trivially in the same way as GMRES(k). A truncation of the orthogonalization of the $\mathbf{A}\mathbf{d}^i$'s leads to the algorithm ORTHOMIN(m) for a given choice of m ($1 \leq m < k$); cf. Figure 5.9.

ORTHOMIN generally requires more arithmetic operations and computer storage per iteration step than GMRES does. In Section 5.11, comparisons between these two algorithms will be described for examples in numerical reservoir simulation.

5.7 BiCGSTAB

The previous three sections dealt with four Krylov subspace algorithms that rely on some form of orthogonalization of the Krylov vectors to obtain an approximate solution. This section considers a family of Krylov subspace algorithms that are instead defined by a biorthogonalization approach due to Lanczos (1952). These algorithms are projection methods that are intrinsically nonorthogonal. They have some appealing properties but are more difficult to analyze theoretically.

BiCGSTAB Algorithm

Given $\mathbf{p}^0 \in \mathbb{R}^M$, set $\mathbf{r}^0 = \mathbf{f} - \mathbf{A}\mathbf{p}^0$ and $\mathbf{d}^0 = \mathbf{r}^0$; $\hat{\mathbf{r}}^0$ arbitrary.
For $k = 1, 2, \ldots,$ compute $\mathbf{p}^k$ and $\mathbf{d}^k$ by

$$\alpha_{k-1} = \frac{(\mathbf{r}^{k-1})^T \hat{\mathbf{r}}^0}{\left(\mathbf{A}\mathbf{d}^{k-1}\right)^T \hat{\mathbf{r}}^0};$$

$$\mathbf{p}_2^{k-1} = \mathbf{r}^{k-1} - \alpha_{k-1}\mathbf{A}\mathbf{d}^{k-1};$$

$$\omega_{k-1} = \frac{\left(\mathbf{A}\mathbf{p}_2^{k-1}\right)^T \mathbf{p}_2^{k-1}}{\left(\mathbf{A}\mathbf{p}_2^{k-1}\right)^T \left(\mathbf{A}\mathbf{p}_2^{k-1}\right)};$$

$$\mathbf{p}^k = \mathbf{p}^{k-1} + \alpha_{k-1}\mathbf{d}^{k-1} + \omega_{k-1}\mathbf{p}_2^{k-1};$$

$$\mathbf{r}^k = \mathbf{p}_2^{k-1} - \omega_{k-1}\mathbf{A}\mathbf{p}_2^{k-1};$$

$$\beta_{k-1} = \frac{(\mathbf{r}^k)^T \hat{\mathbf{r}}^0}{(\mathbf{r}^{k-1})^T \hat{\mathbf{r}}^0} \frac{\alpha_{k-1}}{\omega_{k-1}};$$

$$\mathbf{d}^k = \mathbf{r}^k + \beta_{k-1}\left(\mathbf{d}^{k-1} - \omega_{k-1}\mathbf{A}\mathbf{d}^{k-1}\right).$$

Figure 5.10. *The algorithm BiCGSTAB.*

The earliest such method is the BCG (biconjugate gradient) algorithm (Lanczos, 1952). BCG does not enforce a minimization principle; instead, the kth residual must satisfy the *biorthogonality* condition

$$(\mathbf{r}^k)^T \mathbf{v} = 0 \qquad \forall\, \mathbf{v} \in \widehat{\mathcal{K}}^k,$$

where the Krylov space $\widehat{\mathcal{K}}^k$ of $\mathbf{A}^T$ is defined by

$$\widehat{\mathcal{K}}^k = \operatorname{span}\left(\mathbf{r}^0, \mathbf{A}^T\mathbf{r}^0, \ldots, \left(\mathbf{A}^T\right)^{k-1}\mathbf{r}^0\right).$$

A problem with BCG is that a transpose-vector product is needed, which at best will require additional programming and, at worst, may not be feasible. A remedy for this problem is the CGS (conjugate gradient squared) algorithm (Sonneveld, 1989). CGS replaces the transpose-vector product with an additional matrix-vector product and is based on squaring the residual polynomial. A problem with this approach is that substantial rounding errors can build up. BiCGSTAB (van der Vorst, 1992) was developed to overcome this difficulty and to smooth convergence of CGS; cf. Figure 5.10.

There is no convergence theory for BiCGSTAB. The iteration can break down in the steps computing the coefficients α_{k-1} and β_{k-1}. The cost in storage and in floating point operations per iteration remains bounded in the entire iteration. A single iteration requires four scalar products. In the case where many GMRES iterations are needed and a matrix-vector product is fast, BiCGSTAB can have a much lower average cost per iteration than

GMRES. The reason is that the cost of orthogonalization in the latter algorithm can be much higher than that of the matrix-vector product in BiCGSTAB if the dimension of the Krylov space is large.

5.8 Preconditioned Iterations

To reduce the condition number of matrix $\mathbf{A}$, and thus to improve the performance of the iterative algorithms developed in the previous four sections, one can replace system (5.1) with another system that has the same solution. In practice, all the Krylov subspace algorithms are often useless without *preconditioning*. This section discusses the preconditioned versions of some of the iterative algorithms, particularly of the CG and GMRES algorithms, but without being specific about the particular preconditioners used. The next section will consider the choice of standard preconditioners, and practical preconditioners in numerical reservoir simulation will be discussed in Section 5.10.

The term *preconditioning* was used for the first time by Turing (1948) to reduce the effect of round-off errors on direct algorithms. Its first application to iterative algorithms was presented by Evans (1968) on Chebyshev acceleration of SSOR.

5.8.1 Preconditioned CG

Assume that $\mathbf{A}$ is symmetric positive definite and that a preconditioner $\mathbf{M}$ is available. The *preconditioner* $\mathbf{M}$ is a matrix that approximates $\mathbf{A}$ in some sense (e.g., $\mathbf{M}^{-1}\mathbf{A}$ is close to the identity matrix). We assume that $\mathbf{M}$ is also symmetric positive definite. From a practical point of view, the only requirement for $\mathbf{M}$ is that it is inexpensive to solve the linear system $\mathbf{Mp} = \mathbf{f}$ because preconditioned algorithms require the solution of a linear system with $\mathbf{M}$ as the system matrix at each step. A preconditioned system is of the form

$$\mathbf{M}^{-1}\mathbf{Ap} = \mathbf{M}^{-1}\mathbf{f}. \tag{5.27}$$

In general, $\mathbf{M}^{-1}\mathbf{A}$ is unlikely to be symmetric, and thus CG cannot be directly applied to system (5.27).

When $\mathbf{M}$ possesses a Cholesky factorization:

$$\mathbf{M} = \mathbf{LL}^T,$$

a simple way to preserve symmetry is to split the preconditioner between left and right; i.e.,

$$\mathbf{L}^{-1}\mathbf{AL}^{-T}\mathbf{q} = \mathbf{L}^{-1}\mathbf{f}, \quad \mathbf{p} = \mathbf{L}^{-T}\mathbf{q}, \tag{5.28}$$

which generates a symmetric system. However, it is unnecessary to split M in this way to preserve symmetry. Note that $\mathbf{M}^{-1}\mathbf{A}$ is self-adjoint in the $\mathbf{M}$-inner product:

$$(\mathbf{x}, \mathbf{y})_{\mathbf{M}} = \mathbf{y}^T\mathbf{Mx},$$

because

$$(\mathbf{M}^{-1}\mathbf{Ax}, \mathbf{y})_{\mathbf{M}} = (\mathbf{Ax}, \mathbf{y}) = (\mathbf{x}, \mathbf{M}(\mathbf{M}^{-1}\mathbf{A})\mathbf{y}) = (\mathbf{x}, \mathbf{M}^{-1}\mathbf{Ay})_{\mathbf{M}}.$$

PCG Algorithm

Given $\mathbf{p}^0 \in \mathbb{R}^M$, set $\mathbf{r}^0 = \mathbf{f} - \mathbf{A}\mathbf{p}^0$, $\mathbf{z}^0 = \mathbf{M}^{-1}\mathbf{r}^0$, and $\mathbf{d}^0 = \mathbf{r}^0$.
For $k = 1, 2, \ldots,$ determine $\mathbf{p}^k$ and $\mathbf{d}^k$ by

$$\alpha_{k-1} = \frac{(\mathbf{r}^{k-1})^T \mathbf{z}^{k-1}}{\left(\mathbf{d}^{k-1}\right)^T \mathbf{A}\mathbf{d}^{k-1}};$$

$$\mathbf{p}^k = \mathbf{p}^{k-1} + \alpha_{k-1}\mathbf{d}^{k-1};$$

$$\mathbf{r}^k = \mathbf{r}^{k-1} - \alpha_{k-1}\mathbf{A}\mathbf{d}^{k-1};$$

$$\mathbf{z}^k = \mathbf{M}^{-1}\mathbf{r}^k;$$

$$\beta_{k-1} = \frac{(\mathbf{r}^k)^T \mathbf{z}^k}{(\mathbf{r}^{k-1})^T \mathbf{z}^{k-1}};$$

$$\mathbf{d}^k = \mathbf{z}^k + \beta_{k-1}\mathbf{d}^{k-1}.$$

Figure 5.11. *The algorithm PCG.*

Hence an alternative is to replace the usual Euclidean inner product $(\cdot, \cdot)$ in CG by the $\mathbf{M}$-inner product. In CG, $\mathbf{r}^k = \mathbf{f} - \mathbf{A}\mathbf{p}^k$ denotes the original residual, and in the preconditioned CG, $\mathbf{z}^k = \mathbf{M}^{-1}\mathbf{r}^k$ indicates the residual for the preconditioned system. Also, since $(\mathbf{z}^k, \mathbf{z}^k)_{\mathbf{M}} = (\mathbf{r}^k)^T\mathbf{z}^k$ and $(\mathbf{M}^{-1}\mathbf{A}\mathbf{d}^k, \mathbf{d}^k)_{\mathbf{M}} = (\mathbf{A}\mathbf{d}^k, \mathbf{d}^k)$, the $\mathbf{M}$-inner product does not have to be calculated explicitly. With these observations, the preconditioned CG (PCG) can be defined as in Figure 5.11.

When $\mathbf{M}$ possesses a Cholesky factorization, two options are available, the splitting technique (5.28) and the above PCG. One naturally asks, which one is better? Surprisingly, these two options produce the identical iterates (Saad, 2004).

5.8.2 Preconditioned GMRES

Preconditioning for GMRES and other iterative algorithms for nonsymmetric systems is different from that for CG. There is no concern to preserve symmetry for the preconditioned system. However, there are two different approaches to viewing preconditioning: *left* and *right preconditioning*.

Left preconditioned GMRES

The straightforward application of GMRES to the left preconditioned system (5.27) gives the preconditioned version of GMRES as in Figure 5.12.

Recall that $\mathbf{V}^k = (\mathbf{v}^1, \mathbf{v}^2, \ldots, \mathbf{v}^k)$. The Arnoldi algorithm constructs an orthogonal basis of the left preconditioned Krylov subspace

$$\text{span}\left(\mathbf{p}^0, \mathbf{M}^{-1}\mathbf{A}\mathbf{p}^0, \ldots, (\mathbf{M}^{-1}\mathbf{A})^{k-1}\mathbf{p}^0\right).$$

Left Preconditioned GMRES (k)

Given $\mathbf{p}^0 \in \mathbb{R}^M$, set $\mathbf{r}^0 = \mathbf{M}^{-1}\left(\mathbf{f} - \mathbf{A}\mathbf{p}^0\right)$, $\beta = \|\mathbf{r}^0\|_2$, $\mathbf{v}^1 = \mathbf{r}^0/\beta$.
For the $(k+1) \times k$ matrix $\mathbf{H}^k = (h_{ij})$, set $\mathbf{H}^k = \mathbf{0}$.
For $j = 1, 2, \ldots, k$, compute
$\mathbf{w}^j = \mathbf{M}^{-1}\mathbf{A}\mathbf{v}^j$;
$h_{ij} = (\mathbf{v}^i)^T\mathbf{w}^j$ for $i = 1, 2, \ldots, j$;
$\mathbf{w}^j = \mathbf{w}^j - \sum_{i=1}^{j} h_{ij}\mathbf{v}^i$;
$h_{j+1,j} = \|\mathbf{w}^j\|_2$;
If $h_{j+1,j} = 0$, set $k = j$ and skip the next step;
$\mathbf{v}^{j+1} = \mathbf{w}^j/h_{j+1,j}$.
Determine the minimizer $\mathbf{q}^k$ of $\|\beta\mathbf{e}_1 - \mathbf{H}^k\mathbf{q}^k\|_2$.
Set $\mathbf{p}^k = \mathbf{p}^0 + \mathbf{V}^k\mathbf{q}^k$.
If satisfied, stop; else set $\mathbf{p}^0 = \mathbf{p}^k$ and iterate.

Figure 5.12. *The left preconditioned version of GMRES.*

Right preconditioned GMRES

The right preconditioned GMRES solves a system of the form

$$\mathbf{A}\mathbf{M}^{-1}\underline{\mathbf{q}} = \mathbf{f}, \quad \underline{\mathbf{q}} = \mathbf{M}\mathbf{p}. \tag{5.29}$$

The new variable $\underline{\mathbf{q}}$ does not need to be invoked explicitly. In fact, once the initial residual $\mathbf{r}^0 = \mathbf{f} - \mathbf{A}\mathbf{p}^0 = \mathbf{f} - \mathbf{A}\mathbf{M}^{-1}\underline{\mathbf{q}}^0$ is evaluated, all subsequent vectors of the Krylov subspace can be found without any reference to the $\mathbf{q}$-variables (Saad, 2004). Observe that $\underline{\mathbf{q}}^0$ is not required at all; the initial residual for the preconditioned system can be obtained from $\mathbf{r}^0 = \mathbf{f} - \mathbf{A}\mathbf{p}^0$, which is identical to $\mathbf{f} - \mathbf{A}\mathbf{M}^{-1}\underline{\mathbf{q}}^0$. With this observation, the right preconditioned version of GMRES can be defined as in Figure 5.13.

This time, the Arnoldi algorithm constructs an orthogonal basis of the right preconditioned Krylov subspace

$$\text{span}\left(\mathbf{p}^0, \mathbf{A}\mathbf{M}^{-1}\mathbf{p}^0, \ldots, (\mathbf{A}\mathbf{M}^{-1})^{k-1}\mathbf{p}^0\right).$$

The residual norm is now relative to the initial system, $\mathbf{f} - \mathbf{A}\mathbf{p}^k$, because the algorithm implicitly obtains the residual $\mathbf{r}^k = \mathbf{f} - \mathbf{A}\mathbf{p}^k = \mathbf{f} - \mathbf{A}\mathbf{M}^{-1}\underline{\mathbf{q}}^k$. That is an essential difference between the left and right preconditioned GMRES algorithms. The spectra of the two preconditioned matrices $\mathbf{M}^{-1}\mathbf{A}$ and $\mathbf{A}\mathbf{M}^{-1}$ are the same. Hence their convergence behaviors are expected to be similar, though the eigenvalues do not always govern convergence. Right

Right Preconditioned GMRES (k)

Given $\mathbf{p}^0 \in \mathbb{R}^M$, set $\mathbf{r}^0 = \mathbf{f} - \mathbf{A}\mathbf{p}^0$, $\beta = \|\mathbf{r}^0\|_2$, and $\mathbf{v}^1 = \mathbf{r}^0/\beta$.

For the $(k+1) \times k$ matrix $\mathbf{H}^k = (h_{ij})$, set $\mathbf{H}^k = \mathbf{0}$.

For $j = 1, 2, \ldots, k$, compute

$\mathbf{w}^j = \mathbf{A}\mathbf{M}^{-1}\mathbf{v}^j$;

$h_{ij} = (\mathbf{v}^i)^T \mathbf{w}^j$ for $i = 1, 2, \ldots, j$;

$$\mathbf{w}^j = \mathbf{w}^j - \sum_{i=1}^{j} h_{ij}\mathbf{v}^i;$$

$h_{j+1,j} = \|\mathbf{w}^j\|_2$;

If $h_{j+1,j} = 0$, set $k = j$ and skip the next step;

$\mathbf{v}^{j+1} = \mathbf{w}^j / h_{j+1,j}$.

Determine the minimizer $\mathbf{q}^k$ of $\|\beta\mathbf{e}_1 - \mathbf{H}^k\mathbf{q}^k\|_2$.

Set $\mathbf{p}^k = \mathbf{p}^0 + \mathbf{M}^{-1}\mathbf{V}^k\mathbf{q}^k$.

If satisfied, stop; else set $\mathbf{p}^0 = \mathbf{p}^k$ and iterate.

Figure 5.13. *The right preconditioned version of GMRES.*

preconditioning has been employed as a basis for an algorithm that changes the preconditioner $\mathbf{M}$ as the iteration progresses, i.e., the FGMRES (*flexible GMRES*) algorithm (Saad, 2004), which will be discussed next.

Flexible GMRES

The preconditioner $\mathbf{M}$ has been so far assumed to be fixed; i.e, it does not vary from step to step. In some cases, the matrix $\mathbf{M}$ may not be available; the operation $\mathbf{M}^{-1}\mathbf{p}$ is only the result of some unspecified calculation. $\mathbf{M}$ may not be a constant matrix in such cases. The left and right preconditioned GMRES algorithms will not converge if $\mathbf{M}$ is not fixed; they must be modified to accommodate variations in the preconditioner. In this section, we state a flexible variant of GMRES, *FGMRES* (Saad, 2004).

Suppose that the preconditioner $\mathbf{M}_j$ in the right preconditioned GMRES can change at every step. Then, in the fourth line of the right preconditioned GMRES(k), the vector

$$\mathbf{z}^j = \mathbf{M}_j^{-1}\mathbf{v}^j$$

must be saved. It is now natural to find the solution $\mathbf{p}^k$ in the form

$$\mathbf{p}^k = \mathbf{p}^0 + \mathbf{Z}^k\mathbf{q}^k,$$

where $\mathbf{Z}^k = (\mathbf{z}^1, \mathbf{z}^2, \ldots, \mathbf{z}^k)$ and $\mathbf{q}^k$ is obtained as in the right preconditioned GMRES. With this modification, FGMRES can be defined as in Figure 5.14.

FGMRES (k)

Given $\mathbf{p}^0 \in \mathbb{R}^M$, set $\mathbf{r}^0 = \mathbf{f} - \mathbf{A}\mathbf{p}^0$, $\beta = \|\mathbf{r}^0\|_2$, and $\mathbf{v}^1 = \mathbf{r}^0/\beta$.

For the $(k+1) \times k$ matrix $\mathbf{H}^k = (h_{ij})$, set $\mathbf{H}^k = \mathbf{0}$.

For $j = 1, 2, \ldots, k$, compute

$$\mathbf{z}^j = \mathbf{M}_j^{-1}\mathbf{v}^j;$$

$$\mathbf{w}^j = \mathbf{A}\mathbf{z}^j;$$

$$h_{ij} = (\mathbf{v}^i)^T\mathbf{w}^j \text{ for } i = 1, 2, \ldots, j;$$

$$\mathbf{w}^j = \mathbf{w}^j - \sum_{i=1}^{j} h_{ij}\mathbf{v}^i;$$

$$h_{j+1,j} = \|\mathbf{w}^j\|_2;$$

If $h_{j+1,j} = 0$, set $k = j$ and skip the next step;

$$\mathbf{v}^{j+1} = \mathbf{w}^j / h_{j+1,j}.$$

Determine the minimizer $\mathbf{q}^k$ of $\|\beta\mathbf{e}_1 - \mathbf{H}^k\mathbf{q}^k\|_2$.

Set $\mathbf{p}^k = \mathbf{p}^0 + \mathbf{Z}^k\mathbf{q}^k$.

If satisfied, stop; else set $\mathbf{p}^0 = \mathbf{p}^k$ and iterate.

Figure 5.14. *The flexible GMRES algorithm.*

The major difference between the right preconditioned GMRES and FGMRES is that the vectors $\mathbf{z}^j$ $(j = 1, 2, \ldots, k)$ must be stored and the solution must be updated using these vectors in the latter. If $\mathbf{M}_j = \mathbf{M}$ for $j = 1, 2, \ldots, k$, these two algorithms are mathematically equivalent. Note that the $\mathbf{z}^j$'s can be selected without reference to any preconditioner. This added flexibility may cause FGMRES some problems. In fact, $\mathbf{z}^j$ may be so poorly chosen that a breakdown could occur, such as in the worst case where $\mathbf{z}^j = \mathbf{0}$.

An optimality property similar to that for GMRES (cf. (5.22) or (5.23)) can be shown for FGMRES. Indeed, one can prove that the approximate solution $\mathbf{p}^k$ computed at the kth step of this algorithm minimizes the residual norm $\|\mathbf{f} - \mathbf{A}\mathbf{p}^k\|_2$ over $\mathbf{p}^0 + \text{span}(\mathbf{Z}^k)$ (Saad, 2004).

5.9 Preconditioners

Roughly speaking, a *preconditioner* $\mathbf{M}$ is some form of approximation of the original matrix $\mathbf{A}$ that makes the preconditioned system easier to solve using a given iterative algorithm. One commonly used and easily computable preconditioner is based on *Jacobi preconditioning* where $\mathbf{M}$ is the inverse of the diagonal part of $\mathbf{A}$. One can also utilize other preconditioners that are related to the simple stationary iterative algorithms such as Gauss–Seidel, SOR, and SSOR. In most practical situations in reservoir simulation, these preconditioners may be somewhat useful but should not be expected to have significant effects.

General ILU Factorization

For $i = 2, 3, \ldots, M$,
  For $k = 1, 2, \ldots, i-1$ and $(i, k) \notin Z$,
    $a_{ik} := a_{ik}/a_{kk}$
    For $j = k+1, \ldots, M$ and $(i, j) \notin Z$,
      $a_{ij} := a_{ij} - a_{ik}a_{kj}$.
    End
  End
End

Figure 5.15. *The general ILU factorization.*

Another type of preconditioner is based on an *incomplete Cholesky factorization* of the original matrix $\mathbf{A}$ (Buleev, 1959; Varga, 1960). Such a preconditioner stems from a decomposition of the form $\mathbf{A} = \mathbf{LU} - \mathbf{R}$, where $\mathbf{L}$ and $\mathbf{U}$ have the same nonzero structure as the lower and upper parts of $\mathbf{A}$, respectively, and $\mathbf{R}$ is the *residual* or *error* of the factorization. This incomplete factorization, called ILU(0), is easy and inexpensive to implement. On the other hand, it may generate an approximation that requires the underlying Krylov subspace algorithm to converge in many iterations. To remedy this difficulty, a number of alternative incomplete factorizations have been proposed by allowing some *fill-in* in $\mathbf{L}$ and $\mathbf{U}$. In general, the more accurate the ILU factorization, the faster the resulting preconditioned Krylov subspace algorithm. The preprocessing cost to compute the more accurate $\mathbf{L}$ and $\mathbf{U}$, however, is higher. From the point of view of robustness (e.g., in terms of applicability and reliability), these more accurate factorizations may be needed. In this section, we concentrate on the construction of ILU(0) and its variants.

Consider any sparse matrix $\mathbf{A} = (a_{ij})$. A general ILU factorization algorithm generates a sparse lower triangular matrix $\mathbf{L}$ and a sparse upper triangular matrix $\mathbf{U}$, and so the residual matrix $\mathbf{R} = \mathbf{LU} - \mathbf{A}$ satisfies certain conditions such as having zero entries in some locations. This general algorithm can be obtained by performing Gaussian elimination and dropping certain entries in predetermined nondiagonal positions. The entries to drop at each step can be predetermined statically, by choosing some zero pattern, for example. The sole restriction on the zero pattern is that it should not include diagonal entries. Hence, for any zero-pattern set Z, such as

$$Z \subset \{(i, j) : i \neq j,\ i, j = 1, 2, \ldots, M\},$$

a general ILU factorization takes the form presented in Figure 5.15 (Saad, 2004).

It can be shown (Saad, 2004) that this algorithm produces matrices $\mathbf{L}$ and $\mathbf{U}$ such that $\mathbf{A} = \mathbf{LU} - \mathbf{R}$, where $-\mathbf{R}$ is the matrix of the entries that are dropped during the incomplete elimination process. For $(i, j) \in Z$, an entry r_{ij} of $\mathbf{R}$ equals $-a_{ij}$ computed at the completion of the kth loop in the above algorithm. Otherwise, $r_{ij} = 0$.

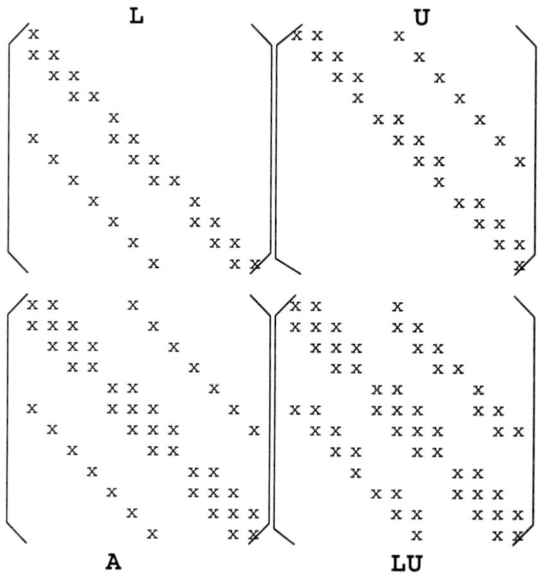

Figure 5.16. *An illustration of ILU*(0).

5.9.1 ILU(0)

The zero-pattern set Z depends on prescribed levels of fill-in or *thresholds*. If **L** and **U** have the same sparsity pattern as **A**, i.e., the zero pattern Z is precisely the zero pattern of **A**, the resulting ILU factorization is indicated by ILU(0). This technique allows no fill-in.

ILU(0) is best illustrated by Figure 5.16. Consider a matrix **A** of the form shown in this figure, any lower triangular matrix **L** that has the same structure as that of the lower part of **A**, and any upper triangular matrix **U** that has the same structure as that of the upper part of **A** (cf. Figure 5.16). If the product **LU** was performed, the resulting matrix would have the pattern displayed in this figure. In general, it is impossible to match the given matrix **A** with this product for any **L** and **U**. This is due to the extra diagonals in the product. The entries in these extra diagonals are termed *fill-in*. If these fill-in entries are dropped, then it is possible to find **L** and **U** such that their product equals **A** in other diagonals. This defines the ILU(0) factorization: the entries of $\mathbf{A} - \mathbf{LU}$ are zero in the locations where $a_{ij} \neq 0$, $i, j = 1, 2, \ldots, M$. That is, with the pattern Z being the zero pattern of **A** (i.e., $Z = Z(\mathbf{A})$), ILU(0) is defined as in Figure 5.17.

5.9.2 ILU(l)

The ILU(0) factorization makes the Krylov subspace algorithms developed in the previous sections very simple and efficient to implement. The accuracy of ILU(0) may be insufficient to generate an adequate rate of convergence for certain realistic problems that arise in numerical reservoir simulation. More accurate ILU factorizations are often needed. These more accurate factorizations ILU(l) differ from ILU(0) by allowing some fill-in.

ILU(0) Factorization

For $i = 2, 3, \ldots, M$,
    For $k = 1, 2, \ldots, i-1$ and $(i, k) \notin Z(\mathbf{A})$,
        $a_{ik} := a_{ik}/a_{kk}$
        For $j = k+1, \ldots, M$ and $(i, j) \notin Z(\mathbf{A})$,
            $a_{ij} := a_{ij} - a_{ik}a_{kj}.$
        End
    End
End

Figure 5.17. *The ILU(0) factorization.*

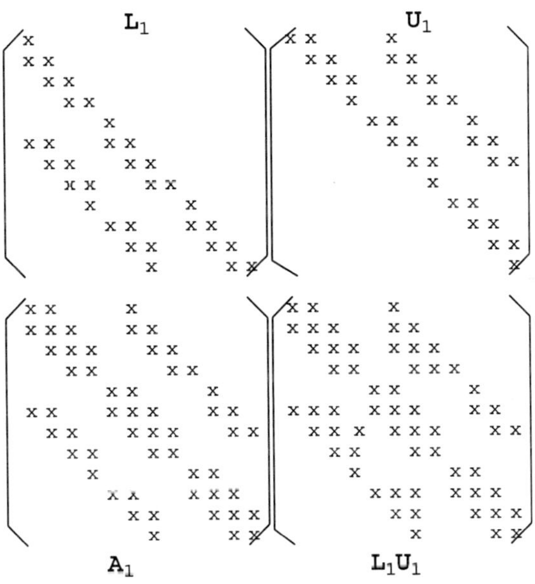

Figure 5.18. *An illustration of ILU(l).*

The idea of ILU(1) is geometrically illustrated with the same example as for ILU(0) in Figure 5.16 (Saad, 2004). That is, ILU(1) comes from taking Z to be the zero pattern of the product $\mathbf{LU}$ of the factors $\mathbf{L}$ and $\mathbf{U}$ obtained from ILU(0); see Figure 5.18. Pretend that the original matrix $\mathbf{A}$ has this "augmented" pattern. In other words, the fill-in locations created in this product belong to the augmented pattern, but their actual values equal zero. The factors $\mathbf{L}_1$ and $\mathbf{U}_1$ of ILU(1) are now obtained by performing an ILU(0) factorization on this augmented pattern matrix. The new product $\mathbf{L}_1\mathbf{U}_1$ has two additional diagonals in the lower and upper locations (cf. Figure 5.18).

ILU(l) Factorization

For all nonzero entries a_{ij}, set $lev_{ij} = 0$
For $i = 2, 3, \ldots, M$,
    For $k = 1, 2, \ldots, i-1$ and $lev_{ik} \le l$,
        $a_{ik} := a_{ik}/a_{kk}$
        $\mathbf{a}_i := \mathbf{a}_i - a_{ik}\mathbf{a}_i$
        Update the levels of fill of the nonzero a_{ij}'s by (5.30)
    End
    Replace any entry in the ith row with $lev_{ij} > l$ by zero
End

Figure 5.19. *The ILU(l) factorization.*

A problem with this illustration is that it does not generalize to general sparse matrices. To extend it, we introduce the notion of *level of fill*. A level of fill is attributed to each matrix entry that occurs in the elimination process. Fill-ins are dropped based on the value of the level of fill. Initially, suppose that a nonzero entry has a level of fill of zero and a zero entry has a level of fill of ∞. Namely, the initial level of fill of an entry a_{ij} of $\mathbf{A}$ is (Saad, 2004)

$$lev_{ij} = \begin{cases} 0 & \text{if } a_{ij} \neq 0 \text{ or } i = j, \\ \infty & \text{otherwise.} \end{cases}$$

Each time this entry is modified according to the general ILU factorization (i.e., by the formula $a_{ij} := a_{ij} - a_{ik}a_{kj}$), its level of fill must be updated:

$$lev_{ij} = \min\{lev_{ij}, lev_{ik} + lev_{kj} + 1\}. \tag{5.30}$$

Note that the level of fill of an entry will never increase during elimination. If $a_{ij} \neq 0$ in the original matrix $\mathbf{A}$, the entry in the (i, j)th location will have a level of fill of zero throughout the elimination process. The introduction of this concept of level of fill yields a natural strategy for dropping entries. In ILU(l), all fill-in entries whose level of fill does not exceed l are kept. Hence the zero pattern for ILU(l) is the set

$$Z_l = \{(i, j) : lev_{ij} > l\},$$

where lev_{ij} is the value of level of fill after all updates in (5.30) have been performed. The case $l = 0$ coincides with the definition of the ILU(0) factorization. In the ILU(l) factorization (cf. Figure 5.19), $\mathbf{a}_i$ indicates the ith row of matrix $\mathbf{A}$.

In the ILU factorization so far, the entries that are dropped during the elimination process have been simply discarded. There are techniques that attempt to reduce the effect of dropping by *compensating* for the discarded entries. A popular technique is to add up all

ILUT Algorithm

For $i = 1, 2, \ldots, M$,
  $\mathbf{w} = \mathbf{a}_i$
  For $k = 1, 2, \ldots, i-1$ and if $w_k \neq 0$,
    $w_k := w_k / a_{kk}$
    Applying a dropping rule to w_k
    If $w_k \neq 0$, then
      $\mathbf{w} = \mathbf{w} - w_k \mathbf{u}_i$
    EndIf
  End
  Applying a dropping rule to the row $\mathbf{w}$
  $l_{ij} = w_j$ for $j = 1, 2, \ldots, i-1$
  $u_{ij} = w_j$ for $j = 1, 2, \ldots, M$
  $\mathbf{w} = \mathbf{0}$
End

Figure 5.20. *The ILUT algorithm.*

the entries that were dropped at the completion of k-loop of the general ILU factorization algorithm. Then this sum is subtracted from the diagonal entry in $\mathbf{U}$. This diagonal compensation technique is referred to as the *modified ILU* (*MILU*) factorization, which will not be considered further.

The ILU(l) factorization algorithm also has a few drawbacks. First, the amounts of fill-in and computational work for obtaining this factorization are not generally predictable for $l > 0$. Second, updating the levels in this algorithm can be very expensive. Third, the level of fill-in for indefinite matrices may not be a good indicator for the size of the entries that are being discarded. In other words, the algorithm may discard large entries. To overcome these drawbacks, a preconditioning technique, known as *ILUT*, is described next.

5.9.3 ILUT

As noted above, the entries that are dropped in the ILU factorization depend only on the structure of matrix $\mathbf{A}$. There are a few alternative algorithms available that are based on dropping entries in the incomplete factorization process according to their magnitude rather than their locations. In these algorithms, the zero-pattern set Z is determined dynamically. One such algorithm is ILUT (ILU with *threshold* (Saad, 2004)).

An ILUT algorithm can be obtained from the general ILU factorization algorithm by applying a set of rules for dropping small entries. Below, applying a dropping rule for an entry will mean replacing this entry by zero if it satisfies a set of criteria. In the next algorithm (cf. Figure 5.20), $\mathbf{w} = (w_1, w_2, \ldots, w_M)$ is a full-length working row that accumulates linear combinations of rows in the elimination, and $\mathbf{u}_i$ represents the ith row of $\mathbf{U}$.

ILU(0) can be treated as a special case of ILUT. The dropping rule for ILU(0) is to drop the entries that are not in locations of the original structure of matrix $\mathbf{A}$. Similar to ILU(l), one can also define the ILUT(l, ϵ) factorization, where ϵ is a *dropping tolerance* used in a dropping criterion. In ILUT(l, ϵ), the following rules are applied:

- In the fifth line of ILUT, an entry w_k is dropped (i.e., replaced by zero) if its magnitude is below the relative tolerance ϵ_i obtained by multiplying ϵ by the norm (e.g., ℓ_2-norm) of the ith row.

- In the tenth line of ILUT, a different dropping rule is used. First, drop again any entry in the row with a magnitude less than ϵ_i. Then, in addition to keeping the diagonal entry, keep only the l largest entries in the $\mathbf{L}$ part of the row and the l largest entries in the $\mathbf{U}$ part of the row.

The second dropping step is to control the number of entries per row. Roughly speaking, the parameter l is used to control memory usage, while ϵ is viewed to reduce computational cost. In many cases, good results are obtained for values of ϵ in the range 10^{-4}–10^{-2}, but an optimal value is strongly problem dependent.

It is well known that ILU preconditioners are not easily parallelizable. The reason is that Gaussian elimination, on which the ILU factorization is based, offers limited scope for parallelization. Furthermore, the forward elimination and backward substitution that form the preconditioning operations are highly sequential in their nature, and parallelization for these operations is difficult.

Preconditioning techniques based on *sparse approximate inverses* have been recently developed (Benson and Frederickson, 1982). The idea of these techniques is that a sparse matrix $\mathbf{M} \approx \mathbf{A}^{-1}$ is explicitly computed and used as a preconditioner for the Krylov subspace algorithms for the solution of (5.1). Their major advantage is that the preconditioning operation can be easily implemented in parallel because it consists only of matrix-vector products. However, like the ILU preconditioning approach, this approach lacks algorithmic scalability (e.g., in terms of operation counts), which has led to the development of a number of variants based on the multigrid method (Hackbusch, 1985; Bramble, 1993), the algebraic multilevel method (Stüben, 1983), and the domain decomposition method (Smith et al., 1996). This class of preconditioning techniques are optimal for linear systems arising from certain partial differential problems (e.g., elliptic or parabolic problems) in the sense that the required number of arithmetic operations is of order $\mathcal{O}(M)$.

5.10 Practical Considerations

Preconditioners can be derived from a knowledge of the original physical problems from which the linear system arises. In the numerical simulation of multiphase flow in reservoirs, for example, the governing partial differential equations involve many distinct variables such as pressure, saturation, and concentration (cf. Chapter 2) and are coupled with injection and production wells (source and sink terms). The system matrix $\mathbf{A}$ for such applications has blocks with different natures. A feasible approach to the construction of a preconditioner is to precondition these blocks differently and separately, using their natures as fully as possible. In this section, we discuss the construction of preconditioners based on this approach.

5.10.1 Decoupling preconditioners

Consider a block representation of system (5.1) in the form

$$\mathbf{A}\mathbf{p} \equiv \begin{pmatrix} \mathbf{A}_{11} & \mathbf{A}_{12} \\ \mathbf{A}_{21} & \mathbf{A}_{22} \end{pmatrix} \begin{pmatrix} \mathbf{p}_1 \\ \mathbf{p}_2 \end{pmatrix} = \begin{pmatrix} \mathbf{f}_1 \\ \mathbf{f}_2 \end{pmatrix}, \tag{5.31}$$

where $\mathbf{p}_1$ and $\mathbf{p}_2$ correspond to the degrees of freedom for two different variables such as pressure and saturation (or concentration). While only two variables are considered, it is straightforward to include more variables. We assume that the off-diagonal block entries responsible for the interaction between these two variables are small compared to the respective entries of the diagonal blocks.

In some situations, the accuracy required for a preconditioner is higher for one variable (e.g., pressure) than for the other (e.g., saturation). This is the case where the coupled system of pressure and saturation equations for two-phase flow (cf. Chapter 7) is solved simultaneously, as the pressure equation causes the most difficulty in the iterative process. If an accurate approximation $\bar{\mathbf{p}}_1$ can be found and an easy-to-invert approximation $\bar{\mathbf{A}}_{22}$ to $\mathbf{A}_{22}$ is available, then $\bar{\mathbf{A}}_{22}^{-1}(\mathbf{f}_2 - \mathbf{A}_{21}\bar{\mathbf{p}}_1)$ is a meaningful approximation to $\mathbf{p}_2$. The choice of $\bar{\mathbf{A}}_{22} = \mathbf{A}_{22}$ implies an exact solution for the second variable $\mathbf{p}_2$; if the stiffness of $\mathbf{A}_{22}$ is less than that of $\mathbf{A}_{11}$, $\mathbf{A}_{22}$ can be replaced by a simple approximation such as the ILU(0) factorization introduced in the previous section.

Assume that an ILU factorization of $\mathbf{A}_{11}$ is given by

$$\mathbf{A}_{11} = \mathbf{LDU} - \mathbf{R},$$

where $\mathbf{L}$, $\mathbf{D}$, and $\mathbf{U}$ are unit lower triangular, diagonal, and unit upper triangular matrices, respectively, and $\mathbf{R}$ is the residual matrix. Then an approximation of $\mathbf{A}$ is given by

$$\begin{aligned} &\begin{pmatrix} \mathbf{LDU} & \mathbf{A}_{12} \\ \mathbf{A}_{21} & \mathbf{A}_{22} \end{pmatrix} \\ &= \begin{pmatrix} \mathbf{LDU} & \mathbf{0} \\ \mathbf{A}_{21} & \mathbf{I} \end{pmatrix} \begin{pmatrix} \mathbf{I} & (\mathbf{LDU})^{-1}\mathbf{A}_{12} \\ \mathbf{0} & \mathbf{A}_{22} - \mathbf{A}_{21}(\mathbf{LDU})^{-1}\mathbf{A}_{12} \end{pmatrix}, \end{aligned} \tag{5.32}$$

where $\mathbf{I}$ is the identity matrix. If $\mathbf{LDU}$ is exact (i.e., Gaussian elimination is used), so is this factorization. When $\mathbf{LDU}$ is an incomplete factorization, the right-hand side of (5.32) can be viewed as an incomplete factorization of $\mathbf{A}$. A problem with this factorization is that $(\mathbf{LDU})^{-1}\mathbf{A}_{12}$ is generally a full matrix (so is $\mathbf{A}_{22} - \mathbf{A}_{21}(\mathbf{LDU})^{-1}\mathbf{A}_{12}$). Thus an approximation to $(\mathbf{LDU})^{-1}\mathbf{A}_{12}$ should be applied. The simplest remedy is the following modification of (5.32):

$$\begin{pmatrix} \mathbf{L} & \mathbf{0} \\ \mathbf{A}_{21} & \mathbf{I} \end{pmatrix} \begin{pmatrix} \mathbf{DU} & \mathbf{DA}_{12} \\ \mathbf{0} & \mathbf{A}_{22} - \mathbf{A}_{21}\mathbf{DA}_{12} \end{pmatrix}. \tag{5.33}$$

This factorization weakens the coupling between the first and second variables. Many preconditioners based on similar approaches have been constructed in reservoir simulation, such as the *constrained pressure residual preconditioner* (Wallis et al., 1985).

5.10.2 COMBINATIVE preconditioners

The assumption that one variable dominates the other is sometimes too restrictive. A preconditioner that can provide a moderate feedback for the interaction between these two variables should be used. An example of such a preconditioner is the two-stage *COMBINATIVE preconditioner* (Behie and Vinsome, 1982). The idea of this approach is to decouple the equation for the first variable and then to find an appropriate preconditioner that provides the feedback:

(1) Solve the equation $\mathbf{A}_{11}\mathbf{p}_1 = \mathbf{f}_1$.

(2) Form the residual

$$\begin{pmatrix} \mathbf{r}_1 \\ \mathbf{r}_2 \end{pmatrix} = \begin{pmatrix} \mathbf{f}_1 \\ \mathbf{f}_2 \end{pmatrix} - \begin{pmatrix} \mathbf{A}_{11} \\ \mathbf{A}_{12} \end{pmatrix} \mathbf{p}_1.$$

(3) Precondition the new residual and update the first variable:

$$\begin{pmatrix} \mathbf{p}_1 \\ \mathbf{p}_2 \end{pmatrix} := \mathbf{M}^{-1} \begin{pmatrix} \mathbf{r}_1 \\ \mathbf{r}_2 \end{pmatrix} - \begin{pmatrix} \mathbf{p}_1 \\ \mathbf{0} \end{pmatrix},$$

where $\mathbf{M}$ is a preconditioner for $\mathbf{A}$ that provides the feedback mentioned. Experience with the construction of COMBINATIVE preconditioners reveals that $\mathbf{M}$ can be chosen to be a rather rough (or weak) preconditioner because its goal is to provide feedback. ILU(0) can serve for this purpose, for example. The combination (that suggests the name COMBINATIVE) of steps (1)–(3) yields a preconditioner for $\mathbf{A}$:

$$\begin{pmatrix} \mathbf{A}_{11}^{-1} & \mathbf{0} \\ \mathbf{0} & \mathbf{0} \end{pmatrix} + \mathbf{M}^{-1}\left(\mathbf{I} - \begin{pmatrix} \mathbf{A}_{11} \\ \mathbf{A}_{12} \end{pmatrix} \mathbf{A}_{11}^{-1}\right). \tag{5.34}$$

The matrix $\mathbf{A}_{11}^{-1}$ may be replaced by a preconditioner for $\mathbf{A}_{11}$ such as an accurate ILU preconditioner.

5.10.3 Bordered systems

The system arising from fully coupled flow and well implicit reservoir simulation is of the form

$$\begin{pmatrix} \mathbf{A}_{11} & \mathbf{A}_{12} \\ \mathbf{A}_{21} & \mathbf{A}_{22} \end{pmatrix} \begin{pmatrix} \mathbf{p} \\ \mathbf{p}_w \end{pmatrix} = \begin{pmatrix} \mathbf{f}_1 \\ \mathbf{f}_2 \end{pmatrix}, \tag{5.35}$$

where $\mathbf{p}_w$ corresponds to the degrees of freedom for the bottom-hole well pressure, $\mathbf{A}_{11}$ and $\mathbf{A}_{22}$ are associated with the flow equations and the well constraint equations, respectively, and $\mathbf{A}_{12}$ and $\mathbf{A}_{21}$ indicate their interaction. Since system (5.35) is identical to (5.31) in form, the approximate factorizations in (5.32)–(5.34) developed for the latter apply to the former.

5.10.4 Choice of initial solutions

The residual matrix $\mathbf{R}$ in the equation $\mathbf{A} = \mathbf{LDU} - \mathbf{R}$ can be taken into account in an approximate fashion. An approach by Gustafsson (1978) modifies $\mathbf{D}$ so that $\mathbf{R}$ has zero row sums. This diagonal modification approach can be used with any order incomplete factorization method (e.g., MILU).

An alternative approach to accounting for the error matrix $\mathbf{R}$ is given by Appleyard et al. (1981). This approach is based on the observation that in most reservoir simulations the sum of entries in the right-hand-side vector $\mathbf{f}$ of (5.1) is equal to the net rate of mass accumulation. If an initial solution $\mathbf{p}^0$ is selected by

$$\mathbf{LDUp}^0 = \mathbf{f}, \tag{5.36}$$

then the sum of the entries in the initial residual

$$\mathbf{r}^0 = \mathbf{f} - \mathbf{Ap}^0 = \mathbf{Rp}^0$$

represents the material balance error. This sum equals zero and will remain zero for all subsequent iterations if the column sums of $\mathbf{R}$ are zero. For a symmetric matrix $\mathbf{A}$, these two approaches produce the same factorizations. For a nonsymmetric $\mathbf{A}$, they are different. The latter approach is based on a physical observation.

The error accounting approach by Appleyard et al. (1981) can be applied to any ILU factorization studied in the previous section. Suppose that an incomplete factorization of $\mathbf{A}$, $\mathbf{LDU}$, is constructed. Instead of changing the factors $\mathbf{L}$, $\mathbf{D}$, and $\mathbf{U}$ via elimination across a row of $\mathbf{A}$, they are constructed via elimination down a column of $\mathbf{A}$. Entries that are to be ignored in the incomplete factorization process (i.e., the error entries) are subtracted from the diagonal entry lying in the same column (instead of the same row). This approach generates an error matrix with zero column sums. If the initial solution $\mathbf{p}^0$ is selected as in (5.36), all subsequent residual sums will be zero.

5.11 Concluding Remarks and Comparisons

Direct and iterative algorithms have been presented in this chapter. The direct algorithms are based on the factorization of the system matrix $\mathbf{A}$ into easily invertible matrices, and are widely employed in many petroleum reservoir codes where reliability is the primary concern. Indeed, direct solvers are very robust, and they tend to require a predictable amount of resources in terms of storage and time. With a state-of-the-art sparse direct solver, it is possible to solve efficiently linear systems of fairly large size in a reasonable amount of time, particularly when the underlying problem is two-dimensional.

Unfortunately, direct algorithms scale poorly with problem size in terms of operation counts and memory requirements, particularly for three-dimensional problems. Three-dimensional multiphase flow simulations lead to linear systems of many millions of equations in as many unknowns. For such simulations, iterative algorithms are the only option available. While the iterative algorithms require less storage and fewer operations than the direct algorithms (particularly when an approximate solution of relatively low accuracy is sought), they do not have the reliability of the latter algorithms. In some applications, they even fail to converge in a reasonable amount of time. Thus preconditioning is necessary, though not always sufficient.

As noted earlier, the linear systems arising in numerical reservoir simulation are sparse, highly nonsymmetric, and indefinite. Three leading iterative algorithms for solving such systems are the CGN, GMRES, and BiCGSTAB algorithms. These three algorithms differ

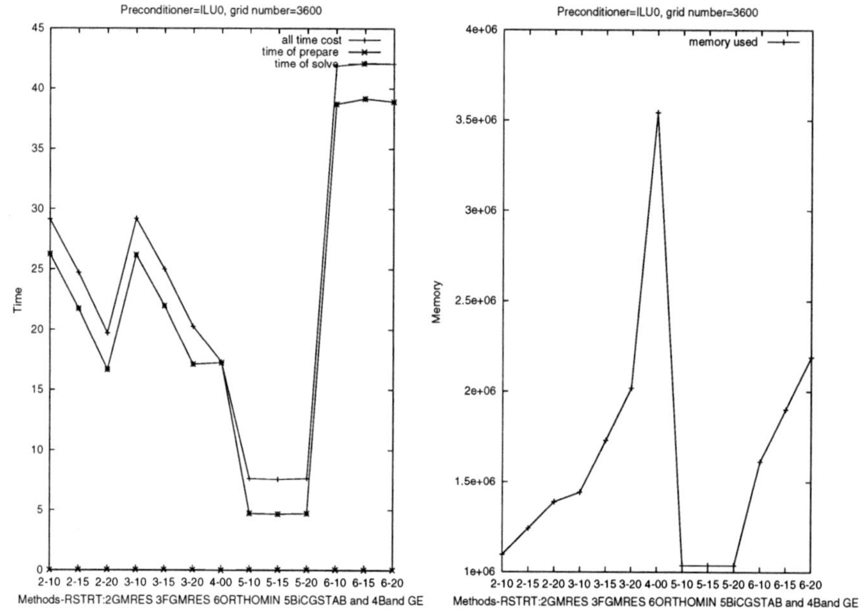

Figure 5.21. *Computational time (sec.) (left); memory (byte) (right).*

fundamentally in their capabilities. As demonstrated by Nachtigal et al. (1992), examples of system matrices can be constructed so that each type of iteration can outperform the others by a factor on the order of $\sqrt{M}$ or M (or even more). Hence, in general, it is very difficult to compare these algorithms for practical reservoir problems. In this section, we just give some indications on their performance for the solution of a linear system that arises from simulation of multiphase flow problems; we present comparisons of Gaussian elimination (the direct banded solver), GMRES (including ORTHOMIN and FGMRES), and BiCGSTAB in terms of computational time and storage memory. This linear system stems from the discretization of the pressure equation for a two-phase flow problem using a standard finite element method, which will be described in detail in Chapter 7. Two cases where the numbers of grid nodes are 3,600 and 10,000 are tested. The preconditioning techniques are based on ILU(l) and ILUT(l). The restart numbers for the iterative algorithms are set to be 10, 15, and 20. The numerical experiments were performed on a Compaq Alpha ES40 workstation with four CPUs, 883 MHZ CPU frequency, and 32 GB RAM (Chen et al., 2002D), and the numerical results are displayed in Figures 5.21–5.28. The numbers 2–6 on the horizontal axes in these figures indicate, respectively, GMRES, FGMRES, banded Gaussian elimination, BiCGSTAB, and ORTHOMIN. From these figures we make the following observations:

- The GMRES, FGMRES, BiCGSTAB, and ORTHOMIN algorithms are much faster than the direct banded Gaussian elimination algorithm, particularly when high-order preconditioners (e.g., ILU(8) and ILUT(10)) are used, and they use much less memory than the latter for large-scale problems such as the one with a grid number of 10,000 under consideration.

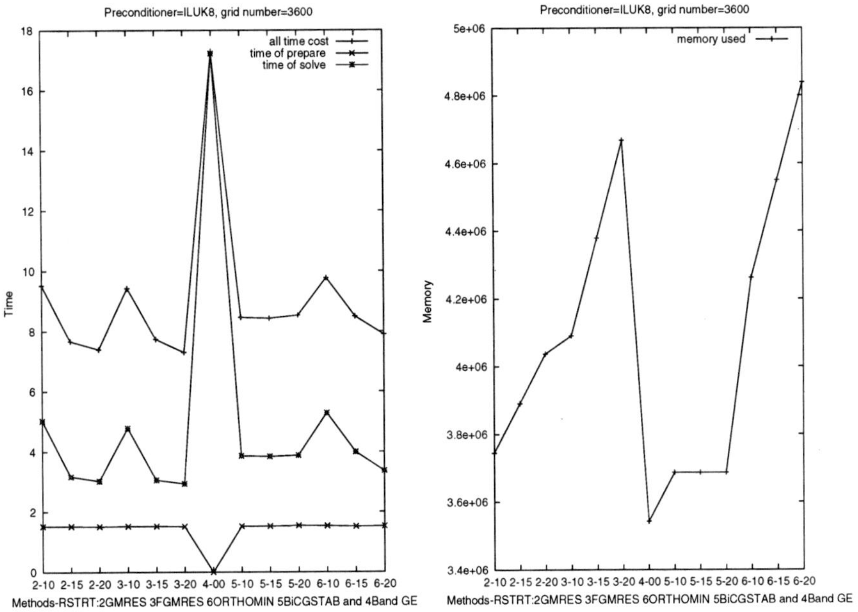

Figure 5.22. *Computational time (sec.) (left); memory (byte) (right).*

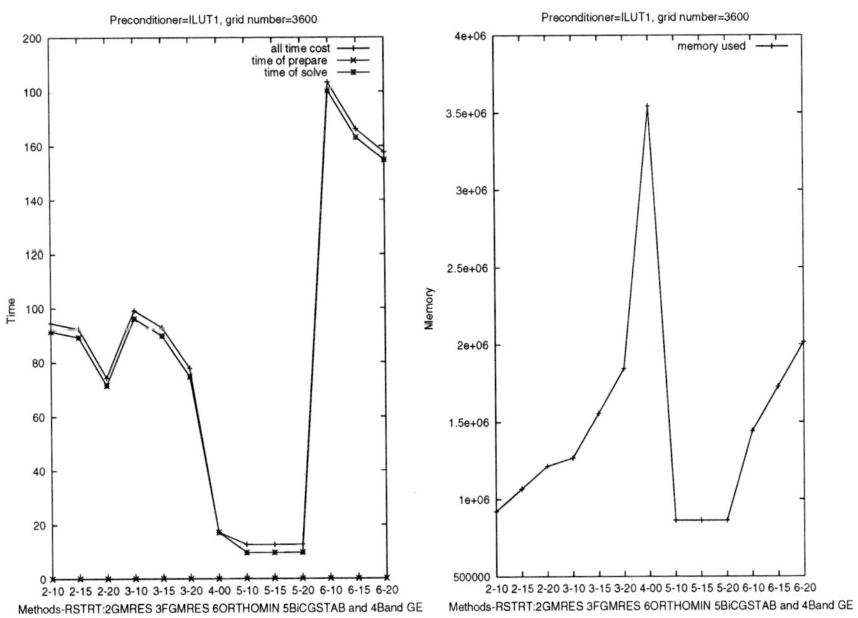

Figure 5.23. *Computational time (sec.) (left); memory (byte) (right).*

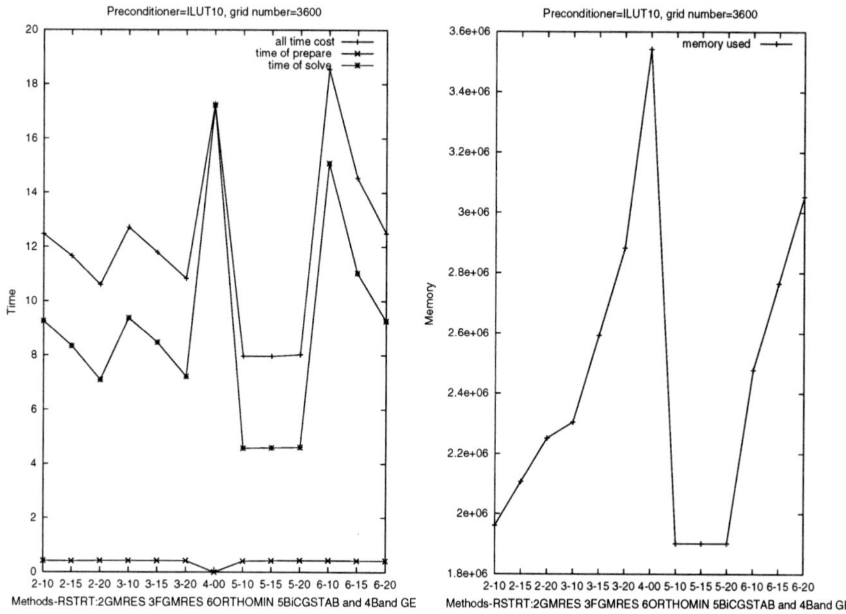

Figure 5.24. *Computational time (sec.) (left); memory (byte) (right).*

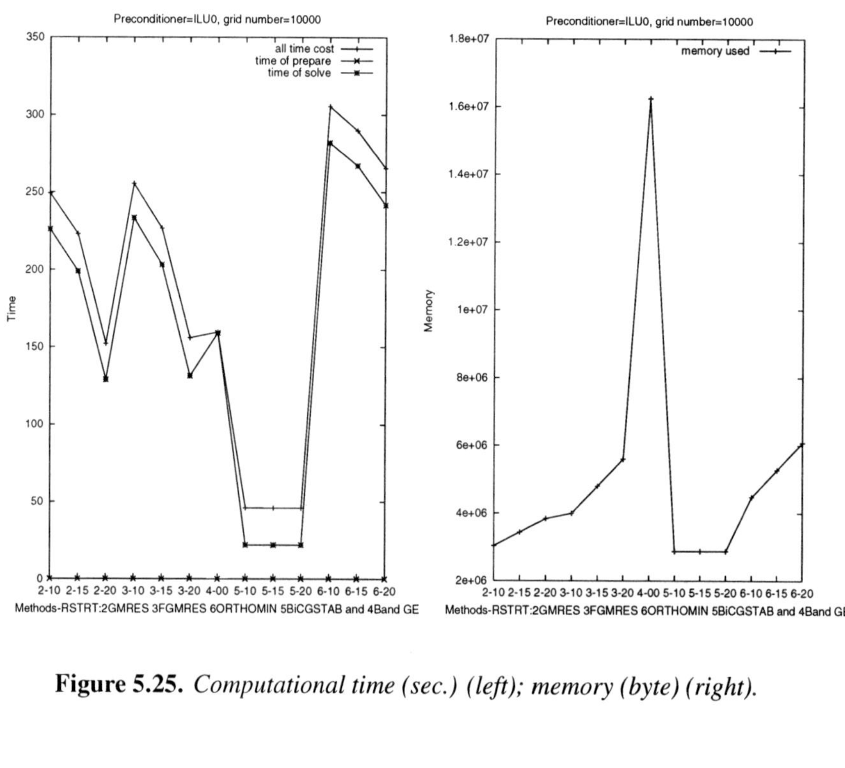

Figure 5.25. *Computational time (sec.) (left); memory (byte) (right).*

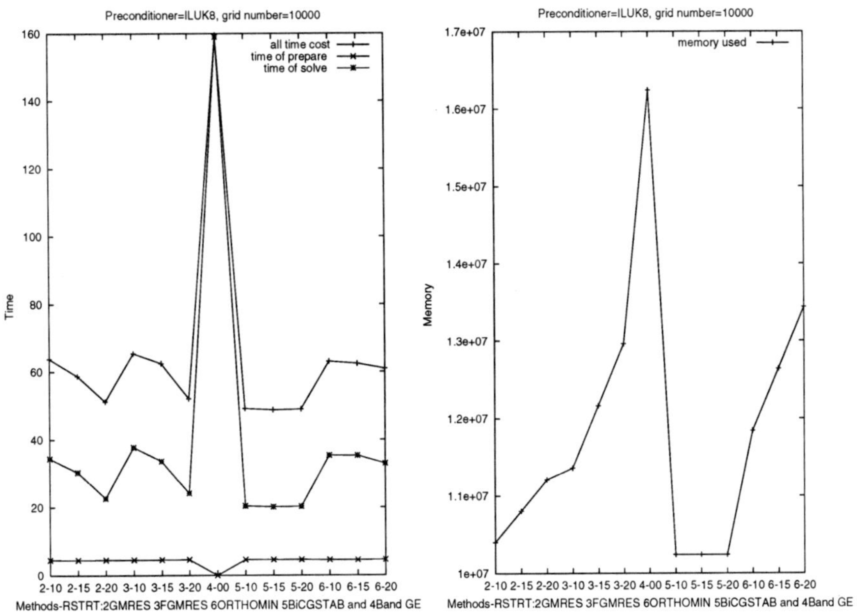

Figure 5.26. *Computational time (sec.) (left); memory (byte) (right).*

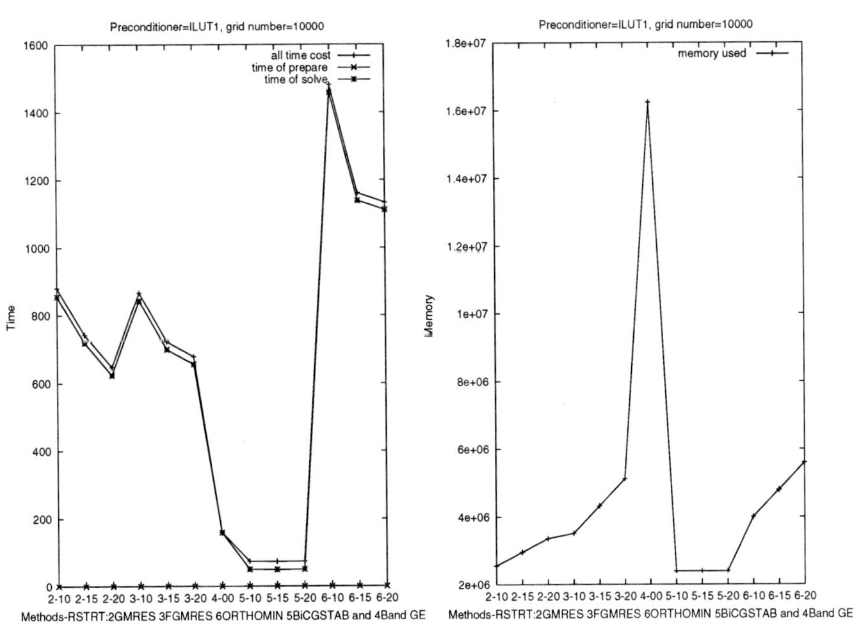

Figure 5.27. *Computational time (sec.) (left); memory (byte) (right).*

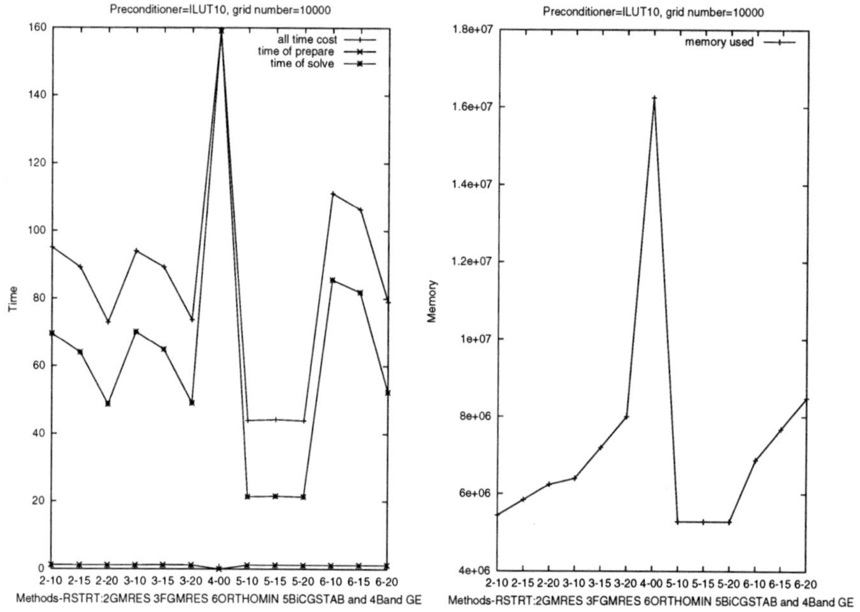

Figure 5.28. *Computational time (sec.) (left); memory (byte) (right).*

- With nearly the same memory requirement, GMRES and FGMRES are better than ORTHOMIN in terms of computational time, when lower-order preconditioners (e.g., ILU(0), ILUT(1)) are employed.

- With nearly the same memory requirement, GMRES, FGMRES, and ORTHOMIN have the same trend with respect to the restart and grid numbers when preconditioners are fixed. Their computational time decreases and storage increases as the restart number increases, for example.

- Of ORTHOMIN, GMRES, FGMRES, and BiCGSTAB with the same preconditioner and nearly the same memory requirement, BiCGSTAB seems the fastest for the problem under consideration.

- With nearly the same memory requirement, ILUT is more efficient than ILU.

- For a fixed linear solver (e.g., GMRES) and a fixed restart number (e.g., 10) for the case with a grid number of 10,000, the higher-order preconditioner ILUT(10) takes as much as 11% of the total CPU time and uses 2.12 times as much memory as that of the lower-order ILUT(1).

Comparisons for more complicated problems, such as for the black oil model (cf. Chapter 8), have been also performed, and observations similar to those made here have been made (Li et al. (2005)).

5.12 Bibliographical Remarks

There are numerous books on Krylov subspace algorithms and their preconditioned versions discussed in this chapter (e.g., Axelsson, 1994; Golub and van Loan, 1996; Saad, 2004). The content of Sections 5.5–5.9 closely follows Saad (2004). The numerical results in Section 5.11 are extracted from the paper by Chen et al. (2002D). Linear algebra routines are available for algorithms of general applicability, such as LAPACK (Anderson et al., 1999), LINPACK (Dongarra et al., 1979), Netlib (Moore et al., 2002), PETSc (Balay et al., 2004), and SPARSKIT (Saad, 1990).

Exercises

5.1. Extend Thomas' algorithm defined in Section 5.1 to a block tridiagonal system with the system matrix $\mathbf{A}$ given in (5.4).

5.2. Show that the number of arithmetic operations in (5.13) for a symmetric matrix $\mathbf{A}$ is asymptotically of order $M^3/6$.

5.3. Prove that the number of operations to factor an $M \times M$ matrix with bandwidth L is $ML^2/2$ (cf. (5.14)).

5.4. Show that if the Arnoldi algorithm does not stop before the kth step, then the vectors $\mathbf{v}^1, \mathbf{v}^2, \ldots, \mathbf{v}^k$ generated by this algorithm form an orthonormal basis for $\mathcal{K}^k$.

5.5. Verify equation (5.24).

5.6. Consider the problem on the unit square $\Omega = (0, 1) \times (0, 1)$:

$$-\frac{\partial^2 p}{\partial x_1^2} - \frac{\partial^2 p}{\partial x_2^2} = q(x_1, x_2), \qquad (x_1, x_2) \in \Omega, \tag{5.37}$$

where q indicates an injector located at $(0.1667, 0.1667)$ or a producer located at $(0.8333, 0.8333)$. A homogeneous Neumann boundary condition (*no-flow boundary condition*) is

$$\frac{\partial p}{\partial \boldsymbol{\nu}} = 0,$$

where $\partial p/\partial \boldsymbol{\nu}$ is the normal derivative and $\boldsymbol{\nu}$ is the outward unit normal to $\Gamma = \partial\Omega$ (the boundary of Ω). (I) Formulate a finite difference scheme for (5.37) similar to scheme (4.20) using a block-centered grid with three equal subintervals in each of the x_1- and x_2-directions. (II) Discretize the Neumann boundary condition using a first-order scheme analogous to (4.14) with $g = 0$. (III) The well term q is evaluated:

$$q_{i,j} = \frac{2\pi}{\ln(r_e/r_w)}(p_{bh} - p_{i,j}) \quad \text{with } (i, j) = (1, 1) \text{ or } (3, 3),$$

where the *wellbore radius* r_w equals 0.001, the *drainage radius* r_e of both wells is given by $r_e = 0.2h$ with h the step size in the x_1- and x_2-directions, and the *wellbore pressure* p_{bh} equals 1.0 at the injector and -1.0 at the producer. Write the finite difference scheme derived in (I) in matrix form $\mathbf{Ap} = \mathbf{q}$ and find matrix $\mathbf{A}$ and vector $\mathbf{q}$. (IV) Use Gaussian elimination as given in Section 5.1 (cf. (5.16) and (5.17)) to solve this system. (V) Use the direct banded solver defined in Section 5.1 to solve the same system. (VI) Use ORTHOMIN defined in Section 5.6 to solve the same system, where the maximum orthogonal number can be 10–25, the iteration can be restarted, and it stops when $\|\mathbf{r}^k\|/\|\mathbf{q}\| \leq 0.00001$. (VII) Use ILU(0) given in Section 5.9.1 as a preconditioner for ORTHOMIN to solve the same system. (VIII) Compare the numerical solutions obtained in (IV)–(VII).

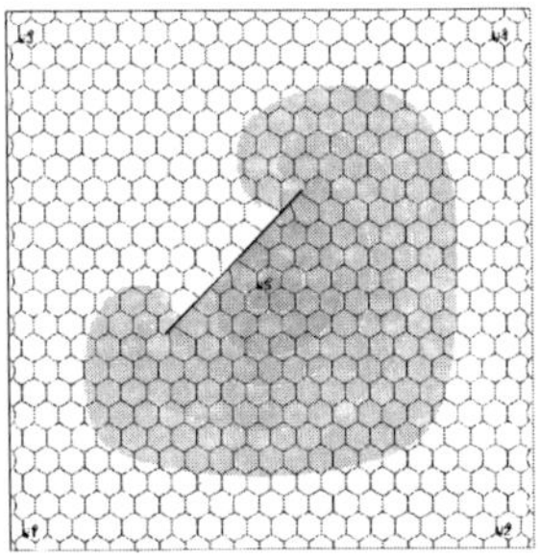

Chapter 6
Single Phase Flow

As noted in Chapter 1, in the very early stage, the reservoir usually contains a single fluid such as oil or gas. Often the pressure at this stage is so high that oil or gas is produced by simple *natural decompression* without any pumping effort at the wells. This stage is referred to as *primary recovery*, and it ends when a pressure equilibrium between the oil or gas field and the atmosphere occurs. The basic differential equations for the flow of a slightly compressible fluid are described in Section 6.1. Then an analytic solution for a one-dimensional radial flow is obtained, and is compared with numerical solutions in Section 6.2. In Section 6.3, finite element methods for general differential equations of single phase flow are presented. Finally, bibliographical information is given in Section 6.4.

6.1 Basic Differential Equations

From Section 2.2.3, the basic differential equation describing the flow of a slightly compressible fluid in a porous medium $\Omega \subset \mathbb{R}^d$ $(1 \leq d \leq 3)$ is (cf. (2.20))

$$\phi \rho c_t \frac{\partial p}{\partial t} = \nabla \cdot \left(\frac{\rho}{\mu} \mathbf{k} \left(\nabla p - \rho \wp \nabla z \right) \right), \tag{6.1}$$

where ϕ and $\mathbf{k}$ are the porosity and absolute permeability tensor of the porous medium; ρ, p, and μ are the density, pressure, and viscosity of the fluid; $\wp$ is the magnitude of the gravitational acceleration; z is the depth; and

$$c_t = c_f + \frac{\phi^o}{\phi} c_R \tag{6.2}$$

is the *total compressibility*, with c_f and c_R the respective compressibility of the fluid and rock and ϕ^o the porosity at a reference pressure p^0. Equation (6.1) is a parabolic equation in p. Existence, uniqueness, and regularity of a solution to this equation can be determined (Chavent and Jaffré, 1986; Friedman, 1982). Its numerical solutions can be also readily performed (cf. Section 6.3).

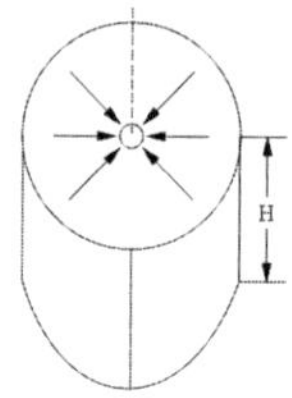

Figure 6.1. *One-dimensional radial flow.*

6.2 One-Dimensional Radial Flow

6.2.1 An analytic solution

In this section, we obtain an *analytic solution* for (6.1) that can be used to check the approximation accuracy for a numerical method for fluid flow in porous media. We assume that Ω is an isotropic medium (cf. Section 2.2.1), and thus $\mathbf{k} = k\mathbf{I}$, where $\mathbf{I}$ is the identity tensor. In cylindrical coordinates (r, θ, x_3), (6.1) takes the form (cf. Exercise 6.1)

$$\begin{aligned}\phi\rho c_t \frac{\partial p}{\partial t} &= \frac{1}{r}\frac{\partial}{\partial r}\left[\frac{r\rho k}{\mu}\left(\frac{\partial p}{\partial r} - \rho\wp\frac{\partial z}{\partial r}\right)\right] \\ &+ \frac{1}{r^2}\frac{\partial}{\partial \theta}\left[\frac{\rho k}{\mu}\left(\frac{\partial p}{\partial \theta} - \rho\wp\frac{\partial z}{\partial \theta}\right)\right] \\ &+ \frac{\partial}{\partial x_3}\left[\frac{\rho k}{\mu}\left(\frac{\partial p}{\partial x_3} - \rho\wp\frac{\partial z}{\partial x_3}\right)\right].\end{aligned} \tag{6.3}$$

We consider a reservoir Ω with an infinite extent in the horizontal direction. Assume that there is an isolated production well (located at $(0, 0, x_3)$) in this reservoir, all its properties are symmetric with respect to the axis of this well, and the reservoir is homogeneous in the vertical direction (cf. Figure 6.1). In addition, if the gravity effect and density change are ignored, (6.3) reduces to

$$\frac{1}{\chi}\frac{\partial p}{\partial t} = \frac{\partial^2 p}{\partial r^2} + \frac{1}{r}\frac{\partial p}{\partial r}, \tag{6.4}$$

where

$$\chi = \frac{k}{\phi\mu c_t}.$$

Thus pressure p is a function of r and t only. That is, the flow is one-dimensional in the radial direction. We find an analytic solution to this one-dimensional equation. Initially, we assume that

$$p(r, 0) = p_0, \qquad 0 \leq r < \infty, \tag{6.5}$$

where p_0 is constant. The boundary conditions are given by

$$\begin{aligned} p(r,t) &= p_0 \qquad && \text{as } r \to \infty,\ t \geq 0, \\ r\frac{\partial p}{\partial r} &= \frac{Q\mu}{2\pi k H} \qquad && \text{as } r \to 0,\ t > 0, \end{aligned} \tag{6.6}$$

where r_w is the radius of the well, Q is a fixed production rate of the well, and H is the thickness of the reservoir.

To solve (6.4), we introduce the *Boltzmann change of variable*

$$y = \frac{r^2}{4t\chi}, \qquad t > 0.$$

Then we see that

$$\begin{aligned}
\frac{\partial p}{\partial r} &= \frac{dp}{dy}\frac{\partial y}{\partial r} = \frac{dp}{dy}\frac{r}{2t\chi}, \\
\frac{\partial^2 p}{\partial r^2} &= \frac{\partial}{\partial r}\left(\frac{dp}{dy}\frac{r}{2t\chi}\right) = \frac{d^2 p}{dy^2}\left(\frac{r}{2t\chi}\right)^2 + \frac{dp}{dy}\frac{1}{2t\chi}, \\
\frac{\partial p}{\partial t} &= \frac{dp}{dy}\frac{\partial y}{\partial t} = -\frac{dp}{dy}\frac{r^2}{4t^2\chi}.
\end{aligned} \tag{6.7}$$

Substituting (6.7) into (6.4) yields

$$y\frac{d^2 p}{dy^2} + (1+y)\frac{dp}{dy} = 0. \tag{6.8}$$

Using the *method of separation of variables*, from (6.8) we obtain (cf. Exercise 6.2)

$$\frac{dp}{dy} = \frac{C}{y}e^{-y}, \tag{6.9}$$

where C is an arbitrary constant. Applying the boundary condition (6.6) to (6.9) gives

$$\frac{dp}{dy} = \frac{Q\mu}{4\pi kH}\frac{e^{-y}}{y}. \tag{6.10}$$

Note that

$$\begin{aligned}
p &= p_0 && \text{when } y = \infty,\ t = 0, \\
p &= p(r,t) && \text{when } y = \frac{r^2}{4t\chi},\ t > 0.
\end{aligned}$$

Integration of (6.10) from $t = 0$ to any t implies

$$p(r,t) = p_0 - \frac{Q\mu}{4\pi kH}\int_{r^2/(4t\chi)}^{\infty} \frac{e^{-y}}{y}\,dy. \tag{6.11}$$

The function $\int_{r^2/(4t\chi)}^{\infty} \frac{e^{-y}}{y}\,dy$ is the *exponential integral function*, and is usually written as

$$\int_{r^2/(4t\chi)}^{\infty} \frac{e^{-y}}{y}\,dy = -Ei\left(-\frac{r^2}{4t\chi}\right) = -Ei(-y).$$

Consequently, it follows from (6.11) that pressure at any r is

$$p(r,t) = p_0 + \frac{Q\mu}{4\pi kH}Ei\left(-\frac{r^2}{4t\chi}\right), \qquad t > 0. \tag{6.12}$$

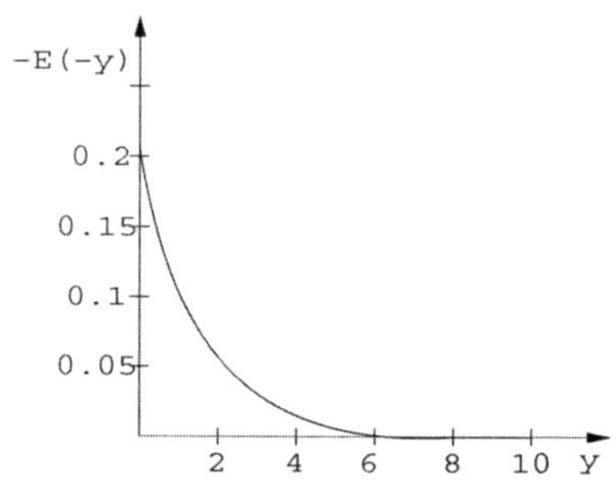

Figure 6.2. *The graph of* $-Ei(-y)$.

The graph of $-Ei(-y)$ in terms of y is displayed in Figure 6.2, which shows that as y increases (r increases or t decreases), $-Ei(-y)$ decreases, so $p(r,t)$ increases and $p_0 - p$ decreases. That is, the farther we are from the well, the larger the pressure but the smaller the pressure drop. The same phenomenon can be observed as t decreases.

If the well starts to operate at $t = t_0$ instead of $t = 0$, the pressure becomes

$$p(r,t) = p_0 + \frac{Q\mu}{4\pi k H} Ei\left(-\frac{r^2}{4(t-t_0)\chi}\right), \qquad t > t_0. \tag{6.13}$$

Similarly, if the well is located at $(x_{1,0}, x_{2,0})$ instead of $(0,0)$, the pressure becomes

$$p(r,t) = p_0 + \frac{Q\mu}{4\pi k H} Ei\left(-\frac{(x_1 - x_{1,0})^2 + (x_2 - x_{2,0})^2}{4t\chi}\right), \qquad t > 0. \tag{6.14}$$

The exponential integral function can be expanded in the series (cf. Exercise 6.3)

$$Ei\left(-\frac{r^2}{4t\chi}\right) = -\ln\left(\frac{4t\chi}{r^2}\right) + 0.5772 - \frac{r^2}{4t\chi} + \frac{1}{4}\left(\frac{r^2}{4t\chi}\right)^2 - \cdots, \qquad t > 0.$$

When $r^2/(4t\chi) < 0.01$, this function can be approximated by

$$Ei\left(-\frac{r^2}{4t\chi}\right) \approx -\ln\left(\frac{4t\chi}{r^2}\right) + 0.5772 = -\ln\left(\frac{2.25t\chi}{r^2}\right),$$

and the resulting approximation error is less than 0.25%. The corresponding simplified analytic solution from (6.12) is

$$p(r,t) \approx p_0 - \frac{Q\mu}{4\pi k H} \ln\left(\frac{2.25t\chi}{r^2}\right). \tag{6.15}$$

At $r = r_w$, $r^2/(4t\chi)$ is small because r_w is small. Then, in a few seconds $r^2/(4t\chi) < 0.01$. Hence (6.15) can be used to find the pressure of the wellbore:

$$p_w(t) = p_0 - \frac{Q\mu}{4\pi k H} \ln\left(\frac{2.25t\chi}{r_w^2}\right). \tag{6.16}$$

Table 6.1. *Parameters for a reservoir.*

Item	Description	Unit	Value
Q_o	Oil production rate	STB/D	300
μ	Oil viscosity	cp	1.06
k	Permeability	md	300
H	Thickness	ft	100
c_o	Oil compressibility (i.e., c_f)	1/psi	0.00001
c_R	Rock compressibility	1/psi	0.000004
ϕ	Porosity	fraction	0.2
p_0	Initial pressure	psia	3,600
p_b	Bubble point pressure	psia	2,000
B_{ob}	Oil formation volume factor at p_b	fraction	1.063
r_w	Radius of wellbore	ft	0.1875
x_{1max}	Length in the x_1-direction	ft	8,100
x_{2max}	Length in the x_2-direction	ft	8,100
h	Length of triangles used in simulation	ft	300
A	Local refinement area near wellbore	ft^2	19,627.7

6.2.2 Numerical comparisons

The simplified analytic solution given in (6.15) can be used to check approximation accuracy for a numerical method. For this, we consider a reservoir with its property parameters given in Table 6.1.

In order to compare with the analytic solution obtained in the previous subsection, we need to change units from the British system to the physics system:

$$
\begin{aligned}
1 \text{ ft} &= 30.48 \text{ cm},\\
1 \text{ day} &= 86,400 \text{ sec}.\\
1 \text{ psi} &= 0.068046 \text{ atm},\\
1 \text{ md} &= 0.001 \text{ darcy},\\
1 \text{ bbl} &= 0.1589873 \times 10^6 \text{ cm}^3.
\end{aligned}
$$

Using these unit transfers, we obtain the following parameters for the analytic solution:

$$
\begin{aligned}
r_w &= 0.1875 \text{ ft} = 5.715 \text{ cm},\\
r_e &= 0.2\sqrt{A} = 0.2\sqrt{19,624.7} = 28.0176 \text{ ft} = 853.98 \text{ cm},\\
k &= 300 \text{ md} = 0.3 \text{ darcy},\\
H &= 100 \text{ ft} = 3,048 \text{ cm},\\
\mu &= 1.06 \text{ cp},\\
c_t &= c_o + c_R = 1.4 \times 10^{-5} \text{ 1/psi} = 2.05743 \times 10^{-4} \text{ 1/atm},\\
\chi &= \frac{k}{\phi \mu c_t} = 6,877.97 \text{ cm}^2/\text{sec}.\\
B_o &= B_{ob}\big(1 - c_o(p_0 - p_b)\big)\\
&= 1.063\big(1 - 10^{-5}(3,600 - 2,000)\big) = 1.04599,
\end{aligned}
$$

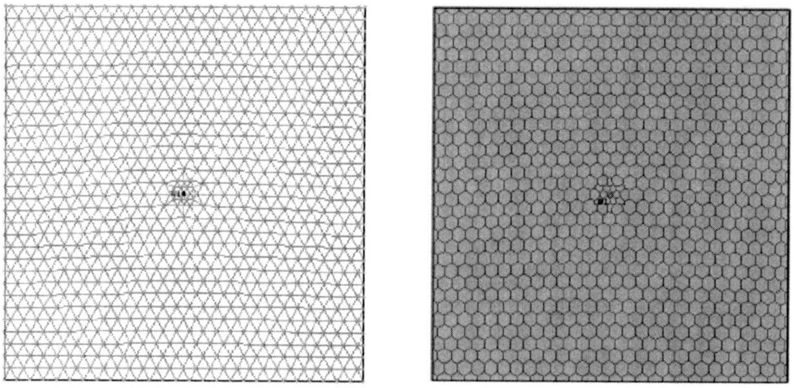

Figure 6.3. *Base triangles and control volumes.*

$$Q = Q_o B_o = 300 \cdot 1.04599 = 313.7976 \text{ RB/D}$$
$$= 313.7976 \cdot 0.1589873 \times 10^6/86,400 = 577.4286 \text{ cm}^3/\text{sec}.$$

where r_e is the *equivalent radius* (cf. Chapter 13). We compare the numerical pressure with the simplified analytic solution at $r = r_w,\ r_e$:

$$p(r,t) = p_0 - \frac{Q\mu}{4\pi kH} \ln\left(\frac{2.25t\chi}{r^2}\right) \qquad \text{when } \frac{r^2}{4t\chi} < 0.01. \tag{6.17}$$

The numerical method is based on the control volume finite element method with piecewise linear functions presented in Section 4.3. Triangles are used to construct the control volumes (cf. Figure 6.3). When this method is used to solve (6.4), we employ local grid refinement near the well to maintain uniform accuracy. The comparisons between the numerical pressure p_h and the analytical pressure p at $r = r_w$ and $r = r_e$ are shown in, respectively, Tables 6.2 and 6.3. From these tables, we see that the numerical solution is very close to the analytical solution. When the size of the triangles used in the numerical experiments is reduced, convergence of the numerical to the analytical solution can be observed.

We now increase the size of the reservoir in the horizontal direction; the length in the x_1- and x_2-directions, x_{1max} and x_{2max}, is increased from 8,100 ft to 13,500 ft. The corresponding numerical and analytical solutions at $r = r_w$ are given in Table 6.4. We see that the difference between these two solutions is now less than 0.01. Thus the numerical solution gets closer to the analytical solution as the size of the reservoir in the horizontal direction becomes larger.

6.3 Finite Element Methods for Single Phase Flow

We now return to the three-dimensional single phase flow equation (6.1). For generality, we write it in a more general form:

Table 6.2. *The pressure comparison at $r = r_w$.*

Time days	Time sec.	$r^2/(4t\chi)$ $\times 10^{-8}$	p_h psia	p psia	$p_h - p$ psia
0.1	8,640	13.740	3,596.32	3,595.92	0.40
0.2	17,280	6.870	3,595.47	3,595.38	0.09
0.3	25,920	4.580	3,595.07	3,595.06	0.01
0.4	34,560	3.435	3,594.80	3,594.84	−0.04
0.5	43,200	2.748	3,594.60	3,594.66	−0.06
0.6	51,840	2.290	3,594.45	3,594.52	−0.07
0.7	60,480	1.963	3,594.31	3,594.40	−0.09
0.8	69,120	1.718	3,594.20	3,594.29	−0.09
0.9	77,760	1.527	3,594.10	3,594.20	−0.10
1.0	86,400	1.374	3,594.01	3,594.12	−0.11
1.5	129,600	0.916	3,593.69	3,593.80	−0.11
2.0	172,800	0.687	3,593.46	3,593.58	−0.12
2.5	216,000	0.550	3,593.28	3,593.40	−0.12
3.0	259,200	0.458	3,593.13	3,593.26	−0.13
4.0	345,600	0.344	3,592.90	3,593.03	−0.13

Table 6.3. *The pressure comparison at $r = r_e$.*

Time days	Time sec.	$r^2/(4t\chi)$ $\times 10^{-4}$	p_h psia	p psia	$p_h - p$ psia
0.1	8,640	30.680	3,588.48	3,588.08	0.40
0.2	17,280	15.340	3,587.63	3,587.54	0.09
0.3	25,920	10.226	3,587.23	3,587.22	0.01
0.4	34,560	7.670	3,586.96	3,587.00	−0.04
0.5	43,200	6.136	3,586.76	3,586.82	−0.06
0.6	51,840	5.113	3,586.61	3,586.68	−0.07
0.7	60,480	4.383	3,586.47	3,586.56	−0.09
0.8	69,120	3.835	3,586.36	3,586.45	−0.09
0.9	77,760	3.409	3,586.26	3,586.36	−0.10
1.0	86,400	3.068	3,586.17	3,586.28	−0.11
1.5	129,600	2.045	3,585.85	3,585.96	−0.11
2.0	172,800	1.534	3,585.62	3,585.74	−0.12
2.5	216,000	1.227	3,585.44	3,585.56	−0.12
3.0	259,200	1.023	3,585.29	3,585.42	−0.13
4.0	345,600	0.767	3,585.06	3,585.19	−0.13

$$
\begin{aligned}
c(p)\frac{\partial p}{\partial t} - \nabla \cdot \big(\mathbf{a}(p)\nabla p\big) &= f(p) && \text{in } \Omega \times J, \\
\mathbf{a}(p)\nabla p \cdot \boldsymbol{\nu} &= 0 && \text{on } \Gamma \times J, \\
p(\cdot, 0) &= p_0 && \text{in } \Omega,
\end{aligned}
\tag{6.18}
$$

where $c(p) = c(\mathbf{x}, t, p)$, $\mathbf{a}(p) = \mathbf{a}(\mathbf{x}, t, p)$, and $f(p) = f(\mathbf{x}, t, p)$ depend on pressure p, $\boldsymbol{\nu}$ is the outward unit normal to the boundary Γ of Ω, the function p_0 is given, and $J = (0, T)$ $(T > 0)$ is the time interval of interest. Various numerical methods were given in Chapter 4 for the solution of a linear version of (6.18). Now is a good time to see how to extend these

Table 6.4. *The pressure comparison at $r = r_w$ for a larger reservoir.*

Time	Time	$r^2/(4t\chi)$	p_h	p	$p_h - p$
days	sec.	$\times 10^{-8}$	psia	psia	psia
0.1	8,640	13.740	3,596.32	3,596.32	0.00
0.2	17,280	6.870	3,595.47	3,595.46	0.01
0.3	25,920	4.580	3,595.07	3,595.06	0.01
0.4	34,560	3.435	3,594.80	3,594.80	0.00
0.5	43,200	2.748	3,594.60	3,594.60	0.00
0.6	51,840	2.290	3,594.45	3,594.44	0.01
0.7	60,480	1.963	3,594.31	3,594.31	0.00
0.8	69,120	1.718	3,594.20	3,594.19	0.01
0.9	77,760	1.527	3,594.10	3,594.09	0.01
1.0	86,400	1.374	3,594.01	3,594.01	0.00
1.5	129,600	0.916	3,593.69	3,593.68	0.01
2.0	172,800	0.687	3,593.46	3,593.45	0.01
2.5	216,000	0.550	3,593.28	3,593.27	0.01
3.0	259,200	0.458	3,593.13	3,593.12	0.01
4.0	345,600	0.344	3,592.90	3,593.90	0.00

methods to this nonlinear equation. As an example, we just consider the standard finite element methods discussed in Section 4.2 for (6.18); similar considerations can be given for other finite element methods (Chen, 2005; also cf. Exercise 6.4 and Section 7.5).

In (6.18), for notational convenience we drop the dependence of these coefficients on $\mathbf{x}$ and t and assume that (6.18) admits a unique solution. Furthermore, we assume that the coefficients $c(p)$, $\mathbf{a}(p)$, and $f(p)$ are *globally Lipschitz continuous* in p; i.e., for some constants C_ξ, they satisfy

$$|\xi(p_1) - \xi(p_2)| \le C_\xi |p_1 - p_2|, \qquad p_1,\ p_2 \in \mathbb{R},\ \xi = c,\ \mathbf{a},\ \text{or } f. \tag{6.19}$$

With $V = H^1(\Omega)$ (cf. Section 4.2), problem (6.18) can be written in the variational form: Find $p : J \to V$ such that

$$\begin{aligned} &\left(c(p)\frac{\partial p}{\partial t}, v\right) + \big(\mathbf{a}(p)\nabla p, \nabla v\big) = \big(f(p), v\big) && \forall v \in V,\ t \in J, \\ &p(\mathbf{x}, 0) = p_0(\mathbf{x}) && \forall \mathbf{x} \in \Omega. \end{aligned} \tag{6.20}$$

Let V_h be a finite element subspace of V (cf. Section 4.2.1). The finite element version of (6.20) is: Find $p_h : J \to V_h$ such that

$$\begin{aligned} &\left(c(p_h)\frac{\partial p_h}{\partial t}, v\right) + \big(\mathbf{a}(p_h)\nabla p_h, \nabla v\big) = \big(f(p_h), v\big) && \forall v \in V_h, \\ &(p_h(\cdot, 0), v) = (p_0, v) && \forall v \in V_h. \end{aligned} \tag{6.21}$$

As for (4.96), after the introduction of basis functions in V_h, (6.21) can be stated in matrix form (cf. Exercise 6.5)

$$\begin{aligned} &\mathbf{C}(\mathbf{p})\frac{d\mathbf{p}}{dt} + \mathbf{A}(\mathbf{p})\mathbf{p} = \mathbf{f}(\mathbf{p}), \qquad t \in J, \\ &\mathbf{B}\mathbf{p}(0) = \mathbf{p}_0. \end{aligned} \tag{6.22}$$

Under the assumption that the coefficient $c(p)$ is bounded below by a positive constant, this nonlinear system of ODEs possesses a unique solution (at least locally). In fact, because of assumption (6.19) on c, $\mathbf{a}$, and f, the solution $\mathbf{p}(t)$ exists for all t. Several approaches for solving (6.22) are discussed next.

6.3.1 Linearization approaches

Let $0 = t^0 < t^1 < t^2 < \cdots < t^N$ be a partition of J, and set $\Delta t^n = t^n - t^{n-1}$, $n = 1, 2, \ldots, N$. The nonlinear system (6.22) can be linearized by allowing the nonlinearities to lag one time step behind. Thus the modified backward Euler method for (6.18) takes the form: Find $p_h^n \in V_h$, $n = 1, 2, \ldots, N$, such that

$$\begin{aligned} \left(c\left(p_h^{n-1}\right) \frac{p_h^n - p_h^{n-1}}{\Delta t^n}, v \right) + \left(\mathbf{a}\left(p_h^{n-1}\right) \nabla p_h^n, \nabla v\right) & \\ = \left(f\left(p_h^{n-1}\right), v\right) \qquad & \forall v \in V_h, \\ \left(p_h^0, v\right) = (p_0, v) \qquad \forall v \in V_h. & \end{aligned} \tag{6.23}$$

In matrix form it is given by

$$\begin{aligned} \mathbf{C}\left(\mathbf{p}^{n-1}\right) \frac{\mathbf{p}^n - \mathbf{p}^{n-1}}{\Delta t^n} + \mathbf{A}\left(\mathbf{p}^{n-1}\right)\mathbf{p}^n &= \mathbf{f}\left(\mathbf{p}^{n-1}\right), \\ \mathbf{B}\mathbf{p}(0) &= \mathbf{p}_0. \end{aligned} \tag{6.24}$$

Note that (6.24) is a system of linear equations in $\mathbf{p}^n$, which can be solved using iterative algorithms discussed in the previous chapter, for example. When V_h is the finite element space of piecewise linear functions, the error $p^n - p_h^n$ $(0 \le n \le N)$ in the $L^2(\Omega)$-norm is asymptotically of order $\mathcal{O}(\Delta t + h^2)$ under appropriate smoothness assumptions on p and for Δt small enough (Thomée, 1984; Chen and Douglas, 1991), where $\Delta t = \max_{1 \le n \le N} \Delta t^n$. We may use the Crank–Nicholson discretization method in (6.23). However, the linearization decreases the order of the time discretization error to $\mathcal{O}(\Delta t)$, giving $\mathcal{O}(\Delta t + h^2)$ overall. This is true for any higher-order time discretization method with the present linearization technique. This drawback can be overcome by using *extrapolation techniques* in the linearization of the coefficients c, $\mathbf{a}$, and f (cf. Section 7.5.3). Combined with an appropriate extrapolation, the Crank–Nicholson method can be shown to produce an error of order $\mathcal{O}((\Delta t)^2)$ in time (Douglas, 1961; Thomée, 1984). On the other hand, higher-order extrapolations generally increase data storage.

6.3.2 Implicit time approximations

We now consider a fully *implicit time approximation* scheme for problem (6.18): Find $p_h^n \in V_h$, $n = 1, 2, \ldots, N$, such that

$$\left(c\left(p_h^n\right)\frac{p_h^n - p_h^{n-1}}{\Delta t^n}, v\right) + \left(\mathbf{a}\left(p_h^n\right)\nabla p_h^n, \nabla v\right)$$
$$= \left(f\left(p_h^n\right), v\right) \qquad \forall v \in V_h, \tag{6.25}$$
$$\left(p_h^0, v\right) = (p_0, v) \qquad \forall v \in V_h.$$

Its matrix form is

$$\mathbf{C}\left(\mathbf{p}^n\right)\frac{\mathbf{p}^n - \mathbf{p}^{n-1}}{\Delta t^n} + \mathbf{A}(\mathbf{p}^n)\mathbf{p}^n = \mathbf{f}(\mathbf{p}^n),$$
$$\mathbf{B}\mathbf{p}(0) = \mathbf{p}_0. \tag{6.26}$$

Now, system (6.26) is a system of nonlinear equations in $\mathbf{p}^n$, which must be solved at each time step via an iteration method. Let us consider *Newton's method* (or the Newton–Raphson method; cf. Chapter 8). Note that the first equation of (6.26) can be rewritten as

$$\left(\mathbf{A}(\mathbf{p}^n) + \frac{1}{\Delta t^n}\mathbf{C}\left(\mathbf{p}^n\right)\right)\mathbf{p}^n - \frac{1}{\Delta t^n}\mathbf{C}\left(\mathbf{p}^n\right)\mathbf{p}^{n-1} - \mathbf{f}(\mathbf{p}^n) = \mathbf{0}.$$

We express this equation as

$$\mathbf{F}(\mathbf{p}^n) = \mathbf{0}. \tag{6.27}$$

Newton's method for (6.27) is

$$\text{Set } \mathbf{v}^0 = \mathbf{p}^{n-1};$$
$$\text{Iterate } \mathbf{v}^k = \mathbf{v}^{k-1} + \mathbf{d}^k, \qquad k = 1, 2, \ldots,$$

where $\mathbf{d}^k$ satisfies the equation

$$\mathbf{G}(\mathbf{v}^{k-1})\mathbf{d}^k = -\mathbf{F}(\mathbf{v}^{k-1})$$

with $\mathbf{G}$ the Jacobian matrix of the vector function $\mathbf{F}$:

$$\mathbf{G} = \left(\frac{\partial F_i}{\partial p_j}\right)_{i,j=1,2,\ldots,M}$$

(recall that M is the dimension of $\mathbf{p}$). If the matrix $\mathbf{G}(\mathbf{p}^n)$ is nonsingular and the second partial derivatives of $\mathbf{F}$ are bounded, Newton's method converges quadratically in a neighborhood of $\mathbf{p}^n$; i.e., there are constants $\epsilon > 0$ and C such that if $|\mathbf{v}^{k-1} - \mathbf{p}^n| \le \epsilon$, then

$$\left|\mathbf{v}^k - \mathbf{p}^n\right| \le C\left|\mathbf{v}^{k-1} - \mathbf{p}^n\right|^2.$$

The main difficulty with Newton's method is to get a sufficiently good initial guess $\mathbf{v}^0$. Once it is obtained, Newton's method converges with very few iterations. This method is a very powerful iteration method for strongly nonlinear problems. There are many variants of Newton's method available in the literature (Ostrowski, 1973; Rheinboldt, 1998). The Crank–Nicholson discretization scheme in time can be also used in (6.25). In the present implicit case, this scheme generates second-order accuracy in time. Numerical experience has revealed that the Crank–Nicholson scheme may not be a good choice for nonlinear parabolic equations because it can be unstable for such equations.

6.3.3 Explicit time approximations

We conclude with a remark about the application of a *forward, explicit time approximation* method to (6.18): Find $p_h^n \in V_h, n = 1, 2, \ldots, N$, such that

$$\left(c\left(p_h^n\right)\frac{p_h^n - p_h^{n-1}}{\Delta t^n}, v\right) + \left(\mathbf{a}\left(p_h^{n-1}\right)\nabla p_h^{n-1}, \nabla v\right) = \left(f\left(p_h^{n-1}\right), v\right) \qquad \forall v \in V_h, \tag{6.28}$$

$$\left(p_h^0, v\right) = (p_0, v) \qquad \forall v \in V_h.$$

In matrix form it is written as

$$\begin{aligned} \mathbf{C}\left(\mathbf{p}^n\right)\frac{\mathbf{p}^n - \mathbf{p}^{n-1}}{\Delta t^n} + \mathbf{A}\left(\mathbf{p}^{n-1}\right)\mathbf{p}^{n-1} &= \mathbf{f}\left(\mathbf{p}^{n-1}\right), \\ \mathbf{B}\mathbf{p}(0) &= \mathbf{p}_0. \end{aligned} \tag{6.29}$$

Note that the only nonlinearity is in matrix $\mathbf{C}$. With an appropriate *mass lumping* (a diagonalization technique; off-diagonal quantities are placed in the right-hand side of (6.29)) in this matrix, the first equation in (6.29) represents M scalar nonlinear equations of the form

$$\mathcal{F}(p_i^n) = 0, \qquad i = 1, 2, \ldots, M. \tag{6.30}$$

Each single equation in (6.30) can be easily solved via any standard method (Ostrowski, 1973; Rheinboldt, 1998).

For the explicit method (6.28) to be stable in the sense defined in Section 4.2.4, a *stability condition* of the following type must be satisfied:

$$\Delta t^n \le Ch^2, \qquad n = 1, 2, \ldots, N, \tag{6.31}$$

where C now depends on c and $\mathbf{a}$ (cf. (4.108)). Unfortunately, this condition on the time steps is very restrictive for long-time integration, as noted earlier.

In summary, we have developed linearization, implicit, and explicit time approximation approaches for numerically solving (6.18). In terms of computational effort, the explicit approach is the simplest at each time step; however, it requires an impractical stability restriction. The linearization approach is more practical, but it reduces the order of accuracy in time for high-order time discretization methods (unless extrapolations are exploited). An efficient and accurate method is the fully implicit approach; the extra cost involved at each time step for this implicit method is usually more than compensated for by the fact that larger time steps may be taken, particularly when Newton's method with a good initial guess is employed. Modified implicit methods such as *semi-implicit methods* (Aziz and Settari, 1979) can be applied; for a given physical problem, the linearization approach should be applied for weak nonlinearity (e.g., the dependence of viscosity μ on pressure p), while the implicit one should be used for strong nonlinearity (e.g., the dependence of density ρ on p); refer to Chen et al. (2000C).

6.4 Bibliographical Remarks

The content of Section 6.3 closely follows Chen (2005). For numerical solutions of problem (6.18) using the discontinuous, mixed, characteristic, and adaptive finite element methods introduced in Chapter 4, the reader can refer to Chen (2005).

Exercises

6.1. Use the cylindrical coordinates (r, θ, x_3) to show that equation (6.1) can be written in the form of (6.3) (recall that $x_1 = r\cos\theta$, $x_2 = r\sin\theta$, and $x_3 = x_3$).

6.2. Prove that equation (6.8) can be reduced to (6.9) using the method of separation of variables.

6.3. Define the *Euler constant* $\gamma = \int_0^\infty e^{-x}\ln x\,dx \approx -5772$. Prove that if $0 < x < 0.1$, the exponential integral function Ei can be approximated:

$$Ei(-x) \approx \ln x + \gamma.$$

(Hint: first show that $Ei(-x) = \gamma + \ln x - x + \frac{1}{2\cdot 2!}x^2 - \frac{1}{3\cdot 3!}x^3 + \cdots$.)

6.4. After introducing appropriate spaces $\mathbf{V}$ and W (cf. Section 4.5.2), write problem (6.18) in a mixed variational formulation.

6.5. After the introduction of basis functions in V_h and of appropriate matrices and vectors, show that system (6.21) can be written as (6.22).

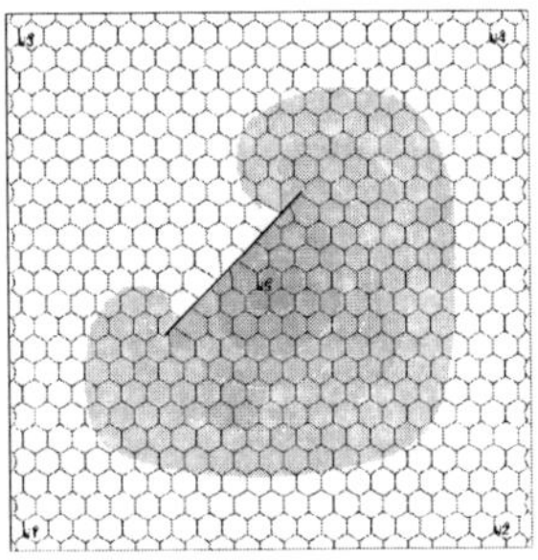

Chapter 7
Two-Phase Flow

As mentioned in Chapter 1, to recover part of the remaining oil after the primary recovery, a fluid (usually water) is injected into some wells (injection wells) while oil is produced through other wells (production wells). This process serves to maintain high reservoir pressure and flow rates. It also displaces some of the oil and pushes it toward the production wells. This stage of oil recovery is called *secondary recovery*. In this recovery stage, if the reservoir pressure is above the *bubble point pressure* of the oil phase, then there is *two-phase immiscible flow*, one phase being water and the other being oil, without mass transfer between the phases. As defined in Chapter 3, a *bubble point* is defined as the state in which the flow system entirely consists of liquids (water and oil), and the pressure at this point is the bubble point pressure.

The basic differential equations for two-phase immiscible flow are described in Section 7.1. A one-dimensional case where an *analytic solution* can be obtained is studied in Section 7.2. In Section 7.3, we consider a solution approach, *IMPES* (implicit pressure-explicit saturation) for solving the differential equations governing two-phase flow, and compare it with a recently introduced approach, an *improved IMPES*. Alternative differential formulations for these differential equations are discussed and compared in Section 7.4. Various finite element methods developed in Chapter 4 are applied to these formulations and compared in Section 7.5. Simulation of miscible displacement of one fluid by another is described in Section 7.6. Finally, bibliographical information is given in Section 7.7.

7.1 Basic Differential Equations

We describe the basic differential equations for two-phase flow in a porous medium Ω. The phase (e.g., water) that wets the porous medium more than the other (e.g., oil) is the *wetting phase* and is indicated by a subscript w. The other phase is termed the *nonwetting phase* and indicated by o. The basic equations can be found in Section 2.3.1; for completeness, we review these equations.

Mass is conserved within each fluid phase:

$$\frac{\partial(\phi\rho_\alpha S_\alpha)}{\partial t} = -\nabla\cdot(\rho_\alpha \mathbf{u}_\alpha) + q_\alpha, \qquad \alpha = w, o, \tag{7.1}$$

where ϕ is the porosity of the porous medium and each phase has its own saturation S_α, density ρ_α, Darcy's velocity $\mathbf{u}_\alpha$, and mass flow rate q_α. Darcy's law for each phase reads:

$$\mathbf{u}_\alpha = -\frac{k_{r\alpha}}{\mu_\alpha}\mathbf{k}\left(\nabla p_\alpha - \rho_\alpha \wp \nabla z\right), \qquad \alpha = w,\, o, \tag{7.2}$$

where $\mathbf{k}$ is the absolute permeability tensor of the porous medium; $k_{r\alpha}$, p_α, and μ_α are the relative permeability, pressure, and viscosity for phase α; $\wp$ is the magnitude of the gravitational acceleration; and z is the depth. The fact that the two fluids jointly fill the voids implies the relation

$$S_w + S_o = 1, \tag{7.3}$$

and the pressure difference between the two phases is given by the capillary pressure

$$p_c(S_w) = p_o - p_w. \tag{7.4}$$

Typical functions of p_c and $k_{r\alpha}$ were given in Chapter 3. Equations (7.1)–(7.4) provide six equations for the six unknowns p_α, $\mathbf{u}_\alpha$, and S_α, $\alpha = w, o$. Alternative differential equations were discussed in Section 2.3.2 and are further discussed in this chapter. The existence, uniqueness, and regularity of a solution to the two-phase flow system were shown under the assumption that the two fluids are incompressible (Chen, 2001; Chen, 2002A).

7.2 One-Dimensional Flow

7.2.1 An analytic solution

As in the treatment of single phase flow, an analytic solution for a simple two-phase flow system is obtained.

Analytic solution before water breakthrough

The *breakthrough time* t_B is an important event in the water-oil displacement; as $t > t_B$, we are producing some of the water being injected. Assume that Ω is a rigid, isotropic medium and that it is homogeneous in the x_2- and x_3-directions (cf. Section 2.2.1). All its properties depend only on x_1. That is, we consider a one-dimensional flow in the x-direction ($x = x_1$). In addition, if the gravity and capillary effects are ignored, the mass conservation equations (7.1) become

$$\begin{aligned}\phi\frac{\partial S_w}{\partial t} + \frac{\partial u_w}{\partial x} &= 0,\\ \phi\frac{\partial S_o}{\partial t} + \frac{\partial u_o}{\partial x} &= 0,\end{aligned} \tag{7.5}$$

and Darcy's law (7.2) simplifies to

$$\begin{aligned}u_w &= -k\frac{k_{rw}(S_w)}{\mu_w}\frac{\partial p}{\partial x},\\ u_o &= -k\frac{k_{ro}(S_o)}{\mu_o}\frac{\partial p}{\partial x}.\end{aligned} \tag{7.6}$$

We introduce the *phase mobilities*

$$\lambda_\alpha(S_\alpha) = \frac{k_{r\alpha}(S_\alpha)}{\mu_\alpha}, \qquad \alpha = w, o,$$

and the *total mobility*

$$\lambda(S_w) = \lambda_w(S_w) + \lambda_o(1 - S_w).$$

The *fractional flow functions* are

$$f_w(S_w) = \frac{\lambda_w(S_w)}{\lambda(S_w)}, \quad f_o(S_w) = \frac{\lambda_o(1 - S_w)}{\lambda(S_w)}.$$

We also define the total velocity

$$u = u_w + u_o. \tag{7.7}$$

Using (7.3) and (7.5), we see that

$$\frac{\partial u}{\partial x} = 0, \tag{7.8}$$

and thus u is independent of x. Because $u_w = f_w(S_w)u$, it follows that

$$\frac{\partial u_w}{\partial x} = f_w \frac{\partial u}{\partial x} + u \frac{d f_w(S_w)}{d S_w} \frac{\partial S_w}{\partial x} = u F_w(S_w) \frac{\partial S_w}{\partial x}, \tag{7.9}$$

where the *distribution function* F_w of saturation is defined by

$$F_w(S_w) = \frac{d f_w(S_w)}{d S_w}.$$

Now, we substitute (7.9) into the first equation of (7.5) to see that

$$\phi \frac{\partial S_w}{\partial t} + u F_w(S_w) \frac{\partial S_w}{\partial x} = 0. \tag{7.10}$$

This equation defines a *characteristic* $x(t)$ along the *interstitial velocity* v by

$$\frac{dx}{dt} = v(x, t) \equiv \frac{u F_w(S_w)}{\phi}. \tag{7.11}$$

Along this characteristic, it follows from (7.10) that S_w is constant; i.e.,

$$\frac{d S_w(x(t), t)}{dt} = \frac{\partial S_w}{\partial x} \frac{dx}{dt} + \frac{\partial S_w}{\partial t} = 0. \tag{7.12}$$

Let A be the cross-sectional area (in the x_2x_3-plane) of Ω, and define the *cumulative liquid production*

$$V(t) = A \int_0^t u \, dt. \tag{7.13}$$

From (7.11), along the characteristic $x(t)$ we see that

$$\int_0^t dx = \frac{F_w(S_w)}{\phi} \int_0^t u \, dt,$$

so, by (7.13),

$$x(S_w, t) = \frac{F_w(S_w)}{\phi A} V(t), \tag{7.14}$$

from which we can find the saturation S_w before water breaks through.

Analytic solution at the water front

Let S_{wf} be the water saturation at the water front, and S_{wc} be the critical saturation (cf. Section 3.1.2). From the *material balance equation*

$$u_w\Big|_{\text{at water front}} = \phi(S_{wf} - S_{wc})\frac{dx}{dt},$$

we have

$$\phi(S_{wf} - S_{wc})\frac{dx}{dt} = f_w u, \tag{7.15}$$

since $u_w = f_w(S_w)u$. Applying (7.11) to (7.15) gives

$$(S_{wf} - S_{wc})F_w = f_w;$$

i.e.,

$$\frac{df_w}{dS_w}(S_{wf}) = \frac{f_w(S_{wf})}{S_{wf} - S_{wc}}. \tag{7.16}$$

Equation (7.16) indicates that the slope of the tangent to the curve of f_w at S_{wf} equals the slope of the secant line through the points $(S_{wf}, f_w(S_{wf}))$ and $(S_{wc}, f_w(S_{wc}))$ (note that $f_w(S_{wc}) = 0$; cf. Section 3.1.2). Thus a graphical method based on this feature can be used to find the water saturation at the water front from (7.16).

Analytic solution after water breakthrough

Let L be the length of Ω in the x-direction and S_{we} be the value of the saturation at $x = L$. At $x = L$, it follows from (7.14) that

$$V(t) = \frac{\phi AL}{F_w(S_{we})}. \tag{7.17}$$

We define the nondimensional cumulative liquid production

$$\bar{V}(t) = \frac{V(t)}{\phi AL}.$$

Then we see that

$$\bar{V}(t) = \frac{1}{F_w(S_{we})}. \tag{7.18}$$

Also, we introduce the *cumulative water production*

$$V_w(t) = \int_{t_B}^{t} f_w \, dV(t) = A\int_{t_B}^{t} u_w \, dt, \tag{7.19}$$

where we recall that t_B is the water breakthrough time (i.e., S_w equals the critical value S_{wc} at $t = t_B$) and where we used the fact that $f_w dV = Au_w dt$ by (7.13). The nondimensional cumulative water production is

$$\bar{V}_w = \frac{V_w}{\phi AL}.$$

It follows from (7.19) and integration by parts that

$$\bar{V}_w = \frac{1}{\phi AL} \int_{t_B}^{t} f_w \, dV(t) = \frac{1}{\phi AL} \left(f_w V - \int_{t_B}^{t} V \, df_w \right),$$

since $f_w(S_{wc}) = 0$. Consequently, by the fact that $df_w = F_w \, dS_w$, we see that

$$\bar{V}_w = \frac{1}{\phi AL} \left(f_w V - \int_{t_B}^{t} V F_w \, dS_w \right).$$

Finally, applying (7.17), we obtain

$$\bar{V}_w = \frac{f_w(S_{we})}{F_w(S_{we})} - (S_{we} - S_{wc}), \tag{7.20}$$

which defines the value of S_{we}.

We can also define the cumulative oil production

$$V_o(t) = \int_{t_B}^{t} f_o \, dV(t) = A \int_{t_B}^{t} u_o \, dt,$$

and the corresponding nondimensional value

$$\bar{V}_o = \frac{V_o}{\phi AL}.$$

Then we derive (cf. Exercise 7.1)

$$\bar{V}_o = \frac{1 - f_w(S_{we})}{F_w(S_{we})} + (S_{we} - S_{wc}) \tag{7.21}$$

and

$$\bar{V} = \bar{V}_w + \bar{V}_o.$$

Either of (7.20) and (7.21) can be utilized to find S_{we}.

7.2.2 An example

We consider an example with $\mu_w = 0.42$ cp, $\mu_o = 15.5$ cp, and the water and oil relative permeabilities given in Table 7.1.

Based on Table 7.1, we can construct the fractional flow function $f_w(S_w)$, which is shown in Figure 7.1. From (7.16) with $S_{wc} = 0.4$, we obtain the water saturation at the water front using the graphical approach defined by (7.16):

$$S_{wf} = 0.5364.$$

Also, by (7.14), we have

$$x(S_w, t) = \frac{F_w(S_w)}{\phi A} V(t).$$

Table 7.1. *Relative permeabilities.*

S_w	k_{rw}	k_{ro}	S_w	k_{rw}	k_{ro}
0.40	0.0000	1.0000	0.62	0.0605	0.2025
0.42	0.0005	0.9025	0.64	0.0720	0.1600
0.44	0.0020	0.8100	0.66	0.0845	0.1225
0.46	0.0045	0.7225	0.68	0.0980	0.0900
0.48	0.0080	0.6400	0.70	0.1125	0.0625
0.50	0.0125	0.5625	0.72	0.1280	0.0400
0.52	0.0180	0.4900	0.74	0.1445	0.0225
0.54	0.0245	0.4225	0.76	0.1620	0.0100
0.56	0.0320	0.3600	0.78	0.1805	0.0025
0.58	0.0405	0.3025	0.80	0.0200	0.0000
0.60	0.0500	0.2500	0.82	0.4500	0.0000

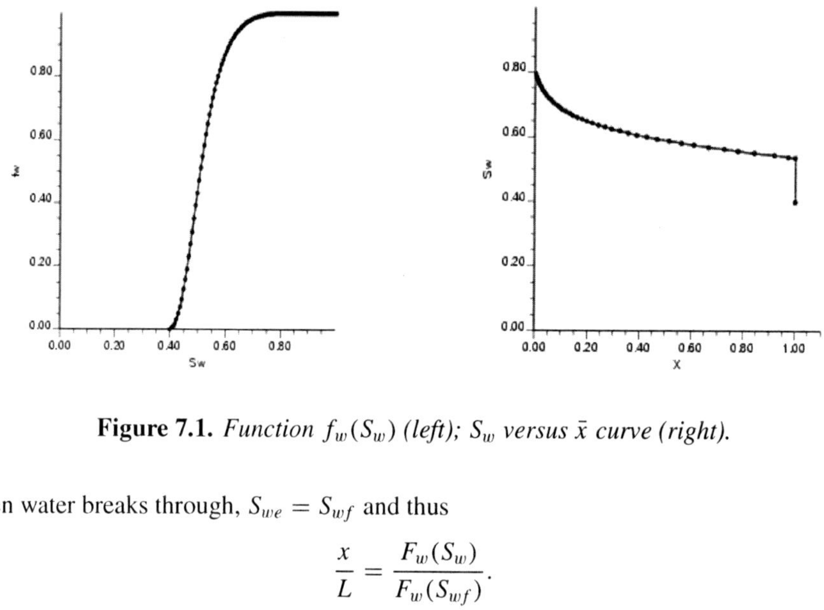

Figure 7.1. *Function $f_w(S_w)$ (left); S_w versus $\bar{x}$ curve (right).*

When water breaks through, $S_{we} = S_{wf}$ and thus

$$\frac{x}{L} = \frac{F_w(S_w)}{F_w(S_{wf})}.$$

Using this equation, the curve of S_w verses $\bar{x}$ is plotted in Figure 7.1, where $\bar{x} = x/L$.

From (7.21) it follows that

$$\bar{V}_o = \frac{1 - f_w(S_{we})}{F_w(S_{we})} + (S_{we} - S_{wc}).$$

Oil recovery is defined by

$$v_o = \frac{\bar{V}_o}{1 - S_{wc}}.$$

The curves of v_o verses the pore volume of water injected and the water cut verses v_o are displayed in Figure 7.2. The *water cut* is defined as $q_w/(q_w + q_o)$, where q_w and q_o are the water and oil production, respectively. In the present example, the water cut equals the fractional flow function f_w since $q_w = f_w(q_w + q_o)$. The curves of v_o verses the pore volume of water injected and the water cut verses v_o indirectly determine S_{we}.

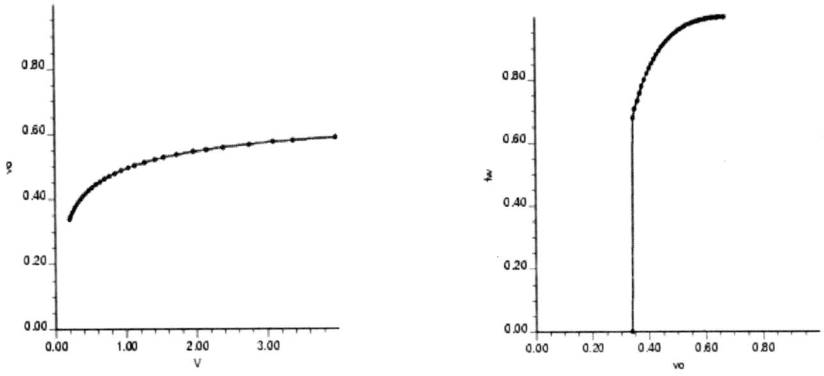

Figure 7.2. *Oil recovery* v_0 *(left); water cut versus* v_o *(right).*

7.3 IMPES and Improved IMPES

Note that the differential equations (7.1)–(7.4) are nonlinear and coupled. There exist a variety of approaches for solving these equations, such as the IMPES, *SS* (simultaneous solution), *sequential*, and *adaptive implicit methods*, as mentioned in Chapter 1. In light of the fact that the IMPES is still popular in the petroleum industry and a very powerful method for solving two-phase flow (particularly for incompressible or slightly compressible fluids), we discuss this solution approach only for this type of flow. Other approaches will be discussed in the next chapter for the black oil model.

7.3.1 Classical IMPES

An IMPES method was originally developed by Sheldon et al. (1959) and Stone and Garder (1961). The basic idea of this classical method for solving (7.1)–(7.4) is to separate the computation of pressure from that of saturation. Namely, the coupled system is split into a pressure equation and a saturation equation, and the pressure and saturation equations are solved using implicit and explicit time approximation approaches, respectively. This method is simple to set up and efficient to implement, and requires less computer memory than other methods such as the SS method (Douglas et al., 1959). However, for it to be stable, this classical method requires very small time steps for the saturation. This requirement is expensive and prohibitive, particularly for long-time integration problems and for small gridblock problems such as coning problems. In this section, we first review the classical IMPES and then introduce an *improved IMPES*. We focus on incompressible flow; compressible flow will be treated in the next chapter.

We use the oil pressure and water saturation as the *primary variables:*

$$p = p_o, \qquad S = S_w. \tag{7.22}$$

Define the total velocity

$$\mathbf{u} = \mathbf{u}_w + \mathbf{u}_o. \tag{7.23}$$

Under the assumption that the fluids are incompressible, we apply (7.3) and (7.23) to (7.1) to see that

$$\nabla \cdot \mathbf{u} = \tilde{q}(p, S) \equiv \tilde{q}_w(p, S) + \tilde{q}_o(p, S), \tag{7.24}$$

and (7.4) and (7.23) to (7.2) to obtain

$$\mathbf{u} = -\mathbf{k}\left[\lambda(S)\nabla p - \lambda_w(S)\nabla p_c - \left(\lambda_w \rho_w + \lambda_o \rho_o\right)\wp \nabla z\right], \tag{7.25}$$

where $\tilde{q}_w = q_w/\rho_w$ and $\tilde{q}_o = q_o/\rho_o$. Substituting (7.25) into (7.24) yields the *pressure equation*

$$-\nabla \cdot \left(\mathbf{k}\lambda \nabla p\right) = \tilde{q} - \nabla \cdot \left(\mathbf{k}(\lambda_w \nabla p_c + (\lambda_w \rho_w + \lambda_o \rho_o)\wp \nabla z)\right). \tag{7.26}$$

The phase velocities $\mathbf{u}_w$ and $\mathbf{u}_o$ are related to the total velocity $\mathbf{u}$ by (cf. Exercise 2.3)

$$\begin{aligned} \mathbf{u}_w &= f_w \mathbf{u} + \mathbf{k}\lambda_o f_w \nabla p_c + \mathbf{k}\lambda_o f_w(\rho_w - \rho_o)\wp \nabla z, \\ \mathbf{u}_o &= f_o \mathbf{u} - \mathbf{k}\lambda_w f_o \nabla p_c + \mathbf{k}\lambda_w f_o(\rho_o - \rho_w)\wp \nabla z. \end{aligned}$$

Similarly, we apply (7.4), (7.23), and (7.25) to (7.1) and (7.2) with $\alpha = w$ to obtain the *saturation equation* (cf. Exercise 7.2)

$$\begin{aligned} \phi \frac{\partial S}{\partial t} + \nabla \cdot \Bigg\{ \mathbf{k} f_w(S)\lambda_o(S)\left(\frac{dp_c}{dS}\nabla S + (\rho_o - \rho_w)\wp \nabla z\right) \\ + f_w(S)\mathbf{u}\Bigg\} = \tilde{q}_w(p, S), \end{aligned} \tag{7.27}$$

where, for notational convenience, we assume that $\phi = \phi(\mathbf{x})$.

Let $J = (0, T]$ $(T > 0)$ be a time interval of interest, and for a positive integer N, let $0 = t^0 < t^1 < \cdots < t^N = T$ be a partition of J. For the pressure computation in the classical IMPES method, the saturation S in (7.26) is supposed to be known, and (7.26) is solved implicitly for p. That is, for each $n = 0, 1, \ldots,$ p^n satisfies

$$-\nabla \cdot \left(\mathbf{k}\lambda(S^n)\nabla p^n\right) = F(p^n, S^n), \tag{7.28}$$

where $F(p, S)$ denotes the right-hand side of (7.26), and S^n is given. It follows from (7.27) that

$$\begin{aligned} \phi \frac{\partial S}{\partial t} = \tilde{q}_w - \nabla \cdot \Bigg\{ \mathbf{k} f_w(S)\lambda_o(S)\left(\frac{dp_c}{dS}\nabla S + (\rho_o - \rho_w)\wp \nabla z\right) \\ + f_w(S)\mathbf{u}\Bigg\}. \end{aligned} \tag{7.29}$$

In the IMPES, (7.29) is explicitly solved for S; i.e., for each $n = 0, 1, 2, \ldots,$ S^{n+1} satisfies

$$\phi \frac{S^{n+1} - S^n}{\Delta t^{n+1}} \approx \phi \left.\frac{\partial S}{\partial t}\right|_{t=t^{n+1}} = G(p^n, \mathbf{u}^n, S^n), \tag{7.30}$$

where $G(p, \mathbf{u}, s)$ represents the right-hand side of (7.29).

The IMPES method goes as follows: After startup, for $n = 0, 1, \ldots,$ we use (7.28) and S^n to evaluate p^n and then (7.25) to evaluate $\mathbf{u}^n$; next, we utilize S^n, p^n, $\mathbf{u}^n$, and (7.30) to compute S^{n+1}. As noted, the time step $\Delta t^n = t^n - t^{n-1}$ must be sufficiently small for this method to be stable (cf. (4.108)).

7.3.2 The seventh SPE project: Horizontal well modeling

We present numerical experiments for the classical IMPES method to check its computational cost and stability. We define the source and sink terms by

$$\tilde{q}_\alpha = \sum_{l,m} q_\alpha^{(l,m)} \delta(\mathbf{x} - \mathbf{x}^{(l,m)}), \qquad \alpha = w, o,$$

where $q_\alpha^{(l,m)}$ indicates the volume of phase α produced or injected per unit time at the lth well and the mth perforated zone, $\mathbf{x}^{(l,m)}$, and δ is the Dirac delta function. Following Peaceman (1991), $q_\alpha^{(l,m)}$ can be defined by

$$q_\alpha^{(l,m)} = \frac{2\pi \rho_\alpha \bar{k} k_{r\alpha} \Delta L^{(l,m)}}{\mu_\alpha \ln\left(r_e^{(l)}/r_w^{(l)}\right)} \left(p_{bh}^{(l)} - p_\alpha - \rho_\alpha \wp \left(z_{bh}^{(l)} - z\right)\right),$$

where $\Delta L^{(l,m)}$ is the length (in the flow direction) of a gridblock (containing the lth well) at the mth perforated zone, $p_{bh}^{(l)}$ is the *bottom hole pressure* at the datum level depth $z_{bh}^{(l)}$, $r_e^{(l)}$ is the equivalent well radius, and $r_w^{(l)}$ is the radius of the lth well. The quantity $\bar{k}$ is some average of $\mathbf{k}$ at the wells (Peaceman, 1991). For a diagonal tensor $\mathbf{k} = \text{diag}(k_{11}, k_{22}, k_{33})$, for example, $\bar{k} = \sqrt{k_{11}k_{22}}$ for a vertical well. In this case, the equivalent radius is calculated by

$$r_e^{(l)} = \frac{0.14\left((k_{22}/k_{11})^{1/2} h_1^2 + (k_{11}/k_{22})^{1/2} h_2^2\right)^{1/2}}{0.5\left((k_{22}/k_{11})^{1/4} + (k_{11}/k_{22})^{1/4}\right)},$$

where h_1 and h_2 are the x_1- and x_2-grid sizes of the gridblock that contains the vertical well. For a horizontal well (e.g., in the x_1-direction), $\bar{k} = \sqrt{k_{22}k_{33}}$ and

$$r_e^{(l)} = \frac{0.14\left((k_{33}/k_{22})^{1/2} h_1^2 + (k_{22}/k_{33})^{1/2} h_3^2\right)^{1/2}}{0.5\left((k_{33}/k_{22})^{1/4} + (k_{22}/k_{33})^{1/4}\right)},$$

where h_3 is the x_3-grid size of the gridblock containing this horizontal well. The treatment of wells will be discussed further in Chapter 13.

The physical data used are taken from the seventh SPE comparative solution project (Nghiem et al., 1991). The reservoir dimensions are

$$\sum_{i=1}^{Nx_1} h_{1,i}, \quad \sum_{j=1}^{Nx_2} h_{2,j}, \quad \text{and} \quad \sum_{k=1}^{Nx_3} h_{3,k},$$

respectively, in the x_1-, x_2-, and x_3-directions, where $Nx_1 = 9$, $Nx_2 = 9$, $Nx_3 = 6$, and (in feet)

$$\begin{aligned}
&h_{1,i} = 300, \quad i = 1, 2, \ldots, 9,\\
&h_{2,1} = h_{2,9} = 620, \quad h_{2,2} = h_{2,8} = 400,\\
&h_{2,3} = h_{2,7} = 200, \quad h_{2,4} = h_{2,6} = 100, \quad h_{2,5} = 60,\\
&h_{3,k} = 20, \quad k = 1, 2, 3, 4,\\
&h_{3,5} = 30, \quad h_{3,6} = 50.
\end{aligned}$$

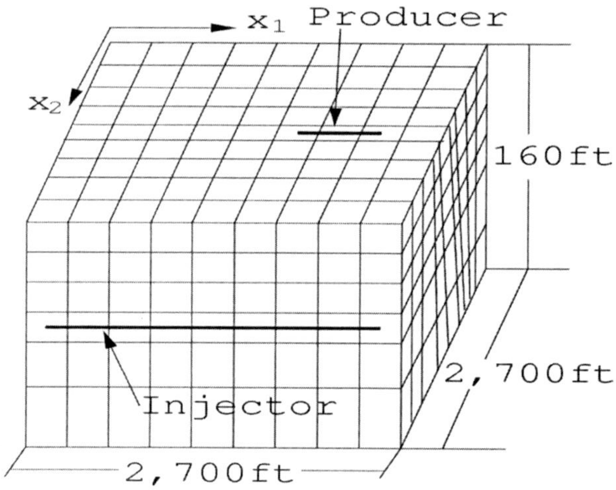

Figure 7.3. *A reservoir.*

Table 7.2. *The relative permeabilities and capillary pressure.*

S	0.22	0.3	0.4	0.5	0.6	0.8	0.9	1
k_{rw}	0	0.07	0.15	0.24	0.33	0.65	0.83	1
k_{ro}	1	0.4	0.125	0.0649	0.0048	0	0	0
p_c (psia)	6.3	3.6	2.7	2.25	1.8	0.9	0.45	0.0

A horizontal oil production well is located in the first layer ($k = 1$) and stretched in gridblocks with $i = 6, 7, 8$ and $j = 5$, and a horizontal water injection well is located in the sixth layer ($k = 6$) and stretched in gridblocks with $i = 1, 2, \ldots, 9$ and $j = 5$. Thus there are two horizontal wells in this experiment (cf. Figure 7.3). The radius of both wells is 2.25 inches. The permeability tensor $\mathbf{k}$ is diagonal with $k_{11} = k_{22} = 300$ md and $k_{33} = 30$ md, and the porosity ϕ is 0.2. The depth z of the centers of the six layers is, respectively, 3,600, 3,620, 3,640, 3,660, 3,685, and 3,725 ft, and the initial water saturation at each layer is 0.289, 0.348, 0.473, 0.649, 0.869, and 1.0. The densities and viscosities are $\rho_o = 0.8975$ g/cm^3, $\rho_w = 0.9814$ g/cm^3, $\mu_o = 0.954$ cp, and $\mu_w = 0.96$ cp. The relative permeability and capillary pressure data are shown in Table 7.2.

Finally, the pressure at the wells is fixed, the datum level depth z_{bh} is 3,600 ft, and the bottom hole pressures p_{bh} for the injection and production wells are, respectively, 3,651.4 and 3,513.6 psia. The final time T is 1,500 days.

As noted, to control the variation of saturation, we need to find a suitable time step Δt^{n+1} before we solve (7.30) for S^{n+1} for each $n = 0, 1, \ldots$. The control strategy is defined as follows: We calculate the maximum value of $\partial S^{n+1}/\partial t$ at all computational nodes, denoted by $(\partial S^{n+1}/\partial t)_{max}$, which is, by (7.30),

$$\left(\frac{\partial S^{n+1}}{\partial t}\right)_{max} = \left(\frac{G(p^n, \mathbf{u}^n, S^n)}{\phi}\right)_{max}.$$

Table 7.3. *The CPU time vs.* DS_{max}.

DS_{max}	0.05	0.02	0.01	0.005	0.002	0.001
N	70	91	86	122	226	432
Pressure-CPU	14.81	19.18	18.13	25.86	47.68	89.49
Saturation-CPU	0.14	0.20	0.19	0.35	0.46	0.88

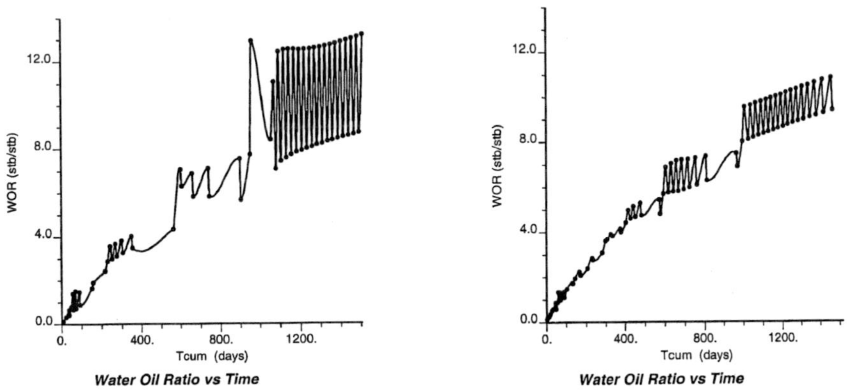

Figure 7.4. $DS_{max} = 0.05$ *(left);* $DS_{max} = 0.02$ *(right).*

Then we apply the following formula to find Δt^{n+1}:

$$\Delta t^{n+1} = \frac{DS_{max}}{\left(\frac{\partial S^{n+1}}{\partial t}\right)_{max}},$$

where DS_{max} is the maximum variation of the saturation to be allowed. Now, we use this time step in (7.30) to obtain S^{n+1}. This approach guarantees that the saturation variation does not exceed DS_{max}. Note that DS_{max} can depend on the time level n.

The mixed finite element method with the Raviart–Thomas–Nédélec space of lowest order over rectangular parallelepipeds in three dimensions is used (cf. Section 4.5.4; also cf. Section 7.5). A *no-flow boundary condition* (homogeneous Neumann boundary condition) is applied. To test stability, we study the curves of the water-oil production ratio (WOR) at the production well verses time (days) in the cases of $DS_{max} = 0.05, 0.02, 0.01, 0.005, 0.002$, and 0.001. The results are displayed in Figures 7.4–7.6. From these figures we see that the WOR does not oscillate only when DS_{max} is smaller than 0.002.

We now check the computational time for the present experiment at $T = 1,500$ days for the six choices of DS_{max}, which is shown in Table 7.3. In this table, the CPU time is in seconds and N (the number of time steps) is such that $t^N = T$. All the computations are carried out on an SGI-O_2 workstation. Table 7.3 shows that the computation of pressure takes far more time than that of saturation.

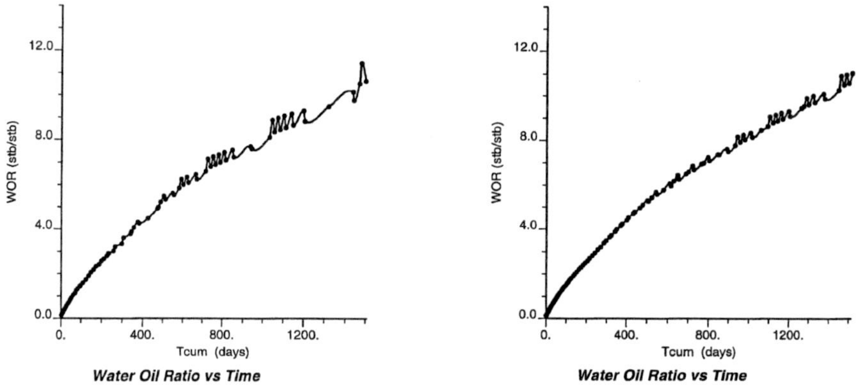

Figure 7.5. $DS_{max} = 0.01$ *(left);* $DS_{max} = 0.005$ *(right).*

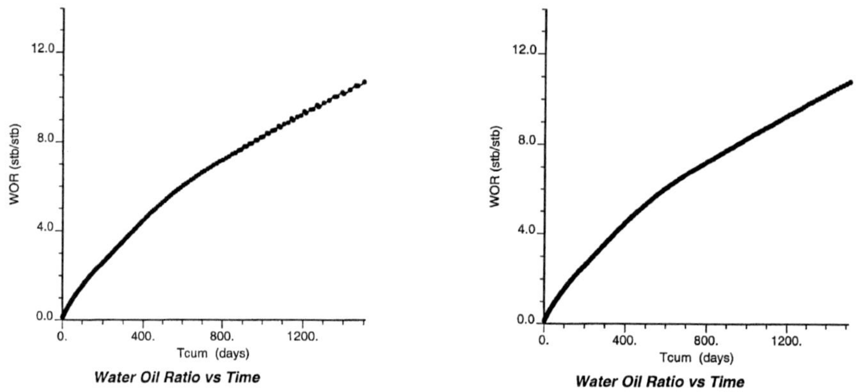

Figure 7.6. $DS_{max} = 0.002$ *(left);* $DS_{max} = 0.001$ *(right).*

7.3.3 Improved IMPES

The improved method

Most of the computational time in the classical IMPES method is spent on the implicit calculation of pressure. It follows from the mechanics of fluid flow in porous media that pressure changes less rapidly in time than saturation. Furthermore, the constraint on time steps is primarily used in the explicit calculation of saturation. For all these reasons, it is appropriate to take a much larger time step for the pressure than for the saturation.

Again, for a positive integer N, let $0 = t^0 < t^1 < \cdots < t^N = T$ be a partition of J into subintervals $J^n = (t^{n-1}, t^n]$, with length $\Delta t_p^n = t^n - t^{n-1}$. This partition is used for pressure. For saturation, each subinterval J^n is divided into sub-subintervals $J^{n,m} = (t^{n-1,m-1}, t^{n-1,m}]$:

$$t^{n-1,m} = t^{n-1} + m\Delta t_p^n / M^n, \qquad m = 1, \ldots, M^n.$$

The length of $J^{n,m}$ is denoted by $\Delta t_S^{n,m} = t^{n-1,m} - t^{n-1,m-1}$, $m = 1, \ldots, M^n, n = 0, 1, \ldots$.

Table 7.4. *The CPU time for the improved IMPES.*

DS_{max}	$M \equiv M^n$	N	Pres-CPU	Satur-CPU	Total CPU
0.01	5	18	3.63	0.28	3.91
0.005	10	12	2.38	0.33	2.71
0.001	50	9	1.76	0.97	2.73

The number of steps, M^n, can depend on n. Below we simply write $t^{n-1,0} = t^{n-1}$ and set $v^{n,m} = v(\cdot, t^{n,m})$.

We denote the right-hand side of (7.25) by $\mathbf{H}(p, S)$. Now, the improved IMPES method is defined: For each $n = 0, 1, \dots$, find p^n such that

$$-\nabla \cdot \left(\mathbf{k}\lambda(S^n)\nabla p^n\right) = F(p^n, S^n) \tag{7.31}$$

and $\mathbf{u}^n$ such that

$$\mathbf{u}^n = \mathbf{H}(p^n, S^n). \tag{7.32}$$

Next, for $m = 1, \dots, M^n$, $n = 0, 1, \dots$, find $S^{n+1,m}$ such that

$$\phi \frac{\partial S^{n+1,m}}{\partial t} = G(p^n, \mathbf{u}^n, S^{n+1,m-1}). \tag{7.33}$$

The time step $\Delta t_S^{n+1,m}$ in (7.33) is chosen as follows: Set

$$\left(\frac{\partial S^{n+1,m}}{\partial t}\right)_{max} = \left(\frac{G(p^n, \mathbf{u}^n, S^{n+1,m-1})}{\phi}\right)_{max}, \tag{7.34}$$

and then calculate

$$\Delta t_S^{n+1,m} = \frac{DS_{max}}{\left(\frac{\partial S^{n+1,m}}{\partial t}\right)_{max}}, \qquad m = 1, \dots, M^n, \ n = 0, 1, \dots. \tag{7.35}$$

Numerical tests

We perform numerical experiments for the improved IMPES method for the same example as in Section 7.3.2. The selection of pressure time steps is automatic, and the total variation of saturation for one pressure time step is fixed at 0.05. We test three values of DS_{max} for the choice of $\Delta t_S^{n+1,m}$, $m = 1, \dots, M^n$, $n = 0, 1, \dots$. The numerical results are reported in Table 7.4, and the WOR curves for these three values are shown in Figure 7.7, where the final time is such that the calculated water cut is up to 98% at the production well.

From Figure 7.7, we see that the WOR curves slightly oscillate when $DS_{max} = 0.01$ and 0.005, and this curve is very smooth when $DS_{max} = 0.001$. From Table 7.4, the total CPU time as $DS_{max} = 0.001$ is 2.73 sec. Also, the ratio of the pressure CPU time to the saturation CPU time is around 1.8:1. This is in sharp contrast with the classical IMPES method, where the total CPU time doubles as DS_{max} is halved and the pressure CPU time is 100 times as great as the saturation CPU time. Furthermore, the total CPU time for the improved IMPES is far less than that for the classical one. For example, at $DS_{max} = 0.001$, the former is 2.73 sec. and the latter is 90.37 sec.

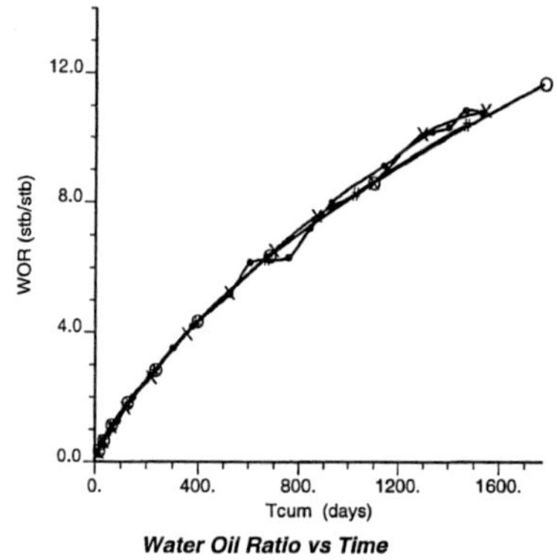

Figure 7.7. $\times = 0.05$, $\bullet = 0.01$, $\circ = 0.001$.

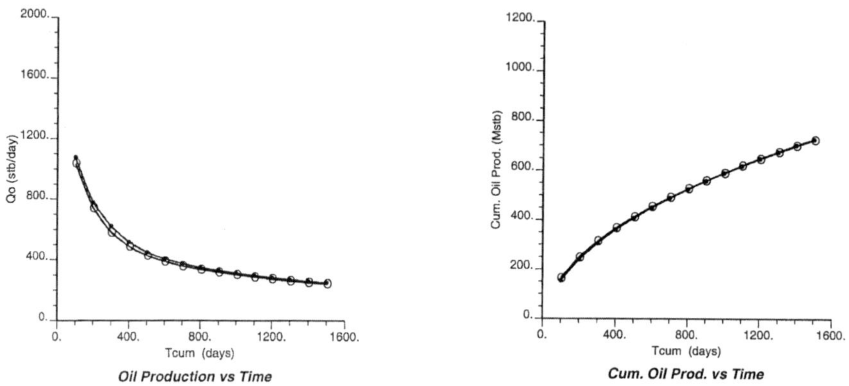

Figure 7.8. $\circ = IMPES$, $\bullet = SS$.

A comparison with SS

To see further the accuracy and efficiency of the improved IMPES method, we compare it with the SS method for the same numerical example. Here the pressure time step is fixed at 100 days, $DS_{max} = 0.001$, and the final time is 1,500 days. The daily oil production rate (verses time), the cumulative oil production, and the WOR curves using these two methods are presented in Figures 7.8 and 7.9 (left). These curves match quite well for these two methods. The total CPU time for the improved IMPES is 5.03 sec., while it is 31.58 sec. for the SS. Thus, for this example, the improved IMPES is 6.3 times as fast as the SS.

Application to a coning problem

The classical IMPES method has not successfully been applied to the solution of a two-phase coning problem. To check its robustness, we apply the improved IMPES method to solve a

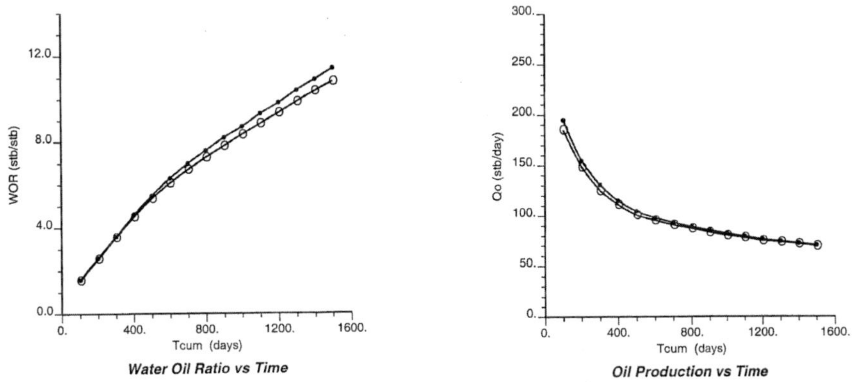

Figure 7.9. ○ = *IMPES*, • = *SS*.

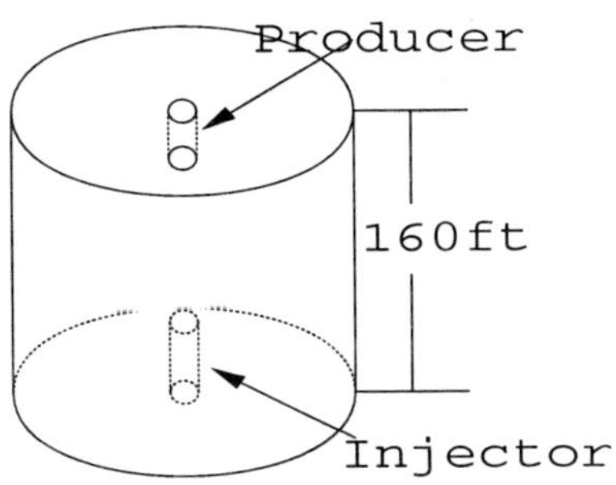

Figure 7.10. *A coning problem.*

problem of this type. Now, the reservoir is a cylindrical domain with its axis parallel to the x_3-axis and its radius equal to 1,343.43 ft. There are two vertical wells located at the center of the reservoir: An oil production well sits vertically in the first layer and a water injection well in the sixth layer (cf. Figure 7.10). Their radius is 0.25 ft. The radii of the innermost to outermost cylinders are, respectively, 4, 8, 16, 32, 64, 128, 256, 512, and 1,343.43 ft. All other data are the same as in the example in Section 7.3.2. The pressure and saturation time steps are the same as in Section 7.3.3. For the present problem, the daily oil production rate, the cumulative oil production, and the WOR curves using the improved IMPES and SS methods are presented in Figures 7.9 (right) and 7.11. Again, the curves match quite well for these two methods. The total CPU time for the former is 2.54 sec., and for the latter is 17.02 sec. Hence this improved IMPES is 6.7 times as fast as the SS for the present coning problem. Also, we point out that the pressure CPU time is 0.39 sec., while the saturation CPU time is 2.15 sec. From this experiment, we see that the improved IMPES method is capable of solving two-phase coning problems.

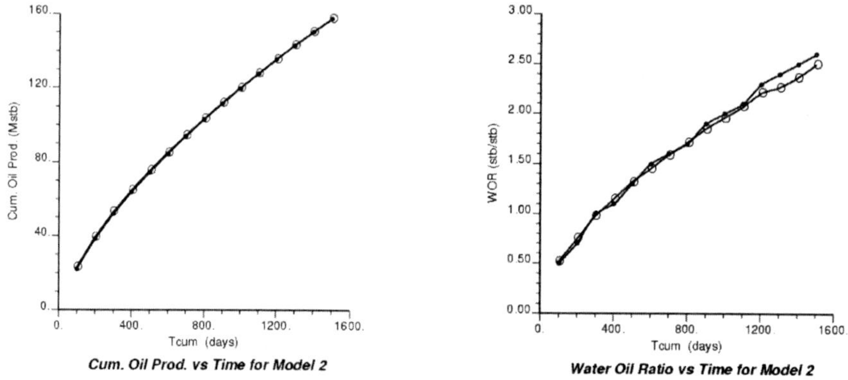

Figure 7.11. ∘ = *IMPES*, • = *SS*.

7.4 Alternative Differential Formulations

Several alternative formulations for the differential equations (7.1)–(7.4) were discussed in Section 2.3.2. We now consider further these alternative formulations and numerically compare their use.

7.4.1 Phase formulation

The phase formulation was used in the previous section. For comparison, we restate this formulation. The oil pressure is employed as the pressure variable:

$$p = p_o. \tag{7.36}$$

The pressure equation consists of the two equations

$$\nabla \cdot \mathbf{u} = \tilde{q} \tag{7.37}$$

and

$$\mathbf{u} = -\mathbf{k}\left(\lambda(S)\nabla p - \lambda_w(S)\nabla p_c - \left(\lambda_w \rho_w + \lambda_o \rho_o\right)\wp \nabla z\right). \tag{7.38}$$

The saturation equation is

$$\begin{aligned}\phi\frac{\partial S}{\partial t} + \nabla \cdot \Bigg\{\mathbf{k} f_w(S)\lambda_o(S)\left(\frac{dp_c}{dS}\nabla S + (\rho_o - \rho_w)\wp\nabla z\right)& \\ + f_w(S)\mathbf{u}\Bigg\} &= \tilde{q}_w(p, S).\end{aligned} \tag{7.39}$$

7.4.2 Weighted formulation

We introduce a pressure that is smoother than the phase pressure:

$$p = S_w p_w + S_o p_o. \tag{7.40}$$

Even if a phase disappears (i.e., either S_w or S_o is zero), there is still a nonzero smooth variable p. Applying the same algebraic manipulations as in deriving the phase formulation, we obtain (cf. Exercise 7.3)

$$\begin{aligned}\mathbf{u} = -\mathbf{k}\big\{&\lambda(S)\nabla p + \big(S\lambda(S) - \lambda_w(S)\big)\nabla p_c + \lambda(S)p_c\nabla S \\ &- \big(\lambda_w\rho_w + \lambda_o\rho_o\big)\wp\nabla z\big\}.\end{aligned} \tag{7.41}$$

Equations (7.37) and (7.39) remain the same.

7.4.3 Global formulation

Note that p_c appears in both (7.38) and (7.41). To remove it, we define a global pressure (Antontsev, 1972; Chavent and Jaffré, 1986):

$$p = p_o - \int^S \left(f_w \frac{dp_c}{dS}\right)(\xi)\, d\xi. \tag{7.42}$$

Using this pressure, the total velocity becomes (cf. Exercise 7.4)

$$\mathbf{u} = -\mathbf{k}\left(\lambda(S)\nabla p - \big(\lambda_w\rho_w + \lambda_o\rho_o\big)\wp\nabla z\right). \tag{7.43}$$

It follows from (7.4) and (7.42) that

$$\lambda\nabla p = \lambda_w\nabla p_w + \lambda_o\nabla p_o,$$

which implies that the global pressure is the pressure that would produce flow of a fluid (with mobility λ) equal to the sum of the flows of fluids w and o. Again, (7.37) and (7.39) remain the same.

The coupling between the pressure and saturation equations in the global formulation is less than that in the phase and weighted formulations, and the nonlinearity is weakened as well. This formulation is most suitable for a mathematical analysis for two-phase flow (Antontsev, 1972; Chavent and Jaffré, 1986; Chen, 2001; Chen, 2002A). When the capillary effect is neglected, the three formulations are the same. In this case, the saturation equation becomes the well-known *Buckley–Leverett equation* (cf. Section 2.3.2).

7.4.4 Numerical comparisons

We perform numerical experiments to compare the three formulations. Since the gravity terms in all the formulations have the same form, we neglect the gravity effect. The reservoir has dimensions 1,000 × 1,000 × 100 ft^3, and the relative permeabilities are

$$k_{rw} = k_{rwmax}\left(\frac{S_w - S_{wc}}{1 - S_{or} - S_{wc}}\right)^2, \qquad k_{ro} = \left(\frac{S_o - S_{or}}{1 - S_{or} - S_{wc}}\right)^2,$$

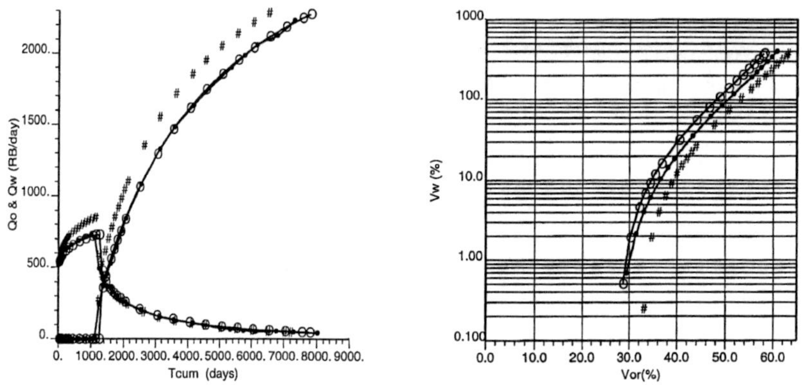

Figure 7.12. *Water (upper) and oil production (lower) (left); characterization curve of displacement (right).* • = *phase formulation,* # = *weighted formulation, and* o = *global formulation.*

where $k_{rwmax} = 0.65$, $S_{wc} = 0.22$, and $S_{or} = 0.2$. The capillary pressure curve is

$$p_c = p_{cmin} - \bar{B} \ln \frac{S - S_{wc}}{1 - S_{wc}},$$

where the constant $\bar{B}$ is the value such that $p_c = p_{cmax}$ as $S = S_{wc}$, $p_{cmin} = 0$ psi, and $p_{cmax} = 70$ psi. Other physical data are chosen as follows:

$$\phi = 0.2, \quad k = 0.1 \text{ darcy}, \quad \mu_w = 0.096 \text{ cp}, \quad \mu_o = 1.14 \text{ cp},$$

where $\mathbf{k} = k\mathbf{I}$. This example is two-dimensional flow in a five spot pattern reservoir. An injection well is located at a corner of the reservoir, and a production well is located at its opposite corner. Water is injected, and oil and/or water is produced. The radius of the two wells is 0.2291667 ft, and the initial saturation equals S_{wc}. Finally, the bottom hole pressure is 3,700 psi at injection and 3,500 psi at production.

In computations, we use the lowest-order Raviart–Thomas mixed finite elements on triangles on a 10×10 grid (triangles are obtained by dividing each rectangle into two triangles in a diagonal direction; cf. Section 4.5.4 or 7.5). The time discretization is based on the backward Euler scheme. The improved IMPES discussed in Section 7.3.3 is employed. A no-flow boundary condition is employed. The oil and water production verses time (in days), the characterization curves of displacement, and the water cut are shown in Figures 7.12 and 7.13. A *characterization curve* is defined as the logarithm of the cumulative water production verses the cumulative oil production. From these figures we see that the results of the global and phase formulations are very close. These results are rather different from those using the weighted formulation. We also checked the CPU times (in seconds) for the three formulations at the final time, $T = 8{,}000$ days, and the results obtained on a Dec Alpha workstation are displayed in seconds in Table 7.5. There is not much difference between the CPU times for this example.

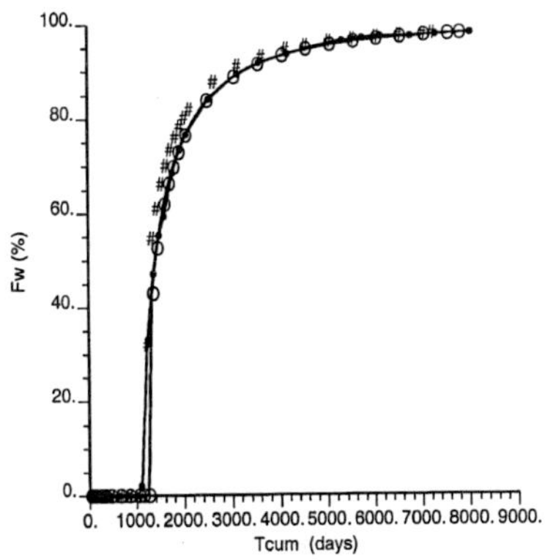

Figure 7.13. *Water cut.* • = *phase formulation, # = weighted formulation, and* o = *global formulation.*

Table 7.5. *CPU times for three formulations.*

	Global	Phase	Weighted
CPU times	33.4748	33.5266	33.6622

7.5 Numerical Methods for Two-Phase Flow

The various discretization methods developed in Chapter 4 are now applied to the solution of the differential equations (7.1)–(7.4) governing two-phase flow in a porous medium $\Omega \subset \mathbb{R}^d$ ($d = 2$ or 3). The standard finite element methods were described for single phase flow in the preceding chapter and can be extended to the present case. Here we discuss the application of the mixed, control volume, and characteristic finite element methods to (7.1)–(7.4). The first two methods are good choices for the pressure equation. Because physical transport dominates diffusive effects in two-phase flow and because the capillary diffusion coefficient in the saturation equation can be zero, it is appropriate to use the characteristic finite element methods to solve this equation.

7.5.1 Mixed finite element methods

As an example, we present mixed finite element methods for the global formulation. Recall that the pressure equation consists of (7.37) and (7.43) in this formulation. The model is completed by specifying boundary and initial conditions. For simplicity, a *no-flow boundary condition* is used for the pressure equation

$$\mathbf{u} \cdot \boldsymbol{\nu} = 0, \qquad \mathbf{x} \in \Gamma, \tag{7.44}$$

where $\boldsymbol{\nu}$ is the outward unit normal to the boundary Γ of Ω. It follows from (7.37) and (7.44) that compatibility to incompressibility of the fluids requires

$$\int_\Omega \tilde{q}\, d\mathbf{x} = 0, \qquad t \geq 0.$$

Set (cf. Section 4.5.2)

$$\mathbf{V} = \{\mathbf{v} \in \mathbf{H}(\text{div}, \Omega) : \mathbf{v} \cdot \boldsymbol{\nu} = 0 \text{ on } \Gamma\}, \quad W = L^2(\Omega).$$

For simplicity, let Ω be a convex polygonal domain. For $0 < h < 1$, let K_h be a regular partition of Ω into elements, say, tetrahedra, rectangular parallelepipeds, or prisms, with maximum mesh size h. Associated with the partition K_h, let $\mathbf{V}_h \times W_h \subset \mathbf{V} \times W$ represent the RT (or RTN), BDM, BDFM, BDDM, or CD mixed finite element spaces (cf. Section 4.5.4). Now, the mixed method for (7.37) and (7.43) is: For $0 \leq n \leq N$, find $\mathbf{u}_h^n \in \mathbf{V}_h$ and $p_h^n \in W_h$ such that

$$\begin{aligned} (\nabla \cdot \mathbf{u}_h^n, w) &= \left(\tilde{q}\left(p_h^n, S_h^n\right), w\right), & w \in W_h, \\ \left(\left(\mathbf{k}\lambda(S_h^n)\right)^{-1}\mathbf{u}_h^n, \mathbf{v}\right) - (p_h^n, \nabla \cdot \mathbf{v}) &= \left(\boldsymbol{\gamma}(S_h^n), \mathbf{v}\right), & \mathbf{v} \in \mathbf{V}_h, \end{aligned} \tag{7.45}$$

where S_h^n is an approximation to S^n (cf. Section 7.5.3) and

$$\boldsymbol{\gamma}(S) = \left(f_w(S)\rho_w + f_o(S)\rho_o\right)\wp\nabla z.$$

Note that system (7.45) is nonlinear, and the various solution approaches (e.g., linearization, implicit time approximation, and explicit time approximation) developed in the preceding chapter for the standard finite element methods can be applied to it in the same fashion.

7.5.2 CVFE methods

Assume that a partition K_h of Ω consists of a set of (open) control volumes V_i:

$$\bar{\Omega} = \bigcup_i \bar{V}_i, \quad V_i \cap V_j = \emptyset, \; i \neq j.$$

(The reader should refer to Section 4.3 for the construction of these control volumes.) On each V_i, integration of (7.37) over V_i and application of the divergence theorem gives

$$\int_{\partial V_i} \mathbf{u} \cdot \boldsymbol{\nu} \, d\ell = \int_{V_i} \tilde{q} \, d\mathbf{x}. \tag{7.46}$$

Substituting (7.43) into this equation yields

$$-\int_{\partial V_i} \lambda(S)\mathbf{k}\nabla p \cdot \boldsymbol{\nu} \, d\ell = \int_{V_i} \tilde{q} \, d\mathbf{x} - \int_{\partial V_i} \left(\lambda_w(S)\rho_w + \lambda_o(S)\rho_o\right)\wp\mathbf{k}\nabla z \cdot \boldsymbol{\nu} \, d\ell. \tag{7.47}$$

Let $M_h \subset H^1(\Omega)$ be a finite element (or function approximation) space associated with the CVFE partition K_h (cf. Section 4.3). Then the CVFE method for the pressure equation reads: For $0 \leq n \leq N$, find $p_h^n \in M_h$ such that

$$\begin{aligned} -\int_{\partial V_i} \lambda(S_h^n)\mathbf{k}\nabla p_h^n \cdot \boldsymbol{\nu} \, d\ell &= \int_{V_i} \tilde{q}\left(p_h^n, S_h^n\right) \, d\mathbf{x} \\ &\quad - \int_{\partial V_i} \left(\lambda_w(S_h^n)\rho_w + \lambda_o(S_h^n)\rho_o\right) \wp\mathbf{k}\nabla z \cdot \boldsymbol{\nu} \, d\ell. \end{aligned} \tag{7.48}$$

The upstream weighting techniques introduced in Section 4.3.4 can be applied to (7.48).

7.5.3 Characteristic finite element methods

As an example, we present the MMOC described in Section 4.6 for the saturation. Introduce

$$\tilde{q}_1(p, S) = \tilde{q}_w(p, S) - \tilde{q}(p, S) f_w(S) + \nabla \cdot (\mathbf{k} f_w(S)\lambda_o(S)(\rho_o - \rho_w)\wp \nabla z).$$

Using (7.37) and (7.39), the saturation equation becomes

$$\phi \frac{\partial S}{\partial t} + \frac{df_w}{dS}\mathbf{u} \cdot \nabla S + \nabla \cdot \left\{ \mathbf{k} f_w(S)\lambda_o(S) \frac{dp_c}{dS} \nabla S \right\} = q_1(p, S). \tag{7.49}$$

Let

$$\mathbf{b}(\mathbf{x}, t) = \frac{df_w}{dS}\mathbf{u}, \quad \psi(\mathbf{x}, t) = \left(\phi^2(\mathbf{x}) + |\mathbf{b}(\mathbf{x}, t)|^2\right)^{1/2},$$

and let the characteristic direction associated with the operator $\phi \frac{\partial}{\partial t} + \mathbf{b} \cdot \nabla$ be denoted by $\boldsymbol{\tau}(\mathbf{x}, t)$, so that

$$\frac{\partial}{\partial \boldsymbol{\tau}} = \frac{\phi(\mathbf{x})}{\psi(\mathbf{x}, t)} \frac{\partial}{\partial t} + \frac{\mathbf{b}(\mathbf{x}, t)}{\psi(\mathbf{x}, t)} \cdot \nabla.$$

Then (7.49) reduces to

$$\psi \frac{\partial S}{\partial \boldsymbol{\tau}} + \nabla \cdot \left\{ \mathbf{k} f_w(S)\lambda_o(S) \frac{dp_c}{dS} \nabla S \right\} = q_1(p, S). \tag{7.50}$$

Note that the characteristic direction $\boldsymbol{\tau}$ depends on the velocity $\mathbf{u}$. Because the saturation step $t^{n-1,m}$ relates to pressure steps by $t^{n-1} < t^{n-1,m} \le t^n$, we need a velocity approximation for (7.50) based on $\mathbf{u}_h^{n-1}$ and earlier values. For this, we utilize a *linear extrapolation approach*: If $n \ge 2$, take the linear extrapolation of $\mathbf{u}_h^{n-2}$ and $\mathbf{u}_h^{n-1}$ determined by

$$E\mathbf{u}_h^{n-1,m} = \left(1 + \frac{t^{n-1,m} - t^{n-1}}{t^{n-1} - t^{n-2}}\right)\mathbf{u}_h^{n-1} - \frac{t^{n-1,m} - t^{n-1}}{t^{n-1} - t^{n-2}}\mathbf{u}_h^{n-2}.$$

For $n = 1$, define

$$E\mathbf{u}_h^{0,m} = \mathbf{u}_h^0.$$

$E\mathbf{u}_h^{n-1,m}$ is first-order accurate in time in the first pressure step and second-order accurate in the later steps.

The MMOC is defined with periodic boundary conditions (cf. Section 4.6). For this reason, we assume that Ω is a rectangular domain, and all functions in (7.50) are spatially Ω-periodic. Let $M_h \subset H^1(\Omega)$ be any finite element space introduced in Section 4.2.1. Then an MMOC procedure for (7.50) is: For each $0 \le n \le N$ and $1 \le m \le M^n$, find $S_h^{n,m} \in M_h$ such that

$$\begin{aligned} &\left(\phi \frac{S_h^{n,m} - \check{S}_h^{n,m-1}}{t^{n,m} - t^{n,m-1}}, w\right) + \left(\mathbf{a}\left(S_h^{n,m-1}\right) \nabla S_h^{n,m}, \nabla w\right) \\ &\quad = \left(\tilde{q}_1\left(p_h^n, S_h^{n,m-1}\right), w\right), \qquad w \in M_h, \end{aligned} \tag{7.51}$$

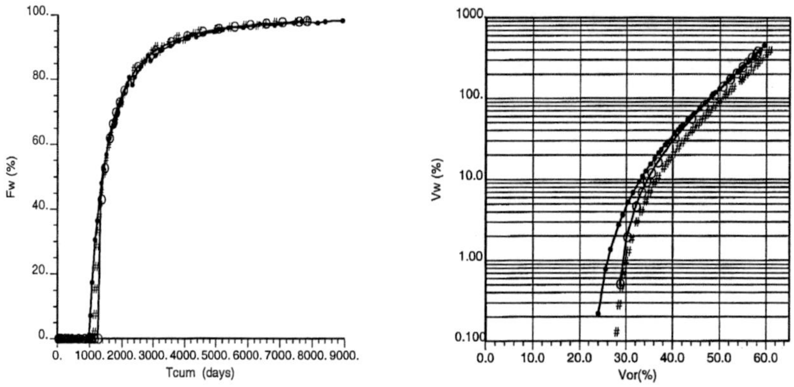

Figure 7.14. *Water cut (left); characterization curve of displacement (right).* • *= finite difference, # = CVFE, and* o *= mixed method.*

where

$$\mathbf{a}(S) = -\mathbf{k} f_w(S)\lambda_o(S)\frac{dp_c}{dS},$$

$$\check{S}_h^{n,m-1} = S_h^{n,m-1}\left(\mathbf{x} - \frac{df_w}{dS}\left(S_h^{n,m-1}\right)\frac{\mathbf{Eu}_h^{n,m}}{\phi(\mathbf{x})}\Delta t^{n,m}, t^{n,m-1}\right)$$

with $\Delta t^{n,m} = t^{n,m} - t^{n,m-1}$. The initial approximate solution S_h^0 can be defined as any appropriate projection of S_0 in M_h (e.g., the L^2-projection of S_0 in M_h). For the improved IMPES approach, the term $(\mathbf{a}(S_h^{n,m-1})\nabla s_h^{n,m}, \nabla w)$ in (7.51) is replaced by $(\mathbf{a}(S_h^{n,m-1})\nabla s_h^{n,m-1}, \nabla w)$.

7.5.4 Comparison between numerical methods

We compare numerically the finite difference, CVFE, and mixed finite element methods for solving the two-phase flow problem described in Section 7.1. To minimize grid orientation effects (cf. Sections 4.1.9 and 4.3.6), the nine-point finite difference method is used, where the partition K_h of Ω is of rectangular type. The CVFE methods are based on linear triangular elements (cf. Section 4.3). Finally, the mixed finite element methods use the lowest-order Raviart–Thomas element on triangles (cf. Section 4.5.4). The grid size is 10×10 (triangles are obtained by dividing each rectangle into two triangles in a diagonal direction). The two-dimensional flow problem in Section 7.4 is employed, and all the physical data are the same. The global formulation and the improved IMPES are utilized. The time discretization is based on the backward Euler scheme.

The numerical results are displayed in Figure 7.14 for the three discretization methods: finite difference, CVFE, and mixed finite element methods. For each of these methods, the pressure and saturation equations are discretized by the same method. The water cut and characterization curve of displacement verses time (in days) are shown in the figure. The numerical results obtained using these three methods match quite well for the present simple two-phase flow problem. Numerical comparisons among the discretization methods will be further performed for more complicated problems in subsequent chapters, such as for the black oil model in the next chapter.

7.6 Miscible Displacement

Miscible displacement was considered in Sections 2.4 and 2.5. Its simulation can be performed using the numerical techniques developed in this chapter for two-phase immiscible flow. The basis for the miscible-immiscible analogy has been long recognized (Lantz, 1970; Chen and Ewing, 1999). To see this analogy, as an example we consider the governing equations for the transport of a component in an incompressible fluid (cf. Section 2.4):

$$\nabla \cdot \mathbf{u} = q,$$
$$\mathbf{u} = -\frac{1}{\mu}\mathbf{k}\left(\nabla p - \rho\wp\nabla z\right), \tag{7.52}$$

and

$$\frac{\partial(\phi c)}{\partial t} + \nabla \cdot \big(c\mathbf{u} - \mathbf{D}(\mathbf{u})\nabla c\big) = \tilde{q}(c), \tag{7.53}$$

where c is the concentration of the component. In form, the pressure equation (7.52) and the concentration equation (7.53) resemble the pressure and saturation equations for two-phase immiscible, incompressible flow, respectively. The concentration equation depends on pressure explicitly through velocity, so mixed finite element methods are a good choice for the discretization of (7.52) (cf. Section 7.5.1). Furthermore, because physical transport dominates diffusive effects in miscible displacement as in two phase flow, characteristic finite element methods are appropriate for numerical solution of (7.53) (cf. Section 7.5.3 and Exercise 7.7). For realistic numerical examples using miscible displacement, the reader should see Todd and Longstaff (1972) and the fifth CSP organized by SPE (Killough and Cossack, 1987). Numerical simulation of miscible displacement processes has been used to show interface instabilities (fingering) due to viscosity and density differences (Homsy, 1987) and heterogeneity of porous media (Ewing et al., 1983).

7.7 Bibliographical Remarks

The content of Sections 7.3, 7.4, and 7.5 closely follows, respectively, Chen et al. (2004A), Chen and Huan (2003), and Chen et al. (2002A). For more details about the data used in the seventh SPE CSP, see Nghiem et al. (1991). For an error analysis of the approximation procedure developed in Section 7.5, the reader can refer to Chen (2005). Finally, for an error analysis of a finite element approximation procedure for the miscible displacement problem addressed in Section 7.6, see Douglas et al. (1983).

Exercises

7.1. Derive equation (7.21).

7.2. Apply equations (7.4), (7.23), and (7.25) to (7.1) and (7.2) with $\alpha = w$ to derive the saturation equation (7.27).

7.3. Show that equation (7.41) defines the total velocity **u** in terms of the weighted pressure p (cf. (7.40)).

7.4. Prove that equation (7.43) defines the total velocity **u** in terms of the global pressure p (cf. (7.42)).

7.5. Use the boundary condition (7.44) and introduce appropriate function spaces to write equations (7.37) and (7.38) in a mixed variational formulation.

7.6. Define a mixed finite element method for the phase formulation of equations (7.37) and (7.38) similar to that for the global formulation developed in Section 7.5.1.

7.7. Develop a mixed finite element approximation procedure for equation (7.52) with the no-flow boundary condition (7.44) as in Section 7.5.1, and a characteristic finite element approximation procedure for equation (7.53) with the periodic boundary condition as in Section 7.5.3.

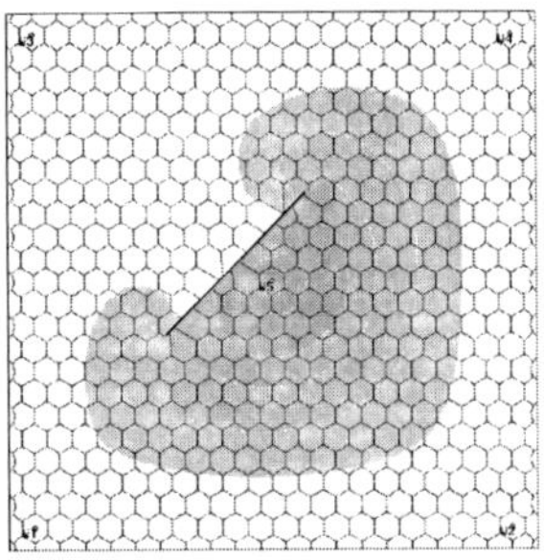

Chapter 8
The Black Oil Model

Recall that in the secondary recovery, if the reservoir pressure drops below the *bubble point pressure*, then oil (more precisely, the hydrocarbon phase) is split into a liquid phase and a gaseous phase at thermodynamical equilibrium. In this case, the flow is of the *black oil type*; the water phase does not exchange mass with the other phases, and the liquid and gaseous phases exchange mass between them. The gas component in this model mainly consists of methane and ethane.

The basic differential equations for the black oil model are reviewed in Section 8.1. The rock and fluid properties are also briefly described there. The Newton–Raphson iteration and three solution techniques (simultaneous solution, sequential, and IMPES) for this model are studied in Section 8.2. Comparisons between these solution techniques are discussed in Section 8.3. An application to a three-phase coning problem is described in Section 8.4. Finally, bibliographical information is given in Section 8.5.

8.1 Basic Differential Equations

8.1.1 The basic equations

The basic differential equations for the black oil model in a porous medium Ω were developed in Section 2.6. For completeness, we review these equations.

We use lower- and uppercase letter subscripts to indicate the three phases—water, oil (i.e., the liquid phase) and gas (i.e., the gaseous phase)—and the three components—water, oil, and gas, respectively. The subscript s represents standard conditions.

Let ϕ and $\mathbf{k}$ denote the porosity and permeability of the porous medium $\Omega \subset \mathbb{R}^3$; S_α, μ_α, p_α, $\mathbf{u}_\alpha$, B_α, and $k_{r\alpha}$ be the saturation, viscosity, pressure, volumetric velocity, formation volume factor, and relative permeability of the α-phase, $\alpha = w, o, g$, respectively; R_{so} be the gas solubility; and $\rho_{\beta s}$ (at standard conditions) and q_β be the density and volumetric rate of the β component, $\beta = W, O, G$. The mass conservation equations on standard volumes are

$$\frac{\partial}{\partial t}\left(\frac{\phi\rho_{Ws}}{B_w}S_w\right) = -\nabla\cdot\left(\frac{\rho_{Ws}}{B_w}\mathbf{u}_w\right) + q_W \tag{8.1}$$

for the water component,

$$\frac{\partial}{\partial t}\left(\frac{\phi\rho_{Os}}{B_o}S_o\right) = -\nabla\cdot\left(\frac{\rho_{Os}}{B_o}\mathbf{u}_o\right) + q_O \tag{8.2}$$

for the oil component, and

$$\frac{\partial}{\partial t}\left[\phi\left(\frac{\rho_{Gs}}{B_g}S_g + \frac{R_{so}\rho_{Gs}}{B_o}S_o\right)\right] = -\nabla\cdot\left(\frac{\rho_{Gs}}{B_g}\mathbf{u}_g + \frac{R_{so}\rho_{Gs}}{B_o}\mathbf{u}_o\right) + q_G \tag{8.3}$$

for the gas component.

Darcy's law for each phase is written in the usual form

$$\mathbf{u}_\alpha = -\frac{k_{r\alpha}}{\mu_\alpha}\mathbf{k}\left(\nabla p_\alpha - \rho_\alpha \wp \nabla z\right), \qquad \alpha = w,\, o,\, g, \tag{8.4}$$

where ρ_α is the mass density of the α-phase, $\wp$ is the magnitude of the gravitational acceleration, and z is the depth. The saturation constraint is

$$S_w + S_o + S_g = 1. \tag{8.5}$$

Finally, the phase pressures are related by capillary pressures

$$p_{cow} = p_o - p_w, \qquad p_{cgo} = p_g - p_o. \tag{8.6}$$

The flow rates are defined by

$$q_W = \frac{q_{Ws}\rho_{Ws}}{B_w}, \quad q_O = \frac{q_{Os}\rho_{Os}}{B_o}, \quad q_G = \frac{q_{Gs}\rho_{Gs}}{B_g} + \frac{q_{Os}R_{so}\rho_{Gs}}{B_o}, \tag{8.7}$$

where q_{Ws}, q_{Os}, and q_{Gs} are the rates at standard conditions. We introduce the potentials

$$\Phi_\alpha = p_\alpha - \rho_\alpha \wp z, \qquad \alpha = w, o, g. \tag{8.8}$$

Moreover, we define the *transmissibility*

$$\mathbf{T}_\alpha = \frac{k_{r\alpha}}{\mu_\alpha B_\alpha}\mathbf{k}, \qquad \alpha = w, o, g. \tag{8.9}$$

Substituting (8.7)–(8.9) into (8.1)–(8.3), neglecting the variation of ρ_α in space, and dividing the resulting equations by ρ_{Ws}, ρ_{Os}, and ρ_{Gs}, respectively, we obtain (cf. Exercise 8.1)

$$\begin{aligned}
&\frac{\partial}{\partial t}\left(\frac{\phi S_w}{B_w}\right) = \nabla\cdot\left(\mathbf{T}_w\nabla\Phi_w\right) + \frac{q_{Ws}}{B_w},\\
&\frac{\partial}{\partial t}\left(\frac{\phi S_o}{B_o}\right) = \nabla\cdot\left(\mathbf{T}_o\nabla\Phi_o\right) + \frac{q_{Os}}{B_o},\\
&\frac{\partial}{\partial t}\left[\phi\left(\frac{S_g}{B_g} + \frac{R_{so}S_o}{B_o}\right)\right]\\
&\quad = \nabla\cdot\left(\mathbf{T}_g\nabla\Phi_g + R_{so}\mathbf{T}_o\nabla\Phi_o\right) + \frac{q_{Gs}}{B_g} + \frac{q_{Os}R_{so}}{B_o}.
\end{aligned} \tag{8.10}$$

The volumetric flow rates at the wells (at standard conditions) are given by (Peaceman, 1991)

$$q_{Ws} = \sum_{j=1}^{N_w} \sum_{m=1}^{M_{wj}} \frac{2\pi \Delta L^{(j,m)}}{\ln(r_e^{(j,m)}/r_w^{(j)})} \frac{\bar{k}k_{rw}}{\mu_w} \left[p_{bh}^{(j)} - p_w - \rho_w \wp (z_{bh}^{(j)} - z)\right] \delta(\mathbf{x} - \mathbf{x}^{(j,m)}),$$

$$q_{Os} = \sum_{j=1}^{N_w} \sum_{m=1}^{M_{wj}} \frac{2\pi \Delta L^{(j,m)}}{\ln(r_e^{(j,m)}/r_w^{(j)})} \frac{\bar{k}k_{ro}}{\mu_o} \left[p_{bh}^{(j)} - p_o - \rho_o \wp (z_{bh}^{(j)} - z)\right] \delta(\mathbf{x} - \mathbf{x}^{(j,m)}),$$

$$q_{Gs} = \sum_{j=1}^{N_w} \sum_{m=1}^{M_{wj}} \frac{2\pi \Delta L^{(j,m)}}{\ln(r_e^{(j,m)}/r_w^{(j)})} \frac{\bar{k}k_{rg}}{\mu_g} \left[p_{bh}^{(j)} - p_g - \rho_g \wp (z_{bh}^{(j)} - z)\right] \delta(\mathbf{x} - \mathbf{x}^{(j,m)}),$$

where $\delta(\mathbf{x})$ is the Dirac delta function, N_w is the total number of wells, $M_{w,j}$ is the total number of perforated zones of the jth well, $\Delta L^{(j,m)}$ and $\mathbf{x}^{(j,m)}$ are the segment length and central location of the mth perforated zone of the jth well, the quantity $\bar{k}$ is an average of $\mathbf{k}$ at the wells (cf. Section 7.3.2 and Chapter 13), $r_w^{(j)}$ denotes the wellbore radius of the jth well, $r_e^{(j,m)}$ is the drainage radius of the jth well at the gridblock in which $\mathbf{x}^{(j,m)}$ is located, and $p_{bh}^{(j)}$ is the bottom hole pressure of the jth well at the well datum $z_{bh}^{(j)}$. The treatment of wells will be further discussed in Chapter 13. Introducing the *well index*

$$WI^{(j,m)} = \frac{2\pi \bar{k} \Delta L^{(j,m)}}{\ln(r_e^{(j,m)}/r_w^{(j)})},$$

the volumetric flow rates at the wells can be written as

$$\begin{aligned} q_{Ws} &= \sum_{j=1}^{N_w} \sum_{m=1}^{M_{wj}} WI^{(j,m)} \frac{k_{rw}}{\mu_w} \left[p_{bh}^{(j)} - p_w - \rho_w \wp (z_{bh}^{(j)} - z)\right] \delta(\mathbf{x} - \mathbf{x}^{(j,m)}), \\ q_{Os} &= \sum_{j=1}^{N_w} \sum_{m=1}^{M_{wj}} WI^{(j,m)} \frac{k_{ro}}{\mu_o} \left[p_{bh}^{(j)} - p_o - \rho_o \wp (z_{bh}^{(j)} - z)\right] \delta(\mathbf{x} - \mathbf{x}^{(j,m)}), \\ q_{Gs} &= \sum_{j=1}^{N_w} \sum_{m=1}^{M_{wj}} WI^{(j,m)} \frac{k_{rg}}{\mu_g} \left[p_{bh}^{(j)} - p_g - \rho_g \wp (z_{bh}^{(j)} - z)\right] \delta(\mathbf{x} - \mathbf{x}^{(j,m)}). \end{aligned} \tag{8.11}$$

Typical expressions of p_{cow}, p_{cgo}, and $k_{r\alpha}$ as functions of S_w and S_g were introduced in Chapter 3. Equations (8.5), (8.6), and (8.10) provide six equations for the six unknowns Φ_α and S_α, $\alpha = w, o, g$. If the bottom hole pressure $p_{bh}^{(j)}$ is not given, the source/sink term defining this pressure introduces one more unknown (i.e., $p_{bh}^{(j)}$). With appropriate boundary and initial conditions, this is a closed differential system for these unknowns. Alternative differential equations can be developed as for two-phase flow in the preceding chapter; they include the phase, weighted, and global pressure formulations (Chen, 2000; also cf. Exercises 8.2–8.7). As an example in this chapter we use the phase formulation.

8.1.2 Rock properties

The rock properties were considered in Chapter 3 for three-phase flow; for completeness, we state them briefly. The oil pressure is one of the primary variables to be used:

$$p = p_o. \tag{8.12}$$

While the capillary pressures are defined in (8.6), for the convenience of programming we usually employ the following definitions:

$$p_{cw} = p_w - p, \quad p_{cg} = p_g - p; \tag{8.13}$$

i.e., $p_{cw} = -p_{cow}$ and $p_{cg} = p_{cgo}$. Moreover, for notational convenience, let $p_{co} = 0$. The capillary pressures p_{cw} and p_{cg} are assumed to be functions of the saturations only (Leverett and Lewis, 1941):

$$p_{cw} = p_{cw}(S_w), \quad p_{cg} = p_{cg}(S_g). \tag{8.14}$$

The relative permeabilities for water and gas are assumed to be of the form

$$\begin{aligned} k_{rw} &= k_{rw}(S_w), \quad k_{row} = k_{row}(S_w), \\ k_{rg} &= k_{rg}(S_g), \quad k_{rog} = k_{rog}(S_g). \end{aligned} \tag{8.15}$$

As an example, Stone's model II for the oil relative permeability is used (cf. Section 3.1.2)

$$\begin{aligned} k_{ro}(S_w, S_g) = k_{rc}\Bigg\{&\left[\frac{k_{row}(S_w)}{k_{rc}} + k_{rw}(S_w)\right]\left[\frac{k_{rog}(S_g)}{k_{rc}} + k_{rg}(S_g)\right] \\ &- k_{rw}(S_w) - k_{rg}(S_g)\Bigg\}, \end{aligned} \tag{8.16}$$

where $k_{rc} = k_{row}(S_{wc})$ and S_{wc} is the critical saturation (cf. Chapter 3). Finally, the porosity ϕ is assumed to have the form

$$\phi = \phi^o\big(1 + c_R(p - p^o)\big), \tag{8.17}$$

where ϕ^o is the porosity at a reference pressure p^o and c_R is the rock compressibility.

8.1.3 Fluid properties

The fluid properties were stated in Chapter 3; we briefly review the definitions of densities and viscosities. The water density ρ_{Ws} at standard conditions is determined using water salinities (cf. Section 3.2.1), while the water phase density ρ_w is determined by

$$\rho_w = \frac{\rho_{Ws}}{B_{wi}}\big(1 + c_w(p - p_o)\big), \tag{8.18}$$

where B_{wi} is the water formation volume factor at the initial formation pressure p_o, and c_w is the water compressibility. The water viscosity μ_w is taken to be constant.

The black oil model involves three phases and three components: water, oil, and gas. The relationship between the phases and components is that the water component is all the water phase with density ρ_w, the oil component exists solely in the oil phase with density

ρ_{Oo}, and the gas component is divided into two parts: one part in the gas phase that is called *free gas* with density ρ_g, and the other part in the oil phase that is termed the *solution gas* with density ρ_{Go}. Thus the oil phase density ρ_o is given by

$$\rho_o = \rho_{Oo} + \rho_{Go}. \tag{8.19}$$

The oil component density ρ_{Oo} is evaluated from

$$\rho_{Oo} = \frac{\rho_{Os}}{B_o}, \tag{8.20}$$

where the oil formation volume factor B_o is

$$B_o = B_{ob}(p_b)\big(1 - c_o(p - p_b)\big), \tag{8.21}$$

with B_{ob} being the formation volume factor at the bubble point pressure p_b and c_o the oil compressibility. The solution gas density ρ_{Go} is computed by

$$\rho_{Go} = \frac{R_{so}\rho_{Gs}}{B_o}. \tag{8.22}$$

The free gas density ρ_g is defined by

$$\rho_g = \frac{\rho_{Gs}}{B_g}, \tag{8.23}$$

where

$$\rho_{Gs} = Y_G \rho_{air}, \quad B_g = \frac{ZT}{p}\frac{p_s}{T_s}, \tag{8.24}$$

with Y_G being the raw gas density (which is unity for air), ρ_{air} the air density, Z the gas deviation factor, T the temperature, and p_s and T_s the formation pressure and temperature at standard conditions.

The oil viscosity μ_o is given by

$$\mu_o = \mu_{ob}(p_b)\big(1 + c_\mu(p - p_b)\big), \tag{8.25}$$

where μ_{ob} is the oil viscosity at p_b and c_μ is the *oil viscosity compressibility*. The gas viscosity μ_g is a function of p:

$$\mu_g = \mu_g(p). \tag{8.26}$$

8.1.4 Phase states

In the secondary recovery, if the reservoir pressure is above the bubble point pressure of the oil phase, the flow is two-phase; if the pressure drops below the bubble point pressure, then the flow is of black oil type. Because of the frequent change in injection and production in a reservoir, the bubble point pressure varies. If all three phases coexist, the reservoir is referred to as being in the *saturated state*. When all gas dissolves into the oil phase, there is no gas phase present (no free gas); i.e., $S_g = 0$. In this case, the reservoir is said to be in the *undersaturated state*. The critical pressure at which the saturated state becomes the undersaturated state or vice versa is the *bubble point pressure*. In the saturated state, $S_g \neq 0$ and $p_b = p$; the densities and viscosities depend only on pressure p:

$$\rho_{Oo}(p) = \frac{\rho_{Os}}{B_{ob}(p)}, \quad \rho_{Go}(p) = \frac{R_{so}(p)\rho_{Gs}}{B_{ob}(p)}, \quad \rho_g(p) = \frac{\rho_{Gs}}{B_g(p)}, \\ \mu_o = \mu_o(p), \qquad \mu_g = \mu_g(p). \tag{8.27}$$

In the undersaturated state, $S_g = 0$ and $p_b < p$. The densities and viscosity in the oil phase depend on both p and p_b:

$$\rho_{Oo}(p, p_b) = \frac{\rho_{Os}}{B_{ob}(p_b)}\big(1 + c_o(p - p_b)\big), \\ \rho_{Go}(p, p_b) = \frac{R_{so}(p_b)\rho_{Gs}}{B_{ob}(p_b)}\big(1 + c_o(p - p_b)\big), \quad \rho_g(p) = \frac{\rho_{Gs}}{B_g(p)}, \\ \mu_o(p, p_b) = \mu_{ob}(p_b)\big(1 + c_\mu(p - p_b)\big), \quad \mu_g = \mu_g(p). \tag{8.28}$$

For numerical solutions of the black oil model, the choice of the primary unknowns depends on the states. In the saturated state, $p = p_o$, S_w, and S_o are the primary unknowns; in the undersaturated state, $p = p_o$, p_b, and S_w are the primary unknowns. Consequently, the initial conditions are either

$$p(\mathbf{x}, 0) = p^0(\mathbf{x}), \quad S_w(\mathbf{x}, 0) = S_w^0(\mathbf{x}), \quad S_o(\mathbf{x}, 0) = S_o^0(\mathbf{x}), \quad \mathbf{x} \in \Omega, \tag{8.29}$$

or

$$p(\mathbf{x}, 0) = p^0(\mathbf{x}), \quad S_w(\mathbf{x}, 0) = S_w^0(\mathbf{x}), \quad p_b(\mathbf{x}, 0) = p_b^0(\mathbf{x}), \quad \mathbf{x} \in \Omega, \tag{8.30}$$

depending on the initial state of a reservoir.

8.2 Solution Techniques

The choice of a solution technique is crucial for a coupled system of differential equations. In this section, we discuss several solution techniques that are currently used in the simulation of multiphase flow. These techniques include *simultaneous solution* (SS), *sequential, implicit pressure-explicit saturation* (IMPES) or *iterative IMPES*, *adaptive implicit*, and *parallel techniques*. IMPES was studied for two-phase flow in the preceding chapter and is further considered for the black oil model.

8.2.1 The Newton–Raphson method

Consider a general system of nonlinear differential equations:

$$\pounds_m\{F_m[\mathbf{p}(\mathbf{x})]\} = f_m(\mathbf{x}), \qquad m = 1, 2, \ldots, M, \quad \mathbf{x} \in \Omega, \tag{8.31}$$

where $\pounds_m$ denotes a linear differential operator, $F_m(\cdot)$ is a nonlinear function, $\mathbf{p} = (p_1, p_2, \ldots, p_M)^T$ is the vector of dependent variables, $\mathbf{f} = (f_1, f_2, \ldots, f_M)^T$ is a given vector, and M is the total number of equations. The *Newton–Raphson iteration* for solving (8.31) establishes an iterative equation system. Taylor's series expansion for $F_m(\mathbf{p} + \delta\mathbf{p})$ is

$$F_m(\mathbf{p} + \delta\mathbf{p}) = F_m(\mathbf{p}) + \nabla F_m(\mathbf{p}) \cdot \delta\mathbf{p} + \mathcal{O}(|\delta\mathbf{p}|^2), \tag{8.32}$$

where $|\delta \mathbf{p}|$ is the Euclidian norm of $\delta \mathbf{p}$. If the higher-order term $\mathcal{O}(|\delta \mathbf{p}|^2)$ (relative to $|\delta \mathbf{p}|$) is truncated, $F_m(\mathbf{p}+\delta \mathbf{p})$ can be approximated as

$$F_m(\mathbf{p}+\delta \mathbf{p}) \approx F_m(\mathbf{p}) + \nabla F_m(\mathbf{p}) \cdot \delta \mathbf{p}. \tag{8.33}$$

If we substitute (8.33) into (8.31), we obtain the iterative equations

$$\pounds_m \left[F_m(\mathbf{p}^l) + \nabla F_m(\mathbf{p}^l) \cdot \delta \mathbf{p}^l \right] = f_m(\mathbf{x}), \qquad m = 1, 2, \ldots, M, \quad \mathbf{x} \in \Omega, \tag{8.34}$$

where $\mathbf{p}^l$ is the lth iterative solution of $\mathbf{p}$ and $\nabla F_m(\mathbf{p}^l)$ is $\nabla F_m(\mathbf{p})$ at $\mathbf{p} = \mathbf{p}^l$, with an initial solution $\mathbf{p}^0$. In the iterative equation system (8.34), the correction vector $\delta \mathbf{p}^l$ is the unknown. This system can be rewritten as

$$\pounds_m \left[\nabla F_m(\mathbf{p}^l) \cdot \delta \mathbf{p}^l \right] = g_m(\mathbf{x}), \qquad m = 1, 2, \ldots, M, \quad \mathbf{x} \in \Omega, \tag{8.35}$$

where $g_m(\mathbf{x}) = f_m(\mathbf{x}) - \pounds_m \left[F_m(\mathbf{p}^l) \right]$, and $F_m(\mathbf{p}^l)$ and $\nabla F_m(\mathbf{p}^l)$ are treated as fixed. Now, (8.35) is a linear system for $\delta \mathbf{p}^l$, and the various numerical methods developed in Chapter 4 can be applied.

A new solution vector $\mathbf{p}^{l+1}$ is obtained by adding the correction vector $\delta \mathbf{p}^l$ to the previous iterative solution vector $\mathbf{p}^l$; i.e.,

$$\mathbf{p}^{l+1} = \mathbf{p}^l + \delta \mathbf{p}^l.$$

This iteration proceeds until the Euclidian norm of $\delta \mathbf{p}^l$ is smaller than a prescribed value.

8.2.2 The SS technique

The most natural solution technique for system (8.10) is to solve the three equations simultaneously, which suggests the SS technique. This technique was initially introduced by Douglas et al. (1959) and is still widely used in reservoir simulation.

Let $n > 0$ (an integer) indicate a time step. For any function v of time, we use $\bar{\delta} v$ to denote the time increment at the nth step:

$$\bar{\delta} v = v^{n+1} - v^n.$$

An implicit time approximation for system (8.10) can be defined as

$$\begin{aligned}
&\frac{1}{\Delta t} \bar{\delta} \left(\frac{\phi S_w}{B_w} \right) = \nabla \cdot \left(\mathbf{T}_w^{n+1} \nabla \Phi_w^{n+1} \right) + \frac{q_{Ws}^{n+1}}{B_w^{n+1}}, \\
&\frac{1}{\Delta t} \bar{\delta} \left(\frac{\phi S_o}{B_o} \right) = \nabla \cdot \left(\mathbf{T}_o^{n+1} \nabla \Phi_o^{n+1} \right) + \frac{q_{Os}^{n+1}}{B_o^{n+1}}, \\
&\frac{1}{\Delta t} \bar{\delta} \left[\phi \left(\frac{S_g}{B_g} + \frac{R_{so} S_o}{B_o} \right) \right] \\
&\quad = \nabla \cdot \left(\mathbf{T}_g^{n+1} \nabla \Phi_g^{n+1} + R_{so}^{n+1} \mathbf{T}_o^{n+1} \nabla \Phi_o^{n+1} \right) + \frac{q_{Gs}^{n+1}}{B_g^{n+1}} + \frac{q_{Os}^{n+1} R_{so}^{n+1}}{B_o^{n+1}},
\end{aligned} \tag{8.36}$$

where $\Delta t = t^{n+1} - t^n$. System (8.36) is nonlinear in the unknowns Φ_α^{n+1} and S_α^{n+1}, $\alpha = w, o, g$, and can be linearized via the Newton–Raphson iteration. For this, we write

$$\Phi_\alpha^{n+1,l+1} = \Phi_\alpha^{n+1,l} + \delta \Phi_\alpha, \quad S_\alpha^{n+1,l+1} = S_\alpha^{n+1,l} + \delta S_\alpha, \quad \alpha = w, o, g,$$

where l denotes the iteration number of Newton–Raphson's iterations and $\delta\Phi_\alpha$ and δS_α represent the increments of the potential and saturation, respectively, in this iteration step (we omit the superscript l in the increments for notational convenience). Note that for any function v of time,

$$v^{n+1} \approx v^{n+1,l+1} = v^{n+1,l} + \delta v,$$

so that

$$\bar{\delta} v \approx v^{n+1,l} - v^n + \delta v.$$

Using this approximation in system (8.36) yields

$$\begin{aligned}
&\frac{1}{\Delta t}\left[\left(\frac{\phi S_w}{B_w}\right)^{n+1,l} - \left(\frac{\phi S_w}{B_w}\right)^{n} + \delta\left(\frac{\phi S_w}{B_w}\right)\right] \\
&\qquad = \nabla\cdot\left(\mathbf{T}_w^{n+1,l+1}\nabla\Phi_w^{n+1,l+1}\right) + \frac{q_{Ws}^{n+1,l+1}}{B_w^{n+1,l+1}}, \\
&\frac{1}{\Delta t}\left[\left(\frac{\phi S_o}{B_o}\right)^{n+1,l} - \left(\frac{\phi S_o}{B_o}\right)^{n} + \delta\left(\frac{\phi S_o}{B_o}\right)\right] \\
&\qquad = \nabla\cdot\left(\mathbf{T}_o^{n+1,l+1}\nabla\Phi_o^{n+1,l+1}\right) + \frac{q_{Os}^{n+1,l+1}}{B_o^{n+1,l+1}}, \\
&\frac{1}{\Delta t}\left\{\left[\phi\left(\frac{S_g}{B_g} + \frac{R_{so}S_o}{B_o}\right)\right]^{n+1,l} - \left[\phi\left(\frac{S_g}{B_g} + \frac{R_{so}S_o}{B_o}\right)\right]^{n}\right. \\
&\qquad\qquad \left. + \delta\left[\phi\left(\frac{S_g}{B_g} + \frac{R_{so}S_o}{B_o}\right)\right]\right\} \\
&= \nabla\cdot\left(\mathbf{T}_g^{n+1,l+1}\nabla\Phi_g^{n+1,l+1} + R_{so}^{n+1,l+1}\mathbf{T}_o^{n+1,l+1}\nabla\Phi_o^{n+1,l+1}\right) \\
&\qquad + \frac{q_{Gs}^{n+1,l+1}}{B_g^{n+1,l+1}} + \frac{q_{Os}^{n+1,l+1}R_{so}^{n+1,l+1}}{B_o^{n+1,l+1}}.
\end{aligned} \tag{8.37}$$

In this system the increments $\delta\Phi_\alpha$ and δS_α are unknowns, $\alpha = w, o, g$. When no ambiguity occurs, we replace $v^{n+1,l+1}$ and $v^{n+1,l}$ by v^{l+1} and v^l, respectively (i.e., the superscript $n+1$ is omitted).

In the saturated state, the primary unknowns are

$$\delta p, \quad \delta S_w, \quad \delta S_o,$$

and in the undersaturated state, they are

$$\delta p, \quad \delta S_w, \quad \delta p_b.$$

In the former case, $\delta S_g = -\delta S_w - \delta S_o$, and in the latter case, $\delta S_g = 0$ and $\delta S_o = -\delta S_w$. Accordingly, the left-hand side of system (8.37) can be expanded as follows. For the water component,

$$\delta\left(\frac{\phi S_w}{B_w}\right) = c_{wp}\delta p + c_{wS_w}\delta S_w, \tag{8.38}$$

where

$$c_{wp} = \phi^o c_R \left(\frac{S_w}{B_w}\right)^l + \left(\phi S_w \frac{dB_w^{-1}}{dp}\right)^l, \quad c_{wS_w} = \left(\frac{\phi}{B_w}\right)^l.$$

For the oil component in the saturated state,

$$\delta\left(\frac{\phi S_o}{B_o}\right) = c_{op}\delta p + c_{oS_o}\delta S_o, \tag{8.39}$$

where

$$c_{op} = \phi^o c_R \left(\frac{S_o}{B_o}\right)^l + \left(\phi S_o \frac{dB_o^{-1}}{dp}\right)^l, \quad c_{oS_o} = \left(\frac{\phi}{B_o}\right)^l,$$

and in the undersaturated state,

$$\delta\left(\frac{\phi S_o}{B_o}\right) = c_{op}\delta p + c_{oS_w}\delta S_w + c_{op_b}\delta p_b, \tag{8.40}$$

where

$$c_{op} = \phi^o c_R \left(\frac{S_o}{B_o}\right)^l + \left(\phi S_o \frac{\partial B_o^{-1}}{\partial p}\right)^l, \quad c_{oS_w} = -\left(\frac{\phi}{B_o}\right)^l, \quad c_{op_b} = \left(\phi S_o \frac{\partial B_o^{-1}}{\partial p_b}\right)^l.$$

For the gas component in the saturated state,

$$\delta\left[\phi\left(\frac{S_g}{B_g} + \frac{R_{so}S_o}{B_o}\right)\right] = c_{gp}\delta p + c_{gS_w}\delta S_w + c_{gS_o}\delta S_o, \tag{8.41}$$

where

$$c_{gp} = \phi^o c_R \left(\frac{S_g}{B_g} + \frac{R_{so}S_o}{B_o}\right)^l + \left[\phi\left(S_g \frac{dB_g^{-1}}{dp} + S_o \frac{d}{dp}\left(\frac{R_{so}}{B_o}\right)\right)\right]^l,$$

$$c_{gS_w} = -\left(\frac{\phi}{B_g}\right)^l, \quad c_{gS_o} = -\left(\frac{\phi}{B_g}\right)^l + \left(\frac{\phi R_{so}}{B_o}\right)^l,$$

and in the undersaturated state,

$$\delta\left[\phi\left(\frac{S_g}{B_g} + \frac{R_{so}S_o}{B_o}\right)\right] = c_{gp}\delta p + c_{gS_w}\delta S_w + c_{gp_b}\delta p_b, \tag{8.42}$$

where

$$c_{gp} = \phi^o c_R \left(\frac{R_{so}S_o}{B_o}\right)^l + \left(\phi S_o \frac{\partial}{\partial p}\left(\frac{R_{so}}{B_o}\right)\right)^l,$$

$$c_{gS_w} = -\left(\frac{\phi R_{so}}{B_o}\right)^l, \quad c_{gp_b} = \left(\phi S_o \frac{\partial}{\partial p_b}\left(\frac{R_{so}}{B_o}\right)\right)^l.$$

The expansion of the right-hand side of system (8.37) in terms of the primary unknowns depends on the solution technique.

In the SS method, the phase potentials are evaluated by

$$\Phi_\alpha^{l+1} = p^{l+1} + p_{c\alpha}^{l+1} - \rho_\alpha^{l+1} \wp z, \qquad \alpha = w, o, g. \tag{8.43}$$

Similarly, the transmissibilities are computed:

$$\mathbf{T}_{\alpha}^{l+1} = \frac{k_{r\alpha}^{l+1}}{\mu_{\alpha}^{l+1} B_{\alpha}^{l+1}} \mathbf{k}, \qquad \alpha = w, o, g, \tag{8.44}$$

where $\mu_{w}^{l+1} = \mu_w$. The flow rates at wells are determined by

$$\begin{aligned}
q_{Ws}^{l+1} &= \sum_{j=1}^{N_w} \sum_{m=1}^{M_{wj}} WI^{(j,m)} \frac{k_{rw}^{l+1}}{\mu_w} \Bigg[\left(p_{bh}^{(j)}\right)^{l+1} - p^{l+1} - p_{cw}^{l+1} \\
&\qquad - \rho_w^{l+1} \wp (z_{bh}^{(j)} - z) \Bigg] \delta(\mathbf{x} - \mathbf{x}^{(j,m)}), \\
q_{Os}^{l+1} &= \sum_{j=1}^{N_w} \sum_{m=1}^{M_{wj}} WI^{(j,m)} \frac{k_{ro}^{l+1}}{\mu_o^{l+1}} \Bigg[\left(p_{bh}^{(j)}\right)^{l+1} - p^{l+1} \\
&\qquad - \rho_o^{l+1} \wp (z_{bh}^{(j)} - z) \Bigg] \delta(\mathbf{x} - \mathbf{x}^{(j,m)}), \\
q_{Gs}^{l+1} &= \sum_{j=1}^{N_w} \sum_{m=1}^{M_{wj}} WI^{(j,m)} \frac{k_{rg}^{l+1}}{\mu_g^{l+1}} \Bigg[\left(p_{bh}^{(j)}\right)^{l+1} - p^{l+1} - p_{cg}^{l+1} \\
&\qquad - \rho_g^{l+1} \wp (z_{bh}^{(j)} - z) \Bigg] \delta(\mathbf{x} - \mathbf{x}^{(j,m)}).
\end{aligned} \tag{8.45}$$

We now expand the potentials, transmissibilities, and flow rates at wells in terms of the primary unknowns: δp, δS_w, and δS_o in the saturated state, and δp, δS_w, and δp_b in the undersaturated state. For the water component,

$$\Phi_w^{l+1} = \Phi_w^l + d_{wp}\delta p + d_{wS_w}\delta S_w, \tag{8.46}$$

where

$$d_{wp} = 1 - \left(\frac{d\rho_w}{dp}\right)^l \wp z, \quad d_{wS_w} = \left(\frac{dp_{cw}}{dS_w}\right)^l.$$

For the oil component in the saturated state,

$$\Phi_o^{l+1} = \Phi_o^l + d_{op}\delta p, \tag{8.47}$$

where

$$d_{op} = 1 - \left(\frac{d\rho_o}{dp}\right)^l \wp z,$$

and in the undersaturated state,

$$\Phi_o^{l+1} = \Phi_o^l + d_{op}\delta p + d_{op_b}\delta p_b, \tag{8.48}$$

where

$$d_{op} = 1 - \left(\frac{\partial \rho_o}{\partial p}\right)^l \wp z, \quad d_{op_b} = -\left(\frac{\partial \rho_o}{\partial p_b}\right)^l \wp z.$$

For the gas component in the saturated state,

$$\Phi_g^{l+1} = \Phi_g^l + d_{gp}\delta p + d_{gS}(\delta S_w + \delta S_o), \tag{8.49}$$

where

$$d_{gp} = 1 - \left(\frac{d\rho_g}{dp}\right)^l \wp z, \quad d_{gS} = -\left(\frac{dp_{cg}}{dS_g}\right)^l.$$

The transmissibilities can be expanded in a similar fashion. For the water component,

$$\mathbf{T}_w^{l+1} = \mathbf{T}_w^l + \mathbf{E}_{wp}\delta p + \mathbf{E}_{wS_w}\delta S_w, \tag{8.50}$$

where

$$\mathbf{E}_{wp} = \left(\frac{k_{rw}}{\mu_w}\frac{dB_w^{-1}}{dp}\right)^l \mathbf{k}, \qquad \mathbf{E}_{wS_w} = \left(\frac{dk_{rw}}{dS_w}\frac{1}{\mu_w B_w}\right)^l \mathbf{k}.$$

For the oil component in the saturated state,

$$\mathbf{T}_o^{l+1} = \mathbf{T}_o^l + \mathbf{E}_{op}\delta p + \mathbf{E}_{oS_w}\delta S_w + \mathbf{E}_{oS_o}\delta S_o, \tag{8.51}$$

where

$$\mathbf{E}_{op} = \left(k_{ro}\frac{d}{dp}\left(\frac{1}{\mu_o B_o}\right)\right)^l \mathbf{k}, \qquad \mathbf{E}_{oS_o} = -\left(\frac{dk_{ro}}{dS_g}\frac{1}{\mu_o B_o}\right)^l \mathbf{k},$$

$$\mathbf{E}_{oS_w} = \left(\left(\frac{dk_{ro}}{dS_w} - \frac{dk_{ro}}{dS_g}\right)\frac{1}{\mu_o B_o}\right)^l \mathbf{k},$$

and in the undersaturated state,

$$\mathbf{T}_o^{l+1} = \mathbf{T}_o^l + \mathbf{E}_{op}\delta p + \mathbf{E}_{oS_w}\delta S_w + \mathbf{E}_{op_b}\delta p_b, \tag{8.52}$$

where

$$\mathbf{E}_{op} = \left(k_{ro}\frac{\partial}{\partial p}\left(\frac{1}{\mu_o B_o}\right)\right)^l \mathbf{k}, \qquad \mathbf{E}_{oS_w} = \left(\frac{dk_{ro}}{dS_w}\frac{1}{\mu_o B_o}\right)^l \mathbf{k},$$

$$\mathbf{E}_{op_b} = \left(k_{ro}\frac{\partial}{\partial p_b}\left(\frac{1}{\mu_o B_o}\right)\right)^l \mathbf{k}.$$

For the gas component in the saturated state,

$$\mathbf{T}_g^{l+1} = \mathbf{T}_g^l + \mathbf{E}_{gp}\delta p + \mathbf{E}_{gS}(\delta S_w + \delta S_o), \tag{8.53}$$

where

$$\mathbf{E}_{gp} = \left(k_{rg}\frac{d}{dp}\left(\frac{1}{\mu_g B_g}\right)\right)^l \mathbf{k}, \qquad \mathbf{E}_{gS} = -\left(\frac{dk_{rg}}{dS_g}\frac{1}{\mu_g B_g}\right)^l \mathbf{k}.$$

The flow rates at wells are expanded in a similar manner. For the water component,

$$q_{Ws}^{l+1} = q_{Ws}^l + \sum_{j=1}^{N_w}\sum_{m=1}^{M_{wj}} WI^{(j,m)}\left[e_{wp}^{(j)}\delta p + e_{wS_w}^{(j)}\delta S_w + e_{wp_{bh}}\delta p_{bh}^{(j)}\right]\delta(\mathbf{x} - \mathbf{x}^{(j,m)}), \tag{8.54}$$

where

$$e_{wp}^{(j)} = -\frac{1}{\mu_w}\left[k_{rw}\left(1 + \frac{d\rho_w}{dp}\wp(z_{bh}^{(j)} - z)\right)\right]^l, \quad e_{wp_{bh}} = \frac{k_{rw}^l}{\mu_w},$$

$$e_{wS_w}^{(j)} = \frac{1}{\mu_w}\left[\frac{dk_{rw}}{dS_w}\left(p_{bh}^{(j)} - p - p_{cw} - \rho_w\wp(z_{bh}^{(j)} - z)\right) - k_{rw}\frac{dp_{cw}}{dS_w}\right]^l.$$

For the oil component in the saturated state,

$$q_{Os}^{l+1} = q_{Os}^{l} + \sum_{j=1}^{N_w}\sum_{m=1}^{M_{wj}} WI^{(j,m)}\Big[e_{op}^{(j)}\delta p + e_{oS_w}^{(j)}\delta S_w + e_{oS_o}^{(j)}\delta S_o + e_{op_{bh}}\delta p_{bh}^{(j)}\Big]\delta(\mathbf{x}-\mathbf{x}^{(j,m)}), \tag{8.55}$$

where

$$\begin{aligned}
e_{op}^{(j)} &= \left\{k_{ro}\left[\frac{d\mu_o^{-1}}{dp}\left(p_{bh}^{(j)} - p - \rho_o\wp(z_{bh}^{(j)} - z)\right) - \frac{1}{\mu_o}\left(1 + \frac{d\rho_o}{dp}\wp(z_{bh}^{(j)} - z)\right)\right]\right\}^l, \\
e_{oS_w}^{(j)} &= \left[\frac{1}{\mu_o}\left(\frac{dk_{ro}}{dS_w} - \frac{dk_{ro}}{dS_g}\right)\left(p_{bh}^{(j)} - p - \rho_o\wp(z_{bh}^{(j)} - z)\right)\right]^l, \\
e_{oS_o}^{(j)} &= -\left[\frac{dk_{ro}}{dS_g}\frac{1}{\mu_o}\left(p_{bh}^{(j)} - p - \rho_o\wp(z_{bh}^{(j)} - z)\right)\right]^l, \quad e_{op_{bh}} = \left(\frac{k_{ro}}{\mu_o}\right)^l,
\end{aligned}$$

and in the undersaturated state,

$$q_{Os}^{l+1} = q_{Os}^{l} + \sum_{j=1}^{N_w}\sum_{m=1}^{M_{wj}} WI^{(j,m)}\Big[e_{op}^{(j)}\delta p + e_{oS_w}^{(j)}\delta S_w + e_{op_b}^{(j)}\delta p_b + e_{op_{bh}}\delta p_{bh}^{(j)}\Big]\delta(\mathbf{x}-\mathbf{x}^{(j,m)}), \tag{8.56}$$

where

$$\begin{aligned}
e_{op}^{(j)} &= \left\{k_{ro}\left[\frac{\partial\mu_o^{-1}}{\partial p}\left(p_{bh}^{(j)} - p - \rho_o\wp(z_{bh}^{(j)} - z)\right) - \frac{1}{\mu_o}\left(1 + \frac{\partial\rho_o}{\partial p}\wp(z_{bh}^{(j)} - z)\right)\right]\right\}^l, \\
e_{oS_w}^{(j)} &= \left[\frac{1}{\mu_o}\frac{dk_{ro}}{dS_w}\left(p_{bh}^{(j)} - p - \rho_o\wp(z_{bh}^{(j)} - z)\right)\right]^l, \quad e_{op_{bh}} = \left(\frac{k_{ro}}{\mu_o}\right)^l, \\
e_{op_b}^{(j)} &= \left\{k_{ro}\left[\frac{\partial\mu_o^{-1}}{\partial p_b}\left(p_{bh}^{(j)} - p - \rho_o\wp(z_{bh}^{(j)} - z)\right) - \frac{1}{\mu_o}\frac{\partial\rho_o}{\partial p_b}\wp(z_{bh}^{(j)} - z)\right]\right\}^l.
\end{aligned}$$

For the gas component in the saturated state,

$$q_{Gs}^{l+1} = q_{Gs}^{l} + \sum_{j=1}^{N_w}\sum_{m=1}^{M_{wj}} WI^{(j,m)}\Big[e_{gp}^{(j)}\delta p + e_{gS}^{(j)}(\delta S_w + \delta S_o) + e_{gp_{bh}}\delta p_{bh}^{(j)}\Big]\delta(\mathbf{x}-\mathbf{x}^{(j,m)}), \tag{8.57}$$

where

$$e_{gp}^{(j)} = \left\{ k_{rg} \left[\frac{d\mu_g^{-1}}{dp} \left(p_{bh}^{(j)} - p - p_{cg} - \rho_g \wp (z_{bh}^{(j)} - z) \right) \right.\right.$$

$$\left.\left. - \frac{1}{\mu_g} \left(1 + \frac{d\rho_g}{dp} \wp (z_{bh}^{(j)} - z) \right) \right] \right\}^l, \quad e_{gp_{bh}} = \left(\frac{k_{rg}}{\mu_g} \right)^l,$$

$$e_{gS}^{(j)} = - \left[\frac{1}{\mu_g} \left(\frac{dk_{rg}}{dS_g} \left(p_{bh}^{(j)} - p - p_{cg} - \rho_g \wp (z_{bh}^{(j)} - z) \right) - k_{rg} \frac{dp_{cg}}{dS_g} \right) \right]^l.$$

Finally, we expand R_{so} and B_α in (8.37), $\alpha = w, o, g$. In the saturated state,

$$R_{so}^{l+1} = R_{so}^l + r_{sp}\delta p, \quad B_\alpha^{l+1} = B_\alpha^l \left(1 - b_{\alpha p}\delta p \right), \quad \alpha = w, o, g, \tag{8.58}$$

where

$$r_{sp} = \left(\frac{dR_{so}}{dp} \right)^l, \quad b_{\alpha p} = - \left(\frac{1}{B_\alpha} \frac{dB_\alpha}{dp} \right)^l, \quad \alpha = w, o, g,$$

and in the undersaturated state,

$$R_{so}^{l+1} = R_{so}^l + r_{sp}\delta p + r_{sp_b}\delta p_b, \quad B_o^{l+1} = B_o^l \left(1 - b_{op}\delta p - b_{op_b}\delta p_b \right), \tag{8.59}$$

where

$$r_{sp} = \left(\frac{\partial R_{so}}{\partial p} \right)^l, \qquad r_{sp_b} = \left(\frac{\partial R_{so}}{\partial p_b} \right)^l,$$

$$b_{op} = - \left(\frac{1}{B_o} \frac{\partial B_o}{\partial p} \right)^l, \qquad b_{op_b} = - \left(\frac{1}{B_o} \frac{\partial B_o}{\partial p_b} \right)^l.$$

Saturated state

Substituting (8.38)–(8.59) into (8.37) leads to a linear system in terms of the primary unknowns. Because the choice of the unknowns depends on the state of a reservoir, we separate the discussion of the saturated state from that of the undersaturated state. In the former case, the primary unknowns are δp, δS_w, and δS_o. For the water component, substituting (8.38), (8.46), (8.50), (8.54), and (8.58) into the first equation of (8.37) and ignoring the higher-order terms in δp and δS_w gives (cf. Exercise 8.8)

$$\begin{aligned} &\frac{1}{\Delta t} \left[\left(\frac{\phi S_w}{B_w} \right)^l - \left(\frac{\phi S_w}{B_w} \right)^n + c_{wp}\delta p + c_{wS_w}\delta S_w \right] \\ &= \nabla \cdot \left(\left(\mathbf{T}_w^l + \mathbf{E}_{wp}\delta p + \mathbf{E}_{wS_w}\delta S_w \right) \nabla \Phi_w^l \right) \\ &+ \nabla \cdot \left(\mathbf{T}_w^l \nabla \left(d_{wp}\delta p \right) \right) + \nabla \cdot \left(\mathbf{T}_w^l \nabla \left(d_{wS_w}\delta S_w \right) \right) \\ &+ \frac{1}{B_w^l} \left\{ q_{Ws}^l + \sum_{j=1}^{N_w} \sum_{m=1}^{M_{wj}} WI^{(j,m)} \left[e_{wp}^{(j)}\delta p + e_{wS_w}^{(j)}\delta S_w \right.\right. \\ &\left.\left. + e_{wp_{bh}}^{(j)}\delta p_{bh}^{(j)} \right] \delta(\mathbf{x} - \mathbf{x}^{(j,m)}) \right\} + \frac{b_{wp}q_{Ws}^l}{B_w^l}\delta p. \end{aligned} \tag{8.60}$$

For the oil component in the saturated state, substituting (8.39), (8.47), (8.51), (8.55), and (8.58) into the second equation of (8.37) gives (cf. Exercise 8.9)

$$\begin{aligned}
&\frac{1}{\Delta t}\left[\left(\frac{\phi S_o}{B_o}\right)^l - \left(\frac{\phi S_o}{B_o}\right)^n + c_{op}\delta p + c_{oS_o}\delta S_o\right] \\
&\quad = \nabla\cdot\left(\left(\mathbf{T}_o^l + \mathbf{E}_{op}\delta p + \mathbf{E}_{oS_w}\delta S_w + \mathbf{E}_{oS_o}\delta S_o\right)\nabla\Phi_o^l\right) \\
&\qquad + \nabla\cdot\left(\mathbf{T}_o^l\nabla\left(d_{op}\delta p\right)\right) \\
&\qquad + \frac{1}{B_o^l}\Bigg\{q_{Os}^l + \sum_{j=1}^{N_w}\sum_{m=1}^{M_{wj}} WI^{(j,m)}\Big[e_{op}^{(j)}\delta p + e_{oS_w}^{(j)}\delta S_w \\
&\qquad\qquad + e_{oS_o}^{(j)}\delta S_o + e_{op_{bh}}^{(j)}\delta p_{bh}^{(j)}\Big]\delta(\mathbf{x}-\mathbf{x}^{(j,m)})\Bigg\} + \frac{b_{op}q_{Os}^l}{B_o^l}\delta p.
\end{aligned} \tag{8.61}$$

For the gas component in the saturated state, substituting (8.41), (8.49), (8.53), (8.57), and (8.58) into the third equation of (8.37) yields (cf. Exercise 8.10)

$$\begin{aligned}
&\frac{1}{\Delta t}\Bigg\{\left[\phi\left(\frac{S_g}{B_g}+\frac{R_{so}S_o}{B_o}\right)\right]^l - \left[\phi\left(\frac{S_g}{B_g}+\frac{R_{so}S_o}{B_o}\right)\right]^n \\
&\qquad + c_{gp}\delta p + c_{gS_w}\delta S_w + c_{gS_o}\delta S_o\Bigg\} \\
&= \nabla\cdot\left(\left(\mathbf{T}_g^l + \mathbf{E}_{gp}\delta p + \mathbf{E}_{gS}\left(\delta S_w + \delta S_o\right)\right)\nabla\Phi_g^l\right) \\
&\quad + \nabla\cdot\left(\mathbf{T}_g^l\nabla\left(d_{gp}\delta p\right)\right) + \nabla\cdot\left(\mathbf{T}_g^l\nabla\left(d_{gS}\left(\delta S_w+\delta S_o\right)\right)\right) \\
&\quad + \nabla\cdot\Bigg[\Bigg(R_{so}^l\left(\mathbf{T}_o^l + \mathbf{E}_{op}\delta p + \mathbf{E}_{oS_w}\delta S_w + \mathbf{E}_{oS_o}\delta S_o\right) \\
&\qquad\qquad + r_{sp}\mathbf{T}_o^l\delta p\Bigg)\nabla\Phi_o^l\Bigg] + \nabla\cdot\left(R_{so}^l\mathbf{T}_o^l\nabla\left(d_{op}\delta p\right)\right) \\
&\quad + \frac{1}{B_g^l}\Bigg\{q_{Gs}^l + \sum_{j=1}^{N_w}\sum_{m=1}^{M_{wj}} WI^{(j,m)}\Big[e_{gp}^{(j)}\delta p + e_{gS}^{(j)}\left(\delta S_w + \delta S_o\right) \\
&\qquad\qquad + e_{gp_{bh}}^{(j)}\delta p_{bh}^{(j)}\Big]\delta(\mathbf{x}-\mathbf{x}^{(j,m)})\Bigg\} + \frac{b_{gp}q_{Gs}^l}{B_g^l}\delta p \\
&\quad + \frac{R_{so}^l}{B_o^l}\Bigg\{q_{Os}^l + \sum_{j=1}^{N_w}\sum_{m=1}^{M_{wj}} WI^{(j,m)}\Big[e_{op}^{(j)}\delta p + e_{oS_w}^{(j)}\delta S_w + e_{oS_o}^{(j)}\delta S_o \\
&\qquad\qquad + e_{op_{bh}}^{(j)}\delta p_{bh}^{(j)}\Big]\delta(\mathbf{x}-\mathbf{x}^{(j,m)})\Bigg\} + \frac{q_{Os}^l}{B_o^l}\left(R_{so}^l b_{op} + r_{sp}\right)\delta p.
\end{aligned} \tag{8.62}$$

At each grid node there are three differential equations (8.60)–(8.62) that must be solved simultaneously in the SS technique. Note that $\delta p_{bh}^{(j)}$ appears in these equations and may be unknown. When the well bottom hole pressure is given at the jth well, $\delta p_{bh}^{(j)} = 0$. When a flow rate is given, $\delta p_{bh}^{(j)}$ is an unknown, and an additional equation is required to supplement (8.60)–(8.62). Thus, in the case of a given flow rate these three equations and the well control equations must be solved simultaneously; see Section 8.2.5 for the well treatment.

Undersaturated state

In the undersaturated state, analogous equations can be obtained for the primary unknowns δp, δS_w, and δp_b. Equation (8.60) for the water component remains the same. For the oil component in the undersaturated state, substituting (8.40), (8.48), (8.52), (8.56), and (8.59) into the second equation of (8.37) produces (cf. Exercise 8.11)

$$\begin{aligned}
&\frac{1}{\Delta t}\left[\left(\frac{\phi S_o}{B_o}\right)^l - \left(\frac{\phi S_o}{B_o}\right)^n + c_{op}\delta p + c_{oS_w}\delta S_w + c_{op_b}\delta p_b\right] \\
&\quad = \nabla\cdot\left(\left(\mathbf{T}_o^l + \mathbf{E}_{op}\delta p + \mathbf{E}_{oS_w}\delta S_w + \mathbf{E}_{op_b}\delta p_b\right)\nabla\Phi_o^l\right) \\
&\qquad + \nabla\cdot\left(\mathbf{T}_o^l\nabla\left(d_{op}\delta p\right)\right) + \nabla\cdot\left(\mathbf{T}_o^l\nabla\left(d_{op_b}\delta p_b\right)\right) \qquad (8.63) \\
&\qquad + \frac{1}{B_o^l}\Bigg\{q_{Os}^l + \sum_{j=1}^{N_w}\sum_{m=1}^{M_{wj}} WI^{(j,m)}\Big[e_{op}^{(j)}\delta p + e_{oS_w}^{(j)}\delta S_w + e_{op_b}^{(j)}\delta p_b \\
&\qquad\qquad + e_{op_{bh}}^{(j)}\delta p_{bh}^{(j)}\Big]\delta(\mathbf{x}-\mathbf{x}^{(j,m)})\Bigg\} + \frac{q_{Os}^l}{B_o^l}\left(b_{op}\delta p + b_{op_b}\delta p_b\right).
\end{aligned}$$

For the gas component in the undersaturated state, substituting (8.42), (8.48), (8.52), (8.56), and (8.59) into the third equation of (8.37) yields (cf. Exercise 8.12)

$$\begin{aligned}
&\frac{1}{\Delta t}\left\{\left(\frac{\phi R_{so} S_o}{B_o}\right)^l - \left(\frac{\phi R_{so} S_o}{B_o}\right)^n + c_{gp}\delta p + c_{gS_w}\delta S_w + c_{gp_b}\delta p_b\right\} \\
&\quad = \nabla\cdot\Bigg[\Bigg(R_{so}^l\left(\mathbf{T}_o^l + \mathbf{E}_{op}\delta p + \mathbf{E}_{oS_w}\delta S_w + \mathbf{E}_{op_b}\delta p_b\right) \\
&\qquad\qquad + \mathbf{T}_o^l\left(r_{sp}\delta p + r_{sp_b}\delta p_b\right)\Bigg)\nabla\Phi_o^l\Bigg] \\
&\quad + \nabla\cdot\left(R_{so}^l\mathbf{T}_o^l\nabla\left(d_{op}\delta p + d_{op_b}\delta p_b\right)\right) \qquad (8.64) \\
&\quad + \frac{R_{so}^l}{B_o^l}\Bigg\{q_{Os}^l + \sum_{j=1}^{N_w}\sum_{m=1}^{M_{wj}} WI^{(j,m)}\Big[e_{op}^{(j)}\delta p + e_{oS_w}^{(j)}\delta S_w \\
&\qquad\qquad + e_{op_b}^{(j)}\delta p_b + e_{op_{bh}}^{(j)}\delta p_{bh}^{(j)}\Big]\delta(\mathbf{x}-\mathbf{x}^{(j,m)})\Bigg\} \\
&\quad + \frac{q_{Os}^l}{B_o^l}\left[\left(R_{so}^l b_{op} + r_{sp}\right)\delta p + \left(R_{so}^l b_{op_b} + r_{sp_b}\right)\delta p_b\right].
\end{aligned}$$

Again, three differential equations (8.60), (8.63), and (8.64) at each grid node, together with the well control equations, must be solved simultaneously.

Termination of the Newton–Raphson iteration

To terminate a Newton–Raphson iteration, some important factors should be considered. First, the iteration number should be smaller than a given maximum number. Second, the iteration values of the unknowns and the right-hand vectors of the linear equation systems (LESs) to be solved are used as a part of the termination condition. The absolute iteration values of the increments of pressure, water saturation, oil saturation (respectively, bubble point pressure), and the bottom hole pressure of wells must be less than their respective allowable maximum limits. Third, from our simulation experience the ratio of the infinite norm of the right-hand side vector of a linear system of equations to the maximum absolute value of the sum of the oil and gas component flow rates of perforated zones of wells must be less than a certain given limit. Mass balance errors are not used as a part of the termination condition of the Newton–Raphson iteration but are monitored during a simulation. *Mass balance* means that the cumulative component mass production equals the initial component mass in place minus the current component mass in place.

Treatment of bubble point problems

It is very important to deal properly with the bubble point problem to control convergence of a Newton–Raphson iteration. The state of a reservoir can change from saturated to undersaturated or vice versa. Determining a proper state during the state transition is the *bubble point problem.* If the bubble point problem can be promptly recognized and reasonable unknowns can be selected for different states of a reservoir, convergence of the Newton–Raphson iteration can be better monitored and sped up.

To handle the bubble point problem properly, we must figure out the *trigger* that causes the transition of states of a reservoir using the state machine (Booch et al., 1998) shown in Figure 8.1. A location in the reservoir can stay in either the saturated state or the undersaturated state. Furthermore, from the lth iteration to the $(l+1)$th iteration in a Newton–Raphson iteration at the $(n+1)$th time step, the location can stay in the same state or transfer to another state. The constraint conditions and triggers are different in different states. In the undersaturated state, the constraint conditions are

$$\begin{aligned} S_w^{n+1,l} + S_o^{n+1,l} &= 1, \\ p^{n+1,l} &> p_b^{n+1,l}. \end{aligned} \tag{8.65}$$

On the other hand, in the saturated state, the constraint conditions are

$$\begin{aligned} S_w^{n+1,l} + S_o^{n+1,l} + S_g^{n+1,l} &= 1, \\ p^{n+1,l} &= p_b^{n+1,l}. \end{aligned} \tag{8.66}$$

The trigger that causes the transition from the undersaturated state to the saturated state is

$$p_b^{n+1,l} + \delta p_b > p^{n+1,l+1}, \tag{8.67}$$

and the trigger that causes the transition from the saturated state to the undersaturated state is

$$S_g^{n+1,l+1} < 0. \tag{8.68}$$

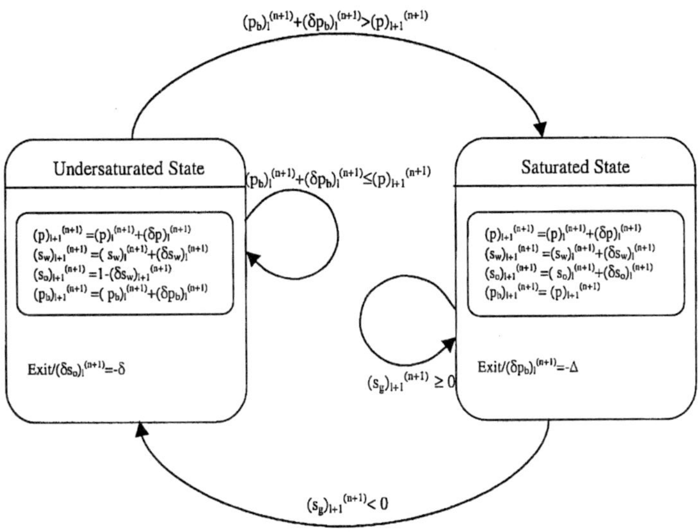

Figure 8.1. *A state machine.*

To deal with the bubble point problem properly, we must check the triggers to determine whether a location in a reservoir stays in the old state or transfers to a new state. Then we let the unknowns satisfy the constraint conditions of the corresponding state. When the reservoir pressure at a location in a reservoir drops below the bubble point pressure, then $(p_b)^{n+1,l} + \delta p_b > p^{n+1,l+1}$, the dissolved gas comes out from the oil phase, and the oil saturation decreases. It triggers the state to transfer from the undersaturated state to the saturated state at this location. In order to enter the new state, δS_o is set with a small negative value so that the gas saturation is greater than zero and the dissolved gas is released. When the reservoir at this location is in the saturated state, the unknowns corresponding to the grid point of the location are updated to satisfy the constraint conditions (8.66). Similarly, if the reservoir pressure at a location increases to the point that all the gas dissolves into the oil phase, then the state changes from the saturated state to the undersaturated state at this location and $S_g^{n+1,l+1} < 0$, which triggers the state to transfer from the saturated state to the undersaturated state. In order to guarantee that the oil phase pressure will be greater than the bubble point pressure in the new state, δp_b is set with a small negative value. After the reservoir at the location enters this new state, the unknowns are updated to meet the constraint conditions (8.65) in the undersaturated state.

8.2.3 The sequential technique

The *sequential solution technique* (MacDonald and Coats, 1970) is similar to the SS technique discussed in the previous subsection. The difference is that the three equations in system (8.37) are now solved separately and *sequentially*.

In the sequential technique, all the saturation functions $k_{rw}, k_{ro}, k_{rg}, p_{cw}$, and p_{cg} use the previous Newton–Raphson iteration values of saturations; i.e., the phase potentials and transmissibilities are

$$\begin{aligned} \Phi_\alpha^{l+1} &= p^{l+1} + p_{c\alpha}^l - \rho_\alpha^{l+1} \wp z, \\ \mathbf{T}_\alpha^{l+1} &= \frac{k_{r\alpha}^l}{\mu_\alpha^{l+1} B_\alpha^{l+1}} \mathbf{k}, \qquad \alpha = w, o, g. \end{aligned} \tag{8.69}$$

The flow rates at wells are given by

$$\begin{aligned} q_{Ws}^{l+1} &= \sum_{j=1}^{N_w} \sum_{m=1}^{M_{wj}} WI^{(j,m)} \frac{k_{rw}^l}{\mu_w} \Big[\left(p_{bh}^{(j)}\right)^{l+1} - p^{l+1} - p_{cw}^l \\ &\qquad - \rho_w^{l+1} \wp \left(z_{bh}^{(j)} - z\right) \Big] \delta(\mathbf{x} - \mathbf{x}^{(j,m)}), \\ q_{Os}^{l+1} &= \sum_{j=1}^{N_w} \sum_{m=1}^{M_{wj}} WI^{(j,m)} \frac{k_{ro}^l}{\mu_o^{l+1}} \Big[\left(p_{bh}^{(j)}\right)^{l+1} - p^{l+1} \\ &\qquad - \rho_o^{l+1} \wp \left(z_{bh}^{(j)} - z\right) \Big] \delta(\mathbf{x} - \mathbf{x}^{(j,m)}), \\ q_{Gs}^{l+1} &= \sum_{j=1}^{N_w} \sum_{m=1}^{M_{wj}} WI^{(j,m)} \frac{k_{rg}^l}{\mu_g^{l+1}} \Big[\left(p_{bh}^{(j)}\right)^{l+1} - p^{l+1} - p_{cg}^l \\ &\qquad - \rho_g^{l+1} \wp \left(z_{bh}^{(j)} - z\right) \Big] \delta(\mathbf{x} - \mathbf{x}^{(j,m)}). \end{aligned} \tag{8.70}$$

Hence the potentials for all three components are expanded:

$$\Phi_\alpha^{l+1} = \Phi_\alpha^l + d_{\alpha p} \delta p, \quad d_{\alpha p} = 1 - \left(\frac{d\rho_\alpha}{dp}\right)^l \wp z, \quad \alpha = w, o, g, \tag{8.71}$$

and the transmissibilities are expanded analogously:

$$\mathbf{T}_\alpha^{l+1} = \mathbf{T}_\alpha^l + \mathbf{E}_{\alpha p} \delta p, \quad \mathbf{E}_{\alpha p} = \left(k_{r\alpha} \frac{d}{dp} \left(\frac{1}{\mu_\alpha B_\alpha} \right) \right)^l \mathbf{k} \tag{8.72}$$

for $\alpha = w, o, g$. The flow rates at wells are expanded. For the water component,

$$q_{Ws}^{l+1} = q_{Ws}^l + \sum_{j=1}^{N_w} \sum_{m=1}^{M_{wj}} WI^{(j,m)} \left[e_{wp}^{(j)} \delta p + e_{wp_{bh}} \delta p_{bh}^{(j)} \right] \delta(\mathbf{x} - \mathbf{x}^{(j,m)}), \tag{8.73}$$

where

$$e_{wp}^{(j)} = -\frac{1}{\mu_w} \left[k_{rw} \left(1 + \frac{d\rho_w}{dp} \wp \left(z_{bh}^{(j)} - z\right) \right) \right]^l, \quad e_{wp_{bh}} = \frac{k_{rw}^l}{\mu_w}.$$

For the oil component,

$$q_{Os}^{l+1} = q_{Os}^{l} + \sum_{j=1}^{N_w} \sum_{m=1}^{M_{wj}} WI^{(j,m)} \left[e_{op}^{(j)} \delta p + e_{op_{bh}} \delta p_{bh}^{(j)} \right] \delta(\mathbf{x} - \mathbf{x}^{(j,m)}), \tag{8.74}$$

where

$$e_{op}^{(j)} = \left\{ k_{ro} \left[\frac{d\mu_o^{-1}}{dp} \left(p_{bh}^{(j)} - p - \rho_o \wp (z_{bh}^{(j)} - z) \right) \right. \right.$$
$$\left. \left. - \frac{1}{\mu_o} \left(1 + \frac{d\rho_o}{dp} \wp (z_{bh}^{(j)} - z) \right) \right] \right\}^l, \quad e_{op_{bh}} = \left(\frac{k_{ro}}{\mu_o} \right)^l.$$

For the gas component,

$$q_{Gs}^{l+1} = q_{Gs}^{l} + \sum_{j=1}^{N_w} \sum_{m=1}^{M_{wj}} WI^{(j,m)} \left[e_{gp}^{(j)} \delta p + e_{gp_{bh}} \delta p_{bh}^{(j)} \right] \delta(\mathbf{x} - \mathbf{x}^{(j,m)}), \tag{8.75}$$

where

$$e_{gp}^{(j)} = \left\{ k_{rg} \left[\frac{d\mu_g^{-1}}{dp} \left(p_{bh}^{(j)} - p - p_{cg} - \rho_g \wp (z_{bh}^{(j)} - z) \right) \right. \right.$$
$$\left. \left. - \frac{1}{\mu_g} \left(1 + \frac{d\rho_g}{dp} \wp (z_{bh}^{(j)} - z) \right) \right] \right\}^l, \quad e_{gp_{bh}} = \left(\frac{k_{rg}}{\mu_g} \right)^l.$$

Equation (8.58) still holds for the sequential technique.

Saturated state

Substituting (8.38)–(8.42) and (8.71)–(8.75) into (8.37) leads to a linear system in terms of the primary unknowns in the sequential technique. For the water component, substituting (8.38), (8.71)–(8.73), and (8.58) into the first equation of (8.37) and ignoring the higher order terms in δp gives (cf. Exercise 8.13)

$$\begin{aligned} &\frac{1}{\Delta t} \left[\left(\frac{\phi S_w}{B_w} \right)^l - \left(\frac{\phi S_w}{B_w} \right)^n + c_{wp} \delta p + c_{wS_w} \delta S_w \right] \\ &= \nabla \cdot \left(\left(\mathbf{T}_w^l + \mathbf{E}_{wp} \delta p \right) \nabla \Phi_w^l \right) + \nabla \cdot \left(\mathbf{T}_w^l \nabla \left(d_{wp} \delta p \right) \right) \\ &+ \frac{1}{B_w^l} \left\{ q_{Ws}^l + \sum_{j=1}^{N_w} \sum_{m=1}^{M_{wj}} WI^{(j,m)} \left[e_{wp}^{(j)} \delta p + e_{wp_{bh}}^{(j)} \delta p_{bh}^{(j)} \right] \right. \\ &\left. \cdot \, \delta(\mathbf{x} - \mathbf{x}^{(j,m)}) \right\} + \frac{b_{wp} q_{Ws}^l}{B_w^l} \delta p. \end{aligned} \tag{8.76}$$

For the oil component, substituting (8.39), (8.71), (8.72), (8.74), and (8.58) into the second equation of (8.37) gives (cf. Exercise 8.14)

$$
\begin{aligned}
&\frac{1}{\Delta t}\left[\left(\frac{\phi S_o}{B_o}\right)^l-\left(\frac{\phi S_o}{B_o}\right)^n+c_{op}\delta p+c_{oS_o}\delta S_o\right]\\
&\quad=\nabla\cdot\left(\left(\mathbf{T}_o^l+\mathbf{E}_{op}\delta p\right)\nabla\Phi_o^l\right)+\nabla\cdot\left(\mathbf{T}_o^l\nabla\left(d_{op}\delta p\right)\right)\\
&\quad+\frac{1}{B_o^l}\left\{q_{Os}^l+\sum_{j=1}^{N_w}\sum_{m=1}^{M_{wj}}WI^{(j,m)}\left[e_{op}^{(j)}\delta p\right.\right.\\
&\qquad\left.\left.+e_{op_{bh}}^{(j)}\delta p_{bh}^{(j)}\right]\delta(\mathbf{x}-\mathbf{x}^{(j,m)})\right\}+\frac{b_{op}q_{Os}^l}{B_o^l}\delta p.
\end{aligned}
\tag{8.77}
$$

For the gas component, substituting (8.41), (8.71), (8.72), (8.75), and (8.58) into the third equation of (8.37) yields (cf. Exercise 8.15)

$$
\begin{aligned}
&\frac{1}{\Delta t}\left\{\left[\phi\left(\frac{S_g}{B_g}+\frac{R_{so}S_o}{B_o}\right)\right]^l-\left[\phi\left(\frac{S_g}{B_g}+\frac{R_{so}S_o}{B_o}\right)\right]^n\right.\\
&\qquad\left.+c_{gp}\delta p+c_{gS_w}\delta S_w+c_{gS_o}\delta S_o\right\}\\
&=\nabla\cdot\left(\left(\mathbf{T}_g^l+\mathbf{E}_{gp}\delta p\right)\nabla\Phi_g^l\right)+\nabla\cdot\left(\mathbf{T}_g^l\nabla\left(d_{gp}\delta p\right)\right)\\
&+\nabla\cdot\left[\left(R_{so}^l\left(\mathbf{T}_o^l+\mathbf{E}_{op}\delta p\right)+r_{sp}\mathbf{T}_o^l\delta p\right)\nabla\Phi_o^l\right]\\
&+\nabla\cdot\left(R_{so}^l\mathbf{T}_o^l\nabla\left(d_{op}\delta p\right)\right)\\
&+\frac{1}{B_g^l}\left\{q_{Gs}^l+\sum_{j=1}^{N_w}\sum_{m=1}^{M_{wj}}WI^{(j,m)}\left[e_{gp}^{(j)}\delta p\right.\right.\\
&\qquad\left.\left.+e_{gp_{bh}}^{(j)}\delta p_{bh}^{(j)}\right]\delta(\mathbf{x}-\mathbf{x}^{(j,m)})\right\}+\frac{b_{gp}q_{Gs}^l}{B_g^l}\delta p\\
&+\frac{R_{so}^l}{B_o^l}\left\{q_{Os}^l+\sum_{j=1}^{N_w}\sum_{m=1}^{M_{wj}}WI^{(j,m)}\left[e_{op}^{(j)}\delta p\right.\right.\\
&\left.\left.+e_{op_{bh}}^{(j)}\delta p_{bh}^{(j)}\right]\delta(\mathbf{x}-\mathbf{x}^{(j,m)})\right\}+\frac{q_{Os}^l}{B_o^l}\left(R_{so}^lb_{op}+r_{sp}\right)\delta p.
\end{aligned}
\tag{8.78}
$$

Equations (8.76)–(8.78) can be also obtained from (8.60)–(8.62) by setting $\delta S_w=0$ and $\delta S_o=0$ in the right-hand sides (cf. Exercise 8.16).

Multiply (8.76)–(8.78) by Δt and write the resulting respective equations as

$$
\begin{aligned}
&c_{wp}\delta p+c_{wS_w}\delta S_w=F_w(\delta p,\delta p_{bh}),\\
&c_{op}\delta p+c_{oS_o}\delta S_o=F_o(\delta p,\delta p_{bh}),\\
&c_{gp}\delta p+c_{gS_w}\delta S_w+c_{gS_o}\delta S_o=F_g(\delta p,\delta p_{bh}).
\end{aligned}
\tag{8.79}
$$

From the first and second equations of (8.79), we see that

$$\begin{aligned} \delta S_w &= \frac{1}{c_{wS_w}} \left(F_w(\delta p, \delta p_{bh}) - c_{wp} \delta p \right), \\ \delta S_o &= \frac{1}{c_{oS_o}} \left(F_o(\delta p, \delta p_{bh}) - c_{op} \delta p \right). \end{aligned} \tag{8.80}$$

Substituting these two equations into the third equation of (8.79) yields

$$\begin{aligned} &\left(c_{gp} - \frac{c_{gS_w} c_{wp}}{c_{wS_w}} - \frac{c_{gS_o} c_{op}}{c_{oS_o}} \right) \delta p \\ &= F_g(\delta p, \delta p_{bh}) - \frac{c_{gS_w}}{c_{wS_w}} F_w(\delta p, \delta p_{bh}) - \frac{c_{gS_o}}{c_{oS_o}} F_o(\delta p, \delta p_{bh}), \end{aligned} \tag{8.81}$$

which is the pressure equation and is solved implicitly in the sequential technique. In the case of a given flow rate at a well, this equation and the well control equations must be solved simultaneously for δp and δp_{bh}.

To compute δS_w and δS_o, we use the same equations as in the SS (cf. (8.60)):

$$\begin{aligned} &\frac{1}{\Delta t} \left[\left(\frac{\phi S_w}{B_w} \right)^l - \left(\frac{\phi S_w}{B_w} \right)^n + c_{wp} \delta p + c_{wS_w} \delta S_w \right] \\ &\quad = \nabla \cdot \left(\left(\mathbf{T}_w^l + \mathbf{E}_{wp} \delta p + \mathbf{E}_{wS_w} \delta S_w \right) \nabla \Phi_w^l \right) \\ &\quad + \nabla \cdot \left(\mathbf{T}_w^l \nabla \left(d_{wp} \delta p \right) \right) + \nabla \cdot \left(\mathbf{T}_w^l \nabla \left(d_{wS_w} \delta S_w \right) \right) \\ &\quad + \frac{1}{B_w^l} \Bigg\{ q_{Ws}^l + \sum_{j=1}^{N_w} \sum_{m=1}^{M_{wj}} WI^{(j,m)} \Big[e_{wp}^{(j)} \delta p + e_{wS_w}^{(j)} \delta S_w \\ &\qquad + e_{wp_{bh}}^{(j)} \delta p_{bh}^{(j)} \Big] \delta(\mathbf{x} - \mathbf{x}^{(j,m)}) \Bigg\} + \frac{b_{wp} q_{Ws}^l}{B_w^l} \delta p \end{aligned} \tag{8.82}$$

and (cf. (8.61))

$$\begin{aligned} &\frac{1}{\Delta t} \left[\left(\frac{\phi S_o}{B_o} \right)^l - \left(\frac{\phi S_o}{B_o} \right)^n + c_{op} \delta p + c_{oS_o} \delta S_o \right] \\ &\quad = \nabla \cdot \left(\left(\mathbf{T}_o^l + \mathbf{E}_{op} \delta p + \mathbf{E}_{oS_w} \delta S_w + \mathbf{E}_{oS_o} \delta S_o \right) \nabla \Phi_o^l \right) \\ &\quad + \nabla \cdot \left(\mathbf{T}_o^l \nabla \left(d_{op} \delta p \right) \right) \\ &\quad + \frac{1}{B_o^l} \Bigg\{ q_{Os}^l + \sum_{j=1}^{N_w} \sum_{m=1}^{M_{wj}} WI^{(j,m)} \Big[e_{op}^{(j)} \delta p + e_{oS_w}^{(j)} \delta S_w \\ &\quad + e_{oS_o}^{(j)} \delta S_o + e_{op_{bh}}^{(j)} \delta p_{bh}^{(j)} \Big] \delta(\mathbf{x} - \mathbf{x}^{(j,m)}) \Bigg\} + \frac{b_{op} q_{Os}^l}{B_o^l} \delta p. \end{aligned} \tag{8.83}$$

Now, equations (8.81)–(8.83) at each grid node are solved sequentially; each equation is solved implicitly.

Undersaturated state

Setting $\delta S_w = 0$ and $\delta p_b = 0$ in the right-hand sides of (8.63) and (8.64), we obtain the pressure and saturation equations for the sequential technique:

$$
\begin{aligned}
&\frac{1}{\Delta t}\left[\left(\frac{\phi S_o}{B_o}\right)^l - \left(\frac{\phi S_o}{B_o}\right)^n + c_{op}\delta p + c_{oS_w}\delta S_w + c_{op_b}\delta p_b\right] \\
&\quad = \nabla\cdot\left(\left(\mathbf{T}_o^l + \mathbf{E}_{op}\delta p\right)\nabla\Phi_o^l\right) + \nabla\cdot\left(\mathbf{T}_o^l\nabla\left(d_{op}\delta p\right)\right) \\
&\quad\quad + \frac{1}{B_o^l}\left\{q_{Os}^l + \sum_{j=1}^{N_w}\sum_{m=1}^{M_{wj}} WI^{(j,m)}\left[e_{op}^{(j)}\delta p\right.\right. \\
&\quad\quad\quad \left.\left. + e_{op_{bh}}^{(j)}\delta p_{bh}^{(j)}\right]\delta(\mathbf{x}-\mathbf{x}^{(j,m)})\right\} + \frac{b_{op}q_{Os}^l}{B_o^l}\delta p
\end{aligned}
\tag{8.84}
$$

and

$$
\begin{aligned}
&\frac{1}{\Delta t}\left\{\left(\frac{\phi R_{so}S_o}{B_o}\right)^l - \left(\frac{\phi R_{so}S_o}{B_o}\right)^n\right. \\
&\quad\quad \left. + c_{gp}\delta p + c_{gS_w}\delta S_w + c_{gp_b}\delta p_b\right\} \\
&= \nabla\cdot\left[\left(R_{so}^l\left(\mathbf{T}_o^l + \mathbf{E}_{op}\delta p\right) + r_{sp}\mathbf{T}_o^l\delta p\right)\nabla\Phi_o^l\right] \\
&\quad + \nabla\cdot\left(R_{so}^l\mathbf{T}_o^l\nabla\left(d_{op}\delta p\right)\right) \\
&\quad + \frac{R_{so}^l}{B_o^l}\left\{q_{Os}^l + \sum_{j=1}^{N_w}\sum_{m=1}^{M_{wj}} WI^{(j,m)}\left[e_{op}^{(j)}\delta p\right.\right. \\
&\quad \left.\left. + e_{op_{bh}}^{(j)}\delta p_{bh}^{(j)}\right]\delta(\mathbf{x}-\mathbf{x}^{(j,m)})\right\} + \frac{q_{Os}^l}{B_o^l}\left(R_{so}^l b_{op} + r_{sp}\right)\delta p.
\end{aligned}
\tag{8.85}
$$

Equation (8.76) for the water component remains the same.

Multiply (8.76), (8.84), and (8.85) by Δt and write the resulting respective equations as

$$
\begin{aligned}
c_{wp}\delta p + c_{wS_w}\delta S_w &= F_w(\delta p, \delta p_{bh}), \\
c_{op}\delta p + c_{oS_w}\delta S_w + c_{op_b}\delta p_b &= F_o(\delta p, \delta p_{bh}), \\
c_{gp}\delta p + c_{gS_w}\delta S_w + c_{gp_b}\delta p_b &= F_g(\delta p, \delta p_{bh}).
\end{aligned}
\tag{8.86}
$$

From the last two equations of (8.86) it follows that

$$
\begin{aligned}
c_{oS_w}\delta S_w + c_{op_b}\delta p_b &= F_o(\delta p, \delta p_{bh}) - c_{op}\delta p, \\
c_{gS_w}\delta S_w + c_{gp_b}\delta p_b &= F_g(\delta p, \delta p_{bh}) - c_{gp}\delta p.
\end{aligned}
\tag{8.87}
$$

Set

$$D = \begin{vmatrix} c_{oS_w} & c_{op_b} \\ c_{gS_w} & c_{gp_b} \end{vmatrix} = c_{oS_w} c_{gp_b} - c_{gS_w} c_{op_b},$$

$$D_S = \begin{vmatrix} F_o(\delta p, \delta p_{bh}) - c_{op}\delta p & c_{op_b} \\ F_g(\delta p, \delta p_{bh}) - c_{gp}\delta p & c_{gp_b} \end{vmatrix}$$

$$= (F_o(\delta p, \delta p_{bh}) - c_{op}\delta p)c_{gp_b} - (F_g(\delta p, \delta p_{bh}) - c_{gp}\delta p)c_{op_b},$$

$$D_p = \begin{vmatrix} c_{oS_w} & F_o(\delta p, \delta p_{bh}) - c_{op}\delta p \\ c_{gS_w} & F_g(\delta p, \delta p_{bh}) - c_{gp}\delta p \end{vmatrix}$$

$$= c_{oS_w}(F_g(\delta p, \delta p_{bh}) - c_{gp}\delta p) - c_{gS_w}(F_o(\delta p, \delta p_{bh}) - c_{op}\delta p).$$

From (8.87) it follows that

$$\delta S_w = \frac{D_S}{D}, \quad \delta p_b = \frac{D_p}{D},$$

which we substitute into the first equation of (8.86) to obtain the pressure equation in the undersaturated state:

$$c_{wp}\delta p + c_{wS_w}\frac{D_S}{D}(\delta p, \delta p_{bh}) = F_w(\delta p, \delta p_{bh}). \tag{8.88}$$

Equation (8.88) is solved implicitly for δp. Equation (8.82) is used to obtain δS_w, and equation (8.64) for the gas component is employed to compute δp_b:

$$\begin{aligned}
&\frac{1}{\Delta t}\left\{\left(\frac{\phi R_{so} S_o}{B_o}\right)^l - \left(\frac{\phi R_{so} S_o}{B_o}\right)^n \right.\\
&\qquad\qquad \left. + c_{gp}\delta p + c_{gS_w}\delta S_w + c_{gp_b}\delta p_b\right\}\\
&= \nabla\cdot\left[\left(R^l_{so}\left(\mathbf{T}^l_o + \mathbf{E}_{op}\delta p + \mathbf{E}_{oS_w}\delta S_w + \mathbf{E}_{op_b}\delta p_b\right)\right.\right.\\
&\qquad\qquad \left.\left. + \mathbf{T}^l_o\left(r_{sp}\delta p + r_{sp_b}\delta p_b\right)\right)\nabla\Phi^l_o\right]\\
&+ \nabla\cdot\left(R^l_{so}\mathbf{T}^l_o\nabla\left(d_{op}\delta p + d_{op_b}\delta p_b\right)\right)\\
&+ \frac{R^l_{so}}{B^l_o}\left\{q^l_{Os} + \sum_{j=1}^{N_w}\sum_{m=1}^{M_{wj}} WI^{(j,m)}\left[e^{(j)}_{op}\delta p + e^{(j)}_{oS_w}\delta S_w\right.\right.\\
&\qquad\qquad \left.\left. + e^{(j)}_{op_b}\delta p_b + e^{(j)}_{op_{bh}}\delta p^{(j)}_{bh}\right]\delta(\mathbf{x} - \mathbf{x}^{(j,m)})\right\}\\
&+ \frac{q^l_{Os}}{B^l_o}\left[\left(R^l_{so}b_{op} + r_{sp}\right)\delta p + \left(R^l_{so}b_{op_b} + r_{sp_b}\right)\delta p_b\right].
\end{aligned} \tag{8.89}$$

Again, there are three equations (8.88), (8.82), and (8.89) at each grid node that are solved implicitly and sequentially.

In summary, the sequential technique has the following features:

- The difference between the SS and sequential techniques is that the three differential equations are solved simultaneously in the SS rather than sequentially at each grid node.

- All the saturation functions k_{rw}, k_{ro}, k_{rg}, p_{cw}, and p_{cg} use the previous Newton–Raphson iteration values of saturations.

- The left-hand sides of the water, oil, and gas component equations are treated in the same fashion as in the SS.

- The equations used to solve for the second and third unknowns are the same for both the SS and sequential techniques.

Selection of time steps

The bubble point problem in the sequential technique can be treated in the same way as in Section 8.2.2 for the SS method. Compared with the SS, the implicitness of the sequential technique is lower. Selecting reasonable time steps is key to controlling convergence of a Newton–Raphson iteration and speeding up a simulation procedure. If the time steps are too small, too much computational time will be consumed; if they are too large, a Newton–Raphson iteration may diverge.

To select suitable time steps, from our experimental experience we have adopted the following empirical rules:

- With a given maximum time step Δt_{max}, the time step Δt should satisfy that $0 < \Delta t \leq \Delta t_{max}$.

- In the saturated state, Δt is bounded by

$$\Delta t \leq \Delta t^n \min \left\{ 3, \frac{(dp)_{max}}{(\delta p)^n_{max}}, \frac{(dS_w)_{max}}{(\delta S_w)^n_{max}}, \frac{(dS_o)_{max}}{(\delta S_o)^n_{max}} \right\}, \tag{8.90}$$

  where Δt^n is the previous time step size; $(dp)_{max}$, $(dS_w)_{max}$, and $(dS_o)_{max}$ are the allowable maximum values of the pressure, water saturation, and oil saturation increments, respectively; and $(\delta p)^n_{max}$, $(\delta S_w)^n_{max}$, $(\delta S_o)^n_{max}$ are the maximum values of these increments at the nth time step. In the undersaturated state, (8.90) becomes

$$\Delta t \leq \Delta t^n \min \left\{ 3, \frac{(dp)_{max}}{(\delta p)^n_{max}}, \frac{(dS_w)_{max}}{(\delta S_w)^n_{max}}, \frac{(dp_b)_{max}}{(\delta p_b)^n_{max}} \right\}, \tag{8.91}$$

  where $(dp_b)_{max}$ is the allowable maximum value of the bubble point pressure increment.

- For a given time period, Δt should guarantee that the simulation time reaches the period time.

With these rules, a time step Δt can be automatically selected. Its choice must also take into account the convergence of a Newton–Raphson iteration. If the number of iterations is larger than a given maximum number when Δt is selected according to these rules, then the selected time step may be too large and must be reduced. First, we reduce Δt by $\Delta t/3$ because of the occurrence of 3 in (8.90) and (8.91). Then the oil phase and bubble point pressures and water and oil saturations at the nth time step are taken as the first iteration values of the Newton–Raphson iteration at the $(n+1)$th time step.

8.2.4 Iterative IMPES

The IMPES algorithm was discussed in the preceding chapter for two-phase flow and is a very useful technique for flow of this type. Particularly, the improved IMPES introduced in Section 7.3.3 is very powerful for solving two-phase flow. We now discuss IMPES for the solution of the black oil model. When IMPES is used within a Newton–Raphson iteration, it is called *iterative IMPES*. In iterative IMPES, only the pressure equation is computed implicitly, and the other two (saturation and bubble point pressure) equations are evaluated explicitly.

In iterative IMPES, all the saturation functions k_{rw}, k_{ro}, k_{rg}, p_{cw}, and p_{cg} are evaluated at the saturation values of the previous time step in a Newton–Raphson iteration, and the fluid formation volume factors and viscosities in the transmissibilities, phase potentials, and well terms are computed using the previous Newton–Raphson iteration values. Thus the phase potentials are

$$\Phi_\alpha^{l+1} = p^{l+1} + p_{c\alpha}^n - \rho_\alpha^l \wp z, \qquad \alpha = w, o, g, \tag{8.92}$$

and the transmissibilities are

$$\mathbf{T}_\alpha^{l+1} = \frac{k_{r\alpha}^n}{\mu_\alpha^l B_\alpha^l}\mathbf{k}, \qquad \alpha = w, o, g. \tag{8.93}$$

Furthermore, the flow rates at the wells are

$$\begin{aligned} q_{Ws}^{l+1} = \sum_{j-1}^{N_w}\sum_{m-1}^{M_{wj}} WI^{(j,m)}\frac{k_{rw}^n}{\mu_w}\Bigg[\left(p_{bh}^{(j)}\right)^{l+1} - p^{l+1} - p_{cw}^n \\ - \rho_w^l \wp\left(z_{bh}^{(j)} - z\right)\Bigg]\delta(\mathbf{x}-\mathbf{x}^{(j,m)}), \\ q_{Os}^{l+1} = \sum_{j=1}^{N_w}\sum_{m=1}^{M_{wj}} WI^{(j,m)}\frac{k_{ro}^n}{\mu_o^l}\Bigg[\left(p_{bh}^{(j)}\right)^{l+1} - p^{l+1} \\ - \rho_o^l \wp\left(z_{bh}^{(j)} - z\right)\Bigg]\delta(\mathbf{x}-\mathbf{x}^{(j,m)}), \\ q_{Gs}^{l+1} = \sum_{j=1}^{N_w}\sum_{m=1}^{M_{wj}} WI^{(j,m)}\frac{k_{rg}^n}{\mu_g^l}\Bigg[\left(p_{bh}^{(j)}\right)^{l+1} - p^{l+1} - p_{cg}^n \\ - \rho_g^l \wp\left(z_{bh}^{(j)} - z\right)\Bigg]\delta(\mathbf{x}-\mathbf{x}^{(j,m)}). \end{aligned} \tag{8.94}$$

Therefore, the potentials for all three components can be expanded,

$$\Phi_\alpha^{l+1} = \Phi_\alpha^l + \delta p, \qquad \alpha = w, o, g, \tag{8.95}$$

and the flow rates at wells are expanded:

$$\begin{aligned}
q_{Ws}^{l+1} &= q_{Ws}^l + \sum_{j=1}^{N_w}\sum_{m=1}^{M_{wj}} WI^{(j,m)} \frac{k_{rw}^n}{\mu_w}\left[\delta p_{bh}^{(j)} - \delta p\right]\delta(\mathbf{x}-\mathbf{x}^{(j,m)}),\\
q_{Os}^{l+1} &= q_{Os}^l + \sum_{j=1}^{N_w}\sum_{m=1}^{M_{wj}} WI^{(j,m)} \frac{k_{ro}^n}{\mu_o^l}\left[\delta p_{bh}^{(j)} - \delta p\right]\delta(\mathbf{x}-\mathbf{x}^{(j,m)}),\\
q_{Gs}^{l+1} &= q_{Gs}^l + \sum_{j=1}^{N_w}\sum_{m=1}^{M_{wj}} WI^{(j,m)} \frac{k_{rg}^n}{\mu_g^l}\left[\delta p_{bh}^{(j)} - \delta p\right]\delta(\mathbf{x}-\mathbf{x}^{(j,m)}).
\end{aligned} \tag{8.96}$$

Saturated state

Substituting (8.38)–(8.42), (8.95), and (8.96) into (8.37) leads to a linear system in terms of the primary unknowns in iterative IMPES. For the water component, substituting (8.38), (8.95), and (8.96) into the first equation of (8.37) and ignoring the higher-order terms in δp gives (cf. Exercise 8.17)

$$\begin{aligned}
&\frac{1}{\Delta t}\left[\left(\frac{\phi S_w}{B_w}\right)^l - \left(\frac{\phi S_w}{B_w}\right)^n + c_{wp}\delta p + c_{wS_w}\delta S_w\right]\\
&= \nabla\cdot\left(\mathbf{T}_w^l \nabla \Phi_w^l\right) + \nabla\cdot\left(\mathbf{T}_w^l \nabla(\delta p)\right)\\
&+ \frac{1}{B_w^l}\left\{q_{Ws}^l + \sum_{j=1}^{N_w}\sum_{m=1}^{M_{wj}} WI^{(j,m)} \frac{k_{rw}^n}{\mu_w}\left[\delta p_{bh}^{(j)} - \delta p\right]\delta(\mathbf{x}-\mathbf{x}^{(j,m)})\right\}.
\end{aligned} \tag{8.97}$$

For the oil component in the saturated state, substituting (8.39), (8.95), and (8.96) into the second equation of (8.37) gives (cf. Exercise 8.18)

$$\begin{aligned}
&\frac{1}{\Delta t}\left[\left(\frac{\phi S_o}{B_o}\right)^l - \left(\frac{\phi S_o}{B_o}\right)^n + c_{op}\delta p + c_{oS_o}\delta S_o\right]\\
&= \nabla\cdot\left(\mathbf{T}_o^l \nabla \Phi_o^l\right) + \nabla\cdot\left(\mathbf{T}_o^l \nabla(\delta p)\right)\\
&+ \frac{1}{B_o^l}\left\{q_{Os}^l + \sum_{j=1}^{N_w}\sum_{m=1}^{M_{wj}} WI^{(j,m)} \frac{k_{ro}^n}{\mu_o^l}\left[\delta p_{bh}^{(j)} - \delta p\right]\delta(\mathbf{x}-\mathbf{x}^{(j,m)})\right\}.
\end{aligned} \tag{8.98}$$

For the gas component in the saturated state, substituting (8.41), (8.95), and (8.96) into the third equation of (8.37) yields (cf. Exercise 8.19)

$$
\begin{aligned}
&\frac{1}{\Delta t}\left\{\left[\phi\left(\frac{S_g}{B_g}+\frac{R_{so}S_o}{B_o}\right)\right]^l-\left[\phi\left(\frac{S_g}{B_g}+\frac{R_{so}S_o}{B_o}\right)\right]^n\right.\\
&\qquad\left.+c_{gp}\delta p+c_{gS_w}\delta S_w+c_{gS_o}\delta S_o\right\}\\
&=\nabla\cdot\left(\mathbf{T}_g^l\nabla\Phi_g^l\right)+\nabla\cdot\left(\mathbf{T}_g^l\nabla(\delta p)\right)\\
&\quad+\nabla\cdot\left(R_{so}^l\mathbf{T}_o^l\nabla\Phi_o^l\right)+\nabla\cdot\left(R_{so}^l\mathbf{T}_o^l\nabla(\delta p)\right)\\
&\quad+\frac{1}{B_g^l}\left\{q_{Gs}^l+\sum_{j=1}^{N_w}\sum_{m=1}^{M_{wj}}WI^{(j,m)}\frac{k_{rg}^n}{\mu_g^l}\left[\delta p_{bh}^{(j)}-\delta p\right]\delta(\mathbf{x}-\mathbf{x}^{(j,m)})\right\}\\
&\quad+\frac{R_{so}^l}{B_o^l}\left\{q_{Os}^l+\sum_{j=1}^{N_w}\sum_{m=1}^{M_{wj}}WI^{(j,m)}\frac{k_{ro}^n}{\mu_o^l}\left[\delta p_{bh}^{(j)}-\delta p\right]\delta(\mathbf{x}-\mathbf{x}^{(j,m)})\right\}.
\end{aligned}
\tag{8.99}
$$

Multiply (8.97)–(8.99) by Δt and write the resulting respective equations as

$$
\begin{aligned}
&c_{wp}\delta p+c_{wS_w}\delta S_w=F_w(\delta p,\delta p_{bh}),\\
&c_{op}\delta p+c_{oS_o}\delta S_o=F_o(\delta p,\delta p_{bh}),\\
&c_{gp}\delta p+c_{gS_w}\delta S_w+c_{gS_o}\delta S_o=F_g(\delta p,\delta p_{bh}).
\end{aligned}
\tag{8.100}
$$

From the first and second equations of (8.100), we see that

$$
\begin{aligned}
\delta S_w&=\frac{1}{c_{wS_w}}\left(F_w(\delta p,\delta p_{bh})-c_{wp}\delta p\right),\\
\delta S_o&=\frac{1}{c_{oS_o}}\left(F_o(\delta p,\delta p_{bh})-c_{op}\delta p\right).
\end{aligned}
\tag{8.101}
$$

Substituting these equations into the third equation of (8.100) yields

$$
\begin{aligned}
&\left(c_{gp}-\frac{c_{gS_w}c_{wp}}{c_{wS_w}}-\frac{c_{gS_o}c_{op}}{c_{oS_o}}\right)\delta p\\
&=F_g(\delta p,\delta p_{bh})-\frac{c_{gS_w}}{c_{wS_w}}F_w(\delta p,\delta p_{bh})-\frac{c_{gS_o}}{c_{oS_o}}F_o(\delta p,\delta p_{bh}),
\end{aligned}
\tag{8.102}
$$

which is the pressure equation and is solved, together with the well control equations, implicitly. After solving for δp and δp_{bh}, they are substituted into (8.101) to compute δS_w and δS_o.

Undersaturated state

Substituting (8.40), (8.42), (8.95), and (8.96) into (8.37), the equations in the undersaturated state can be similarly obtained (cf. Exercise 8.20):

$$\begin{aligned}
&\frac{1}{\Delta t}\left[\left(\frac{\phi S_o}{B_o}\right)^l - \left(\frac{\phi S_o}{B_o}\right)^n + c_{op}\delta p + c_{oS_w}\delta S_w + c_{op_b}\delta p_b\right] \\
&\quad = \nabla\cdot\left(\mathbf{T}_o^l \nabla \Phi_o^l\right) + \nabla\cdot\left(\mathbf{T}_o^l \nabla(\delta p)\right) \\
&\qquad + \frac{1}{B_o^l}\left\{q_{Os}^l + \sum_{j=1}^{N_w}\sum_{m=1}^{M_{wj}} WI^{(j,m)}\frac{k_{ro}^n}{\mu_o^l}\left[\delta p_{bh}^{(j)} - \delta p\right]\delta(\mathbf{x}-\mathbf{x}^{(j,m)})\right\}
\end{aligned} \tag{8.103}$$

and

$$\begin{aligned}
&\frac{1}{\Delta t}\left\{\left(\frac{\phi R_{so} S_o}{B_o}\right)^l - \left(\frac{\phi R_{so} S_o}{B_o}\right)^n + c_{gp}\delta p + c_{gS_w}\delta S_w + c_{gp_b}\delta p_b\right\} \\
&\quad = \nabla\cdot\left(R_{so}^l \mathbf{T}_o^l \nabla \Phi_o^l\right) + \nabla\cdot\left(R_{so}^l \mathbf{T}_o^l \nabla(\delta p)\right) \\
&\qquad + \frac{R_{so}^l}{B_o^l}\left\{q_{Os}^l + \sum_{j=1}^{N_w}\sum_{m=1}^{M_{wj}} WI^{(j,m)}\frac{k_{ro}^n}{\mu_o^l}\left[\delta p_{bh}^{(j)} - \delta p\right]\delta(\mathbf{x}-\mathbf{x}^{(j,m)})\right\}.
\end{aligned} \tag{8.104}$$

Equation (8.97) for the water component is unchanged.

Multiply (8.97), (8.103), and (8.104) by Δt and write the resulting respective equations as

$$\begin{aligned}
c_{wp}\delta p + c_{wS_w}\delta S_w &= F_w(\delta p, \delta p_{bh}), \\
c_{op}\delta p + c_{oS_w}\delta S_w + c_{op_b}\delta p_b &= F_o(\delta p, \delta p_{bh}), \\
c_{gp}\delta p + c_{gS_w}\delta S_w + c_{gp_b}\delta p_b &= F_g(\delta p, \delta p_{bh}).
\end{aligned} \tag{8.105}$$

From the last two equations of (8.105) it follows that

$$\begin{aligned}
c_{oS_w}\delta S_w + c_{op_b}\delta p_b &= F_o(\delta p, \delta p_{bh}) - c_{op}\delta p, \\
c_{gS_w}\delta S_w + c_{gp_b}\delta p_b &= F_g(\delta p, \delta p_{bh}) - c_{gp}\delta p.
\end{aligned} \tag{8.106}$$

Define the determinants

$$\begin{aligned}
D &= \begin{vmatrix} c_{oS_w} & c_{op_b} \\ c_{gS_w} & c_{gp_b} \end{vmatrix} = c_{oS_w}c_{gp_b} - c_{gS_w}c_{op_b}, \\
D_S &= \begin{vmatrix} F_o(\delta p, \delta p_{bh}) - c_{op}\delta p & c_{op_b} \\ F_g(\delta p, \delta p_{bh}) - c_{gp}\delta p & c_{gp_b} \end{vmatrix} \\
&= (F_o(\delta p, \delta p_{bh}) - c_{op}\delta p)c_{gp_b} - (F_g(\delta p, \delta p_{bh}) - c_{gp}\delta p)c_{op_b}, \\
D_p &= \begin{vmatrix} c_{oS_w} & F_o(\delta p, \delta p_{bh}) - c_{op}\delta p \\ c_{gS_w} & F_g(\delta p, \delta p_{bh}) - c_{gp}\delta p \end{vmatrix} \\
&= c_{oS_w}(F_g(\delta p, \delta p_{bh}) - c_{gp}\delta p) - c_{gS_w}(F_o(\delta p, \delta p_{bh}) - c_{op}\delta p).
\end{aligned}$$

It follows from (8.106) that

$$\delta S_w = \frac{D_S}{D}, \quad \delta p_b = \frac{D_p}{D}, \tag{8.107}$$

which we substitute into the first equation of (8.105) to obtain the pressure equation in the undersaturated state:

$$c_{wp}\delta p + c_{wS_w}\frac{D_S}{D}(\delta p, \delta p_{bh}) = F_w(\delta p). \tag{8.108}$$

Equation (8.108) and the well control equations are solved implicitly for δp and δp_{bh}. After their computation, they are substituted into (8.107) to obtain δS_w and δp_b.

In summary, iterative IMPES has the following features:

- The difference between iterative IMPES and classical IMPES is that the iterative version is used within each Newton–Raphson iteration loop, while the classical one is exploited before a Newton–Raphson iteration.
- All the saturation functions k_{rw}, k_{ro}, k_{rg}, p_{cw}, and p_{cg} use the previous time step values of saturations in a Newton–Raphson iteration.
- The fluid formation volume factors and viscosities in the transmissibilities, phase potentials, and well terms are computed using the previous Newton–Raphson iteration values.
- The left-hand sides of the water, oil, and gas component equations are treated in the same fashion as in the SS.
- The unknown pressure is obtained implicitly, and the other two unknowns are obtained explicitly.

As in the sequential technique, the saturation functions k_{rw}, k_{ro}, k_{rg}, p_{cw}, and p_{cg} may use the previous Newton–Raphson iteration values of saturations, instead of the previous time step values of saturation. The bubble point problem in the iterative IMPES can be treated in the same manner as in the SS, and the time steps can be controlled in a similar way as in the sequential technique. The improved IMPES developed in the preceding chapter for two-phase flow can be extended to iterative IMPES for the black oil model. In particular, the time steps can be different for pressure than for saturations.

8.2.5 Well coupling

Various *well constraints* need to be taken into account. Two kinds of well constraints are used for an injection well.

Either the well bottom hole pressure p_{bh} is given, or a phase injection rate is fixed. In the former case,

$$p_{bh}^{(j)} = P_{bh}^{(j)}, \tag{8.109}$$

where j is the number of the well with this kind of well control and $P_{bh}^{(j)}$ is the given bottom hole pressure at this well. In this case,

$$\delta p_{bh}^{(j)} = 0. \tag{8.110}$$

In the latter case, it follows from (8.11) that the injection rate controls for water and gas injection wells are, respectively,

$$Q_{Ws}^{(j)} = \sum_{m=1}^{M_{wj}} WI^{(j,m)} \frac{k_{rwmax}}{\mu_w} \left[p_{bh}^{(j)} - p_w - \rho_w \wp (z_{bh}^{(j)} - z) \right] \delta(\mathbf{x} - \mathbf{x}^{(j,m)}) \tag{8.111}$$

and

$$Q_{Gs}^{(j)} = \sum_{m=1}^{M_{wj}} WI^{(j,m)} \frac{k_{rgmax}}{\mu_g} \left[p_{bh}^{(j)} - p_g - \rho_g \wp (z_{bh}^{(j)} - z) \right] \delta(\mathbf{x} - \mathbf{x}^{(j,m)}), \tag{8.112}$$

where $Q_{Ws}^{(j)}$ and $Q_{Gs}^{(j)}$ are the given water and gas injection rates, respectively, at the jth well and $k_{r\alpha max}$ is the maximum relative permeability of the α-phase, $\alpha = w, g$. A Newton–Raphson iteration can be used to solve the well control equations (8.111) and (8.112). For example, in the SS technique, it follows from (8.54) and (8.57) that the iteration applied to (8.111) and (8.112) gives

$$\begin{aligned} Q_{Ws}^{(j)} = (q_{Ws}^{(j)})^l + \sum_{m=1}^{M_{wj}} WI^{(j,m)} \Big[& e_{wp}^{(j)} \delta p + e_{wS_w}^{(j)} \delta S_w \\ & + e_{wp_{bh}} \delta p_{bh}^{(j)} \Big] \delta(\mathbf{x} - \mathbf{x}^{(j,m)}) \end{aligned} \tag{8.113}$$

and

$$\begin{aligned} Q_{Gs}^{(j)} = (q_{Gs}^{(j)})^l + \sum_{m=1}^{M_{wj}} WI^{(j,m)} \Big[& e_{gp}^{(j)} \delta p + e_{gS}^{(j)} (\delta S_w + \delta S_o) \\ & + e_{gp_{bh}} \delta p_{bh}^{(j)} \Big] \delta(\mathbf{x} - \mathbf{x}^{(j,m)}), \end{aligned} \tag{8.114}$$

where $Q_{Ws}^{(j)} = (q_{Ws}^{(j)})^{l+1}$, $Q_{Gs}^{(j)} = (q_{Gs}^{(j)})^{l+1}$, and the coefficients in these two equations are determined as in (8.54) and (8.57). For the sequential and iterative IMPES techniques, the flow rate terms can be expanded as in Sections 8.2.3 and 8.2.4.

For a production well, there are three kinds of well constraints: a constant bottom hole pressure, a constant total liquid production rate, and a constant total flow rate. The constant bottom hole pressure constraint has the form (8.109), and thus (8.110) holds. The constant total liquid production rate control takes the form

$$\begin{aligned} Q_{Ls}^{(j)} = & \sum_{m=1}^{M_{wj}} WI^{(j,m)} \frac{k_{rw}}{\mu_w} \left[p_{bh}^{(j)} - p_w - \rho_w \wp (z_{bh}^{(j)} - z) \right] \delta(\mathbf{x} - \mathbf{x}^{(j,m)}) \\ & + \sum_{m=1}^{M_{wj}} WI^{(j,m)} \frac{k_{ro}}{\mu_o} \left[p_{bh}^{(j)} - p_o - \rho_o \wp (z_{bh}^{(j)} - z) \right] \delta(\mathbf{x} - \mathbf{x}^{(j,m)}), \end{aligned} \tag{8.115}$$

where $Q_{Ls}^{(j)}$ is the given total liquid production rate at the jth well. The *water cut*, defined as the ratio of water production to the sum of water and oil production, at a perforated zone of a well with this kind of well constraint must be less than a certain limit; over this limit, that perforated zone must be shut down. The constant total flow rate control can be defined similarly; in this case, gas production is added.

In the SS technique, using (8.54) and (8.55) in the saturated state, a Newton–Raphson iteration applied to (8.115) gives

$$\begin{aligned} Q_{Ls}^{(j)} = (q_{Ls}^{(j)})^l &+ \sum_{m=1}^{M_{wj}} WI^{(j,m)} \Big[e_{wp}^{(j)} \delta p + e_{wS_w}^{(j)} \delta S_w \\ &\qquad + e_{wp_{bh}} \delta p_{bh}^{(j)} \Big] \delta(\mathbf{x} - \mathbf{x}^{(j,m)}) \\ &+ \sum_{m=1}^{M_{wj}} WI^{(j,m)} \Big[e_{op}^{(j)} \delta p + e_{oS_w}^{(j)} \delta S_w + e_{oS_o}^{(j)} \delta S_o \\ &\qquad + e_{op_{bh}} \delta p_{bh}^{(j)} \Big] \delta(\mathbf{x} - \mathbf{x}^{(j,m)}), \end{aligned} \tag{8.116}$$

where $Q_{Ls}^{(j)} = (q_{Ls}^{(j)})^{l+1}$ is fixed and the coefficients in this equation can be determined as in (8.54) and (8.55). In the SS in the undersaturated state, the sequential technique, and iterative IMPES, $Q_{Ls}^{(j)}$ can be expanded as in Sections 8.2.2–8.2.4.

8.2.6 The adaptive implicit and other techniques

An *adaptive implicit technique* was introduced in reservoir simulation by Thomas and Thurnau (1983). The principal idea of this technique is to seek an efficient middle ground between the IMPES (or sequential) and SS techniques. That is, at a given time step, the expensive SS technique is confined to those gridblocks that require it, while on the remaining gridblocks the IMPES technique is implemented. In this technique, pressure is computed implicitly everywhere in a porous medium (as in the IMPES, sequential, and SS techniques), but the computation of saturation is implicit in selected gridblocks and explicit elsewhere. This division into implicit and explicit gridblocks may be different from one time step to the next. The principal issue in implementation of this technique is a *switching criterion* that determines whether the saturation equation should be considered implicit or explicit.

In the original work (Thomas and Thurnau, 1983), the switching criterion is based on solution variable changes (as in local grid refinement; cf. Section 4.7). When a change at an IMPES gridblock exceeds a specified threshold value, the gridblock switches to the SS treatment. This criterion has the drawback that although instability leads to large solution changes, small changes do not guarantee stability. This drawback has led to the development of other criteria such as those based on eigenvalues (Fung et al., 1989) and hyperbolic equation stability analysis (i.e., the well-known Courant–Friedrichs–Lewy (CFL) stability analysis; cf. Section 4.1.8).

The adaptive implicit technique has been exclusively used in the finite difference simulation of reservoirs. Its application to the finite element method is not promulgated in the literature. The finite difference method is defined locally on grid points, and thus the CFL switching criterion can be easily analyzed in terms of local grid step sizes. However, the finite element method is defined globally on a whole domain, and hence how a switching criterion can be defined is not so clear. Thus we do not discuss this solution technique in this book.

Research on parallel computation in reservoir simulation was extensively carried out in the late 1980s, particularly due to the introduction of shared and distributed memory computers. For example, Scott et al. (1987) presented a multiple instruction multiple data (MIMD) approach to reservoir simulation, and Chien et al. (1987) described parallel processing on distributed memory machines. Several methods are available in the literature for parallelization of reservoir codes. Most of them are based on message passing techniques such as PVM (parallel virtual machine) and MPI (message passing interface) and *domain decomposition* methods. In most parallel approaches, a reservoir is split into a number of subdomains, and a processor is assigned to each subdomain problem (Killough and Wheeler, 1987); the *Schur complement* method can be used to solve interface problems (Smith et al., 1996). Parallel computing will be further discussed in Chapter 14.

Parallel algorithms have been used in the SS (Mayer, 1989), IMPES (Rutledge et al., 1991), and adaptive implicit (Verdière et al., 1999) solution techniques for various multiphase flows. That is, in each of these solution techniques, both the pressure and saturation equations are solved in a parallel fashion. Benchmark computations have indicated that linear (or nearly linear) speedup in CPU time can be obtained with an increasing number of processors. The parallel idea can be also used as a solution technique for multiphase flow. In the IMPES, sequential, and SS techniques, the pressure and saturation equations are solved either separately or simultaneously on the same processor. However, these two equations can be solved in parallel; i.e., their solution can be assigned to different processors at the same time point. This idea seems very useful for multicomponent, multiphase flow where the equations for different components (or phases) can be assigned to different processors. This research direction is yet to be investigated.

8.3 Comparisons between Solution Techniques

This section presents comparative results of the SS, sequential, and iterative IMPES solution techniques for the black oil model for both saturated and undersaturated reservoirs. For an undersaturated reservoir, the nonlinearity of the model's governing equations caused by the high compressibility and low viscosity of the gas component is relatively weaker than that for a saturated reservoir. In addition, there is no bubble point problem for undersaturated reservoirs. Since the sequential and iterative IMPES techniques have lower implicitness, they may be applicable to an undersaturated reservoir, but not to a saturated reservoir. We test the three solution techniques for both types of reservoirs.

8.3.1 An undersaturated reservoir

The simulation model comes from a development scheme design for water flooding of an oil field. The dimensions of the oil field are 6,890 ft $\times$ 6,726 ft $\times$ 4,227 ft. It has four geological layers with an irregularly shaped boundary, top, and base, and has reservoir temperature 165.2° F. The absolute permeability and compressibility of rock and the thickness of the layers vary in space. The water, oil, and oil viscosity compressibilities are 3.1×10^{-6}, 3.1×10^{-6}, and 0 psi^{-1}, respectively. The stock-tank densities for oil and water are, respectively, 60.68 and 62.43 lbm/ft^3. The gas specific gravity at standard conditions (expressed as the ratio of the molecular weight of the gas to the molecular weight of air) is 0.5615. The

Table 8.1. *PVT property data.*

p (psia)	B_o (RB/STB)	μ_o (cp)	R_{so} (SCF/STB)	B_w (RB/STB)	μ_w (cp)	Z	μ_g (cp)
87.02	1.0057	52.8	6.74	1.022	0.42	0.993	0.0151
435.11	1.0208	37.6	39.19	1.022	0.42	0.966	0.0141
870.23	1.0415	26.3	83.66	1.022	0.42	0.936	0.0132
1305.34	1.0632	19.7	130.25	1.022	0.42	0.913	0.0141
1624.42	1.0795	15.5	165.63	1.022	0.42	0.898	0.0151

Table 8.2. *Saturation function data for a water-oil system.*

S_w	k_{rw}	k_{row}	p_{cow} (psi)
0.2400	0.000	1.000	2.4656
0.3050	0.001	0.809	1.1603
0.3266	0.002	0.707	0.8702
0.3483	0.004	0.606	0.5802
0.3699	0.007	0.513	0.3916
0.3915	0.010	0.421	0.2321
0.4131	0.014	0.349	0.1450
0.5000	0.037	0.260	0.0725
0.6000	0.087	0.200	0.0435
0.7000	0.155	0.150	0.0232
0.8000	0.230	0.100	0.0000
0.9000	0.400	0.000	0.0000
1.0000	1.000	0.000	0.0000

Table 8.3. *Saturation function data for a gas-oil system.*

S_g	k_{rg}	k_{rog}	p_{cgo} (psi)
0.00	0.000	1.0000	0.0
0.04	0.000	0.4910	0.0
0.10	0.001	0.2990	0.0
0.20	0.003	0.1200	0.0
0.22	0.007	0.1030	0.0
0.29	0.015	0.0400	0.0
0.33	0.030	0.0210	0.0
0.37	0.065	0.0087	0.0
0.40	0.131	0.0021	0.0
0.46	0.250	0.0000	0.0
0.76	1.000	0.0000	0.0

depths to the gas/oil contact (GOC) and water/oil contact (WOC) are 3,666 ft and 4,593 ft, respectively. The reservoir is initially at capillary/gravity equilibrium with a pressure of 1,624 psia at depth 3,684 ft. The capillary pressures at the GOC and WOC are zero. Other PVT and rock data are given in Tables 8.1–8.3, where Z is the gas deviation factor (cf. Chapter 3 or Section 8.1.3).

There are 50 oil production wells and 20 water injection wells. They perforate all the layers (above the WOC). The wellbore radius of each well is 0.25 ft. The well controls can be the bottom hole pressure, water injection rate, oil production rate, and liquid production rate controls with a water cut limit of 0.95.

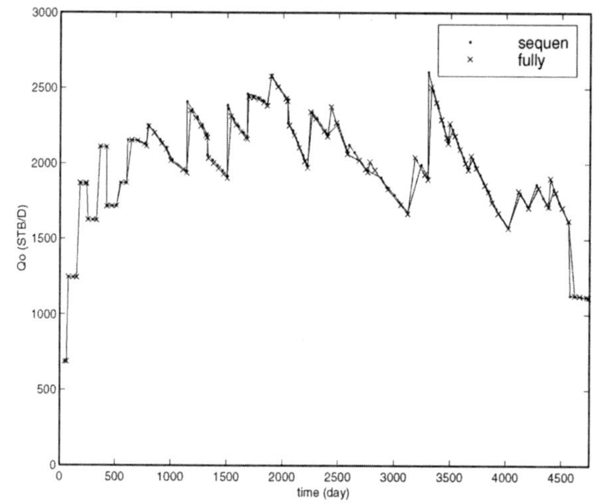

Figure 8.2. *Oil production rate of an undersaturated reservoir.*

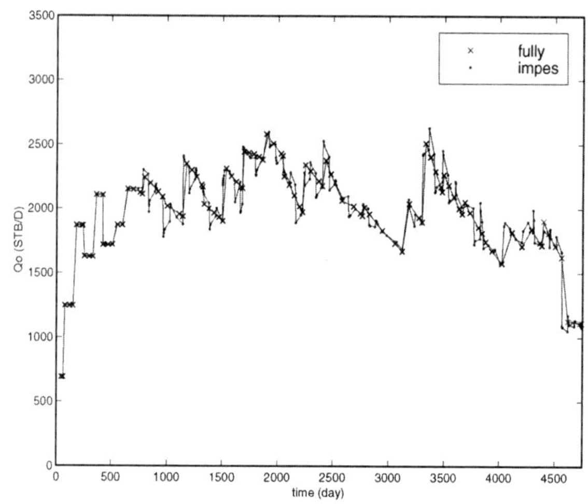

Figure 8.3. *Oil production rate of an undersaturated reservoir.*

Due to the layer structure in the vertical direction of this reservoir, we divide its domain into hexagonal prisms, i.e., hexagons in the x_1x_2-plane and rectangles in the x_3-coordinate direction, as shown in Figure 4.36. The number of control volumes is 2,088 $\times$ 4 (4 is the number of layers). The CVFE method with linear elements is used for the discretization of the governing equations (cf. Section 4.3). We run the simulator with $(dp)_{max} = 300$ psia, $(dS_w)_{max} = 0.05$, and $(dp_b)_{max} = 300$ psia (cf. Section 8.2.3) and stop running at 4,740 days for all three solution techniques. The ORTHOMIN algorithm with incomplete LU factorization preconditioners (cf. Chapter 5) is used to solve the LESs. The plots of the oil production rate, water cut, and oil recovery for this reservoir are shown in Figures 8.2–8.7,

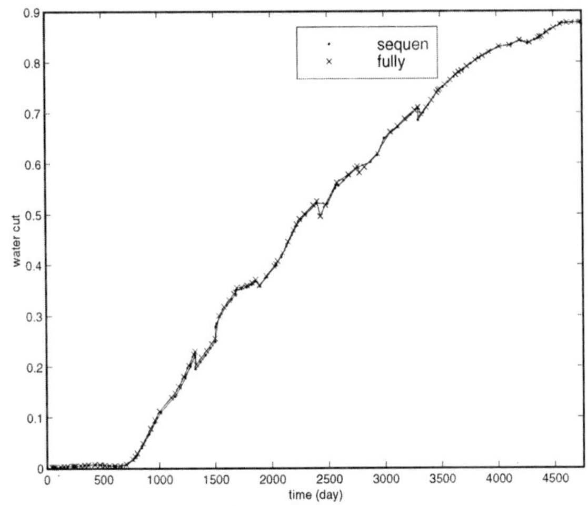

Figure 8.4. *Water cut of an undersaturated reservoir.*

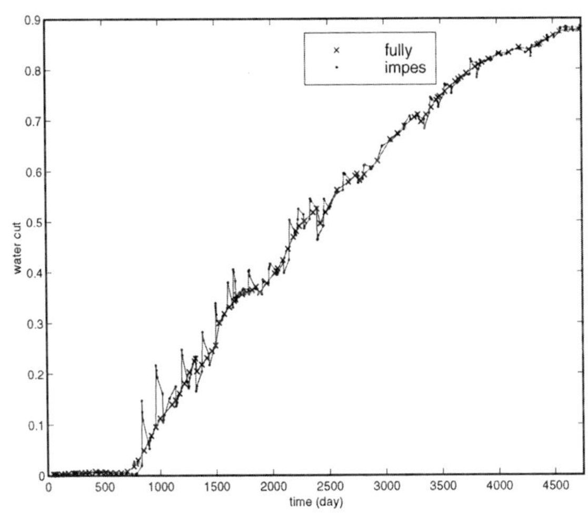

Figure 8.5. *Water cut of an undersaturated reservoir.*

where the SS is indicated by "fully." The comparative results of memory space and computational cost for the solution techniques are shown in Table 8.4. The results obtained from the sequential technique match those from the SS very well, but there are oscillations in the results from the IMPES, as seen in Figures 8.3 and 8.5. These oscillations can be made to disappear by reducing time steps to such an extent that the simulation process advances slowly. The memory used by the sequential and IMPES techniques to solve the LESs is as little as 20.01% of that of the SS, as a result of the size reduction of the LESs. The CPU time used by the sequential technique to solve the LESs is just 12.06% of that by the SS, and the total CPU time by the sequential technique is only 23.89% of that by the SS.

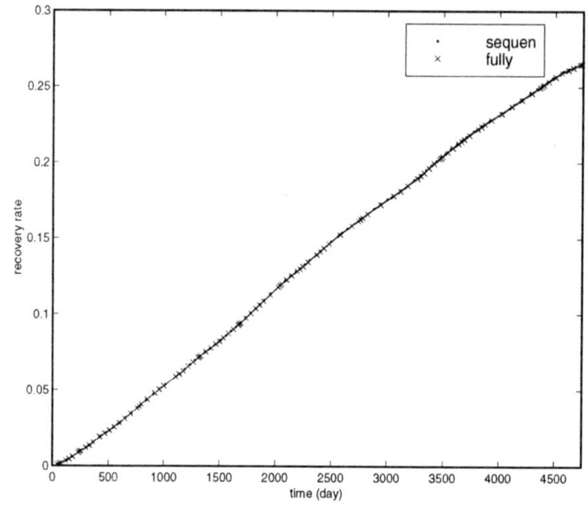

Figure 8.6. *Oil recovery of an undersaturated reservoir.*

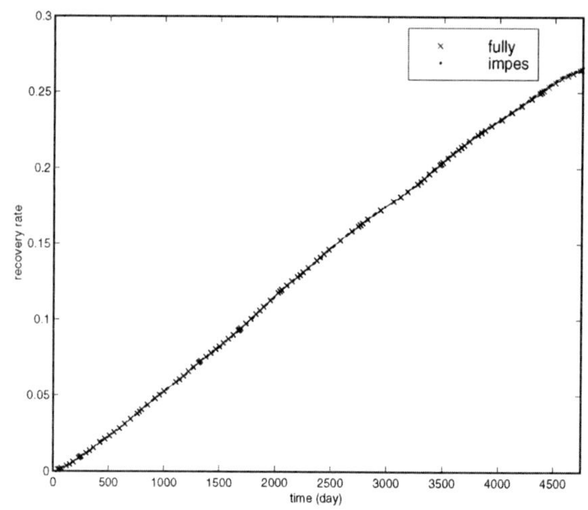

Figure 8.7. *Oil recovery of an undersaturated reservoir.*

From Table 8.4, we see that 83.55% of the computational cost is spent on solving the LESs, and other calculations just take 549.39 seconds in the SS. But in the sequential technique, only 42.17% of the total CPU time is spent on solving the LESs, and the CPU time taken by other calculations is only 87.89 seconds less than that used by the SS. Therefore, the primary reason that the sequential technique is faster than the SS is that using the sequential technique can greatly reduce the computational cost for the solution of the LESs. Also, note that the sequential technique requires less memory. These remarks also apply to the iterative IMPES technique.

Table 8.4. *Comparison among the SS, sequential, and iterative IMPES techniques for an undersaturated reservoir.*

Solution technique	SS	Sequential	IMPES
Memory for LES solver (MB)	18.099264	3.621892	3.621892
Total memory (MB)	26.326132	11.84876	11.84876
CPU time for LES (sec.)	2790.80	336.55	543.17
Total CPU time (sec.)	3340.19	798.05	1518.15
Number of time steps	30	30	30

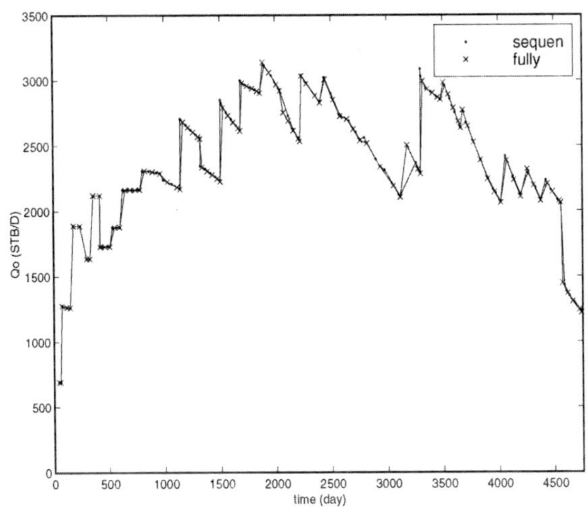

Figure 8.8. *Oil production rate for water flooding of a saturated reservoir.*

8.3.2 A saturated reservoir

Because the IMPES technique is clearly not a good choice even for undersaturated reservoirs, we do not test this technique for saturated reservoirs. Two cases are designed to compare the SS and sequential techniques for simulation of a saturated reservoir. In the first case, we simply raise the initial bubble point pressure of the oil field described in the above example to 1,642 psia so that we initially have a saturated reservoir. For the second case, we change a production well, which is located at an upper part of this field and is shut down at 510 days, into a gas injection well to improve oil recovery with an upper limit of GOR (gas-oil ratio) 0.2 MSCF/RB after 600 days. Its injection rate is 500 MSCF/day.

For these two cases, we run the simulator with the same control parameters as those in the above example, using both the SS and sequential techniques. The computational results are shown in Figures 8.8–8.16. The memory and computational cost for both techniques are given for cases 1 and 2 in Tables 8.5 and 8.6, respectively. From Figures 8.8–8.11, we see that the oil production rate, GOR, water cut, and oil recovery obtained from these two techniques match very well for the first case. In this case, the total CPU time taken by the sequential technique increases to 34.60% of that for the SS. However, for the second case, although the oil production rate, water cut, and oil recovery from the sequential technique still match

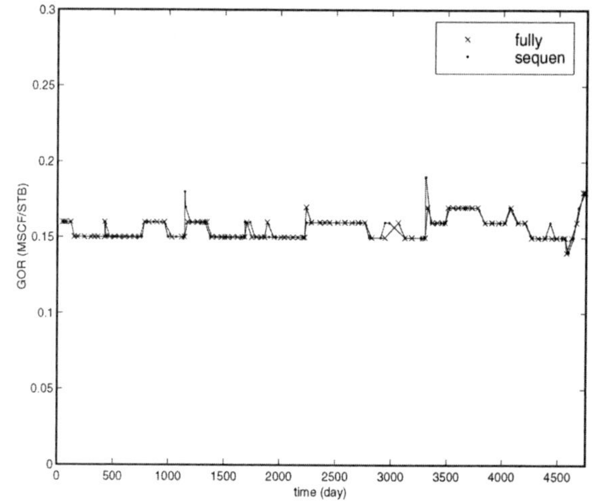

Figure 8.9. *GOR for water flooding of a saturated reservoir.*

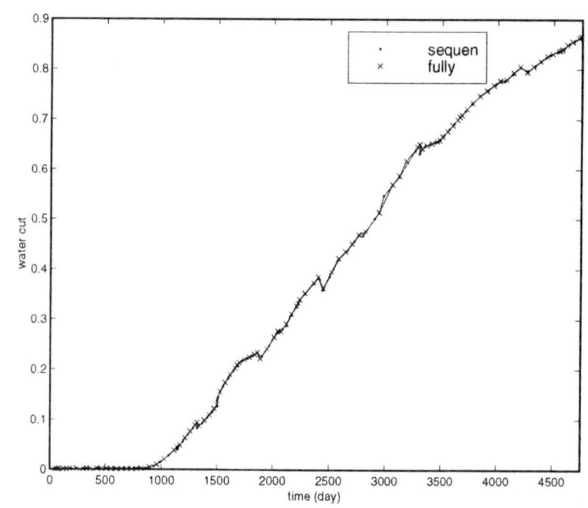

Figure 8.10. *Water cut for water flooding of a saturated reservoir.*

those from the SS, there is a deviation between the GORs for these two techniques after 3,700 days (cf. Figure 8.14). Also, in this case, the CPU time for the sequential technique to solve the LESs is 18.22% of that for the SS, and the total computational time for the former becomes 40.78% of that for the latter. The number of Newton–Raphson's iterations taken by the sequential technique is 10 more than that for the SS.

The nonlinearity caused by the free gas and the bubble point problem is the main reason for these phenomena. The free gas has a large compressibility, compared with water and oil. It makes a great contribution to the flow term in the governing equation of the

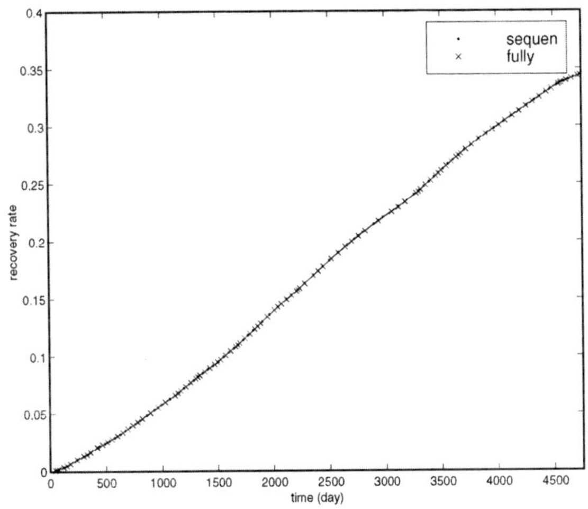

Figure 8.11. *Oil recovery for water flooding of a saturated reservoir.*

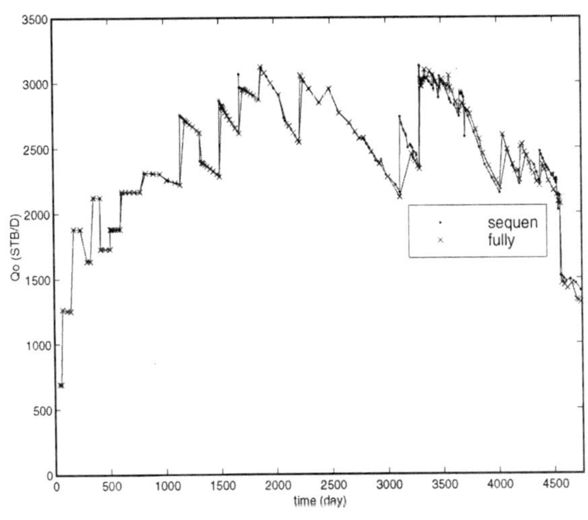

Figure 8.12. *Oil production rate for gas injection of a saturated reservoir.*

gas component for a grid point of a reservoir in the saturated state. If the contribution is ignored by the sequential technique to obtain a pressure equation, it will introduce a large approximation error into the resulting pressure equation. Particularly, this may lead to divergence of the Newton–Raphson iteration at a bubble point. For a saturated reservoir, the state at a location may transfer from the saturated state to the undersaturated state. At a bubble point, if the pressure is not correct, inappropriate PVT data of oil will be used and the Newton–Raphson iteration will approach an incorrect value. In the first case, the free gas comes from the dissolved gas in the reservoir, and the GOR is just 0.15, which is rather low.

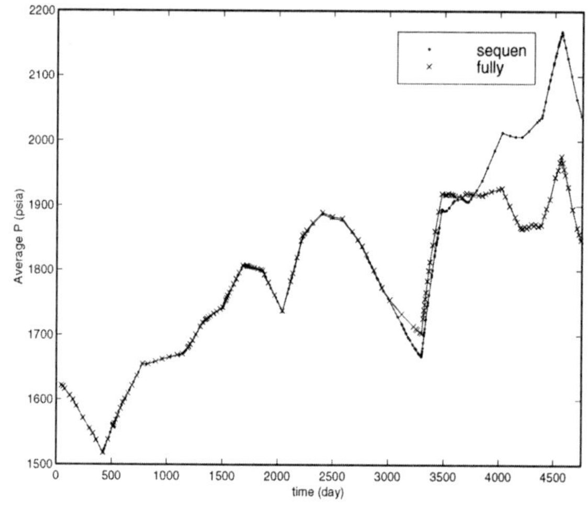

Figure 8.13. *Average reservoir pressure for gas injection of a saturated reservoir.*

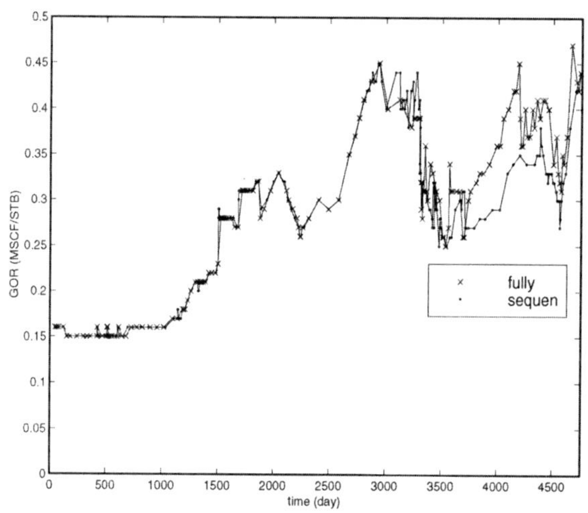

Figure 8.14. *GOR for gas injection of a saturated reservoir.*

The nonlinearity caused by the free gas is weak. The approximation error for the pressure equation introduced by the sequential technique is small. Therefore, its convergence rate is high. However, in the second case, a great amount of free gas is injected into the reservoir. The nonlinearity caused by the free gas is strong. After 3,700 days, the oil production rate drops quickly, the pressure obtained from the sequential technique is higher than the real value because it ignores the nonlinearity caused by the free gas and water (cf. Figure 8.13), and at this pressure more free gas dissolves into oil and leads to deviation of the GOR from its correct value.

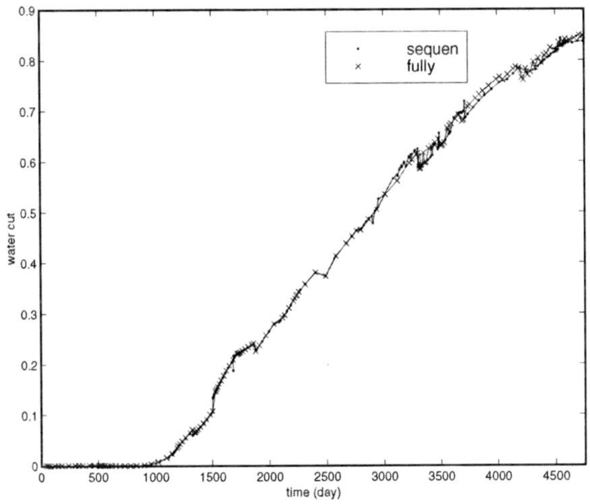

Figure 8.15. *Water cut for gas injection of a saturated reservoir.*

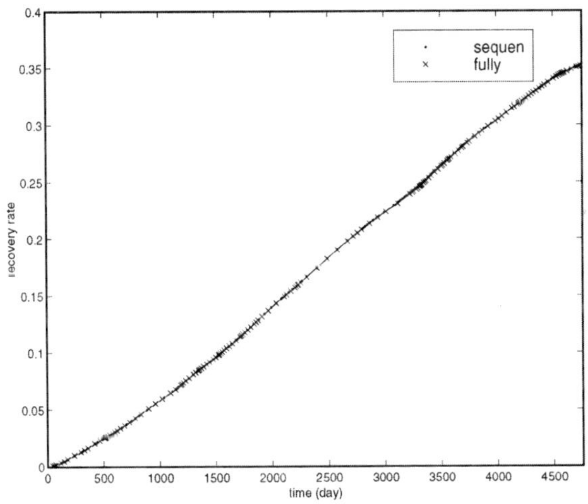

Figure 8.16. *Oil recovery for gas injection of a saturated reservoir.*

Table 8.5. *Comparison between the SS and sequential techniques for water flooding of a saturated reservoir in case* 1.

Solution technique	SS	Sequential
Memory for LES solver (MB)	18.099264	3.621892
Total memory (MB)	26.326132	11.84876
CPU time for LES (sec.)	2485.88	518.58
Total CPU time (sec.)	3067.10	1061.09
Number of time steps	30	30
Number of Newton iterations	104	146

Table 8.6. *Comparison between the SS and sequential techniques for gas injection of a saturated reservoir in case 2.*

Solution technique	SS	Sequential
Memory for LES solver (MB)	18.099264	3.621892
Total memory (MB)	26.326132	11.84876
CPU time for LES (sec.)	5008.60	912.93
Total CPU time (sec.)	5869.08	2393.66
Number of time steps	30	30
Number of Newton iterations	137	147

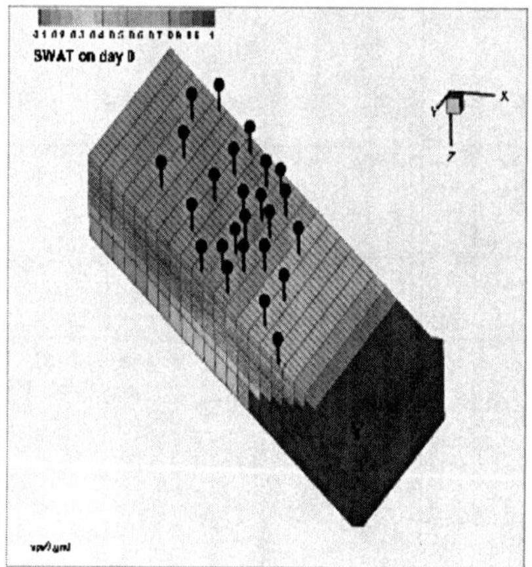

Figure 8.17. *The reservoir of the ninth CSP problem.*

8.3.3 The ninth SPE project: Black oil simulation

The benchmark problem of the ninth comparative solution project (CSP) (Killough, 1995) is challenging because, first, the permeability of the reservoir is generated from geostatistical modeling, which can lead to a strong heterogeneity; second, the water-oil capillary pressure has a discontinuity at a water saturation of 0.35, which may cause divergence of a Newton–Raphson iteration; third, the capillary pressure has a tail that does not extend to the water saturation 1.0 (cf. Figure 8.19 later).

A grid of rectangular parallelepipeds for the reservoir under consideration is given in Figure 8.17. Its dimensions are $7{,}200 \times 7{,}500 \times 359$ ft^3. The depth to cell (1,1,1) of this rectangular grid is 9,000 ft. It has a dip in the x_1-direction of 10 degrees. The GOC and WOC are located at, respectively, 8,800 ft and 9,950 ft. The reservoir has 15 layers.

The values of porosity and thickness for each layer and of oil and gas PVT property data are based on the second CSP (Weinstein et al., 1986; also see Section 8.4), and are given in Tables 8.7 and 8.8. The gas specific gravity equals 0.92. The gas-oil saturation functions given in Table 8.9 are also taken from the second CSP. The relative permeabilities and capillary pressure for a water-oil system are shown in Figures 8.18 and 8.19.

Table 8.7. *Reservoir description.*

Layer	Thickness (ft)	Porosity
1	20	0.087
2	15	0.097
3	26	0.111
4	15	0.160
5	16	0.130
6	14	0.170
7	8	0.170
8	8	0.080
9	18	0.140
10	12	0.130
11	19	0.120
12	18	0.105
13	20	0.120
14	50	0.116
15	100	0.157

Table 8.8. *PVT property data.*

p (psia)	B_o (RB/STB)	μ_o (cp)	R_{so} (SCF/STB)	Z	μ_g (cp)
14.7	1.000	1.20	0	0.9999	0.0125
400	1.0120	1.17	165	0.8369	0.0130
800	1.0255	1.14	335	0.8370	0.0135
1200	1.0380	1.11	500	0.8341	0.0140
1600	1.0150	1.08	665	0.8341	0.0145
2000	1.0630	1.06	828	0.8370	0.0150
2400	1.0750	1.03	985	0.8341	0.0155
2800	1.0870	1.00	1130	0.8341	0.0160
3200	1.0985	0.98	1270	0.8398	0.0165
3600	1.1100	0.95	1390	0.8299	0.0170
4000	1.1200	0.94	1500	0.8300	0.0175

Table 8.9. *Saturation function data for a gas-oil system.*

S_g	k_{rg}	k_{rog}	p_{cgo} (psi)
0.0	0.0	1.0	0.0
0.04	0.0	0.60	0.2
0.10	0.0220	0.33	0.5
0.20	0.1000	0.10	1.0
0.30	0.2400	0.02	1.5
0.40	0.3400	0.0	2.0
0.50	0.4200	0.0	2.5
0.60	0.5000	0.0	3.0
0.70	0.8125	0.0	3.5
0.88	1.0	0.0	3.9

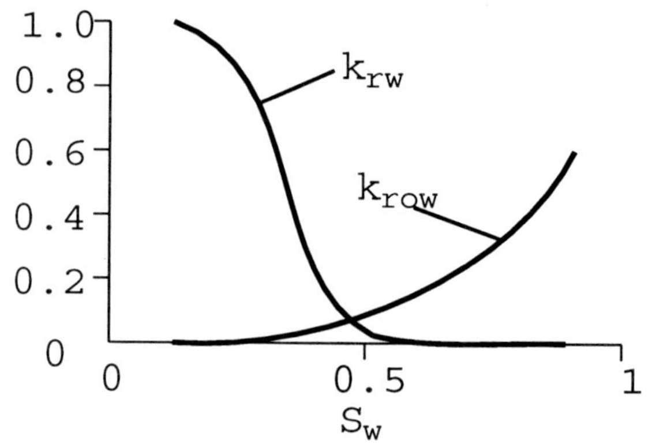

Figure 8.18. *Water-oil relative permeabilities.*

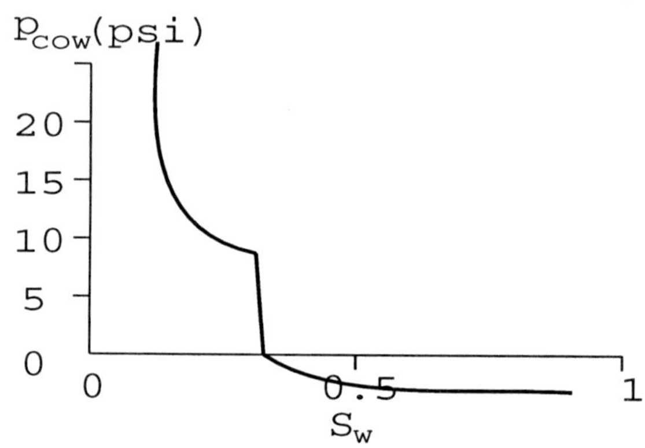

Figure 8.19. *Water-oil capillary presure.*

In the initial state, the reservoir reaches equilibrium with an initial reservoir pressure of 3,600 psia at 9,035 ft and with a reservoir temperature of 100° F. The bubble point pressure of oil is 3,600 psia. At 1,000 psi above the bubble point pressure p_b, B_o is 0.999 times the value of B_o at p_b. The density of the stock-tank oil is 0.7296 gm/cc. The oil pressure gradient is approximately 0.3902 psi/ft at 3,600 psia. The stock-tank density of water is 1.0095 gm/cc, with a water formation volume factor B_w at 3,600 psia of 1.0034 RB/STB yielding a water gradient of approximately 0.436 psi/ft. The rock compressibility is 1.0×10^{-6} 1/psi. The Stone II model is used for calculating the relative permeability of the oil phase when three phases coexist (cf. Chapter 3).

There are one water injector and 25 producers, whose wellbore radii are 0.50 ft. Their locations are shown in Figure 8.17. The injector is perforated at layers 11–15, and the producers are perforated at layers 2–4. The water injection rate is 5,000 STB/D with a maximum bottom hole pressure of 4,000 psia. Initially, the oil production rate of the producers is set to 1,500 STB/D. They are reduced to 100 STB/D at 300 days. Then they

Figure 8.20. *Gas saturation at* 50 *days.*

are raised to 1,500 STB/D until the end of the simulation at 900 days. The reference depths of all wells are 9,110 ft.

For this problem, the CVFA method presented in Section 4.3.5 is used for the space discretization. To check the accuracy, stability, and convergence of this method, we compare its results in the sequential and SS techniques with those generated by the 9-point finite difference (FD) method in the SS and by VIP-EXECUTIVE, which is a three-dimensional, three-phase finite difference reservoir simulator developed originally by the firm J. S. Nolen and Associates (now part of Western ATLAS Software). For the CVFA method, we use hexagonal prisms (hexagons in the x_1x_2-plane and rectangles in the x_3-coordinate direction, cf. Figure 4.36) as base gridblocks since the reservoir considered has a layer structure. In order for the wells to be located at the destination positions, the base gridblocks are adjusted with the techniques of corner point correction and local grid refinement (cf. Section 13.4.4). The total number of gridblocks is 765×15, where 15 is the number of layers. The ORTHOMIN iterative algorithm is used to solve the LESs, and incomplete LU(0) factorizations are used as preconditioners (cf. Chapter 5). The maximum saturation and pressure changes during the computational processes are set to 0.05 and 150 psi, respectively, for the SS, while the maximum saturation change for the sequential technique is set to 0.02 to control convergence.

Figure 8.20 shows the gas saturation distribution of the first layer at 50 days, where S_g is in one of the intervals [0, .02], (.02, .04], (.04, .06], and (.06, .08] represented by dark to light colors. The gas saturation distribution is quite unusual. It is caused by the strong heterogeneity of the reservoir, whose permeability has a lognormal distribution. Figures 8.21–8.27 are the comparative results. The results from the CVFA in the sequential technique are closer to those from the FD method in the SS than to those from VIP-EXECUTIVE. The reason may be that there are minor differences between our simulator and VIP-EXECUTIVE in the treatment of the well models, linearization of conservation equations, time step control, iteration control, or type of grids used. From these plots, we see that the reservoir pressures

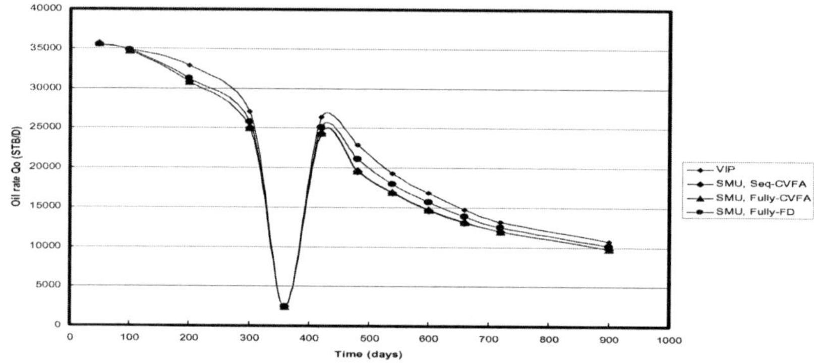

Figure 8.21. *Comparison of oil production rates.*

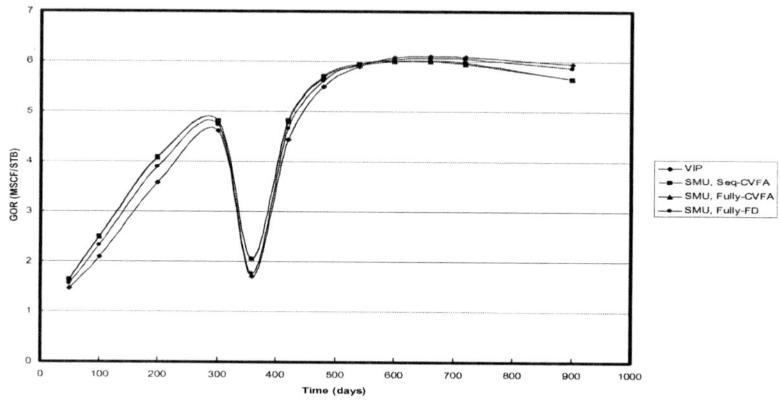

Figure 8.22. *Comparison of GORs versus time.*

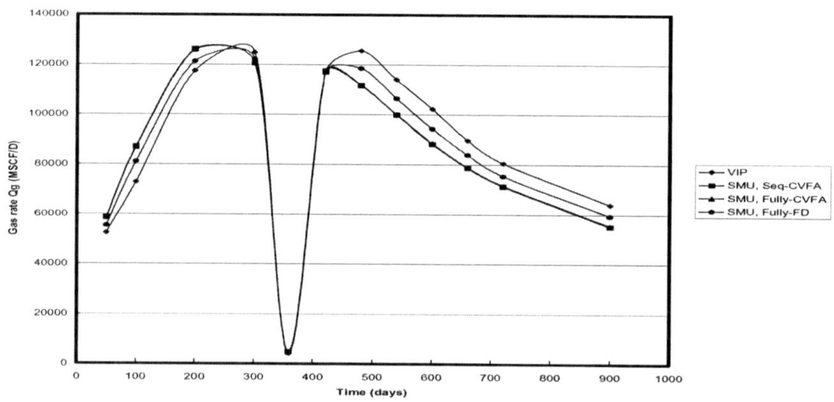

Figure 8.23. *Comparison of field gas rates.*

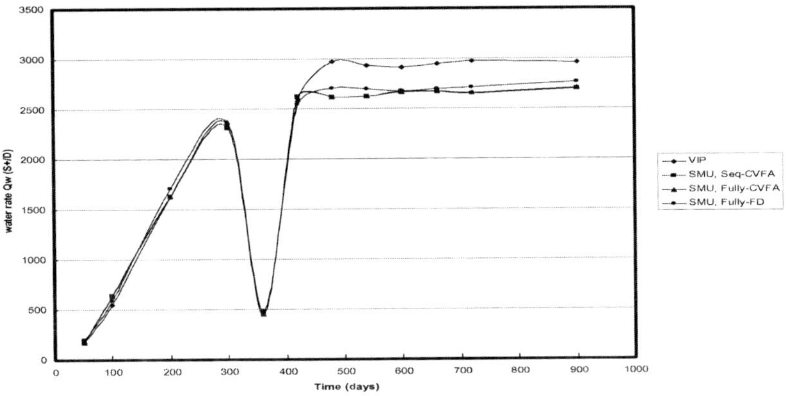

Figure 8.24. *Comparison of field water rates.*

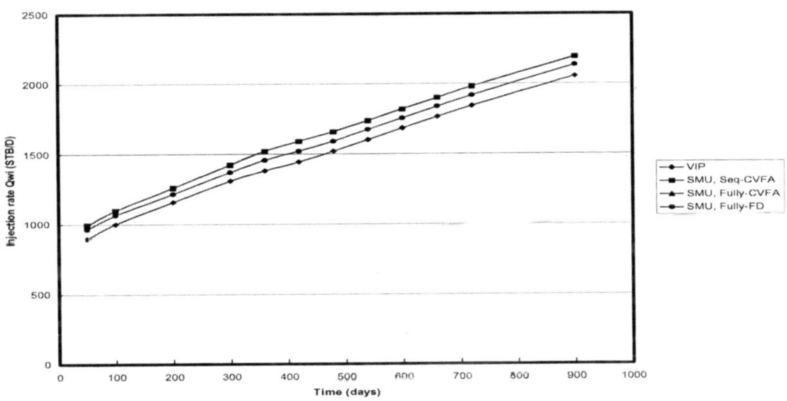

Figure 8.25. *Comparison of injected water rates.*

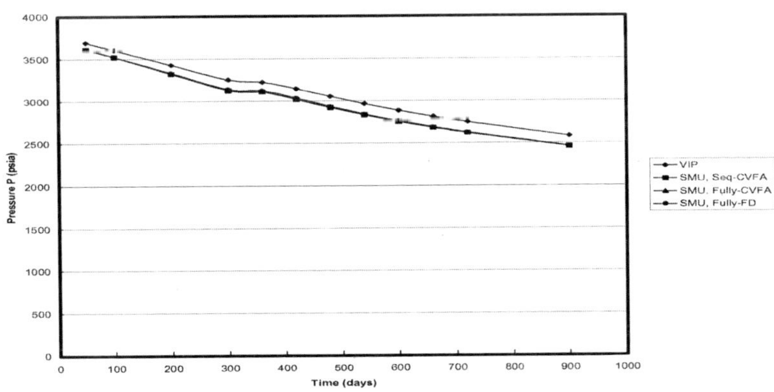

Figure 8.26. *Comparison of average reservoir pressures.*

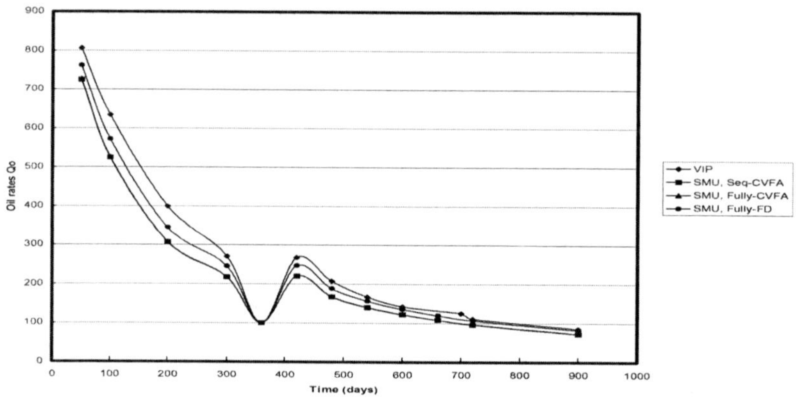

Figure 8.27. *Comparison of oil rates for well* 21.

Table 8.10. *Comparison of computational cost between the SS and sequential techniques for the ninth CSP problem.*

Solution technique	SS	Sequential
CPU time for LESs (sec.)	4141.92	1174.48
Total CPU time (sec.)	5172.76	2819.80
Number of time steps	119	179

match perfectly between the CVFA and FD methods, as shown in Figure 8.26; there exist slight differences for other quantities. Since this benchmark problem has a very strong heterogeneity generated by geostatistical modeling, the unstructured grids used in the CVFA can more accurately describe the heterogeneity of the reservoir, which is reflected in the production rates. Table 8.10 shows that the sequential technique just takes 28% of the CPU time of the SS to solve the linear equations. The total CPU time is smaller by 45.5%.

8.3.4 Remarks on numerical experiments

We have applied the SS, sequential, and iterative IMPES solution techniques to black oil reservoir simulation. The FD, CVFE, and CVFA methods have been employed for the discretization of the governing equations of the black oil model. Field-scale simulation models of an oil reservoir have been used to test these solution schemes for both the saturated and undersaturated states of this reservoir.

From the numerical experiment results, we can draw the following conclusions for black oil reservoir simulation:

- The iterative IMPES technique is not a good choice for this type of simulation.
- The SS technique is the most stable and robust, but it has the highest memory and computational costs.
- The sequential technique is convergent and stable for an undersaturated reservoir, and it can significantly reduce memory and computational cost compared with the SS.

For a saturated reservoir the accuracy of the sequential scheme depends on whether free gas is injected. For no gas injection, this scheme is convergent and accurate and can reduce computational cost. But, for gas injection, the pressures and GORs obtained from this technique differ from those from the SS, even though it seems convergent.

- For the ninth SPE CSP benchmark problem, the results from the SS and sequential techniques match very well.

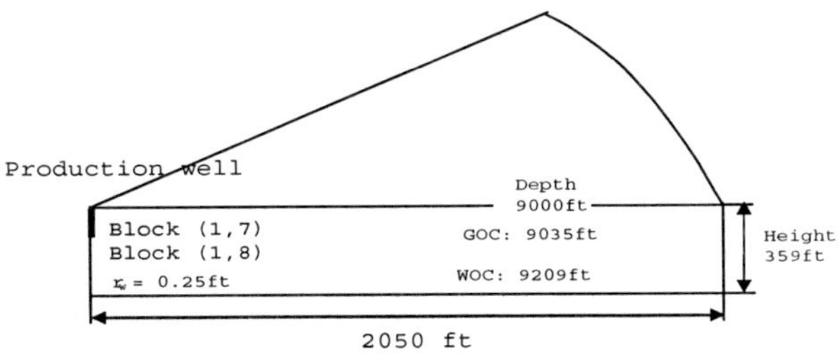

Figure 8.28. *Cross-sectional view of the second SPE CSP reservoir.*

8.4 The Second SPE Project: Coning Problems

This section deals with a three-phase coning problem. The *coning problem* is caused by a large gradient of a phase potential in the axial direction of a well (Fanchi, 2001). In the initial stage of a recovery process of a reservoir, the equal-potential surface has the shape of a semisphere with an infinite radius, and the gradient of the potential on the surface is zero everywhere. After a producer is perforated, this gradient is no longer zero. In the axial direction of the well, it reaches a highest value because of production. This results in a change of shape of the equal-potential surface. It changes gradually into a cone, with the top of the cone toward the perforated zones of the producer. Therefore, the water and/or gas fronts gradually reach the perforated zones of the producer. Near the wellbore, the saturations and pressure change very rapidly during the formation of water and/or gas coning, which may cause unstability of a reservoir simulator.

The second SPE CSP (Weinstein et al., 1986) was used to test the stability of reservoir simulators to deal with a coning problem. A cross-sectional view of the reservoir is seen in Figure 8.28. The reservoir dimensions, permeabilities, and porosities are presented in Table 8.11, where k_h ($= k_{11} = k_{22}$) and k_v ($= k_{33}$) denote the horizontal and vertical permeabilities, respectively. The radial extent of the reservoir is 2,050 ft. In the radial direction, 10 blocks are used. Their boundaries are at 2.00, 4.32, 9.33, 20.17, 43.56, 94.11, 203.32, 439.24, 948.92, and 2,050 ft, respectively. There are 15 vertical layers. The depth to the top of formation is 9,000 ft. The pore, water, oil, and oil viscosity compressibilities are 4×10^{-6}, 4×10^{-6}, 3×10^{-6}, and 0 psi^{-1}, respectively. The stock-tank densities for oil and water are 45.0 and 63.02 lbm/ft^3. The gas density at standard conditions is 0.0702 lbm/ft^3. The depths to the GOC, which is the interface between the gas zone and the oil

Table 8.11. *Reservoir description.*

Layer	Thickness (ft)	k_h (md)	k_v (md)	Porosity
1	20	35.000	3.500	0.087
2	15	47.500	4.750	0.097
3	26	148.000	14.800	0.111
4	15	202.000	20.200	0.160
5	16	90.000	9.000	0.130
6	14	418.500	41.850	0.170
7	8	775.000	77.500	0.170
8	8	60.000	6.000	0.080
9	18	682.000	68.200	0.140
10	12	472.000	47.200	0.130
11	19	125.000	12.500	0.120
12	18	300.000	30.000	0.105
13	20	137.000	13.750	0.120
14	50	191.000	19.100	0.116
15	100	350.000	35.000	0.157

Table 8.12. *Saturation function data for a water-oil system.*

S_w	k_{rw}	k_{row}	p_{cow} (psi)
0.22	0.0	1.0	7.0
0.30	0.07	0.4000	4.0
0.40	0.15	0.1250	3.0
0.50	0.24	0.0649	2.5
0.60	0.33	0.0048	2.0
0.80	0.65	0.0	1.0
0.90	0.83	0.0	0.5
1.00	1.0	0.0	0.0

zone, and WOC, which is the interface between the water zone and the oil zone, are 9,035 and 9,209 ft, respectively. The reservoir is initially at capillary/gravity equilibrium with a pressure of 3,600 psia at the GOC. The capillary pressures at the GOC and WOC are zero. The single well at the center of the radial system is completely perforated at the 7th and 8th layers, has the wellbore radius 0.25 ft, and has a minimum bottom hole pressure of 3,000 psia. The saturation function data and PVT property data are presented in Tables 8.9, 8.12, and 8.13, and the well production schedule is shown in Table 8.14.

To model the radial flow pattern of this single well, we use a hybrid grid to present the reservoir (cf. Figure 8.29) and apply the CVFA method to discretize the governing equations (cf. Section 4.3.5). The center blocks are cylinders, and other blocks are obtained by uniformly partitioning in the angular direction. The total number of gridblocks is $(18 \times 9 + 1) \times 15$, where 15 is the number of layers. The radial sizes of gridblocks are the same as those given in the problem statement. The drainage radius of the center gridblocks, which are cylindrical gridblocks, is

$$r_e = \sqrt{r_w r_1},$$

where r_1 indicates the radius of the center block and r_w is the wellbore radius. To choose appropriate time steps, the maximum saturation change per step is set to 0.05.

Table 8.13. *PVT property data.*

p (psia)	B_o (RB/STB)	μ_o (cp)	R_{so} (SCF/STB)	B_w (RB/STB)	μ_w (cp)	B_g (RB/STB)	μ_g (cp)
400	1.0120	1.17	165	1.01303	0.96	5.90	0.0130
800	1.0255	1.14	335	1.01182	0.96	2.95	0.0135
1200	1.0380	1.11	500	1.01061	0.96	1.96	0.0140
1600	1.0150	1.08	665	1.00940	0.96	1.47	0.0145
2000	1.0630	1.06	828	1.00820	0.96	1.18	0.0150
2400	1.0750	1.03	985	1.00700	0.96	0.98	0.0155
2800	1.0870	1.00	1130	1.00580	0.96	0.84	0.0160
3200	1.0985	0.98	1270	1.00460	0.96	0.74	0.0165
3600	1.1100	0.95	1390	1.00341	0.96	0.65	0.0170
4000	1.1200	0.94	1500	1.00222	0.96	0.59	0.0175
4400	1.1300	0.92	1600	1.00103	0.96	0.54	0.0180
4800	1.1400	0.91	1676	0.99985	0.96	0.49	0.0185
5200	1.1480	0.90	1750	0.99866	0.96	0.45	0.0190
5600	1.1550	0.89	1810	0.99749	0.96	0.42	0.0195

Table 8.14. *Production schedule.*

Period number	Period time (day)	Oil production rate (STB/D)
1	1–10	1,000
2	10–50	100
3	50–720	1,000
4	720–900	100

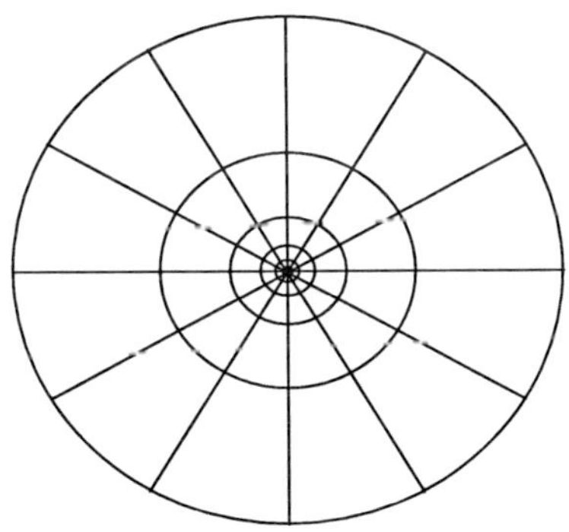

Figure 8.29. *Cross-sectional view of the grid system.*

We compare the CVFA and FD methods for this three-phase coning problem; the CVFE uses grids based on triangles or tetrahedra that cannot accurately model the cylindrical boundary. The FD method in an (r, z)-coordinate system is adopted. The total number of gridblocks is 10×15.

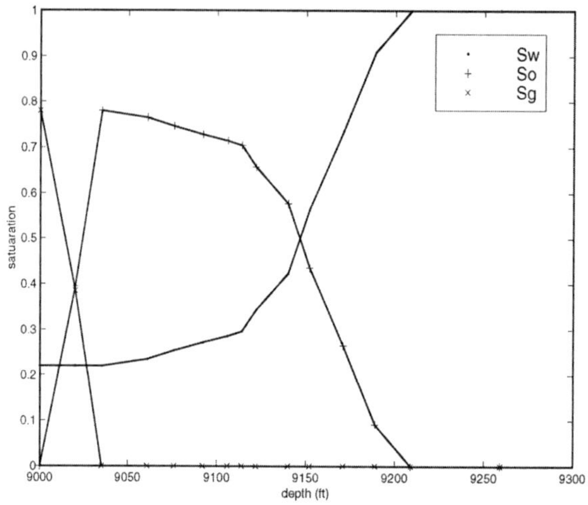

Figure 8.30. *Initial saturation distribution.*

Table 8.15. *Initial fluids in place and time on decline.*

Method	Oil (10^6STB)	Water (10^6STB)	Gas (10^6STB)	Time on decline (day)
FD	28.87	73.98	47.13	230
CVFA	28.89	73.96	47.08	220

Figure 8.30 shows the plot of initial saturations versus depth. The gas saturation drops to zero if the depth is greater than 9,035 ft, which is consistent with the positions of the given GOC and WOC. Also, the initial saturations satisfy the constraint (8.5). Table 8.15 shows the initial fluids in place. Figures 8.31–8.35 give plots of the oil production rate, water cut, GOR, bottom hole pressure, and *pressure drawdown* (p(1,7)-bhp, a decline in well pressure with time due to production), all versus time for the CVFA and FD methods, where (1,7) is the first radial gridblock and the 7th layer. There are slight differences between the two methods for the quantities shown in these figures.

To check the stability of the CVFA methods for stronger coning, we design three cases A, B, and C by changing the ratio of the vertical permeability to the horizontal permeability k_v/k_h from 0.1 to 0.5 for case A, changing $Q_{o,max}$ (the maximum oil production rate) from 1,000 STB/D to 2,000 STB/D for case B, and changing $Q_{o,max}$ from 1,000 STB/D to 3,000 STB/D for case C based on the original data. Figures 8.36–8.40 are the oil production rate, water cut, GOR, bottom hole pressure, and pressure drawdown at block (1,7) all versus time for these cases. We can see that water and gas coning becomes more serious if k_v/k_h changes to 0.5; the transients become significant if the maximum oil production rate is doubled or tripled. However, no oscillations occur.

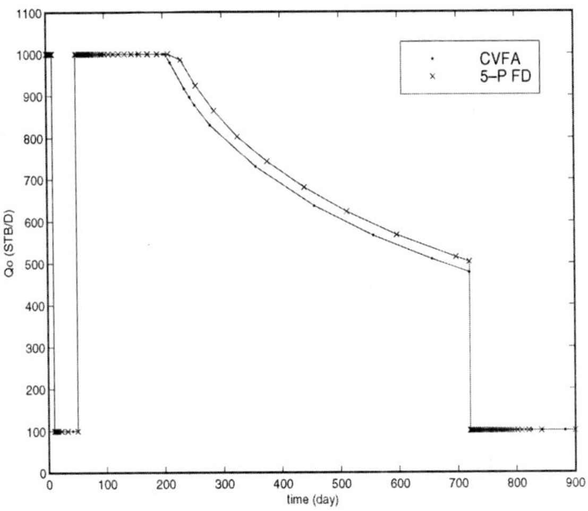

Figure 8.31. *Oil production rate versus time.*

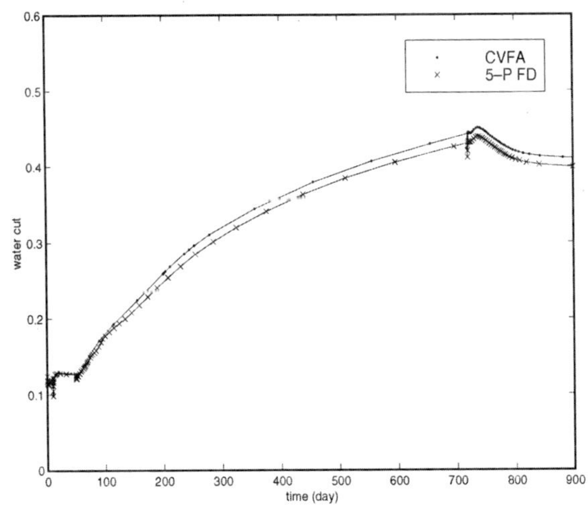

Figure 8.32. *Water cut versus time.*

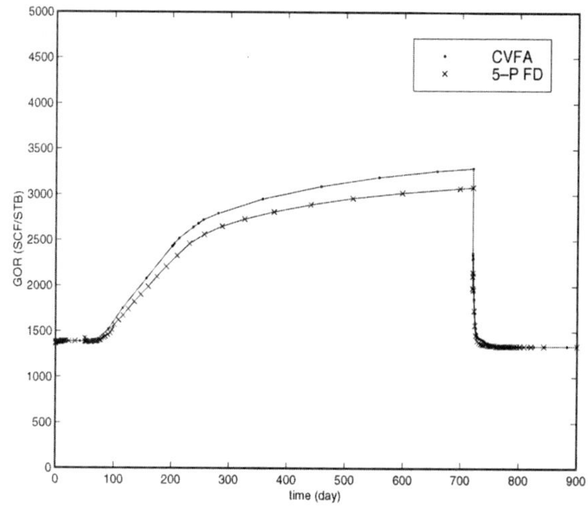

Figure 8.33. *GOR versus time.*

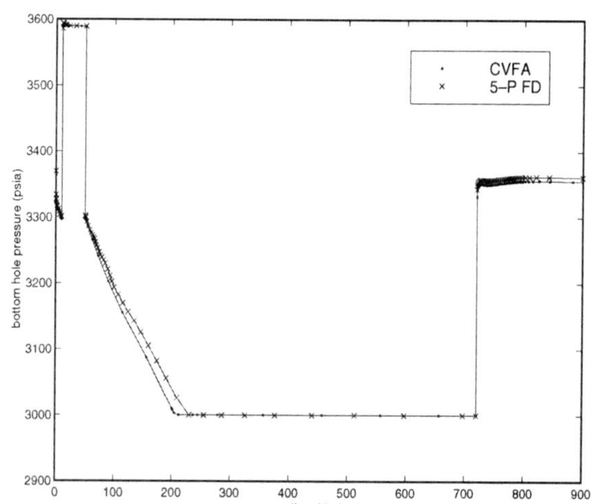

Figure 8.34. *Bottom hole pressure versus time.*

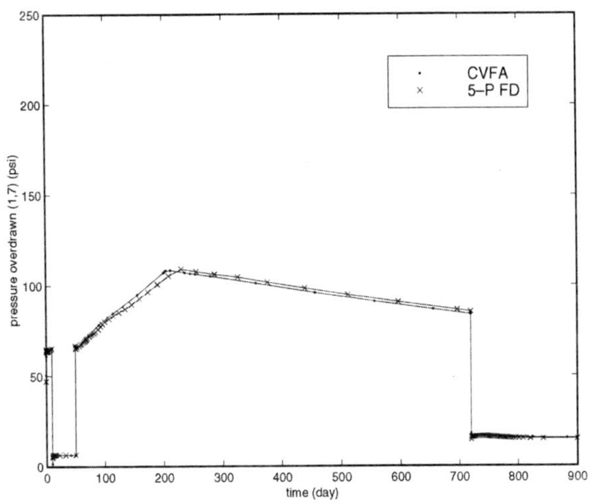

Figure 8.35. *Pressure drawdown* (1,7) *versus time.*

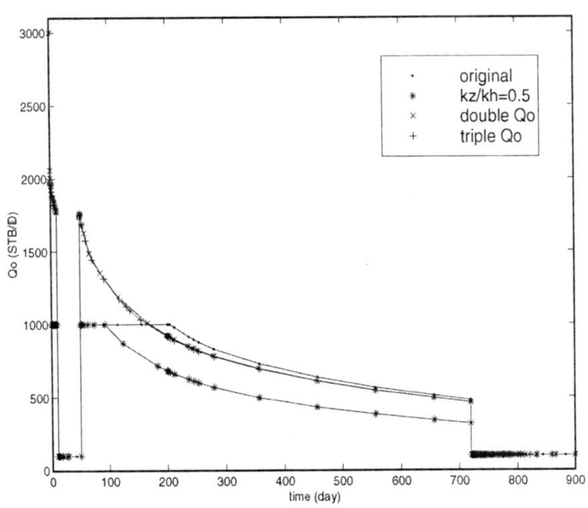

Figure 8.36. *Oil production rate for different parameters.*

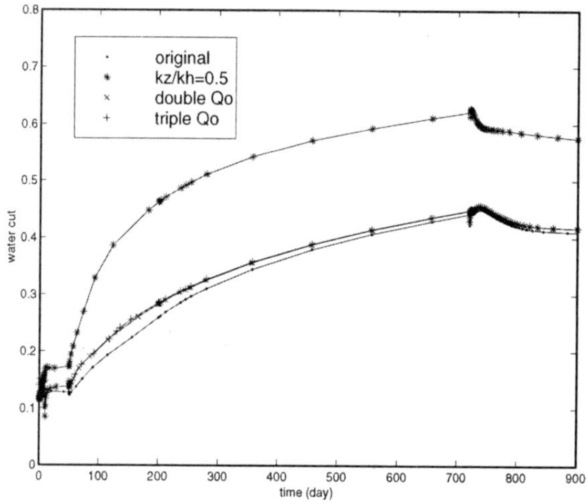

Figure 8.37. *Water cut for different parameters.*

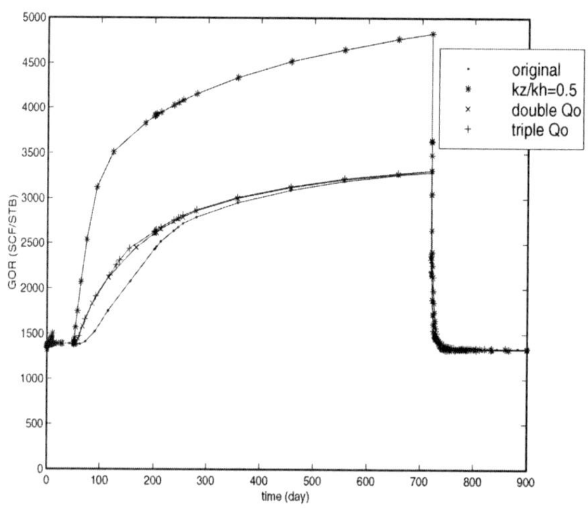

Figure 8.38. *GOR for different parameters.*

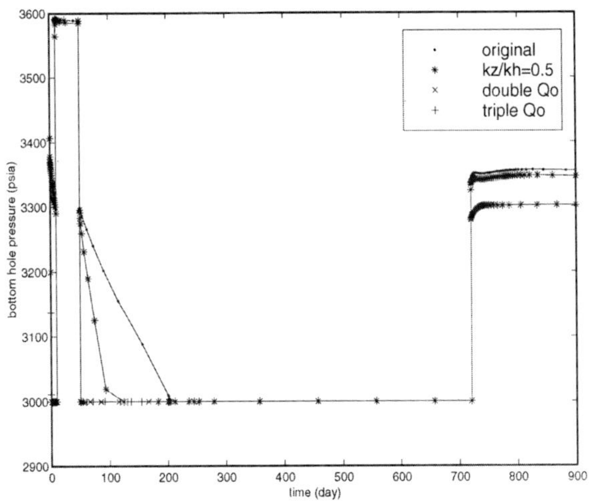

Figure 8.39. *Bottom hole pressure for different parameters.*

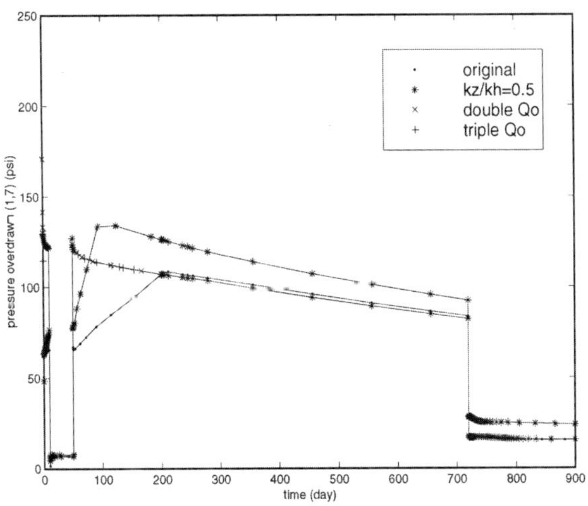

Figure 8.40. *Pressure overdrawn* (1,7) *for different parameters.*

8.5 Bibliographical Remarks

The numerical results reported in Sections 8.3 and 8.4 are taken from Li et al. (2003A; 2004A; 2004B). For more numerical results, the reader should refer to these papers. For more information about the data used in the second and ninth SPE CSPs, see Weinstein et al. (1986) and Killough (1995), respectively.

Exercises

8.1. Derive equation (8.10). (Hint: Substitute (8.7)–(8.9) into (8.1)–(8.3) and neglect the variation of ρ_α with respect to space.)

8.2. The phase, weighted, and global pressure formulations were developed for two-phase flow in Chapter 7 and can be extended to the black oil model under consideration. This and the next five exercises are devoted to the development of these formulations. If necessary, the reader can refer to Chen (2000) for their derivation. Recall the mass conservation equations on standard volumes

$$\begin{aligned}
&\phi\frac{\partial}{\partial t}\left(\frac{S_w}{B_w}\right) = -\nabla\cdot\left(\frac{1}{B_w}\mathbf{u}_w\right) + \tilde{q}_W,\\
&\phi\frac{\partial}{\partial t}\left(\frac{S_o}{B_o}\right) = -\nabla\cdot\left(\frac{1}{B_o}\mathbf{u}_o\right) + \tilde{q}_O,\\
&\phi\frac{\partial}{\partial t}\left(\frac{S_g}{B_g}+\frac{R_{so}S_o}{B_o}\right) = -\nabla\cdot\left(\frac{1}{B_g}\mathbf{u}_g+\frac{R_{so}}{B_o}\mathbf{u}_o\right) + \tilde{q}_G,
\end{aligned} \tag{8.117}$$

Darcy's law

$$\mathbf{u}_\alpha = -\frac{k_{r\alpha}}{\mu_\alpha}\mathbf{k}\left(\nabla p_\alpha - \rho_\alpha\wp\nabla z\right), \qquad \alpha = w,\, o,\, g, \tag{8.118}$$

and the saturation and pressure constraints

$$\begin{aligned}
&S_w + S_o + S_g = 1,\\
&p_{c\alpha} = p_\alpha - p_o, \qquad \alpha = w,\, o,\, g,
\end{aligned} \tag{8.119}$$

where $\tilde{q}_\beta = q_\beta/\rho_{\beta s}$, $\beta = W, O, G$. Recall the phase mobility functions

$$\lambda_\alpha = k_{r\alpha}/\mu_\alpha, \qquad \alpha = w, o, g,$$

the total mobility

$$\lambda = \sum_{\alpha=w}^{g}\lambda_\alpha,$$

and the fractional flow functions

$$f_\alpha = \lambda_\alpha/\lambda, \qquad \alpha = w, o, g.$$

We use the oil phase pressure as the pressure variable in this exercise

$$p = p_o, \tag{8.120}$$

and the total velocity

$$\mathbf{u} = \sum_{\beta=w}^{g} \mathbf{u}_\beta. \tag{8.121}$$

Show that equations (8.117)–(8.119) can be written as

$$\mathbf{u} = -\mathbf{k}\lambda\left(\nabla p - G_\lambda + \sum_\beta f_\beta \nabla p_{c\beta}\right),$$

$$\nabla\cdot\mathbf{u} = \sum_\beta B_\beta\left(\tilde{q}_\beta - \phi S_\beta \frac{\partial}{\partial t}\left(\frac{1}{B_\beta}\right) - \mathbf{u}_\beta\cdot\nabla\left(\frac{1}{B_\beta}\right)\right)$$
$$-B_g\left(R_{so}\tilde{q}_o + \frac{\phi S_o}{B_o}\frac{\partial R_{so}}{\partial t} + \frac{1}{B_o}\mathbf{u}_o\cdot\nabla R_{so}\right),$$

and

$$\phi\frac{\partial S_\alpha}{\partial t} + \nabla\cdot\mathbf{u}_\alpha = B_\alpha\left(\tilde{q}_\alpha - \phi S_\alpha\frac{\partial}{\partial t}\left(\frac{1}{B_\alpha}\right) - \mathbf{u}_\alpha\cdot\nabla\left(\frac{1}{B_\alpha}\right)\right),$$
$$\mathbf{u}_\alpha = f_\alpha\mathbf{u} + \mathbf{k}f_\alpha\sum_\beta \lambda_\beta\big(\nabla(p_{c\beta} - p_{c\alpha}) - (\rho_\beta - \rho_\alpha)\wp\nabla z\big)$$

for $\alpha = w, o$, where

$$\sum_\beta = \sum_{\beta=w}^{g}, \qquad G_\lambda = \wp\nabla z\sum_\beta f_\beta\rho_\beta.$$

8.3. Note that in the above phase formulation the quadratic terms in the velocities $\mathbf{u}_\alpha$ appear. To remove them, we modify the definition of the total velocity. Toward that end, set

$$\lambda_w = \frac{k_{rw}}{B_w\mu_w}, \quad \lambda_o = \frac{1+R_{so}}{B_o\mu_o}k_{ro}, \quad \lambda_g = \frac{k_{rg}}{B_g\mu_g}, \quad \lambda = \sum_\beta \lambda_\beta,$$

and

$$f_\alpha = \lambda_\alpha/\lambda, \qquad \alpha = w, o, g.$$

The pressure variable is defined as in (8.120), but the total velocity is modified to

$$\mathbf{u} = \sum_\beta \frac{1}{B_\beta}\mathbf{u}_\beta + \frac{R_{so}}{B_o}\mathbf{u}_o. \tag{8.122}$$

Prove that the pressure and saturation equations now become

$$\mathbf{u} = -\mathbf{k}\lambda\left(\nabla p - G_\lambda + \sum_\beta f_\beta\nabla p_{c\beta}\right),$$

$$\phi\frac{\partial}{\partial t}\left(\sum_\beta \frac{S_\beta}{B_\beta} + \frac{S_o R_{so}}{B_o}\right) + \nabla\cdot\mathbf{u} = \sum_\beta \tilde{q}_\beta,$$

and

$$\phi \frac{\partial}{\partial t}\left(\frac{S_\alpha}{B_\alpha}\right) + \nabla \cdot \left(\frac{1}{B_\alpha}\mathbf{u}_\alpha\right) = \tilde{q}_\alpha, \qquad \alpha = w, o,$$

where

$$\mathbf{u}_o = \frac{B_o}{1+R_{so}}\left\{ f_o \mathbf{u} + \mathbf{k} f_o \sum_\beta \lambda_\beta \big(\nabla p_{c\beta} - (\rho_\beta - \rho_o)\wp \nabla z\big)\right\},$$
$$\mathbf{u}_w = B_w \left\{ f_w \mathbf{u} + \mathbf{k} f_w \sum_\beta \lambda_\beta \big(\nabla (p_{c\beta} - p_{cw}) - (\rho_\beta - \rho_w)\wp \nabla z\big)\right\}.$$

8.4. We now define a smoother pressure than the phase pressure, i.e., the weighted fluid pressure

$$p = \sum_\alpha S_\alpha p_\alpha.$$

The phase pressures are

$$p_\alpha = p + p_{c\alpha} - \sum_\beta S_\beta p_{c\beta}, \qquad \alpha = w,\ o,\ g.$$

With λ_α, λ, f_α, and the modified total velocity defined as in Exercise 8.3, show that the pressure equation is

$$\mathbf{u} = -\mathbf{k}\lambda\left(\nabla p - G_\lambda + \sum_\beta f_\beta \nabla p_{c\beta} - \sum_\beta \nabla (S_\beta p_{c\beta})\right),$$
$$\phi \frac{\partial}{\partial t}\left(\sum_\beta \frac{S_\beta}{B_\beta} + \frac{S_o R_{so}}{B_o}\right) + \nabla \cdot \mathbf{u} = \sum_\beta \tilde{q}_\beta,$$

and the saturation equations are the same as in Exercise 8.3.

8.5. To define a *global pressure*, we assume that the fractional flow functions f_α depend solely on the saturations S_w and S_g (for pressure-dependent functions f_α, see the next exercise) and there exists a function $(S_w, S_g) \longmapsto p_c(S_w, S_g)$ such that

$$\nabla p_c = f_w \nabla p_{cw} + f_g \nabla p_{cg}.$$

This is true if and only if the following equations are satisfied (cf. Exercise 2.7):

$$\frac{\partial p_c}{\partial S_w} = f_w \frac{\partial p_{cw}}{\partial S_w} + f_g \frac{\partial p_{cg}}{\partial S_w}, \qquad \frac{\partial p_c}{\partial S_g} = f_w \frac{\partial p_{cw}}{\partial S_g} + f_g \frac{\partial p_{cg}}{\partial S_g}. \tag{8.123}$$

A necessary and sufficient condition for existence of a function p_c satisfying (8.123) is (cf. Exercise 2.7)

$$\frac{\partial f_w}{\partial S_g}\frac{\partial p_{cw}}{\partial S_w} + \frac{\partial f_g}{\partial S_g}\frac{\partial p_{cg}}{\partial S_w} = \frac{\partial f_w}{\partial S_w}\frac{\partial p_{cw}}{\partial S_g} + \frac{\partial f_g}{\partial S_w}\frac{\partial p_{cg}}{\partial S_g}. \tag{8.124}$$

When the condition (8.124) is satisfied,

$$p_c(S_w, S_g) = \int_1^{S_w} \left\{ f_w(\xi, 0)\frac{\partial p_{cw}}{\partial S_w}(\xi, 0) + f_g(\xi, 0)\frac{\partial p_{cg}}{\partial S_w}(\xi, 0) \right\} d\xi$$
$$+ \int_0^{S_g} \left\{ f_w(S_w, \xi)\frac{\partial p_{cw}}{\partial S_g}(S_w, \xi) + f_g(S_w, \xi)\frac{\partial p_{cg}}{\partial S_g}(S_w, \xi) \right\} d\xi,$$

where we assume that the integrals are well defined. Define the global pressure and the total velocity

$$p = p_o + p_c, \quad \mathbf{u} = \sum_\beta \mathbf{u}_\beta.$$

Show that equations (8.117)–(8.119) can be written as

$$\mathbf{u} = -\mathbf{k}\lambda(\nabla p - G_\lambda),$$
$$\nabla \cdot \mathbf{u} = \sum_\beta B_\beta \left(\tilde{q}_\beta - \phi S_\beta \frac{\partial}{\partial t}\left(\frac{1}{B_\beta}\right) - \mathbf{u}_\beta \cdot \nabla \left(\frac{1}{B_\beta}\right) \right)$$
$$- B_g \left(R_{so}\tilde{q}_o + \frac{\phi S_o}{B_o}\frac{\partial R_{so}}{\partial t} + \frac{1}{B_o}\mathbf{u}_o \cdot \nabla R_{so} \right),$$

and

$$\phi \frac{\partial S_\alpha}{\partial t} + \nabla \cdot \mathbf{u}_\alpha = B_\alpha \left(\tilde{q}_\alpha - \phi S_\alpha \frac{\partial}{\partial t}\left(\frac{1}{B_\alpha}\right) - \mathbf{u}_\alpha \cdot \nabla \left(\frac{1}{B_\alpha}\right) \right),$$
$$\mathbf{u}_\alpha = f_\alpha \mathbf{u} + \mathbf{k}\lambda_\alpha \big(\nabla(p_c - p_{c\alpha}) - \delta_\alpha\big)$$

for $\alpha = w, o$, where

$$\delta_\alpha = \big(f_\beta(\rho_\beta - \rho_\alpha) + f_\gamma(\rho_\gamma - \rho_\alpha)\big)\wp\nabla z,$$
$$\alpha, \beta, \gamma = w, o, g, \ \alpha \neq \beta, \beta \neq \gamma, \gamma \neq \alpha.$$

8.6. To combine the modified total velocity and global pressure concepts, we assume that the solubility factor R_{so}, the formation factors B_α, and the viscosity functions μ_α depend only on their respective phase pressure. Furthermore, to derive the global pressure p, we assume that these functions essentially depend on p. The second assumption ignores the error caused by calculating them for the α-phase at p instead of p_α. The third assumption is that there exists a function $(S_w, S_g, p) \longmapsto p_c(S_w, S_g, p)$ satisfying

$$\nabla p_c = f_w \nabla p_{cw} + f_g \nabla p_{cg} + \frac{\partial p_c}{\partial p}\nabla p,$$

where f_α is defined as in Exercise 8.3. A necessary and sufficient condition for existence of a function p_c satisfying such a condition is (8.124), where p is treated as a parameter. Under this condition,

$$p_c(S_w, S_g, p) = \int_1^{S_w} \left\{ f_w(\xi, 0, p)\frac{\partial p_{cw}}{\partial S_w}(\xi, 0) + f_g(\xi, 0, p)\frac{\partial p_{cg}}{\partial S_w}(\xi, 0) \right\} d\xi$$
$$+ \int_0^{S_g} \left\{ f_w(S_w, \xi, p)\frac{\partial p_{cw}}{\partial S_g}(S_w, \xi) + f_g(S_w, \xi, p)\frac{\partial p_{cg}}{\partial S_g}(S_w, \xi) \right\} d\xi.$$

With this definition, $p = p_o + p_c$, and λ_α, λ, f_α, and the modified total velocity defined as in Exercise 8.3 prove that equations (8.117)–(8.119) can be written as

$$\mathbf{u} = -\mathbf{k}\lambda(\omega\nabla p - G_\lambda),$$

$$\phi\frac{\partial}{\partial t}\left(\sum_\beta \frac{S_\beta}{B_\beta} + \frac{S_o R_{so}}{B_o}\right) + \nabla\cdot\mathbf{u} = \sum_\beta \tilde{q}_\beta,$$

and

$$\phi\frac{\partial}{\partial t}\left(\frac{S_\alpha}{B_\alpha}\right) + \nabla\cdot\left(\frac{1}{B_\alpha}\mathbf{u}_\alpha\right) = \tilde{q}_\alpha, \qquad \alpha = w, o,$$

where

$$\mathbf{u}_o = \frac{B_o}{1 + R_{so}}\left\{\omega^{-1} f_o\mathbf{u} + \mathbf{k}\lambda_o\big(\nabla p_c - \delta_o\big) - \omega^{-1}\frac{\partial p_c}{\partial p}G_\lambda\right\},$$

$$\mathbf{u}_w = B_w\left\{\omega^{-1} f_w\mathbf{u} + \mathbf{k}\lambda_w\big(\nabla(p_c - p_{cw}) - \delta_w\big) - \omega^{-1}\frac{\partial p_c}{\partial p}G_\lambda\right\},$$

and

$$\omega(s_w, s_g, p) = 1 - \frac{\partial p_c}{\partial p}.$$

8.7. The global pressure formulation in Exercises 8.5 and 8.6 requires the *total differential condition* (8.124) on the shape of three-phase relative permeability and capillary pressure functions. We now introduce a *pseudoglobal pressure* formulation, which does not require such a condition. As an example, consider this formulation with the total velocity defined in Exercise 8.2. Assume that the capillary pressures satisfy the condition

$$p_{cw} = p_{cw}(S_w), \quad p_{cg} = p_{cg}(S_g). \tag{8.125}$$

Then introduce the mean values

$$\widehat{f_w}(S_w) = \frac{1}{1 - S_w}\int_0^{1-S_w} f_w(S_w, \zeta)\, d\zeta,$$

$$\widehat{f_g}(S_g) = \frac{1}{1 - S_g}\int_0^{1-S_g} f_g(\zeta, S_g)\, d\zeta,$$

and the pseudoglobal pressure

$$p = p_o + \int_{S_{wc}}^{S_w} \widehat{f_w}(\zeta)\frac{dp_{cw}(\zeta)}{dS_w}\, d\zeta + \int_{S_{gc}}^{S_g} \widehat{f_g}(\zeta)\frac{dp_{cg}(\zeta)}{dS_g}\, d\zeta,$$

where S_{wc} and S_{gc} are such that $p_{cw}(S_{wc}) = 0$ and $p_{cg}(S_{gc}) = 0$. With this pressure and the total velocity defined as in (8.121), show that the pressure equation is

$$\mathbf{u} = -\mathbf{k}\lambda\left\{\nabla p - G_\lambda + \sum_\alpha \left(f_\alpha - \widehat{f}_\alpha\right)\frac{dp_{c\alpha}}{dS_\alpha}\nabla S_\alpha\right\},$$

$$\nabla\cdot\mathbf{u} = \sum_\beta B_\beta\left(\tilde{q}_\beta - \phi S_\beta\frac{\partial}{\partial t}\left(\frac{1}{B_\beta}\right) - \mathbf{u}_\beta\cdot\nabla\left(\frac{1}{B_\beta}\right)\right)$$
$$- B_g\left(R_{so}\tilde{q}_o + \frac{\phi S_o}{B_o}\frac{\partial R_{so}}{\partial t} + \frac{1}{B_o}\mathbf{u}_o\cdot\nabla R_{so}\right),$$

and the saturation equations are as in Exercise 8.2.

8.8. Derive equation (8.60) by substituting (8.38), (8.46), (8.50), (8.54), and (8.58) into the first equation of (8.37) and ignoring the higher-order terms in δp and δS_w.

8.9. Show equation (8.61) by substituting (8.39), (8.47), (8.51), (8.55), and (8.58) into the second equation of (8.37) and ignoring the higher-order terms in δp, δS_w, and δS_o.

8.10. Prove equation (8.62) by substituting (8.41), (8.49), (8.53), (8.57), and (8.58) into the third equation of (8.37) and ignoring the higher-order terms in δp, δS_w, and δS_o.

8.11. Derive equation (8.63) by substituting (8.40), (8.48), (8.52), (8.56), and (8.59) into the second equation of (8.37) and ignoring the higher-order terms in δp, δS_w, and δp_b.

8.12. Show equation (8.64) by substituting (8.42), (8.48), (8.52), (8.56), and (8.59) into the third equation of (8.37) and ignoring the higher-order terms in δp, δS_w, and δp_b.

8.13. Prove equation (8.76) by substituting (8.38), (8.71)–(8.73), and (8.58) into the first equation of (8.37) and ignoring the higher-order terms in δp.

8.14. Derive equation (8.77) by substituting (8.39), (8.71), (8.72), (8.74), and (8.58) into the second equation of (8.37) and ignoring the higher-order terms in δp.

8.15. Show equation (8.78) by substituting (8.41), (8.71), (8.72), (8.75), and (8.58) into the third equation of (8.37) and ignoring the higher-order terms in δp.

8.16. Prove that equations (8.76)–(8.78) can be obtained from equations (8.60)–(8.62) by setting $\delta S_w = 0$ and $\delta S_o = 0$ in the right-hand sides of the latter equations.

8.17. Derive equation (8.97) by substituting (8.38), (8.95), and (8.96) into the first equation of (8.37) and ignoring the higher-order terms in δp.

8.18. Show equation (8.98) by substituting (8.39), (8.95), and (8.96) into the second equation of (8.37) and ignoring the higher-order terms in δp.

8.19. Prove equation (8.99) by substituting (8.41), (8.95), and (8.96) into the third equation of (8.37) and ignoring the higher-order terms in δp.

8.20. Derive equation (8.103) by substituting (8.40), (8.42), (8.95), and (8.96) into (8.37) and ignoring the higher-order terms in δp.

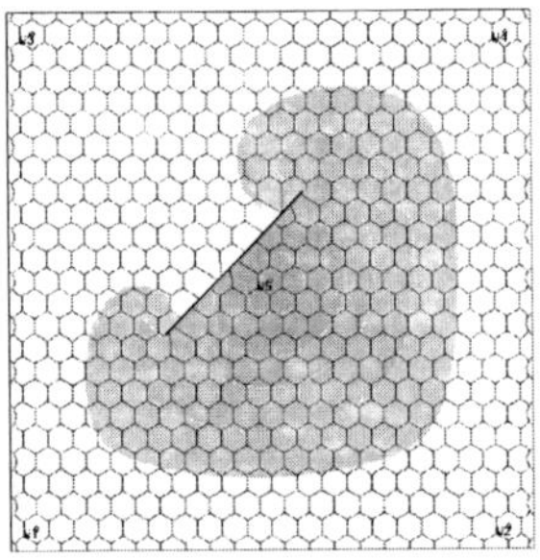

Chapter 9
The Compositional Model

Recall that to recover some of the hydrocarbons after water flooding, several enhanced recovery techniques are used. These involve complex chemical and thermal effects and are termed *tertiary recovery* or *enhanced recovery*. There are many different variations of enhanced recovery techniques. One of the main objectives of these techniques is to achieve miscibility and thus eliminate residual oil saturation. Miscibility can be achieved by increasing temperature (e.g., in situ combustion) or by injecting other chemical species such as CO_2. A typical flow in enhanced recovery is the *compositional flow*, where only the number of chemical species is a priori given, and the number of phases and the composition of each phase in terms of the given species depend on the thermodynamic conditions and the overall concentration of each species.

The governing equations for the compositional model are stated in Section 9.1. The Peng–Robinson equation of state is also briefly reviewed there. The iterative IMPES solution technique developed for the black oil model in Chapter 8 is further studied for the compositional model in Section 9.2. In Section 9.3, the solution of equilibrium relations that describe the mass distribution of chemical species among the fluid phases is discussed in detail. Numerical results based on the third CSP organized by the SPE are reported in Section 9.4. Finally, bibliographical information is given in Section 9.5.

9.1 Basic Differential Equations

9.1.1 The basic equations

The basic equations for the compositional model in a porous medium Ω were described in Section 2.8. For completeness, we review these equations. We describe a compositional model under the assumptions that the flow process is isothermal (i.e., constant temperature), the components form at most three phases (e.g., water, oil, and gas), there is no mass interchange between the water phase and the hydrocarbon phases (i.e., the oil and gas phases), and diffusive effects are neglected.

Let ϕ and $\mathbf{k}$ denote the porosity and permeability of the porous medium $\Omega \subset \mathbb{R}^3$, and let S_α, μ_α, p_α, $\mathbf{u}_\alpha$, and $k_{r\alpha}$ be the saturation, viscosity, pressure, volumetric velocity, and relative permeability, respectively, of the α-phase, $\alpha = w, o, g$. Also, let ξ_{io} and ξ_{ig} represent the molar densities of component i in the oil (liquid) and gas (vapor) phases, respectively, $i = 1, 2, \ldots, N_c$, where N_c is the number of components. The molar density of phase α is given by

$$\xi_\alpha = \sum_{i=1}^{N_c} \xi_{i\alpha}, \qquad \alpha = o,\ g. \tag{9.1}$$

The mole fraction of component i in phase α is then

$$x_{i\alpha} = \xi_{i\alpha}/\xi_\alpha, \qquad i = 1, 2, \ldots, N_c,\ \alpha = o, g. \tag{9.2}$$

The total mass is conserved for each component:

$$\begin{aligned}
&\frac{\partial(\phi\xi_w S_w)}{\partial t} + \nabla \cdot (\xi_w \mathbf{u}_w) = q_w, \\
&\frac{\partial(\phi[x_{io}\xi_o S_o + x_{ig}\xi_g S_g])}{\partial t} + \nabla \cdot (x_{io}\xi_o \mathbf{u}_o + x_{ig}\xi_g \mathbf{u}_g) \\
&\qquad = x_{io}q_o + x_{ig}q_g, \qquad i = 1, 2, \ldots, N_c,
\end{aligned} \tag{9.3}$$

where ξ_w is the molar density of water (that is the water mass density ρ_w for the present model) and q_α stands for the flow rate of phase α at wells. In (9.3), the volumetric velocity $\mathbf{u}_\alpha$ is given by Darcy's law:

$$\mathbf{u}_\alpha = -\frac{k_{r\alpha}}{\mu_\alpha}\mathbf{k}\left(\nabla p_\alpha - \rho_\alpha \wp \nabla z\right), \qquad \alpha = w,\ o,\ g, \tag{9.4}$$

where ρ_α is the mass density of the α-phase, $\wp$ is the magnitude of the gravitational acceleration, and z is the depth. The mass density ρ_α is related to the molar density ξ_w by (2.93). The fluid viscosity $\mu_\alpha(p_\alpha, T, x_{1\alpha}, x_{2\alpha}, \ldots, x_{N_c\alpha})$ can be calculated from pressure, temperature, and compositions (Lohrenz et al., 1964).

In addition to the differential equations (9.3) and (9.4), there are also algebraic constraints. The mole fraction balance implies that

$$\sum_{i=1}^{N_c} x_{io} = 1, \quad \sum_{i=1}^{N_c} x_{ig} = 1. \tag{9.5}$$

In the transport process, the saturation constraint reads

$$S_w + S_o + S_g = 1. \tag{9.6}$$

Finally, the phase pressures are related by capillary pressures:

$$p_{cow} = p_o - p_w, \quad p_{cgo} = p_g - p_o. \tag{9.7}$$

Mass interchange between phases is characterized by the variation of mass distribution of each component in the oil and gas phases. As usual, these two phases are assumed to be in the phase equilibrium state at every moment. This is physically reasonable since mass interchange between phases occurs much faster than the flow of porous media fluids. Consequently, the distribution of each hydrocarbon component into the two phases is subject to the condition of *stable thermodynamic equilibrium*, which is given by minimizing the Gibbs free energy of the compositional system (Bear, 1972; Chen et al., 2000):

$$f_{io}(p_o, x_{1o}, x_{2o}, \ldots, x_{N_c o}) = f_{ig}(p_g, x_{1g}, x_{2g}, \ldots, x_{N_c g}), \tag{9.8}$$

where f_{io} and f_{ig} are the fugacity functions of the ith component in the oil and gas phases, respectively, $i = 1, 2, \ldots, N_c$.

Equations (9.3)–(9.8) provide $2N_c+9$ independent relations, differential or algebraic, for the $2N_c+9$ dependent variables: $x_{io}, x_{ig}, \mathbf{u}_\alpha, p_\alpha$, and $S_\alpha, \alpha = w, o, g, i = 1, 2, \ldots, N_c$. With appropriate boundary and initial conditions, this is a closed differential system for these unknowns.

9.1.2 Equations of state

The rock properties reviewed in Section 8.1.2 for the black oil model also apply to the compositional model. In particular, for convenience of programming, we define

$$p_{cw} = p_w - p_o, \qquad p_{cg} = p_g - p_o; \tag{9.9}$$

i.e., $p_{cw} = -p_{cow}$ and $p_{cg} = p_{cgo}$. Moreover, for notational convenience, let $p_{co} = 0$.

Several equations of state (EOSs) were introduced in Section 3.2.5 for the definition of the fugacity functions f_{io} and f_{ig}, including the Redlich–Kwong, Redlich–Kwong–Soave, and Peng–Robinson EOSs. Here we briefly review the most frequently used Peng–Robinson EOS (Peng and Robinson, 1976; Coats, 1980).

The mixing principle for the Peng–Robinson equation of state is

$$a_\alpha = \sum_{i=1}^{N_c}\sum_{j=1}^{N_c} x_{i\alpha}x_{j\alpha}(1-\kappa_{ij})\sqrt{a_i a_j},$$
$$b_\alpha = \sum_{i=1}^{N_c} x_{i\alpha}b_i, \qquad \alpha = o, g,$$

where κ_{ij} is a *binary interaction* parameter between components i and j, and a_i and b_i are empirical factors for the pure component i. The interaction parameters account for molecular interactions between two unlike molecules. By definition, κ_{ij} is zero when i and j represent the same component, small when i and j represent components that do not differ much (e.g., when components i and j are both alkanes), and large when i and j represent components that are substantially different. Ideally, κ_{ij} depends on pressure and temperature and only on the identities of components i and j (Zudkevitch and Joffe, 1970; Whitson, 1982).

The factors a_i and b_i can be computed from

$$a_i = \Omega_{ia}\alpha_i \frac{R^2 T_{ic}^2}{p_{ic}}, \qquad b_i = \Omega_{ib}\frac{R\, T_{ic}}{p_{ic}},$$

where R is the universal gas constant, T is the temperature, T_{ic} and p_{ic} are the critical temperature and pressure, the EOS parameters Ω_{ia} and Ω_{ib} are given by

$$\begin{aligned}
&\Omega_{ia} = 0.45724, \quad \Omega_{ib} = 0.077796,\\
&\alpha_i = \left(1 - \lambda_i\left[1 - \sqrt{T/T_{ic}}\right]\right)^2,\\
&\lambda_i = 0.37464 + 1.5423\omega_i - 0.26992\omega_i^2,
\end{aligned}$$

and ω_i is the *acentric factor* for components i. The acentric factors roughly express the deviation of the shape of a molecule from a sphere (Reid et al., 1977). Define

$$A_\alpha = \frac{a_\alpha p_\alpha}{R^2 T^2}, \qquad B_\alpha = \frac{b_\alpha p_\alpha}{R\, T}, \qquad \alpha = o, g, \tag{9.10}$$

where the pressure p_α is given by the Peng–Robinson two-parameter EOS

$$p_\alpha = \frac{RT}{V_\alpha - b_\alpha} - \frac{a_\alpha(T)}{V_\alpha(V_\alpha + b_\alpha) + b_\alpha(V_\alpha - b_\alpha)} \tag{9.11}$$

with V_α being the molar volume of phase α. Introduce the compressibility factor

$$Z_\alpha = \frac{p_\alpha V_\alpha}{R\, T}, \qquad \alpha = o, g. \tag{9.12}$$

Equation (9.11) can be expressed as a cubic equation in Z_α:

$$\begin{aligned}
Z_\alpha^3 - (1 - B_\alpha)Z_\alpha^2 + (A_\alpha - 2B_\alpha - 3B_\alpha^2)Z_\alpha&\\
- (A_\alpha B_\alpha - B_\alpha^2 - B_\alpha^3) = 0.&
\end{aligned} \tag{9.13}$$

The correct choice of the root of (9.13) will be discussed in Section 9.3.4. Now, for $i = 1, 2, \ldots, N_c$ and $\alpha = o, g$, the fugacity coefficient $\varphi_{i\alpha}$ of component i in the mixture can be obtained from

$$\begin{aligned}
\ln \varphi_{i\alpha} = {}& \frac{b_i}{b_\alpha}(Z_\alpha - 1) - \ln(Z_\alpha - B_\alpha)\\
& - \frac{A_\alpha}{2\sqrt{2}B_\alpha}\left(\frac{2}{a_\alpha}\sum_{j=1}^{N_c} x_{j\alpha}(1 - \kappa_{ij})\sqrt{a_i a_j} - \frac{b_i}{b_\alpha}\right)\\
& \cdot \ln\left(\frac{Z_\alpha + (1 + \sqrt{2})B_\alpha}{Z_\alpha - (1 - \sqrt{2})B_\alpha}\right).
\end{aligned} \tag{9.14}$$

Finally, the fugacity of component i is

$$f_{i\alpha} = p_\alpha x_{i\alpha} \varphi_{i\alpha}, \qquad i = 1, 2, \ldots, N_c, \ \alpha = o, g. \tag{9.15}$$

The mass distribution of each hydrocarbon component into the fluid (oil) and vapor (gas) phases is given by the thermodynamic equilibrium relation (9.8).

9.2 Solution Techniques

The choice of a solution technique is crucial for a coupled system of partial differential equations. In the preceding chapter, we discussed several solution techniques that are currently used in the numerical solution of the black oil model. These techniques include the iterative IMPES, sequential, SS, and adaptive implicit techniques. They can be also employed for the numerical simulation of the compositional model. However, a typical compositional simulator includes about a dozen chemical components; the SS would be a very expensive technique for this type of flow, even with today's computing power. The iterative IMPES and sequential techniques are widely used and are thus studied here. As an example, we develop iterative IMPES for the compositional model. An extension from this technique to the sequential technique can be carried out as in the preceding chapter for the black oil model.

9.2.1 Choice of primary variables

Equations (9.3)–(9.8) form a strongly coupled system of time-dependent, nonlinear differential equations and algebraic constraints. While there are $2N_c + 9$ equations for the same number of dependent variables, this system can be written in terms of $2N_c + 2$ primary variables, and other variables can be expressed as functions of them. These primary variables must be carefully chosen so that the main physical properties inherent in the governing equations and constraints are preserved, the nonlinearity and coupling between the equations is weakened, and efficient numerical methods for the solution of the resulting system can be devised.

To simplify the expressions in (9.3), we introduce the potentials

$$\Phi_\alpha = p_\alpha - \rho_\alpha \wp z, \qquad \alpha = w, o, g. \tag{9.16}$$

Also, we use the *total mass variable* F of the hydrocarbon system (Nolen, 1973; Young and Stephenson, 1983)

$$F = \xi_o S_o + \xi_g S_g, \tag{9.17}$$

and the *mass fractions* of oil and gas in this system,

$$L = \frac{\xi_o S_o}{F}, \qquad V = \frac{\xi_g S_g}{F}. \tag{9.18}$$

Note that

$$L + V = 1.$$

Next, instead of exploiting the individual mole fractions, we use the *total mole fraction* of the components in the hydrocarbon system

$$z_i = Lx_{io} + (1 - L)x_{ig}, \qquad i = 1, 2, \ldots, N_c. \tag{9.19}$$

Then we see, using (9.5), (9.17), and (9.18), that

$$\sum_{i=1}^{N_c} z_i = 1 \tag{9.20}$$

and

$$x_{io}\xi_o S_o + x_{ig}\xi_g S_g = F z_i, \qquad i = 1, 2, \ldots, N_c. \tag{9.21}$$

Consequently, applying (9.4) and (9.16), the second equation in (9.3) becomes (cf. Exercise 9.1)

$$\begin{aligned} &\frac{\partial(\phi F z_i)}{\partial t} - \nabla \cdot \left(\mathbf{k}\left[\frac{x_{io}\xi_o k_{ro}}{\mu_o}\nabla\Phi_o + \frac{x_{ig}\xi_g k_{rg}}{\mu_g}\nabla\Phi_g\right]\right) \\ &\qquad = x_{io}q_o + x_{ig}q_g, \qquad i = 1, 2, \ldots, N_c. \end{aligned} \tag{9.22}$$

Adding equations (9.22) over i and exploiting (9.5) and (9.20) gives

$$\frac{\partial(\phi F)}{\partial t} - \nabla \cdot \left(\mathbf{k}\left[\frac{\xi_o k_{ro}}{\mu_o}\nabla\Phi_o + \frac{\xi_g k_{rg}}{\mu_g}\nabla\Phi_g\right]\right) = q_o + q_g. \tag{9.23}$$

Equation (9.22) is the individual flow equation for the ith component (say, $i = 1, 2, \ldots, N_c - 1$) and (9.23) is the global hydrocarbon flow equation.

To simplify the differential equations further, we define the transmissibilities

$$\begin{aligned} \mathbf{T}_\alpha &= \frac{\xi_\alpha k_{r\alpha}}{\mu_\alpha}\mathbf{k}, \qquad && \alpha = w, o, g, \\ \mathbf{T}_{i\alpha} &= \frac{x_{i\alpha}\xi_\alpha k_{r\alpha}}{\mu_\alpha}\mathbf{k}, \qquad && \alpha = o, g,\ i = 1, 2, \ldots, N_c. \end{aligned} \tag{9.24}$$

We now summarize the equations needed in iterative IMPES. The equilibrium relation (9.8) is recast as

$$\begin{aligned} f_{io}(p_o, x_{1o}, x_{2o}, \ldots, x_{N_c o}) &= f_{ig}(p_o + p_{cg}, x_{1g}, x_{2g}, \ldots, x_{N_c g}), \\ &\qquad i = 1, 2, \ldots, N_c. \end{aligned} \tag{9.25}$$

Using (9.24), equation (9.22) becomes

$$\begin{aligned} \frac{\partial(\phi F z_i)}{\partial t} &= \nabla \cdot (\mathbf{T}_{io}\nabla\Phi_o + \mathbf{T}_{ig}\nabla\Phi_g) + x_{io}q_o + x_{ig}q_g, \\ &\qquad i = 1, 2, \ldots, N_c - 1. \end{aligned} \tag{9.26}$$

Similarly, it follows from (9.23) that

$$\frac{\partial(\phi F)}{\partial t} = \nabla \cdot (\mathbf{T}_o\nabla\Phi_o + \mathbf{T}_g\nabla\Phi_g) + q_o + q_g. \tag{9.27}$$

Next, applying the first equation of (9.3) and (9.24) yields

$$\frac{\partial(\phi\xi_w S_w)}{\partial t} = \nabla\cdot(\mathbf{T}_w\nabla\Phi_w) + q_w. \tag{9.28}$$

Finally, using (9.17) and (9.18), the saturation state equation (9.6) becomes

$$F\left(\frac{L}{\xi_o} + \frac{1-L}{\xi_g}\right) + S = 1. \tag{9.29}$$

The differential system consists of the $2N_c + 2$ equations (9.25)–(9.29) for the $2N_c + 2$ primary unknowns: x_{io} (or x_{ig}), L (or V), z_i, F, $S = S_w$, and $p = p_o$, $i = 1, 2, \ldots, N_c - 1$.

9.2.2 Iterative IMPES

Let $n > 0$ (an integer) indicate a time step. For any function v of time, we use $\bar{\delta}v$ to denote the time increment at the nth step:

$$\bar{\delta}v = v^{n+1} - v^n.$$

A time approximation at the $(n+1)$th level for the system of equations (9.25)–(9.29) is

$$\begin{aligned}
&f_{io}(p_o^{n+1}, x_{1o}^{n+1}, x_{2o}^{n+1}, \ldots, x_{N_co}^{n+1}) \\
&\quad = f_{ig}(p_g^{n+1}, x_{1g}^{n+1}, x_{2g}^{n+1}, \ldots, x_{N_cg}^{n+1}), \qquad i = 1, 2, \ldots, N_c, \\
&\frac{1}{\Delta t}\bar{\delta}(\phi F z_i) = \nabla\cdot(\mathbf{T}_{io}^n\nabla\Phi_o^{n+1} + \mathbf{T}_{ig}^n\nabla\Phi_g^{n+1}) \\
&\qquad\qquad + x_{io}^{n+1}q_o^n + x_{ig}^{n+1}q_g^n, \qquad i = 1, 2, \ldots, N_c - 1, \\
&\frac{1}{\Delta t}\bar{\delta}(\phi F) = \nabla\cdot(\mathbf{T}_o^n\nabla\Phi_o^{n+1} + \mathbf{T}_g^n\nabla\Phi_g^{n+1}) + q_o^n + q_g^n, \\
&\frac{1}{\Delta t}\bar{\delta}(\phi\xi_w S) = \nabla\cdot(\mathbf{T}_w^n\nabla\Phi_w^{n+1}) + q_w^n, \\
&\left[F\left(\frac{L}{\xi_o} + \frac{1-L}{\xi_g}\right) + S\right]^{n+1} = 1,
\end{aligned} \tag{9.30}$$

where $\Delta t = t^{n+1} - t^n$. Note that the transmissibilities and well terms in (9.30) are evaluated at the previous time level.

System (9.30) is nonlinear in the primary unknowns, and can be linearized via the Newton–Raphson iteration introduced in Section 8.2.1. For a generic function v of time, we use the iteration

$$v^{n+1,l+1} = v^{n+1,l} + \delta v,$$

where l refers to the iteration number of Newton–Raphson's iterations and δv represents the increment in this iteration step. When no ambiguity occurs, we will replace $v^{n+1,l+1}$ and $v^{n+1,l}$ by v^{l+1} and v^l, respectively (i.e., the superscript $n+1$ is omitted). Observe that

$$v^{n+1} \approx v^{l+1} = v^l + \delta v,$$

so

$$\bar{\delta} v \approx v^l - v^n + \delta v.$$

Using this approximation in system (9.30) gives

$$\begin{aligned}
& f_{io}(p_o^{l+1}, x_{1o}^{l+1}, x_{2o}^{l+1}, \ldots, x_{N_co}^{l+1}) \\
& \quad = f_{ig}(p_g^{l+1}, x_{1g}^{l+1}, x_{2g}^{l+1}, \ldots, x_{N_cg}^{l+1}), \qquad i = 1, 2, \ldots, N_c, \\
& \frac{1}{\Delta t}\left[(\phi F z_i)^l - (\phi F z_i)^n + \delta(\phi F z_i)\right] \\
& \quad = \nabla \cdot (\mathbf{T}_{io}^n \nabla \Phi_o^{l+1} + \mathbf{T}_{ig}^n \nabla \Phi_g^{l+1}) + x_{io}^{l+1} q_o^n + x_{ig}^{l+1} q_g^n, \\
& \qquad\qquad\qquad\qquad\qquad\qquad i = 1, 2, \ldots, N_c - 1, \\
& \frac{1}{\Delta t}\left[(\phi F)^l - (\phi F)^n + \delta(\phi F)\right] \\
& \quad = \nabla \cdot (\mathbf{T}_o^n \nabla \Phi_o^{l+1} + \mathbf{T}_g^n \nabla \Phi_g^{l+1}) + q_o^n + q_g^n, \\
& \frac{1}{\Delta t}\left[(\phi \xi_w S)^l - (\phi \xi_w S)^n + \delta(\phi \xi_w S)\right] = \nabla \cdot (\mathbf{T}_w^n \nabla \Phi_w^{l+1}) + q_w^n, \\
& \left[F\left(\frac{L}{\xi_o} + \frac{1-L}{\xi_g}\right) + S\right]^{l+1} = 1.
\end{aligned} \tag{9.31}$$

We expand the potentials and transmissibilities in terms of the primary unknowns. Toward that end, we must identify these unknowns. If the gas phase dominates in the hydrocarbon system (e.g., $L < 0.5$), the primary unknowns will be x_{io}, L, z_i, F, S, and p, $i = 1, 2, \ldots, N_c - 1$, which is the $L - X$ *iteration type* in compositional modeling. If the oil phase dominates (e.g., $L \geq 0.5$), the primary unknowns will be x_{ig}, V, z_i, F, S, and p, $i = 1, 2, \ldots, N_c - 1$, which corresponds to the $V - Y$ *iteration type*. As an example, we illustrate how to expand the potentials and transmissibilities in terms of δx_{io}, δL, δz_i, δF, δS, and δp, $i = 1, 2, \ldots, N_c - 1$; a similar expansion can be performed for the $V - Y$ iteration type.

For the ith component flow equation,

$$\delta(\phi F z_i) = c_{ip}\delta p + c_{iF}\delta F + c_{iz}\delta z_i, \qquad i = 1, 2, \ldots, N_c - 1, \tag{9.32}$$

where

$$c_{ip} = \phi^o c_R (F z_i)^l, \quad c_{iF} = (\phi z_i)^l, \quad c_{iz} = (\phi F)^l,$$

with ϕ^o being the porosity at a reference pressure p^o and c_R the rock compressibility. For the global hydrocarbon flow equation,

$$\delta(\phi F) = c_p \delta p + c_F \delta F, \tag{9.33}$$

where

$$c_p = \phi^o c_R F^l, \quad c_F = \phi^l.$$

For the water flow equation,

$$\delta(\phi \xi_w S) = c_{wp} \delta p + c_{wS} \delta S, \tag{9.34}$$

where

$$c_{wp} = \phi^o c_R (\xi_w S)^l + \left(\phi \frac{d\xi_w}{dp} S\right)^l, \quad c_{wS} = (\phi \xi_w)^l.$$

In iterative IMPES, all the saturation functions (k_{rw}, k_{ro}, k_{rg}, p_{cw}, and p_{cg}), densities, and viscosities are evaluated at the saturation values of the previous time step in the Newton–Raphson iteration. The phase potentials are calculated by

$$\Phi_\alpha^{l+1} = p^{l+1} + p_{c\alpha}^n - \rho_\alpha^n \wp z, \qquad \alpha = w, o, g, \tag{9.35}$$

and the transmissibilities by

$$\begin{aligned} \mathbf{T}_\alpha^n &= \frac{\xi_\alpha^n k_{r\alpha}^n}{\mu_\alpha^n} \mathbf{k}, & \alpha &= w, o, g, \\ \mathbf{T}_{i\alpha}^n &= \frac{x_{i\alpha}^n \xi_\alpha^n k_{r\alpha}^n}{\mu_\alpha^n} \mathbf{k}, & \alpha &= o, g, \ i = 1, 2, \dots, N_c. \end{aligned} \tag{9.36}$$

It follows from (9.35) that

$$\Phi_\alpha^{l+1} = \Phi_\alpha^l + \delta p, \qquad \alpha = w, o, g. \tag{9.37}$$

We now expand each of the equations in system (9.31). For this, we replace the derivatives in x_{ig} by those in the primary variables, $i = 1, 2, \dots, N_c$. Applying relation (9.19), we see that

$$\frac{\partial x_{ig}}{\partial x_{io}} = \frac{L}{L-1}, \quad \frac{\partial x_{ig}}{\partial z_i} = \frac{1}{1-L},$$
$$\frac{\partial x_{ig}}{\partial L} = \frac{x_{io} - x_{ig}}{L-1}, \qquad i = 1, 2, \dots, N_c.$$

Consequently, the chain rule implies

$$\begin{aligned} \frac{\partial}{\partial x_{io}} &= \frac{\partial x_{ig}}{\partial x_{io}} \frac{\partial}{\partial x_{ig}} = \frac{L}{L-1} \frac{\partial}{\partial x_{ig}}, \\ \frac{\partial}{\partial z_i} &= \frac{\partial x_{ig}}{\partial z_i} \frac{\partial}{\partial x_{ig}} = \frac{1}{1-L} \frac{\partial}{\partial x_{ig}}, \\ \frac{\partial}{\partial L} &= \frac{\partial x_{ig}}{\partial L} \frac{\partial}{\partial x_{ig}} = \frac{x_{io} - x_{ig}}{L-1} \frac{\partial}{\partial x_{ig}}. \end{aligned}$$

Thus, after using (9.5) and (9.20) to eliminate $x_{N_c o}$ and z_{N_c}, the first equation in (9.31) can be expanded:

$$\begin{aligned}
&\sum_{j=1}^{N_c-1}\left\{\left(\frac{\partial f_{io}}{\partial x_{jo}}\right)^l - \left(\frac{\partial f_{io}}{\partial x_{N_c o}}\right)^l + \frac{L^l}{1-L^l}\left[\left(\frac{\partial f_{ig}}{\partial x_{jg}}\right)^l - \left(\frac{\partial f_{ig}}{\partial x_{N_c g}}\right)^l\right]\right\}\delta x_{jo} \\
&\quad + \frac{1}{1-L^l}\sum_{j=1}^{N_c}\left(\frac{\partial f_{ig}}{\partial x_{jg}}(x_{jo}-x_{jg})\right)^l \delta L \\
&= f_{ig}^l - f_{io}^l + \left[\left(\frac{\partial f_{ig}}{\partial p}\right)^l - \left(\frac{\partial f_{io}}{\partial p}\right)^l\right]\delta p \\
&\quad + \frac{1}{1-L^l}\sum_{j=1}^{N_c-1}\left[\left(\frac{\partial f_{ig}}{\partial x_{jg}}\right)^l - \left(\frac{\partial f_{ig}}{\partial x_{N_c g}}\right)^l\right]\delta z_j,
\end{aligned} \tag{9.38}$$

where, for $i = 1, 2, \ldots, N_c$,

$$f_{io}^l = f_{io}(p_o^l, x_{1o}^l, x_{2o}^l, \ldots, x_{N_c o}^l), \qquad f_{ig}^l = f_{ig}(p_g^l, x_{1g}^l, x_{2g}^l, \ldots, x_{N_c g}^l).$$

The linear equation (9.38) is used to solve for $(\delta x_{1o}, \delta x_{2o}, \ldots, \delta x_{(N_c-1)o}, \delta L)$ in terms of $(\delta z_1, \delta z_2, \ldots, \delta z_{N_c-1}, \delta p)$.

Next, applying (9.32) and (9.37), from the second equation in (9.31) it follows that, for $i = 1, 2, \ldots, N_c - 1$,

$$\begin{aligned}
&\frac{1}{\Delta t}\left[(\phi F z_i)^l - (\phi F z_i)^n + c_{ip}\delta p + c_{iF}\delta F + c_{iz}\delta z_i\right] \\
&\quad = \nabla\cdot(\mathbf{T}_{io}^n \nabla\Phi_o^l + \mathbf{T}_{ig}^n \nabla\Phi_g^l) + \nabla\cdot\left((\mathbf{T}_{io}^n + \mathbf{T}_{ig}^n)\nabla(\delta p)\right) \\
&\quad + \left(x_{io}^l + \delta x_{io}\right) q_o(\delta p) + \left(x_{ig}^l + \delta x_{ig}\right) q_g(\delta p).
\end{aligned} \tag{9.39}$$

Equation (9.39) is solved for $(\delta z_1, \delta z_2, \ldots, \delta z_{N_c-1})$ in terms of $(\delta F, \delta p)$. Similarly, from the third equation in (9.31) we see that

$$\begin{aligned}
&\frac{1}{\Delta t}\left[(\phi F)^l - (\phi F)^n + c_p\delta p + c_F\delta F\right] \\
&\quad = \nabla\cdot(\mathbf{T}_o^n \nabla\Phi_o^l + \mathbf{T}_g^n \nabla\Phi_g^l) + \nabla\cdot\left((\mathbf{T}_o^n + \mathbf{T}_g^n)\nabla(\delta p)\right) \\
&\quad + q_o(\delta p) + q_g(\delta p),
\end{aligned} \tag{9.40}$$

which gives δF in terms of δp. From the fourth equation in (9.31), (9.34), and (9.37), we have

$$\frac{1}{\Delta t}\left[(\phi\xi_w S)^l - (\phi\xi_w S)^n + c_{wp}\delta p + c_{ws}\delta S\right] = \nabla\cdot(\mathbf{T}_w^n\nabla\Phi_w^l) + \nabla\cdot\left(\mathbf{T}_w^n\nabla(\delta p)\right) + q_w(\delta p). \tag{9.41}$$

Equation (9.41) gives δS in terms of δp.

It follows from (9.12) that

$$\frac{1}{\xi_\alpha} = \frac{Z_\alpha(p_\alpha, x_{1\alpha}, x_{2\alpha}, \ldots, x_{N_c\alpha})RT}{p_\alpha}, \qquad \alpha = o, g.$$

Applying (9.5) and (9.20), it follows from the last equation in (9.31) that

$$\begin{aligned}
&\left(\frac{FLRT}{p}\right)^l \sum_{j=1}^{N_c-1}\left\{\left(\frac{\partial Z_o}{\partial x_{jo}}\right)^l - \left(\frac{\partial Z_o}{\partial x_{N_co}}\right)^l \right.\\
&\qquad\qquad \left. - \left[\left(\frac{\partial Z_g}{\partial x_{jg}}\right)^l - \left(\frac{\partial Z_g}{\partial x_{N_cg}}\right)^l\right]\right\}\delta x_{jo}\\
&+\left(\frac{FRT}{p}\right)^l\left[Z_o - Z_g - \sum_{j=1}^{N_c}\left(\frac{\partial Z_g}{\partial x_{jg}}(x_{jo}-x_{jg})\right)^l\right]\delta L\\
&+\left(\frac{FRT}{p}\right)^l\sum_{j=1}^{N_c-1}\left\{\left(\frac{\partial Z_g}{\partial x_{jg}}\right)^l - \left(\frac{\partial Z_g}{\partial x_{N_cg}}\right)^l\right\}\delta z_j\\
&+\left(\frac{RT}{p}(LZ_o + (1-L)Z_g)\right)^l\delta F + \delta S\\
&+\left(\frac{FRT}{p}\left[L\frac{\partial Z_o}{\partial p} - \frac{LZ_o}{p} + (1-L)\frac{\partial Z_g}{\partial p} - \frac{(1-L)Z_g}{p}\right]\right)^l\delta p\\
&= 1 - \left(F\left[\frac{L}{\xi_o} + \frac{1-L}{\xi_g}\right] + S\right)^l.
\end{aligned} \tag{9.42}$$

After substituting δx_{jo}, δL, δz_j, δF, and δS, $j = 1, 2, \ldots, N_c - 1$, into (9.42) using (9.38)–(9.41), the resulting equation becomes the pressure equation, which, together with the well control equations (cf. Chapter 8), is implicitly solved for δp. After δp is obtained, (9.41), (9.40), (9.39), and (9.38) are solved explicitly for δS, δF, $(\delta z_1, \delta z_2, \ldots, \delta z_{N_c-1})$, and $(\delta x_{1o}, \delta x_{2o}, \ldots, \delta x_{(N_c-1)o}, \delta L)$, respectively. The numerical methods introduced in Chapter 4 can be applied to the discretization of (9.38)–(9.42) in space.

In summary, the iterative IMPES for the compositional model has following features:

- The difference between iterative IMPES and classical IMPES is that the iterative one is used within each Newton–Raphson iteration loop, while the classical one is utilized outside the Newton–Raphson iteration.

- The saturation constraint equation is used to solve implicitly for pressure p.
- The equilibrium relation is solved for $(x_{1o}, x_{2o}, \ldots, x_{(N_c-1)o}, L)$.
- The hydrocarbon component flow equations are used to obtain $(z_1, z_2, \ldots, z_{N_c-1})$ explicitly.
- The global hydrocarbon flow equation is exploited to solve explicitly for F.
- The water flow equation is explicitly solved for S.
- Relation (9.19) generates $(x_{1g}, x_{2g}, \ldots, x_{N_c g})$.

As in the sequential technique for the black oil model, the saturation functions k_{rw}, k_{ro}, k_{rg}, p_{cw}, and p_{cg} can use the previous Newton–Raphson iteration values of saturations, instead of the previous time step values of saturations.

9.3 Solution of Equilibrium Relations

We discuss the solution of the thermodynamic equilibrium relation (9.25), which describes the mass distribution of each component in the oil and gas phases. As an example, we concentrate on the Peng–Robinson equation of state.

9.3.1 Successive substitution method

The *successive substitution method* is often employed to find an initial guess for the computation of the thermodynamic equilibrium relation (9.38) in the Newton–Raphson flash calculation discussed in the next subsection. The *equilibrium flash vaporization ratio* for component i is defined by

$$K_i = \frac{x_{ig}}{x_{io}}, \qquad i = 1, 2, \ldots, N_c, \tag{9.43}$$

where the quantity K_i is the K-value of component i. If the iterative IMPES in the previous section is used (i.e., the capillary pressure p_{cg} is evaluated at the previous time step value of saturations in the Newton–Raphson iteration), it follows from (9.15) that

$$f_{i\alpha} = p x_{i\alpha} \varphi_{i\alpha}, \qquad i = 1, 2, \ldots, N_c, \ \alpha = o, g. \tag{9.44}$$

Then, using (9.8), we see that

$$x_{io}\varphi_{io} = x_{ig}\varphi_{ig}, \qquad i = 1, 2, \ldots, N_c.$$

Thus, by (9.43), we have

$$K_i = \frac{\varphi_{io}}{\varphi_{ig}}, \qquad i = 1, 2, \ldots, N_c, \tag{9.45}$$

where the fugacity coefficients φ_{io} and φ_{ig} are defined in (9.14).

A *flash calculation* is an instant phase equilibrium:

$$\begin{aligned} &\text{Given } p,\ T,\ \text{and } z_i; \\ &\text{Find } L \text{ (or } V),\ x_{io},\ \text{and } x_{ig}, \qquad i = 1, 2, \ldots, N_c. \end{aligned}$$

It follows from (9.19) and (9.43) that

$$
\begin{aligned}
x_{io} &= \frac{z_i}{L + (1 - L)K_i}, \qquad i = 1, 2, \ldots, N_c, \\
\sum_{i=1}^{N_c} \frac{z_i(1 - K_i)}{L + (1 - L)K_i} &= 0.
\end{aligned}
\tag{9.46}
$$

Based on (9.46), we introduce the following successive substitution method for the flash calculation:

Initially, K_i is evaluated by the empirical formula

$$
K_i = \frac{1}{p_{ir}} \exp\left(5.3727(1 + \omega_i)\left[1 - \frac{1}{T_{ir}}\right]\right), \quad p_{ir} = \frac{p}{p_{ic}}, \; T_{ir} = \frac{T}{T_{ic}};
$$

(F1) Given K_i and z_i, find L by

$$
\sum_{i=1}^{N_c} \frac{z_i(1 - K_i)}{L + (1 - L)K_i} = 0;
$$

(F2) Find x_{io} and x_{ig} by

$$
x_{io} = \frac{z_i}{L + (1 - L)K_i}, \qquad x_{ig} = K_i x_{io}, \quad i = 1, 2, \ldots, N_c;
$$

(F3) Calculate K_i and z_i by

$$
K_i = \frac{\varphi_{io}}{\varphi_{ig}}, \qquad z_i = L x_{io} + (1 - L) x_{ig}, \quad i = 1, 2, \ldots, N_c;
$$

Return to (F1) and iterate until the convergence of the values K_i.

In general, convergence of this successive substitution method is very slow. However, it can be used as an initialization for the Newton–Raphson flash iteration discussed below.

9.3.2 Newton–Raphson's flash calculation

Introduce the notation

$$
\begin{aligned}
G_{ij} &= \left(\frac{\partial f_{io}}{\partial x_{jo}}\right)^l - \left(\frac{\partial f_{io}}{\partial x_{N_c o}}\right)^l + \frac{L^l}{1 - L^l}\left[\left(\frac{\partial f_{ig}}{\partial x_{jg}}\right)^l - \left(\frac{\partial f_{ig}}{\partial x_{N_c g}}\right)^l\right], \\
G_{iN_c} &= \frac{1}{1 - L^l} \sum_{j=1}^{N_c} \left(\frac{\partial f_{ig}}{\partial x_{jg}} (x_{jo} - x_{jg})\right)^l, \\
H_i(\delta p, \delta z_1, \delta z_2, \ldots, \delta z_{N_c - 1}) &= f_{ig}^l - f_{io}^l + \left[\left(\frac{\partial f_{ig}}{\partial p}\right)^l - \left(\frac{\partial f_{io}}{\partial p}\right)^l\right] \delta p \\
&\quad + \frac{1}{1 - L^l} \sum_{j=1}^{N_c - 1} \left[\left(\frac{\partial f_{ig}}{\partial x_{jg}}\right)^l - \left(\frac{\partial f_{ig}}{\partial x_{N_c g}}\right)^l\right] \delta z_j
\end{aligned}
$$

for $i = 1, 2, \ldots, N_c$, $j = 1, 2, \ldots, N_c - 1$. Then (9.38) can be written in matrix form

$$\begin{pmatrix} G_{11} & G_{12} & \cdots & G_{1,N_c-1} & G_{1,N_c} \\ G_{21} & G_{22} & \cdots & G_{2,N_c-1} & G_{2,N_c} \\ \vdots & \vdots & \vdots & \vdots & \vdots \\ G_{N_c-1,1} & G_{N_c-1,2} & \cdots & G_{N_c-1,N_c-1} & G_{N_c-1,N_c} \\ G_{N_c,1} & G_{N_c,2} & \cdots & G_{N_c,N_c-1} & G_{N_c,N_c} \end{pmatrix} \begin{pmatrix} \delta x_{1o} \\ \delta x_{2o} \\ \vdots \\ \delta x_{(N_c-1)o} \\ \delta L \end{pmatrix} = \begin{pmatrix} H_1 \\ H_2 \\ \vdots \\ H_{N_c-1} \\ H_{N_c} \end{pmatrix}. \tag{9.47}$$

This system gives $(\delta x_{1o}, \delta x_{2o}, \ldots, \delta x_{(N_c-1)o}, \delta L)$ in terms of δz_i, $i = 1, 2, \ldots, N_c - 1$, and δp.

We point out the difference between the successive substitution method and the Newton–Raphson iteration in the flash calculation.

- The former method is easier to implement and is more reliable, even near a critical point. However, its convergence is usually slower; it may take over 1,000 iterations near the critical point.
- The latter method is faster. But it needs a good initial guess for x_{io} and L, $i = 1, 2, \ldots, N_c$; moreover, this method may not converge near a critical point.
- These two methods can be combined. For example, the former is used to find a good initial guess for the latter. Also, in places where the latter is difficult to converge, the former can be utilized instead.

9.3.3 Derivatives of fugacity coefficients

We calculate the partial derivatives involved in the Jacobian coefficient matrix of (9.47). First, by (9.44), for $i, j = 1, 2, \ldots, N_c$, $\alpha = o, g$,

$$\frac{\partial f_{i\alpha}}{\partial p} = x_{i\alpha}\varphi_{i\alpha} + p x_{i\alpha}\frac{\partial \varphi_{i\alpha}}{\partial p}, \quad \frac{\partial f_{i\alpha}}{\partial x_{j\alpha}} = p\frac{\partial x_{i\alpha}}{\partial x_{j\alpha}}\varphi_{i\alpha} + p x_{i\alpha}\frac{\partial \varphi_{i\alpha}}{\partial x_{j\alpha}},$$

where

$$\frac{\partial x_{i\alpha}}{\partial x_{j\alpha}} = \begin{cases} 1 & \text{if } i = j, \\ 0 & \text{if } i \neq j. \end{cases}$$

So it suffices to find the derivatives of $\varphi_{i\alpha}$, which is defined by (9.14), $i = 1, 2, \ldots, N_c$, $\alpha = o, g$.

It follows from (9.10) that

$$\frac{\partial A_\alpha}{\partial p} = \frac{a_\alpha}{R^2T^2}, \quad \frac{\partial B_\alpha}{\partial p} = \frac{b_\alpha}{RT}, \quad \alpha = o, g. \tag{9.48}$$

Differentiating both sides of (9.14) gives

$$\begin{aligned}\frac{1}{\varphi_{i\alpha}}\frac{\partial\varphi_{i\alpha}}{\partial p} &= \frac{b_i}{b_\alpha}\frac{\partial Z_\alpha}{\partial p} - \frac{1}{Z_\alpha - B_\alpha}\left(\frac{\partial Z_\alpha}{\partial p} - \frac{B_\alpha}{p}\right)\\ &\quad - \frac{A_\alpha}{2\sqrt{2}B_\alpha}\left(\frac{2}{a_\alpha}\sum_{j=1}^{N_c} x_{j\alpha}(1-\kappa_{ij})\sqrt{a_i a_j} - \frac{b_i}{b_\alpha}\right)\\ &\quad \cdot 2B_\alpha\left(\frac{Z_\alpha}{p} - \frac{\partial Z_\alpha}{\partial p}\right)\Big/\left(Z_\alpha^2 + 2\sqrt{2}Z_\alpha B_\alpha + B_\alpha^2\right).\end{aligned} \tag{9.49}$$

Similarly, we can obtain $\partial\varphi_{i\alpha}/\partial x_{j\alpha}$ using the expressions (cf. Exercise 9.2)

$$\begin{aligned}\frac{\partial A_\alpha}{\partial x_{j\alpha}} &= \frac{p}{R^2T^2}\frac{\partial a_\alpha}{\partial x_{j\alpha}}, & \frac{\partial B_\alpha}{\partial x_{j\alpha}} &= \frac{p}{RT}\frac{\partial b_\alpha}{\partial x_{j\alpha}},\\ \frac{\partial a_\alpha}{\partial x_{j\alpha}} &= 2\sum_{i=1}^{N_c} x_{i\alpha}(1-\kappa_{ij})\sqrt{a_i a_j}, & \frac{\partial b_\alpha}{\partial x_{j\alpha}} &= b_j\end{aligned} \tag{9.50}$$

for $i, j = 1, 2, \ldots, N_c$, $\alpha = o, g$.

The Z-factors, Z_α ($\alpha = o, g$), are determined by (9.13), which can be differentiated to find their derivatives. Implicit differentiation on (9.13) yields

$$\begin{aligned}\frac{\partial Z_\alpha}{\partial p} = &-\left\{\frac{\partial B_\alpha}{\partial p}Z_\alpha^2 + \left(\frac{\partial A_\alpha}{\partial p} - 2\left[1+3B_\alpha\right]\frac{\partial B_\alpha}{\partial p}\right)Z_\alpha\right.\\ &\left. - \left(\frac{\partial A_\alpha}{\partial p}B_\alpha + \left[A_\alpha - 2B_\alpha - 3B_\alpha^2\right]\frac{\partial B_\alpha}{\partial p}\right)\right\}\\ &\Big/\left(2Z_\alpha^2 - 2(1-B_\alpha)Z_\alpha + (A_\alpha - 2B_\alpha - 3B_\alpha^2)\right).\end{aligned} \tag{9.51}$$

Consequently, substituting (9.48) into (9.51) gives $\partial Z_\alpha/\partial p$. A similar argument, together with (9.50), gives the derivatives $\partial Z_\alpha/\partial x_{j\alpha}$ (cf. Exercise 9.3), $j = 1, 2, \ldots, N_c$.

9.3.4 Solution of Peng–Robinson's cubic equation

The Peng–Robinson cubic equation (9.13) has the form

$$\mathcal{Z}^3 + B\mathcal{Z}^2 + C\mathcal{Z} + D = 0 \tag{9.52}$$

with given inputs B, C, and D. Before discussing the solution of this equation, we consider a simpler cubic equation:

$$X^3 + PX + Q = 0. \tag{9.53}$$

With

$$\Delta = \left(\frac{Q}{2}\right)^2 + \left(\frac{P}{3}\right)^3,$$

equation (9.53) has three roots (cf. Exercise 9.4)

$$X_1 = \sqrt[3]{-\frac{Q}{2} + \sqrt{\Delta}} + \sqrt[3]{-\frac{Q}{2} - \sqrt{\Delta}},$$
$$X_2 = \omega\sqrt[3]{-\frac{Q}{2} + \sqrt{\Delta}} + \omega^2\sqrt[3]{-\frac{Q}{2} - \sqrt{\Delta}},$$
$$X_3 = \omega^2\sqrt[3]{-\frac{Q}{2} + \sqrt{\Delta}} + \omega\sqrt[3]{-\frac{Q}{2} - \sqrt{\Delta}},$$

where

$$\omega = \frac{-1 + i\sqrt{3}}{2}, \quad \omega^2 = \frac{-1 - i\sqrt{3}}{2}, \quad i^2 = -1.$$

Note that (cf. Exercise 9.5)

$$X_1 + X_2 + X_3 = 0, \quad \frac{1}{X_1} + \frac{1}{X_2} + \frac{1}{X_3} = -\frac{P}{Q}, \quad X_1X_2X_3 = -Q. \tag{9.54}$$

If $\Delta > 0$, (9.53) has only one real root X_1. If $P = Q = 0$, there is solely the trivial solution $X_1 = X_2 = X_3 = 0$. When $\Delta \leq 0$, there are three real roots given by

$$\begin{aligned} &X_1 = 2\sqrt[3]{\mathcal{R}}\cos\theta, \quad X_2 = 2\sqrt[3]{\mathcal{R}}\cos\left(\frac{2\pi}{3} + \theta\right), \\ &X_3 = 2\sqrt[3]{\mathcal{R}}\cos\left(\frac{4\pi}{3} + \theta\right), \end{aligned} \tag{9.55}$$

where

$$\mathcal{R} = \sqrt{-\left(\frac{P}{3}\right)^3}, \quad \theta = \frac{1}{3}\arccos\left(-\frac{Q}{2\mathcal{R}}\right).$$

To solve (9.52), set $\mathcal{Z} = X - \frac{B}{3}$. Then (9.52) is converted into (9.53) with (cf. Exercise 9.6)

$$P = -\frac{B^2}{3} + C, \quad Q = \frac{2B^3}{27} - \frac{BC}{3} + D.$$

Thus the roots of (9.52) are

$$\mathcal{Z}_1 = X_1 - \frac{B}{3}, \quad \mathcal{Z}_2 = X_2 - \frac{B}{3}, \quad \mathcal{Z}_3 = X_3 - \frac{B}{3}. \tag{9.56}$$

If $\mathcal{Z}_1$ is the sole real root, it is selected. In the case where there are three real roots, say,

$$\mathcal{Z}_1 > \mathcal{Z}_2 > \mathcal{Z}_3,$$

we select $\mathcal{Z}_1$ if the vapor (gas) phase dominates. If the liquid (oil) phase dominates, we select $\mathcal{Z}_1$ when $\mathcal{Z}_2 \leq 0$; select $\mathcal{Z}_2$ when $\mathcal{Z}_2 > 0$ and $\mathcal{Z}_3 \leq 0$; select $\mathcal{Z}_3$ when $\mathcal{Z}_3 > 0$.

9.3.5 Practical considerations

We point out a few practical issues in programming the solution of equilibrium relations.

Iteration switch

As noted, depending on the size of L, different variables, either x_{io} and L or x_{ig} and V, should be used in the flash calculation, $i = 1, 2, \ldots, N_c$. If the gas phase dominates in the hydrocarbon system (e.g., $L < 0.5$), the primary unknowns will be x_{io} and L. If the oil phase dominates (e.g., $L \geq 0.5$), the primary unknowns will be x_{ig} and V. This choice can improve solution accuracy and convergence speed. For example, as L gets close to one, the flash calculation may not converge. In this case, the primary unknown needs to be switched to V. In programming, the switch of iterations should be done automatically.

Determination of bubble points

The following system of $N_c + 1$ equations are solved simultaneously for finding the *bubble point pressure* p and the compositions x_{ig} by an Newton–Raphson iteration ($i = 1, 2, \ldots, N_c$):

$$\begin{aligned} &z_i \varphi_{io}(p, x_{1o}, x_{2o}, \ldots, x_{N_c o}) = x_{ig} \varphi_{ig}(p, x_{1g}, x_{2g}, \ldots, x_{N_c g}), \\ &\sum_{i=1}^{N_c} x_{ig} = 1. \end{aligned} \tag{9.57}$$

In the late steps of the iteration (e.g., after ten iterations), the second equation in (9.57) can be replaced by

$$\sum_{i=1}^{N_c} \frac{\varphi_{io}}{\varphi_{ig}} z_i = 1 \tag{9.58}$$

to speedup convergence. In the Newton–Raphson iteration, if the successive values of pressure change less than a certain value (e.g., 0.01 psi), then this iteration is considered to have converged. We consider that it fails to converge if more than 30 iterations are required or if $|z_i - x_{ig}| < 0.001|z_i|$. In the latter case, the successive substitution method can be used to obtain p and x_{ig}, $i = 1, 2, \ldots, N_c$. A trivial solution occurs when $x_{ig} = z_i$ for any value of p, indicating that a *dew point* occurs.

Determination of dew points

The *dew point pressure* p and the compositions x_{io} satisfy the system of $N_c + 1$ equations ($i = 1, 2, \ldots, N_c$):

$$\begin{aligned} &x_{io} \varphi_{io}(p, x_{1o}, x_{2o}, \ldots, x_{N_c o}) = z_i \varphi_{ig}(p, x_{1g}, x_{2g}, \ldots, x_{N_c g}), \\ &\sum_{i=1}^{N_c} x_{io} = 1. \end{aligned} \tag{9.59}$$

Again, after about ten Newton–Raphson's iterations, the second equation in (9.59) is replaced by

$$\sum_{i=1}^{N_c} \frac{\varphi_{ig}}{\varphi_{io}} z_i - 1. \tag{9.60}$$

Table 9.1. *Reservoir grid data.*

$Nx_1 = Nx_2 = 9$, $Nx_3 = 4$; $h_1 = h_2 = 293.3$ ft
$h_3 = 30, 30, 50, 50$ ft; Datum=7,500 ft. (subsurface)
Porosity: 0.13 (at initial reservoir pressure)
Gas-water contact: 7,500 ft; S_w at contact: 1.0
p_{cgw} at contact: 0.0 psi; initial pressure at contact: 3,550 psia
Water density at contact: 63.0 lb/ft^3; c_w=3.0E-6 psi^{-1}
Formation water viscosity: 0.78 cp; Rock comp.: 4.0E-6 psi^{-1}

Table 9.2. *Reservoir model description.*

Layer	Thickness (ft)	k_h (md)	k_v (md)	Depth to center (ft)
1	30	130	13	7,330
2	30	40	4	7,360
3	50	20	2	7,400
4	50	150	15	7,450

Using the same guidelines as in the treatment of bubble points, if the successive values of pressure in the iteration process change less than 0.01 psi, this iteration is considered to have converged. We consider that the convergence fails if more than 30 iterations are required or if $|z_i - x_{io}| < 0.001|z_i|$. In the latter case, the successive substitution method can be used to obtain p and x_{io}, $i = 1, 2, \ldots, N_c$. A trivial solution occurs when $x_{io} = z_i$ for any value of p, indicating that a *bubble point* occurs.

9.4 The Third SPE Project: Compositional Flow

The simulation problem is chosen from the benchmark problem of the third CSP (Kenyon and Behie, 1987). Nine companies participated in this comparative project. It is a study of gas cycling in a rich retrograte condensate reservoir. Two prediction cases are considered. The first case is gas cycling with constant sales gas removal, and the second case is cycling with some gas sales deferral to enhance pressure maintenance in the early life of the reservoir. The specification of the reservoir model is presented in Tables 9.1–9.5, where k_h $(= k_{11} = k_{22})$ and k_v $(= k_{33})$ denote the horizontal and vertical permeabilities, respectively. A reservoir grid with $9 \times 9 \times 4$ is shown in Figure 9.1, and it is diagonally symmetrical, indicating that it would be possible to simulate half of this reservoir. We chose to model the full reservoir. Also, the reservoir layers are homogeneous and have a constant porosity, but there are permeability and thickness variations between layers, a factor leading to unequal sweepout. The two-well pattern is arbitrary and is employed to allow for some retrograde condensation without significant revaporization by recycling gas to simulate what occurs in sweep-inaccessable parts of a real reservoir.

The CVFE method with linear elements introduced in Section 4.3 is used for the discretization of the governing equations for the compositional model. Due to the layer structure in the vertical direction of the reservoir under consideration, we divide its domain into hexagonal prisms, i.e., hexagons in the horizontal plane and rectangles in the vertical

Table 9.3. *Production, injection, and sales data.*

Production	Location: $i = j = 7$; perforations: $k = 3, 4$; radius = 1 ft; rate: 6,200 MSCF/D (gas rate); min p_{bh}, 500 psi
Injection	Location: $i = j = 1$; perforations: $k = 1, 2$; radius = 1 ft; rate: separator rate-sales rate; max p_{bh}: 4,000 psi
Sales rate for case 1	Constant sales rate to blowdown: $0 < t < 10$ yr, 1,500 MSCF/D; $t > 10$ yr, all produced gas to sales
Sales rate for case 2	Deferred sales: $0 < t < 5$ yr, 500 MSCF/D; $5 < t < 10$ yr, 2,500 MSCF/D; $t > 10$ yr, all produced gas to sales

Table 9.4. *Saturation function data.*

Phase saturation	k_{rg}	k_{ro}	k_{rw}	p_{cgw} (psi)	p_{cgo} (psi)
0.00	0.00	0.00	0.00	> 50	0
0.04	0.005	0.00	0.00	> 50	0
0.08	0.013	0.00	0.00	> 50	0
0.12	0.026	0.00	0.00	> 50	0
0.16	0.040	0.00	0.00	50	0
0.20	0.058	0.00	0.002	32	0
0.24	0.078	0.00	0.010	21	0
0.28	0.100	0.005	0.020	15.5	0
0.32	0.126	0.012	0.033	12.0	0
0.36	0.156	0.024	0.049	9.2	0
0.40	0.187	0.040	0.066	7.0	0
0.44	0.222	0.060	0.090	5.3	0
0.48	0.260	0.082	0.119	4.2	0
0.52	0.300	0.112	0.150	3.4	0
0.56	0.348	0.150	0.186	2.7	0
0.60	0.400	0.196	0.227	2.1	0
0.64	0.450	0.250	0.277	1.7	0
0.68	0.505	0.315	0.330	1.3	0
0.72	0.562	0.400	0.390	1.0	0
0.76	0.620	0.513	0.462	0.7	0
0.80	0.680	0.650	0.540	0.5	0
0.84	0.740	0.800	0.620	0.4	0
0.88	—	—	0.710	0.3	0
0.92	—	—	0.800	0.2	0
0.96	—	—	0.900	0.1	0
1.00	—	—	1.000	0.0	0

Table 9.5. *Separator pressures and temperatures.*

Separator	Pressure (psia)	Temperature (°F)
Primary*	815	80
Primary	315	80
Second stage	65	80
Stock tank	14.7	60

*Primary separation at 815 psia until reservoir pressure (at datum) falls below 2,500 psia; then switch to primary separation at 315 psia.

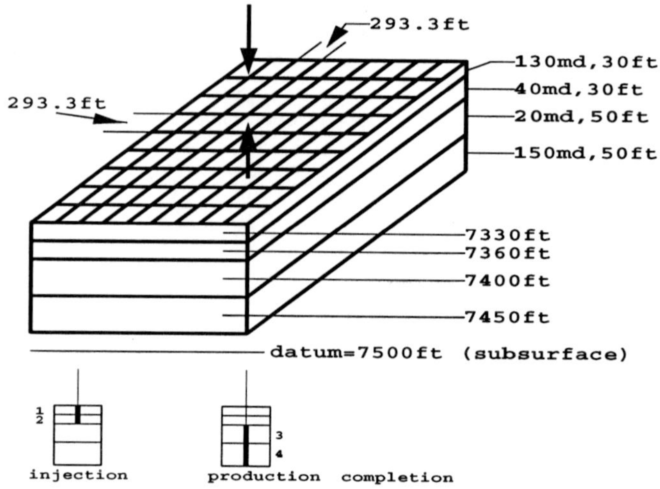

Figure 9.1. *A reservoir domain.*

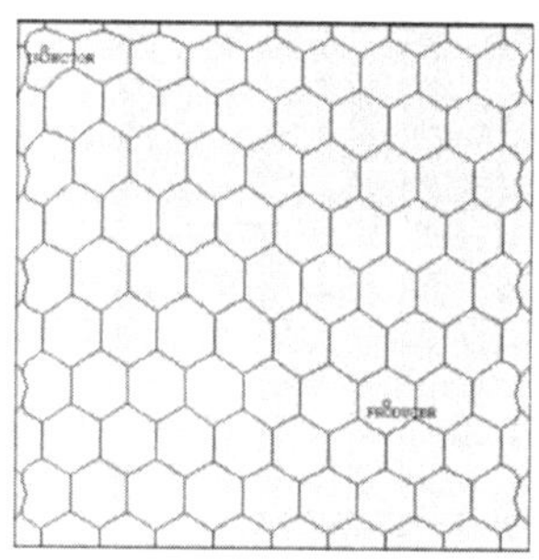

Figure 9.2. *A planar view of the grid.*

direction, as seen in Figure 4.36; also see Figure 9.2 for a planar view of the grid. The initial conditions, the location of the gas-water contact, and the capillary pressure data produce a water-gas transition zone extending to the pay zones. However, the very small compressibility and water volume make water quite insignificant for the present problem. Relative permeability data are used under the assumption that the phase relative permeability function depends only on its own phase saturation. Oil is immobile to 24% saturation, and k_{rg} is reduced from 0.74 to 0.4 as condensate builds to this saturation with irreducible water present.

Production is separator gas rate controlled. Liquid production through multistage separation is to be predicted. The separator train is given, and the primary separator pressure depends on reservoir pressure as shown in Table 9.5. Sales gas is removed from the bulked separator gas, and the remaining gas is recycled. Volumetrically, the two cases under consideration provide for exactly the same amount of recycling gas to be reinjected over the cycling period (10 years), but more gas is recycled in the critical early years in the second case. Blowdown (all gas to sales) starts at the end of the tenth year of cycling, and simulations are run up to 15 years or 1,000 psi average reservoir pressure, whichever occurs

Table 9.6. *Mole fractions of the reservoir fluids.*

Component	Mol percent
Carbon dioxide (CO_2)	1.21
Nitrogen (N_2)	1.94
Methane (C_1)	65.99
Ethane (C_2)	8.69
Propane (C_3)	5.91
Iso-butane (IC_4)	2.39
N-butane (NC_4)	2.78
Iso-pentane (IC_5)	1.57
N-pentane (NC_5)	1.12
Hexanes (C_6)	1.81
Heptanes plus (C_{7+})*	6.59

*Properties of heptanes plus: specific gravity at 60°F = 0.774;
API gravity at 60°F = 51.4; molecular weight=140.
Computed separator gas gravity (air=1.0)=0.736.
Computed gross heating value for separator gas=1,216 Btu
per cubic foot of dry gas at 14.65 psia and 60°F.
Primary separator gas/separator liquid ratio
=4,812 SCF/bbl at 72°F and 2,000 psig.

Table 9.7. *Pressure volume relations of reservoir fluid at* 200° *F.*

Pressure (psig)	Relative volume	Deviation factor Z
6,000	0.8045	1.129
5,500	0.8268	1.063
5,000	0.8530	0.998
4,500	0.8856	0.933
4,000	0.9284	0.869
3,600	0.9745	0.822
3,428 (dew point)	1.0000	0.803*
3,400	1.0043	
3,350	1.0142	
3,200	1.0468	
3,000	1.0997	
2,800	1.1644	
2,400	1.3412	
2,000	1.6113	
1,600	2.0412	
1,300	2.5542	
1,030	3.2925	
836	4.1393	

*Gas expansion factor=1.295 MSCF/bbl.

first. The simulations are initialized at pressure about 100 psi above the dew point pressure 3,443 psia.

The entire compositional simulation study is divided into two steps:

- A PVT phase behavior study to obtain accurate EOS parameters and prediction results.
- A reservoir simulation study of the compositional flow using the CVFE.

Table 9.8. *Hydrocarbon analysis of lean gas sample.*

Component*	Mol percent	GPM
Hydrogen sulfids	Nil	
Carbon dioxide (CO_2)	Nil	
Nitrogen (N_2)	Nil	
Methane (C_1)	94.69	
Ethane (C_2)	5.27	1.401
Propane (C_3)	0.05	0.014
Butanes plus (C_{4+})	Nil	
Total	100.00	1.415

*Computed gas gravity (air=1.0)=0.58.
Computed gross heating value =1,216 Btu
per cubic foot of dry gas at 14.65 psia and 60°F.

Table 9.9. *Pressure volume relations of mixture No.* 1 *at* 200° *F.*

Pressure (psig)	Relative volume*	Liquid volume (percent of saturated volume)
6,000	0.9115	
5,502	0.9387	
5,000	0.9719	
4,500	1.0135	
4,000	1.0687	
3,800	1.0965	
3,700	1.1116	
3,650	1.1203	
3,635 (dew point)	1.1224	0.0
3,600	1.1298	0.3
3,500	1.1508	1.7
3,300	1.1969	6.8
3,000	1.2918	12.8

*Relative volumes and liquid volume percents are all based on original hydrocarbon pore volume at 3,428 psig and 200°F.

Table 9.10. *Pressure volume relations of mixture No.* 2 *at* 200° *F.*

Pressure (psig)	Relative volume	Liquid volume (percent of saturated volume)
6,000	1.1294	
5,500	1.1686	
5,000	1.2162	
4,500	1.2767	
4,300	1.3064	
4,100	1.3385	
4,050	1.3479	
4,015 (dew point)	1.3542	0.0
3,950	1.3667	0.1
3,800	1.3992	0.5
3,400	1.5115	4.5
3,000	1.6709	9.4

Table 9.11. *Pressure volume relations of mixture No.* 3 *at* 200° *F.*

Pressure (psig)	Relative volume	Liquid volume (percent of saturated volume)
6,000	1.6865	
5,600	1.7413	
5,300	1.7884	
5,100	1.8233	
5,000	1.8422	
4,950	1.8519	
4,900	1.8620	
4,800	1.8827	
4,700	1.9043	
4,610 (dew point)	1.9248	
4,500	1.9512	0.1
4,200	2.0360	0.3
3,900	2.1378	0.6
3,500	2.3193	2.1
3,000	2.6348	6.0

Table 9.12. *Pressure volume relations of mixture No.* 4 *at* 200° *F.*

Pressure (psig)	Relative volume	Liquid volume (percent of saturated volume)
6,000	2.2435	
5,500	2.3454	
5,000	2.4704	
4,880 (dew point)	2.5043	0.0
4,800	2.5288	Trace
4,600	2.5946	0.1
4,400	2.6709	0.3
4,000	2.8478	0.7
3,500	3.1570	1.4
3,000	3.5976	3.6

9.4.1 PVT phase behavior study

PVT data

The measured PVT data are shown in Tables 9.6–9.16. These data include hydrocarbon sample analysis, constant composition expansion data, constant volume depletion data, and swelling data of four mixtures of reservoir gas with lean gas. Table 9.6 gives the mole fractions of the reservoir fluids. Table 9.7 describes the constant composition expansion data, and the computed Z-factors at and above the dew point pressure. Tables 9.8–9.12 show data for the swelling tests of reservoir gas with lean gas. Table 9.8 gives the lean gas composition. Note that it is virtually free of C_{3+} fractions. This contrasts with the separator gas recycled in the reservoir problem, which has about 10% of C_{3+}. Hence matching the swelling data is more significant for recycling with gas plant residue gas than for typical separator gas compositions. Tables 9.9–9.12 indicate the pressure-volume data for expansions at 200° F for four mixtures (with the respective mole fractions: 0.1271, 0.3046, 0.5384, and 0.6538)

Table 9.13. *Retrograde condensation during gas depletion at* 200° *F.*

Pressure (psig)	Retrograde liquid volume (percent of hydrocarbon pore space)
3,428 (dew point)	0.0
3,400	0.9
3,350	2.7
3,200	8.1
3,000 (first depletion level)	15.0
2,400	19.9
1,800	19.2
1,200	17.1
700	15.2
0	10.2

Table 9.14. *Computed cumulative recovery during depletion.*

	Reservoir pressure (psig)						
Cumulative recovery per MMSCF of original fluid	Initial in place	3,428	3,000	2,400	1,800	1,200	700
Well stream (MSCF)	1,000	0	90.95	247.02	420.26	596.87	740.19
Normal temp. separation* Stock tank liquid (B)	131.00	0	7.35	14.83	20.43	25.14	29.25
Primary separator gas (MSCF)	750.46	0	74.75	211.89	369.22	530.64	666.19
Second stage gas (MSCF)	107.05	0	7.25	16.07	23.76	31.45	32.92
Stock tank gas (MSCF)	27.25	0	2.02	4.70	7.15	9.69	11.67
Total "plant products" in primary separator sas (Gallons) Propane (C_3)	801	0	85	249	443	654	876
Butanes (total C_4)	492	0	54	613	295	440	617
Pentanes plus (C_{5+})	206	0	22	67	120	176	255
Total "plant products" in 2nd stage gas (gallons) Propane (C_3)	496	0	35	80	119	161	168
Butanes (total C_4)	394	0	30	69	106	146	153
Pentanes plus (C_{5+})	164	0	12	29	45	62	65
Total plant products in well stream (gallons) Propane (C_3)	1,617	0	141	374	629	900	1,146
Butanes (total C_4)	1,648	0	137	352	580	821	1,049
Pentanes plus (C_{5+})	5,464	0	321	678	973	1,240	1,488

*Primary separator at 800 psig and 80° F reduced to 300 psig and 80° F for reservoir pressure below 1,200 psig; second stage at 50 psig and 80° F; stock tonk at 0 psig and 60° F.

of lean gas with reservoir gas. Liquid dropout data are shown for each of the expansions. Table 9.13 gives retrograde condensation during gas depletion (constant volume depletion) of the original reservoir fluids. Table 9.14 indicates the computed yields of separator and gas plant products, and Table 9.15 shows compositions of equilibrium gas during constant volume depletion. We use these data to match the surface volumes generated by reservoir gas processed in the multistage separators. Table 9.16 gives the results of the swelling

Table 9.15. *Hydrocarbon analysis of produced well stream-Mol percent: Depletion study at 200° F.*

	Reservoir pressure (psig)						
Component	3,428	3,000	2,400	1,800	1,200	700	700*
Carbon dioxide (CO_2)	1.21	1.24	1.27	1.31	1.33	1.32	0.44
Nitrogen (N_2)	1.94	2.13	2.24	2.27	2.20	2.03	0.14
Methane (C_1)	65.99	69.78	72.72	73.98	73.68	71.36	12.80
Ethane (C_2)	8.69	8.66	8.63	8.79	9.12	9.66	5.27
Propane (C_3)	5.91	5.67	5.46	5.38	5.61	6.27	7.12
Iso-butane (IC_4)	2.39	2.20	2.01	1.93	2.01	2.40	4.44
N-butane (NC_4)	2.78	2.54	2.31	2.18	2.27	2.60	5.96
Iso-pentane (IC_5)	1.57	1.39	1.20	1.09	1.09	1.23	4.76
N-pentane (NC_5)	1.12	0.96	0.82	0.73	0.72	0.84	3.74
Hexanes (C_6)	1.81	1.43	1.08	0.88	0.83	1.02	8.46
Heptanes (C_7)	1.44	1.06	0.73	0.55	0.49	0.60	8.09
Octanes (C_8)	1.50	1.06	0.66	0.44	0.34	0.40	9.72
Nonanes (C_9)	1.05	0.69	0.40	0.25	0.18	0.16	7.46
Decanes (C_{10})	0.73	0.43	0.22	0.12	0.08	0.07	5.58
Undecanes (C_{11})	0.49	0.26	0.12	0.06	0.03	0.02	3.96
Dodecanes plus (C_{12+})	1.38	0.50	0.13	0.04	0.02	0.02	12.06
Total	100.00	100.00	100.00	100.00	100.00	100.00	100.00
Molecular weight of heptanes plus (C_{7+})	140	127	118	111	106	105	148
Specific gravity of heptanes plus (C_{7+})	0.774	0.761	0.752	0.745	0.740	0.739	0.781
Deviation Z-factor Equilibrium gas	0.803	0.798	0.802	0.830	0.877	0.924	
Two phase	0.803	0.774	0.748	0.730	0.703	0.642	
Well stream produced-Cumulative percent of initial	0.00	9.095	24.702	42.026	59.687	74.019	
GPM from smooth compositions Propane plus (C_{3+})	8.729	6.598	5.159	4.485	4.407	5.043	
Butanes plus (C_{4+})	7.112	5.046	3.665	3.013	2.872	3.328	
Pentanes plus (C_{5+})	5.464	3.535	2.287	1.702	1.507	1.732	

*Equilibrium liquid phase, representing 10.762% of original well stream.

Table 9.16. *Solubility and swelling test at* 200° *F (injection gas-lean gas).*

Mixture number	Cumul. gas injected (SCF/bbl)(1)	Cumul. gas injected (Mol fraction)(2)	Swollen volume(3)	Dew point pressure (psig)
0*	0.0	0.0000	1.0000	3,428
1	190	0.1271	1.1224	3,635
2	572	0.3046	1.3542	4,015
3	1,523	0.5384	1.9248	4,610
4	2,467	0.6538	2.5043	4,880

*Original reservoir fluid.
(1) SCF/bbl is the cumulative cubic feet of injection gas at 14.65 psia and 60° F per barrel of original reservoir fluid at 3,428 psig and 200° F.
(2) Mol fraction is cumulative mols of injection gas per total mols of indicated mixture.
(3) Swollen volume is barrels of indicated mixture at its dew point pressure and 200° F per barrel of original reservoir fluid at 3,428 psig and 200° F.

Table 9.17. *HC_1, HC_2, and HC_3.*

Component	Mole fraction	Molecular weights	Specific gravity
HC_1	0.05011	118.44	0.74985
HC_2	0.01340	193.95	0.81023
HC_3	0.00238	295.30	0.86651

Table 9.18. *Pseudogrouping of components.*

Pseudocomponent	P_1	P_2	P_3	P_4	P_5	P_6	P_7
Natural component	C_1, N_2	C_2, CO_2	C_3, C_4	C_5, C_6	HC_1	HC_2	HC_3
Mole fraction	0.6793	0.0990	0.1108	0.0450	0.05011	0.0134	0.00238
Molecular weights	16.38	31.77	50.64	77.78	118.44	193.95	295.30

experiments of reservoir gas with lean gas for the four samples. Note that the dew point pressure increases by approximately 50% for lean gas additions of 2,467 SCF/bbl for a total gas content of about 8,000–9,000 SCF/STB.

PVT study for matching the PVT data

The PVT study includes:

- splitting C_{7+},
- pseudogrouping,
- constant composition expansion and constant volume depletion,
- swelling tests,
- critical parameters at the formation and separator conditions for compositional modeling.

The heavy C_{7+} component is split into three components, HC_1, HC_2, and HC_3, to enhance the accuracy of PVT data matching. The mole fractions, molecular weights, and specific gravity of these components are stated in Table 9.17.

We use a pseudogrouping approach to group components. The purpose of pseudogrouping is to reduce the number of components involved in compositional modeling. These pseudocomponents are described in Table 9.18.

Detailed matches of the PVT data are displayed in Figures 9.3–9.6. Figure 9.3 shows pressure-volume data in constant composition expansion of the reservoir gas at 200° F. Figure 9.4 indicates retrograde condensate during constant volume depletion. Liquid yield by multistage surface separation in reservoir gas produced by constant volume depletion is displayed in Figure 9.5. The results of swelling of reservoir gas with increasing the dew point pressure of injected lean gas are given in Figure 9.6. There is a very good agreement between the laboratory and computed PVT data.

Finally, Tables 9.19–9.22 give a summary for the characterization data and binary interaction coefficients of the components at the formation and separator conditions.

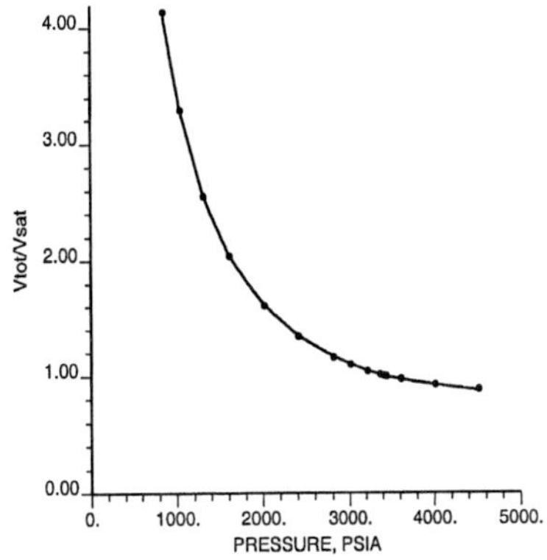

Figure 9.3. *Pressure-volume relation of reservoir fluid at* 200° *F: Constant composition expansion (cf. Table* 9.7*); laboratory data (dotted) and computed data (solid).*

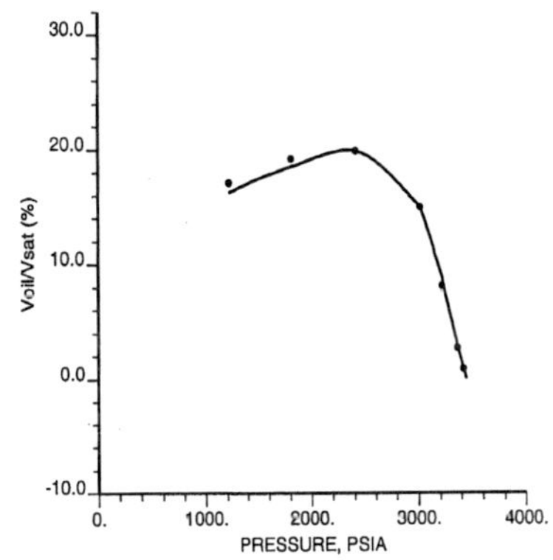

Figure 9.4. *Retrograde condensate during constant volume gas depletion at* 200° *F (cf. Table* 9.13*); laboratory data (dotted) and computed data (solid).*

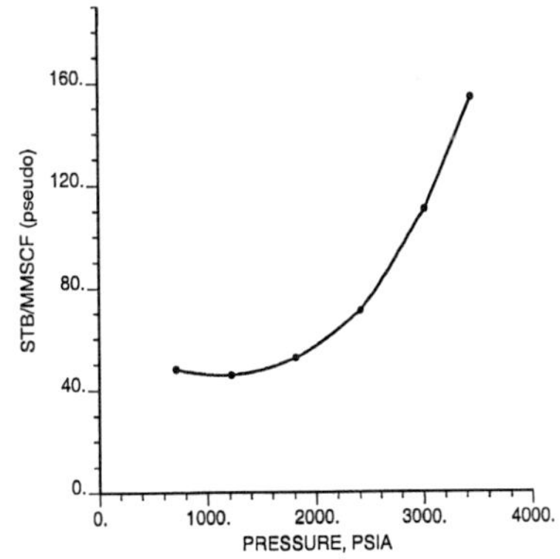

Figure 9.5. *Three-stage separator yield during constant volume gas depletion at* 200° *F (cf. Table* 9.14*); laboratory data (dotted) and computed data (solid).*

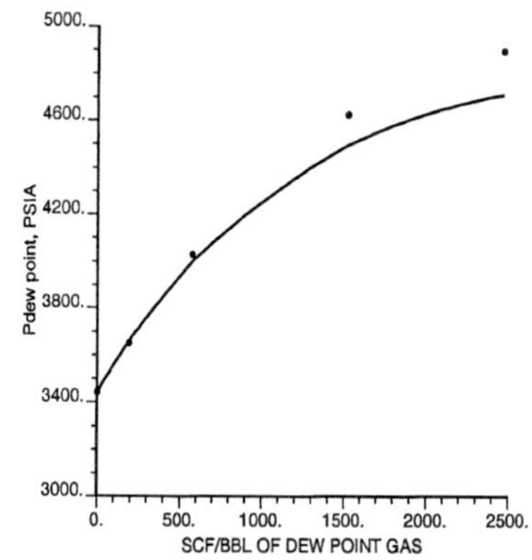

Figure 9.6. *Dew point pressure versus cumulative gas injected during swelling with lean gas at 200° F (cf. Table 9.16); laboratory data (dotted) and computed data (solid).*

Table 9.19. *Characterization data of components at the formation conditions.*

Pseudo-components	Z_c	P_c (psia)	T_c (°F)	Molecular weight	Acentric ω	Ω_a	Ω_b
P_1	0.28968	667.96	−119.11	16.38	0.00891	0.34477208	0.06328161
P_2	0.28385	753.82	90.01	31.77	0.11352	0.52197368	0.09982480
P_3	0.27532	586.26	252.71	50.64	0.17113	0.51497212	0.10747888
P_4	0.26699	469.59	413.50	77.78	0.26910	0.41916871	0.09345540
P_5	0.27164	410.14	605.99	118.44	0.34196	0.48594317	0.07486045
P_6	0.23907	260.33	795.11	193.95	0.51730	0.57058309	0.10120595
P_7	0.22216	183.92	988.26	295.30	0.72755	0.45723552	0.07779607

Table 9.20. *Binary interaction coefficients at the formation conditions.*

Components	P_1	P_2	P_3	P_4	P_5	P_6	P_7
P_1	0.0						
P_2	0.000622	0.0					
P_3	−0.002471	−0.001540	0.0				
P_4	0.011418	0.010046	0.002246	0.0			
P_5	−0.028367	0.010046	0.002246	0.0	0.0		
P_6	−0.100000	0.010046	0.002246	0.0	0.0	0.0	
P_7	0.206868	0.010046	0.002246	0.0	0.0	0.0	0.0

Table 9.21. *Characterization data of components at the separator conditions.*

Pseudo-components	Z_c	P_c (psia)	T_c (°F)	Molecular weight	Acentric ω	Ω_a	Ω_b
P_1	0.28968	667.96	−119.11	16.38	0.00891	0.50202385	0.09960379
P_2	0.28385	753.82	90.01	31.77	0.11352	0.45532152	0.08975547
P_3	0.27532	586.26	252.71	50.64	0.17113	0.46923415	0.08221724
P_4	0.26699	469.59	413.50	77.78	0.26910	0.58758251	0.08178213
P_5	0.27164	410.14	605.99	118.44	0.34196	0.55567652	0.06715680
P_6	0.23907	260.33	795.11	193.95	0.51730	0.49997263	0.07695341
P_7	0.22216	183.92	988.26	295.30	0.72755	0.45723552	0.07779607

Table 9.22. *Binary interaction coefficients at the separator conditions.*

Components	P_1	P_2	P_3	P_4	P_5	P_6	P_7
P_1	0.0						
P_2	0.000622	0.0					
P_3	−0.002471	−0.001540	0.0				
P_4	0.011418	0.010046	0.002246	0.0			
P_5	0.117508	0.010046	0.002246	0.0	0.0		
P_6	0.149871	0.010046	0.002246	0.0	0.0	0.0	
P_7	0.112452	0.010046	0.002246	0.0	0.0	0.0	0.0

Table 9.23. *The initial fluids in-place.*

Wet gas (BSCF)	Dry gas (BSCF)	Stock tank oil (MMSTB)
25.774	23.246	3.450

9.4.2 Reservoir simulation study

The initial fluids in-place using multistage separation are given in Table 9.23. Simulation results for the compositional model considered are given in Figures 9.7–9.13. The time step size used in iterative IMPES is about 30 days (in the first few time steps, it is smaller). The compositional simulator uses the ORTHOMIN Krylov subspace algorithm, with incomplete LU factorization preconditioners (cf. Chapter 5), as the linear solver.

As noted earlier, the first case is gas cycling with constant sales gas removal, while the second case is cycling with some gas sales deferral to enhance pressure maintenance in the early life of the reservoir. The total sales gas removal is the same for the two cases; the difference lies in the way sales gas is removed in the first ten years (cf. Table 9.3). For a gas condensate reservoir, decreasing the occurrence of retrograde condensate phenomena leads to less loss of heavy hydrocarbon components and more production of oil.

Stock-tank oil rates for the first and second cases and the corresponding cumulative liquid production for these cases at the final simulation time of 15 years are shown in Figures 9.7–9.10. Incremental stock-tank oil produced by gas-sales deferral (the second case minus the first), and oil saturations are given in Figures 9.11–9.13. Primary separator switchout occurs late in the cycling phase (10 years). The predicted surface oil rate is closely correlated with the liquid yield predictions shown in Figure 9.5.

Figure 9.11 gives the incremental stock-tank oil produced by gas-sales deferral. At the peak of this curve (at the eighth year), the cumulative stock-tank oil produced by the second case is 182 MSTB more than that from the first case (i.e, 9.76% increase). At the final production time (the 15th year), the increase is down to 159 MSTB (6.65%). This phenomenon can be understood from the observation that after injection of recycle gas stops, liquid production is due to depletion only, and the heavy end fractions vaporize into the vapor phase and are produced.

Figures 9.12 and 9.13 give the oil saturation at the gridblock (7,7,4) for these two cases, respectively. From these two figures, we see that the oil saturation in the second case is smaller than that in the first case. This shows that the retrograde condensate phenomenon in the second case occurs less than that in the first.

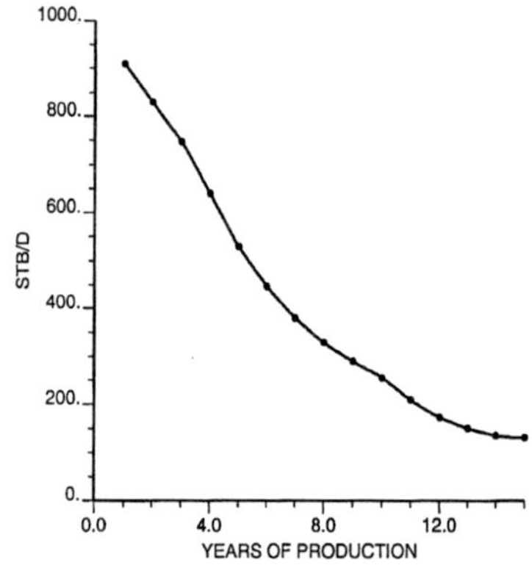

Figure 9.7. *Stock-tank oil production rate in case* 1.

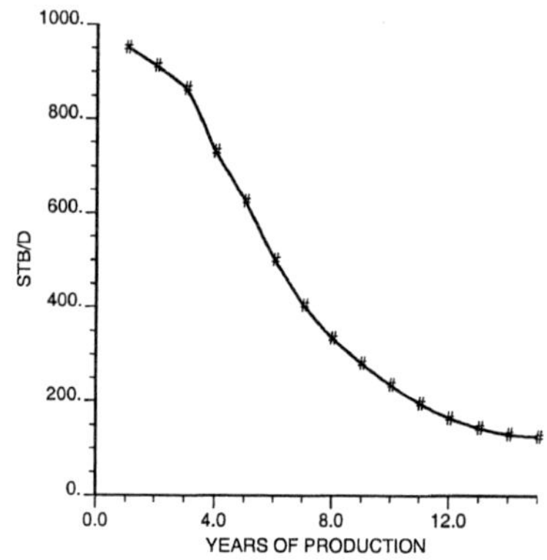

Figure 9.8. *Stock-tank oil production rate in case* 2.

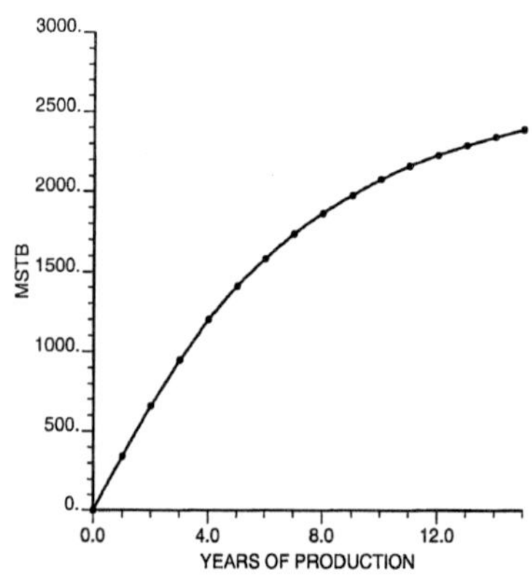

Figure 9.9. *Cumulative stock-tank oil production in case* 1.

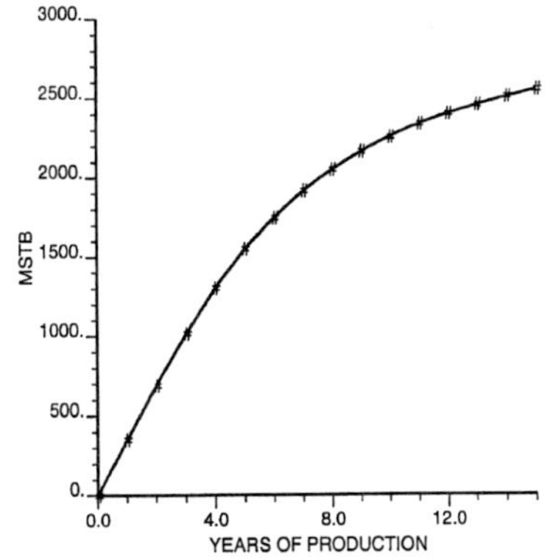

Figure 9.10. *Cumulative stock-tank oil production in case* 2.

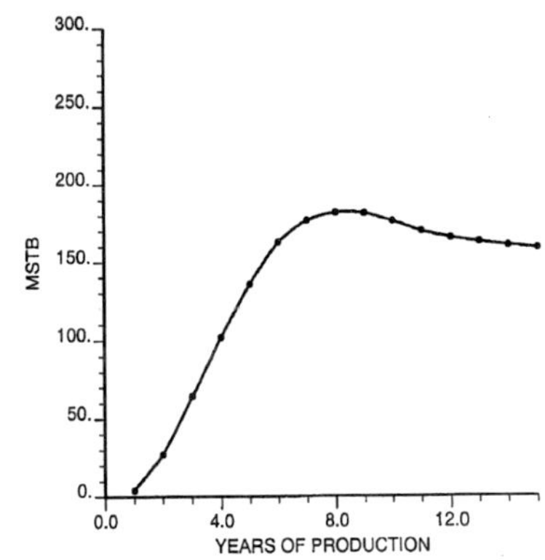

Figure 9.11. *Incremental stock-tank oil produced by gas-sales deferral (case* 2 *minus case* 1*).*

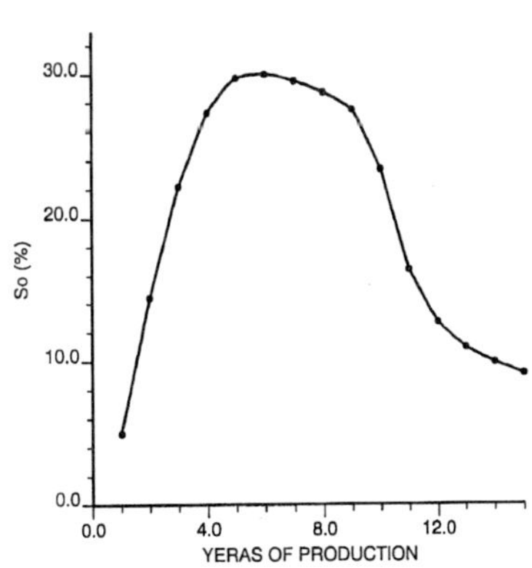

Figure 9.12. *Oil saturation in grid block* (7,7,4) *in case* 1.

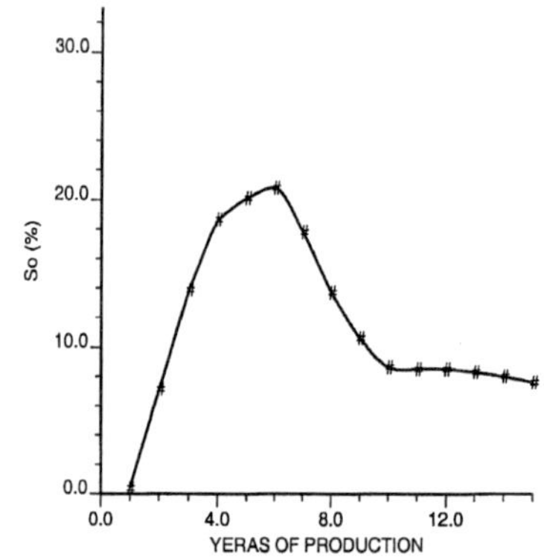

Figure 9.13. *Oil saturation in grid block* (7,7,4) *in case* 2.

Compared with those prepared by the nine companies (Kenyon and Behie, 1987), the numerical results in Figures 9.7–9.13 show that the numerical scheme here performs very well. The stock-tank oil rate and corresponding cumulative production are close to the respective averaged values of those provided by nine companies (Kenyon and Behie, 1987). In the numerical scheme for the compositional simulation here, the treatment of crossing "bubble points" and "dew points" in a Newton–Raphson iteration is very accurate, which leads to a very accurate computation of Jacobian matrices when the flow changes from three-phase to two-phase or vice versa. The scheme here also utilizes an accurate postprocessing technique for checking consistency of the solution variables (F, L) with the natural variables (S_o, S_g) after the Newton–Raphson iteration.

9.4.3 Computational remarks

We have applied an iterative IMPES solution technique to the numerical simulation of three-dimensional, three-phase, multicomponent compositional flow in porous media. The CVFE method with linear elements was employed for discretizing the governing equations of this compositional model. Numerical experiments were presented for the benchmark problem of the third CSP and showed that the iterative IMPES technique performs very well for this problem of a moderate size. To simulate accurately the process of recycle gas injection in a gas condensate reservoir using a compositional model, from our experience the following factors are very important:

- Through a PVT data match of the retrograde condensate curve during constant volume depletion, one can predict accurately the change of the reservoir oil saturation during a pressure decrease.

- Through a PVT data match of swelling tests, one sees that the increase of the dew point pressure after injection of recycle gas can lead to the transfer of heavy hydrocarbon components in the thermodynamic equilibrium from the liquid phase to the vapor phase and to the production of these components at production wells, thus increasing production.

- In compositional simulations, it is necessary to input two sets of critical PVT data; one for high pressure used for simulation of a reservoir flow process, and the other for lower pressure used for simulation of a separator process. The efficiency of enhanced oil recovery depends on the accuracy of the separator simulation.

The simulations in this section were performed on an SGI Power Indigo with 1 GB RAM, and the CPU time for the present compositional problem at the final time of 15 years is about 39 seconds.

9.5 Bibliographical Remarks

The choice of primary variables made in Section 9.2.1 follows Nolen (1973) and Young and Stephenson (1983). The numerical results reported in Section 9.4 are taken from Chen et al. (2006), which contains additional numerical results. More information about the data used in the third SPE CSP can be found in Kenyon and Behie (1987).

Exercises

9.1. Derive equation (9.22) using the second equation in (9.3) and equations (9.4), (9.16), and (9.21) and neglecting the variation of ρ_α with respect to space.

9.2. For the Newton–Raphson flash calculation introduced in Section 9.3.2, evaluate $\partial \varphi_{i\alpha}/\partial x_{j\alpha}$, $i,j = 1,2,\ldots,N_c$, $\alpha = o,g$.

9.3. For the Newton–Raphson flash calculation introduced in Section 9.3.2, compute $\partial Z_\alpha/\partial x_{j\alpha}$, $j = 1,2,\ldots,N_c$, $\alpha = o,g$.

9.4. Given the cubic equation

$$X^3 + PX + Q = 0,$$

show that its three roots are

$$X_1 = \sqrt[3]{-\frac{Q}{2} + \sqrt{\left(\frac{Q}{2}\right)^2 + \left(\frac{P}{3}\right)^3}} + \sqrt[3]{-\frac{Q}{2} - \sqrt{\left(\frac{Q}{2}\right)^2 + \left(\frac{P}{3}\right)^3}},$$

$$X_2 = \omega\sqrt[3]{-\frac{Q}{2} + \sqrt{\left(\frac{Q}{2}\right)^2 + \left(\frac{P}{3}\right)^3}} + \omega^2\sqrt[3]{-\frac{Q}{2} - \sqrt{\left(\frac{Q}{2}\right)^2 + \left(\frac{P}{3}\right)^3}},$$

$$X_3 = \omega^2\sqrt[3]{-\frac{Q}{2} + \sqrt{\left(\frac{Q}{2}\right)^2 + \left(\frac{P}{3}\right)^3}} + \omega\sqrt[3]{-\frac{Q}{2} - \sqrt{\left(\frac{Q}{2}\right)^2 + \left(\frac{P}{3}\right)^3}}.$$

9.5. Let X_1, X_2, and X_3 be the three roots in Exercise 9.4. Prove that they satisfy equations (9.54).

9.6. Defining $\mathcal{Z} = X - \frac{B}{3}$, show that equation (9.52) can be transformed into equation (9.53) with

$$P = -\frac{B^2}{3} + C, \quad Q = \frac{2B^3}{27} - \frac{BC}{3} + D.$$

9.7. Prove that the three roots $\mathcal{Z}_1$, $\mathcal{Z}_2$, and $\mathcal{Z}_3$ of equation (9.52) satisfy

$$\mathcal{Z}_1 + \mathcal{Z}_2 + \mathcal{Z}_3 = -B, \quad \frac{1}{\mathcal{Z}_1} + \frac{1}{\mathcal{Z}_2} + \frac{1}{\mathcal{Z}_3} = -\frac{C}{D}, \quad \mathcal{Z}_1\mathcal{Z}_2\mathcal{Z}_3 = -D.$$

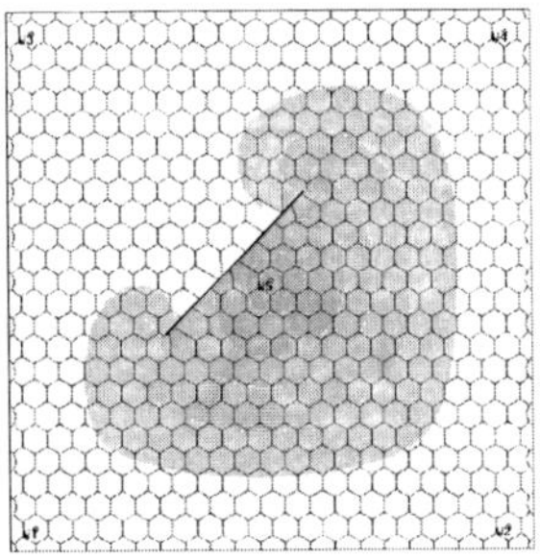

Chapter 10
Nonisothermal Flow

Isothermal flows were considered in Chapters 6–9; we now discuss numerical simulation of *nonisothermal flow* in a petroleum reservoir. Thermal methods, particularly *steam drive* and *soak*, make up a very large share of the enhanced oil recovery (EOR) projects in the petroleum industry and have experienced rapid growth since the early 1970s. Steam methods recently accounted for nearly 80% of the EOR oil in USA (Lake, 1989). Thermal flooding has been commercially successful for the past 40 years.

Thermal methods rely on several displacement mechanisms to recover oil, such as viscosity reduction, distillation, miscible displacement, thermal expansion, wettability changes, cracking, and lowered oil-water interfacial tension. For many applications, most important is the reduction of crude viscosity with increasing temperature. Four basic approaches to achieve this mechanism are hot water flooding, steam soak, steam drive, and in situ combustion. In a steam soak (stimulation or huff'n puff), for example, steam is introduced into a well, and then the well is returned to production after a brief shut-in period.

The basic differential equations for nonisothermal flow are reviewed in Section 10.1. The rock and fluid properties are also briefly stated there. The SS technique developed for the black oil model in Chapter 8 is extended to the nonisothermal flow in Section 10.2. Numerical results based on the fourth CSP organized by the SPE are reported in Section 10.3. Finally, bibliographical information is given in Section 10.4.

10.1 Basic Differential Equations

The governing equations for nonisothermal flow in a porous medium Ω were described in Section 2.9. The mass conservation equations and Darcy's laws are the same as for the compositional model discussed in Chapter 9; an additional energy conservation equation is required. For the convenience of the reader, we review these equations.

The governing equations are based on the displacement mechanisms of thermal methods: (a) reduction of crude viscosity with increasing temperature, (b) change of relative permeabilities for greater oil displacement, (c) vaporization of connate water and of a portion of crudes for miscible displacement of light components, and (d) high temperatures of fluids and rock to maintain high reservoir pressure. They can model the following important physical factors and processes:

- viscosity, gravity, and capillary forces,
- heat conduction and convection processes,
- heat losses to overburden and underburden of a reservoir,
- mass transfer between phases,
- effects of temperature on the physical property parameters of oil, gas, and water,
- rock compression and expansion.

10.1.1 The basic equations

We assume that the chemical components form at most three phases (e.g., water, oil, and gas), N_c chemical components may exist in all three phases, and diffusive effects are neglected.

Let ϕ and $\mathbf{k}$ denote the porosity and permeability of a porous medium $\Omega \subset \mathbb{R}^3$, and let S_α, μ_α, p_α, $\mathbf{u}_\alpha$, and $k_{r\alpha}$ be the saturation, viscosity, pressure, volumetric velocity, and relative permeability, respectively, of the α-phase, $\alpha = w, o, g$. Also, let $\xi_{i\alpha}$ represent the molar density of component i in the α-phase, $i = 1, 2, \ldots, N_c$, $\alpha = w, o, g$. The molar density of phase α is given by

$$\xi_\alpha = \sum_{i=1}^{N_c} \xi_{i\alpha}, \qquad \alpha = w,\ o,\ g. \tag{10.1}$$

The mole fraction of component i in phase α is then defined by

$$x_{i\alpha} = \xi_{i\alpha}/\xi_\alpha, \qquad i = 1, 2, \ldots, N_c,\ \alpha = w,\ o,\ g. \tag{10.2}$$

The total mass is conserved for each component:

$$\begin{aligned} &\frac{\partial}{\partial t} \sum_{\alpha=w}^{g} x_{i\alpha}\xi_\alpha S_\alpha + \nabla \cdot \sum_{\alpha=w}^{g} x_{i\alpha}\xi_\alpha \mathbf{u}_\alpha \\ &\qquad = \sum_{\alpha=w}^{g} x_{i\alpha} q_\alpha, \qquad i = 1, \ldots, N_c, \end{aligned} \tag{10.3}$$

where q_α stands for the flow rate of phase α at the wells. In (10.3), the volumetric velocity $\mathbf{u}_\alpha$ is given by Darcy's law:

$$\mathbf{u}_\alpha = -\frac{k_{r\alpha}}{\mu_\alpha}\mathbf{k}\left(\nabla p_\alpha - \rho_\alpha \wp \nabla z\right), \qquad \alpha = w,\ o,\ g, \tag{10.4}$$

where ρ_α is the mass density of the α-phase, $\wp$ is the magnitude of the gravitational acceleration, and z is the depth. The energy conservation equation takes the form

$$\begin{aligned} &\frac{\partial}{\partial t}\left(\phi \sum_{\alpha=w}^{g} \rho_\alpha S_\alpha U_\alpha + (1-\phi)\rho_s C_s T\right) \\ &\quad + \nabla \cdot \sum_{\alpha=w}^{g} \rho_\alpha \mathbf{u}_\alpha H_\alpha - \nabla \cdot (k_T \nabla T) = q_c - q_L, \end{aligned} \tag{10.5}$$

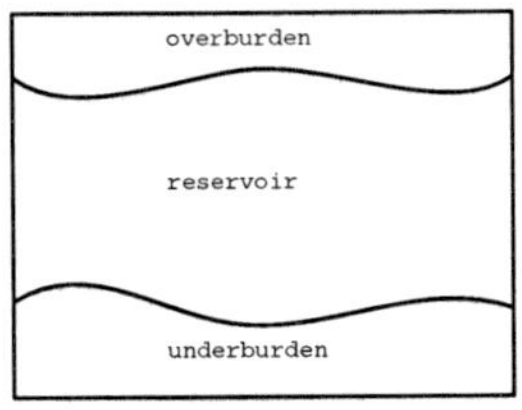

Figure 10.1. *Reservoir, overburden, and underburden.*

where T is the temperature, U_α and H_α are the specific internal energy and enthalpy of the α-phase (per unit mass), ρ_s and C_s are the density and the specific heat capacity of the solid, k_T represents the total thermal conductivity, q_c denotes the heat source item, and q_L indicates the heat loss to overburden and underburden. In (10.5), the *specific internal energy* U_α and *enthalpy* H_α of phase α can be computed from

$$U_\alpha = C_{V\alpha} T, \quad H_\alpha = C_{p\alpha} T,$$

where $C_{V\alpha}$ and $C_{p\alpha}$ represent the *heat capacities* of phase α at constant volume and constant pressure, respectively.

In addition to the differential equations (10.3)–(10.5), there are also algebraic constraints. The mole fraction balance implies

$$\sum_{i=1}^{N_c} x_{i\alpha} = 1, \qquad \alpha = w,\ o,\ g. \tag{10.6}$$

In the transport process, the saturation constraint reads

$$S_w + S_o + S_g = 1. \tag{10.7}$$

Finally, the phase pressures are related by capillary pressures

$$p_{cow} = p_o - p_w, \quad p_{cgo} = p_g - p_o. \tag{10.8}$$

The equilibrium relations describing the mass distribution of hydrocarbon components into the phases are given by

$$\begin{aligned} f_{iw}(p_w, T, x_{1w}, x_{2w}, \dots, x_{N_c w}) &= f_{io}(p_o, T, x_{1o}, x_{2o}, \dots, x_{N_c o}), \\ f_{io}(p_o, T, x_{1o}, x_{2o}, \dots, x_{N_c o}) &= f_{ig}(p_g, T, x_{1g}, x_{2g}, \dots, x_{N_c g}), \end{aligned} \tag{10.9}$$

where $f_{i\alpha}$ is the fugacity function of the ith component in the α-phase (cf. Section 3.2.5), $i = 1, 2, \dots, N_c$, $\alpha = w, o, g$.

In thermal methods, heat is lost to the adjacent strata of a reservoir or the *overburden* and *underburden*, which is included in the term q_L of (10.5). We assume that the overburden and underburden extend to infinity along both the positive and negative x_3-axis (the vertical direction); see Figure 10.1. If the overburden and underburden are impermeable, heat is transferred entirely through conduction. With all fluid velocities and convective fluxes being zero, the energy conservation equation (10.5) reduces to

$$\frac{\partial}{\partial t}\left(\rho_{ob} C_{p,ob} T_{ob}\right) = \nabla \cdot (k_{ob} \nabla T_{ob}), \tag{10.10}$$

where the subscript *ob* indicates that the variables are associated with the overburden and $C_{p,ob}$ is the heat capacity at constant pressure. The initial condition is the original temperature $T_{ob,0}$ of the overburden:

$$T_{ob}(\mathbf{x}, 0) = T_{ob,0}(\mathbf{x}).$$

The boundary condition at the top of the reservoir is

$$T_{ob}(x_1, x_2, x_3, t) = T(x_1, x_2, x_3, t).$$

At $x_3 = \infty$, T_{ob} is fixed:

$$T_{ob}(x_1, x_2, \infty, t) = T_\infty.$$

On other boundaries, we use the impervious boundary condition

$$k_{ob} \nabla T_{ob} \cdot \boldsymbol{\nu} = 0,$$

where $\boldsymbol{\nu}$ represents the outward unit normal to these boundaries. Now, the rate of heat loss to the overburden can be calculated by $k_{ob}\nabla T_{ob} \cdot \boldsymbol{\nu}$, where $\boldsymbol{\nu}$ is the unit normal to the interface between the overburden and reservoir (pointing to the overburden). For the underburden, the heat conduction equation is

$$\frac{\partial}{\partial t}\left(\rho_{ub} C_{p,ub} T_{ub}\right) = \nabla \cdot (k_{ub} \nabla T_{ub}), \tag{10.11}$$

and similar initial and boundary conditions can be developed as for the overburden.

Equations (10.3)–(10.9) provide $3N_c + 10$ independent relations, differential or algebraic, for the $3N_c + 10$ dependent variables: $x_{i\alpha}$, $\mathbf{u}_\alpha$, p_α, T, and S_α, $\alpha = w, o, g$, $i = 1, 2, \ldots, N_c$. If (10.10) and (10.11) are included, two more unknowns T_{ob} and T_{ub} are added. With proper initial and boundary conditions, this is a closed differential system for these unknowns.

10.1.2 Rock properties

The *rock properties* for nonisothermal flow are similar to those for the isothermal black oil and compositional models, but now these properties depend on temperature (cf. Section 3.3). In particular, the capillary pressures are of the form

$$p_{cw}(S_w, T) = p_w - p_o, \qquad p_{cg}(S_g, T) = p_g - p_o, \tag{10.12}$$

where $p_{cw} = -p_{cow}$ and $p_{cg} = p_{cgo}$. For notational convenience, we set $p_{co} = 0$. Similarly, the relative permeabilities for water, oil, and gas are

$$\begin{aligned} k_{rw} &= k_{rw}(S_w, T), & k_{row} &= k_{row}(S_w, T), \\ k_{rg} &= k_{rg}(S_g, T), & k_{rog} &= k_{rog}(S_g, T), \\ k_{ro} &= k_{ro}(S_w, S_g, T). \end{aligned} \tag{10.13}$$

Stone's models (cf. Section 3.1.2) can be adapted for the oil relative permeability k_{ro}, for example.

As an example, the relative permeability functions k_{rw} and k_{row} for a water-oil system can be defined by

$$
\begin{aligned}
k_{rw} &= k_{rwro}(T)\left(\frac{S_w - S_{wir}(T)}{1 - S_{orw}(T) - S_{wir}(T)}\right)^{nw}, \\
k_{row} &= k_{rocw}(T)\left(\frac{1 - S_w - S_{orw}(T)}{1 - S_{orw}(T) - S_{wc}(T)}\right)^{now},
\end{aligned}
\tag{10.14}
$$

and for a gas-oil system, k_{rg} and k_{rog} by

$$
\begin{aligned}
k_{rg} &= k_{rgro}(T)\left(\frac{S_g - S_{gr}^*}{1 - S_{wc}(T) - S_{oinit} - S_{gr}^*}\right)^{ng}, \\
k_{rog} &= k_{rocw}(T)\left(\frac{1 - S_g - S_{wc}(T) - S_{org}(T)}{1 - S_{wc}(T) - S_{org}(T)}\right)^{nog},
\end{aligned}
\tag{10.15}
$$

where nw, now, ng, and nog are nonnegative real numbers measured in the laboratory; S_{wc}, S_{wir}, S_{orw}, S_{org}, and S_{gr}^* are the connate water saturation, irreducible water saturation, residual oil saturation in the water-oil system, residual oil saturation in the gas oil system, and residual gas saturation; k_{rwro}, k_{rocw}, and k_{rgro} are the water relative permeability at the residual oil saturation for the water-oil system, the oil relative permeability at the connate water saturation, and the gas relative permeability at the residual oil saturation for the gas-oil system, respectively; and S_{oinit} is the initial oil saturation in the gas-oil system. Finally, for the rock properties, one must consider the thermal conductivity and heat capacity of the reservoir, overburden, and underburden.

10.1.3 Fluid properties

The equations of state discussed in Section 3.2.5 can be used to define the fugacity functions $f_{i\alpha}$ in (10.9). Because of complexity of nonisothermal flow, however, an *equilibrium K-value approach* is often used to describe the equilibrium relations (cf. Section 3.2.5):

$$x_{iw} = K_{iw}(p, T)x_{io}, \quad x_{ig} = K_{ig}(p, T)x_{io}, \quad i = 1, 2, \ldots, N_c. \tag{10.16}$$

One example of evaluating the K-values $K_{i\alpha}$ uses the empirical formula

$$K_{i\alpha} = \left(\kappa_{i\alpha}^1 + \frac{\kappa_{i\alpha}^2}{p} + \kappa_{i\alpha}^3\, p\right)\exp\left(-\frac{\kappa_{i\alpha}^4}{T - \kappa_{i\alpha}^5}\right), \tag{10.17}$$

where the constants $\kappa_{i\alpha}^j$ are obtained in the laboratory, $i = 1, 2, \ldots, N_c$, $j = 1, 2, 3, 4, 5$, $\alpha = w, g$, and p and T are pressure and temperature. For notational convenience, we use $K_{io} = 1, i = 1, 2, \ldots, N_c$.

Water properties

Physical properties of water and steam, such as density, internal energy, enthalpy, and viscosity, can be found from a *water-steam table* (Lake, 1989). Such a table is given in terms of the independent variables: pressure and temperature. In the case where all three phases coexist, a reservoir is in the saturated state. In this case, there is free gas; pressure and temperature are related, and only one of them is employed as an independent variable.

Oil properties

While any number of hydrocarbon components can be treated in the differential system describing the nonisothermal multiphase multicomponent flow considered in this chapter, computational work and time significantly increase as the number of components increases. It is often computationally convenient (or necessary) to group several similar chemical components into one mathematical component (cf. Section 9.4.1). In this way, only a few components (or *pseudocomponents*) are simulated in practical applications.

The oil phase is a mixture of hydrocarbon components, and these components range from the lightest component, methane (CH_4), to the heaviest component, bitumen. One way to reduce the number of components is to introduce pseudocomponents, as noted. According to the compositions of each pseudocomponent, one can deduce its physical properties, such as its pseudomolecular weight (which may not be a constant), critical pressure and temperature, compressibility, density, viscosity, thermal expansion coefficient, and specific heat. These properties are functions of pressure and temperature.

The most important property is the oil and gas phase viscosity dependence on temperature:

$$\mu_{io} = \exp\left(a_1 T^{b_1}\right) + c_1, \quad \mu_{ig} = a_2 T^{b_2},$$

where T is in absolute degrees, a_1, b_1, c_1, a_2, and b_2 are empirical parameters that can be measured in the laboratory, and μ_{io} and μ_{ig} are the viscosities of the ith component in the oil and gas phases, respectively.

10.2 Solution Techniques

In simulation of nonisothermal flow, three parts must be treated: the oil reservoir, overburden, and underburden. Because of the weak coupling between the reservoir and the overburden and underburden, the equations in these three parts can be decoupled; that is, they are solved in a sequential manner. In the reservoir domain, the IMPES, sequential, and SS techniques introduced for the black oil model in Chapter 8 can be applied. For the nonisothermal flow, because there exist strong nonlinearity and coupling in the governing equations, pressure and temperature vary greatly, and mass and energy transfer frequently between the oil and gas phases, the SS technique should be used for the reservoir system. The heat conduction equations for overburden and underburden are simple enough that a fully implicit scheme in time can be employed for their solution.

10.2.1 Choice of primary variables

As discussed earlier, (10.3)–(10.9) form a strongly coupled system of time-dependent nonlinear differential equations and algebraic constraints for $3N_c + 10$ unknowns. Although there are the same number of equations for these dependent variables, the entire system can be rewritten in terms of certain primary variables, with other variables being obtained from these variables.

Undersaturated state

As discussed in Section 8.1.4, if all three phases coexist, a reservoir is in the saturated state. When all the gas dissolves into the oil phase (i.e., there is no free gas; $S_g = 0$), the reservoir is in the undersaturated state. The choice of primary unknowns depends on the state of a reservoir.

We introduce the potentials

$$\Phi_\alpha = p_\alpha - \rho_\alpha \wp z, \qquad \alpha = w, o, g. \tag{10.18}$$

Also, we define the transmissibilities

$$\begin{aligned} \mathbf{T}_\alpha &= \frac{\rho_\alpha k_{r\alpha}}{\mu_\alpha}\mathbf{k}, \\ \mathbf{T}_{i\alpha} &= \frac{x_{i\alpha}\xi_\alpha k_{r\alpha}}{\mu_\alpha}\mathbf{k}, \qquad i = 1, 2, \ldots, N_c, \ \alpha = w, o, g. \end{aligned} \tag{10.19}$$

Moreover, we use the total mole fraction

$$x_i = \sum_{\alpha=w}^{g} x_{i\alpha}, \qquad i = 1, 2, \ldots, N_c. \tag{10.20}$$

Using (10.16), equation (10.20) becomes

$$x_{io} = \frac{1}{K_{iwog}(p, T)} x_i, \qquad i = 1, 2, \ldots, N_c, \tag{10.21}$$

where $K_{iwog}(p, T) = K_{iw} + 1 + K_{ig}$. As a result, we see that

$$x_{iw} = \frac{K_{iw}}{K_{iwog}} x_i, \quad x_{ig} = \frac{K_{ig}}{K_{iwog}} x_i, \quad i = 1, 2, \ldots, N_c. \tag{10.22}$$

Thus x_i should be used as a primary unknown, $i = 1, 2, \ldots, N_c$. Due to equation (10.6), only $N_c - 2$ unknowns are independent. Consequently, in the undersaturated state, $(p, S, x_1, x_2, \ldots, x_{N_c-2}, T)$ are chosen as the primary unknowns, where $p = p_o$ and $S = S_w$. The differential system for these unknowns consists of the N_c component mass conservation equations (cf. Exercise 10.1)

$$\frac{\partial(\phi F_i x_i)}{\partial t} = \sum_{\alpha=w}^{g} \nabla \cdot (\mathbf{T}_{i\alpha} \nabla \Phi_\alpha) + \sum_{\alpha=w}^{g} x_{i\alpha} q_\alpha, \qquad i = 1, 2, \ldots, N_c, \tag{10.23}$$

and the energy conservation equation

$$\frac{\partial}{\partial t}\left(\phi \sum_{\alpha=w}^{g} \rho_\alpha S_\alpha C_{V\alpha} T + (1-\phi)\rho_s C_s T\right) - \nabla \cdot \sum_{\alpha=w}^{g} C_{p\alpha} T \mathbf{T}_\alpha \nabla \Phi_\alpha - \nabla \cdot (k_T \nabla T) = q_c - q_L, \tag{10.24}$$

where

$$F_i = \sum_{\alpha=w}^{g} \frac{K_{i\alpha}}{K_{iwog}} \xi_\alpha S_\alpha .$$

Saturated state

In the saturated state, there is free gas. Pressure p and temperature T are related; their relationship may be given through a saturated steam table. Thus only one can be used as a primary unknown. In this case, we choose the primary unknowns $(p, S_w, S_o, x_1, x_2, \ldots, x_{N_c-2})$, where $p = p_o$. The system of differential equations is composed of the N_c component mass conservation equations (10.23) and the energy conservation equation (10.24).

10.2.2 The SS technique

Let $n > 0$ (an integer) indicate a time step. For any function v of time, we use $\bar{\delta} v$ to denote the forward time increment: $\bar{\delta} v = v^{n+1} - v^n$. A time approximation for the system of equations (10.23) and (10.24) can be defined ($i = 1, 2, \ldots, N_c$) as

$$\begin{aligned}
&\frac{1}{\Delta t}\bar{\delta}\left(\phi F_i x_i\right) = \sum_{\alpha=w}^{g} \nabla \cdot \left(\mathbf{T}_{i\alpha}^{n+1} \nabla \Phi_\alpha^{n+1}\right) + \sum_{\alpha=w}^{g} x_{i\alpha}^{n+1} q_\alpha^{n+1}, \\
&\frac{1}{\Delta t}\bar{\delta}\left(\phi \sum_{\alpha=w}^{g} \rho_\alpha S_\alpha C_{V\alpha} T + (1-\phi)\rho_s C_s T\right) \\
&\quad - \nabla \cdot \sum_{\alpha=w}^{g} \tilde{\mathbf{T}}_\alpha^{n+1} \nabla \Phi_\alpha^{n+1} - \nabla \cdot \left(k_T^{n+1} \nabla T^{n+1}\right) \\
&\qquad = q_c^{n+1} - q_L^{n+1},
\end{aligned} \tag{10.25}$$

where $\Delta t = t^{n+1} - t^n$ and $\tilde{\mathbf{T}}_\alpha^{n+1} = C_{p\alpha}^{n+1} T^{n+1} \mathbf{T}_\alpha^{n+1}$.

Since system (10.25) is nonlinear in the primary unknowns, it can be linearized via the Newton–Raphson iteration introduced in Section 8.2.1. For a generic function v of time, set

$$v^{n+1,l+1} = v^{n+1,l} + \delta v,$$

where l refers to the Newton–Raphson iteration number and δv represents the increment in this iteration step. When no ambiguity occurs, we replace $v^{n+1,l+1}$ and $v^{n+1,l}$ by v^{l+1} and v^l, respectively (i.e., the superscript $n+1$ is omitted). Note that $v^{n+1} \approx v^{l+1} = v^l + \delta v$, so $\bar{\delta} v \approx v^l - v^n + \delta v$. Applying this approximation to system (10.25) yields, for $i = 1, 2, \ldots, N_c$,

$$
\begin{aligned}
&\frac{1}{\Delta t}\left[(\phi F_i x_i)^l - (\phi F_i x_i)^n + \delta\left(\phi F_i x_i\right)\right] \\
&\qquad = \sum_{\alpha=w}^{g} \nabla \cdot \left(\mathbf{T}_{i\alpha}^{l+1} \nabla \Phi_\alpha^{l+1}\right) + \sum_{\alpha=w}^{g} x_{i\alpha}^{l+1} q_\alpha^{l+1}, \\
&\qquad \frac{1}{\Delta t}\left[\left(\phi \sum_{\alpha=w}^{g} \rho_\alpha S_\alpha C_{V\alpha} T + (1-\phi)\rho_s C_s T\right)^l \right. \\
&\qquad\qquad - \left(\phi \sum_{\alpha=w}^{g} \rho_\alpha S_\alpha C_{V\alpha} T + (1-\phi)\rho_s C_s T\right)^n \\
&\qquad\qquad \left. + \delta\left(\phi \sum_{\alpha=w}^{g} \rho_\alpha S_\alpha C_{V\alpha} T + (1-\phi)\rho_s C_s T\right)\right] \\
&\qquad - \nabla \cdot \sum_{\alpha=w}^{g} \tilde{\mathbf{T}}_\alpha^{l+1} \nabla \Phi_\alpha^{l+1} - \nabla \cdot \left(k_T^{l+1} \nabla T^{l+1}\right) = q_c^{l+1} - q_L^{l+1}.
\end{aligned}
\tag{10.26}
$$

Undersaturated state

We expand the left- and right-hand sides of the equations in system (10.26) in terms of the primary variables. Recall that the capillary pressures $p_{c\alpha}$ and relative permeabilities $k_{r\alpha}$ are known functions of saturation and temperature, and the viscosities μ_α, molar densities ξ_α, and mass densities ρ_α are functions of their respective phase pressure, compositions, and temperature, $\alpha = w, o, g$.

For the ith component flow equation,

$$
\begin{aligned}
\delta(\phi F_i x_i) &= c_{ip}\delta p + c_{iS}\delta S + \sum_{j=1}^{N_c-2} c_{ix_j}\delta x_j + c_{iT}\delta T, \\
& i = 1, 2, \ldots, N_c,
\end{aligned}
\tag{10.27}
$$

where

$$
c_{i\xi} = \left(\frac{\partial\left(\phi F_i x_i\right)}{\partial \xi}\right)^l, \qquad \xi = p,\ S,\ x_j,\ T.
$$

For the energy conservation equation,

$$
\begin{aligned}
&\delta\left(\phi \sum_{\alpha=w}^{o} \rho_\alpha S_\alpha C_{V\alpha} T + (1-\phi)\rho_s C_s T\right) \\
&\qquad = c_{Ep}\delta p + c_{ES}\delta S + \sum_{j=1}^{N_c-2} c_{Ex_j}\delta x_j + c_{ET}\delta T,
\end{aligned}
\tag{10.28}
$$

where

$$
c_{E\xi} = \left(\frac{\partial}{\partial \xi}\left(\phi \sum_{\alpha=w}^{o} \rho_\alpha S_\alpha C_{V\alpha} T + (1-\phi)\rho_s C_s T\right)\right)^l, \qquad \xi = p,\ S,\ x_j,\ T.
$$

In the undersaturated state, $\delta S_o = -\delta S$ and $\delta S_g = 0$.

In the SS technique, the potentials and transmissibilities are evaluated by

$$\Phi_\alpha^{l+1} = p^{l+1} + p_{c\alpha}^{l+1} - \rho_\alpha^{l+1} \wp z, \qquad \alpha = w, o, g,$$

and

$$\mathbf{T}_\alpha^{l+1} = \frac{\rho_\alpha^{l+1} k_{r\alpha}^{l+1}}{\mu_\alpha^{l+1}} \mathbf{k},$$

$$\mathbf{T}_{i\alpha}^{l+1} = \frac{x_{i\alpha}^{l+1} \xi_\alpha^{l+1} k_{r\alpha}^{l+1}}{\mu_\alpha^{l+1}} \mathbf{k}, \qquad i = 1, 2, \ldots, N_c, \ \alpha = w, o, g.$$

Consequently, we see that

$$\Phi_\alpha^{l+1} = \Phi_\alpha^l + d_{\alpha p}\delta p + d_{\alpha S}\delta S + \sum_{j=1}^{N_c-2} d_{\alpha x_j}\delta x_j + d_{\alpha T}\delta T, \tag{10.29}$$

where

$$d_{\alpha\xi} = \left(\frac{\partial \Phi_\alpha}{\partial \xi}\right)^l, \qquad \xi = p,\ S,\ x_j,\ T,\ \alpha = w, o, g.$$

Similarly, the transmissibilities are expanded:

$$\begin{aligned}\tilde{\mathbf{T}}_\alpha^{l+1} &= \tilde{\mathbf{T}}_\alpha^l + \tilde{\mathbf{E}}_{\alpha p}\delta p + \tilde{\mathbf{E}}_{\alpha S}\delta S + \sum_{j=1}^{N_c-2} \tilde{\mathbf{E}}_{\alpha x_j}\delta x_j + \tilde{\mathbf{E}}_{\alpha T}\delta T,\\ \mathbf{T}_{i\alpha}^{l+1} &= \mathbf{T}_{i\alpha}^l + \mathbf{E}_{i\alpha p}\delta p + \mathbf{E}_{i\alpha S}\delta S + \sum_{j=1}^{N_c-2} \mathbf{E}_{i\alpha x_j}\delta x_j + \mathbf{E}_{i\alpha T}\delta T,\end{aligned} \tag{10.30}$$

where, for $i = 1, 2, \ldots, N_c, \alpha = w, o, g,$

$$\tilde{\mathbf{E}}_{\alpha\xi} = \left(\frac{\partial \tilde{\mathbf{T}}_\alpha}{\partial \xi}\right)^l, \quad \mathbf{E}_{i\alpha\xi} = \left(\frac{\partial \mathbf{T}_{i\alpha}}{\partial \xi}\right)^l, \quad \xi = p,\ S,\ x_j,\ T.$$

The source/sink terms q_α^{l+1} can be expanded as for the black oil model in Chapter 8:

$$\begin{aligned}&\sum_{\alpha=w}^{g} x_{i\alpha}^{l+1} q_\alpha^{l+1}\\ &= \sum_{\alpha=w}^{g}\left(x_{i\alpha}^l q_\alpha^l + q_\alpha(\delta p, \delta p_{bh}, \delta S, \delta x_1, \delta x_2, \ldots, \delta x_{N_c-2}, \delta T)\right),\end{aligned} \tag{10.31}$$

where p_{bh} is the well bottom hole pressure. If this pressure is given, then $\delta p_{bh} = 0$.

Substituting (10.27)–(10.31) into (10.26) and neglecting higher-order terms of the increments yield the differential system in the increments of the primary unknowns at the

$(l+1)$th Newton–Raphson iteration of the $(n+1)$ time level in the undersaturated case (cf. Exercise 10.2), $i = 1, 2, \ldots, N_c$,

$$
\begin{aligned}
&\frac{1}{\Delta t}\left[(\phi F_i x_i)^l - (\phi F_i x_i)^n + c_{ip}\delta p + c_{iS}\delta S + \sum_{j=1}^{N_c-2} c_{ix_j}\delta x_j + c_{iT}\delta T\right] \\
&= \sum_{\alpha=w}^{g} \nabla\cdot\left\{\mathbf{T}^l_{i\alpha}\nabla\left(\Phi^l_\alpha + d_{\alpha p}\delta p + d_{\alpha S}\delta S + \sum_{j=1}^{N_c-2} d_{\alpha x_j}\delta x_j + d_{\alpha T}\delta T\right)\right.\\
&\qquad\left. + \left(\mathbf{E}_{i\alpha p}\delta p + \mathbf{E}_{i\alpha S}\delta S + \sum_{j=1}^{N_c-2}\mathbf{E}_{i\alpha x_j}\delta x_j + \mathbf{E}_{i\alpha T}\delta T\right)\nabla\Phi^l_\alpha\right\}\\
&\quad + \sum_{\alpha=w}^{g}\left(x^l_{i\alpha}q^l_\alpha + q_\alpha(\delta p, \delta p_{bh}, \delta S, \delta x_1, \delta x_2, \ldots, \delta x_{N_c-2}, \delta T)\right),
\end{aligned}
$$

$$
\begin{aligned}
&\frac{1}{\Delta t}\left[\left(\phi\sum_{\alpha=w}^{g}\rho_\alpha S_\alpha C_{V\alpha}T + (1-\phi)\rho_s C_s T\right)^l\right.\\
&\quad - \left(\phi\sum_{\alpha=w}^{g}\rho_\alpha S_\alpha C_{V\alpha}T + (1-\phi)\rho_s C_s T\right)^n\\
&\quad\left. + c_{Ep}\delta p + c_{ES}\delta S + \sum_{j=1}^{N_c-2} c_{Ex_j}\delta x_j + c_{ET}\delta T\right]\\
&\quad - \nabla\cdot\sum_{\alpha=w}^{g}\left\{\tilde{\mathbf{T}}^l_\alpha\nabla\left(\Phi^l_\alpha + d_{\alpha p}\delta p + d_{\alpha S}\delta S + \sum_{j=1}^{N_c-2} d_{\alpha x_j}\delta x_j + d_{\alpha T}\delta T\right)\right.\\
&\qquad\left. + \left(\tilde{\mathbf{E}}_{\alpha p}\delta p + \tilde{\mathbf{E}}_{\alpha S}\delta S + \sum_{j=1}^{N_c-2}\tilde{\mathbf{E}}_{\alpha x_j}\delta x_j + \tilde{\mathbf{E}}_{\alpha T}\delta T\right)\nabla\Phi^l_\alpha\right\}\\
&\quad - \nabla\cdot\left(k_T^{l+1}\nabla(T^l + \delta T)\right) = q_c^{l+1} - q_L^{l+1}.
\end{aligned}
\tag{10.32}
$$

This system is linear in the increments of the primary variables. The Newton Raphson iterations are constrained by maximum changes in these variables over the iterations, and an automatic time step size is determined by the maximum changes over the time step (cf. Section 8.2.2).

Saturated state

In the saturated state, the primary unknowns are p, S_w, S_o, and x_i, $i = 1, 2, \ldots, N_c - 2$. Thus, in this case, for the ith component flow equation ($i = 1, 2, \ldots, N_c$),

$$
\delta(\phi F_i x_i) = c_{ip}\delta p + c_{iS_w}\delta S_w + c_{iS_o}\delta S_o + \sum_{j=1}^{N_c-2} c_{ix_j}\delta x_j, \tag{10.33}
$$

where

$$
c_{i\xi} = \left(\frac{\partial(\phi F_i x_i)}{\partial \xi}\right)^l, \qquad \xi = p,\ S_w,\ S_o,\ x_j.
$$

Analogously, for the energy conservation equation,

$$\begin{aligned} \delta\left(\phi \sum_{\alpha=w}^{g} \rho_\alpha S_\alpha C_{V\alpha} T + (1-\phi)\rho_s C_s T\right) \\ = c_{Ep}\delta p + c_{ES_w}\delta S_w + c_{ES_o}\delta S_o + \sum_{j=1}^{N_c-2} c_{Ex_j}\delta x_j, \end{aligned} \tag{10.34}$$

where, for $\xi = p,\ S_w,\ S_o,\ x_j$,

$$c_{E\xi} = \left(\frac{\partial}{\partial \xi}\left(\phi \sum_{\alpha=w}^{g} \rho_\alpha S_\alpha C_{V\alpha} T + (1-\phi)\rho_s C_s T\right)\right)^l .$$

In the saturated state, $\delta S_g = -\delta S_w - \delta S_o$. The potentials are expanded:

$$\Phi_\alpha^{l+1} = \Phi_\alpha^l + d_{\alpha p}\delta p + d_{\alpha S_w}\delta S_w + d_{\alpha S_o}\delta S_o + \sum_{j=1}^{N_c-2} d_{\alpha x_j}\delta x_j, \tag{10.35}$$

where

$$d_{\alpha\xi} = \left(\frac{\partial \Phi_\alpha}{\partial \xi}\right)^l, \qquad \xi = p,\ S_w,\ S_o,\ x_j,\ \alpha = w, o, g.$$

The transmissibilities are evaluated by

$$\begin{aligned} \tilde{\mathbf{T}}_\alpha^{l+1} &= \tilde{\mathbf{T}}_\alpha^l + \tilde{\mathbf{E}}_{\alpha p}\delta p + \tilde{\mathbf{E}}_{\alpha S_w}\delta S_w + \tilde{\mathbf{E}}_{\alpha S_o}\delta S_o + \sum_{j=1}^{N_c-2} \tilde{\mathbf{E}}_{\alpha x_j}\delta x_j, \\ \mathbf{T}_{i\alpha}^{l+1} &= \mathbf{T}_{i\alpha}^l + \mathbf{E}_{i\alpha p}\delta p + \mathbf{E}_{i\alpha S_w}\delta S_w + \mathbf{E}_{i\alpha S_o}\delta S_o + \sum_{j=1}^{N_c-2} \mathbf{E}_{i\alpha x_j}\delta x_j, \end{aligned} \tag{10.36}$$

where, for $i = 1, 2, \ldots, N_c$, $\alpha = w, o, g$,

$$\tilde{\mathbf{E}}_{\alpha\xi} = \left(\frac{\partial \tilde{\mathbf{T}}_\alpha}{\partial \xi}\right)^l, \quad \mathbf{E}_{i\alpha\xi} = \left(\frac{\partial \mathbf{T}_{i\alpha}}{\partial \xi}\right)^l, \quad \xi = p,\ S_w,\ S_o,\ x_j.$$

Finally, the source/sink terms q_α^{l+1} have the form

$$\begin{aligned} &\sum_{\alpha=w}^{g} x_{i\alpha}^{l+1} q_\alpha^{l+1} \\ &= \sum_{\alpha=w}^{g} \left(x_{i\alpha}^l q_\alpha^l + q_\alpha(\delta p, \delta p_{bh}, \delta S_w, \delta S_o, \delta x_1, \delta x_2, \ldots, \delta x_{N_c-2})\right). \end{aligned} \tag{10.37}$$

Substituting (10.33)–(10.37) into (10.26) gives the differential system in the increments of the primary unknowns at the $(l+1)$th Newton–Raphson iteration of the $(n+1)$

time level in the saturated state (cf. Exercise 10.3):

$$
\begin{aligned}
&\frac{1}{\Delta t}\left[(\phi F_i x_i)^l - (\phi F_i x_i)^n + c_{ip}\delta p + c_{iS_w}\delta S_w + c_{iS_o}\delta S_o + \sum_{j=1}^{N_c-2} c_{ix_j}\delta x_j\right] \\
&= \sum_{\alpha=w}^{g} \nabla\cdot\left\{\mathbf{T}_{i\alpha}^l \nabla\left(\Phi_\alpha^l + d_{\alpha p}\delta p + d_{\alpha S_w}\delta S_w + d_{\alpha S_o}\delta S_o + \sum_{j=1}^{N_c-2} d_{\alpha x_j}\delta x_j\right)\right. \\
&\qquad \left. + \left(\mathbf{E}_{i\alpha p}\delta p + \mathbf{E}_{i\alpha S_w}\delta S_w + \mathbf{E}_{i\alpha S_o}\delta S_o + \sum_{j=1}^{N_c-2} \mathbf{E}_{i\alpha x_j}\delta x_j\right)\nabla\Phi_\alpha^l\right\} \\
&\quad + \sum_{\alpha=w}^{g}\left(x_{i\alpha}^l q_\alpha^l + q_\alpha(\delta p, \delta p_{bh}, \delta S_w, \delta S_o, \delta x_1, \delta x_2, \ldots, \delta x_{N_c-2})\right),
\end{aligned}
\tag{10.38}
$$

for $i = 1, 2, \ldots, N_c$, and

$$
\begin{aligned}
&\frac{1}{\Delta t}\left[\left(\phi\sum_{\alpha=w}^{g}\rho_\alpha S_\alpha C_{V\alpha}T + (1-\phi)\rho_s C_s T\right)^l\right. \\
&\qquad - \left(\phi\sum_{\alpha=w}^{g}\rho_\alpha S_\alpha C_{V\alpha}T + (1-\phi)\rho_s C_s T\right)^n \\
&\qquad \left. + c_{Ep}\delta p + c_{ES_w}\delta S_w + c_{ES_o}\delta S_o + \sum_{j=1}^{N_c-2} c_{Ex_j}\delta x_j\right] \\
&\quad - \nabla\cdot\sum_{\alpha=w}^{g}\left\{\tilde{\mathbf{T}}_\alpha^l \nabla\left(\Phi_\alpha^l + d_{\alpha p}\delta p + d_{\alpha S_w}\delta S_w + d_{\alpha S_o}\delta S_o + \sum_{j=1}^{N_c-2} d_{\alpha x_j}\delta x_j\right)\right. \\
&\qquad \left. + \left(\tilde{\mathbf{E}}_{\alpha p}\delta p + \tilde{\mathbf{E}}_{\alpha S_w}\delta S_w + \tilde{\mathbf{E}}_{\alpha S_o}\delta S_o + \sum_{j=1}^{N_c-2} \tilde{\mathbf{E}}_{\alpha x_j}\delta x_j\right)\nabla\Phi_\alpha^l\right\} \\
&\quad - \nabla\cdot\left(k_T^{l+1}\nabla(T^l + c_{Tp}\delta p)\right) = q_c^{l+1} - q_L^{l+1},
\end{aligned}
\tag{10.39}
$$

where $c_{Tp} = (dT/dp)^l$. Again, these equations are linear in the increments of the primary variables.

10.3 The Fourth SPE Project: Steam Injection

The experimental problems are chosen from the benchmark problems of the fourth CSP (Aziz et al., 1985). Six companies participated in the comparative project. Two related steam injection problems were numerically studied. The first deals with cyclic steam injection in a nondistillable petroleum reservoir with two-dimensional radial cross-sectional grids, and the second with nondistillable oil displacement by steam in an inverted nine-spot pattern by considering one-eighth of the full pattern (see Figure 10.2). Standard conditions for these problems are 14.7 psia and 60° F. The problems were chosen to exercise features of the models that are important in practical applications, though they may not represent a real field analysis.

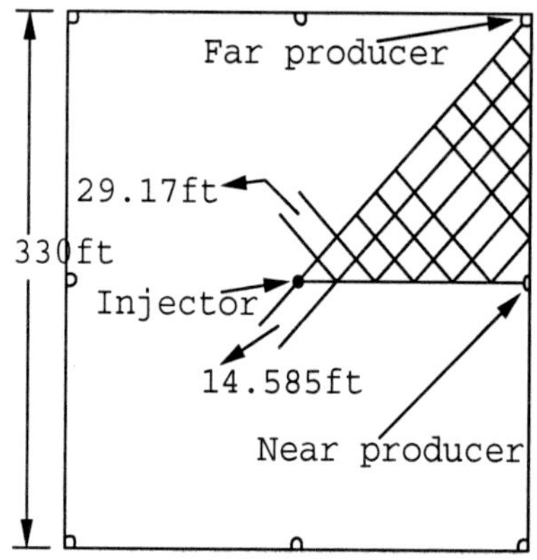

Figure 10.2. *Element of symmetry in an inverted nine-spot.*

Table 10.1. *Rock properties.*

k_h starting with the top layer: 2,000, 500, 1,000, and 2,000 md
k_v: 50% of k_h
Porosity: 0.3 for all layers
Thermal conductivity: 24 Btu/(ft.-day-°F)
Heat capacity: 35 Btu/(ft^3 of rock-°F)
Effective rock compressibility: 5.0E − 4 psi^{-1}

Table 10.2. *Oil properties.*

Density at standard conditions: 60.68 lb/ft^3
Compressibility: 5.0E − 6 psi^{-1}
Molecular weight: 600
Thermal expansion coefficient: 3.8E − 4 1/R
Specific heat: 0.5 Btu/(lb.-R)

10.3.1 The first problem

The aim is to simulate cyclic steam injection in a two-dimensional reservoir (closed system) with four layers. The rock properties are stated in Table 10.1, where k_h $(= k_{11} = k_{22})$ and k_v $(= k_{33})$ denote the horizontal and vertical permeabilities, respectively, and the thermal conductivity and heat capacity are given for the reservoir, overburden, and underburden. Water is assumed to be pure water with standard properties. Oil properties are listed in Table 10.2, and the viscosity dependence on temperature is given in Table 10.3. The capillary pressures are zero. The relative permeability functions are defined by (10.14) and (10.15) with the data $nw = 2.5$, $now = nog = 2$, $ng = 1.5$, $S_{wc} = S_{wir} = 0.45$, $S_{orw} = 0.15$, $S_{org} = 0.1$, $S^*_{gr} = 0.06$, $k_{rwro} = 0.1$, $k_{rocw} = 0.4$, and $k_{rgro} = 0.2$. The initial conditions are presented in Table 10.4, where pressure is distributed according to the gravity head.

The computational grid is cylindrical with 13 grid points in the radial direction. The well radius is 0.3 ft, and the exterior radius is 263.0 ft. The block boundaries in the radial

Table 10.3. *Oil viscosity dependence on temperature.*

Temp (°F)	75	100	150	200	250	300	350	500
Viscosity (cp)	5,780	1,389	187	47	17.4	8.5	5.2	2.5

Table 10.4. *Initial conditions.*

Oil saturation: 0.55
Water saturation: 0.45
Reservoir temperature: 125° F
Pressure at the center of the top layer: 75 psia

direction are at 0.30, 3.0, 13.0, 23.0, 33.0, 43.0, 53.0, 63.0, 73.0, 83.0, 93.0, 103.0, 143.0, and 263.0 ft, and the block boundaries in the vertical direction are at 0.0 (top of pay), 10.0, 30.0, 55.0, and 80.0 ft. The depth to the top of pay is 1,500 ft subsea.

The spatial discretization scheme of the simulation is based on the Raviart–Thomas mixed finite element method on rectangles (cf. Section 4.5.4). Upstream-weighted interblock flow and injection and production terms are included. The linear system of algebraic equations is solved by the ORTHOMIN (orthogonal minimum residual) iterative algorithm, with incomplete LU factorization preconditioners (cf. Chapter 5).

Finally, the operating conditions are summarized. All layers are open to flow during injection and production (a zero skin factor; cf. Chapter 13). The energy content of the injected steam is based on 0.7 quality and 450° F. Steam quality at bottom hole conditions is fixed at 0.7. Three cycles are simulated: each cycle is 365 days with injection for 10 days followed by a 7-day soak period, and the cycle is completed with 348 days of production. Steam is injected at capacity subject to the following conditions: the maximum bottom hole pressure is 1,000 psia at the center of the top layer, and the maximum injection rate is 1,000 STB/day. The production capacity is subject to the following constraints: the minimum bottom hole pressure is 17 psia at the center of the top layer, and the maximum production rate is 1,000 STB/day of liquids.

Figures 10.3 and 10.4 show the cumulative oil production and oil production rates, respectively. Compared with the results presented by Aziz et al. (1985), the two quantities in Figures 10.3 and 10.4 are closer to the respective averaged values of those provided by the six companies.

10.3.2 The second problem

The objective is to simulate one-eighth of an inverted nine-spot pattern via symmetry. The total pattern area is 2.5 acres. The rock and fluid properties, relative permeability data, and initial conditions are the same as for the first problem. The grid dimensions are $9 \times 5 \times 4$ (uniform in the horizontal direction). The radius of all wells is 0.3 ft.

The operating conditions are as follows: injection occurs only in the bottom layer, and production occurs from all four layers. Steam conditions are the same as in the first problem. Steam is injected at capacity subject to the following conditions: the maximum bottom hole pressure is 1,000 psia at the center of the bottom layer, and the maximum injection rate is 1,000 STB/day on a full-well basis. The production capacity is subject to

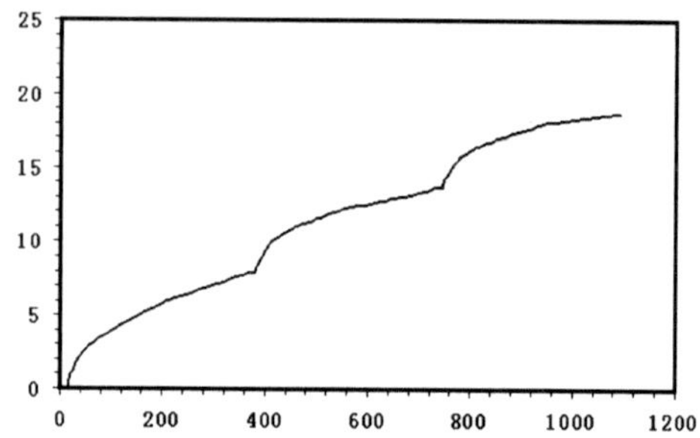

Figure 10.3. *Cumulative oil production (MSTB) versus time (days).*

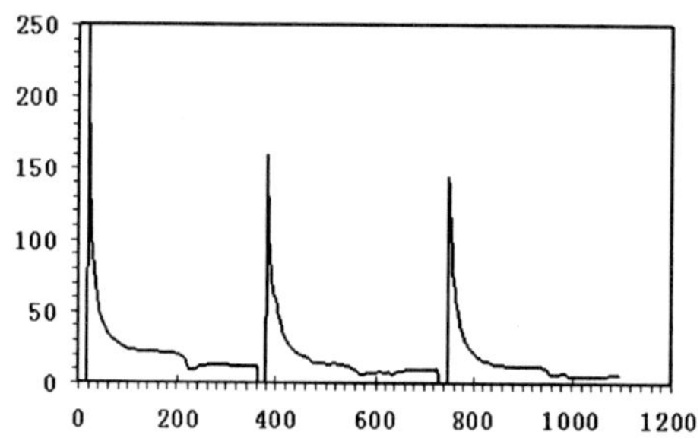

Figure 10.4. *Oil production rate (STB/day).*

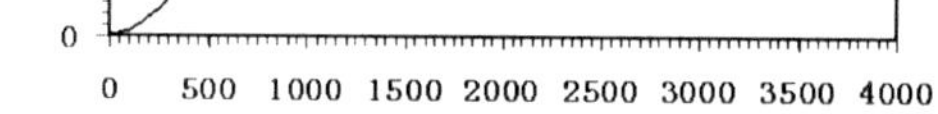

Figure 10.5. *Cumulative oil production for the full pattern (MSTB versus days).*

the constraints: the minimum bottom hole pressure is 17 psia at the center of the top layer, the maximum production rate is 1,000 STB/day of liquids, and the maximum steam rate is 10 STB/day. The simulation time is 10 years of injection and production.

Figures 10.5–10.7 indicate the cumulative oil production for the full pattern, the oil production rate for the far producer, and the oil production rate for the near producer, respectively. All well data presented are on a full-well basis, and the pattern results are for the full pattern consisting of four quarter (far) producers and four half (near) producers. Again, compared with the results presented by Aziz et al. (1985), the three quantities are closer to the respective mean values of those provided by the six companies.

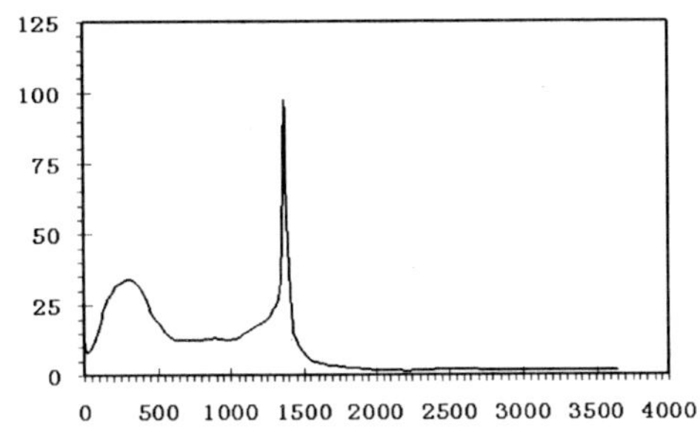

Figure 10.6. *Oil production rate for the far producer (STB/day).*

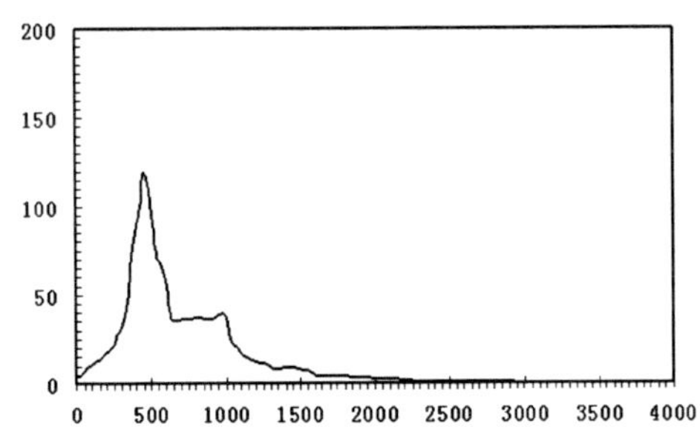

Figure 10.7. *Oil production rate for the near producer (STB/day).*

10.4 Bibliographical Remarks

Most of the content in this chapter is taken from Chen and Yu (2006). More details about the data used in the fourth SPE CSP can be found in Aziz et al. (1985).

Exercises

10.1. Derive equation (10.23) using equations (10.3), (10.4), and (10.18)–(10.22) and ignoring the variation of the mass densities ρ_α in space, $\alpha = w, o, g$.

10.2. Derive system (10.32) by substituting (10.27)–(10.31) into (10.26) and neglecting the higher-order terms of the increments of the primary unknowns ($\delta p, \delta S, \delta x_1, \delta x_2, \ldots, \delta x_{N_c-2}, \delta T$).

10.3. Derive equations (10.38) and (10.39) by substituting (10.33)–(10.37) into (10.26) and neglecting the higher-order terms of the increments of the primary unknowns ($\delta p, \delta S_w, \delta S_o, \delta x_1, \delta x_2, \ldots, \delta x_{N_c-2}$).

10.4. As an example, the SS technique was developed for the solution of the nonisothermal flow governing equations in Section 10.2. Develop a sequential solution technique for these equations in a manner similar to that given in Section 8.2.3.

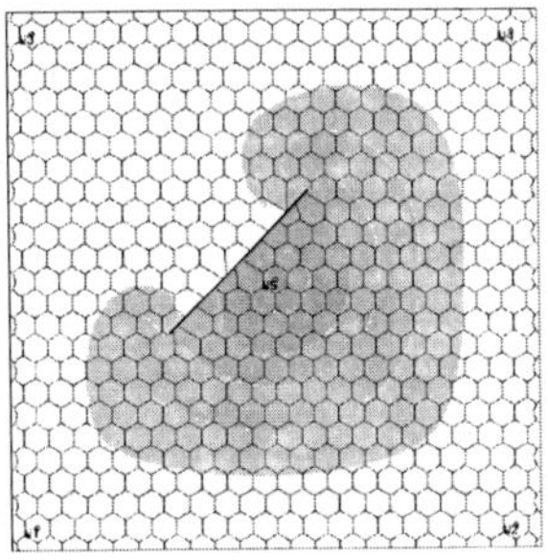

Chapter 11
Chemical Flooding

Enhanced oil recovery (EOR) is achieved by injecting materials that are not normally present in a petroleum reservoir. An important approach in EOR is *chemical flooding*: for example, alkaline, surfactant, polymer, and foam (*ASP+foam*) flooding. The injection of these chemical species reduces fluid mobility to improve the sweep efficiency of the reservoir, i.e., increases the volume of the permeable medium contacted at any given time. While chemical flooding in the petroleum industry has a larger scale of oil recovery efficiency than water flooding, it is far more technical, costly, and risky. The displacement mechanisms in this type of flooding involve interfacial tension lowering, capillary desaturation, chemical synergetic effects, and mobility control, and the flow and transport model describes such physicochemical phenomena as dispersion, diffusion, adsorption, chemical reaction, and in situ generation of surfactant from acidic crude oil.

We develop and study a multicomponent multiphase model for ASP+foam flooding. This model describes synergetic effects in the form of an interfacial tension function, the foam flow resistance in the function of surfactant and oil concentrations, capillary pressure, permeability, gas-liquid ratio, and gas velocity, and the phase behavior in terms of equations of state (EOSs). The balance equations are the mass balance equation for each chemical species, the aqueous phase pressure equation, and the energy balance equation. The major physical variables modeled are density, viscosity, velocity-dependent dispersion, molecular diffusion, adsorption, interfacial tension, relative permeability, capillary pressure, capillary trapping, cation exchange, and polymer and gel properties such as permeability reduction, inaccessible pore volume, and non-Newtonian rheology (Pope and Nelson, 1978). *Phase mobilization* is described through entrapped phase saturation and relative permeability dependence on the trapping number. *Chemical reactions* include aqueous electrolyte chemistry, precipitation/dissolution of minerals, ion-exchange reactions with the matrix (the geochemical option), reactions of acidic components of oil with the bases in the aqueous solution, and polymer reactions with cross-linking agents to form gel (Bhuyan et al., 1991).

The basic differential equations governing chemical flooding were described in Section 2.10 and are reviewed in Section 11.1. Then, in Sections 11.2–11.5, respectively, we describe the mathematical formulations for alkaline, polymer, surfactant, and foam displacement mechanisms. The rock and fluid properties are stated in Section 11.6. A numerical solution scheme is briefly presented in Section 11.7. Numerical results are reported in Sections 11.8 and 11.9. Finally, bibliographical information is given in Section 11.10.

11.1 Basic Differential Equations

The basic equations for a chemical flooding compositional model in a porous medium Ω were developed in Section 2.10. The governing differential equations for the chemical compositional model consist of a mass conservation equation for each chemical component, an energy equation, Darcy's law, and an overall mass conservation or continuity equation for pressure. These equations are developed under the following assumptions: local thermodynamic equilibrium, immobile solid phase, Fickian dispersion, ideal mixing, slightly compressible soil and fluids, and Darcy's law.

We consider the general case where N_c chemical components form N_p phases. Let ϕ and $\mathbf{k}$ denote the porosity and permeability of a porous medium $\Omega \subset \mathbb{R}^3$, and let ρ_α, S_α, μ_α, p_α, $\mathbf{u}_\alpha$, and $k_{r\alpha}$ be the density, saturation, viscosity, pressure, volumetric velocity, and relative permeability, respectively, of the α-phase, $\alpha = 1, 2, \ldots, N_p$. The mass conservation for component i is expressed in terms of the overall concentration of this component per unit pore volume:

$$\frac{\partial}{\partial t}(\phi \tilde{c}_i \rho_i) = -\nabla \cdot \left(\sum_{\alpha=1}^{N_p} \rho_i \left[c_{i\alpha} \mathbf{u}_\alpha - \mathbf{D}_{i\alpha} \nabla c_{i\alpha} \right] \right) + q_i, \tag{11.1}$$

for $i = 1, 2, \ldots, N_c$, where the overall concentration $\tilde{c}_i$ is the sum over all phases, including the adsorbed phases:

$$\tilde{c}_i = \left(1 - \sum_{j=1}^{N_{cv}} \hat{c}_j \right) \sum_{\alpha=1}^{N_p} S_\alpha c_{i\alpha} + \hat{c}_i, \qquad i = 1, 2, \ldots, N_c, \tag{11.2}$$

N_{cv} is the total number of volume-occupying components (such as water, oil, surfactant, and air); $\hat{c}_i$, ρ_i, and q_i are the adsorbed concentration, mass density, and source/sink term of component i; and $c_{i\alpha}$ and $\mathbf{D}_{i\alpha}$ are the concentration and diffusion-dispersion tensor, respectively, of component i in phase α.

The density ρ_i is related to a reference phase pressure p_r by

$$C_i = \frac{1}{\rho_i} \frac{\partial \rho_i}{\partial p_r} \bigg|_T$$

at a fixed temperature T, where C_i is the compressibility of component i. For a slightly compressible fluid, ρ_i is (cf. (2.13))

$$\rho_i = \rho_i^o \left(1 + C_i^o (p_r - p_r^o) \right), \tag{11.3}$$

where C_i^o and ρ_i^o are the constant compressibility and the density at the reference pressure p_r^o, respectively.

The diffusion-dispersion tensor $\mathbf{D}_{i\alpha}$ for multiphase flow is (cf. Section 2.4)

$$\mathbf{D}_{i\alpha}(\mathbf{u}_\alpha) = \phi \left\{ S_\alpha d_{i\alpha} \mathbf{I} + |\mathbf{u}_\alpha| \left(d_{l\alpha} \mathbf{E}(\mathbf{u}_\alpha) + d_{t\alpha} \mathbf{E}^\perp(\mathbf{u}_\alpha) \right) \right\}, \tag{11.4}$$

where $d_{i\alpha}$ is the molecular diffusion coefficient of component i in phase α; $d_{l\alpha}$ and $d_{t\alpha}$ are, respectively, the longitudinal and transverse dispersion coefficients of phase α; $|\mathbf{u}_\alpha|$ is the Euclidean norm of $\mathbf{u}_\alpha = (u_{1\alpha}, u_{2\alpha}, u_{3\alpha})$, $|\mathbf{u}_\alpha| = \sqrt{u_{1\alpha}^2 + u_{2\alpha}^2 + u_{3\alpha}^2}$; $\mathbf{E}(\mathbf{u}_\alpha)$ is the orthogonal projection along the velocity,

$$\mathbf{E}(\mathbf{u}_\alpha) = \frac{1}{|\mathbf{u}_\alpha|^2} \begin{pmatrix} u_{1\alpha}^2 & u_{1\alpha}u_{2\alpha} & u_{1\alpha}u_{3\alpha} \\ u_{2\alpha}u_{1\alpha} & u_{2\alpha}^2 & u_{2\alpha}u_{3\alpha} \\ u_{3\alpha}u_{1\alpha} & u_{3\alpha}u_{2\alpha} & u_{3\alpha}^2 \end{pmatrix};$$

$\mathbf{E}^\perp(\mathbf{u}_\alpha) = \mathbf{I} - \mathbf{E}(\mathbf{u}_\alpha)$; and $\mathbf{I}$ is the identity matrix, $i = 1, 2, \ldots, N_c$, $\alpha = 1, 2, \ldots, N_p$. The source/sink term q_i combines all rates for component i and is expressed as

$$q_i = \phi \sum_{\alpha=1}^{N_p} S_\alpha r_{i\alpha} + (1 - \phi) r_{is} + \tilde{q}_i, \tag{11.5}$$

where $r_{i\alpha}$ and r_{is} are the reaction rates of component i in the α fluid phase and rock phase, respectively, and $\tilde{q}_i$ is the injection/production rate of the same component per bulk volume. The volumetric velocity $\mathbf{u}_\alpha$ is given by Darcy's law

$$\mathbf{u}_\alpha = -\frac{1}{\mu_\alpha} \mathbf{k} k_{r\alpha} (\nabla p_\alpha - \rho_\alpha \wp \nabla z), \qquad \alpha = 1, 2, \ldots, N_p, \tag{11.6}$$

where $\wp$ is the magnitude of the gravitational acceleration and z is the depth.

The energy conservation equation reads

$$\begin{aligned} &\frac{\partial}{\partial t}\left(\phi \sum_{\alpha=1}^{N_p} \rho_\alpha S_\alpha U_\alpha + (1-\phi)\rho_s C_s T\right) \\ &\quad + \nabla \cdot \sum_{\alpha=1}^{N_p} \rho_\alpha \mathbf{u}_\alpha H_\alpha - \nabla \cdot (k_T \nabla T) = q_c - q_L, \end{aligned} \tag{11.7}$$

where T is the temperature, U_α and H_α are the specific internal energy and the enthalpy of the α-phase (per unit mass), ρ_s and C_s are the density and the specific heat capacity of the solid, k_T represents the total thermal conductivity, q_c denotes the heat source item, and q_L indicates the heat loss to overburden and underburden (cf. Chapter 10). In (11.7), the specific internal energy U_α and the enthalpy H_α of phase α can be computed from

$$U_\alpha = C_{V\alpha} T, \qquad H_\alpha = C_{p\alpha} T,$$

where $C_{V\alpha}$ and $C_{p\alpha}$ respectively represent the heat capacities of phase α at constant volume and constant pressure.

In the numerical simulation of chemical flooding, a pressure equation for the aqueous phase (e.g., phase 1) is obtained by an overall mass balance on the volume-occupying components. Other phase pressures are evaluated using the capillary pressure functions:

$$p_{c\alpha 1} = p_\alpha - p_1, \qquad \alpha = 1, 2, \ldots, N_p, \tag{11.8}$$

where $p_{c11} = 0$ for convenience. Introduce the phase mobility

$$\lambda_\alpha = \frac{k_{r\alpha}}{\mu_\alpha} \sum_{i=1}^{N_{cv}} \rho_i c_{i\alpha}, \qquad \alpha = 1, 2, \ldots, N_p,$$

and the total mobility

$$\lambda = \sum_{\alpha=1}^{N_p} \lambda_\alpha .$$

Note that

$$\sum_{i=1}^{N_{cv}} \rho_i \mathbf{D}_{i\alpha} \nabla c_{i\alpha} = \mathbf{0}, \quad \sum_{i=1}^{N_{cv}} r_{i\alpha} = \sum_{i=1}^{N_{cv}} r_{is} = 0, \quad \alpha = 1, 2, \ldots, N_p .$$

Now, by adding (11.1) over i, $i = 1, 2, \ldots, N_{cv}$, we obtain the pressure equation (cf. Exercise 11.1)

$$\phi c_t \frac{\partial p_1}{\partial t} - \nabla (\lambda \mathbf{k} \nabla p_1) = \nabla \cdot \sum_{\alpha=1}^{N_p} \lambda_\alpha \mathbf{k} \left(\nabla p_{c\alpha 1} - \rho_\alpha \wp \nabla z \right) + \sum_{i=1}^{N_{cv}} \tilde{q}_i , \tag{11.9}$$

where the total compressibility c_t is

$$c_t = \frac{1}{\phi} \frac{\partial}{\partial p_1} \sum_{i=1}^{N_{cv}} \phi \tilde{c}_i \rho_i .$$

Assume that the rock compressibility c_R at the reference pressure p_r^0 is (cf. (2.16))

$$\phi = \phi^o \left(1 + c_R (p_r - p_r^o) \right), \tag{11.10}$$

where ϕ^o is the porosity at p_r^o. With $p_r = p_1$, using (11.3) and (11.10), we have

$$\phi \tilde{c}_i \rho_i = \phi^o \tilde{c}_i \rho_i^o \left(1 + (c_R + C_i^0)(p_1 - p_1^o) + c_R C_i^0 (p_1 - p_1^o)^2 \right).$$

Neglecting the higher-order term (due to the slight compressibility of rock and fluid phases), this equation becomes

$$\phi \tilde{c}_i \rho_i \approx \phi^o \tilde{c}_i \rho_i^o \left(1 + (c_R + C_i^0)(p_1 - p_1^o) \right). \tag{11.11}$$

Applying (11.11), the total compressibility c_t simplifies to

$$c_t = \frac{\phi^o}{\phi} \sum_{i=1}^{N_{cv}} \tilde{c}_i \rho_i^o \left(c_R + C_i^0 \right). \tag{11.12}$$

There are more dependent variables than there are differential and algebraic relations; there are formally $N_c + N_{cv} + N_c N_p + 3N_p + 1$ dependent variables: c_i, $\hat{c}_j$, $c_{i\alpha}$, T, $\mathbf{u}_\alpha$, p_α, and S_α, $\alpha = 1, 2, \ldots, N_p$, $i = 1, 2, \ldots, N_c$, $j = 1, 2, \ldots, N_{cv}$. Equations (11.1) and (11.6)–(11.9) provide $N_c + 2N_p$ independent relations, differential or algebraic; the additional $N_{cv} + N_c N_p + N_p + 1$ relations are given by the following constraints:

$$
\begin{aligned}
&\sum_{\alpha=1}^{N_p} S_\alpha = 1 \qquad \text{(a saturation constraint)},\\
&\sum_{i=1}^{N_{cv}} c_{i\alpha} = 1 \qquad (N_p \text{ phase concentration constraints}),\\
&c_i = \sum_{\alpha=1}^{N_p} S_\alpha c_{i\alpha} \qquad (N_c \text{ component concentration constraints}),\\
&\hat{c}_j = \hat{c}_j(c_1, c_2, \ldots, c_{N_c}) \qquad (N_{cv} \text{ adsorption constraints}),\\
&f_{i\alpha}(p_\alpha, T, c_{1\alpha}, \ldots, c_{N_c\alpha}) = f_{i\beta}(p_\beta, T, c_{1\beta}, \ldots, c_{N_c\beta})\\
&\qquad\qquad (N_c(N_p - 1) \text{ phase equilibrium relations}),
\end{aligned}
\tag{11.13}
$$

where $f_{i\alpha}$ is the fugacity function of the ith component in the α-phase. For a general compositional flow, several equations of state were developed to define the fugacity functions $f_{i\alpha}$, such as the Redlich–Kwong, Redlich–Kwong–Soave, and Peng–Robinson equations of state (cf. Section 3.2.5). For each individual chemical flooding considered in this chapter, the phase behavior model will be described in one of the following four sections.

As an example, the phases are numbered in the order water (aqueous), oil (oleic), microemulsion, and gas (air), and the components in the order water, oil, surfactant, polymer, chloride, calcium, alcohol, and gas (air).

11.2 Surfactant Flooding

Due to strong surface tension, a large amount of oil is trapped in small pores and cannot be washed out by water flooding. Surfactants can be injected to create low interfacial tension to reduce capillary forces and thus mobilize trapped oil. *Surfactants* have a greater role in EOR than lowering interfacial tension. They can be employed to alter wettability, stabilize dispersions, lower bulk-phase viscosity, and promote emulsification and entrainment.

The *surfactant phase behavior* in the water, oil, and surfactant system involves up to five volumetric components (water, oil, surfactant, and two alcohols) that form three pseudocomponents in solution. For simplicity, only three components (water, oil, and surfactant) are considered. Salinity and divalent cation concentrations strongly affect phase behavior. At low salinity, an excess oil phase that is essentially pure oil and a microemulsion phase that contains water plus electrolytes, some solubilized oil, and surfactant coexist. The tie lines (distribution curves) at low salinity have negative slope (cf. Figure 11.1 (left)). This kind of phase environment is referred to as *type II*(-) or *Winsor type I* (Winsor, 1954). At high salinity, an excess water phase and a microemulsion phase that contains some solubilized water and most of the oil and surfactant coexist. This kind of phase environment is termed *type II*(+) (cf. Figure 11.1 (right)). At intermediate salinity, excess water and oil phases and a microemulsion phase whose composition is represented by an invariant point coexist. Such a three-phase environment is called *type III* or *Winsor type III* (cf. Figure 11.2). The water, oil, and surfactant phase behavior model can be represented as a function of *effective salinity* once the binodal curve and tie lines (distribution curves) are given.

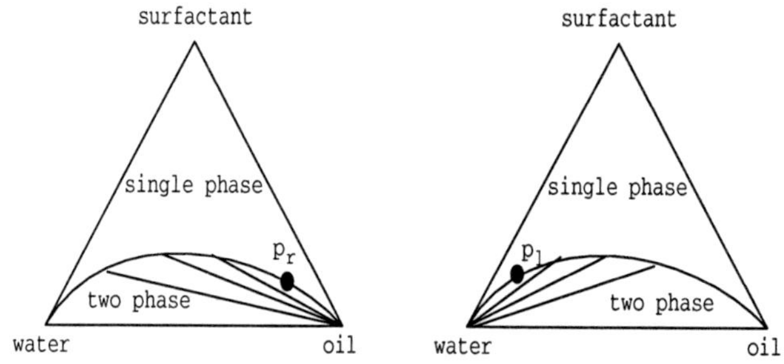

Figure 11.1. *Schematic plot of type II(-) (left); schematic plot of type II(+) (right).*

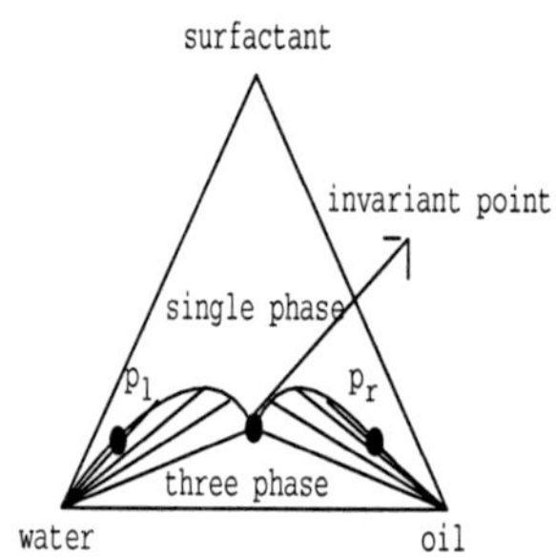

Figure 11.2. *Schematic plot of type III.*

11.2.1 Effective salinity

The *effective salinity* increases with the divalent cations bound to micelles (an aggregate (or cluster) of surface molecules; see Glover et al., 1979; Hirasaki, 1982), decreases as temperature increases for anionic surfactants, and increases as temperature increases for nonionic surfactants:

$$c_{SE} = c_{51}(1 - \beta_6 f_6^s)^{-1} \left[1 + \beta_T (T - T^o)\right]^{-1}, \tag{11.14}$$

where c_{51} is the aqueous phase anion concentration, β_6 is an effective salinity positive constant for calcium, $f_6^s = c_6^s / c_3^m$ is the fraction of the total divalent cations bound to surfactant micelles, β_T is a temperature coefficient, and T^o is a reference temperature. The effective salinities at which the three equilibrium phases form or disappear are called the lower and upper limits of effective salinity, c_{SEL} and c_{SEU}.

11.2.2 Binodal curves

The formulation of the *binodal curve* using *Hand's rule* is the same in all phase environments. This rule is based on the empirical observation that equilibrium phase concentration ratios are straight lines on a log-log scale. The *ternary diagram* for a type II(-) environment with the equilibrium phases numbered 2 and 3 and the corresponding Hand plot is shown in

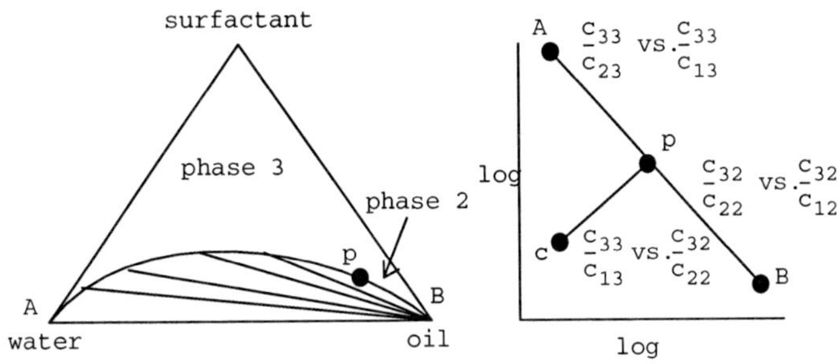

Figure 11.3. *Correspondence between ternary diagram and Hand plot.*

Figure 11.3. Hand's rule (Hand, 1939) to formulate the binodal curve reads

$$\frac{c_{3\alpha}}{c_{2\alpha}} = A\left(\frac{c_{3\alpha}}{c_{1\alpha}}\right)^{B} \qquad \text{for } \alpha = 1, 2, \text{ or } 3, \tag{11.15}$$

where the parameters A and B are empirical. $B = -1$ for the symmetric binodal curve. In this case, all phase concentrations are computed explicitly in terms of oil concentrations $c_{2\alpha}$:

$$\begin{aligned} c_{3\alpha} &= \frac{1}{2}\left(-Ac_{2\alpha} + \sqrt{(Ac_{2\alpha})^2 + 4Ac_{2\alpha}(1 - c_{2\alpha})}\right), \\ c_{1\alpha} &= 1 - c_{2\alpha} - c_{3\alpha} \qquad \text{for } \alpha = 1, 2, \text{ or } 3. \end{aligned} \tag{11.16}$$

The parameter A is related to the height of the binodal curve:

$$A_m = \left(\frac{2c_{3max,m}}{1 - c_{3max,m}}\right)^2, \qquad m = 0, 1, 2, \tag{11.17}$$

where $m = 0$, 1, and 2 are associated with low, optimal, and high salinities, and the height $c_{3max,m}$ is determined by a linear function of temperature:

$$c_{3max,m} = H_{BNC,m} + H_{BNT,m}(T - T^o), \qquad m = 0, 1, 2,$$

with the input parameters $H_{BNC,m}$ and $H_{BNT,m}$. Then A is linearly interpolated as

$$A = (A_0 - A_1)\left(1 - \frac{c_{SE}}{c_{SEOP}}\right) + A_1 \quad \text{if } c_{SE} \le c_{SEOP},$$

$$A = (A_2 - A_1)\left(\frac{c_{SE}}{c_{SEOP}} - 1\right) + A_1 \quad \text{if } c_{SE} > c_{SEOP},$$

where c_{SEOP} denotes the optimum effective salinity.

11.2.3 Tie lines for two phases

For both types II(-) and II(+), the phase behavior involves only two phases below the binodal curve. *Tie lines* are the lines joining the compositions of the equilibrium phases:

$$\frac{c_{3\alpha}}{c_{2\alpha}} = E\left(\frac{c_{33}}{c_{13}}\right)^{F}, \tag{11.18}$$

where $\alpha = 1$ for type II(+) and $\alpha = 2$ for type II(-). If the data for tie lines are not available, we set $F = -1/B$. For the symmetric binodal curve ($B = -1$), $F = 1$. Because the plait point is on both the binodal curve and tie line, it follows that

$$E = \frac{c_{1P}}{c_{2P}} = \frac{1 - c_{2P} - c_{3P}}{c_{2P}},$$

which, together with an application of the binodal curve to the plait point, gives

$$E = \frac{1}{c_{2P}}\left(1 - c_{2P} - \frac{1}{2}\left[-Ac_{2P} + \sqrt{(Ac_{2P})^2 + 4Ac_{2P}(1 - c_{2P})}\right]\right), \tag{11.19}$$

where c_{2P} is the oil concentration at the plait point and is an input parameter for both types II(-) and II(+). Note that c_{1P} and c_{3P} are the water and surfactant concentrations, respectively, at the plait point.

11.2.4 Tie lines for three phases

The computation of phase compositions for the three-phase region of type III is performed under the assumption that the excess oleic and aqueous phases are pure. The microemulsion phase composition is defined by the coordinates of an invariant point (M), which are evaluated as a function of effective salinity:

$$c_{2M} = \frac{c_{SE} - c_{SEL}}{c_{SEU} - c_{SEL}}. \tag{11.20}$$

The concentrations c_{1M} and c_{3M} are calculated by substituting c_{2M} into (11.16).

11.2.5 Phase saturations

In the presence of surfactant, the *phase saturations* in the saturated zone are computed using the phase concentrations and overall component concentrations:

$$\sum_{\alpha=1}^{3} S_\alpha = 1, \quad c_i = \sum_{\alpha=1}^{3} S_\alpha c_{i\alpha}, \quad i = 1, 2, 3. \tag{11.21}$$

11.2.6 Interfacial tension

The water/oil (σ_{ow}) and water/air *interfacial tensions* (σ_{aw}) are assumed to be constants. The models for computing microemulsion/oil (σ_{23}) and microemulsion/water (σ_{13}) interfacial tensions are based on Healy and Reed's model (Healy and Reed, 1974):

$$\begin{aligned}
\log_{10}\sigma_{\alpha 3} &= \log_{10} F_\alpha + G_{\alpha 2} + \frac{G_{\alpha 1}}{1 + G_{\alpha 3}R_{\alpha 3}} \quad \text{if } R_{\alpha 3} \geq 1,\\
\log_{10}\sigma_{\alpha 3} &= \log_{10} F_\alpha + (1 - R_{\alpha 3})\log_{10}\sigma_{ow}\\
&\quad + R_{\alpha 3}\left(G_{\alpha 2} + \frac{G_{\alpha 1}}{1 + G_{\alpha 3}}\right) \quad \text{if } R_{\alpha 3} < 1,
\end{aligned} \tag{11.22}$$

where the $G_{\alpha i}$'s are input parameters ($i = 1, 2$), $R_{\alpha 3} = c_{\alpha 3}/c_{33}$ is the solubilization ratio, and the correction factor F_α guarantees that the interfacial tension at the plait point is zero (Hirasaki, 1981):

$$F_\alpha = \frac{1 - e^{-\sqrt{con_\alpha}}}{1 - e^{-\sqrt{2}}}, \qquad con_\alpha = \sum_{i=1}^{3} (c_{i\alpha} - c_{i3})^2, \qquad \alpha = 1, 2.$$

Other models such as Huh's model (Huh, 1979) can be also used to calculate σ_{13} and σ_{23}. In the absence of surfactant or if the surfactant concentration is below the *critical micelle concentration*, these interfacial tensions are equal to σ_{ow}, which will be discussed below.

11.2.7 Interfacial tension without mass transfer

While injection of surfactants with high concentration greatly improves oil recovery, it can be very expensive. In most applications, the concentration of surfactants used is below the critical micelle concentration. In this case, the system of water, oil, and surfactant does not involve mass transfer between phases. As a result, the entire system is composed of only an aqueous phase containing all the surfactant, electrolytes, and dissolved oil at the water solubility limit and a pure excess oil phase. Such a system is called a *sparse system* without mass interchange. An ASP+foam displacement mechanism for this type of system is accomplished through the synergetic effect of water, oil, surfactant, and alkaline. This effect is described by the interfacial tension function

$$\sigma_{ow} = \sigma_{ow}(c_S, c_A),$$

where σ_{ow} is the interfacial tension between the aqueous and oil phases and c_S and c_A are the concentrations of surfactant and alkaline, respectively. This function is obtained via experiment.

11.2.8 Trapping numbers

A displacement mechanism in EOR is the mobilization of a trapped organic phase due to reduced interfacial tension resulting from the injection of surfactants (Brown et al., 1994). *Buoyancy forces* also affect the mobilization of the trapped phase and can be defined by the bond number (Morrow and Songkran, 1982). The *bond* and *capillary numbers* are two dimensionless numbers; the former represents gravity/capillary forces, and the latter represents viscous/capillary forces. Traditionally, the *capillary number* (Lake, 1989) is defined by

$$N_{c\alpha} = \frac{|\mathbf{k} \cdot \nabla \Phi_\beta|}{\sigma_{\alpha\beta}}, \qquad \alpha, \ \beta = 1, 2, \ldots, N_p, \tag{11.23}$$

where α and β are the displaced and displacing fluids and the potentials Φ_β are

$$\Phi_\beta = p_\beta - \rho_\beta \wp z, \qquad \beta = 1, 2, \ldots, N_p. \tag{11.24}$$

The bond number is

$$N_{B\alpha} = \frac{k \wp (\rho_\alpha - \rho_\beta)}{\sigma_{\alpha\beta}}, \qquad \alpha, \ \beta = 1, 2, \ldots, N_p, \tag{11.25}$$

where k is such that $\mathbf{k} = k\mathbf{I}$.

11.2.9 Relative permeabilities

Residual saturations are related to the trapping numbers by

$$S_{\alpha r} = \min\left\{S_\alpha, S_{\alpha r}^H + \frac{S_{\alpha r}^L - S_{\alpha r}^H}{1 + C_\alpha N_{c\alpha}}\right\}, \qquad \alpha = 1, 2, \ldots, N_p,$$

where C_α is a positive input parameter based on the experimental observation of the relation between the residual saturations and the trapping number, and $S_{\alpha r}^L$ and $S_{\alpha r}^H$ are the input residual saturations for phase α at low and high trapping numbers, respectively. This correlation was obtained based on experimental data for n-decane (Delshad et al., 2000).

The *relative permeability* curves change as the residual saturations change at high trapping numbers due to detrapping, which can be accounted for by the expressions

$$k_{r\alpha} = k_{r\alpha}^0 (S_{n\alpha})^{n_\alpha}, \qquad \alpha = 1, 2, \ldots, N_p,$$

where $S_{n\alpha}$ is the normalized saturation of phase α

$$S_{n\alpha} = (S_\alpha - S_{\alpha r}) \Bigg/ \left(1 - \sum_{\alpha=1}^{N_p} S_{\alpha r}\right), \qquad \alpha = 1, 2, \ldots, N_p.$$

The endpoints and exponents in the relative permeability functions are evaluated as a linear interpolation between the given input values at low and high trapping numbers $\left(k_{r\alpha}^L, k_{r\alpha}^H, n_\alpha^L, n_\alpha^H\right)$:

$$k_{r\alpha}^0 = k_{r\alpha}^L + \frac{S_{\beta r}^L - S_{\beta r}}{S_{\beta r}^L - S_{\beta r}^H}\left(k_{r\alpha}^H - k_{r\alpha}^L\right),$$

$$n_\alpha = n_\alpha^L + \frac{S_{\beta r}^L - S_{\beta r}}{S_{\beta r}^L - S_{\beta r}^H}\left(n_\alpha^H - n_\alpha^L\right), \qquad \alpha, \beta = 1, 2, \ldots, N_p.$$

11.3 Alkaline Flooding

Oil recovery mechanisms in *alkaline* or *high-pH flooding* have been attributed to many mechanisms (de Zabala et al., 1982), such as interfacial tension lowering, emulsion formation, and wettability. In surfactant flooding, the surfactant is injected, whereas in high-pH flooding, it is generated in situ. Alkaline and acidic hydrocarbon species in crude oil react to generate the surfactant. Also, interactions of alkaline chemicals and permeable media minerals can cause excessive retardation in the propagation of these chemicals through the media. The physicochemical phenomena in high-pH flooding are described through a *chemical reaction equilibrium model* (Bhuyan et al., 1991). The reaction chemistry in this model includes aqueous electrolyte chemistry, precipitation/dissolution of minerals, ion-exchange reactions with the matrix (the geochemical option), and reactions of acidic components of oil with the bases in the aqueous solution. This model can be utilized to compute the chemical composition of the reservoir rock and fluids in the presence of chemical reactions among the injected chemical species and the reservoir rock and fluids.

11.3.1 Basic assumptions

The reaction equilibrium model is established under the following assumptions (Delshad et al., 2000):

- All reactions attain local thermodynamic equilibrium.
- No redox reaction exists.
- Temperature, pressure, and volume changes resulting from chemical reactions are negligibly small. In particular, the reservoir is isothermal.
- Activity coefficients of all reactive species are unity so that molar concentrations replace activities in reaction equilibrium computations.
- Water present in any phase always has the same chemical composition and is in equilibrium with matrix minerals.
- Supersaturation of aqueous species is not allowed.
- The active acid species in the crude oil can be represented collectively by a single pseudoacid component. This pseudocomponent is highly soluble in oil, and it partitions between water and oil with a constant partition coefficient.

11.3.2 Mathematical formulations of reaction equilibria

Assume that the reactive system is composed of N_F fluid species, N_S solid species, N_I matrix-adsorbed cations, and N_M micelle-associated cations all made up of N independent elements. Then there exist $N_F + N_S + N_I + N_M$ unknown equilibria concentrations for which the same number of independent equations are needed.

Mass balance equations

The N elemental mass balance equations are

$$c_r^t = \sum_{j=1}^{N_F} h_{rj} c_j + \sum_{k=1}^{N_S} g_{rk} \hat{c}_k + \sum_{i=1}^{N_I} f_{ri} \bar{c}_i + \sum_{m=1}^{N_M} e_{rm} \check{c}_m, \tag{11.26}$$

for $r = 1, 2, \ldots, N$, where c_r^t is the total concentration of element r; c_j, $\hat{c}_k$, $\bar{c}_i$, and $\check{c}_m$ are the concentrations of the jth fluid species, the kth solid species, the ith matrix-adsorbed cation, and the mth micelle-associated cation, respectively; and h_{rj}, g_{rk}, f_{ri}, and e_{rm} are the reaction coefficients of the rth element in the respective species and cations. *Electrical neutrality* in the bulk fluid phase gives an additional equation

$$\sum_{j=1}^{N_F} Z_j c_j + \sum_{m=1}^{N_M} \check{Z}_m \check{c}_m = 0, \tag{11.27}$$

where Z_j and $\check{Z}_m$ are the *electroneutrality coefficients* of the jth fluid species and the mth micelle-associated cation, respectively. Equation (11.27) is a linear combination of the mass balance equations given in (11.26). Thus this equation is not independent but can be used to replace any of the elemental mass balance equations.

Aqueous reaction equilibrium relations

From the N_F fluid chemical species, N independent elements can be arbitrarily selected so that the concentrations of the remaining $N_F - N$ fluid species are expressed in terms of the concentrations of the independent ones via equilibrium relations of the form

$$c_r = k_r^{eq} \prod_{j=1}^{N} c_j^{w_{rj}}, \qquad r = N+1, N+2, \ldots, N_F, \tag{11.28}$$

where k_r^{eq} and w_{rj} are the reaction equilibrium constants and exponents, respectively.

Solubility product constraints

For each solid species, there is a *solubility product constraint*

$$k_k^{sp} \geq \prod_{j=1}^{N} c_j^{w_{kj}}, \qquad k = 1, 2, \ldots, N_S, \tag{11.29}$$

where the solubility product constants k_k^{sp} are defined in terms of the concentrations of the independent chemical species only. If a solid is not present, the corresponding solubility product constraint is the inequality in (11.29); if the solid is present, it is an equality.

Ion exchange equilibria on matrix substrate

For each substrate allowing exchange among N_I cations, there exists an *electroneutrality condition*

$$Q_v = \sum_{i=1}^{N_I} \bar{Z}_i \bar{c}_i, \tag{11.30}$$

where Q_v is the cation exchange capacity on matrix surface and $\bar{Z}_i$ is the electroneutrality coefficient of the ith matrix-adsorbed cation.

In addition, for these N_I adsorbed cations, there are $N_I - 1$ independent *exchange equilibria relations* of the form

$$k_s^{ex} = \prod_{j=1}^{N} c_j^{y_{sj}} \prod_{i=1}^{N_I} \bar{c}_i^{x_{si}}, \qquad s = 1, 2, \ldots, N_I - 1, \tag{11.31}$$

where k_s^{ex} is the exchange equilibrium constant on matrix surface, and x_{si} and y_{sj} are equilibrium exponents.

Ion exchange equilibrium with micelles

For N_M cations associated with surfactant micelles, there are $N_M - 1$ *cation exchange* (on micelle) *equilibria relations*:

$$k_q^{exm} = \prod_{j=1}^{N} c_j^{y_{qj}} \prod_{j=1}^{N_M} \check{c}_m^{x_{qm}}, \qquad q = 1, 2, \ldots, N_M - 1, \tag{11.32}$$

where k_q^{exm} is the exchange equilibrium constant on micelle surfaces.

It has been observed that an electrostatic association model, where the mass action equilibrium "constants" are really functions of the total anionic surfactant concentration, adequately describes these ion exchange equilibrium relations (Hirasaki, 1982). These equilibrium "constants" are modified to

$$k_q^{exm} = \beta_q^{exm}\left(c_{A^-} + c_{S^-}\right), \qquad q = 1, 2, \ldots, N_M - 1,$$

where c_{A^-} and c_{S^-} are the concentrations of the surfactants generated in situ and injected, respectively. They are determined by the *electroneutrality condition* for the micelles as a whole:

$$c_{A^-} + c_{S^-} = \sum_{m=1}^{N_M} \check{Z}_m \check{c}_m. \tag{11.33}$$

In summary, there are N mass balance equations (11.26), $N_F - N$ aqueous reaction equilibrium relations (11.28), N_S solubility product constraints (11.29), 1 matrix surface electroneutrality condition (11.30), $N_I - 1$ cation exchange (on the matrix surface) equilibrium relations (11.31), $N_M - 1$ cation exchange (on micelle) equilibrium relations (11.32), and 1 electroneutrality condition for the micelles (11.33), giving a total number $N_F + N_S + N_I + N_M$ of independent equations to compute the equilibrium concentrations of N_F fluid species, N_S solid species, N_I matrix-adsorbed cations, and N_M cations adsorbed on the micelle surfaces. An iterative method such as the Newton–Raphson iteration (cf. Section 8.2.1) can be used to solve this set of nonlinear equations.

11.4 Polymer Flooding

In general, *polymer flooding* is economic only if the water flooding mobility ratio is high, the reservoir is highly heterogeneous, or both. In a polymer flooding procedure, polymer is added to water to decrease its mobility. The resulting increase in viscosity, together with a decrease in the aqueous phase permeability, leads to a lower mobility ratio, which increases the efficiency of water flooding through larger volumetric sweep efficiency and a lower swept zone oil saturation.

11.4.1 Viscosity

At a certain shear rate the *polymer solution viscosity* is a function of salinity and polymer concentration (Flory, 1953):

$$\mu_P^0 = \mu_w\left(1 + \left[a_{P1}c_{4\alpha} + a_{P2}c_{4\alpha}^2 + a_{P3}c_{4\alpha}^3\right] c_{SEP}^{b_P}\right), \qquad \alpha = 1 \text{ or } 3, \tag{11.34}$$

where $c_{4\alpha}$ is the polymer concentration in water or microemulsion, μ_w is the water viscosity, c_{SEP} is the effective salinity for polymer, and a_{P1}, a_{P2}, a_{P3}, and b_P are input parameters. The constant b_P determines how the polymer viscosity depends on salinity.

The reduction in the polymer solution viscosity as a function of shear rate γ' is modeled by Meter's relation (Meter and Bird, 1964)

$$\mu_P = \mu_w + \frac{\mu_P^0 - \mu_w}{1 + (\gamma'/\gamma'_{1/2})^{n_M - 1}}, \tag{11.35}$$

where n_M is an empirical coefficient and $\gamma'_{1/2}$ is the shear rate at which $\mu_P = (\mu_P^0 + \mu_w)/2$. When (11.35) is applied to flow in porous media, μ_P is often called the *apparent viscosity*, and the shear rate is an equivalent shear rate γ'_{eq}. The in situ shear rate for phase α is calculated using the modified Blake–Kozeny capillary bundle equation for multiphase flow (Sorbie, 1991):

$$\gamma'_{eq,\alpha} = \frac{\gamma'_c |\mathbf{u}_\alpha|}{\sqrt{\bar{k} k_{r\alpha} \phi S_\alpha}}, \tag{11.36}$$

where $\gamma'_c = 3.97C$ sec.$^{-1}$, C is the shear rate coefficient used to account for nonideal effects such as slip at pore walls (Wreath et al., 1990), and $\bar{k}$ is the average permeability

$$\bar{k} = \left(\frac{1}{k_{11}} \left(\frac{u_{1\alpha}}{|\mathbf{u}_\alpha|} \right)^2 + \frac{1}{k_{22}} \left(\frac{u_{2\alpha}}{|\mathbf{u}_\alpha|} \right)^2 + \frac{1}{k_{33}} \left(\frac{u_{3\alpha}}{|\mathbf{u}_\alpha|} \right)^2 \right)^{-1}$$

with $\mathbf{u}_\alpha = (u_{1\alpha}, u_{2\alpha}, u_{3\alpha})$ and $\mathbf{k} = \text{diag}(k_{11}, k_{22}, k_{33})$.

11.4.2 Permeability reduction

Polymer reduces both the effective permeability of porous media and the mobility of displacing fluids. The *permeability reduction* is described by a *reduction factor* R_k:

$$R_k = \frac{k_w}{k_P}, \tag{11.37}$$

where k_w and k_P are the effective permeabilities of water and polymer. The mobility change due to the combined effect of increased viscosity and reduced permeability is the *resistance factor* R_r:

$$R_r = \frac{R_k \mu_P}{\mu_w}. \tag{11.38}$$

The effect of permeability reduction persists even after the polymer solution has gone through the porous media. This effect is described by the *residual resistance factor* R_{rr}:

$$R_{rr} = \frac{\lambda_P}{\tilde{\lambda}_P}, \tag{11.39}$$

where λ_P and $\tilde{\lambda}_P$ are the mobilities before and after polymer solution, respectively.

11.4.3 Inaccessible pore volume

The reduction in porosity due to inaccessible or excluded pores because of the great size of polymer molecules is termed the *inaccessible pore volume*. The result is that polymer moves more quickly than water. This effect can be incorporated by multiplying the porosity in the polymer conservation by the input parameter of effective porous volume.

11.5 Foam Flooding

Foam flooding uses surfactants to reduce gas-phase mobility through formation of stable gas-liquid foams. Interfacial tension lowering is not a significant mechanism. Gas-liquid foams offer an alternative to polymers for providing mobility control in micellar flooding. In contrast to individual foam flooding, ASP+foam flooding generates foams of smaller sizes. For an initially oil-wet porous medium, these foams can enter the small pores that are not reached with water flooding, thus mobilizing the residual oil there. In addition, because of a low interfacial tension between oil and the ASP+foam system, this type of flooding can effectively displace the residual oil trapped on the rock surface after water flooding.

Foams flowing in porous media can drastically reduce the mobility of a gas phase. This is illustrated in the following relation:

$$k_{rg}^f = \frac{k_{rg}}{R_s R_u}, \tag{11.40}$$

where k_{rg} and k_{rg}^f are the gas relative permeabilities before and after the formation of foams, and R_s and R_u are independent *gas mobility reduction factors*. R_s depends on the oil phase saturation, surfactant, permeability, and capillary force, while R_u is related to the gas velocity and gas-liquid ratio. They can be determined using (11.41)–(11.43) below.

11.5.1 Critical oil saturation

The presence of crude oil is not favorable to formation of foams, mainly due to the fact that the oil-water surface tension is lower than the gas-water surface tension. When these two surfaces coexist in a reservoir, the surface energy changes in the decreasing direction of surface tension so that foaming agents move from the gas-water surface to the oil-water surface. Then foams will lose the protection of a surfactant film and quickly break. Consequently, in ASP+foam flooding, there is a *critical oil saturation* S_{oc}. When S_o is greater than S_{oc}, foams do not form; otherwise, foams can form.

11.5.2 Critical surfactant concentration

Foams are dispersions of gas bubbles in liquids. Such dispersions are normally quite unstable and break up in less than a second. However, if surfactants are added to the liquids, stability is greatly improved so that some foams can persist. If the concentration of the surfactants used as foaming agents is too low, foams do not form. Only when the concentration is higher than a *critical concentration* c_s^c are foams present.

11.5.3 Critical capillary force

The capillary force in the reservoir rock plays an important role in formation of foams. Only when this capillary force is small enough do foams form. When bubbles move through small pore throats, capillary pressure decreases as the bubble sizes increase, and then the pressure gradient in the liquids causes the liquids to enter these throats from the surrounding areas. If the capillary pressure is small enough, the liquids will fully fill the throats, which will cause large bubbles to split up into smaller bubbles. Hence the formation of foams in this type of mechanism requires the application of a sufficiently small capillary force. In general, for a reservoir there is a *critical capillary force* p_c^* such that the property of foams changes dramatically in a small neighborhood $(p_c^* - \epsilon, p_c^* + \epsilon)$ of p_c^*, where ϵ is a positive constant. When the capillary pressure p_c satisfies $p_c > p_c^* + \epsilon$, foams do not form; when $p_c < p_c^* - \epsilon$, the strength of the foams formed is very strong. If p_c is a function of the water phase saturation S_w,

$$p_c = p_c(S_w),$$

a corresponding critical S_{wc} can be obtained from this function.

11.5.4 Oil relative permeability effects

In core flow experiments, as ASP+foam species are injected, liquid production decreases in high-permeability zones of the core; it increases in its low-permeability zones. This indicates that foams have a preference for the blocking of the high-permeability zones.

According to the discussions given so far in this section, a function of the *mobility reduction factor* R_s can be defined as follows:

$$R_s = 1 \quad \text{if } S_o > S_{oc} \text{ or } c_s < c_s^c, \tag{11.41}$$

and if both $S_o \le S_{oc}$ and $c_s \ge c_s^c$,

$$R_s = \begin{cases} 1, & S_w \le S_{wc} - \epsilon, \\ \left(1 + (R_{\max} - 1)\left(\dfrac{S_w - S_{wc} + \epsilon}{2\epsilon}\right)\right)\left(1 + \dfrac{k}{\bar{k}}\right)^2, & \\ \qquad\qquad S_{wc} - \epsilon < S_w < S_{wc} + \epsilon, & \\ R_{\max}\left(1 + \dfrac{k}{\bar{k}}\right)^2, & S_w \ge S_{wc} + \epsilon, \end{cases} \tag{11.42}$$

where $R_{\max}$ is an experimentally determined constant and $\bar{k}$ is the weighted average of the permeability k with the effective thickness of each layer as a weight.

11.5.5 Gas-liquid ratio effects

In ASP+foam flooding, there is an optimum gas-liquid ratio R_{gl}^* under which the strength of foams is the greatest and oil recovery is the most efficient. If the *gas-liquid ratio* R_{gl} is higher or lower than R_{gl}^*, the strength of foams will weaken and so will oil recovery efficiency.

11.5.6 Gas velocity effects

The strength of foams also depends on the *gas velocity* u_g. The lower the gas velocity, the stronger the foam strength. The effect of R_{gl} and u_g on the *mobility reduction factor* R_u

can be modeled by the function

$$R_u = \begin{cases} \left(u_g/u_g^o\right)^{\sigma-1} & \text{if } R_{gl} \leq R_{gl}^*, \\ \left(u_g/u_g^o\right)^{\sigma-1} R_{gl}^{-\omega} & \text{if } R_{gl} > R_{gl}^*, \end{cases} \tag{11.43}$$

where u_g^o is a reference gas velocity and σ and ω are experimentally determined constants.

11.6 Rock and Fluid Properties

In ASP+foam flooding, very complex physical and chemical phenomena can occur between the reservoir rock and fluids, such as adsorption, cation exchange, and the change of phase specific weights and viscosities with compositions.

11.6.1 Adsorption

Surfactant

Surfactant adsorption has been the subject of extensive study for many decades and is now quite well understood. In general, the surfactant adsorption isotherm is very complicated (Somasundaran and Hanna, 1977; Scamehorn et al., 1982). This is particularly true when the surfactant is not isomerically pure and the substrate is not a pure mineral. However, it has been believed that a *Langmuir-type isotherm* can be used to capture the essential features of the surfactant adsorption in simulating oil recovery (Camilleri et al., 1987). This type of isotherm describes the adsorption level of surfactant that takes into account salinity, surfactant concentration, and rock permeability. The adsorbed concentration of surfactant is described by

$$\hat{c}_i = \min\left\{\tilde{c}_i, \frac{a_i(\tilde{c}_i - \hat{c}_i)}{1 + b_i(\tilde{c}_i - \hat{c}_i)}\right\}, \tag{11.44}$$

where $i = 3$ (for surfactant) and b_i is a constant. The minimum is taken to ensure that adsorption is not greater than the total surfactant concentration. Adsorption increases linearly with effective salinity and decreases as permeability increases:

$$a_i = (a_{i1} + a_{i2}c_{SE})\sqrt{\frac{k^o}{k}},$$

where c_{SE} is the effective salinity, a_{i1} and a_{i2} are constants, k is the permeability, and k^o is a reference permeability. The reference permeability is the permeability at which the input adsorption parameters are specified. The ratio a_i/b_i represents the maximum level of adsorbed surfactant, and b_i controls the curvature of the isotherm.

From our experience with petroleum applications, in many situations the Langmuir-type isotherm is not valid; the adsorbed surfactant concentration curve must be remeasured in the laboratory. According to our laboratory experiments (Chen et al., 2007), the adsorbed concentration $\hat{c}_i^o$ at a reference value pH_r of pH may be calculated by its relation to the surfactant concentration c_i:

$$\hat{c}_i^o = \hat{c}_i^o(c_i).$$

The adsorbed concentration $\hat{c}_i$ varies with pH:

$$\hat{c}_i = \left(1 - \frac{a_i(\mathrm{pH} - \mathrm{pH}_r)}{\mathrm{pH}_{\max} - \mathrm{pH}_r}\right)\hat{c}_i^o,$$

where $\mathrm{pH}_{\max}$ is the maximum value of pH and a_i is an experimental constant.

Polymer

The retention of polymer in a porous medium is due both to adsorption onto the solid surface and to trapping within small pores. *Polymer retention* is analogous to that of surfactant, slows down the polymer velocity, and depletes the polymer slug. *Polymer adsorption* is given by (11.44) with the parameter a_i specified by

$$a_i = (a_{i1} + a_{i2}c_{SEP})\sqrt{\frac{k^o}{k}},$$

where $i = 4$ (for polymer) and c_{SEP} is the effective salinity for polymer:

$$c_{SEP} = \frac{c_{51} + (\beta_P - 1)c_{61}}{c_{11}}$$

with c_{51}, c_{61}, and c_{11} being the anion, calcium, and water concentrations in the aqueous phase and β_P an input parameter measured in the laboratory.

11.6.2 Phase-specific weights

Phase-specific weights ($\gamma_\alpha = \rho_\alpha \wp$) are functions of pressure and composition:

$$\begin{aligned} \gamma_\alpha &= c_{1\alpha}\gamma_{1\alpha} + c_{2\alpha}\gamma_{2\alpha} + c_{3\alpha}\gamma_{3\alpha} + 0.02533c_{5\alpha} \\ &\quad - 0.001299c_{6\alpha} + c_{8\alpha}\gamma_{8\alpha}, \qquad \alpha = 1, 2, \ldots, N_p, \end{aligned} \tag{11.45}$$

where $\gamma_{i\alpha} = \gamma_i^o[1 + C_i^o(p_\alpha - p_r^o)]$ and γ_i^o is the specific weight of component i at a reference pressure p_r^o.

11.6.3 Phase viscosities

The *liquid phase viscosities* are expressed in terms of pure component viscosities and the phase concentrations of the organic, water, and surfactant:

$$\begin{aligned} \mu_\alpha &= c_{1\alpha}\mu_w e^{\beta_1(c_{2\alpha}+c_{3\alpha})} + c_{2\alpha}\mu_o e^{\beta_2(c_{1\alpha}+c_{3\alpha})} \\ &\quad + c_{3\alpha}\beta_3 e^{\beta_4 c_{1\alpha}+\beta_5 c_{2\alpha}} \qquad \text{for } \alpha = 1, 2, \text{ or } 3, \end{aligned} \tag{11.46}$$

where the parameters β_i are determined by matching laboratory microemulsion viscosities at several compositions. In the absence of surfactant and polymer, the water and oil phase viscosities reduce to pure water and oil viscosities μ_w and μ_o. When polymer is present, μ_w is replaced by the polymer viscosity μ_P defined by (11.35).

The following exponential expressions can be used to calculate viscosities as functions of temperature:

$$\mu_i = \mu_i^o \exp\left(b_i\left[\frac{1}{T} - \frac{1}{T^o}\right]\right), \qquad i = \text{water, oil, or gas}, \tag{11.47}$$

where μ_i^o is the viscosity at a reference temperature T^o and b_i is an input parameter. The viscosity of air is a linear function of pressure:

$$\mu_a = \mu_a^o + \mu_a^s(p_r - p_r^o), \tag{11.48}$$

where μ_a^o is the air viscosity at a reference pressure p_r^o and μ_a^s is the slope (rate of change) of the air viscosity vs. pressure.

11.6.4 Cation exchange

An incompatibility in the electrolyte composition of the initial and injected fluids saturating a porous medium leads to *cation exchange*. Cation exchange affects the transport of ions in solution and thus can influence the optimum salinity, surfactant phase behavior, and surfactant adsorption (Pope et al., 1978; Fountain, 1992). The type and concentration of cations involved in exchanges also have an effect on the permeability (Fetter, 1993). Cations exist in the form of free ions, adsorbed on clay surfaces, and associated with either surfactant micelles or adsorbed surfactant. Hirasaki's model (Hirasaki, 1982) can be used to describe the cation exchange: The mass action equations for the exchange of calcium ($i = 6$) and sodium ($i = 12$) on clay and surfactant are

$$\frac{\left(c_{12}^s\right)^2}{c_6^s} = \beta^s c_3^m \frac{\left(c_{12}^f\right)^2}{c_6^f}, \qquad \frac{\left(c_{12}^a\right)^2}{c_6^a} = \beta^a Q_v \frac{\left(c_{12}^f\right)^2}{c_6^f}, \tag{11.49}$$

where the superscripts f, a, and s indicate free cations, adsorbed cations on clay, and adsorbed cations on micelles, respectively; β^s and β^a are the ion exchange constants for clay and surfactant; c_3^m is the concentration of surfactant in meq/ml; and Q_v is the cation exchange capacity of the mineral. Electrical neutrality and mass conservation are required to close the system of ion exchange equations:

$$\begin{aligned} c_5 &= c_{12}^f + c_6^f, \\ c_6 &= c_6^f + c_6^s + c_6^a, \\ c_3 &= c_6^s + c_{12}^s, \\ Q_v &= c_6^a + c_{12}^a, \\ c_5 - c_6 &= c_{12}^f + c_{12}^s + c_{12}^a. \end{aligned} \tag{11.50}$$

All concentrations in these equations are given in meq/ml of water. The molar volume concentration of surfactant is evaluated from

$$c_3^m = \frac{1{,}000 c_3}{c_1 M_3}, \tag{11.51}$$

where M_3 is the equivalent weight of surfactant. The cation exchange equations (11.49)–(11.51) are solved for the six unknowns c_6^a, c_{12}^a, c_6^f, c_{12}^f, c_6^s, and c_{12}^s using the Newton–Raphson iteration (cf. Section 8.2.1).

11.7 Numerical Methods

The various numerical methods developed in Chapter 4 and the solution techniques described in Chapter 8 can be applied for the numerical solution of the governing equations for chemical flooding. For the numerical results presented in the next section, the temporal discretization is based on the backward Euler scheme, while the spatial discretization is based on the Raviart–Thomas mixed finite element method on rectangular parallelepipeds (cf. Section 4.5.4). The solution technique used is sequential and is evolved from the IMPEC (i.e., implicit in pressure and explicit in composition; cf. Section 8.2.4) technique developed by Delshad et al. (2000) for a compositional simulator of chemical flooding. Because of the explicitness for the solution of compositions, the size of time steps must be restricted to stabilize the overall procedure. In contrast, the sequential technique (cf. Section 8.2.3) solves both the pressure and compositions implicitly, and relaxes the time step restriction. The Newton–Raphson iterations for each of the pressure and composition equations are constrained by maximum changes in these variables over the iteration (cf. Section 8.2.3), and an automatic choice of time step sizes is determined by these maximum changes over the time step. Upstream-weighted interblock flow (e.g., for mobilities) and injection/production terms are included. The linear system of algebraic equations is solved by the ORTHOMIN iterative algorithm, with incomplete LU factorization preconditioners (cf. Chapter 5).

Both an implicit scheme in time for each of the pressure and composition equations and an implicit bottom hole pressure treatment add stability and preserve user-specified rates and constraints. In fact, for the numerical tests carried out in the next section, we have observed that the sequential technique is approximately four times faster than the IMPEC.

The sequential solution technique proceeds in the following order:

1. Solve the pressure equation implicitly.
2. Solve the transport system implicitly for the overall concentration of each component.
3. Use the chemical reaction equilibrium model to obtain the effective salinities.
4. Utilize a flash calculation to obtain the phase saturations and the concentrations of components in each phase.
5. Compute the interfacial tensions, trapping numbers, residual phase saturations, relative permeabilities, phase densities, viscosities, mobility reduction factors, etc.
6. Go back to step 1 to repeat this procedure until a final state is reached.

11.8 Numerical Results

The chemical compositional model developed in Sections 11.1–11.6 is applied to three experiments: a chemical flow without mass transfer between phases, a laboratory sandstone core, and an ASP+foam displacement problem with mass transfer. The purpose of the first experiment is to show that this chemical model is reliable and practical. Because there is no analytical solution available for the chemical compositional problem under consideration, the second experiment is used to compare numerical and laboratory results. The third experimental problem is more realistic than the first, and is exploited to study oil recovery

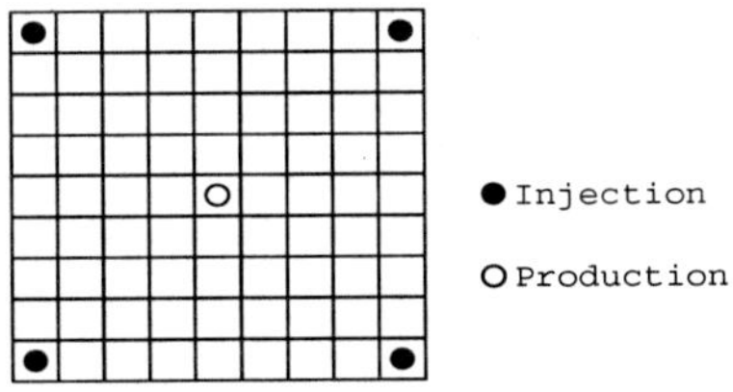

Figure 11.4. *A five-spot pattern.*

efficiency using different development methods, the oil displacement mechanisms, and the effects of different factors on ASP+foam flooding. Numerical simulation can be employed to conduct mechanism study, feasibility evaluation, pilot plan optimization, and performance prediction for chemical flooding to improve oil recovery efficiency and reduce operational costs.

11.8.1 Example 1

This is a typical five-spot pattern problem with four injection wells and one production well (cf. Figure 11.4). The distance between the injection and production wells is 250 m. The number of horizontal grids is 9×9 with a spatial grid size of 44.19 m. The temporal step size is of the order of several days. There are two layers in the vertical direction; the effective thickness of each layer is 3 m. The permeabilities in the first and second layers are 800 and 1,500 md, respectively, and the porosity is 0.26. The initial water saturation is 0.45, and an injection rate of 0.19 PV/D is used. *Water cut* (WC) is defined as the ratio of water production to the sum of water and oil production.

There are three types of injections: water, polymer, and ASP flooding. The injection modes are the following:

- Water flooding: Water is injected until WC = 98%.
- Polymer flooding: 0.05 PV water is injected, followed by polymer (1,000 ppm in solution) injection until the total injection reaches 0.38 PV, and then water is injected again until WC – 98%.
- ASP flooding: 0.05 PV water is injected, followed by ASP injection with 0.3% surfactant, polymer with 1,000 ppm in solution, and 2.0wt% NaOH until the total injection reaches 0.38 PV, and then water is injected again until WC = 98%.

The active function table of interfacial tension used in this simulation is given in Table 11.1. The recovery rates of the second (polymer flooding) and third (ASP flooding) types of injections are 23% and 32% OIP (oil in place), respectively. The WC curves for different injection methods are presented in Figure 11.5. Figure 11.6 displays the residual oil saturation for the first layer using the polymer and ASP flooding, respectively, when WC equals 98%. Figure 11.5 shows that WC decreases to 79.85% and 66.56% from the highest value 92.34% for the second and third types, respectively, and that the third type

Table 11.1. *The active function table of interfacial tension.*

Alkaline \ Surfactant	0	0.001	0.002	0.003	0.004	0.005
0	20	0.9	0.2	0.12	0.07	0.04
0.5%	0.758	0.017	0.004	0.00019	0.00015	0.00010
1.0%	0.173	0.011	0.001	0.00009	0.00004	0.00003
1.5%	0.073	0.006	0.0007	0.00005	0.00003	0.00002
2.0%	0.03	0.002	0.0003	0.00002	0.00002	0.00001
3.0%	0.06	0.008	0.0007	0.00012	0.00010	0.00005

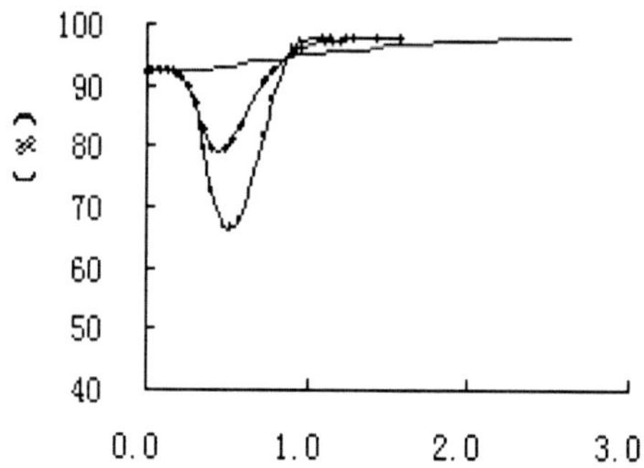

Figure 11.5. *Water cut versus injected PV (water: top, polymer: middle, and ASP: bottom).*

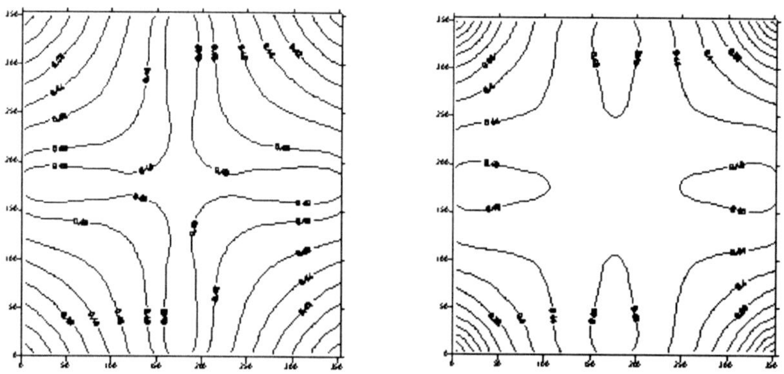

Figure 11.6. *Polymer flooding (left); ASP flooding (right).*

reduces the residual oil saturation much more dramatically than does the second type. These observations are in good agreement with physical intuition, and indicate that the chemical simulator is practical. While quite a coarse grid is utilized, an observation similar to that in Figure 11.5 has been made for refined grids.

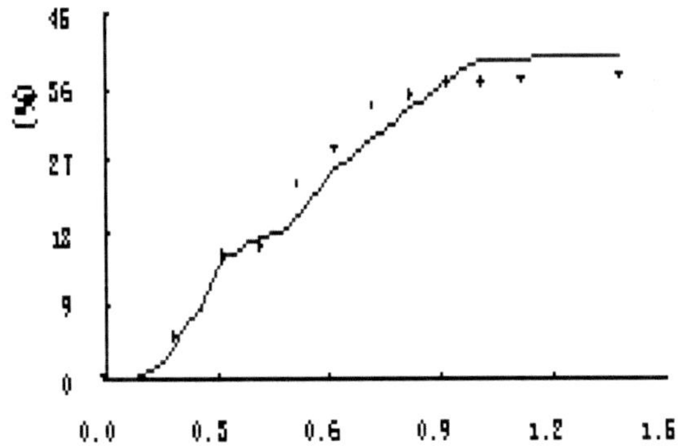

Figure 11.7. *Oil recovery versus injected PV (numerical: solid and laboratory: dotted).*

11.8.2 Example 2

To test the accuracy of the chemical compositional simulator, we compare numerical and laboratory results for a *core flow experiment*. It is a sandstone core, inhomogeneous in the horizontal direction. The dimensions of this core are $30 \times 4.5 \times 4.5$ cm^3; it has three layers, each having a thickness of 1.5 cm. The average permeability of each layer is 1,000 md, with a variation of 0.72. The porosity is 0.26, and water flooding has reached the stage of WC $= 98\%$.

There are primary and secondary injections. In the primary, ASP consists of ORS41 with a concentration of 0.3%, 1.0wt% NaOH solution, and polymer 1275A with 2,000 ppm in solution; in the secondary injection, ASP is composed of ORS41 with a concentration of 0.05%, 1.0wt% NaOH solution, and polymer 1275A with 1,800 ppm in solution. These injections are alternating equal-sized injections of (natural) gas and liquids, with 0.05 PV injected in each cycle. In the primary injection, the gas and liquids are injected 0.3 PV each; in the secondary, they are injected 0.1 PV each. After these two injections, there is a protection period. In this period, 0.05 PV polymer 1275A with 800 ppm in solution is first injected, then 0.15 PV polymer 1275A with 500 ppm in solution is injected, and water is finally injected. The oil recovery rates (relative to the current OIP) obtained using the numerical simulation and laboratory experiment for this problem are shown in Figure 11.7, and the corresponding WCs are presented in Figure 11.8. These two figures show that the numerical and laboratory results match.

We remark that while the differential equations in Section 11.1 were derived for slightly compressible fluids, they apply to the gas injection experiments in this section. Gas injection is studied in the context of ASP+foam flooding. In this type of flooding, on one hand, the polymer viscosity is quite large, and, on the other hand, due to the presence of surfactants and foams, the emulsive phenomenon is significant. As a result, the viscosity of formed emulsions is large and their mobility is low. Therefore, in the entire ASP+foam flooding process, the oil reservoir considered is at a very high pressure. Under such a high pressure, most of the gas flow is in the form of foam, and its volume does not change much.

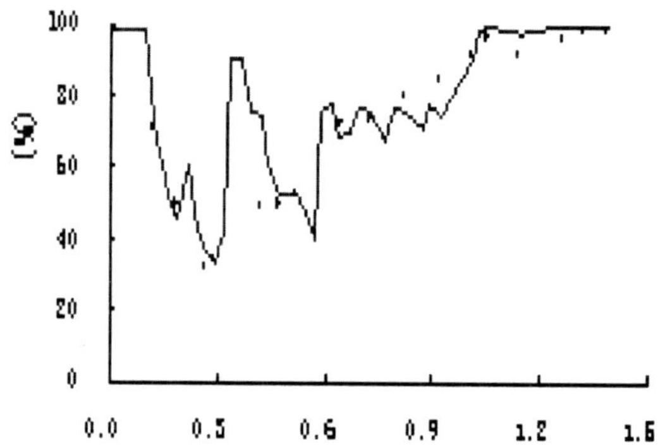

Figure 11.8. *Water cut versus injected PV (numerical: solid and laboratory: dotted).*

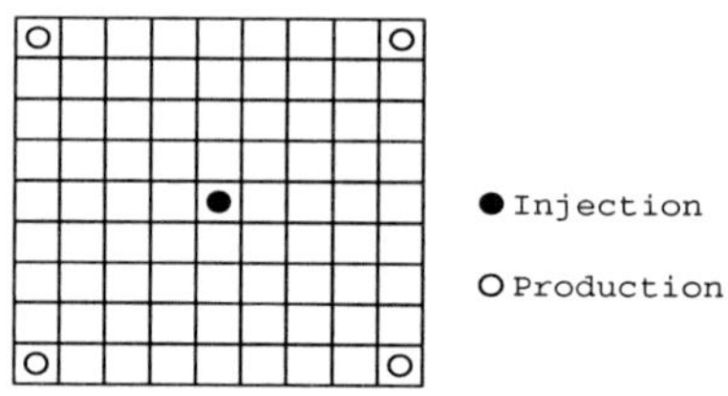

Figure 11.9. *Another five-spot pattern.*

11.8.3 Example 3

This example is more realistic than the first. We use the chemical compositional model to study oil recovery efficiency using different development methods, the oil displacement mechanisms, and the effects of different factors on ASP+foam flooding.

The model

This is another five-spot pattern problem with one injection well and four production wells, and the distance between the injection and production wells is 250 m (cf. Figure 11.9). There are three vertical layers, each having a thickness of 2 m. The average permeability of the first, second, and third layers is 154, 560, and 2,421 md, respectively, with a variation of 0.72 on each layer. The porosity is 0.26, and the initial water saturation is 0.26. The number of grids used is $9 \times 9 \times 3$, and the horizontal grid size is 44.1942 m. The injection rate is 0.19 PV/D.

Oil recovery study

The chemical compositional simulator is applied to four different injection methods: water, polymer, ASP, and ASP+foam flooding. These four injection procedures are the following:

- Water flooding: Water is injected until WC = 98%.

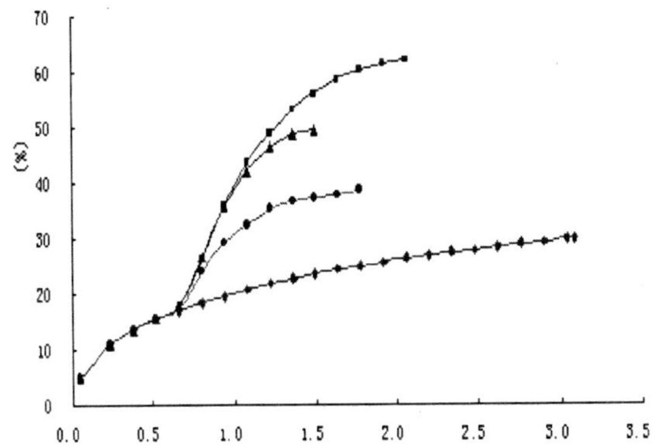

Figure 11.10. *Oil recovery versus injected PV (from bottom to top: water, polymer, ASP, and ASP + foam).*

- Polymer flooding: Water is injected until $S_w = 0.915$, followed by polymer (1,000 ppm in solution) injection until the total injection reaches 0.57 PV, and then water is injected again until WC = 98%.
- ASP flooding: Water is injected until $S_w = 0.915$, followed by 0.015 PV polymer (1,000 ppm in solution) injection in a protection period, then ASP with 0.3% surfactant, 1.0wt% NaOH, and polymer with 1,000 ppm in solution is injected until the total injection reaches 0.57 PV, and finally water is injected again until WC = 98%.
- ASP+foam flooding: Water is injected until $S_w = 0.915$, followed by 0.015 PV polymer (1,000 ppm in solution) injection in the protection period, then ASP+foam is injected with a simultaneous injection of gas and liquids, where the gas-liquid ratio is 1 : 1 and ASP+foam consists of 0.3% surfactant, 1.0wt% NaOH, and polymer with 1,000 ppm in solution, until the total injection reaches 0.57 PV, and finally water is injected again until WC = 98%.

The oil recovery rates using these four injection methods are shown in Figure 11.10. It seems from this figure that ASP+foam flooding is the most efficient.

Displacement mechanism study

As discussed in Section 11.5, in ASP+foam flooding for an initially oil-wet porous medium, because of a change of foam mobility resistance, ASP+foams enter small pores that are not reached by water flooding and displace a large amount of residual oil there. Hence this type of flooding increases the efficiency of water flooding through larger volumetric sweep efficiency and a lower swept zone oil saturation.

Improving larger volumetric sweep efficiency is the ultimate goal of ASP+foam flooding in order to increase oil recovery from water, gas, or steam flooding in a petroleum reservoir. The improvement of this sweep efficiency heavily depends on the blocking capacity of foams in a porous medium. Numerical simulation is a useful approach in studying the

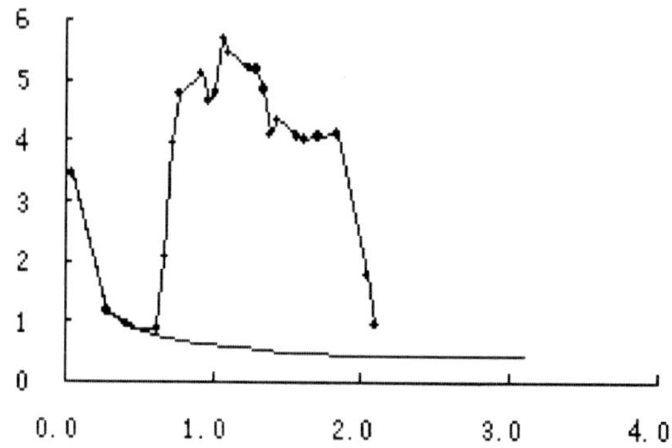

Figure 11.11. *Liquid production (m^3) versus injected PV (water: bottom, and ASP+foam: top).*

mobility of ASP+foams in different permeability zones of the medium to determine the blocking role of foams.

In water flooding for a highly heterogeneous porous medium, most liquid is produced from high-permeability zones, while a small amount of liquid is produced from low-permeability zones. When foams are injected, they first enter the high-permeability zones. As they are continually injected, they soon play a blocking role in these zones so that the mobility resistance there increases, and then they gradually move to the low-permeability zones. That is why a larger volume can be swept by this type of flooding.

We simulate water and ASP+foam flooding for the present problem. These two floodings and their injection slugs are the following:

- Water flooding: Water is injected until WC = 98%.
- ASP+foam flooding: Water is injected until $S_w = 0.915$, followed by 0.015 PV polymer (1,000 ppm in solution) injection in a protection period, then ASP+foam is injected with a simultaneous injection of gas and liquids, where the gas-liquid ratio is 1 : 1 and ASP+foam consists of 0.3% surfactant, 1.0wt% NaOH, and polymer with 1,000 ppm in solution, until the total injection reaches 0.57 PV, and finally water is injected again until WC = 98%.

The oil recovery rates of water and ASP+foam flooding are, respectively, 29.86% and 62.06% for the model problem considered. Obviously, the second form of flooding is far more efficient. Figures 11.11–11.13 give the liquid production in three different layers (high, intermediate, and low permeability layers) for these two floodings. It is clear from these figures that most liquid is produced from the high-permeability layer, and less is produced from other two layers in water flooding. In ASP+foam flooding, foams can effectively block the high-permeability layer so that liquid production decreases in this layer and increases in the intermediate- and low-permeability layers. In addition, liquid production increases more in the intermediate-permeability layer than at low permeability. These observations agree with the displacement mechanism theory that a larger volume is swept by ASP+foam flooding.

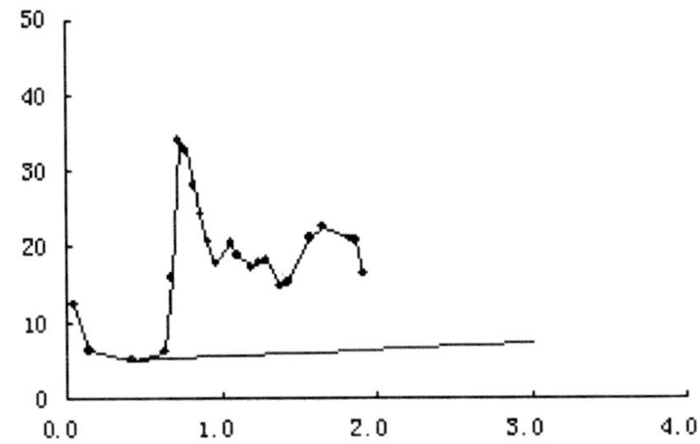

Figure 11.12. *Liquid production (m^3) versus injected PV (water: bottom, and ASP+foam: top).*

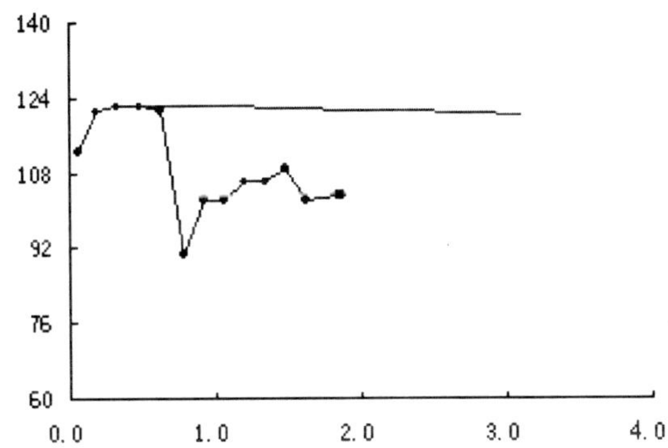

Figure 11.13. *Liquid production (m^3) versus injected PV (water: top, and ASP+foam: bottom).*

Effects of different factors

Many factors affect oil recovery of ASP+foam flooding. Here we numerically study two: the gas-liquid ratio and different injection methods.

(i) Gas-liquid ratio effect

In ASP+foam flooding, the *gas-liquid ratios* are now set to 1 : 1, 3 : 1, and 5 : 1. The oil recovery rates are given in Figure 11.14 for these three cases. It follows from this figure that the ratio 3 : 1 appears better. This ratio generates good quality foams, which can effectively enter and block the high-permeability layer so that more displacing fluids can reach the intermediate- and low-permeability layers, and thus larger volumetric sweep efficiency can be obtained.

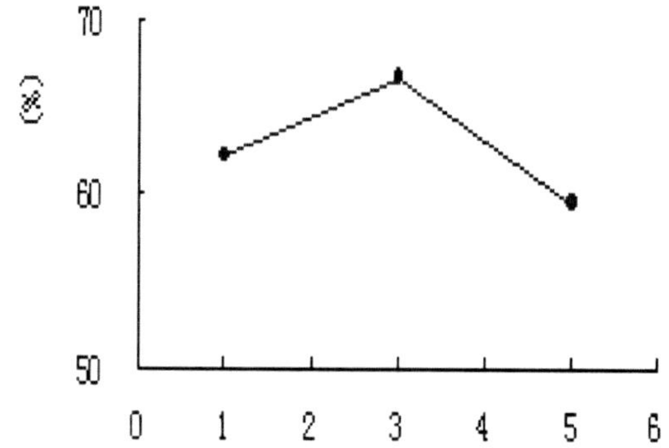

Figure 11.14. *Oil recovery versus different gas-liquid ratios.*

(ii) Gas and liquid injection effect

The gas and liquid injections can be alternating or simultaneous. In addition, in alternating injection, the injection frequency (or cycles) can be different. Different injection methods have different effects on oil recovery.

The gas-liquid ratio is fixed at 3 : 1. We study three injection methods: alternating injection with a low frequency, alternating injection with a high frequency, and simultaneous injection.

- Alternating with a low frequency: 0.095 PV ASP is injected, followed by 0.032 PV gas injection, then they are alternatingly injected until a cumulative ASP reaches 0.57 PV, and finally water is injected again until WC = 98%.
- Alternating with a high frequency: 0.0475 PV ASP is injected, followed by 0.0158 PV gas injection, then they are alternatingly injected until a cumulative ASP reaches 0.57 PV, and finally water is injected again until WC = 98%.
- Simultaneous injection: Gas and liquids are simultaneously injected until a cumulative ASP reaches 0.57 PV, and then water is injected again until WC = 98%.

The recovery rates for these three injection methods are displayed in Figure 11.15. The numerical simulation shows that simultaneous injection is more efficient than the alternating method. For the alternating method, high frequency produces more than does low frequency.

11.9 Application to a Real Oilfield

In this section, the chemical compositional model is used for the numerical study and development prediction of a real oilfield. This oilfield is located in Asia and has been operating since 1963.

11.9.1 Background

This oilfield is large, but the area under study is 0.39 km^2, and the depth to its center is 935 m. Its porous volume is 64.05×10^4 m^3, and the initial OIP is 35.92×10^4 t. The initial pressure

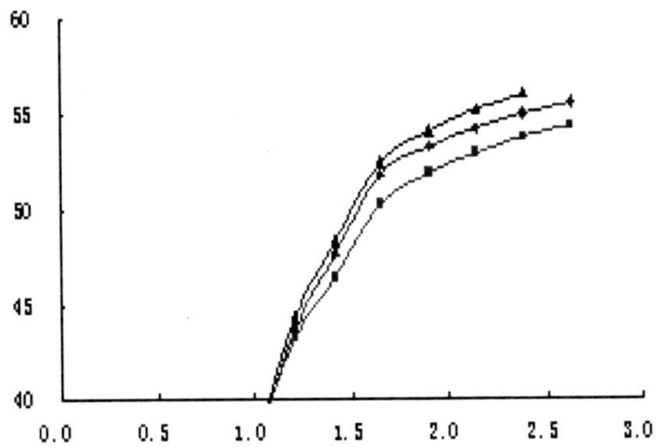

Figure 11.15. *Oil recovery versus injected PV (alternating with low frequency: bottom, alternating with high frequency: middle, and simultaneous: top).*

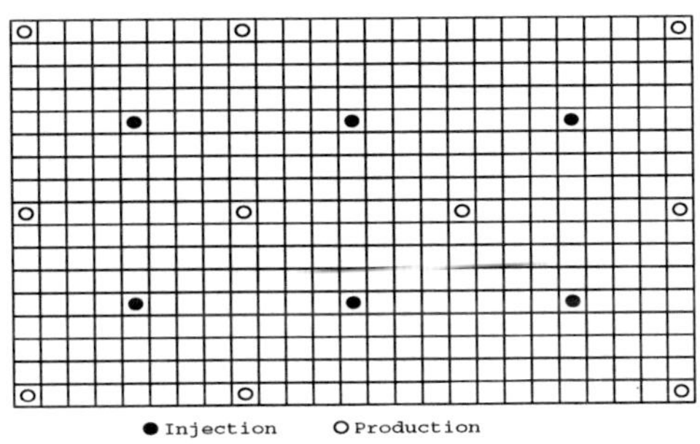

Figure 11.16. *The experimental area.*

of the reservoir is 10.5 Mpa. There are 16 wells; 6 are injection wells, and 10 are production wells. The average distance between the injection wells is 250 m, and the average distance between the injection and production wells is 176 m (cf. Figure 11.16). The two central production wells are the major producers, while other production wells are observatory. The control area, average effective thickness, porous volume, and initial OIP of the two major producers are 0.125 km^2, 6.8 m, 22.44×10^4 m^3, and 12.58×10^4 t, respectively. From March 1989 to September 1993, there were 36 periods of alternating water-gas injections. The cumulative gas injection is $4{,}938 \times 10^4$ m^3 (in standard conditions), i.e., 0.24 PV; the cumulative water injection is 66.92×10^4 m^3, i.e., 0.48 PV.

11.9.2 The numerical model

To simulate this model problem, the injection and production wells are rearranged as in Figure 11.16. A no-flow boundary condition is used. The reservoir has six layers, and the grid dimensions are $25 \times 17 \times 6$. The x_1- and x_2-spatial grid sizes are 31.304 m and 30.829 m, respectively.

Table 11.2. *The reservoir data.*

	Effective thickness (m)	Permeability (μ^2m)	Porosity (%)	Depth (m)
1st layer	0–2.8	0.04–0.378	0.235–0.257	912–950
2nd layer	0–1.4	0.039–0.417	0.235–0.257	914–952
3rd layer	0–2.8	0.04–0.596	0.235–0.257	920–953
4th layer	0.2–2.6	0.039–0.493	0.235–0.257	922–956
5th layer	0.5–2.2	0.039–0.543	0.235–0.257	924–958
6th layer	0–4.1	0.039–0.543	0.235–0.257	926–960

The effective thickness, permeability, porosity, and depth of the grid points where the wells are located were obtained from measurements and are given in Table 11.2. The data for other grid points are interpolated using the well grid points' data. The water saturation before ASP+foam flooding is not known. This saturation at well grid points can be measured using injection, liquid production, and WC data provided by the wells. A WAG (*water-alternating-gas*) test was used to show that 13.88×10^4 m^3 of the injected gas is present in the reservoir before ASP+foam flooding. Since the gas injection region has a pore volume of 139.2×10^4 m^3, the ratio of these two numbers is 9.97%, which can be treated as a reference saturation of the remaining gas.

The physicochemical properties of chemical agents and foams used in this example are obtained from laboratory measurements combined with core flow experiments as in the second example of the previous section. The major properties of foams are that the critical water saturation equals 0.37, the critical concentration of surfactant is 0.0015, the critical oil saturation is 0.25, and the optimal gas-liquid ratio is 3 : 1. The active function of interfacial tension is given in Table 11.1.

11.9.3 Numerical history matching

The numerical experiment involves the water flooding period of January 1–February 24, 1997, the pre-ASP flooding period of February 25–March 26, 1997, the major gas-liquid injection period of March 27, 1997–August 5, 1999, the secondary foam injection period of August 6, 1999–November 16, 2000, and the polymer (800 mg/L in solution) injection period of November 17, 2000–June 30, 2001. The gas and liquids are injected alternatingly. The injection modes are the following:

- Pre-ASP flooding: 0.02 PV ASP is first injected: 0.3% ORS41, 1.2wt% NaOH, and 15,000 (in thousand molecular weights) polymer with 1,200 mg/L in solution.

- Major ASP flooding: 0.55 PV ASP is injected, with 0.3% ORS41, 1.2wt% NaOH, and 15,000 (in thousand molecular weights) polymer with 1,200 mg/L in solution.

- Secondary ASP flooding: 0.3 PV ASP is injected: 0.1% ORS41, 1.2wt % NaOH, and 15,000 (in thousand molecular weights) polymer and natural gas with 1,200 mg/L in polymer solution.

- Protection period: 0.1 PV polymer with 800 mg/L in solution is injected.

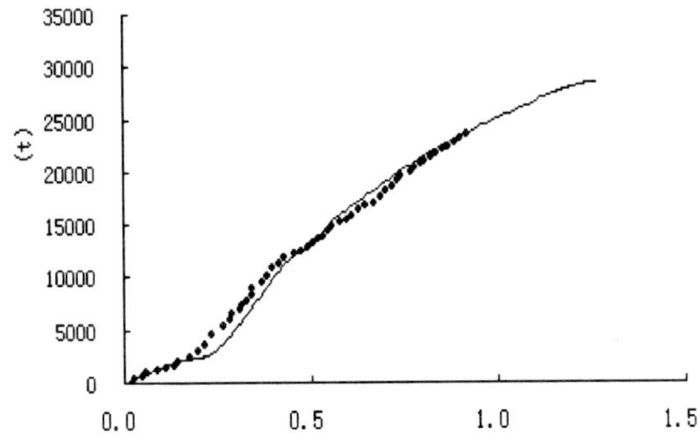

Figure 11.17. *Cumulative oil production versus injected PV (numerical: solid and actual: dotted).*

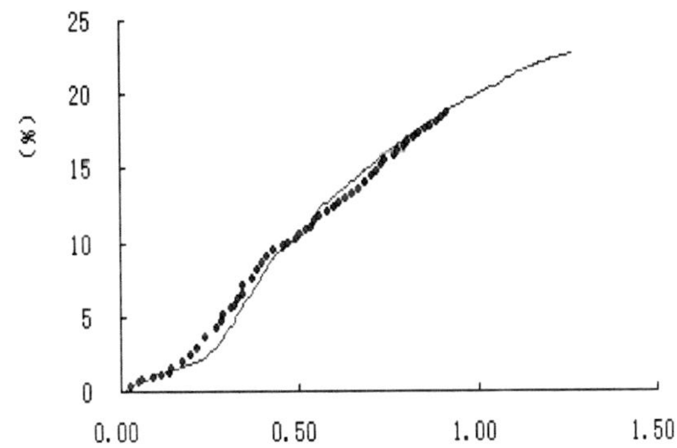

Figure 11.18. *Oil recovery versus injected PV (numerical: solid and actual: dotted).*

The two central production wells are the major producers, so we do *history matching* (cf. Section 14.2) for only these two producers. The history matching covers the period of January 1, 1997–June 30, 2001 from water flooding to the protection period of polymer injection. The matched variables include the daily oil and water production and WC. History matching is performed through an adjustment of relative permeabilities and other physical data. The matches between actual and numerical results for the matched variables are shown in Figures 11.17–11.22 for the injected PV in the range 0–0.97 PV. The cumulative oil production for the same period is given in Table 11.3.

The relative error for WC match is 4.48%. From Figures 11.17–11.22 and Table 11.3, we can see that other variables (daily oil and water productions, cumulative oil production, and recovery rate) also match.

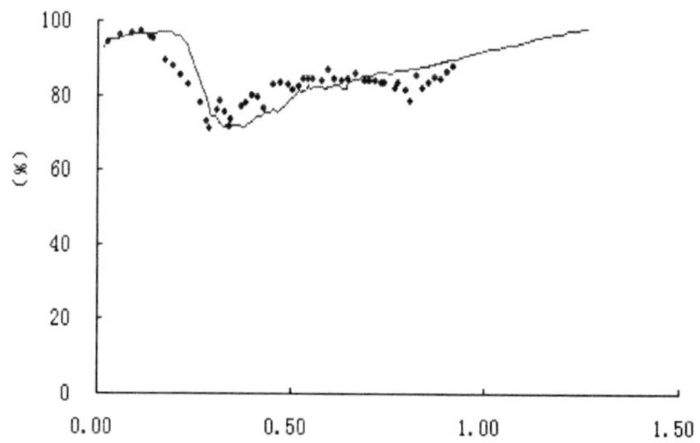

Figure 11.19. *Water cut versus injected PV (numerical: solid and actual: dotted).*

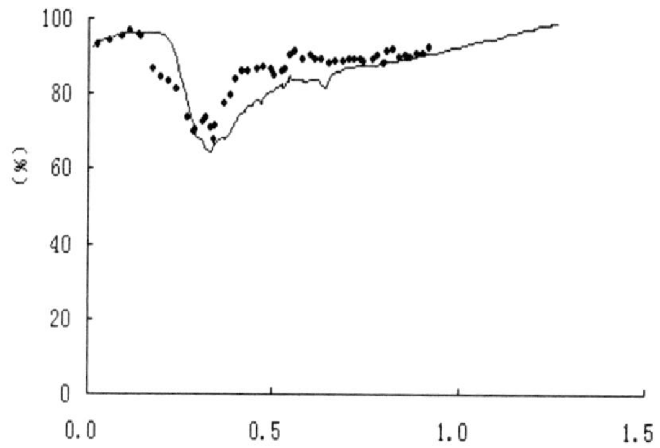

Figure 11.20. *Water cut versus injected PV (numerical: solid and actual: dotted).*

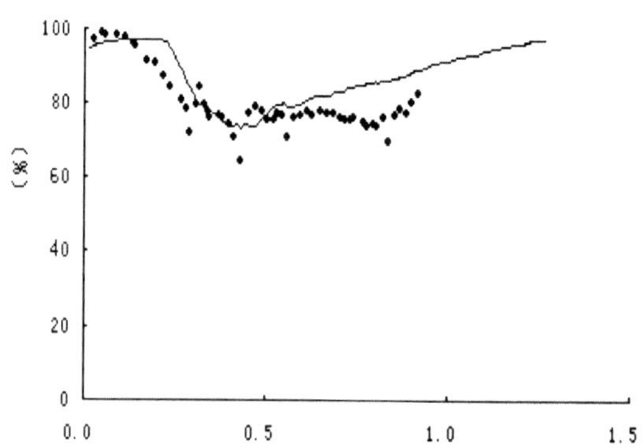

Figure 11.21. *Water cut versus injected PV (numerical: solid and actual: dotted).*

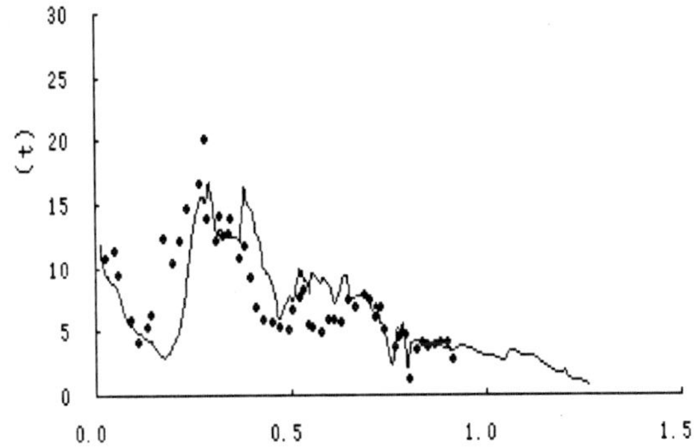

Figure 11.22. *Instantaneous oil production versus injected PV (numerical: solid and actual: dotted).*

Table 11.3. *The history matching of cumulative oil production.*

	Cumulative production (t)	Recovery rates (%)
Actual	23,435	18.63
Numerical	23,647	18.80

11.9.4 Predictions

We can employ the history matching–based adjusted model to predict the development and production of the experimental region using ASP+foam flooding. The prediction is made until WC reaches 98%. The prediction for the two central producers is 28,603 t for the cumulative oil production, 22.74% for the recovery rate for the predicted time period, 67.36% for the recovery rate for the entire simulation time, and 1.27 PV for the injected PV (in the whole experimental oilfield). The predicted results are displayed in Figures 11.17–11.22, where the injected PV is in the range 0.97–1.27 PV.

11.9.5 Assessment of different development methods

An advantage of numerical reservoir simulation is its ability to assess different development methods for a petroleum reservoir in order to choose a robust and reliable method, increase oil and/or gas recovery, and achieve greater economic efficiency. For the present experiment, we compare three different development methods: water flooding, ASP+foam flooding with a protection period of polymer injection, and ASP+foam flooding without this protection period (i.e., water is further injected after the secondary ASP+foam flooding). The cumulative oil production and oil recovery rate for the predicted time period (January 1, 1997–June 30, 2001) are given in Table 11.4 for the two central producers. It is clear that it is very difficult to recover the remaining oil using water flooding alone. ASP+foam flooding recovers much more. Furthermore, the ASP+foam flooding with a protection period recovers even more. This implies that the second development project is the most efficient among the three projects.

Table 11.4. *The assessment of different development methods.*

	Cumulative production (t)	Recovery rates (%)
Water flooding	4,029	3.20
ASP+foam with protection	28,603	22.74
ASP+foam without protection	27,022	21.48

11.10 Bibliographical Remarks

Most of the content in this chapter is taken from Chen et al. (2007). The presentation in Sections 11.2–11.4 follows Delshad et al. (2000).

Exercises

11.1. Derive equation (11.9) by adding equations (11.1) over i, $i = 1, 2, \ldots, N_{cv}$, and using equations (11.6) and (11.8).

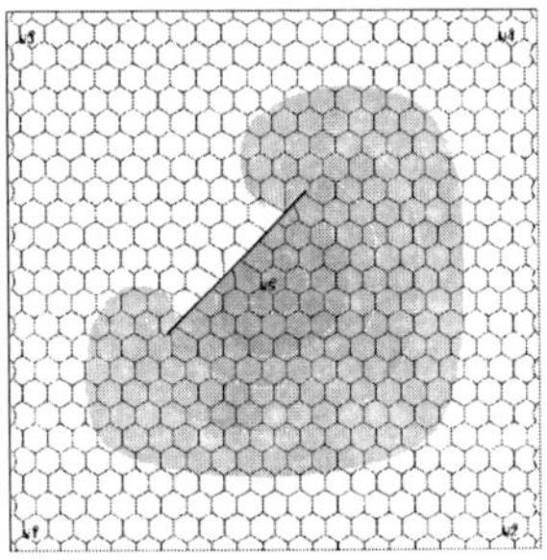

Chapter 12
Flows in Fractured Porous Media

A *fractured porous medium* has throughout its extent a system of interconnected fractures dividing the medium into a series of essentially disjoint blocks of porous rock, called the *matrix blocks* (cf. Figure 2.2). It has two main length scales of interest: the microscopic scale of the fracture thickness (about 10^{-4} m) and the macroscopic scale of the average distance between fracture planes, i.e., the size of the matrix blocks (about 0.1–1 m). Since the entire porous medium is about 10^3–10^4 m across, flow can be mathematically simulated only in some averaged sense. The concept of *dual porosity* (and *dual porosity/permeability*) has been utilized to model the flow of fluids on its various scales (Pirson, 1953; Barenblatt et al., 1960; Warren and Root, 1963; Kazemi, 1969). In this concept, the fracture system is treated as a porous structure distinct from the usual porous structure of the matrix itself. The fracture system is highly permeable, but can store very little fluid, while the matrix has the opposite characteristics. When developing a dual porosity model, it is critical to treat the *flow transfer terms* between the fracture and matrix systems.

There are two approaches to treating a matrix-fracture flow transfer term. In the first approach (known as the Warren–Root approach; cf. Section 2.11.2), this term for a particular fluid phase is directly related to a *shape factor*, the fluid mobility, and the potential difference between these two systems, and the capillary pressure, gravity, and viscous forces are properly incorporated into this term. Here this approach will be reviewed. Moreover, the inclusion of a pressure gradient across a matrix block in this term in a general fashion is also studied. The other approach is to treat the flow transfer term explicitly through boundary conditions on the matrix blocks. This approach avoids the introduction of the ad hoc parameters (e.g., the shape factor and a characteristic length) in the first approach, and is more general. However, the second approach appears to apply only to a dual porosity model, not to a dual porosity/permeability model.

The formulation of the mass balance equation for each fluid phase in a fractured porous medium follows that for an ordinary medium with an additional matrix-fracture transfer term. The *two overlapping continua*, fractures and matrix blocks, are allowed to coexist and interact with each other. Furthermore, there are matrix-matrix connections. In this case, a dual porosity/permeability model is required for the fractured porous medium.

If the matrix blocks act only as a source term to the fracture system and there is no matrix-matrix connection, a dual porosity (and single permeability) model is applied.

The governing equations that describe fluid flow in a fractured porous medium are developed in Section 12.1. The matrix-fracture transfer terms for the dual porosity and dual porosity/permeability models are also introduced in this section. Numerical results based on the sixth CSP organized by the SPE are reported in Section 12.2. Finally, bibliographical information is given in Section 12.3.

12.1 Flow Equations

Dual porosity/permeability models were developed for single phase and compositional flows in fractured porous media in Sections 2.2.6 and 2.11, respectively. To be specific to the application presented here, the fluid flow equations considered are based on a three-component, three-phase black oil model (cf. Section 2.6 or Chapter 8).

To reduce confusion, we distinguish carefully between phases and components. We use lower- and uppercase letter subscripts to denote the phases and components, respectively. Furthermore, a subscript f is used to denote fracture variables.

12.1.1 Dual porosity/permeability models

Let ϕ and $\mathbf{k}$ denote the porosity and permeability of a matrix system, and let S_α, μ_α, p_α, $\mathbf{u}_\alpha$, ρ_α, and $k_{r\alpha}$ be the saturation, viscosity, pressure, volumetric velocity, density, and relative permeability of the α-phase, $\alpha = w, o, g$, respectively. Because of mass interchange between the oil and gas phases, mass is not conserved within each phase, but rather the total mass of each component must be conserved. Thus, for the matrix system, the mass balance equations are

$$\frac{\partial(\phi\rho_w S_w)}{\partial t} = -\nabla\cdot(\rho_w\mathbf{u}_w) - q_{Wm} \tag{12.1}$$

for the water component,

$$\frac{\partial(\phi\rho_{Oo} S_o)}{\partial t} = -\nabla\cdot(\rho_{Oo}\mathbf{u}_o) - q_{Oom} \tag{12.2}$$

for the oil component, and

$$\frac{\partial}{\partial t}\big(\phi(\rho_{Go}S_o + \rho_g S_g)\big) = -\nabla\cdot(\rho_{Go}\mathbf{u}_o + \rho_g\mathbf{u}_g) - (q_{Gom} + q_{Gm}) \tag{12.3}$$

for the gas component, where ρ_{Oo} and ρ_{Go} indicate the partial densities of the oil and gas components in the oil phase, respectively, and q_{Wm}, q_{Oom}, q_{Gom}, and q_{Gm} represent the matrix-fracture transfer terms. Equation (12.3) implies that the gas component may exist in both the oil and gas phases.

Darcy's law for each phase is written in the usual form

$$\mathbf{u}_\alpha = -\frac{k_{r\alpha}}{\mu_\alpha}\mathbf{k}\left(\nabla p_\alpha - \rho_\alpha\wp\nabla z\right), \qquad \alpha = w,\ o,\ g, \tag{12.4}$$

where $\wp$ is the magnitude of the gravitational acceleration and z is the depth. The saturation constraint reads

$$S_w + S_o + S_g = 1. \tag{12.5}$$

Finally, the phase pressures are related by capillary pressures

$$p_{cow} = p_o - p_w, \quad p_{cgo} = p_g - p_o. \tag{12.6}$$

For the fracture system, the mass balance equations are

$$\begin{aligned} \frac{\partial(\phi\rho_w S_w)_f}{\partial t} &= -\nabla\cdot(\rho_w\mathbf{u}_w)_f + q_{Wm} + q_W, \\ \frac{\partial(\phi\rho_{Oo} S_o)_f}{\partial t} &= -\nabla\cdot(\rho_{Oo}\mathbf{u}_o)_f + q_{Oom} + q_{Oo}, \\ \frac{\partial}{\partial t}\left(\phi(\rho_{Go}S_o + \rho_g S_g)\right)_f &= -\nabla\cdot(\rho_{Go}\mathbf{u}_o + \rho_g\mathbf{u}_g)_f \\ &\quad + (q_{Gom} + q_{Gm}) + (q_{Go} + q_G), \end{aligned} \tag{12.7}$$

where q_W, q_{Oo}, q_{Go}, and q_G denote the external sources and sinks. We have assumed that these external terms interact only with the fracture system. This is reasonable since the flow is much faster in this system than in the matrix blocks. Equations (12.4)–(12.6) remain valid for the fracture quantities.

The matrix-fracture transfer terms for the dual porosity/permeability model can be defined using the concept of Warren and Root (1963) and Kazemi (1969). The transfer term for a particular component is directly related to a shape factor σ, the fluid mobility, and the potential difference between the fracture and matrix systems. The capillary pressure, gravity, and viscous forces must be properly incorporated into this term. Furthermore, the contributions from a pressure gradient across each matrix block and the molecular diffusion rate for each component must be also included. For the brevity of presentation, we neglect the diffusion rate, and discuss the contribution from the pressure gradient.

The treatment of a pressure gradient across a block is based on the following observation: for an oil matrix block surrounded with water in the fractures, we see that

$$\Delta p_w = 0, \quad \Delta p_o = \wp(\rho_w - \rho_o).$$

Analogously, for an oil block surrounded with gas fractures and a gas block surrounded with water fractures, we see, respectively, that

$$\Delta p_g = 0, \quad \Delta p_o = \wp(\rho_o - \rho_g),$$

and

$$\Delta p_w = 0, \quad \Delta p_g = \wp(\rho_w - \rho_g).$$

In general, we introduce the global fluid density in the fractures

$$\rho_f = S_{w,f}\rho_w + S_{o,f}\rho_o + S_{g,f}\rho_g,$$

and define the pressure gradients

$$\Delta p_\alpha = \wp\left|\rho_f - \rho_\alpha\right|, \qquad \alpha = w, o, g.$$

Now, the *transfer terms* that include the contributions from the capillary pressure, gravity, and viscous forces, and the pressure gradients across matrix blocks are defined by

$$\begin{aligned} q_{Wm} &= T_m \frac{k_{rw}\rho_w}{\mu_w}\left(\Phi_w - \Phi_{w,f} + L_c \Delta p_w\right), \\ q_{Oom} &= T_m \frac{k_{ro}\rho_{Oo}}{\mu_o}\left(\Phi_o - \Phi_{o,f} + L_c \Delta p_o\right), \\ q_{Gom} &= T_m \frac{k_{ro}\rho_{Go}}{\mu_o}\left(\Phi_o - \Phi_{o,f} + L_c \Delta p_o\right), \\ q_{Gm} &= T_m \frac{k_{rg}\rho_g}{\mu_g}\left(\Phi_g - \Phi_{g,f} + L_c \Delta p_g\right), \end{aligned} \tag{12.8}$$

where Φ_α is the phase potential

$$\Phi_\alpha = p_\alpha - \rho_\alpha \wp z, \qquad \alpha = w, o, g,$$

L_c is the characteristic length for the matrix-fracture flow, and T_m is the *matrix-fracture transmissibility*

$$T_m = k\sigma \left(\frac{1}{l_{x_1}^2} + \frac{1}{l_{x_2}^2} + \frac{1}{l_{x_3}^2} \right)$$

with σ the shape factor and l_{x_1}, l_{x_2}, and l_{x_3} the matrix block dimensions (Kazemi, 1969; Coats, 1989). When the matrix permeability $\mathbf{k}$ is a tensor and different in the three coordinate directions, the matrix-fracture transmissibility is modified to

$$T_m = \sigma \left(\frac{k_{11}}{l_{x_1}^2} + \frac{k_{22}}{l_{x_2}^2} + \frac{k_{33}}{l_{x_3}^2} \right), \qquad \mathbf{k} = \mathrm{diag}(k_{11}, k_{22}, k_{33}).$$

12.1.2 Dual porosity models

For the derivation of a *dual porosity model*, we assume that fluids do not flow directly from one matrix block to another. Rather, they first flow into the fractures, and then they flow into another block or remain in the fractures. This is reasonable since fluids flow more rapidly in the fractures than in the matrix. Therefore, the matrix blocks act as source terms to the fracture system, and there is no matrix-matrix connection for the dual porosity model. In this case, there are two approaches for deriving this model: the first is as in Section 12.1.1, and the second is to be defined in Section 12.1.2 (ii) below.

(i) The Warren–Root approach

In this approach, the mass balance equations in the matrix become

$$\begin{aligned} &\frac{\partial(\phi \rho_w S_w)}{\partial t} = -q_{Wm}, \\ &\frac{\partial(\phi \rho_{Oo} S_o)}{\partial t} = -q_{Oom}, \\ &\frac{\partial}{\partial t}\left(\phi\left(\rho_{Go} S_o + \rho_g S_g\right)\right) = -\left(q_{Gom} + q_{gm}\right), \end{aligned} \tag{12.9}$$

where q_{Wm}, q_{Oom}, q_{Gom}, and q_{gm} are given by (12.8). The fracture equations are the same as in Section 12.1.1.

(ii) The boundary conditions approach

For a dual porosity model, the matrix-fracture transfer terms can be modeled explicitly through boundary conditions on the matrix blocks, following Pirson (1953) and Barenblatt et al. (1960). Let the matrix system be composed of disjoint blocks $\{\Omega_i\}$. On each block $\{\Omega_i\}$, the following mass balance equations hold:

$$\begin{aligned}
&\frac{\partial(\phi\rho_w S_w)}{\partial t} = -\nabla\cdot(\rho_w \mathbf{u}_w),\\
&\frac{\partial(\phi\rho_{Oo} S_o)}{\partial t} = -\nabla\cdot(\rho_{Oo} \mathbf{u}_o),\\
&\frac{\partial}{\partial t}\big(\phi(\rho_{Go} S_o + \rho_g S_g)\big) = -\nabla\cdot(\rho_{Go}\mathbf{u}_o + \rho_g \mathbf{u}_g).
\end{aligned} \tag{12.10}$$

The total mass of water leaving the ith matrix block Ω_i per unit time is

$$\int_{\partial\Omega_i} \rho_w \mathbf{u}_w \cdot \boldsymbol{\nu}\, d\ell,$$

where $\boldsymbol{\nu}$ is the outward unit normal to the surface $\partial\Omega_i$ of Ω_i. The divergence theorem and the first equation of (12.10) imply

$$\int_{\partial\Omega_i} \rho_w \mathbf{u}_w \cdot \boldsymbol{\nu}\, d\ell = \int_{\Omega_i} \nabla\cdot(\rho_w \mathbf{u}_w) d\mathbf{x} = -\int_{\Omega_i} \frac{\partial(\phi\rho_w S_w)}{\partial t} d\mathbf{x}. \tag{12.11}$$

Now, we define q_{Wm} by

$$q_{Wm} = -\sum_i \chi_i(\mathbf{x}) \frac{1}{|\Omega_i|} \int_{\Omega_i} \frac{\partial(\phi\rho_w S_w)}{\partial t} d\mathbf{x}, \tag{12.12}$$

where $|\Omega_i|$ denotes the volume of Ω_i and $\chi_i(\mathbf{x})$ is its characteristic function, i.e.,

$$\chi_i(\mathbf{x}) = \begin{cases} 1 & \text{if } \mathbf{x} \in \Omega_i,\\ 0 & \text{otherwise.} \end{cases}$$

Similarly, q_{Oom} and $q_{Gom} + q_{Gm}$ are (cf. Exercise 12.1)

$$q_{Oom} = -\sum_i \chi_i(\mathbf{x}) \frac{1}{|\Omega_i|} \int_{\Omega_i} \frac{\partial(\phi\rho_{Oo} S_o)}{\partial t} d\mathbf{x} \tag{12.13}$$

and

$$q_{Gom} + q_{Gm} = -\sum_i \chi_i(\mathbf{x}) \frac{1}{|\Omega_i|} \int_{\Omega_i} \frac{\partial\big(\phi(\rho_{Go} S_o + \rho_g S_g)\big)}{\partial t} d\mathbf{x}. \tag{12.14}$$

This approach for defining the transfer terms avoids the introduction of the ad hoc parameters (e.g., the shape factor and characteristic length).

With the definition of q_{Wm}, q_{Oom}, and $q_{Gom} + q_{Gm}$, boundary conditions on the surface of each matrix block can be imposed in a general fashion, and gravitational forces and pressure gradient effects across the block can be incorporated into these conditions (cf. Sections 2.2.6 and 2.11). We define the *phase pseudopotential* as

$$\Phi'_\alpha(p_\alpha) = \int_{p_\alpha^o}^{p_\alpha} \frac{1}{\rho_\alpha(\xi)\wp} \, d\xi - z, \tag{12.15}$$

where p_α^o is some reference pressure, $\alpha = w, o, g$. The inverse of this integral is denoted $\psi'_\alpha(\cdot)$. Now, the boundary conditions for (12.10) on the surface $\partial\Omega_i$ of each matrix block Ω_i are

$$\Phi'_\alpha(p_\alpha) = \Phi'_{\alpha,f}(p_{\alpha,f}) - \Phi_\alpha^o \quad \text{on } \partial\Omega_i, \ \alpha = w, o, g, \tag{12.16}$$

where, for a given $\Phi'_{\alpha,f}$, Φ_α^o is a pseudopotential reference value on each block Ω_i determined by

$$\frac{1}{|\Omega_i|} \int_{\Omega_i} (\phi \rho_\alpha) \left(\psi'_\alpha (\Phi'_{\alpha,f} - \Phi_\alpha^o + x_3) \right) d\mathbf{x} = (\phi \rho_\alpha)(p_{\alpha,f}). \tag{12.17}$$

If we assume that $\partial \rho_\alpha / \partial p_\alpha \geq 0$, (12.17) is solvable for Φ_α^o. (For incompressible α-phase fluid, we set $\Phi_\alpha^o = 0$.)

This model implies that the fracture system, being highly permeable, quickly comes into phase equilibrium locally on the fracture spacing scale. This equilibrium is defined in terms of the *phase pseudopotentials* and is reflected in the matrix equations through the boundary conditions (12.16).

12.2 The Sixth SPE Project: Dual Porosity Simulation

The experimental problems are chosen from the benchmark problems of the sixth CSP (Thomas et al., 1983; Firoozabadi-Thomas, 1990). Ten organizations participated in the comparative project. In these problems, various aspects of the physics of multiphase flow in fractured petroleum reservoirs are examined. The question of a fracture capillary pressure and its influence on reservoir performance is addressed by including zero and nonzero gas-oil capillary pressures in the fractures. The nonzero capillary pressure is not based on actual measurements, but is intended as a parameter for sensitivity studies. The variation of gas-oil interfacial tension with pressure is also incorporated. The gas-oil capillary pressure is directly related to the interfacial tension, and thus this pressure should be adjusted according to the ratio of the interfacial tensions at pressure and at the pressure at which the capillary pressures are specified.

The example under consideration is cross sectional, and is designed to simulate depletion, gas injection, and water injection in fractured petroleum reservoirs. Table 12.1 states the basic physical and fluid property data, Table 12.2 shows the reservoir layer description, Table 12.3 gives the matrix block shape factors, Tables 12.4 and 12.5 indicate the fracture and rock data (relative permeabilities and capillary pressures), and Tables 12.6 and 12.7 represent the oil and gas PVT data, where B_o and B_g are the oil and gas formation volume factors, R_{so} is the gas solubility factor, and c_μ is the oil viscosity compressibility. In all the experiments, the injector is located at $i = 1$, and the producer is located at $i = 10$. The input data for each experiment are given below.

Table 12.1. *Basic physical and fluid data.*

$k = 1$ (md), $\phi = 0.29$, $\phi_f = 0.01$
$Nx_1 = 10$, $Nx_2 = 1$, $Nx_3 = 5$
$h_1 = 200$, $h_2 = 1000$, $h_3 = 50$ (ft)
z-direction transmissibility: multiply computed values by 0.1
Initial pressure: 6014.7 (psia), saturation pressure: 5559.7 (psia)
Water viscosity: 0.35 (cp), water compressibility: 3.5×10^{-6} (psi^{-1})
Water formation volume factor: 1.07 (psig)
Rock and oil compressibility: 3.5×10^{-6}, 1.2×10^{-5} (psi^{-1})
Temperature: 200° F, datum: 13400 (ft), depth to the top: 13400 (ft)
Densities of stock tank oil and water: 0.81918 and 1.0412 (gm/cc)
Gas specific gravity at standard conditions: 0.7595
Rate $= \frac{k_r PI}{B\mu}$, Δp in psi, μ in cp, B in RB/STB, and rate in STB/D

Table 12.2. *Reservoir layer description.*

Layer	k_f (md)	Block height (ft)	$PI(\frac{RB\ cp}{D\ psi})$
1	10	25	1
2	10	25	1
3	90	5	9
4	20	10	2
5	20	10	2

Table 12.3. *Matrix block shape factors.*

Block size (ft)	Water-oil (ft^{-2})	Gas-oil (ft^{-2})
5	1.00	0.08
10	0.25	0.02
25	0.04	0.0032

Table 12.4. *Fracture rock data.*

S_w	k_{rw}	k_{row}	p_{cow}
0.0	0.0	1.0	0.0
1.0	1.0	0.0	0.0

S_g	k_{rg}	k_{rog}	p_{cgo}
0.0	0.0	1.0	0.0375
0.1	0.1	0.9	0.0425
0.2	0.2	0.8	0.0475
0.3	0.3	0.7	0.0575
0.4	0.4	0.6	0.0725
0.5	0.5	0.5	0.0880
0.7	0.7	0.3	0.1260
1.0	1.0	0.0	0.1930

Table 12.5. *Matrix rock data.*

S_w	k_{rw}	k_{row}	p_{cow}
0.2	0.0	1.0	1.0
0.25	0.005	0.860	0.5
0.30	0.010	0.723	0.3
0.35	0.020	0.600	0.15
0.40	0.030	0.492	0.0
0.45	0.045	0.392	−0.2
0.50	0.060	0.304	−1.2
0.60	0.110	0.154	−4.0
0.70	0.180	0.042	−10.0
0.75	0.230	0.000	−40.0
1.0	1.0	0.0	−100.0
S_g	k_{rg}	k_{rog}	p_{cgo}
0.0	0.0	1.0	0.075
0.1	0.015	0.70	0.085
0.2	0.050	0.45	0.095
0.3	0.103	0.25	0.115
0.4	0.190	0.11	0.145
0.5	0.310	0.028	0.255
0.55	0.420	0.0	0.386
0.6	0.553	0.0	1.0
0.8	1.0	0.0	100.0

Table 12.6. *Oil PVT data.*

p_b (psia)	R_{so} (SCF/STB)	μ_o (cp)	c_μ (psi^{-1})	B_o (RB/STB)
1688.7	367	0.529	0.0000325	1.3001
2045.7	447	0.487	0.0000353	1.3359
2544.7	564	0.436	0.0000394	1.3891
3005.7	679	0.397	0.0000433	1.4425
3567.7	832	0.351	0.0000490	1.5141
4124.7	1000	0.310	0.0000550	1.5938
4558.7	1143	0.278	0.0000619	1.6630
4949.7	1285	0.248	0.0000694	1.7315
5269.7	1413	0.229	0.0000751	1.7953
5559.7	1530	0.210	0.0000819	1.8540
7014.7	2259	0.109	0.0001578	2.1978

Depletion. Depletion runs are performed to a maximum of ten years or whenever production is less than 1 STB/D. The producer has a maximum rate of 500 STB/D, and it is constrained by a maximum drawdown of 100 psi. This well is perforated only in the bottom layer. Two cases are studied: zero and nonzero fracture capillary pressures. The nonzero capillary data are reported in Table 12.4. These data are given at the bubble point pressure p_b of 5,545 psig and have been adjusted for the effect of pressure on interfacial tension.

Gas injection. In this experiment 90% of the gas produced from the previous time step is reinjected. The injector is perforated in layers 1–3. The producer is perforated in layers 4 and 5, and is constrained by a maximum drawdown of 100 psi. A maximum rate

Table 12.7. *Gas PVT data.*

p_g (psia)	μ_g (cp)	B_g (RB/STB)	σ_1 (dyne/cm)*
1688.7	0.0162	1.98	6.0
2045.7	0.0171	1.62	4.7
2544.7	0.0184	1.30	3.3
3005.7	0.0197	1.11	2.2
3567.7	0.0213	0.959	1.28
4124.7	0.0230	0.855	0.72
4558.7	0.0244	0.795	0.444
4949.7	0.0255	0.751	0.255
5269.7	0.0265	0.720	0.155
5559.7	0.0274	0.696	0.090
7014.7	0.0330	0.600	0.050

$^*\sigma_1 = IFT(p)/IFT(p_{ref})$, $p_{cgo}(S_g) = p_{cgo,ref}(S_g)\sigma_1$.

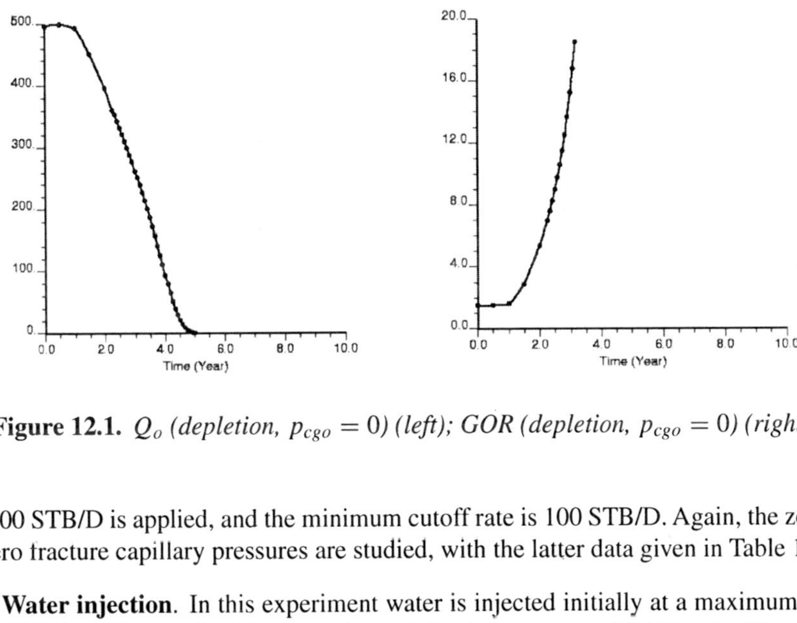

Figure 12.1. *Q_o (depletion, $p_{cgo} = 0$) (left); GOR (depletion, $p_{cgo} = 0$) (right).*

of 1,000 STB/D is applied, and the minimum cutoff rate is 100 STB/D. Again, the zero and nonzero fracture capillary pressures are studied, with the latter data given in Table 12.4.

Water injection. In this experiment water is injected initially at a maximum rate of 1,750 STB/D and constrained by a maximum injection pressure of 6,100 psig. The production rate is set at 1,000 STB/D of the total fluid (water and oil). The injector is perforated in layers 1–4, and the producer is perforated in layers 1–3. The final time of runs is 20 years.

For the numerical results presented here, the temporal discretization is based on the backward Euler scheme, and the spatial discretization is based on the Raviart–Thomas–Nédélec mixed finite element method on rectangular parallelepipeds (cf. Section 4.5.4). We use the simultaneous fully implicit solution technique (cf. Section 8.2.2). The Warren–Root approach is used to model the matrix-fracture flow transfer terms.

Numerical results are reported for the oil production rate (Q_o in STB/D) and gas-oil ratio (GOR in SCF/STB) versus time (years) in the first two studies (depletion and gas injection), and for the oil production rate and water cut (percent) in the water injection study. The results are shown in Figures 12.1–12.5, where the zero and nonzero fracture capillary

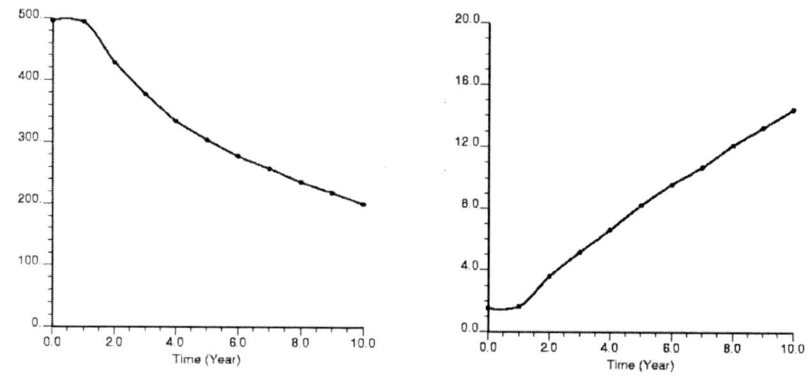

Figure 12.2. *Q_o (depletion, $p_{cgo} \neq 0$) (left); GOR (depletion, $p_{cgo} \neq 0$) (right).*

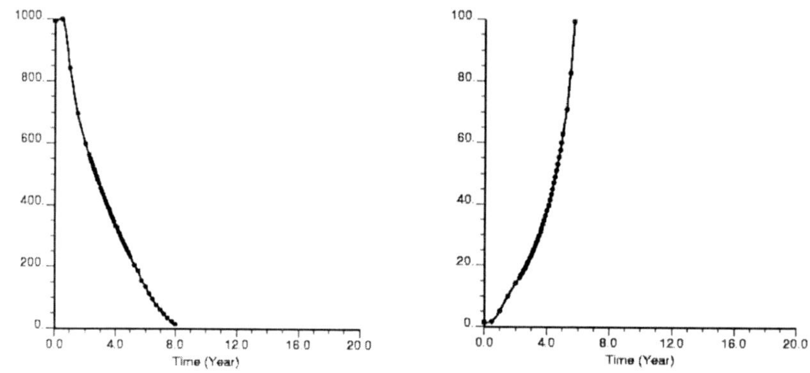

Figure 12.3. *Q_o (gas recycling, $p_{cgo} = 0$) (left); GOR (gas recycling, $p_{cgo} = 0$) (right).*

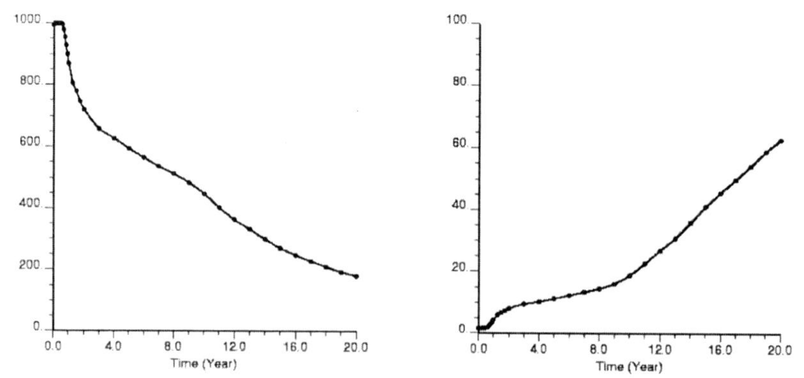

Figure 12.4. *Q_o (gas recycling, $p_{cgo} \neq 0$) (left); GOR (gas recycling, $p_{cgo} \neq 0$) (right).*

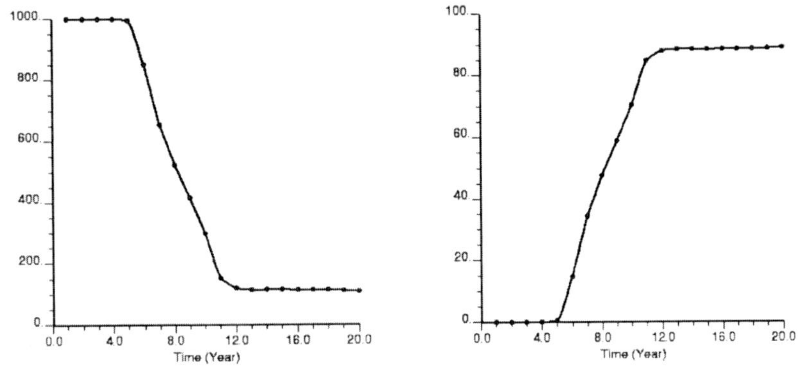

Figure 12.5. *Q_o (water flooding) (left); water cut (water flooding) (right).*

pressure cases are illustrated. A comparison of these two cases indicates that capillary continuity has a major influence on the numerical results. The reason is that in the depletion study, for example, when the capillary pressure force is stronger than the gravity drainage force, the oil flow from the matrix blocks decreases since interfacial tension increases with a decrease in pressure. Note that there is a stable water cut curve after the 10th year. This occurs because the entire fracture system contains water after the 10th year; the major flow exchange mechanism between the matrix and fractures depends on imbibition (minus the value of p_{cow}) with a small flow rate for a long time.

12.3 Bibliographical Remarks

The content in this chapter is taken from Huan et al. (2005). For more information about the data used in the sixth SPE CSP, see Firoozabadi and Thomas (1990).

Exercises

12.1. Derive the matrix-fracture transfer terms q_{Oom} and $q_{Gom} + q_{Gm}$ in equations (12.13) and (12.14) for the dual porosity model of the black oil model.

12.2. Develop a dual porosity/permeability model for the volatile model (cf. Section 2.7) using an approach similar to that for the black oil model in Section 12.1.1.

12.3. Develop a dual porosity model for the volatile model (cf. Section 2.7) using an approach similar to that for the black oil model in Section 12.1.2 (i).

12.4. Develop a dual porosity model for the volatile model (cf. Section 2.7) using an approach similar to that for the black oil model in Section 12.1.2 (ii).

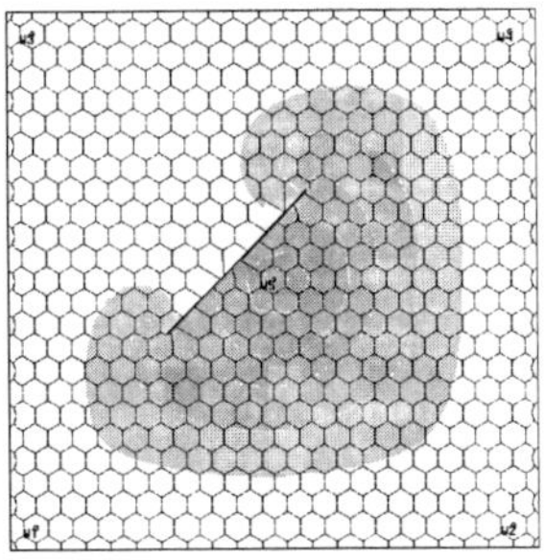

Chapter 13
Well Modeling

Numerical simulation of fluid flows in petroleum reservoirs must account for the presence of wells. The pressure at a gridblock that contains a well is different from the average pressure in that block and different from the *flowing bottom hole pressure* for the well (Peaceman, 1977A). The difficulty in modeling wells in a field-scale numerical simulation is that the region where pressure gradients are the largest is closest to a well and is far smaller than the spatial size of gridblocks. Using local grid refinement around the well can alleviate this problem but can lead to an impratical restriction on time step sizes in the numerical simulation (cf. Section 4.2.4). The fundamental task in modeling wells is to model flows into the wellbore accurately and to develop accurate well equations that allow the computation of the bottom hole pressure when a production or injection rate is given, or the computation of the rate when this pressure is known. In this chapter, we develop well flow equations for numerical simulation of fluid flows in petroleum reservoirs using finite difference methods (Section 13.2), standard finite element methods (Section 13.3), control volume finite element methods (Section 13.4), and mixed finite element methods (Section 13.5). The development of these well equations requires the use of analytical formulas (Section 13.1). Various well controls and constraints are discussed in Section 13.6. Numerical results based on the seventh CSP organized by the SPE are presented in Section 13.7. Bibliographical information is given in Section 13.8.

13.1 Analytical Formulas

The derivation of well flow equations is based on a basic assumption that the flow is *radial* in a neighborhood of the well (cf. Section 6.2.1), and requires the use of *analytical formulas* for radial flow. These formulas are known only in simplified flow situations. Thus we consider single phase incompressible flow in isotropic reservoirs. Furthermore, we focus on steady-state flow; an unsteady-state single phase flow was described in Section 6.2. In the steady state case, the mass conservation equation is (cf. (2.1) and (2.10))

$$\nabla \cdot (\rho \mathbf{u}) = q\delta, \tag{13.1}$$

where ρ and $\mathbf{u}$ are the density and volumetric velocity, respectively, of the fluid; δ is the Dirac delta function representing a well placed at the origin, for example; and q is the mass production/injection at this well. Darcy's law without the gravity term is (cf. (2.4))

$$\mathbf{u} = -\frac{1}{\mu}\mathbf{k}\nabla p, \tag{13.2}$$

where $\mathbf{k}$ is the absolute permeability tensor of the reservoir and p and μ are the fluid pressure and viscosity, respectively.

To obtain an *analytical solution* for (13.1) and (13.2), we assume the following:

- The flow is two-dimensional in the x_1- and x_2-directions (i.e., it is homogeneous in the x_3-direction, and gravity is neglected).
- The reservoir is homogeneous and isotropic; i.e., $\mathbf{k} = k\mathbf{I}$ and k is a constant (cf. Section 2.2.1).
- The viscosity μ and density ρ are constant.
- The flow is radial in a small neighborhood of the well.

With the last assumption, near the well the velocity $\mathbf{u}$ has the form

$$\mathbf{u}(r, \theta) = u(r)(\cos\theta, \sin\theta),$$

where (r, θ) is the polar coordinate system. Since the well is placed at the origin, substitution of this velocity into (13.1) gives (cf. Exercise 13.1)

$$\frac{du}{dr} + \frac{1}{r}u = 0, \qquad r > 0, \tag{13.3}$$

whose solution is $u = C/r$ (cf. Exercise 13.2). The constant C is proportional to q. Note that q represents the mass production/injection. Hence, when the well is an injector, for example, for any small neighborhood B of the origin (a small circle) q is the mass flux

$$q = h_3 \int_B \rho\mathbf{u} \cdot \boldsymbol{\nu}\, da(\mathbf{x}) = 2\pi\rho h_3 C; \quad \text{i.e.,} \quad C = \frac{q}{2\pi\rho h_3},$$

where $\boldsymbol{\nu}$ is the outward unit normal to B and h_3 is the reservoir thickness (or the height of the gridblock containing the well). Consequently, we obtain

$$\mathbf{u} = \frac{q}{2\pi\rho h_3 r}(\cos\theta, \sin\theta). \tag{13.4}$$

Substituting (13.4) into (13.2), taking a dot product of the resulting equation with $\boldsymbol{\nu} = (1, 0)$, and integrating from $(r^o, 0)$ to $(r, 0)$, we obtain (cf. Exercise 13.3)

$$p(r) = p(r^o) - \frac{\mu q}{2\pi\rho k h_3}\ln\left(\frac{r}{r^o}\right), \tag{13.5}$$

where $(r^o, 0)$ is a reference point (e.g., r^o is the well radius r_w). Equation (13.5) is the *analytical flow model* near the well, on which the development of well equations for various numerical methods is based in the next four sections.

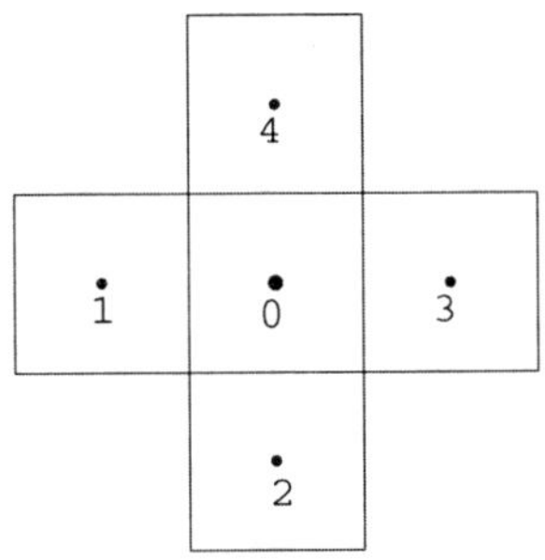

Figure 13.1. *A cell-centered finite difference on a square grid.*

13.2 Finite Difference Methods

The first comprehensive study of well equations was by Peaceman (1977A) for cell-centered finite difference methods on square grids for single phase flow. Peaceman's study gave a proper interpretation of a well-block pressure, and indicated how it relates to the flowing bottom hole pressure. The importance of his study is that the computed block pressure is associated with the steady-state pressure for the actual well at an *equivalent radius* r_e. For a square grid with a grid size h, Peaceman derived a formula for r_e by three different approaches: (1) analytically by assuming that the pressure in the blocks adjacent to the well block is computed exactly by the radial flow model, obtaining $r_e = 0.208h$, (2) numerically by solving the pressure equation on a sequence of grids, deriving $r_e = 0.2h$, and (3) by solving exactly the system of difference equations and using the equation for the pressure drop between the injector and producer in a repeated five-spot pattern problem, finding $r_e = 0.1987h$. From these approaches, he concluded that $r_e \approx 0.2h$. In this chapter, the first approach is adapted not only for finite difference methods but also for finite element methods.

13.2.1 Square grids

For a square grid K_h, we solve (13.1) and (13.2) in the case where the well is located in the center of a grid cell. The adjacent cells are enumerated as in Figure 13.1. Application of a five-point stencil scheme (cf. Section 4.1) to (13.1) and (13.2) gives

$$\frac{\rho k h_3}{\mu}(4p_0 - p_1 - p_2 - p_3 - p_4) = q. \tag{13.6}$$

Using the symmetry of the solution p, i.e., $p_1 = p_2 = p_3 = p_4$, we see that

$$\frac{\rho k h_3}{\mu}(p_0 - p_1) = \frac{q}{4}. \tag{13.7}$$

We assume that the pressure at the adjacent cells is computed accurately. In particular, this means that the analytical well model derived in the previous section can be an accurate approximation in cell 1. Thus, if a bottom hole pressure p_{bh} is given, then it follows from

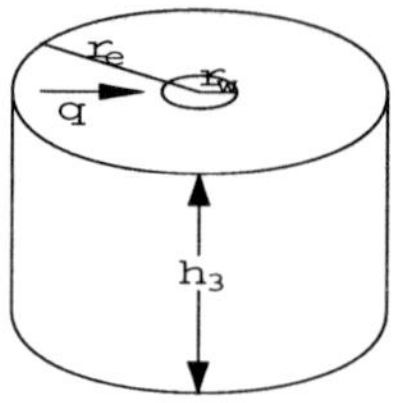

Figure 13.2. *Radial flow.*

(13.5) that

$$p_1 = p_{bh} - \frac{\mu q}{2\pi\rho k h_3} \ln\left(\frac{r_1}{r_w}\right), \tag{13.8}$$

where we recall that r_w is the well radius and $r_1 = h$. Inserting (13.8) into (13.7) yields

$$\begin{aligned} p_0 &= p_{bh} - \frac{\mu q}{2\pi\rho k h_3} \ln\left(\frac{h}{r_w}\right) + \frac{q\mu}{4\rho k} \\ &= p_{bh} + \frac{\mu q}{2\pi\rho k h_3} \left(\ln\left(\frac{r_w}{h}\right) + \frac{\pi}{2}\right) \\ &= p_{bh} + \frac{\mu q}{2\pi\rho k h_3} \ln\left(\frac{r_w}{\alpha_1 h}\right), \end{aligned}$$

where $\alpha_1 = e^{-\pi/2} = 0.20788\ldots$. This is exactly Peaceman's well model:

$$q = \frac{2\pi\rho k h_3}{\mu \ln(r_e/r_w)} (p_{bh} - p), \tag{13.9}$$

where the *equivalent radius* equals $r_e = \alpha_1 h = 0.20788h$ and $p = p_0$ (cf. Figure 13.2). The equivalent radius is the radius at which the steady-state flowing pressure for the actual well equals the numerically computed pressure for the well cell. When the well is a producer, q is

$$q = \frac{2\pi\rho k h_3}{\mu \ln(r_e/r_w)} (p - p_{bh}). \tag{13.10}$$

13.2.2 Extensions

(i) Extension to anisotropic media

The above well model needs be extended in various directions, including to rectangular grids and incorporating gravity force effects, anisotropic reservoirs, skin effects, horizontal wells, and multiphase flows. Here we consider an extension of the model in (13.9) to the first four effects. The gravitational effects must be treated on the same footing as pressure gradient effects. The *skin factor* s_k is a dimensionless number and accounts for the effect resulting from formation damage caused by drilling. With these effects for single phase flow for an anisotropic permeability $\mathbf{k} = \mathrm{diag}(k_{11}, k_{22}, k_{33})$, the well model is extended to

$$q = \frac{2\pi\rho h_3 \sqrt{k_{11}k_{22}}}{\mu\left(\ln(r_e/r_w) + s_k\right)} \left(p_{bh} - p - \rho\wp(z_{bh} - z)\right), \tag{13.11}$$

where $\wp$ is the magnitude of the gravitational acceleration, z is the depth, and z_{bh} is the well datum level depth. The factor $\sqrt{k_{11}k_{22}}$ comes from the coordinate transformation: $x_1' = x_1/\sqrt{k_{11}}$ and $x_2' = x_2/\sqrt{k_{22}}$ (cf. Section 4.3.2).

In the nonsquare grid and anisotropic medium case, the equivalent radius r_e is (see Peaceman, 1983)

$$r_e = \frac{0.14\left((k_{22}/k_{11})^{1/2}h_1^2 + (k_{11}/k_{22})^{1/2}h_2^2\right)^{1/2}}{0.5\left((k_{22}/k_{11})^{1/4} + (k_{11}/k_{22})^{1/4}\right)}, \tag{13.12}$$

where h_1 and h_2 are the x_1- and x_2-grid sizes of the gridblock that contains the vertical well. The *well index* is defined by

$$WI = \frac{2\pi h_3\sqrt{k_{11}k_{22}}}{\ln(r_e/r_w) + s_k}. \tag{13.13}$$

(ii) Extension to horizontal wells

Horizontal wells in either the x_1- or the x_2-coordinate direction use the same well model equations as vertical ones. Only the parameters related to the direction of the wellbore need be modified. The well index for a horizontal well parallel to the x_1-direction is calculated as follows:

$$WI = \frac{2\pi h_1\sqrt{k_{22}k_{33}}}{\ln(r_e/r_w) + s_k}; \tag{13.14}$$

if the well is parallel to the x_2-direction, it is

$$WI = \frac{2\pi h_2\sqrt{k_{11}k_{33}}}{\ln(r_e/r_w) + s_k}. \tag{13.15}$$

Accordingly, in the x_1-direction the equivalent radius r_e is

$$r_e = \frac{0.14\left((k_{33}/k_{22})^{1/2}h_2^2 + (k_{22}/k_{33})^{1/2}h_3^2\right)^{1/2}}{0.5\left((k_{33}/k_{22})^{1/4} + (k_{22}/k_{33})^{1/4}\right)}, \tag{13.16}$$

and in the x_2-direction,

$$r_e = \frac{0.14\left((k_{33}/k_{11})^{1/2}h_1^2 + (k_{11}/k_{33})^{1/2}h_3^2\right)^{1/2}}{0.5\left((k_{33}/k_{11})^{1/4} + (k_{11}/k_{33})^{1/4}\right)}. \tag{13.17}$$

A well in an arbitrary direction (i.e., a *slanted well*) cannot be easily modeled via finite difference methods. It will be discussed in Section 13.4.

(iii) Extension to multiphase flow

The vertical well equations derived for single phase flow can be extended to multiphase flow, e.g., to a flow system of water, oil, and gas:

$$q_\alpha = \frac{2\pi h_3\sqrt{k_{11}k_{22}}}{\ln(r_e/r_w) + s_k}\frac{\rho_\alpha k_{r\alpha}}{\mu_\alpha}\left(p_{bh} - p_\alpha - \rho_\alpha\wp(z_{bh} - z)\right), \tag{13.18}$$

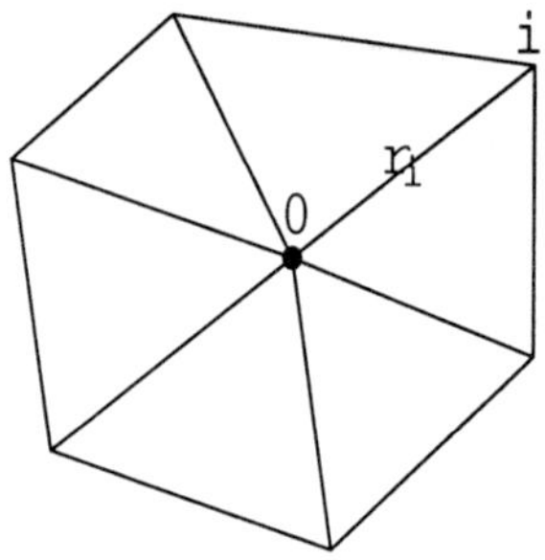

Figure 13.3. *Support* Ω_0 *of* φ_0.

where ρ_α, $k_{r\alpha}$, and p_α are the density, relative permeability, and pressure of phase α, respectively, $\alpha = w, o, g$. Note that the definitions of the well index WI and equivalent radius r_e remain the same. A similar extension to horizontal wells for multiphase flow is possible (cf. Exercise 13.4).

13.3 Standard Finite Element Methods

The well equations derived in the context of finite differences can be extended to finite elements. For finite difference methods, the pressure at the well cell is numerically computed, and the pressure at the adjacent cells is computed using the analytical formula (13.5). This approach is also employed in the context of finite elements. Again, we concentrate on two-dimensional flow.

13.3.1 Triangular finite elements

For simplicity, consider the case where the finite element space V_h is the space of piecewise linear polynomials associated with a triangulation K_h (cf. Section 4.2). Let $\varphi_0 \in V_h$ be the basis function at node $\mathbf{x}_0$ where the well is located, and Ω_0 be the support of φ_0 (cf. Figure 13.3). Then, using (13.1) and (13.2), we see that

$$\frac{k\rho h_3}{\mu} \sum_{K \subset \Omega_0} \int_K \nabla p \cdot \nabla \varphi_0 \, d\mathbf{x} = q. \tag{13.19}$$

Since $p = \sum_i \varphi_i p_i$ on Ω_0, it follows from (13.19) that

$$\frac{k\rho h_3}{\mu} \sum_{K \subset \Omega_0} \sum_i \left(\int_K \nabla \varphi_i \cdot \nabla \varphi_0 \, d\mathbf{x} \right) p_i = q. \tag{13.20}$$

Using the same argument as in Section 4.3, this equation becomes

$$-\frac{k\rho h_3}{\mu} \sum_i T_{0i}(p_i - p_0) = q, \tag{13.21}$$

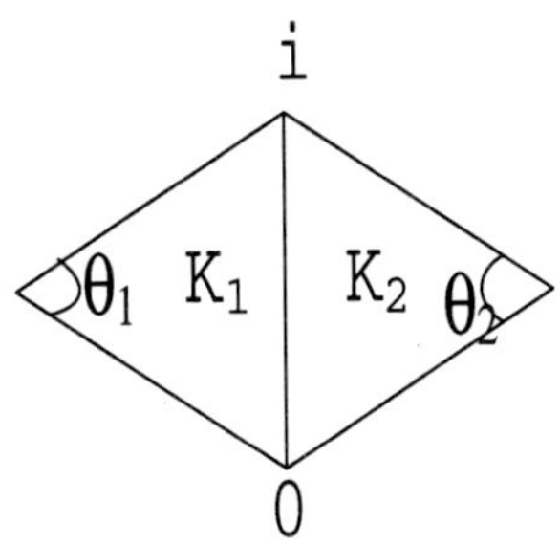

Figure 13.4. *Two adjacent triangles.*

where the *transmissibility coefficient* T_{0i} is (cf. Figure 13.4 and Exercise 13.5)

$$T_{0i} = -\sum_{l=1}^{2} (|K|\nabla\varphi_i \cdot \nabla\varphi_0)\Big|_{K_l} = \sum_{l=1}^{2} \frac{\cot\theta_l}{2}. \tag{13.22}$$

System (13.21) is the linear system of algebraic equations arising from the finite element discretization of (13.1) and (13.2) at node $\mathbf{x}_0$.

At an adjacent node $\mathbf{x}_i$, the analytic model in (13.5) is used to find the pressure

$$p_i = p_{bh} - \frac{\mu q}{2\pi\rho k h_3} \ln\left(\frac{r_i}{r_w}\right), \tag{13.23}$$

where r_i is the distance between $\mathbf{x}_i$ and $\mathbf{x}_0$. Substituting (13.23) into (13.21) gives the well model equation (cf. Exercise 13.6)

$$q = \frac{2\pi\rho k h_3}{\mu \ln(r_e/r_w)}(p_{bh} - p), \tag{13.24}$$

where $p = p_0$ and the *equivalent radius* r_e equals

$$r_e = \exp\left(\left[\sum_i T_{0i} \ln r_i - 2\pi\right] \Big/ \sum_i T_{0i}\right). \tag{13.25}$$

We consider an example where the support of φ_0 is as shown in Figure 13.5. In this case (cf. Exercise 13.7),

$$T_{01} = T_{02} = T_{04} = T_{05} = 1, \quad T_{03} = T_{06} = 0, \tag{13.26}$$

and

$$r_e = he^{-\pi/2} = 0.20788\ldots. \tag{13.27}$$

The radius is exactly the same as that in the finite difference method. This is not surprising because the finite element method is a five-point stencil scheme for the case in Figure 13.5 (cf. Section 4.2.1).

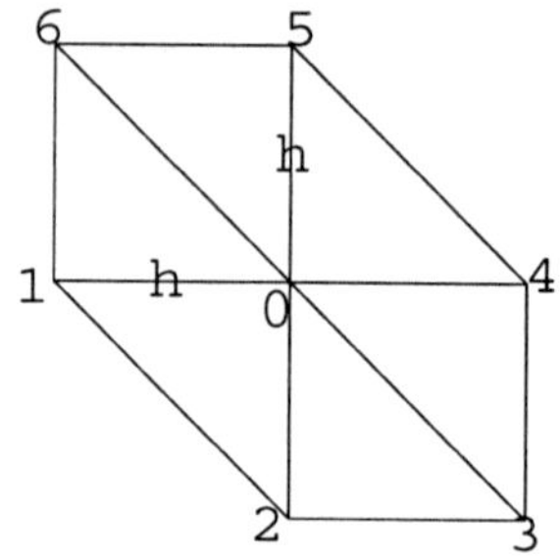

Figure 13.5. *An example of a triangulation near the well.*

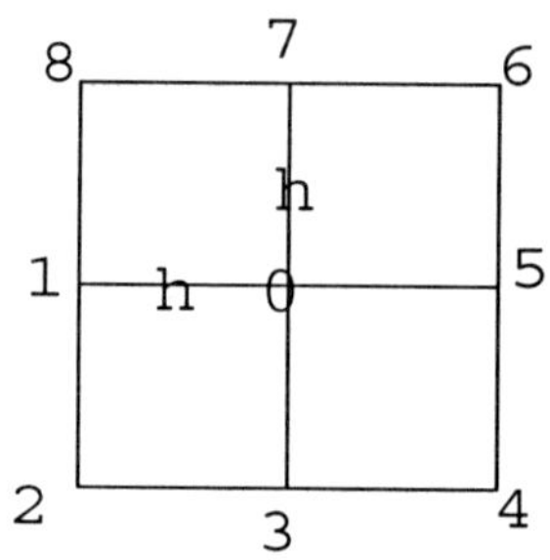

Figure 13.6. *Support Ω_0 for the bilinear finite element.*

13.3.2 Rectangular finite elements

Again, for brevity of presentation, we consider the simplest rectangular finite element, the bilinear finite element (cf. Section 4.2.1). As an example, let the support of φ_0 be given as in Figure 13.6. In this case, (13.19) remains valid. Because of the symmetry assumption of radial flow, $p_1 = p_3 = p_5 = p_7$ and $p_2 = p_4 = p_6 = p_8$. Consequently, it follows from (13.19), with Ω_0 as in Figure 13.6, that (cf. Exercise 13.8)

$$\frac{4}{3}\frac{k\rho h_3}{\mu}(2p_0 - p_1 - p_2) = q. \tag{13.28}$$

Using the analytic model (13.5), we see that

$$\begin{aligned} p_1 &= p_{bh} - \frac{\mu q}{2\pi \rho k h_3} \ln\left(\frac{h}{r_w}\right), \\ p_2 &= p_{bh} - \frac{\mu q}{2\pi \rho k h_3} \ln\left(\frac{\sqrt{2}h}{r_w}\right). \end{aligned} \tag{13.29}$$

Combining (13.28) and (13.29) yields the well model (13.24) with *equivalent radius*

$$r_e = 2^{1/4} e^{-3\pi/4} h. \tag{13.30}$$

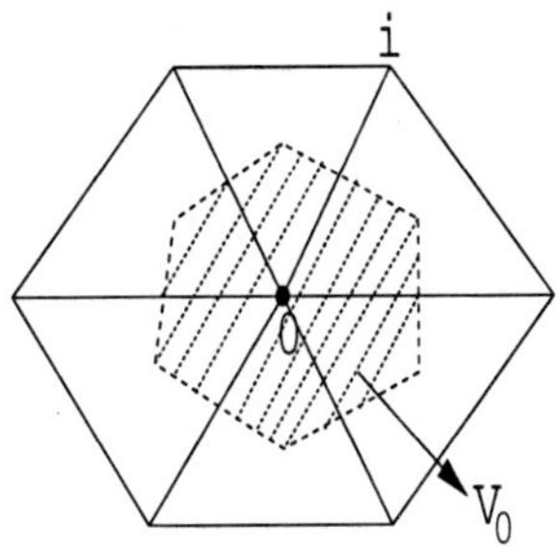

Figure 13.7. *A control volume V_0 for the linear finite element.*

13.4 Control Volume Finite Element Methods

13.4.1 Well model equations

For the control volume finite element (CVFE) method based on the triangular linear elements (cf. Section 4.3), the well model equation (13.24) and the equivalent radius r_e defined in (13.25) remain the same since the linear system arising from this method is the same as that from the standard finite element method using piecewise linear functions (cf. (4.120)). For the CVFE, node $\mathbf{x}_0$ is now the center of a control volume; i.e., the well is now located at a center (cf. Figure 13.7), instead of at a vertex as in the standard finite element method. In practice, the *equivalent radius* r_e for the CVFE can be computed using a simpler formula (Chen et al., 2002C)

$$r_e = \sqrt{\frac{|V_0|}{\pi}}, \tag{13.31}$$

where $|V_0|$ is the area of the control volume V_0 that contains the well (cf. Figure 13.7). The derivation of (13.31) is based on the following principle: $|V_0|$ is approximately the area of a circle with radius r_e that contains the well, and the mean value of pressure on V_0 is approximately the pressure on this circle (Chen et al., 2002C).

13.4.2 Horizontal wells

The well model derived for a vertical well using finite elements can be generalized to include the following effects: gravity forces, anisotropic reservoirs, skin factors, horizontal wells, and multiphase flows. These generalizations can be performed in the same fashion as in the finite difference case; here we focus on the modeling of horizontal wells.

Because of the intrinsic flexibility of finite element grids, the flow pattern near a horizontal well in an arbitrary direction can be modeled accurately, particularly when local grid refinement is used. If the horizontal well passes through a triangle, this triangle needs to be refined: (1) if it passes through two edges of the triangle, we can make the intersections to be the vertices of smaller triangles (or centers of control volumes) by properly adjusting the midpoints of the two edges (cf. Figure 13.8); (2) if it passes through a vertex of the triangle, the local refinement can be done as in Figure 13.9 by connecting the well-edge intersection with the two midpoints of the other edges. The feature of this approach is

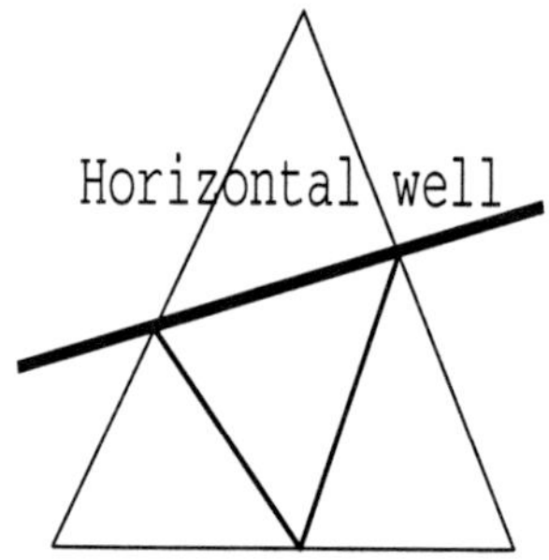

Figure 13.8. *A horizontal well passes through two edges.*

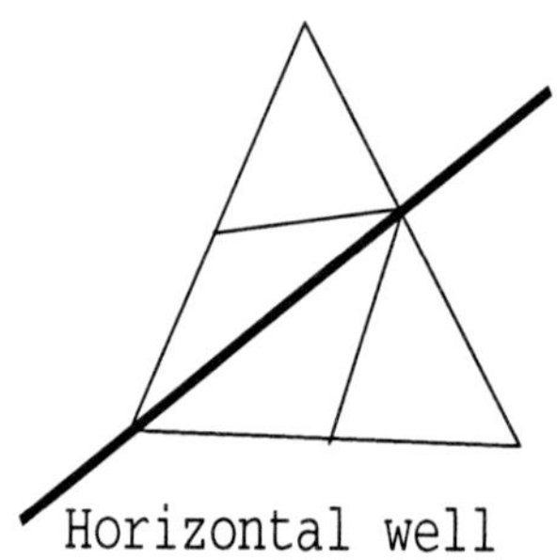

Figure 13.9. *A horizontal well passes through a vertex.*

that the horizontal well contains only triangle vertices (cf. Figure 13.10) or control volume centers (cf. Figure 13.11).

For the CVFE, the well model equation for a horizontal well in an arbitrary direction is derived in an analogous fashion to (13.24):

$$q = \frac{2\pi \rho k \Delta L}{\mu \ln(r_e/r_w)}(p_{bh} - p), \tag{13.32}$$

where ΔL is the diameter of the control volume (that contains the well) in the well direction and the equivalent radius r_e can be defined as in (13.25). For the latter, using a similar principle as for (13.31), a simpler definition is (Chen et al., 2002C)

$$r_e = \sqrt{\frac{|V_0| h_3}{\pi \Delta L}}, \tag{13.33}$$

where h_3 is the x_3-spatial grid size of the block that contains the well. An extension of (13.32) to multiphase flow was given in (8.11).

13.4.3 Treatment of faults

Faults in a petroleum reservoir can be treated in a manner similar to horizontal wells by adjusting the midpoints of edges and the barycenters of triangles in order for them to be on the faults (cf. Figure 13.12). In the present case, only the form and areas of control volumes

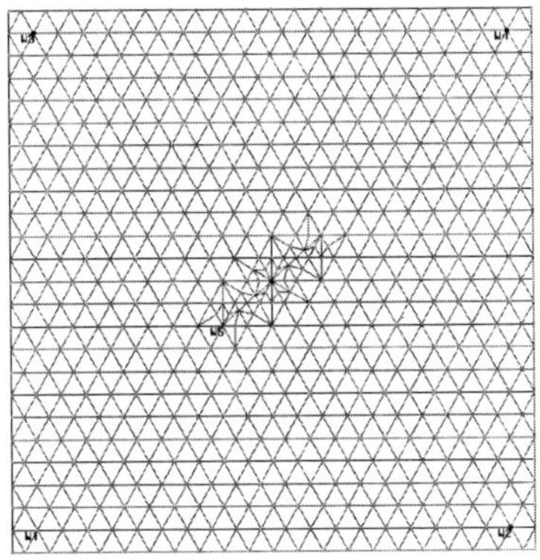

Figure 13.10. *A horizontal well for the triangular case.*

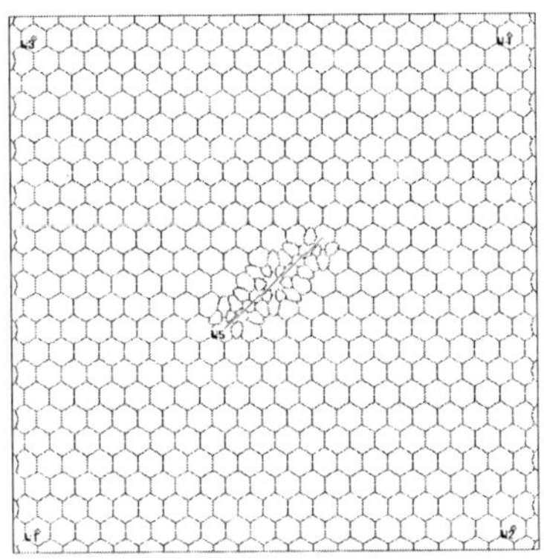

Figure 13.11. *A horizontal well for the CVFE case.*

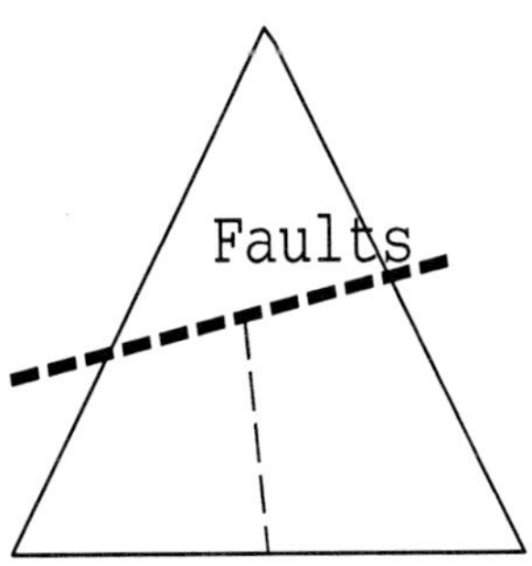

Figure 13.12. *Treatment of faults.*

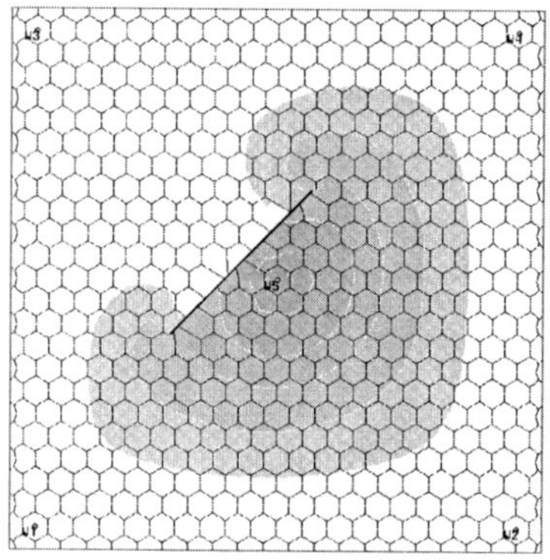

Figure 13.13. *An example of flow around faults.*

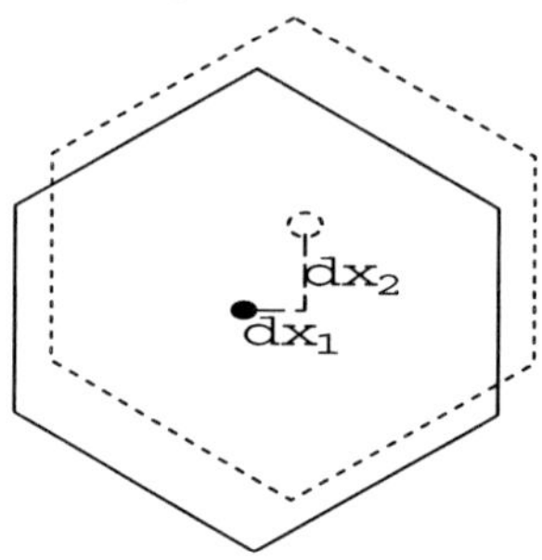

Figure 13.14. *Corner point technique.*

need be changed; nothing else is altered. The transmissibility between two points across a fault in a control volume is set to zero. This approach is easy to implement and practical. A numerical example is shown in Figure 13.13.

13.4.4 Corner point techniques

A *corner point technique* can be used for the finite difference method discussed in Section 13.2 to adjust the locations of gridblocks (Collins et al., 1991). When a vertical well is not located in the center of a rectangle, the vertices of the rectangle must be adjusted (as well as the vertices of other rectangles to preserve the grid orthogonality). This corner point technique can be also applied to the CVFE. We locate the centers of the control volumes that contain vertical wells, find the discrepancies in the x_1- and x_2-directions between these centers and the centers of the wells (cf. Figure 13.14), and use the values of these discrepancies to adjust the location of all control volumes except those that are adjacent to the boundary of a reservoir or contain horizontal wells or faults. Note that grid orthogonality is not required for the CVFE grids.

For the value p_1, we use the well equation (13.5):

$$p_1 = p_{bh} - \frac{\mu q}{2\pi \rho k h_3} \ln\left(\frac{r_1}{r_w}\right), \tag{13.40}$$

where $r_1 = 2\sqrt{2}h/3$ is the distance from the well to the barycenter of the triangle K_1. Substituting (13.40) into (13.39) generates the well model equation (13.9) with *equivalent radius*

$$r_e = \frac{2\sqrt{2}h}{3} e^{-\pi/6}. \tag{13.41}$$

13.6 Well Constraints

Well constraints must be taken into account for numerical simulation of petroleum reservoirs (cf. Section 8.2.5). We restrict the discussion to vertical wells for a multiphase flow system that consists of water, oil, and gas. For an injection well, there are two types of well constraints: either the well bottom hole pressure p_{bh} is given, or a phase injection rate is fixed. In the former case,

$$p_{bh} = P_{bh}, \tag{13.42}$$

where P_{bh} is the given bottom hole pressure at the well, and the phase injection rate is calculated according to formula (13.18). In the latter case, the injection rate control for a water injection well is

$$\frac{2\pi h_3 \sqrt{k_{11}k_{22}}}{\ln(r_e/r_w) + s_k} \frac{\rho_w k_{rwmax}}{\mu_w} (p_{bh} - p_w - \rho_w \wp (z_{bh} - z)) = Q_w, \tag{13.43}$$

where Q_w is a given water injection rate and k_{rwmax} represents the maximum relative permeability of the water phase. In this case, p_{bh} is an unknown and is obtained from (13.43), which is coupled to the flow equations (cf. Section 8.2.5). An analogous control equation holds when a gas injection rate at the well is prescribed.

For a production well, there are three types of well constraints: a fixed bottom hole pressure, a given total liquid production rate, and a given total flow rate. The bottom hole pressure constraint has the form (13.42). The total liquid production rate control is

$$\begin{aligned} \frac{2\pi h_3 \sqrt{k_{11}k_{22}}}{\ln(r_e/r_w) + s_k} \Bigg\{ & \frac{\rho_w k_{rw}}{\mu_w} (p_{bh} - p_w - \rho_w \wp (z_{bh} - z)) \\ & + \frac{\rho_o k_{ro}}{\mu_o} (p_{bh} - p_o - \rho_o \wp (z_{bh} - z)) \Bigg\} = Q_L, \end{aligned} \tag{13.44}$$

where Q_L denotes the given total liquid production rate at the well. The *water cut*, defined as the ratio of water production to the sum of water and oil production, at a perforated zone of the well with this type of well constraint must be less than a certain limit; over this limit, the perforated zone must be shut down in practice. The constant total flow rate control can be defined similarly; in this case, gas production is added.

13.7 The Seventh SPE Project: Horizontal Well Modeling

This benchmark problem deals with production from a horizontal well in a thin reservoir where coning tendencies are strong. It was used for comparing different approaches for modeling horizontal wells in reservoir simulation and studying the effect of well lengths and production rates of horizontal wells on oil recovery (Nghiem et al., 1991).

The dimensions of the reservoir are $2{,}700 \times 2{,}700 \times 160$ ft^3, as seen in Figure 13.16. The reservoir and initial data are shown in Tables 13.1 and 13.2, where k_h $(= k_{11} = k_{22})$ and k_v $(= k_{33})$ represent the horizontal and vertical permeabilities, respectively. The initial bubble point pressure is the same as the initial oil pressure. The fluid property data are given in Table 13.3, and the relative permeability and capillary pressure data are listed in Tables 13.4 and 13.5.

The reservoir has six layers, whose dimensions are given in Table 13.1. The producer is a horizontal well drilled on the top layer. Its entire length is open to flow. Two lengths of the producer are considered: 900 ft and 2,100 ft, as seen in Figure 13.16. Its well constranit

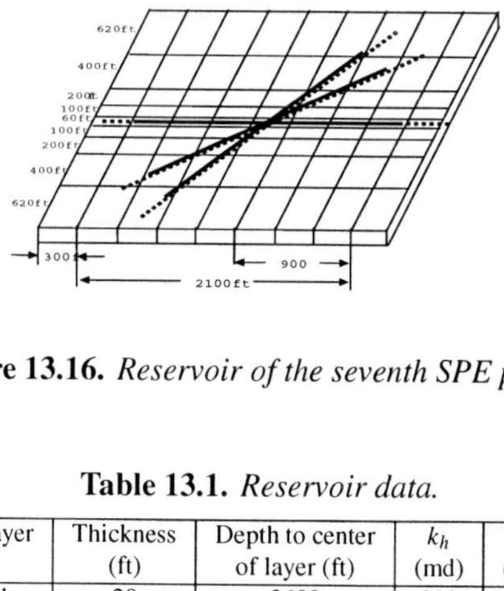

Figure 13.16. *Reservoir of the seventh SPE project.*

Table 13.1. *Reservoir data.*

Layer	Thickness (ft)	Depth to center of layer (ft)	k_h (md)	k_v (md)
1	20	3600	300	30
2	20	3620	300	30
3	20	3640	300	30
4	20	3660	300	30
5	30	3685	300	30
6	50	3725	300	30

Table 13.2. *Reservoir initial data.*

Layer	p_o (psia)	S_o	S_w
1	3600	0.711	0.289
2	3608	0.652	0.348
3	3616	0.527	0.473
4	3623	0.351	0.649
5	3633	0.131	0.869
6 (bottom)	3650	0.000	1.000

Table 13.3. *Fluid property data.*

p (psia)	R_{so} (SCF/STB)	B_o (RB/STB)	B_g (RB/SCF)	μ_o (cp)	μ_g (cp)
400	165	1.0120	0.00590	1.17	0.0130
800	335	1.0255	0.00295	1.14	0.0135
1200	500	1.0380	0.00196	1.11	0.0140
1600	665	1.0510	0.00147	1.08	0.0145
2000	828	1.0630	0.00118	1.06	0.0150
2400	985	1.0750	0.00098	1.03	0.0155
2800	1130	1.0870	0.00084	1.00	0.0160
3200	1270	1.0985	0.00074	0.98	0.0165
3600	1390	1.1100	0.00065	0.95	0.0170
4000	1500	1.1200	0.00059	0.94	0.0175
4400	1600	1.1300	0.00054	0.92	0.0180
4800	1676	1.1400	0.00049	0.91	0.0185
5200	1750	1.1480	0.00045	0.90	0.0190
5600	1810	1.1550	0.00042	0.89	0.0195

Table 13.4. *Saturation function data for water/oil.*

S_w	k_{rw}	k_{row}	p_{cow}
0.22	0.0	1.0	6.30
0.30	0.07	0.4000	3.60
0.40	0.15	0.1250	2.70
0.50	0.24	0.0649	2.25
0.60	0.33	0.0048	1.80
0.80	0.65	0.0	0.90
0.90	0.83	0.0	0.45
1.00	1.0	0.0	0.00

Table 13.5. *Saturation function data for gas/oil.*

S_g	k_{rg}	k_{rog}	p_{cgo}
0.00	0.0000	1.0	0.0
0.04	0.0000	0.60	0.2
0.10	0.0220	0.33	0.5
0.20	0.1000	0.10	1.0
0.30	0.2400	0.02	1.5
0.40	0.3400	0.0	2.0
0.50	0.4200	0.0	2.5
0.60	0.5000	0.0	3.0
0.70	0.8125	0.0	3.5
0.78	1.0	0.0	3.9

is the constant liquid rate. This horizontal well has an internal diameter of 0.1875 ft. The injector is also a horizontal well located in the bottom layer. Its whole length is 2,700 ft. The producer and injector are on the same plane. Three kinds of well directions are considered: 0°, 45°, and 60° (cf. Figure 13.16). Two well controls are assumed for the injector: constant bottom hole pressure and constant water injection rate. Twelve well schemes are designed for runs of this reservoir simulator, as shown in Table 13.6. The last four schemes are newly

Table 13.6. *Producer/injector schemes.*

Case	Direction of well (C)	Producer length (ft)	Liquid production rate (STB/D)	Water injection scheme
1a	0	900	3000	$p = 3700$ psia
1b	0	2100	3000	$p = 3700$ psia
2a	0	900	6000	$p = 3700$ psia
2b	0	2100	6000	$p = 3700$ psia
3a	0	900	9000	$p = 3700$ psia
3b	0	2100	9000	$p = 3700$ psia
4a	0	900	9000	$Q_w = 6000$ STB/D
4b	0	2100	9000	$Q_w = 6000$ STB/D
5a	45	900	9000	$Q_w = 6000$ STB/D
5b	45	2100	9000	$Q_w = 6000$STB/D
6a	30	900	9000	$Q_w = 6000$ STB/D
6b	30	2100	9000	$Q_w = 6000$ STB/D

designed. The simulation time is 1,500 days. We report the oil production rate, cumulative oil production, water-oil ratio (WOR), water production rate, cumulative water production, gas-oil ratio (GOR), cumulative gas production, and bottom hole pressure. The last six cases are different from the first six in the reservoir permeability, well constraint of the injector, and well directions. The reservoir permeability in the last six cases is 10 times that in the first six. The well constraints of the injector for the first six cases are constant bottom hole pressure, and for the last six are a constant water injection rate. The well direction of the last four cases has a positive angle.

For the spatial discretization, we use the CVFE and finite difference (FD) methods and compare the computational results for the first eight cases. We use only the CVFE for the last four cases because the FD cannot easily model the horizontal wells for these cases. In the simulation models using the CVFE, hexagonal prisms are used to represent the reservoir (cf. Figure 4.36). The distance between two neighboring grid points of the base grid in the x_1x_2-plane is 300 ft. For the models using the FD, rectangular parallelepiped gridblocks are used to represent the reservoir. The dimensions of gridblocks are shown in Figure 13.16 and Table 13.1.

The convergence control parameters used for the first six cases are $(\delta t)_{max} = 50$ days, $(\delta p)_{max} = 200$ psia, $(\delta S_w)_{max} = 0.05$, and $(\delta S_g)_{max} = 0.05$ (cf. Section 8.2.3). Since the injector has a fixed bottom hole pressure of 3,700 psia for these cases, very little free gas is released. The iterative computational processes are stable for the chosen parameters. Figures 13.17–13.28 give the computational results in these cases. Tables 13.7 and 13.8 compare the cumulative oil production and bottom hole pressure of the producer at 1,500 days. It can be seen that the computational results obtained using the CVFE and FD methods approach the average values of those of the organizations involved in the seventh CSP. From Figures 13.17–13.28, we see that increasing lengths of horizontal wells can reduce coning tendencies. As a result, the oil production rates increase, WORs decrease, and the water production rates decrease. The numerical results obtained from the CVFE match those from the FD.

Cases 4a and 4b use the well constraint of constant water injection rate. The bottom hole pressure of the injector can dramatically drop, and then a large volume of free gas

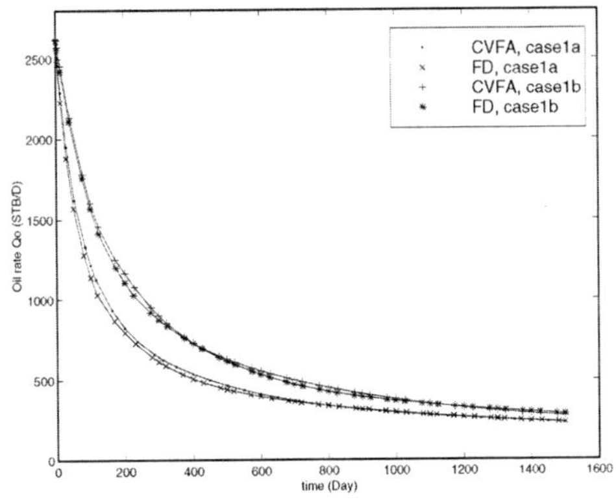

Figure 13.17. *Oil production rates of cases* 1a *and* 1b.

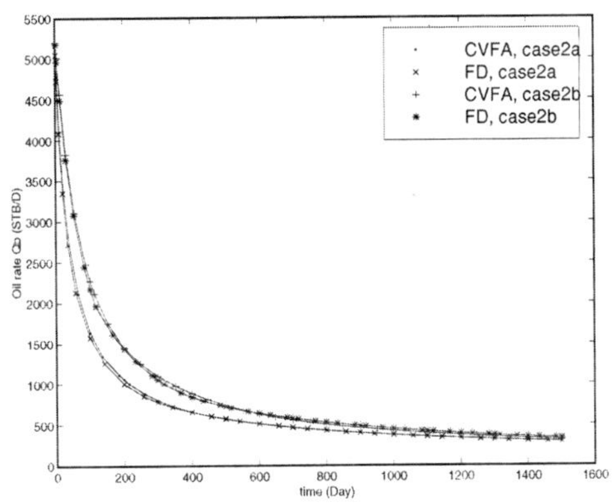

Figure 13.18. *Oil production rates of cases* 2a *and* 2b.

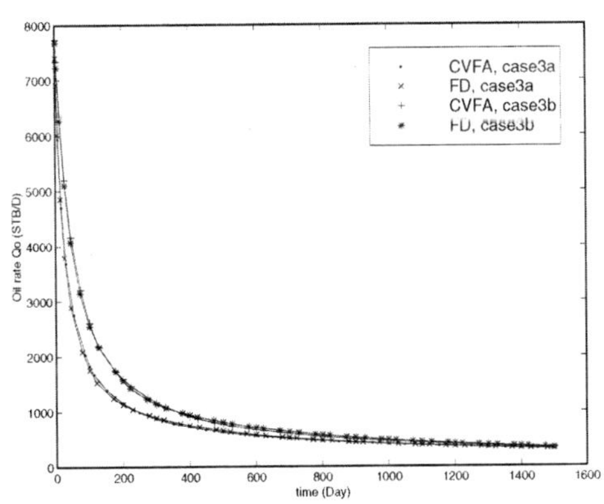

Figure 13.19. *Oil production rates of cases* 3a *and* 3b.

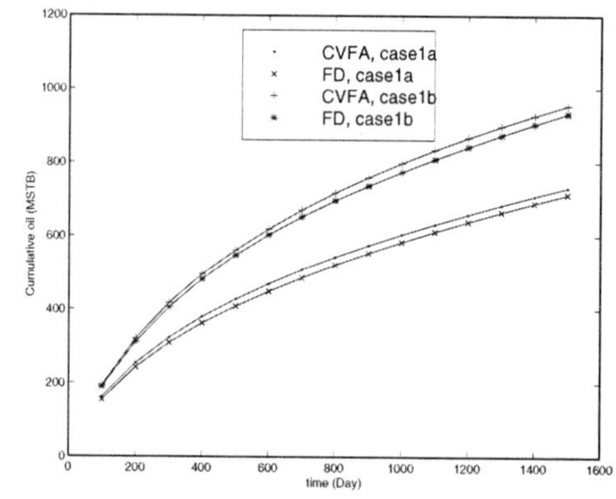

Figure 13.20. *Cumulative oil production of cases* 1a *and* 1b.

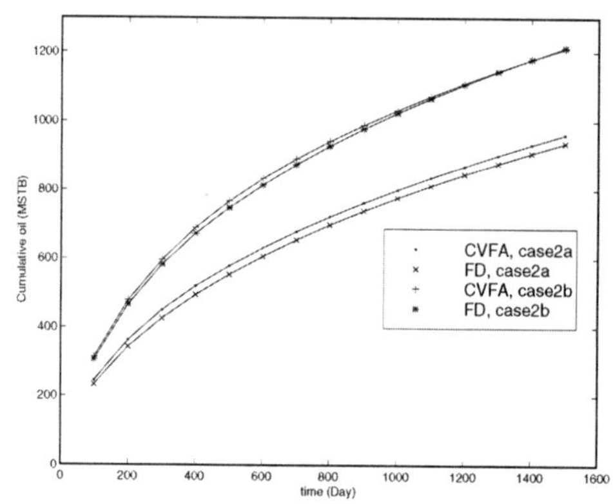

Figure 13.21. *Cumulative oil production of cases* 2a *and* 2b.

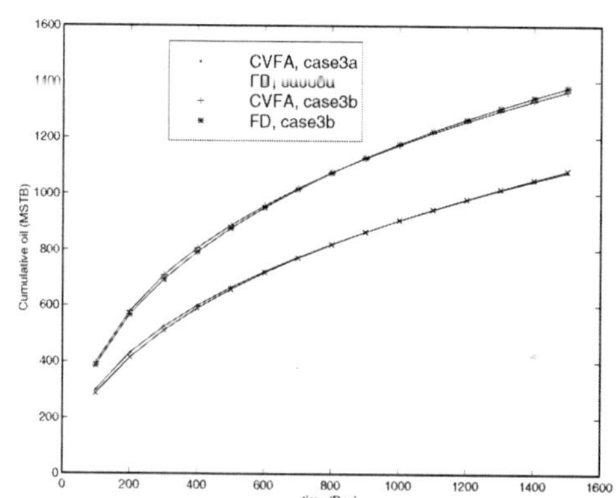

Figure 13.22. *Cumulative oil production of cases* 3a *and* 3b.

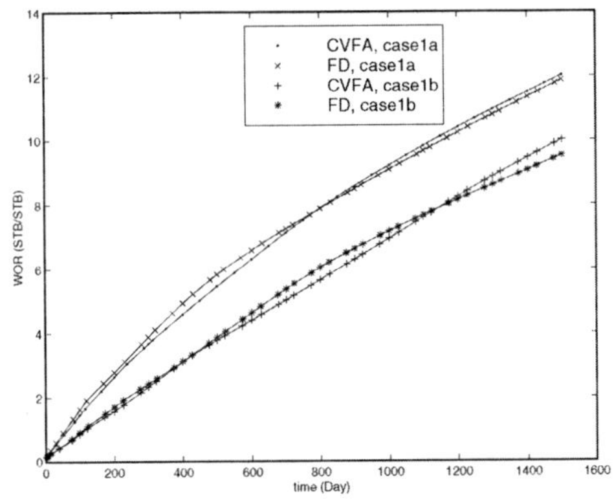

Figure 13.23. *WORs of cases* 1a *and* 1b.

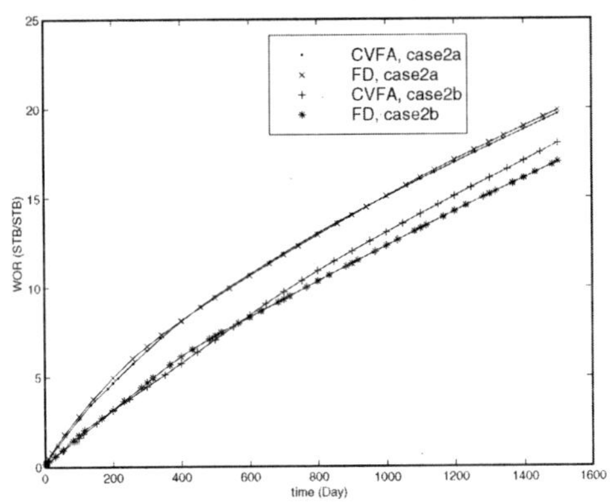

Figure 13.24. *WORs of cases* 2a *and* 2b.

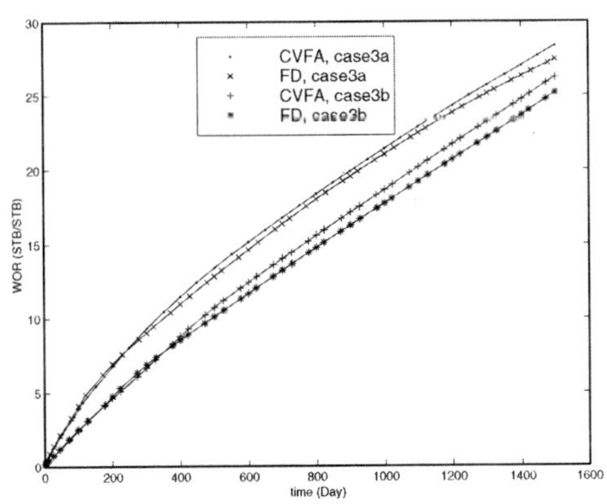

Figure 13.25. *WORs of cases* 3a *and* 3b.

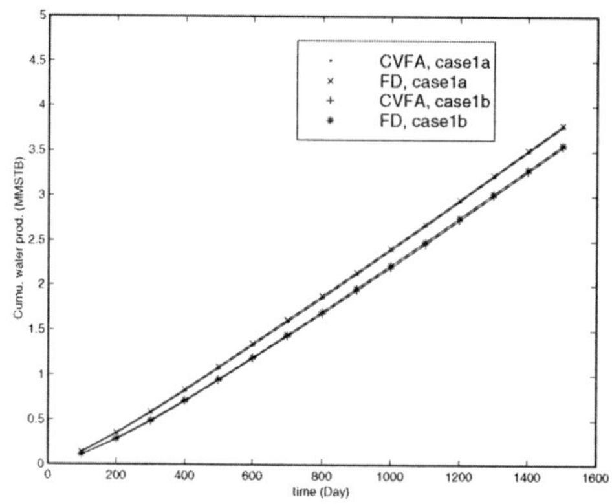

Figure 13.26. *Cumulative water production of cases* 1a *and* 1b.

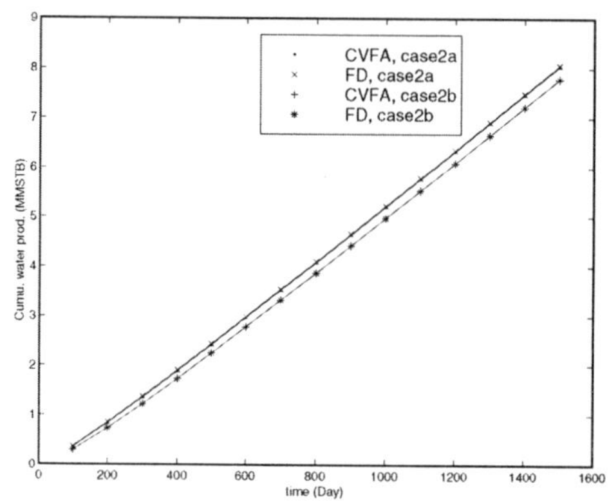

Figure 13.27. *Cumulative water production of cases* 2a *and* 2b.

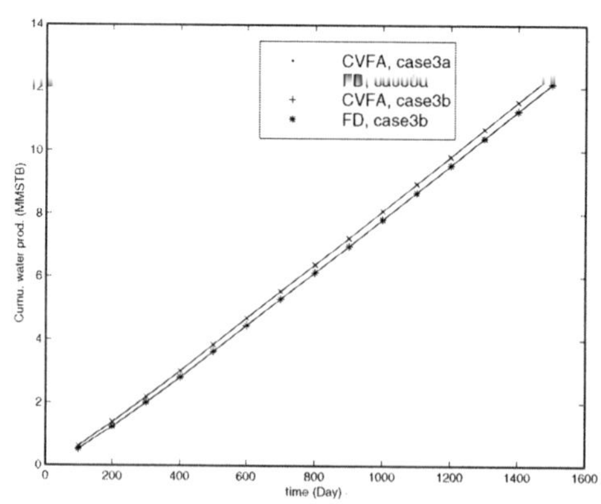

Figure 13.28. *Cumulative water production of cases* 3a *and* 3b.

Table 13.7. *Cumulative oil production in MSTB at* 1,500 *days.*

Participants	1a	1b	2a	2b	3a	3b	4a	4b
ARTEP	747.2	951.7	976.4	1221.0	1096.4	1318.5	740.8	902.8
Chevron	741.0	929.4	958.1	1181.6	1066.0	1274.8	665.3	797.7
CMG	753.6	960.1	983.6	1230.3	1106.1	1330.2	709.0	850.6
ECL	757.2	951.0	1034.2	1251.0	1229.1	1444.8	696.7	827.4
ERC	683.5	870.2	900.3	1106.1	1031.4	1222.3	672.0	788.4
HOT	765.0	961.9	1045.9	1263.7	1247.0	1466.8	714.0	877.6
INTECH	723.3	957.5	949.6	1241.5	1103.2	1414.7	754.4	890.4
JNOC	717.4	951.3	931.6	1245.9	1084.4	1412.7	660.6	843.9
Marathon	722.9	964.3	941.5	1257.1	1096.0	1436.7	781.7	895.8
Philips	750.9	956.8	980.5	1227.1	1103.5	1325.0	712.0	959.7
RSRC	678.7	916.7	877.9	1177.8	1017.1	1333.2	620.5	801.5
Shell	749.0	954.8	978.4	1224.6	1100.0	1322.4	733.5	884.1
Stanford	742.0	943.9	968.7	1211.8	1043.7	1305.6	331.0	457.6
TDC	766.2	989.4	989.4	1210.0	1105.0	1279.2	854.4	933.6
Mean	735.6	946.4	965.4	1217.8	1101.1	1349.1	688.4	829.4
Stand. devia.	27.4	26.7	45.2	41.0	64.7	73.5	117.0	115.4
SMU (CVFE)	731.8	954.9	961.6	1211.6	1077.7	1364.7	657.4	792.3
SMU (FD)	713.1	932.9	936.5	1213.9	1082.5	1377.7	645.0	779.3

Table 13.8. *Bottom hole pressure in psia at* 1,500 *days.*

Participants	1a	1b	2a	2b	3a	3b
ARTEP	3466.76	3575.78	3236.68	3470.49	3002.20	3364.74
Chevron	3464.77	3576.10	3239.19	3464.42	3012.13	3356.08
CMG	3446.32	3558.33	3210.46	3454.76	2970.39	3345.85
ECL	3485.03	3569.71	3326.22	3490.41	3170.46	3412.53
ERC	3439.96	3562.14	3199.89	3453.11	2949.06	3343.41
HOT	3511.65	3582.92	3382.08	3250.19	3256.18	3459.89
INTECH	3530.00	3601.00	3382.00	3541.00	3221.00	3479.00
JNOC	3471.72	3589.29	3251.86	3491.07	3020.84	3405.28
Marathon	3493.24	3593.85	3295.26	3509.80	3085.07	3433.56
Philips	3449.40	3572.40	3203.40	3460.20	2953.20	3351.90
RSRC	3567.80	3610.90	3444.10	3575.30	3318.90	3530.30
Shell	3448.75	3571.38	3201.16	3456.91	2948.98	3345.16
Stanford	3454.64	3572.29	3216.69	3464.30	2977.69	3359.93
TDC	3438.21	3544.40	3203.95	3452.69	2959.80	3343.16
Mean	3476.30	3577.18	3270.92	3486.04	3060.42	3395.06
Stand. devia.	37.96	17.45	81.54	37.87	127.79	60.28
SMU (CVFE)	3482.48	3587.02	3269.03	3496.32	3043.11	3391.13
SMU (FD)	3434.74	3579.45	3171.97	3458.44	2903.07	3353.75

appears. Consequently, if the relatively large convergence control parameters chosen for cases 1–3 are used, the iterative processes of simulation may not be stable for cases 4a and 4b (cf. Section 8.3.2). Hence we run the simulator with stricter convergence control parameters for these two cases (cf. Tables 13.9 and 13.10).

Table 13.9. *Convergence control parameters of cases* 4a *and* 4b.

Case	Method	$(\delta t)_{max}$ (day)	$(\delta p)_{max}$ (psia)	$(\delta S_w)_{max}$	$(\delta S_g)_{max}$
4a	CVFE	20	150	0.01	0.01
4a	FD	20	100	0.01	0.01
4b	CVFE	50	150	0.02	0.02
4b	FD	50	200	0.05	0.05
5a	CVFE	20	150	0.005	0.005
5b	CVFE	20	100	0.01	0.01
6a	CVFE	20	100	0.005	0.005
6b	CVFE	20	100	0.01	0.01

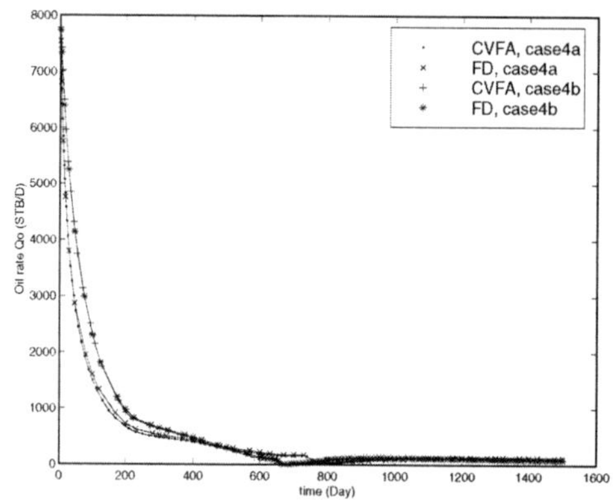

Figure 13.29. *Oil production rates of cases* 4a *and* 4b.

As in cases 1a–3b, increasing the horizontal well lengths in cases 4a and 4b also reduces coning tendencies. Figures 13.29 and 13.30 show the oil production rate and the cumulative oil production, respectively, for these two cases. Figures 13.31 and 13.32 give the water production rate and the cumulative water production. A comparison between cases 4a and 4b indicates that the oil production rate and the cumulative oil production increase, and the water production rate and the cumulative water production decrease, for case 4b. The water production in case 4b drops to its minimum value at about 690 days. But the drop of this production in case 4a has a delay; it drops to the minimum value at about 800 days because of a stronger coning tendency in this case. Figure 13.35 shows the bottom hole pressure of the producer. Accordingly, the bottom hole pressures drop to the minimum value 1,500 psia at 800 and 690 days for cases 4a and 4b, respectively. The drop in water production is caused by the reservoir pressure drop. Since the well constraint of the injector is not the constant bottom hole pressure, the reservoir pressure cannot continue to hold above the bubble point pressure. When it drops below the bubble point pressure, free gas appears. If the reservoir pressure drops to the minimum bottom hole pressure at a perforated zone of the producer, no liquid will be produced at that zone. Therefore, the water production rate decreases. After a certain time, the reservoir pressure goes up; increasing the pressure difference between the reservoir and wellbore increases the water production

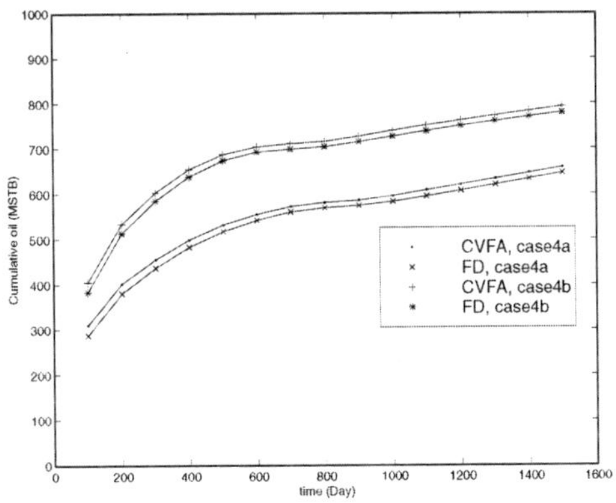

Figure 13.30. *Cumulative oil production of cases* 4a *and* 4b.

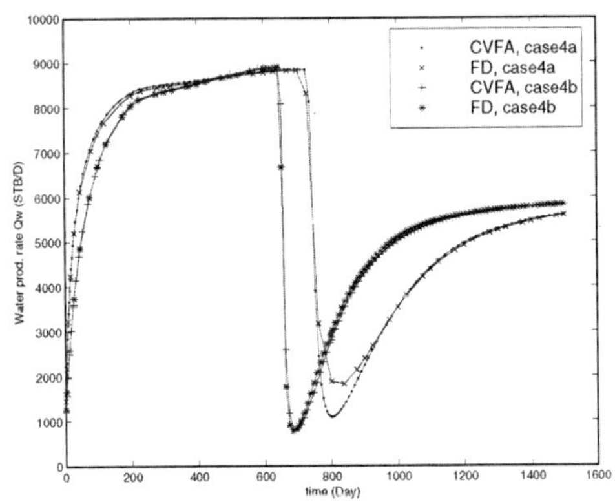

Figure 13.31. *Water production rates of cases* 4a *and* 4b.

rate. The GOR and cumulative gas production are given in Figures 13.33 and 13.34. They show that a large volume of free gas is produced.

Cases 5 and 6 are designed to test the modeling of horizontal wells in arbitrary directions. From Table 13.6, we see that these two cases are different from case 4 only in the direction of wells. Figures 13.36 and 13.37 are the oil production rate and the cumulative oil production. The oil production in cases 4a, 5a, and 6a is quite close to equivalent, and the same is true for cases 4b, 5b, and 6b (cf. Table 13.11). The cumulative oil production in cases 5a and 6a is comparable but is different from that in case 4a. This observation for the cumulative oil production can be also seen for cases 4b, 5b, and 6b. Figures 13.38–13.43 show the water production rate, cumulative water production, WOR, GOR, cumulative gas production, and bottom hole pressure. Figure 13.44 shows water saturation distribution in case 4a. All these figures show that the results in cases 5 and 6 are close to each other and are slightly different from those in case 4. This phenomenon is due to the different directions of the wells in these cases. Although cases 4, 5, and 6 have the same well lengths, injection rates, and production rates, the well locations in cases 5 and 6 are closer.

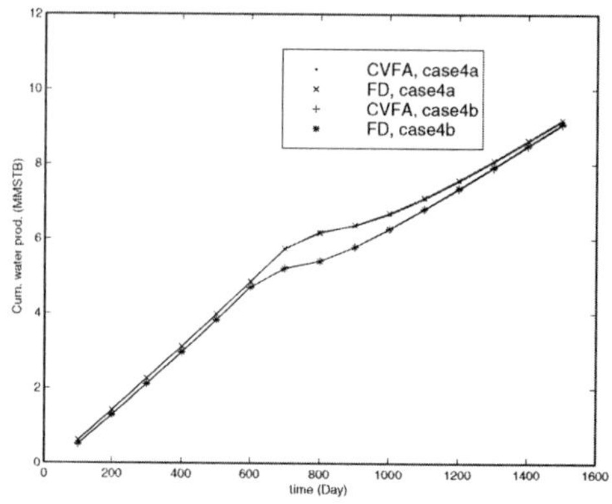

Figure 13.32. *Cumulative water production of cases* 4a *and* 4b.

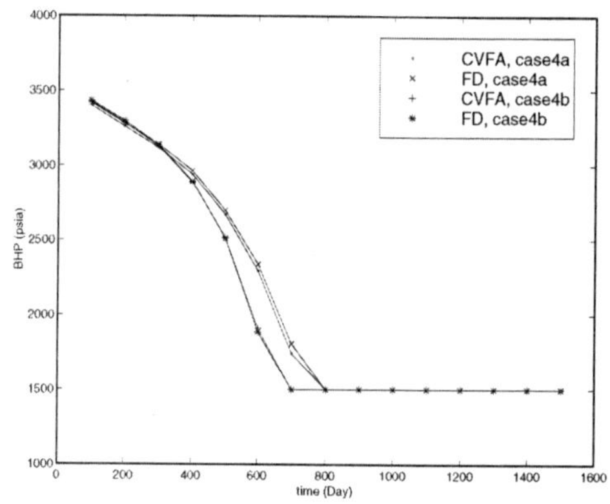

Figure 13.33. *Bottom hole pressures of the producer for cases* 4a–4b.

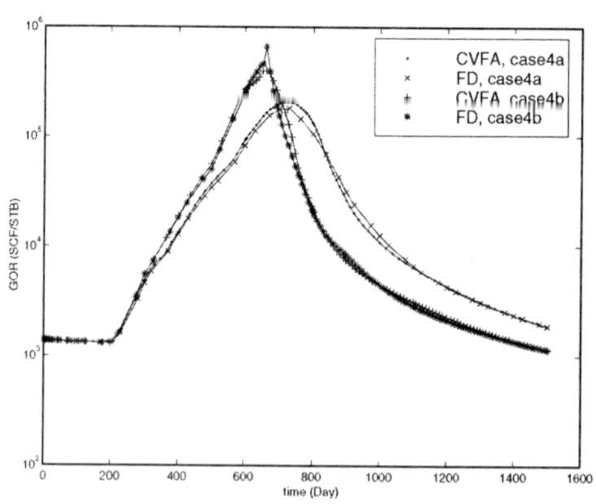

Figure 13.34. *GORs of cases* 4a *and* 4b.

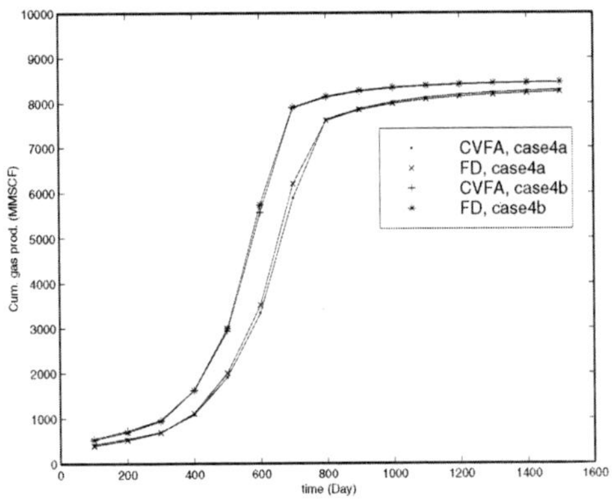

Figure 13.35. *Cumulative gas production of cases* 4a *and* 4b.

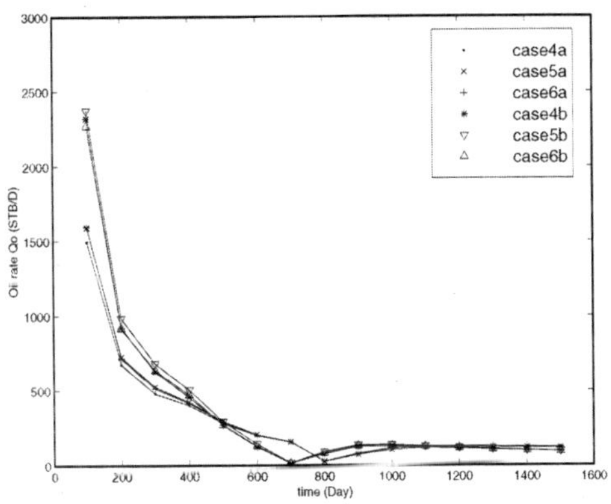

Figure 13.36. *Oil production rates of cases* 4a–6b.

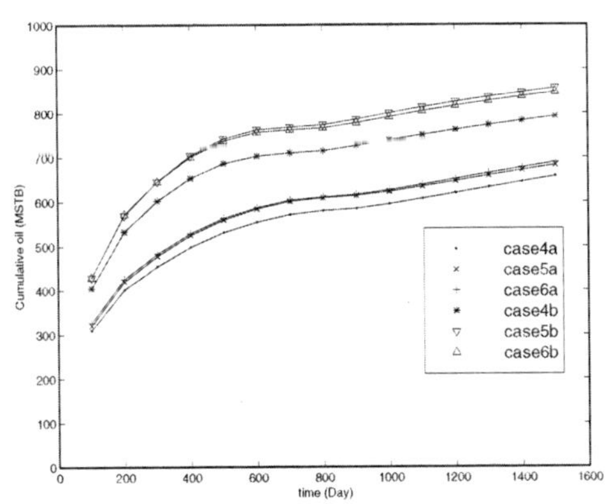

Figure 13.37. *Cumulative oil production of cases* 4a–6b.

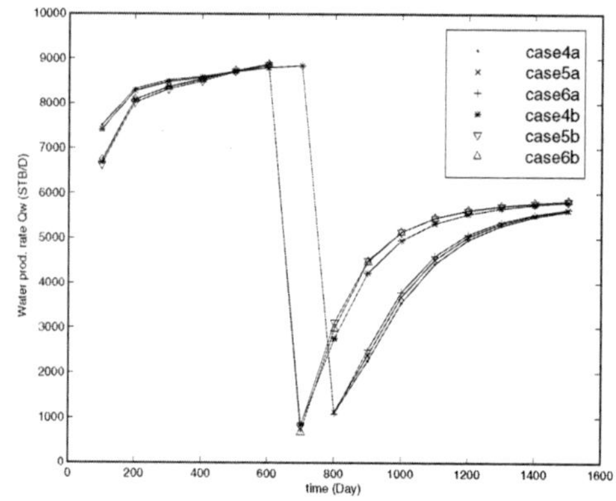

Figure 13.38. *Water production rates of cases* 4a–6b.

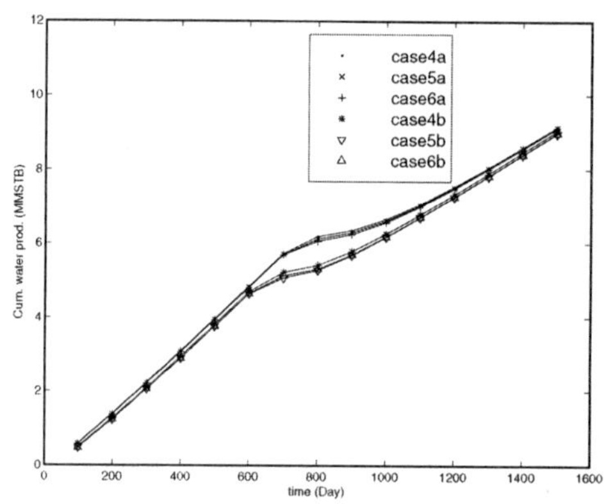

Figure 13.39. *Cumulative water production of cases* 4a–6b.

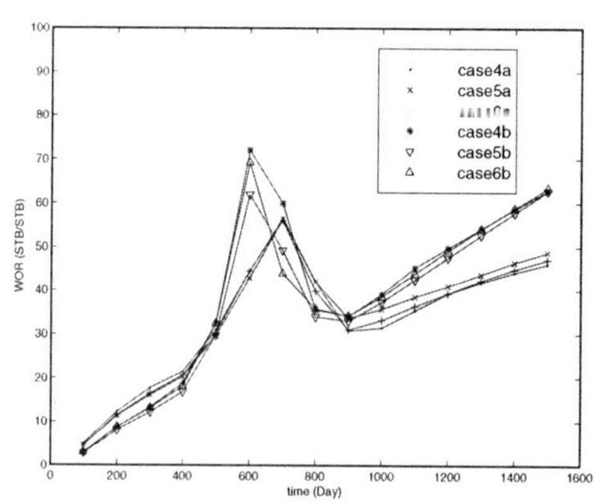

Figure 13.40. *WORs of cases* 4a–6b.

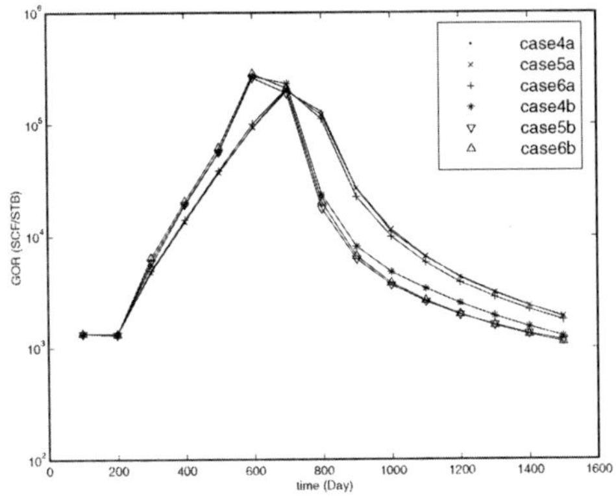

Figure 13.41. *GORs of cases* 4a–6b.

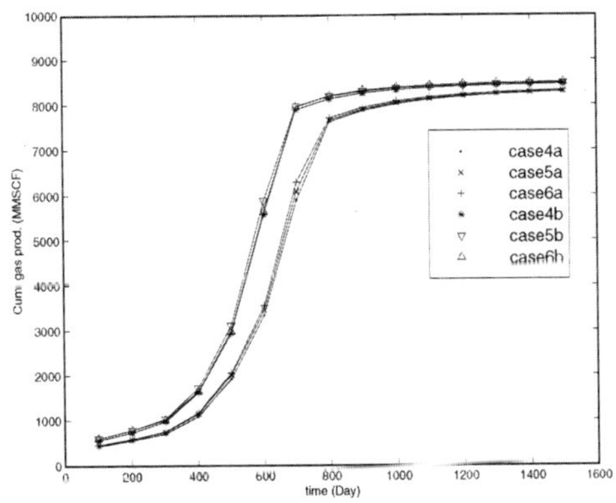

Figure 13.42. *Cumulative gas production of cases* 4a–6b.

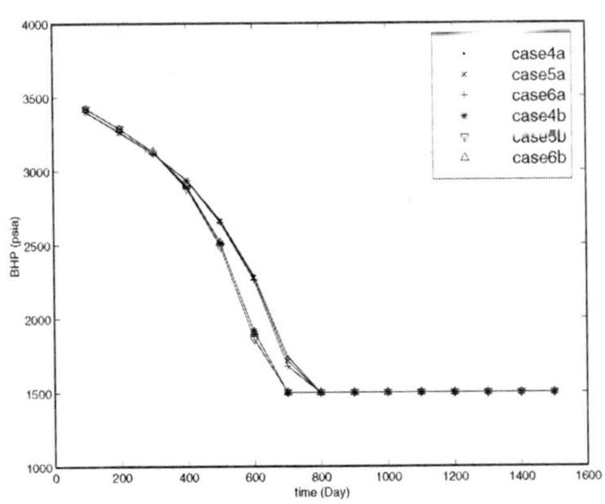

Figure 13.43. *Bottom hole pressure of cases* 4a–6b.

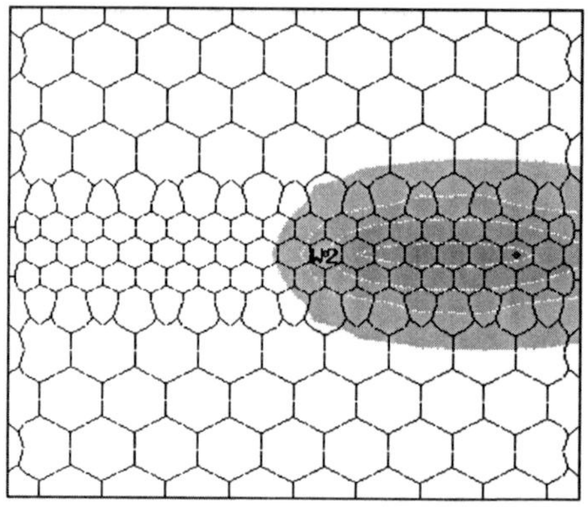

Figure 13.44. *Water saturation of case* 4a.

Table 13.10. *Time steps and Newton's iterations.*

Participants	1a	1b	2a	2b	3a	3b	4a	4b
ARTEP	39 104	39 94	45 120	39 100	47 124	42 107	50 186	49 171
Chevron	36 84	21 63	36 96	23 78	37 120	24 92	66 247	45 246
CMG	24 58	23 61	25 62	25 76	25 61	25 66	31 135	33 154
ECL	23 55	21 51	23 64	23 56	23 65	22 57	35 102	34 103
ERC	26 39	25 38	24 42	27 43	24 51	25 45	149 459	343 943
HOT	17 23	17 23	17 24	17 24	17 27	17 26	102 256	96 182
INTECH	31 92	31 106	33 105	31 104	34 105	33 114	82 392	72 356
JNOC	22 53	21 48	23 57	22 53	24 57	22 53	48 130	47 134
Marathon	155 221	155 192	161 291	157 233	165 346	157 253	288 898	252 961
Philips	47 57	46 50	47 66	47 56	47 70	47 60	47 104	50 101
RSRC	58 58	36 36	158 161	44 45	182 197	71 72	1732 1733	1264 1264
Shell	42 114	42 109	45 123	43 121	42 120	43 125	55 180	47 155
Stanford	20 55	19 44	22 55	20 50	22 60	21 57	49 265	43 116
TDC	318 2093	96 189	632 4441	272 1796	951 6986	421 2882	901 7326	541 3366
SMU (CVFE)	38 121	48 149	37 121	36 114	37 119	49 158	128 431	125 397
SMU (FD)	49 153	48 150	35 114	49 155	50 159	49 167	49 197	118 387

Table 13.11. *Simulation results of cases* 4a–6b *at* 1,500 *days.*

Case	Oil prod. rate (STB/D)	Gas prod. rate (MCF/D)	Water prod. rate (STB/D)	Oil recovery (%)	Water cut (%)
4a	121.55	227.0	5606.73	5.765	97.88
4b	92.41	117.0	5808.17	6.948	98.43
5a	115.35	218.0	5627.49	5.987	97.99
5b	92.77	108.0	5825.35	7.511	98.43
6a	119.68	210.0	5642.36	6.043	97.92
6b	91.94	103.0	5832.61	7.427	98.45

13.8 Bibliographical Remarks

The well model equations for vertical and horizontal wells in the context of finite differences were derived by Peaceman (1977A; 1991). These equations for the CVFE were developed by Chen et al. (2002C). The presentation of the well equation for the triangular mixed method in Section 13.5.2 follows Ewing et al. (1999). For more information on the data used in the seventh SPE CSP, refer to Nghiem et al. (1991). The content of Section 13.7 is taken from Li et al. (2003B).

Exercises

13.1. Derive equation (13.3).

13.2. Find the solution of equation (13.3).

13.3. Derive equation (13.5).

13.4. Extend the horizontal well models developed for single phase flow in Section 13.2.2 (ii) to the multiphase flow considered in Section 13.2.2 (iii).

13.5. Derive equation (13.21) from equation (13.20), with the transmissibility coefficient T_{0i} given in (13.22).

13.6. Derive the well equation (13.24) by substituting equation (13.23) into (13.21).

13.7. For Figure 13.5, check equations (13.26) and (13.27).

13.8. For the bilinear finite element (cf. Section 4.2.1), derive equation (13.28) from (13.19), with Ω_0 given as in Figure 13.6.

13.9. Verify the definition of φ_3 in equation (13.35).

13.10. Derive expressions for the basis functions φ_4 and φ_5 in the triangle K_1 of Figure 13.15, and then prove the orthogonality relation (13.36).

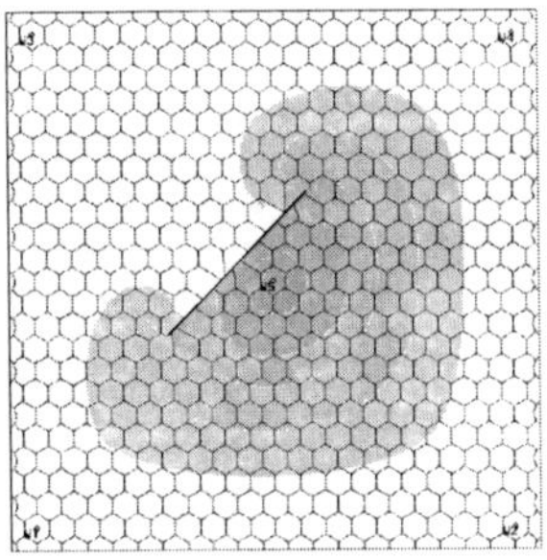

Chapter 14
Special Topics

We briefly discusse some practical issues that must be addressed at certain stages in a petroleum reservoir simulation. These issues include upscaling, history matching, parallel computing, oil recovery optimization, and surface network systems. The presentation is brief and is intended to give the reader an idea of the steps involved and decisions that must be made. A detailed treatment of each topic is beyond the scope of this book. An overview of upscaling is given in Section 14.1. In Section 14.2, history matching is described. The major ingredients in reservoir parallel computing are discussed in Section 14.3. Recovery optimization and surface network systems are presented in Sections 14.4 and 14.5, respectively. Finally, bibliographical information is given in Section 14.6.

14.1 Upscaling

In recent years *upscaling* has become increasingly important for converting highly detailed geological models to computational grids. These geological models usually require fine-scale descriptions of reservoir porosity and permeability on grids of tens of millions of cells to honor the known and inferred statistics of these reservoir properties. The geological grids of this order are far too fine to be used as simulation grids. Even with today's computing power, most of the full-field reservoir models are of the order of 100,000 cells, a factor of 100 less than the geological grids. Upscaling has been developed to bridge the gap between these two scales. Given a fine reservoir description scale and a simulation grid, an *upscaling algorithm* is designed to obtain suitable values for the porosity, permeability, and other property data for use in the coarse grid simulation. Many upscaling methods have been developed, such as pressure-solver (Begg et al., 1989), renormalization (King, 1989), effective medium (King, 1989), power law averaging (Deutsch, 1989), harmonic/arithmetic mean, local averaging (Whitaker, 1986), and homogenization (Amaziane et al., 1991); see the reviews of upscaling and pseudoization techniques by Christie (1996) and Barker and Thibeau (1997), for example. Here we briefly mention a few of these methods.

14.1.1 Single phase flow

For single phase flow, the aim of upscaling is to preserve the gross features of flow on the simulation grid. An algorithm is needed to compute an *effective permeability*, which will result in the same total flow of the fluid through the coarse homogeneous grid as that obtained from the fine heterogeneous grid.

In the *pressure-solver method* (Begg et al., 1989), for example, we set up a single phase flow computation with specific boundary conditions and then ask what value of effective permeability generates the same flow rate as the fine-scale computation. The results obtained depend on the assumptions made, particularly with regard to the boundary conditions. If no-flow boundary conditions are used, a diagonal effective permeability tensor can be derived and entered directly into a reservoir simulator. Alternatively, if periodic boundary conditions are employed, a full effective permeability tensor can be obtained (White and Horne, 1987).

The *renormalization method* (King, 1989) offers a faster but less accurate method for computing an effective permeability. It yields effective permeabilities close to a direct solution of the pressure equation and allows a rapid computation of these permeabilities from very large systems. This method works by breaking a large problem down into a hierarchy of manageable problems.

14.1.2 Two-phase flow

For two-phase flow, it is generally believed that upscaling of the absolute permeability alone is not enough to capture the effects of heterogeneity on two-phase fluid simulation (Muggeridge, 1991; Durlofsky et al., 1994), particularly when the correlation length of the heterogeneity not represented on the flow simulation grid is significant compared with the well spacing. A multiphase upscaling technique must be used. The most obvious technique is the use of *pseudorelative permeabilities*, i.e., pseudos (Lake et al., 1990). The role of pseudorelative permeabilities is to determine the flow rate of each fluid phase out of a gridblock. They relate the flow rate to the pressure gradients between the gridblock and its neighbors, given the average saturation in each gridblock. Both the flow rate and the pressure gradient depend on the details of the saturation distribution within the gridblock. Hence, to obtain a pseudorelative permeability curve, it is necessary to determine the saturation distribution within the block for any given average saturation (Barker and Thibeau, 1997); see the review papers by Christie (1996) and Barker and Thibeau (1997) for the generation of pseudorelative permeabilities.

14.1.3 Limitations in upscaling

A major limitation in upscaling is that it usually gives an answer without any indication of whether the assumptions made in obtaining the answer hold. No rigorous theory exists behind the upscaling process. Furthermore, some factors give rise to a concern about whether the upscaled values are good approximations; these include large-aspect-ratio gridblocks, significant transport at an angle to the grid lines, and upscaled gridblocks close in size to a correlation length of a heterogeneous reservoir. Compared with single phase upscaling, multiphase upscaling is far less developed and understood. The tenth SPE CSP was presented to compare different upscaling methods for two problems (Christie and Blunt, 2001); nine participants took part.

14.2 History Matching

A fundamental task of the reservoir engineer is to predict future production rates for a given reservoir or a specific well. Over the years, reservoir engineers have developed various techniques to accomplish this task. The techniques range from a simple decline curve analysis to the sophisticated multidimensional, multiphase reservoir simulators that we have developed in this book. Whether a simple or sophisticated technique is employed, the basic idea in predicting production rates is first to compute the rates for a time period for which the engineer already has production information. If the computed rates match the actual rates, the computation is assumed to be correct and can be then used to make future predictions. If the computed rates do not match the actual production data, some of the model parameters (e.g., porosity, permeability, etc.) must be modified and the computation must be repeated. Sometimes, this trial-and-error process must be repeated in a number of iterations to obtain a set of usable model parameters. The process of modifying these parameters to match the computed rates with the actual observed rates is called *history matching*.

For a given production schedule, the matching data usually are (1) observed gas-oil ratios (GORs) and water-oil ratios (WORs); (2) observed average pressures (shut-in pressures) or pressures at observation wells; (3) observed flowing well pressures; and (4) observed oil production rates.

The process of history matching is time consuming and extremely difficult. It often represents a large portion of the cost of a petroleum reservoir study. History matching can be done manually or automatically by adjusting model parameters through the above-mentioned trial-and-error procedure. The general approach in manual history matching is to modify the parameters that have the largest uncertainty and also the largest effect on the solution. The sensitivity of the solution to some of the parameters is often established during the history matching process itself. To the best of our knowledge, general guidelines for manual history matching do not exist. However, the following hints may be useful (Aziz and Settari, 1979; Mattax and Dalton, 1990):

- The match of average pressures is influenced by fluid volumes in-place, the size of the aquifer, and the degree of communication between the reservoir and the aquifer. Moreover, a poor match of GORs and WORs can also cause a bad match for the average pressures.

- Pressure drawdown primarily depends on horizontal permeabilities and skin factors.

- GORs and WORs are mainly affected by pressure drawdown, but also by the position of fluid contacts and the thickness of the transition zone (which depends on capillary forces). The shape of GOR and WOR curves after breakthough depends on the relative permeability curves; the breakthough time primarily depends on the endpoints of the latter curves, i.e., the effective permeabilities with only one of the phases flowing.

- Breakthough time is less frequently matched. In fact, matching breakthough times is one of the toughest tasks.

Manual history matching requires a great deal of experience and depends heavily on personal judgment. In recent years, considerable research efforts have been devoted to the development of automatic history matching techniques. While the need for incorporating professional experience is not eliminated with the latter matching, it does have the potential to save significant amounts of time and manpower and to provide more accurate estimates on the model parameters. It generally uses *inverse simulation* that involves output least squares algorithms. These algorithms are based on minimizing an objective functional (cost function), i.e., a quadratic function of the differences between observed and predicted measurements. Gradient-based algorithms are then used to speed up the process of parameter estimation. Constraints and a priori information (via Bayesian estimation) are added to restrict the dimension of the parameter spaces. Finally, sophisticated search algorithms involving trust region methods are employed for the constrained optimization problem. Thus the automatic history matching process becomes a mathematical minimization problem. Reservoir history matching problems are generally characterized by a very large number of unknown parameters. Consequently, the efficiency of numerical minimization algorithms is a primary concern. In addition, these problems are typically ill-conditioned; many quite different sets of parameter estimates may yield nearly identical matches to the data (Ewing et al., 1994). Because of these concerns, much research is yet to be done, and at the current stage of development automatic history matching is of limited use for practical problems.

14.3 Parallel Computing

The rapid development of parallel computers can overcome the limitations of problem size and space resolution for reservoir simulation associated with single-processor machines. In the past decade, the total number of gridblocks employed in a typical reservoir simulator has increased from thousands to millions. This is particularly due to the advent of the most prevalent type of parallel computers, distributed-memory machines, which have hundreds to thousands of processors. Research on *parallel computation* in reservoir simulation was extensively carried out in the late 1980s. There exist parallel black oil, compositional, and thermal reservoir simulators (Briens at al., 1997; Killough at al., 1997; Ma and Chen, 2004). Parallel commercial reservoir simulators are also available, such as Parallel-VIP (Landmark Graphics Corporation) and Eclipse Parallel (Schlumberger Software).

Because 70–90% of the computational time is spent on the assembly and solution of linear systems of algebraic equations, a prevailing strategy in reservoir simulation is to parallelize only this part, i.e., the linear solver part. However, this strategy may not be effective. First, the model scale is limited by the size of accessible memory of the CPU. This difficulty becomes prominent in a parallel environment with a PC or workstation cluster. Also, most preconditioners for linear solvers used in reservoir simulation are based on incomplete LU factorization (cf. Chapter 5), which is by its nature a sequential process. While various techniques, such as the use of parallel approximate inverses, have been introduced to parallelize these preconditioners, additional computations are needed. Thus, to really improve the efficiency of a simulation code, a global parallel scheme must be employed. In a global parallel computation, the use of domain decomposition methods, data communication, load balancing, and time step size control must be addressed.

14.3.1 Domain decomposition

The *domain decomposition* method is a technique for solving a partial differential problem based on a decomposition of the spatial domain of the problem into a number of smaller domains (Chan and Mathew, 1994). In general, this method can be classified as either an overlapping method or a nonoverlapping method. The overlapping method is generally easier to describe and to implement. It is also easier to achieve an optimal convergence rate using this method, and it is often more robust. But, in comparison to the nonoverlapping method, additional work is needed in the overlapped regions. Furthermore, if the coefficients of a differential problem are discontinuous across interboundaries, the extended subdomains have discontinuous coefficients, which makes their solution problematic. On the other hand, the nonoverlapping method requires a solution of interface problems at all interfaces of the subdomains.

14.3.2 Load balancing

In parallel computing, one should try to distribute the work load equally on all processors. In practice, it is difficult to achieve a load balance close to the optimum. Fortunately, in reservoir simulation, there are several guidelines for distributing the work load. First, the gridblocks should be evenly distributed among the processors with not only approximately the same number of internal blocks, but also roughly the same number of external blocks per processor. Second, if natural faults exist in a reservoir, these faults should be used as the interboundaries between subdomains. Some of the PVT and rock property data are discontinuous across the faults, and there should be no data communication across them. Third, all the subdomains should contain the same number of wells. The well operating schemes must be also taken into account for *load balancing*. A well can be an injector or producer. In the thermal modeling (cf. Chapter 10), for example, a well can be both, and the injection, production, and shut-in periods must be considered in distributing the work load. Among these three guidelines, the last should be respected the most.

14.3.3 Data communication

There exist standard procedures for message passing that allow data communication between different processors such as MPI (message passing interface) and PVM (parallel virtual machine). Message passing between processors is an essential component of parallel computing. It can take two forms: blocking (synchronous) and nonblocking (asynchronous). Which form is to be used depends on the characteristics of data to be transferred. In reservoir simulation, according to their time-variant characteristics, the communication data are divided into three basic types, *static data*, *slow transient data*, and *fast transient data*. The data describing the geometric model of a reservoir and rock property parameters are the static data. Essentially, these data do not change in the simulation. At a time step in the iteration process, the values of pressure, temperature, and saturation are the slow transient data. These data need to be recorded at certain times to restart a computation. All others are fast transient data. In particular, those that are frequently transferred over the overlapping regions are of this type. In practice, a blocking communication mode is used to transfer the static and slow transient data, and a nonblocking communication mode is adopted to transfer the fast transient data to reduce communication overhead and improve communication efficiency.

14.3.4 Time step size and communication time control

In parallel computation, the time step sizes on different subdomains can be different. To ensure that the well data of all production periods can be safely loaded and that a simulation process is stable and accurate on each processor, the step size Δt_i^n on the ith subdomain Ω_i can be chosen using an adaptive control strategy developed in Section 7.3.2, for example, that possesses the desired properties, $i = 1, 2, \ldots, N$, where N is the number of the subdomains.

To synchronize the computational processes on different processors and to pass messages efficiently between processors at certain times, the nth communication time is controlled as follows:

1. predict the communication time t_i^n for the ith subdomain, $i = 1, 2, \ldots, N$;

2. determine the nth synchronic communication time t^n by

$$t^n = \min\{t_1^n, t_2^n, \ldots, t_N^n\};$$

3. find the nth communication time t_i^n for the ith subdomain: $t_i^n = t^n$.

While the minimum time level approach is recommended here, we point out that the maximum and weighted time level approaches can be also utilized. From our experience, when a domain decomposition approximately achieves a load balance, these three approaches do not differ much. The approach adopted here generates the most accurate solution.

14.4 Oil Recovery Optimization

Enhanced oil recovery techniques have received considerable attention in recent years. The techniques involve the injection of large amounts of rather expensive fluids into oil-bearing reservoir formations (cf. Chapter 1). Commercial application of any enhanced oil recovery process relies on economic projections that show a decent return on the investment. Because of high chemical costs, it is extremely important to optimize enhanced oil recovery processes to generate the greatest recovery at the lowest chemical injection cost. Optimal control histories or operating strategies are needed to maximize the economic value of enhanced oil recovery techniques. The determination of these strategies is one of the key elements in the successful usage of these oil recovery techniques.

A proper treatment of the economic aspects of the enhanced oil recovery process is crucial because it is the major factor that controls applicability. Most oilfields can use this type of technique to significantly improve recovery efficiencies. However, expenses mainly with in-field drilling and injected chemical costs severely limit its applicability. First, candidate reservoirs must be selected from a preliminary screening, and then precise economic evaluations are obtained using accurate technical predictions such as history matching (cf. Section 14.2). Finally, injection policies must be evaluated to maximize the profitability of the project.

Optimization objectives can be expressed as a *performance index* to be extremized. If a profit index is employed, a maximum is desired. The controls associated with enhanced oil recovery processes are the physical state histories of the injected fluids. Thus the *optimization problem* in enhanced oil recovery is to determine the injection policies that lead to a maximum in the profitability index, subject to the differential equality constraints that describe the system dynamics. For more information on this subject, the reader may consult Ramirez (1987) for application of *optimal control theory* to the determination of optimal operating strategies in the petroleum industry.

14.5 Surface Network Systems

Well production rates and bottom hole pressures must be determined simultaneously from a reservoir, production wells, and a surface network system. Any change in a gathering network affects individual rates of production wells. A *gathering network* consists of pipes, valves, and fittings to connect wellheads to a separation section. The production rate of any well can be accurately computed only from the intersection of the *inflow performance curve* (determined by a reservoir model) and the *outflow performance curve* (defined by well tubing/casing and surface pipeline network models).

Models of multiphase flow in well tubing and surface network devices (e.g., pipelines and valves) must be added to a reservoir model for an integrated full field simulation. Thus the simultaneous simulation of multiphase flow in the reservoir, well tubing, and surface pipeline network system consists of the following models:

- a wellbore model that describes the fluid flow from the reservoir to the wellbores of production wells,
- a well tubing model that governs the flow from the wellbores to the wellheads,
- a surface facility model that determines the flow in the surface pipeline network system.

The reservoir and wellbore models define the inflow performance curve, and the well tubing and surface facility models define the outflow performance curve for each production well. The production rate and bottom hole pressure of the well are computed from the interaction of these two curves.

The reservoir and wellbore models have already been described in detail. Here we briefly touch on the well tubing and surface facility models that utilize models of *flow devices*, *links*, and *nodes*.

14.5.1 Hydraulic models of flow devices

Hydraulic models of multiphase flow in surface network devices (tubing strings, pipelines, valves, etc.) are basic elements of the surface pipeline network system. They are the building blocks for the surface network system. Each flow device has an inlet and an outlet (cf. Figure 14.1). In steady-state flow, a hydraulic model of the device determines the inlet pressure of this device as a function of the outlet pressure and flow rates of hydrocarbon components involved. There are two basic approaches for determining this function: *analytical steady-state modeling* (Beggs, 1991) and *hydraulic look-up tables* (VIP-Executive, 1994). The former approach is widely used in the petroleum industry for the simulation of multiphase flow in well tubing, pipelines, and valves. The latter approach tabulates the inlet pressure of a flow device in terms of the outlet pressure and flow rates. The tabular approach requires

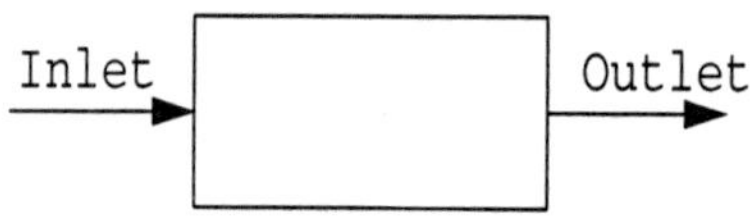

Figure 14.1. *A flow device model.*

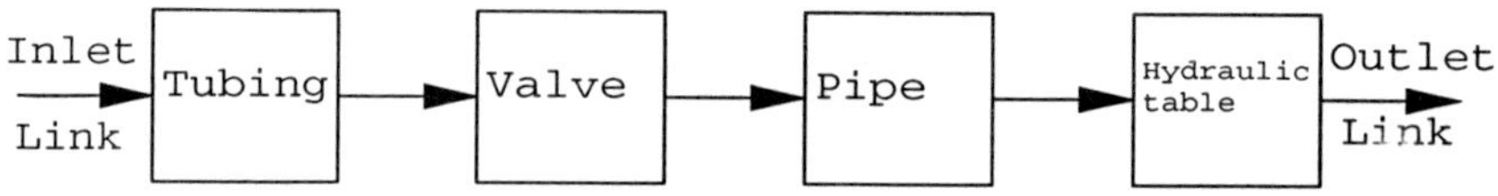

Figure 14.2. *A link example.*

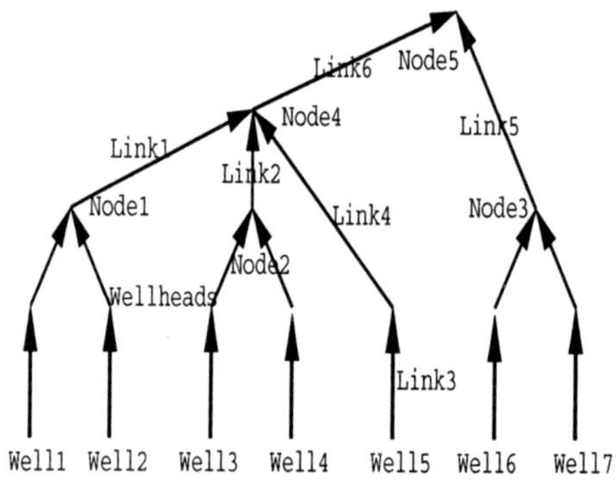

Figure 14.3. *A surface pipeline network system.*

significantly less CPU time than the analytical approach, but it has obvious disadvantages: a preprocessor package is required to establish the tables, and significant computer memory is needed to store a large number of hydraulic tables for each of the sophisticated reservoir models discussed in Chapters 9–12.

14.5.2 Models of links and nodes

A *link* simulates multiphase flow in well tubing, connections between wellheads and nodes of the surface pipeline network system, and the connections between nodes. Each link has only an inlet and an outlet (cf. Figure 14.2). The link shown in Figure 14.2 is composed of four flow devices: tubing, valve, pipe, and hydraulic table.

A *node* is a junction of several links. Each node can have any number of input links but only one output link. Production wells can be connected to any node. Figure 14.3 shows an example of a surface pipeline network with seven production wells. Five nodes are presented, and links are employed for flow simulation in well tubing and node connections.

Pressure equations are formulated for each link. They determine the inlet pressure of the link in terms of the outlet pressure and mass rates of hydrocarbon components in this link. Mass conservation equations for the hydrocarbon components are formulated for each node in the surface network system. These equations state that the mass rate of each component at the outlet of a node equals the sum of mass rates of the component at the outlets of all nodes connected to this node (Litvak and Darlow, 1995).

In summary, the simultaneous simulation of multiphase flow in a reservoir, well tubing, and surface pipeline network system consists of a solution of a reservoir model in gridblocks, a solution of a wellbore model in production wells, and a solution of a well tubing and surface facility model in links and nodes. The solution of these three subsystems can be performed either simultaneously (in a fully coupled fashion) or sequentially (in a decoupled fashion) (Litvak and Darlow, 1995).

14.6 Bibliographical Remarks

In this chapter, five special topics in reservoir simulation have been briefly studied: upscaling, history matching, parallel computing, oil recovery optimization, and surface network systems. Upscaling remains a hot research topic, with much research to be done. For recent work on this topic, the reader can consult the review papers by Christie (1996) and Barker and Thibeau (1997) and the tenth SPE CSP (Christie and Blunt, 2001). The process of history matching is sometimes frustrating. There has been considerable research effort devoted to automating this process (see the recent biannual SPE numerical simulation proceedings). Due to the advent of powerful parallel computers, parallel computing technologies have been massively applied to reservoir simulation, particularly since the 1990s. Ideally, the speedup of CPU times in terms of the numbers of processors can be superlinear (Briens at al., 1997; Killough at al., 1997; Ma and Chen, 2004). Compared with the devotion of time and manpower to numerical reservoir simulation, less research effort has been devoted to oil recovery optimization. The book by Ramirez (1987) is a good starting point in this area. Finally, research on the simultaneous simulation of multiphase flow in a reservoir, well tubing, and surface pipeline network system needs more attention. The approach in Section 14.5 to this topic follows Litvak and Darlow (1995).

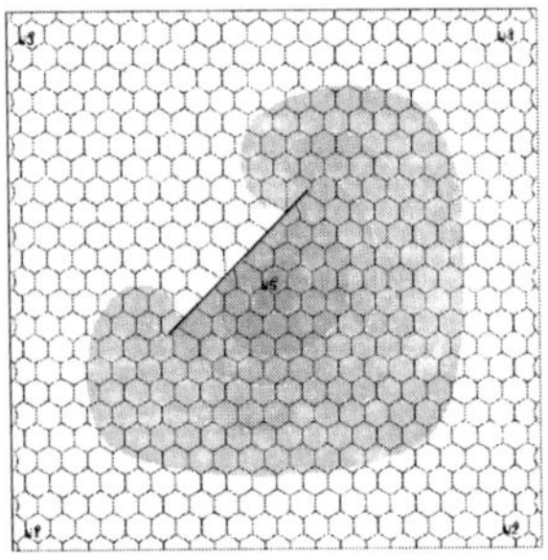

Chapter 15
Nomenclature

15.1 English Abbreviations

ASP	Alkaline, surfactant and polymer
BCG	Biconjugate gradient
BDDF	Brezzi–Douglas–Durán–Fortin
BDFM	Brezzi–Douglas–Fortin–Marini
BDM	Brezzi–Douglas–Marini
BiCGSTAB	Biconjugate gradient stabilized
CD	Chen–Douglas
CG	Conjugate gradient
CGN	CG applied to normal equations
CGS	Conjugate gradient squared
CSP	Comparative solution project
CVFA	Control volume function approximation
CVFE	Control volume finite element
DG	Discontinuous Galerkin
ELLAM	Eulerian–Lagrangian localized adjoint method
EOR	Enhanced oil recovery
EOS	Equation of state
erfc	Complementary error function
FD	Finite difference
FGMRES	Flexible generalized minimum residual
GMRES	Generalized minimum residual
GOC	Gas/oil contact
GOR	Gas-oil ratio
ILU	Incomplete LU factorization
ILUT	Incomplete LU factorization with threshold
IMPES	Implicit pressure-explicit saturation
LES	Linear equation system
MMOC	Modified method of characteristics

OIP	Oil in place
ORTHOMIN	Orthogonal minimum residual
PCG	Preconditioned conjugate gradient
PI	Production index
RT	Raviart–Thomas
RTN	Raviart–Thomas–Nédélec
SDG	Stabilized discontinuous Galerkin
SOR	Successive overrelaxation
SPE	Society of Petroleum Engineers
SS	Simultaneous solution
SSOR	Symmetric successive overrelaxation
WAG	Water-alternating-gas
WC	Water cut
WOC	Water/oil contact
WOR	Water-oil production ratio

15.2 Subscripts

f	Fluid phase or fracture quantity
g	Gas phase
i	Component or coordinate index
o	Oil phase
s	Standard conditions or solid phase
t	Total quantity
w	Water phase
α	Phase index

15.3 Base Quantities

Symbol	Base quantities	Unit
L	Length	m
M	Mass	kg
T	Temperature	K
t	Time	s (or sec.)

15.4 English Symbols

Symbol	Quantity	Unit
A	Area	L^2
A_{PI}	Oil gravity	°API
a	Acceleration	L/t^2
B	Formation volume factor	

B_α	Formation volume factor of phase α	
C_i	Compressibility of component i	Lt^2/M
C_s	Specific heat capacity	$L^2/(Tt^2)$
$C_{V\alpha}$	Heat capacity of phase α at constant volume	$L^2/(Tt^2)$
$C_{p\alpha}$	Heat capacity of phase α at constant pressure	$L^2/(Tt^2)$
$C_{p,ob}$	Overburn heat capacity at constant pressure	$L^2/(Tt^2)$
c	Mass concentration	
c_f	Fluid compressibility	Lt^2/M
c_g	Gas compressibility	Lt^2/M
c_o	Oil compressibility	Lt^2/M
c_w	Water compressibility	Lt^2/M
c_i	Concentration of component i	
$c_{i\alpha}$	Concentration of component i in phase α	
c_R	Rock compressibility	Lt^2/M
c_{SE}	Effective salinity	
c_{SEL}	Lower limit of effective salinity	
c_{SEU}	Upper limit of effective salinity	
c_{SEOP}	Optimum effective salinity	
c_s^c	Critical surfactant concentration	
c_t	Total compressibility	Lt^2/M
c_μ	Oil viscosity compressibility	Lt^2/M
$\tilde{c}_i$	Overall concentration of component i	
$\hat{c}_i$	Adsorbed concentration of component i	
$\hat{c}_i^o$	Reference adsorbed concentration	
$\mathbf{D}$	Diffusion/dispersion tensor	L^2/t
$\mathbf{D}_{i\alpha}$	Diffusion/dispersion of component i in phase α	L^2/t
Dx_i	Grid size in x_i-direction	L
$\mathbf{d}_{i\alpha}$	Diffusive flux of component i in phase α	$M/(L^2t)$
d_m	Molecular diffusion	L^2/t
d_l	Longitudinal dispersion	L
d_t	Transverse dispersion	L
$d\ell$	Line or surface integral sign	L (or L^2)
$d\mathbf{x}$	Area or volume integral sign	L^2 (or L^3)
$\mathbf{E}$	Energy flux	M/t^3
E	Energy	L^2M/t^2
$E_i(\cdot)$	Exponential integral function	
$\mathbf{E}(\mathbf{u})$	Orthogonal projection along $\mathbf{u}$	
$\mathbf{E}^\perp(\mathbf{u})$	Complement of $\mathbf{E}(\mathbf{u})$	$\mathbf{I}-\mathbf{E}(\mathbf{u})$
F	Force	LM/t^2
F	Total mass variable	M/L^3
$F_\alpha(\cdot)$	Distribution function of phase α	
$f_{i\alpha}$	Fugacity function of component i in phase α	$M/(Lt^2)$
f_α	Fractional flow function of phase α	
G	Young modulus	
H_α	Enthalpy of α-phase	L^2/t^2
H	Reservoir thickness	L

h_3	Reservoir thickness	L
h_3	Height of the gridblock containing a well	L
h_i	Grid size in x_i-direction	L
$\mathbf{I}$	Identity tensor	
I	An interval in space	L
$J = (0, T]$	Time interval of interest	t
J^n	Subinterval in time $(t^{n-1}, t^n]$	t
$\|K\|$	Area or volume of set K	L^2 (L^3)
K_i	Equilibrium K-value of component i	
$\mathbf{k}$	Permeability tensor	L^2
k_{ii}	Permeability in x_i-direction	L^2
$\bar{\mathbf{k}}$	Certain average of $\mathbf{k}$	L^2
k_h	Horizontal permeability	L^2
k_v	Vertical permeability	L^2
$\mathbf{k}_\alpha$	Effective permeability of phase α	L^2
$k_{r\alpha}$	Relative permeability of phase α	
k_{row}	Relative permeability of oil-water system	
k_{rog}	Relative permeability of oil-gas system	
k_{rc}	Value of k_{row} at S_{wc}	$k_{row}(S_{wc})$
k_{ij}^f	Forward chemical rate	$M/(L^3t)$
k_{ij}^r	Reverse chemical rate	$M/(L^3t)$
k_T	Thermal conductivity	$ML/(Tt^3)$
k_{ob}	Thermal conductivity of overburden	$ML/(Tt^3)$
L	Mass fraction of oil	
L_c	Characteristic length	L
L_i	Chemical loss rate of component i	$M/(L^3t)$
l_{x_i}	Matrix block dimension in x_i-direction	L
$N_{B\alpha}$	Bond number	
N_c	Number of components	
$N_{c\alpha}$	Capillary number	
N_p	Number of phases	
N_{cv}	Total number of volume-occupying components	
N_w	Number of wells	
Nx_i	Number of gridblocks in x_i-direction	
p	Pressure	$M/(Lt^2)$
p_b	Bubble point pressure	$M/(Lt^2)$
p_α	Pressure of phase α	$M/(Lt^2)$
p_{pc}	Pseudocritical pressure	$M/(Lt^2)$
p_c	Capillary pressure	$M/(Lt^2)$
p_c^*	Critical capillary pressure	
p_{ic}	Critical pressure of component i	$M/(Lt^2)$
$p_{c\alpha 1}$	Capillary pressure	$M/(Lt^2)$
p_{cow}	Capillary pressure $p_o - p_w$	$M/(Lt^2)$
p_{cgo}	Capillary pressure $p_g - p_o$	$M/(Lt^2)$
p_{cw}	Capillary pressure $= -p_{cow}$	$M/(Lt^2)$
p_{cg}	Capillary pressure $= p_{cgo}$	$M/(Lt^2)$

p_{cb}	Threshold pressure	$M/(Lt^2)$
p_{bh}	Bottom hole pressure	$M/(Lt^2)$
P_{bh}	Given bottom hole pressure	$M/(Lt^2)$
p^o	Reference pressure	$M/(Lt^2)$
p_r	Reference phase pressure	$M/(Lt^2)$
Q	Production rate	L^3/t
Q_α	Production rate of phase α	L^3/t
Q_i	Chemical reaction rate of component i	$M/(L^3t)$
$\mathbf{q}_c$	Heat conduction flux	M/t^3
$\mathbf{q}_r$	Heat radiation flux	M/t^3
q	Source/sink	$M/(L^3t)$
q_{ext}	External source/sink	$M/(L^3t)$
q_{Gm}	Matrix-fracture transfer term for gas	$M/(L^3t)$
q_{Gom}	Matrix-fracture transfer term for gas in oil	$M/(L^3t)$
q_{Oom}	Matrix-fracture transfer term for oil in oil	$M/(L^3t)$
q_{Wm}	Matrix-fracture transfer term for water	$M/(L^3t)$
q_{mf}	Matrix-fracture transfer term	$M/(L^3t)$
q_α	Source/sink of phase α	$M/(L^3t)$
$q^{(i)}$	Production/injection rate at well i	L^3/t
q_H	Enthalpy source term	$M/(Lt^3)$
q_L	Heat loss	$M/(Lt^3)$
R	Universal gas constant	$R \approx 0.8205$
R_{gl}	Gas-liquid ratio	
R_k	Permeability reduction factor	
R_r	Resistance factor	
R_{rr}	Residual resistance factor	
R_s (R_u)	Gas mobility reduction factor	
R_{so}	Dissolved gas-oil ratio	
R_v	Oil volatility in gas	
r_e	Equivalent radius	L
r_w	Wellbore radius	L
$r_{i\alpha}$	Reaction rate of component i in phase α	$M/(L^3t)$
r_{is}	Reaction rate of component i in solid	$M/(L^3t)$
S_α	Saturation of phase α	
$S_{n\alpha}$	Normalized saturation of phase α	
S_{oc}	Critical oil saturation	
S_{wc}	Critical water saturation	
S_{nc}	Residual saturation	
S_{or}	Residual oil saturation	
$S_{\alpha r}$	Residual saturation of phase α	
S_{wf}	Water saturation at water front	
s_k	Skin factor	
$\mathbf{T}_\alpha$	Transmissibility of phase α	L^3t/M
T	Temperature	T
T_c	Critical temperature	T
T_m	Matrix-fracture transmissibility	

T_{ob}	Temperature of overburden	T
T_{pc}	Pseudocritical temperature	T
t_B	Water breakthrough time	t
U	Specific internal energy	L^2/t^2
U_α	Specific internal energy of phase α	L^2/t^2
$\mathbf{u}$	Darcy' velocity (u_1, u_2, u_3)	L/t
$\|\mathbf{u}\|$	Euclidean norm of $\mathbf{u}$	L/t
$\mathbf{u}_\alpha$	Velocity of phase α	L/t
V	Volume	L^3
$\|V_0\|$	Area or volume of set V_0	L^2 (L^3)
V	Mass fraction of gas	
W	Molecular weight	M/mole
W_i	Molecular weight of component i	M/mole
WI	Well index	L^3
$\mathbf{w}$	Displacement of fluid	L
$\mathbf{w}_s$	Displacement of solid	L
$\mathbf{x}$	Spatial variable (x_1, x_2, x_3)	L
$\mathbf{x}^{(i)}$	Well location	L
$x_{i\alpha}$	Mole fraction of component i in phase α	
Y_G	Raw gas gravity	
Z	Gas compressibility or deviation factor	
Z_α	Compressibility factor of phase α	
z	Depth	L
z_i	Total mole fraction	
z_{bh}	Datum level depth	L

15.5 Greek Symbols

Symbol	Quantity	Unit
$\bar{\alpha}$	Dimension factor	
β	Inertial or turbulence factor	
κ_{ij}	Binary interaction parameter	
Ω	Solution domain	L^3
Ω_i	ith matrix block	L^3
Ω_0	Support of φ_0	L^3
$\partial\Omega$	Boundary of Ω	L^2
$\partial/\partial t$	Time derivative	t^{-1}
$\partial/\partial x_i$	Spatial derivative	L^{-1}
∇	Gradient operator	L^{-1}
$\nabla\cdot$	Divergence operator	L^{-1}
Δ	Laplacian operator	L^{-2}
ΔL	Well length in a gridblock	L
Δp_α	Pressure gradient across a matrix block	$M/(L^2t^2)$
Φ	Potential	$M/(Lt^2)$
Φ_α	Potential of phase α	$M/(Lt^2)$

Φ^o	Reference potential	M/(Lt^2)
Φ'	Pseudopotential	L
ϕ	Porosity	
ϕ^o	Reference porosity	
$\varphi_{i\alpha}$	Fugacity coefficient of component i in phase α	
ψ	Pseudopressure	M/(Lt^3)
ψ'	Inverse function of (2.7)	L
μ	Viscosity	M/(Lt)
μ_α	Viscosity of phase α	M/(Lt)
μ_P	Polymer viscosity	M/(Lt)
ρ	Density	M/L^3
ρ_f	Fluid density	M/L^3
ρ_f	Global fluid density in fractures	M/L^3
ρ_α	Density of phase α	M/L^3
ρ^o	Reference density	M/L^3
ρ_t	Total mass density	M/L^3
ρ_{Go}	Partial density of gas component in oil	M/L^3
ρ_{Oo}	Partial density of oil component in oil	M/L^3
ρ_{ob}	Density of overburden	M/L^3
$\wp$	Gravitational acceleration	L/t^2
δ	Dirac delta function	1/L^3
δv^l	Increment of v at lth Newton–Raphson	
$\bar{\delta} v^n$	Time increment of v at nth step	
σ	Matrix shape factor	1/L^2
σ	Surface tension	M/t^2
σ_{aw}	Water/air interfacial tension	M/t^2
σ_{ow}	Water/oil interfacial tension	M/t^2
σ_{23}	Microemulsion/oil interfacial tension	M/t^2
σ_{13}	Microemulsion/water interfacial tension	M/t^2
$\boldsymbol{\sigma}$	Stress tensor	M/(Lt^2)
$\boldsymbol{\epsilon}_s$	Strain tensor	M/(Lt^2)
ν	Poisson ratio	
$\boldsymbol{\nu}$	Outward unit normal	
$\chi_i(\cdot)$	Characteristic function	
λ_α	Mobility of phase α	L^{3}t/M
λ	Total mobility	L^{3}t/M
$\xi_{i\alpha}$	Molar density of component i in phase α	mole/L^3
ξ_α	Molar density of phase α	mole/L^3
θ	Contact angle	
ω_i	Acentric factor of components i	
γ_α	Phase specific weight	M/(L^{2}t^2)
$\gamma_{i\alpha}$	Specific weight of component i in phase α	M/(L^{2}t^2)
γ_i^o	Reference specific weight of component i	M/(L^{2}t^2)

15.6 Generic Symbols Used in Chapters 4 and 5

Symbol	Definition
$\mathbf{A}$	Coefficient matrix of a system (stiffness matrix)
$\mathbf{a}$	Diffusion coefficient
$\mathbf{a}_{har}$	Harmonic average of $\mathbf{a}$
a	Diffusion coefficient
$a(\cdot,\cdot)$	Bilinear form
$a_h(\cdot,\cdot)$	Mesh-dependent bilinear form
$a_K(\cdot,\cdot)$	Restriction of $a(\cdot,\cdot)$ on K
a_{ij}	Entries of $\mathbf{A}$
a_{ij}^K	Restriction of a_{ij} on K (element)
a_{har}	Harmonic average of a
a_{num}	Numerical dispersion
$\mathbf{B}$	Mass matrix
$\mathbf{B}_1$	Matrix in an affine mapping
B_{r+1}	Equals $\lambda_1\lambda_2\lambda_3 P_{r-2}$
$B_{r+2,r}$	Refer to Section 4.5.4
$b(\cdot,\cdot)$	Bilinear form
$\mathbf{b}$	Convection or advection coefficient vector
b	Convection or advection coefficient
c	Reaction coefficient
$\mathbf{C}$	Coefficient matrix associated with time
d	Dimension number ($d = 1, 2$, or 3)
e_{ik}	kth edge on ∂V_i
$\mathcal{E}_h^D$	Set of edges on Γ_D
$\mathcal{E}_h^N$	Set of edges on Γ_N
$\mathcal{E}_h^o$	Set of internal edges in K_h
$\mathcal{E}_h^b$	Set of edges on Γ
$\mathcal{E}_h$	Set of edges of the partition K_h
F	Functional or total potential energy
F_i	Equals $\int_{V_i} f(\mathbf{x})\, d\mathbf{x}$
$\mathbf{F}$	Mapping
f	Right-hand function or load
$\mathbf{f}$	Right-hand vector of a system
f_K	Local mean value on K
f_i	ith entry of $\mathbf{f}$
f_α	Fractional flow function
$\mathbf{G}$	Jacobian matrix or a mapping
g	Boundary datum
g_e	Local mean value on e
$\mathbf{H}^k$	$(k+1)\times k$ upper Hessenberg matrix
h	Mesh or grid size
h_e	Length of edge e
h_i	Grid size in x_i-direction

h_i'	Grid size in x_i-direction
h_i''	Grid size in x_i-direction
h_k	Mesh size at the kth level
h_K	Diameter of K (element)
I	Interval in $\mathbb{R}$
I_i	Subintervals
$\mathbf{I}$	Identity matrix or operator
$\check{I}_i(t)$	Trace back of I_i to time t
$\mathcal{I}_i^n$	Space-time region following characteristics
$\bar{i}$	Equals $\sqrt{-1}$
J	Time interval of interest ($J = (0, T]$)
J^n	nth subinterval of time (t^{n-1}, t^n)
K	Element (triangle, rectangle, etc.)
$\|K\|$	Area or volume of K
$\hat{K}$	Reference element
K_h	Triangulation (partition)
$\check{K}(t)$	Trace back of K to time t
$\mathcal{K}^n$	Space-time region following characteristics
$\mathcal{K}^k$	kth Krylov space of $\mathbf{A}$
$\hat{\mathcal{K}}^k$	kth Krylov space of $\mathbf{A}^T$
$L(\cdot)$	Linear functional
$L_-(\cdot)$	Linear functional for symmetric DG
$L_+(\cdot)$	Linear functional for nonsymmetric DG
L_h	Space of Lagrange multipliers
L_i	Bandwidth of ith row of a matrix
L	Bandwidth of a matrix
$\mathbf{L}$	Lower triangular matrix
$\mathcal{L}$	Linear operator
l_{ij}	Elements of $\mathbf{L}$
M	Number of grid points (nodes)
$\mathbf{M}$	Coefficient matrix arising from mixed methods
$\mathbf{m}_c$	Centroid of an element
$\mathbf{m}_i$	Vertices of elements
$\mathbf{m}_{ij}$	Midpoint of an edge
$\mathbf{m}_0$	Centroid of an element
$\mathcal{N}_h$	Set of vertices in K_h
p	Primary unknown
$\mathbf{p}$	Unknown vector of a system
p_h	Approximate solution of p
p_0	Initial datum
$\check{p}_h^{n-1}$	Value of p_h at $\left(\check{x}_n, t^{n-1}\right)$: $p_h\left(\check{x}_n, t^{n-1}\right)$
$\tilde{p}_h$	Interpolant of p_h
$\mathbf{p}^k$	kth iterate
p_{ij}^n	Value of p at $(x_{1,i}, x_{2,j}, t^n)$
P_r	Set of polynomials of total degree $\leq r$
$P_{l,r}$	Set of polynomials defined on prisms

p_α	Pressure of α-phase
Q_r	Set of polynomials of degree $\leq r$ in each variable
$Q_{l,r}$	Set of polynomials of degree l in x_1 and r in x_2
$\mathbf{Q}$	Upper triangular matrix
R	Reaction coefficient
$\mathbb{R}^d$	Euclidean space, $d = 1, 2, 3$
R_i^n	Truncation error
$\mathbf{r}^k$	kth residual vector
$\mathcal{R}_K$	Residual a posteriori estimator
S	Computer storage
t	Time variable
$\check{t}$	See Section 4.6.2
t^n	nth time step
T	Final time
T_{ij}	Transmissibility between nodes i and j
$\mathbf{u}$	Equals $-\mathbf{a}\nabla p$ or $-\nabla p$
$\mathbf{U}$	Unknown vector for u or $\mathbf{u}$
u_{ij}	Elements of $\mathbf{U}$
v_-	Left-hand limit notation
v_+	Right-hand limit notation
V	Linear vector space
V'	Dual space to V
V_h	Finite element space
$\mathbf{V}$	Vector space in a pair of mixed spaces
$\mathbf{V}_h$	Vector space in a pair of mixed finite element spaces
$\mathbf{V}_h(K)$	Restriction of $\mathbf{V}_h$ on K
V_i	Control volume
$\mathbf{V}^k$	Orthogonal projector
w_i	Integration weight
W	Computer work
W	Scalar space in a pair of mixed spaces
W_h	Scalar space in a pair of mixed finite element spaces
$W_h(K)$	Restriction of W_h on K
x	Independent variable in $\mathbb{R}$
$\mathbf{x}$	Independent variable in $\mathbb{R}^d$: $\mathbf{x} = (x_1, x_2, \ldots, x_d)$
$\check{x}_n$	Foot of a characteristic corresponding to x at t^n
$\mathbf{Z}(2, M)$	Coordinate matrix of nodes
$\mathcal{Z}(3, \mathcal{M})$	Matrix of node numbers
Z^n	Maximum error $\sup_i\{\lvert z_i^n \rvert\}$
z_i^n	Error $P_i^n - p_i^n$
$\mathrm{cond}(\mathbf{A})$	Condition number of matrix $\mathbf{A}$
$\mathbb{R}$	Set of real numbers
Ω	Open set in $\mathbb{R}^d$ ($d = 2$ or 3)
$\bar{\Omega}$	Closure of Ω
Ω_e	Union of elements with common edge e
Ω_K	Union of elements adjacent to K

Ω_i	Set of neighboring nodes of $\mathbf{m}_i$
$\Omega_{\mathbf{m}}$	Union of elements with common vertex $\mathbf{m}$
Γ	Boundary of Ω ($\partial\Omega$)
Γ_-	Inflow boundary of Γ
Γ_+	Outflow boundary of Γ
Γ_D	Dirichlet boundary of Γ
Γ_N	Neumann boundary of Γ
∂K	Boundary of K
∂K_-	Inflow part of ∂K
∂K_+	Outflow part of ∂K
∇	Gradient operator
$\nabla\cdot$	Divergence operator (div)
Δ	Laplacian operator
Δ^2	Biharmonic operator ($\Delta\Delta$)
Δt	Time step size
Δt^n	Time step size at the nth step
$\frac{\partial}{\partial x_i}$	Partial derivative with respect to x_i
$\frac{\partial}{\partial t}$	Partial derivative with respect to t (time)
$\frac{\partial}{\partial \boldsymbol{\nu}}$	Normal derivative
$\frac{\partial}{\partial \mathbf{t}}$	Tangential derivative
$\frac{\partial}{\partial \boldsymbol{\tau}}$	Directional derivative along characteristics
$\frac{D}{Dt}$	Material derivative
D^α	Partial derivative notation
$C^\infty(\Omega)$	Space of functions infinitely differentiable
$\mathcal{D}(\Omega)$	Subset of $C^\infty(\Omega)$ having compact support in Ω
$C_0^\infty(\Omega)$	Same as $\mathcal{D}(\Omega)$
$\text{diam}(K)$	Diameter of K
$L^1_{loc}(\Omega)$	Integrable functions on any compact set inside Ω
$L^q(\Omega)$	Lebesgue space
$W^{r,q}(\Omega)$	Sobolev spaces
$W_0^{r,q}(\Omega)$	Completion of $\mathcal{D}(\Omega)$ with respect to $\|\cdot\|_{W^{r,q}(\Omega)}$
$\|\cdot\|$	Norm
$\|\cdot\|_h$	Norm on a nonconforming space
$\|\cdot\|_{L^q(\Omega)}$	Norm of $L^q(\Omega)$
$\|\cdot\|_{W^{r,q}(\Omega)}$	Norm of $W^{r,q}(\Omega)$
$\lvert\cdot\rvert_{W^{r,q}(\Omega)}$	Seminorm of $W^{r,q}(\Omega)$
$\lvert\mathbf{u}\rvert$	Euclidean norm $\sqrt{u_1^2+u_2^2+\cdots+u_d^2}$
$(\cdot,\cdot)$	Inner product
$H^r(\Omega)$	Same as $W^{r,2}(\Omega)$
$H_0^r(\Omega)$	Same as $W_0^{r,2}(\Omega)$
$H^l(K_h)$	Piecewise smooth space
$\mathbf{H}(\text{div}, \Omega)$	Divergence space

$\boldsymbol{\beta}$	Convection or advection coefficient
α	multi-index (a d-tuple): $\alpha = (\alpha_1, \alpha_2, \ldots, \alpha_d)$
β_1	Measure of smallest angle over $K \in K_h$
β_2	Quasi-uniform triangulation constant
ϵ_i^n	Perturbation error
Σ_K	Set of degrees of freedom
γ	Amplification factor
γ_k^n	Magnitude (expansion coefficient) of ϵ_i^n
π_h	Interpolation operator
π_K	Restriction of π_h on element K
Π_h	Projection operator
$\delta(\mathbf{x} - \mathbf{x}^{(l)})$	Dirac delta function at $\mathbf{x}^{(l)}$
ρ_K	Diameter of largest circle inscribed in K
$\boldsymbol{\nu}$	Outward unit normal
ϕ	Time differentiation term coefficient
φ, $\boldsymbol{\varphi}$	Interstitial velocity
φ_i	Basis function of V_h
φ_{ik}^j	Basis function in CVFA
ψ_i	Basis function of W_h
$\boldsymbol{\varphi}_i$	Basis function of $\mathbf{V}_h$
λ_d	Lagrange multipliers
λ_i	Barycentric coordinates ($i = 1, 2, 3$)
λ_{ij}^{up}	Upstream weighted coefficient
τ, $\boldsymbol{\tau}$	Characteristic direction
$[\![\cdot]\!]$	Jump operator notation
$\{\!\{ \cdot \}\!\}$	Averaging operator notation
$\det(\cdot)$	Determinant of a matrix
$\sigma(\mathbf{A})$	Spectrum of $\mathbf{A}$

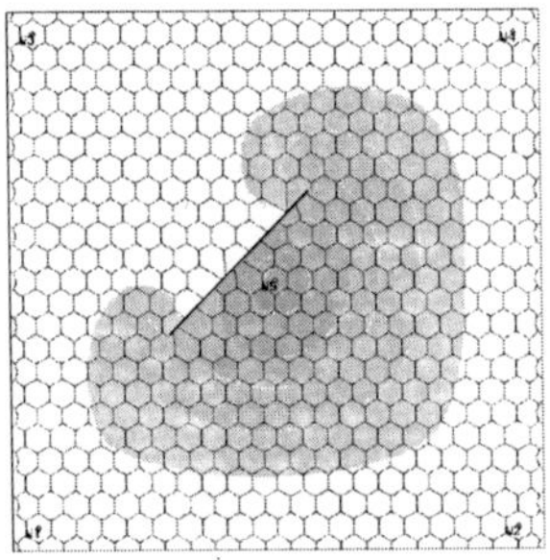

Chapter 16
Units

16.1 Unit Abbreviations

API	American Petroleum Institute
atm	Atmosphere
bbl	Reservoir barrel
Btu	British thermal unit
°C	Degrees Celsius
cc	Cubic centimeter or cubic content
cm	Centimeter
cp	Centipoise
D	Day
dyn	Dyne
°F	Degrees Fahrenheit
ft	Foot
g	Gram
gm	Gram
hr	Hour
J	Joule
K	Kelvin
kg	Kilogram
lb	Pound
lbm	Pound-mole
m	Meter
md	Millidarcy
mg/L	Milligram/liter
mol	Mole
N	Newton
Pa	Pascal
ppm	Parts per million
psi	Pounds per square inch

psia	Pounds per square inch absolute
psig	Pounds per square inch gauge
PV	Pore volume
R	Rankine
RB	Reservoir barrel
s (or sec.)	Second
SCF	Standard cubic feet
SCM	Standard cubic meter
STB	Standard barrel
t	Ton
yr	Year

16.2 Unit Conversions

Length

1 m = 100 cm = 1,000 mm = 3.28084 ft = 39.3701 in
1 ft = 0.30480 m = 30.4800 cm = 3,048 mm = 12 in
1 km = 0.621388 mile

Area

1 m^2 = 10,000 cm^2 = 1,000,000 mm^2 = 10.7639 ft^2 = 1,550.0 in^2
1 ha = 10,000 m^2 = 2.47105 acres
1 $mile^2$ (section) = 2.58985 km^2 = 258.985 ha = 639.965 acres
1 acre = 43,560 ft^2 = 0.404686 ha = 4,046.86 m^2

Volume (capacity)

1 m^3 = 1,000 L = 1,000 dm^3 = 35.3147 ft^3 = 6.28981 bbl
1 L = 1 dm^3 = 0.001 m^3 = 1,000 cm^3 = 0.0353147 ft^3 = 61.0237 in^3
1 ft^3 = 0.0283168 m^3 = 28.3168 L
1 bbl (API) = 0.158987 m^3 = 158.987 L = 5.61458 ft^3

Mass

1 kg = 2.20460 lbm = 1,000 g
1 lbm = 0.453597 kg = 453.597 g
1 t = 1,000 kg = 2,204.60 lbm

Density

1 kg/m^3 = 0.001 g/cm^3 = 0.001 t/m^3 = 0.0624273 lbm/ft^3
1 lbm/ft^3 = 16.0186 kg/m^3 = 0.0160186 g/cm^3
1 g/cm^3 = 1,000 kg/m^3 = 1 t/m^3 = 1 kg/L = 62.4273 lbm/ft^3

Force

1 N = 10^5 dyn = 0.102 kgf = 0.225 lbf
1 kgf = 9.81 N = 9.81 × 10^5 dyn = 2.205 lbf
1 lbf = 4.45 N = 0.454 kgf

Pressure

1 MPa = 10^6 Pa = 9.86923 atm = 10.1972 at = 145.038 psi
1 atm = 0.101325 MPa = 1.03323 at = 14.6959 psi
1 psi = 0.00689476 MPa = 6.89476 kPa = 0.0680460 atm = 0.0703072 at

Temperature

°C = (°F-32)/1.8

K = °C + 273.16

°F = 1.8(°C) + 32

R = °F + 459.67

K = R/1.8

Viscosity

1 mPa·s = 1 cp (dynamic) = 10^{-3} Pa·s

1 mm²/s = 1 cSt = 1.08 × 10^{-5} ft²/s (kinematic)

Permeability

1 $\mu m^2 = 10^{-12}$ m² = 1.01325 darcy = 1.01325 × 10^3 md

1 md = 10^{-3} darcy = 9.86923 × 10^{-16} m² = 9.86923 × 10^{-4} μm^2

1 $\mu m^2 \approx$ 1 darcy = 1,000 md

Surface tension

1 mN/m = 1 dyn/cm

Work, energy, power

1 J = 9.47813 × 10^{-4} Btu

1 Btu = 1,055.06 J

Heat transfer coefficient

1 kJ/(m·day·K) = 1.60996 Btu/(ft·day·°F)

1 Btu/(ft·day·°F) = 6.23067 kJ/(m·day·K)

Specific heat

1 J/(kg·K) = 2.38846 × 10^{-4} Btu/(lb·°F)

1 Btu/(lb·°F) = 4.1868 × 10^3 J/(kg·K)

Some special units

γ_o (oil specific gravity) = 141.5/(131.5 +° API)

1 SCF/STB (gas-oil ratio) = 0.17811 m³/m³ (standard)

1 m³/m³ = 5.6146 SCF/STB

1 psi/ft (pressure gradient) = 0.223248 atm/m = 0.0226206 Mpa/m

16.3 SI and Other Metric Systems

Quantity	Symbol	SI units Base	SI Practical	Mixed units Base	Mixed Practical	Mixed British
Oil production	q_o	m^3/s	m^3/D	cm^3/s	m^3/D	bbl/D
Water injection	q_w	m^3/s	m^3/D	cm^3/s	m^3/D	bbl/D
Critical production	q_c	m^3/s	m^3/D	cm^3/s	m^3/D	bbl/D
Cross section	A	m^2	m^2	cm^2	m^2	ft^2
Permeability	k	m^2	μm^2	darcy	md	md
Effective depth	h	m	m	cm	m	ft
Perforated depth	ΔL	m	m	cm	m	ft
Reservoir length	L	m	m	cm	m	ft
Capillary radius	r_c	m	μm	cm	μm	μin
Wellbore radius	r_w	m	m	cm	m	ft
Drainage radius	r_e	m	m	cm	m	ft
Oil volume factor	B_o					
Oil viscosity	μ_o	Pa·s	MPa·s	cp	cp	cp
Pressure difference	Δp	Pa	MPa	atm	atm	psi
Initial difference	p_0	Pa	MPa	atm	atm	psi
Wellbore pressure	p_{bh}	Pa	MPa	atm	atm	psi
Capillary pressure	p_c	Pa	MPa	$dyne/cm^2$	atm	psi
Oil density	ρ_o	kg/m^3	g/cm^3	g/cm^3	g/cm^3	lbm/ft^3
Water density	ρ_w	kg/m^3	g/cm^3	g/cm^3	g/cm^3	lbm/ft^3
Production time	t	s	h	s	h	h
Saturation	S	%	%	%	%	%
Skin factor	s_k					
Oil compressibility	c_o	Pa^{-1}	MPa^{-1}	atm^{-1}	atm^{-1}	psi^{-1}
Rock compressibility	c_R	Pa^{-1}	MPa^{-1}	atm^{-1}	atm^{-1}	psi^{-1}
Well controlled reserve	N	m^3	m^3	cm^3	m^3	bbl
Turbulence factor	β	m^{-1}	m^{-1}	cm^{-1}	m^{-1}	ft^{-1}
Porosity	ϕ	%	%	%	%	%
Surface tension	σ	N/m	mN/m	dyne/cm	dyne/cm	dyne/cm
Contact angle	θ	$^\circ$	$^\circ$	$^\circ$	$^\circ$	$^\circ$
Energy	E	J	J	J	J	Btu
Heat transfer coeff.	k_T	$\frac{W}{m \cdot K}$	$\frac{kJ}{m \cdot D \cdot K}$	$\frac{W}{m \cdot K}$	$\frac{kJ}{m \cdot D \cdot K}$	$\frac{Btu}{ft \cdot D \cdot {}^\circ F}$
Specific heat	U	$\frac{J}{kg \cdot K}$	$\frac{J}{kg \cdot K}$	$\frac{J}{kg \cdot K}$	$\frac{J}{kg \cdot K}$	$\frac{Btu}{lb \cdot {}^\circ F}$

Bibliography

R. A. Adams (1975), Sobolev Spaces, Academic Press, New York.

P. Allen and J. Allen (1990), Basin Analysis: Principles and Applications, Blackwell Scientific Publications, Oxford, England.

B. Amaziane, A. Bourgeat, and J. Koebbe (1991), Numerical simulation and homogenization of two-phase flow in heterogeneous porous media, *Transp. Porous Media* **6**, 519–539.

E. Anderson, Z. Bai, C. Bischof, J. Demmel, J. Dongarra, J. DuCroz, A. Greenbaum, S. Hammarling, A. McKenney, S. Ostrouchov, and D. Sorensen (1999), LAPACK Users' Guide, 3rd ed., SIAM, Philadelphia, PA.

S. N. Antontsev (1972), On the solvability of boundary value problems for degenerate two-phase porous flow equations, *Dinamika Sploshn. Sredy* **10**, 28–53, in Russian.

J. R. Appleyard, I. M. Cheshire, and R. K. Pollard (1981), Special techniques for fully implicit simulators, in Proc. European Symp. Enhanced Oil Recovery, Bournemouth, England, 395–408.

T. Arbogast (1993), Gravitational forces in dual-porosity systems I. Model derivation by homogenization, *Transp. Porous Media* **13**, 179–203.

T. Arbogast and M. F. Wheeler (1995), A characteristics-mixed finite element for advection-dominated transport problems, *SIAM J. Numer. Anal.* **32**, 404 – 424.

D. N. Arnold, D. Boffi, and R. Falk (2005), Quadrilateral H(div) finite elements, *SIAM J. Numer. Anal.* **42**, 2429–2451.

W. E. Arnoldi (1951), The principle of minimized iteration in the solution of the matrix eigenvalue problem, *Quart. Appl. Math.* **9**, 17–29.

F. E. Ashford (1969), Computed relative permeability drainage and imbibition, SPE 2582, 44th Annual Meeting, Denver.

O. Axelsson (1994), Iterative Solution Methods, Cambridge University Press, Cambridge, England.

K. Aziz, B. Ramesh, and P. T. Woo (1985), Fourth SPE comparative solution project: A comparison of steam injection simulators, SPE 13510, SPE Reservoir Simulation Symposium, Dallas, Texas, 441– 454.

K. Aziz and A. Settari (1979), Petroleum Reservoir Simulation, Applied Science Publishers Ltd., London.

I. Babuška and M. R. Dorr (1981), Error estimates for the combined h and p versions of the finite element method, *Numer. Math.* **37**, 257–277.

I. Babuška, A. Miller, and M. Vogelius (1983), Adaptive methods and error estimation for elliptic problems of structural mechanics, in Adaptive Computational Methods for Partial Differential Equations, I. Babuška et al., eds., SIAM, Philadelphia, 35–56.

I. Babuška and W. C. Rheinboldt (1978A), Error estimates for adaptive finite element computations, *SIAM J. Numer. Anal.* **15**, 736–754.

I. Babuška and W. C. Rheinboldt (1978B), A-posteriori error estimates for the finite element method, *Internat. J. Numer. Methods Engrg.* **12**, 1597–1615.

S. Balay, K. Buschelman, V. Eijkhout, W. D. Gropp, D. Kaushik, M. G. Knepley, L. C. McInnes, B. F. Smith, and H. Zhang (2004), PETSc Users Manual, ANL-95/11-Revision 2.1.5, Argonne National Laboratory, Argonne, IL.

R. E. Bank (1990), PLTMG: A Software Package for Solving Elliptic Partial Differential Equations, User's Guide 6.0, SIAM, Philadelphia.

R. E. Bank, A. H. Sherman, and A. Weiser (1983), Refinement algorithms and data structures for regular local mesh refinement, in Scientific Computing, R. Stepleman et al., eds., North–Holland, Amsterdam, New York, Oxford, 3–17.

R. E. Bank and A. Weiser (1985), Some a posteriori error estimators for elliptic partial differential equations, *Math. Comp.* **44**, 283–301.

G. I. Barenblatt, Iu. P. Zheltov, and I. N. Kochina (1960), Basic concepts in the theory of seepage of homogeneous liquids in fissured rocks [strata], *Prikl. Mat. Mekh.* **24**, 852–864.

J. W. Barker and S. Thibeau (1997), A critical review of the use of pseudo relative permeabilities for upscaling, *SPE Reservoir Engineering J.*, May, 138–143.

J. W. Barrett and K. W. Morton (1984), Approximate symmetrization and Petrov–Galerkin methods for diffusion-convection problems, *Comput. Methods Appl. Mech. Engrg.* **45**, 97–122.

J. Bear (1972), Dynamics of Fluids in Porous Media, Dover, New York.

J. Bear and Y. Bachmat (1991), Introduction to Modeling of Transport Phenomena in Porous Media, Kluwer Academic Publishers, Dordrecht, The Netherlands.

H. D. Beggs (1991), Production Optimization Using Nodal Analysis, OGCI Publication, Oil and Gas Consultants International, Tulsa, OK.

H. D. Beggs and J. R. Robinson (1975), Estimating the viscosity of crude oil systems, *JPT Forum*, September, 1140–1141.

S. H. Beggs, R. R. Carter, and P. Dranfield (1989), Assigning effective values to simulator gridblock parameters for heterogeneous reservoirs, *SPE Reservoir Engineering J.*, Nov., 455–465.

G. Behie and P. Vinsome (1982), Block iterative methods for fully implicit reservoir simulation, *Soc. Pet. Eng. J.*, July, 658–668.

M. W. Benson and P. O. Frederickson (1982), Iterative solution of large sparse linear systems arising in certain multidimensional approximation problems, *Util. Math.* **22**, 127–148.

P. Bedrikovetsky, D. Marchesin, and P. R. Ballin (1996), Mathematical model for immiscible displacement honoring hysteresis, SPE 36132, in Fourth Latin American and Caribbean Petroleum Engineering Conference, Port-of-Spain, Trinidad & Tabago, 557–573.

D. Bhuyan, G. A. Pope, and L. W. Lake (1991), Simulation of high-pH coreflood experiments using a compositional chemical flood simulator, SPE 21029, Proc. 11th SPE Int. Symp. Oilfield Chemistry, Anaheim, CA.

M. A. Biot (1955), Theory of elasticity and consolidation for a porous anisotropic solid, *J. Appl. Phys.* **26**, 182–185.

R. B. Bird, W. E. Stewart, and E. N. Lightfoot (1960), Transport Phenomena, John Wiley & Sons, New York.

G. Booch, J. Rumbaugh, and I. Jacobson (1998), The Unified Modelling Language User Guid, Addison–Wesley, Boston.

D. Braess (1997), Finite Elements, Theory, Fast Solvers, and Applications in Solid Mechanics, Cambridge University Press, Cambridge, England.

J. H. Bramble (1993), Multigrid Methods, Pitman Research Notes in Math., Vol. 294, Longman, London.

S. C. Brenner and L. R. Scott (1994), The Mathematical Theory of Finite Element Methods, Springer-Verlag, New York.

F. Brezzi, J. Douglas, Jr., R. Durán, and M. Fortin (1987A), Mixed finite elements for second order elliptic problems in three variables, *Numer. Math.* **51**, 237–250.

F. Brezzi, J. Douglas, Jr., M. Fortin, and L. D. Marini (1987B), Efficient rectangular mixed finite elements in two and three space variables, *RAIRO Modèl. Math. Anal. Numér.* **21**, 581–604.

F. Brezzi, J. Douglas, Jr., and L. D. Marini (1985), Two families of mixed finite elements for second order elliptic problems, *Numer. Math.* **47**, 217–235.

F. Brezzi and M. Fortin (1991), Mixed and Hybrid Finite Element Methods, Springer-Verlag, New York.

F. J. L. Briens, C. H. Wu, and J. Gazdag (1997), Compositional reservoir simulation in parallel supercomputing environment, SPE 21214, The 11th SPE Symp. on Reserv. Simul., Anaheim, CA.

A. Brooks and T. J. Hughes (1982), Streamline upwind Petrov–Galerkin formulations for convection dominated flows with particular emphasis on the incompressible Navier–Stokes equations, *Comput. Methods Appl. Mech. Engrg.* **32**, 199–259.

R. H. Brooks and A. T. Corey (1964), Hydraulic properties of porous media, Hydrology Papers, Colorado State University, Fort Collins, CO, 1964.

C. L. Brown, G. A. Pope, L. M. Abriola, and K. Sepehrnoori (1994), Simulation of surfactant enhanced aquifer remediation, *Water Resources Research* **30**, 2959–2977.

N. I. Buleev (1959), A numerical method for the solution of two-dimensional and three-dimensional equations of diffusion, *Math. Sb.* **51**, 227–238.

D. A. Calhoun, Jr., M. Lewis, Jr., and R. C. Newman (1949), Experiments on the capillary properties of porous solids, *Trans. SPE AIME* **186**, 180–196.

D. Camilleri, A. Fil, G. A. Pope, B. A. Rouse, and K. Sepehrnoori (1987), Comparison of an improved compositional micellar/polymer simulator with laboratory core floods, *SPE Reservoir Engineering J.*, Nov., 441–451.

J. M. Campbell and R. A. Campbell (1985), Applications of the SI metric system: Part 2—The basic units, *J. Pet. Tech.*, Oct., 1801–1805; *Trans. AIME* **279**.

N. L. Carr, R. Kobayashi, and D. B. Burrows (1954), Viscosity of hydrocarbon gases under pressure, *Trans. AIME* **191**, 264–272.

M. A. Celia, T. F. Russell, I. Herrera, and R. E. Ewing (1990), An Eulerian Lagrangian localized adjoint method for the advection-diffusion equation, *Advances in Water Resources* **13**, 187–206.

T. F. Chan and T. P. Mathew (1994), Domain decomposition algorithms, *Acta Numer.*, 61–143.

G. Chavent and J. Jaffré (1986), Mathematical Models and Finite Elements for Reservoir Simulation, North–Holland, Amsterdam.

H. Chen, Z. Chen, G. Huan, and Z. Wang (2002A), Mixed discontinuous FE methods and their applications to two-phase flow in porous media, in Fluid Flows and Transport in Porous Media, Mathematical and Numerical Treatment, Contemp. Math., Vol. 295, Z. Chen and R. Ewing, eds., AMS, Providence, RI, 113–126.

Z. Chen (2000), Formulations and numerical methods of the black oil model in porous media, *SIAM J. Numer. Anal.* **38**, 489–514.

Z. Chen (2001), Degenerate two-phase incompressible flow I: Existence, uniqueness and regularity of a weak solution, *J. Differential Equations* **171**, 203–232.

Z. Chen (2002A), Degenerate two-phase incompressible flow II: Regularity, stability and stabilization, *J. Differential Equations* **186**, 345–376.

Z. Chen (2002B), Characteristic mixed discontinuous finite element methods for advection-dominated diffusion problems, *Comput. Methods Appl. Mech. Engrg.* **191**, 2509–2538.

Z. Chen (2005), Finite Element Methods and Their Applications, Springer-Verlag, Heidelberg, New York.

Z. Chen and J. Douglas, Jr. (1989), Prismatic mixed finite elements for second order elliptic problems, *Calcolo* **26**, 135–148.

Z. Chen and J. Douglas, Jr. (1991), Approximation of coefficients in hybrid and mixed methods for nonliner parabolic problems, *Mat. Apl. Comput.* **10**, 137–160.

Z. Chen, M. Espedal, and R. E. Ewing (1995), Continuous-time finite element analysis of multiphase flow in groundwater hydrology, *Appl. Math.* **40**, 203–226.

Z. Chen and R. E. Ewing (1997A), Fully discrete finite element analysis of multiphase flow in groundwater hydrology, *SIAM J. Numer. Anal.* **34**, 2228–2253.

Z. Chen and R. E. Ewing (1997B), Comparison of various formulations of three-phase flow in porous media, *J. Comput. Phys.* **132**, 362–373.

Z. Chen and R. Ewing (1999), Mathematical analysis for reservoir models, *SIAM J. Math. Anal.* **30**, 431–453.

Z. Chen, R. E. Ewing, H. Lu, S. L. Lyons, S. Maliassov, M. B. Ray, and T. Sun (2002B), Integrated two-dimensional modeling of fluid flow and compaction in a sedimentary basin, *Comput. Geosci.* **6**, 545–564.

Z. Chen, R. E. Ewing, and Z.-C. Shi (2000A), Numerical Treatment of Multiphase Flows in Porous Media, Lecture Notes in Phys., Vol. 552, Springer-Verlag, Heidelberg.

Z. Chen and G. Huan (2003), Numerical experiments with various formulations for two phase flow in petroleum reservoirs, *Transp. Porous Media* **51**, 89–102.

Z. Chen, G. Huan, and B. Li (2002C), Modeling 2D and 3D horizontal wells using CVFA, *Commun. Math. Sci.* **1**, 30–44.

Z. Chen, G. Huan, and B. Li (2004A), An improved IMPES method for two-phase flow in porous media, *Transp. Porous Media* **54**, 361–376.

Z. Chen, G. Huan, and H. Wang (2006) Computer simulation of gas cycling of retrograde condensate reservoirs using an unstructured CVFE compositional model, *Computing* **78**, 31–53.

Z. Chen, B. Li, G. Huan, D. Espin, H. Klie, and S. Buitrago (2002D), Control volume function approximation method for the black oil model, in Proceedings of the 2nd Meeting on Reservoir Simulation, Universidad Argentina de la Empresa, Buenos Aires, Argentina, CD-ROM.

Z. Chen, S. Lyons, and G. Qin (2000B), Derivation of the Forchheimer law via homogenization, *Transp. Porous Media* **44**, 325–335.

Z. Chen, S. Lyons, and G. Qin (2004B), The mechanical behavior of poroelastic media saturated with a Newtonian fluid derived via homogenization, *Internat. J. Numer. Anal. Modeling* **1**, 75–98.

Z. Chen, Y. Ma, and G. Chen (2007), A sequential numerical chemical compositional simulator. *Transport in Porous Media* **68**, 389–411.

Z. Chen, G. Qin, and R. E. Ewing (2000C), Analysis of a compositional model for fluid flow in porous media, *SIAM J. Appl. Math.* **60**, 747–777.

Z. Chen and X. Yu (2006), Implementation of mixed methods as finite difference methods and applications to nonisothermal multiphase flow in porous media, *Journal of Computational Mathematics* **24**, 281–294.

J. N. Chew and C. A. Connally (1959), A viscosity correlation for gas-saturated crude oils, *Trans. SPE AIME* **216**, 23–25.

M. C. H. Chien, M. L. Wasserman, H. E. Yardumian, and E. Y. Chung (1987), The use of vectorization and parallel processing for reservoir simulation, SPE 16025, The Ninth SPE Symp. on Reservoir Simulation, San Antonio, TX.

I. Christie, D. F. Griffiths, and A. R. Mitchell (1976), Finite element methods for second order differential equations with significant first derivatives, *Internat. J. Numer. Methods Engrg.* **10**, 1389–1396.

M. A. Christie (1996), Upscaling for reservoir simulation, *J. Pet. Tech.*, Nov., 1004–1010.

M. A. Christie and M. J. Blunt (2001), Tenth comparative solution project: A comparison of upscaling techniques, SPE 66599, The 16th SPE Symp. on Reservoir Simulation, Houston, TX.

P. G. Ciarlet (1978), The Finite Element Method for Elliptic Problems, North–Holland, Amsterdam.

P. G. Ciarlet and P.-A. Raviart (1972), The combined effect of curved boundaries and numerical integration in isoparametric finite element methods, in The Mathematical Foundations of the Finite Element Method with Applications to Partial Differential Equations, A. K. Aziz, ed., Academic Press, New York, pp. 409–474.

K. H. Coats (1980), An equation of state compositional model, *Soc. Pet. Eng. J.* **20**, 363–376.

K. H. Coats (1989), Implicit compositional simulation of single-porosity and dual-porosity reservoirs, SPE 18427, SPE Symposium on Reservoir Simulation, Houston, Texas.

D. A. Collins, L. X. Nghiem, R. Sharma, R. Agarwal, and K. Jha (1991), Field scale simulation of horizontal wells with hybrid grids, SPE 21218, The 11th SPE Symposium on Reservoir Simulation, Anaheim, California.

R. E. Collins (1961), Flow of Fluids through Porous Materials, van Nostrand Reinhold, New York.

A. T. Corey (1986), Mechanics of Immiscible Fluids in Porous Media, Water Resources Publications, Littleton, CO.

A. T. Corey, C. H. Rathjens, J. H. Henderson, and M. R. Wyllie (1956), Three-phase relative permeability, *Trans. SPE AIME* **207**, 349–351.

R. Courant (1943), Variational methods for the solution of problems of equilibrium and vibrations, *Bull. Amer. Math. Soc.* **49**, 1–23.

B. C. Craft and M. F. Hawkins (1991), Applied Petroleum Reservoir Engineering, 2nd ed., revised by R. E. Terry, Prentice–Hall, Englewood Cliffs, NJ.

H. Darcy (1856), Les Fontaines Publiques de la Ville de Dijon, Victor Dalmond, Paris.

E. F. D'Azevedo and R. B. Simpson (1989), On optimal interpolation triangle incidences, *SIAM J. Sci. Statist. Comput.* **10**, 1063–1075.

M. Delshad and G. A. Pope (1989), Comparison of the three-phase oil relative permeability models, *Transp. Porous Media* **4**, 59–83.

M. Delshad, G. A. Pope, and K. Sepehrnoori (2000), UTCHEM Version-9.0, Technical Documentation, Center for Petroleum and Geosystems Engineering, The University of Texas at Austin, TX.

J. R. Dempsey (1965), Computer routine treats gas viscosity as a variable, *Oil and Gas Journal*, August 16, 11–17.

P. Deuflhard, P. Leinen, and H. Yserentant (1989), Concepts of an adaptive hierarchical finite element code, *IMPACT Comput. Sci. Engrg.* **1**, 3–35.

C. V. Deutsch (1989), Calculating effective absolute permeability in sandstone/shale sequences, *SPE Research Engineering*, September, 343–350.

J. C. Diaz, R. E. Ewing, R. W. Jones, A. E. McDonald, I. M. Uhler, and D. U. von Rosenberg (1984), Self-adaptive local grid-refinement for time-dependent, two-dimensional simulation, in Finite Elements in Fluids, Vol. VI, Wiley, New York, 479–484.

J. Dongarra, J. R. Bunch, C. B. Moler, and G. W. Stewart (1979), LINPACK Users' Guide, SIAM, Philadelphia.

J. Douglas, Jr. (1961), A survey of numerical methods for parabolic differential equations, in *Advances in Computers*, Vol. 2, F. L. Alt, ed., Academic Press, New York, 1–54.

J. Douglas, Jr., and T. Arbogast (1990), Dual-porosity models for flow in naturally fractured reservoirs, in Dynamics of Fluids in Hierarchical Porous Media, J. H. Cushman, ed., 177–221.

J. Douglas, Jr., R. E. Ewing, and M. Wheeler (1983), The approximation of the pressure by a mixed method in the simulation of miscible displacement, *RAIRO Anal. Numér.* **17**, 17–33.

J. Douglas, Jr., F. Furtado, and F. Pereira (1997), On the numerical simulation of water flooding of heterogeneous petroleum reservoirs, *Comput. Geosci.* **1**, 155–190.

J. Douglas, Jr., D. W. Peaceman, and H. H. Rachford, Jr. (1959), A method for calculating multi-dimensional immiscible displacement, *Trans. SPE AIME* **216**, 297–306.

J. Douglas, Jr., and T. F. Russell (1982), Numerical methods for convection-dominated diffusion problems based on combining the method of characteristics with finite element or finite difference procedures, *SIAM J. Numer. Anal.* **19**, 871–885.

P. M. Dranchuk, R. A. Purvis, and D. B. Robinson (1974), Computer calculation of nature gas compressibility factors using the standing and K-correlation, Institute of Petroleum Technical Series, No. IP 74–008.

J. Dupuit (1863), Estudes Theoriques et Pratiques sur le Mouvement des Eaux, Dunod, Paris.

L. J. Durlofsky, R. C. Jones, and W. J. Milliken (1994), A new method for the scale up of displacement processes in heterogeneous reservoirs, in Proc. Third European Conference on the Mathematics of Oil Recovery, Roros, Norway.

N. S. Espedal and R. E. Ewing (1987), Characteristic Petrov–Galerkin subdomain methods for two phase immiscible flow, *Comput. Methods Appl. Mech. Engrg.* **64**, 113–135.

D. J. Evans (1968), The use of pre-conditioning in iterative methods for solving linear systems with symmetric positive definite matrices, *J. Inst. Math. Its. Appl.* **4**, 295–315.

R. E. Ewing (1983), The Mathematics of Reservoir Simulation, SIAM, Philadelphia.

R. E. Ewing, T. F. Russell, and M. F. Wheeler (1983), Simulation of miscible displacement using mixed methods and a modified method of characteristics, SPE 12241, Reservoir Simulation Symposium, San Francisco, CA.

R. E. Ewing (1986), Efficient adaptive procedures for fluid flow applications, *Comput. Methods Appl. Mech. Engrg.* **55**, 89–103.

R. E. Ewing, R. D. Lazarov, S. L. Lyons, D. V. Papavassiliou, J. E. Pasciak, and G. Qin (1999), Numerical well model for non-Darcy flow through isotropic porous media, *Comput. Geosci.* **3**, 185–204.

R. E. Ewing, M. S. Pilant, J. G. Wade, and A. T. Watson (1994), Estimating parameters in scientific computation: A survey of experience from oil and groundwater modeling, *IEEE Comput. Sci. Engrg.* **1**, 19–31.

G. H. Fancher and J. A. Lewis (1933), Flow of simple fluids through porous materials, *Ind. Eng. Chem. Fund.* **24**, 1139–1147.

J. R. Fanchi (2001), Principles of Applied Reservoir Simulation, 2nd ed., Gulf Professional Publishing, Houston, TX.

C. W. Fetter (1993), Contaminant Hydrology, Macmillan Publishing Company, New York.

A. Firoozabadi and L. K. Thomas (1990), Sixth comparative solution project: Dual porosity simulators, *J. Pet. Tech.* **42**, 710–715.

P. J. Flory (1953), Principles of Polymer Chemistry, Cornell University Press, Ithaca, NY.

P. Forchheimer (1901), Wasserbewegung durch Boden, Z. Ver. Deut. Ing. **45**, 1782–1788 (in German).

P. A. Forsyth (1991), A control volume finite element approach to NAPL groundwater contamination, *SIAM J. Sci. Statist. Comput.* **12**, 1029–1057.

J. C. Fountain (1992), Field test of surfactant flooding: Mobility control of dense nonaqueous phase liquids, in Transport and Remediation of Subsurface Contaminants, D. A. Sabatini and R. C. Knox, eds., American Chemical Society Symposium Series 491, ACS, Washington, DC, Chapter 15.

B. Fraeijs de Veubeke (1965), Displacement and equilibrium models in the finite element method, in Stress Analysis, O. C. Zienkiewicz and G. Holister, eds., John Wiley & Sons, New York.

A. Friedman (1982), Variational Principles and Free-Boundary Problems, John Wiley & Sons, New York.

L. S. Fung, D. A. Collins, and L. X. Nghiem (1989), An adaptive-implicit switching criterion based on numerical stability analysis, *SPE Reservoir Engineering J.*, February, 45–51.

A. O. Garder, D. W. Peaceman, and A. L. Pozzi (1964), Numerical calculations of multidimensional miscible displacement by the method of characteristics, *Soc. Pet. Eng. J.* **4**, 26–36.

C. J. Glover, M. C. Puerto, J. M. Maerker, and E. I. Sandvik (1979), Surfactant phase behavior and retention in porous media, *Soc. Pet. Eng. J.* **19**, 183–193.

G. H. Golub and C. F. van Loan (1996), Matrix Computations, Johns Hopkins University Press, Baltimore, London.

I. Gustafsson (1978), A class of first order factorization methods, *BIT* **18**, 142–156.

W. Hackbusch (1985), Multigrid Methods and Applications, Springer-Verlag, Berlin, Heidelberg, New York.

D. B. Hand (1939), Dineric distribution I. The distribution of a consolute liquid between two immiscible liquids, *J. Phys. Chem.* **34**, 1961–2000.

G. M. Homsy (1987), Viscous fingering in porous media, *Ann. Rev. Fluid Mech.* **19**, 271–280.

R. N. Healy and R. L. Reed (1974), Physicochemical aspects of microemulsion flooding, *Soc. Pet. Eng. J.* **14**, 291–501.

B. Heinrich (1987), Finite Difference Methods on Irregular Networks, Birkhäuser, Basel, Boston, Stuttgart.

K. Hellan (1967), Analysis of elastic plates in flexure by a simplified finite element method, *Acta Polytechnica Scandinavia*, Civil Engineering Series, Trondheim **46**, 1–19.

R. Helmig (1997), Multiphase Flow and Transport Process in the Subsurface, Springer-Verlag, Heidelberg.

L. R. Herrmann (1967), Finite element bending analysis for plates, *J. Eng. Mech. Div. ASCE* **93**, 13–26.

M. R. Hestenes and E. L. Stiefel (1952), Methods of conjugate gradients for solving linear systems, *J. Res. National Bureau of Standards* **49**, Section B, 409–436.

N. J. Higham (1996), Accuracy and Stability of Numerical Algorithms, SIAM, Philadelphia.

G. J. Hirasaki (1981), Application of the theory of multicomponent, multiphase displacement to three-component, two-phase surfactant flooding, *Soc. Pet. Eng. J.*, 191–204.

G. J. Hirasaki (1982), Interpretation of the change in optimal salinity with overall surfactant concentration, *Soc. Pet. Eng. J.*, December, 971–982.

M. K. Hubbert (1956), Darcy's law and the field equations of the flow of underground fluids, *Trans. SPE AIME* **207**, 222–239.

C. Huh (1979), Interfacial tension and solubilizing ability of a microemulsion phase that coexists with oil and brine, *J. Colloid Interface Sci.* **71**, 408–428.

G. Huan, R. E. Ewing, G. Qin, and Z. Chen (2005), Numerical Simulation for Fractured Porous Media, in High Performance Computing and Applications, Lecture Notes in Computational Science and Engineering, Z. Chen et al., eds., Springer-Verlag, Heidelberg, 47–58.

R. H. Jacoby and V. J. Berry (1957), A method for predicting depletion performance of a reservoir producing volatile crude oil, *Trans. SPE AIME* **208**, 27–38.

B. Joe (1986), Delaunay triangular meshes in convex polygons, *SIAM J. Sci. Statist. Comput.* **7**, 514–539.

C. Johnson (1994), Numerical Solutions of Partial Differential Equations by the Finite Element Method, Cambridge University Press, Cambridge, England.

H. Kazemi (1969), Pressure transient analysis of naturally fractured reservoirs with uniform fracture distribution, *Soc. Pet. Eng. J.* **9**, 451–462.

D. E. Kenyon and A. Behie (1987), Third SPE comparative solution project: Gas cycling of retrograde condensate reservoirs, *J. Pet. Tech.* **39**, 981–998.

J. E. Killough (1995), Ninth SPE comparative solution project: A reexamination of black-oil simulation, SPE 29110, The 13th SPE Symposium on Reservoir Simulation, San Antonio, TX.

J. E. Killough and C. Cossack (1987), Fifth comparative solution project: Evaluation of miscible flood simulators, SPE 16000, The ninth SPE Symp. on Reservoir Simulation, San Antonio, TX.

J. E. Killough, D. Camilleri, B. L. Darlow, and J. A. Foster (1997), A parallel reservoir simulator based on local grid refinement, SPE 37978, The 1997 Reservoir Simulation, Dallas, TX.

J. E. Killough and M. F. Wheeler (1987), Parallel iterative linear equation solvers: An investigation of domain decomposition algorithms for reservoir simulation, SPE 16021, The ninth SPE Symp. on Reservoir Simulation, San Antonio, TX.

P. R. King (1989), The use of renormalization for calculating effective permeability, *Transp. Porous Media* **4**, 37–44.

P. M. Knupp and J. L. Lage (1995), Generalization of the Forchheimer-extended Darcy flow model to the tensor permeability case via a variational principle, *J. Fluid Mech.* **299**, 97–104.

Yu. Kuznetsov (1969), Some Problems in the Theory and Applications of Iterative Methods, Ph.D. Thesis, Novosibirsk, Russia, 1969.

L. W. Lake (1989), Enhanced Oil Recovery, Prentice–Hall, Englewood Cliffs, NJ.

L. W. Lake, E. Kasap, and M. Shook (1990), Pseudofunctions–The key to practical use of reservoir description, in North Sea Oil and Gas Reservoirs–II, A. T. Buller, et al., eds., Graham and Trotman, London.

C. Lanczos (1952), Solution of systems of linear equations by minimized iterations, *J. Res. National Bureau of Standards* **49**, 33–53.

R. B. Lantz (1970), Rigorous calculation of miscible displacement using immiscible reservoir simulators, *Trans. SPE AIME* **251**, 313–320.

P. A. Lemonnier (1979), Improvement of reservoir simulation by a triangular discontinuous finite element method, SPE 8249, The 1979 Annual Fall Technical Conference and Exhibition of SPE of AIME, Las Vegas.

I. Lerche (1990), Basin Analysis, Quantitative Methods, Vol. 1, Academic Press, San Diego, CA.

F. W. Letniowski (1992), Three-dimensional Delaunay triangulations for finite element approximations to a second-order diffusion operator, *SIAM J. Sci. Statist. Comput.* **13**, 765–770.

M. C. Leverett and W. B. Lewis (1941), Steady flow of gas-oil-water mixtures through unconsolidated sands, *Trans. SPE AIME* **142**, 107–116.

B. Li, Z. Chen, and G. Huan (2003A), Control volume function approximation methods and their applications to modeling porous media flow I: The two-phase flow, *Adv. Water Resources* **26**, 435–444.

B. Li, Z. Chen, and G. Huan (2003B), Modeling horizontal wells with the CVFA method in black oil reservoir simulations, in Current Trends in Scientific Computing, Z. Chen, R. Glowinski, and K. Li, eds., Contemporary Mathematics, Vol. 329, AMS, Providence, RI, 209–236.

B. Li, Z. Chen, and G. Huan (2004A), Control volume function approximation methods and their applications to modeling porous media flow II: The black oil model, *Adv. Water Resources* **27**, 99–120.

B. Li, Z. Chen, and G. Huan (2004B), Comparison of solution schemes for black oil reservoir simulations with unstructured grids, *Comput. Methods Appl. Mech. Engrg.* **193**, 319–355.

W. Li, Z. Chen, R. Ewing, G. Huan, and B. Li (2005), Comparison of the GMRES and ORTHOMIN for the black oil model in porous media, *Internat. J. Numer. Methods Fluids* **48**, 501–519.

M. L. Litvak and B. L. Darlow (1995), Surface network and well tubinghead pressure constraints in compositional simulation, SPE 29125, The 13th SPE Symposium on Reservoir Simulation, San Antonio, TX.

J. Lohrenz, B. G. Bray, and C. R. Clark (1964), Calculating viscosities of reservoir fluids from their compositions, *J. Pet. Tech.*, 1171–1176; Trans. AIME **23**.

Y. Ma and Z. Chen (2004), Parallel computation for reservoir thermal simulation of multicomponent and multiphase fluid flow, *J. Comput. Phys.* **201**, 224–237.

I. F. MacDonald, M. S. El-Sayed, K. Mow, and F. A. L. Dullien (1979), Flow through porous media: The Ergun equation revisited, *Indust. Chem. Fundam.* **18**, 199–208.

R. C. MacDonald and K. H. Coats (1970), Methods for numerical simulation of water and gas coning, *Trans. SPE AIME* **249**, 425–436.

C. C. Mattax and R. L. Dalton (1990), Reservoir Simulation, SPE Monograph, vol. 13, Society of Petroleum Engineers, Richardson, TX.

D. F. Mayer (1989), Application of reservoir simulation models to a new parallel computing system, SPE 19121, SPE Petroleum Computer Conference, San Antonio, TX.

D. N. Meehan (1980A), Estimating water viscosity at reservoir conditions, *Petroleum Engineers*, July, 117–118.

D. N. Meehan (1980B), A correlation for water compressibility, *Petroleum Engineers*, November, 125–126.

C. C. Mei and J.-L. Auriault (1991), The effect of the weak inertia on flow through a porous medium, *J. Fluid Mech.* **222**, 647–663.

D. M. Meter and R. B. Bird (1964), Tube flow of non-Newtonian polymer solutions, Parts I and II—laminar flow and rheological models, *AICHE J* (Nov.), 878–881, 1143–1150.

K. Miller and R. N. Miller (1981), Moving finite elements I, *SIAM J. Numer. Anal.* **18**, 1019–1032.

S. Moore, A. Baker, J. Dongarra, C. Halloy, and C. Ng (2002), Active Netlib: An active mathematical software collection for inquiry-based computational science and engineering education, *J. Digital Information* **2**, Special Issue on Interactivity in Digital Libraries.

N. R. Morrow (1970), Irreducible wetting-phase saturations in porous media, *Chem. Eng. Sci.* **25**, 1799–1816.

N. R. Morrow and B. Songkran (1982), Surface Phenomena in Enhanced Oil Recovery, D. O. Shah, ed., Plenum Press, New York.

Y. Mualem (1976), A new model for predicting the hydrolic conductivity of porous media, *Water Resour. Research* **12**, 513–522.

A. H. Muggeridge (1991), Generation of effective relative permeabilities from detailed simulation of flow in heterogeneous porous media, in Reservoir Characterization II, L. W. Lake et al., eds., Academic Press, San Diego, CA, 197–225.

M. Muskat (1937), Flow of Homogeneous Fluids Through Porous Media, McGraw–Hill, New York.

M. Muskat (1949), Physical Principles of Oil Production, McGraw–Hill, New York.

N. M. Nachtigal, S. C. Reddy, and L. N. Trefethen (1992), How fast are nonsymmetric matrix iterations?, *SIAM J. Matrix Anal. Appl.* **13**, 778–795.

J. C. Nédélec (1980), Mixed finite elements in $\mathbb{R}^3$, *Numer. Math.* **35**, 315–341.

S. P. Neuman (1981), An Eulerian Lagrangian numerical scheme for the dispersion-convection equation using conjugate-time grids, *J. Comput. Phys.* **41**, 270–294.

L. S. Nghiem, D. A. Collins, and R. Sharma (1991), Seventh SPE comparative solution project: Modeling of horizontal wells in reservoir simulation, SPE 21221, The 11th SPE Symposium on Reservoir Simulation, Anaheim, CA.

J. S. Nolen (1973), Numerical Simulation of Compositional Phenomena in Petroleum Reservoirs, Reprint Series, SPE, Dallas, **11**, 268–284.

D. Numbere, W. E. Brigham, and M. B. Standing (1977), Correlations for physical properties of petroleum reservoir brines, Petroleum Research Institute, Stanford University, Stanford, CA, 8–16.

A. S. Odeh (1981), Comparison of solutions to a three-dimensional black-oil reservoir problem, *J. Pet. Tech.*, January, 1981, 13–25.

J. T. Oden, L. Demkowicz, W. Rachowicz, and T. A. Westermann (1989), Toward a universal $h - p$ adaptive finite element strategy, Part 2. A posteriori error estimation, *Comput. Methods Appl. Mech. Engrg.* **77**, 113–180.

E. R. Oliveira (1971), Optimization of finite element solutions, in Proc. Third Conference on Matrix Methods in Structural Mechanics, Wright-Patterson Air Force Base, OH.

E. S. Oran and J. P. Boris (2001), Numerical Simulation of Reactive Flow, 2nd ed., Cambridge University Press, Cambridge, England.

A. M. Ostrowski (1973), Solution of Equations in Euclidean and Banach Spaces, 3rd ed., Academic Press, New York.

W. W. Owens and D. L. Archer (1971), The effect of rock wettability on oil-water relative permeability relations, *Trans. SPE AIME* **251**, 873–878.

D. W. Peaceman (1977A), Interpretation of well-block pressures in numerical reservoir simulation, SPE 6893, The 52nd Annual Fall Technical Conference and Exhibition, Denver, CO.

D. W. Peaceman (1977B), Fundamentals of Numerical Reservoir Simulation, Elsevier, New York.

D. W. Peaceman (1983), Interpretation of well-block pressures in numerical reservoir simulation with non-square grid blocks and anisotropic permeability, *Soc. Pet. Eng. J.*, June, 531–543.

D. W. Peaceman (1991), Presentation of a horizontal well in numerical reservoir simulation, SPE 21217, The 11th SPE Symposium on Reservoir Simulation, Ananheim, CA.

A. Peneloux, E. Rauzy, and R. Freze (1982), A consistent correction for Redlich-Kwong-Soave volumes, *Fluid Phase Equilibria* **8**, 7-23.

D.-Y. Peng and D. B. Robinson (1976), A new two-constant equation of state, *Industrial and Engineering Chemistry Fundamentals* **15**, 59–64.

O. Pironneau (1982), On the transport-diffusion algorithm and its application to the Navier-Stokes equations, *Numer. Math.* **38**, 309–332.

S. J. Pirson (1953), Performance of fractured oil reservoirs, *Bull. Amer. Assoc. Petroleum Geologists* **37**, 232–244.

P. Ya. Polubarinova-Kochina (1962), Theory of Ground Water Movement, translated by J. M. Roger de Wiest, Princeton University Press, Princeton, NJ.

G. A. Pope, L. W. Lake, and F. G. Helfferich (1978), Cation exchange in chemical flooding: Part I—Basic theory without dispersion, *Soc. Pet. Eng. J.* **18**, 418–434.

G. A. Pope and R. C. Nelson (1978), A chemical flooding compositional simulator, *Soc. Pet. Eng. J.* **18**, 339–354.

S. W. Poston, S. C. Ysrael, A. K. Hossain, E. F. Montgomery, and H. J. Ramey, Jr. (1970), The effect of temperature on irreducible water saturation and relative permeability of unconsolidated sands, *Trans. SPE AIME* **249**, 171–180.

H. S. Price and K. H. Coats (1974), Direct methods in reservoir simulation, *Trans. SPE AIME* **257**, 295–308.

P. Quandalle (1993), The eighth comparative solution project: Gridding techniques in reservoir simulation, SPE 25263, The 12th SPE Symp. of Reservoir Simulation, New Orleans, LA.

A. Quarteroni and A. Valli (1997), Numerical Approximation of Partial Differential Equations, Lecture Notes in Comp. Math., vol. 23, Springer-Verlag, Heidelberg.

W. F. Ramirez (1987), Application of Optimal Control Theory to Enhanced Oil Recovery, Elsevier, New York.

R. Raviart and J.-M. Thomas (1977), A mixed finite element method for second order elliptic problems, Lecture Notes in Math., Vol. 606, Springer-Verlag, Berlin, 292–315.

W. H. Reed and T. R. Hill (1973), Triangular mesh methods for the neutron transport equation, Technical Report, LA-UR-73-479, Los Alamos Scientific Laboratory, Los Alamos, NV.

R. C. Reid, J. M. Prausnitz, and T. K. Sherwood (1977), The Properties of Gases and Liquids, 3rd ed., McGraw–Hill, New York.

W. C. Rheinboldt (1998), Methods for Solving Systems of Nonlinear Equations, 2nd ed., SIAM, Philadelphia.

W. C. Rheinboldt and C. Mesztenyi (1980), On a data structure for adaptive finite element mesh refinement, *ACM Trans. Math. Software* **6**, 166–187.

L. A. Richards (1931), Capillary conduction of liquids through porous media, *Physics* **1**, 318–333.

R. D. Richtmyer and K. W. Morton (1967), Difference Methods for Initial Value Problems, Interscience, New York.

M. C. Rivara (1984A), Algorithms for refining triangular grids suitable for adaptive and multigrid techniques, *Internat. J. Numer. Methods Engrg.* **20**, 745–756.

M. C. Rivara (1984B), Design and data structure of fully adaptive, multigrid, finite element software, *ACM Trans. Math. Software* **10**, 242–264.

T. F. Russell (1990), Eulerian-Lagrangian localized adjoint methods for advection-dominated problems, in Numerical Analysis, Pitman Res. Notes Math. Series, Vol. 228, D. F. Griffiths and G. A. Watson, eds., Longman Scientific and Technical, Harlow, England, 206–228.

T. F. Russell and R. V. Trujillo (1990), Eulerian-Lagrangian Localized Adjoint methods with Variable Coefficients in Multiple Dimensions, Gambolati et al., eds., Comp. Meth. in Surface Hydrology, Springer-Verlag, Berlin.

T. F. Russell and M. F. Wheeler (1983), Finite element and finite difference methods for continuous flows in porous media, in The Mathematics of Reservoir Simulation, R. E. Ewing, ed., SIAM, Philadelphia, 35–106.

J. M. Rutledge, D. R. Jones, W. H. Chen, and E. Y. Chung (1991), The use of a massively parallel SIMD computer for reservoir simulation, SPE 21213, The seventh SPE Symp. on Reservoir Simulation, Anaheim, CA.

Y. Saad (1990), SPARSKIT: A basic tool kit for sparse matrix computations, Technical Report 90-20, Research Institute for Advanced Computer Science, NASA Ames Research Center, Moffet Field, CA.

Y. Saad (2004), Iterative Methods for Sparse Linear Systems, 2nd ed., SIAM, Philadelphia.

Y. Saad and M. H. Schultz (1986), GMRES: A generalized minimal residual algorithm for solving nonsymmetric linear systems, *SIAM J. Sci. Statist. Comput.* **7**, 856–869.

A. E. Scheidegger (1974), The Physics of Flow Through Porous Media, 3rd ed., University of Toronto Press, Toronto.

L. L. Schumaker (1981), Spline Functions: Basic Theory, John Wiley & Sons, New York.

J. F. Scamehorn, R. S. Schechter, and W. H. Wade (1982), Adsorption of surfactants on mineral oxide surface from aqueous solutions, I: Isometrically pure anionic surfactants, *J. Colloid and Inferface Science* **85**, 463–477.

S. L. Scott, R. L. Wainwright, and R. Raghavan (1987), Application of parallel (MIMD) computers to reservoir simulation, SPE 16020, The Ninth SPE Symp. on Reservoir Simulation, San Antonio, TX.

J. W. Sheldon, B. Zondek, and W.T. Cardwell (1959), One-dimensional, incompressible, non-capillary, two-phase fluid flow in a porous medium, *Trans. SPE AIME* **216**, 290–296.

N. D. Shutler (1969), Numerical three-phase simulation of the linear steamflood process, *Trans. SPE AIME* **246**, 232–246.

B. Smith, P. Bjorstad, and W. Gropp (1996), Domain Decomposition, Parallel Multilevel Methods for Elliptic Partial Differential Equations, Cambridge University Press, Cambridge, England.

R. W. Snell (1962), Three-phase relative permeability in unconstituted sand, *J. Inst. Petrol.* **84**, 80–88.

P. Somasundaran and H. S. Hanna (1977), Physico-chemical aspects of adsorption at solid/liquid inferfaces, in Improved Oil Recovery by Surfactant and Polymer Flooding, D. O. Shah and R. S. Schechter, eds., Academic Press, New York, 205–274.

P. Sonneveld (1989), CGS, A fast Lanczos-type solver for nonsymmetric linear systems, *SIAM J. Sci. Statist. Comput.* **10**, 36–52.

K. S. Sorbie (1991), Polymer-Improved Oil Recovery, CRC Press, Boca Raton, FL.

M. B. Standing (1977), Volumetric and phase behavior of oil field hydrocarbon systems, SPE Reprint, 8th ed., 124–125.

E. L. Stiefel (1958), Kernel polynomials in linear algebra and their applications, *U.S. National Bureau of Standards, Applied Mathematics Series* **49**, 1–14.

H. L. Stone (1970), Probability model for estimating three-phase relative permeability, *Trans. SPE AIME* **249**, 214–218.

H. L. Stone (1973), Estimation of three-phase relative permeability and residual oil data, *J. Can. Petrol. Technol.* **12**, 53–61.

H. L. Stone and A. O. Garder, Jr. (1961), Analysis of gas-cap or dissolved-gas reservoirs, *Trans. SPE AIME* **222**, 92–104.

G. Strang and G. J. Fix (1973), An Analysis of the Finite Element Method, Prentice–Hall, Englewood Cliffs, NJ.

K. Stüben (1983), Algebraic multigrid (AMG): Experiences and comparisons, *Appl. Math. Comput.* **13**, 419–440.

B. A. Szabo (1986), Mesh design for the p-version of the finite element method, *Comput. Methods Appl. Mech. Engrg.* **55**, 86–104.

G. W. Thomas and D. H. Thurnau (1983), Reservoir simulation using an adaptive implicit method, *Soc. Pet. Eng. J.*, October, 759–768.

J. W. Thomas (1995), Numerical Partial Differential Equations, Finite Difference Methods, Springer-Verlag, New York.

L. K. Thomas, T. N. Dixon, and R. G. Pierson (1983), Fractured reservoir simulation, *Soc. Pet. Eng. J.*, February, 42–54.

V. Thomée (1984), Galerkin Finite Element Methods for Parabolic Problems, Lecture Notes in Math., Vol. 1054, Springer-Verlag, Berlin.

M. R. Todd and W. J. Longstaff (1972), The development, testing and application of a numerical simulator for predicting miscible flood performance, *Trans. SPE AIME* **253**, 874–882.

A. M. Turing (1948), Rounding-off errors in matrix processes, *Quart. J. Mech. Appl. Math.* **1**, 287–292.

R. S. Varga (1960), Factorization and normalized iterative methods, in Boundary Problems in Differential Equations, University of Wisconsin Press, Madison, WI, 121–142.

H. A. van der Vorst (1992), BI-CGSTAB: A fast and smoothly converging variant of BI-CG for the solution of nonsymmetric linear systems, *SIAM J. Sci Statist. Comput.* **13**, 631–644.

M. Th. van Genuchten (1980), A closed form equation for predicting the hydraulic conductivity in soils, *Soil Sci. Soc. Arm. J.* **44**, 892–898.

M. Vasquez and H. D. Beggs (1980), Correlations for fluid physical property predictions, *J. Pet. Tech.*, June, 968–970.

S. Verdière, L. H. Quettier, P. Samier, and A. M. Thompson (1999), Applications of a parallel simulator to industrial test cases, SPE 51887, The 15th SPE Symp. on Reservoir Simulation, Houston, TX.

R. Verfürth (1996), A Review of a Posteriori Error Estimation and Adaptive Mesh-Refinement Techniques, Wiley/Teubner, Chichester, Stuttgart.

P. K. W. Vinsome (1976), Orthomin, An iterative method for solving sparse sets of simultaneous linear equations, in Proc. of Fourth Symposium on Reservoir simulations, Society of Petroleum Engineers of AIME, 149–157.

VIP-Executive (1994), VIP-Executive Technical Reference, Western Atlas International, Houston, TX.

J. W. Wallis, R. P. Kendall, and T. E. Little (1985), Constrained residual acceleration of conjugate residual methods, SPE 13536, The SPE Reservoir Simulation Symp., Dallas, TX.

H. Wang (2000), An optimal-order error estimate for an ELLAM scheme for two-dimensional linear advection-diffusion equations, *SIAM J. Numer. Anal.* **37**, 1338–1368.

H. F. Wang and M. P. Anderson (1982), Introduction to Groundwater Modeling, Finite Difference and Finite Element Methods, W. H. Freeman, San Francisco.

J. Wang and T. Mathew (1994), Mixed finite element methods over quadrilaterals, in the Proceedings of the Third International Conference on Advances in Numerical Methods and Applications, I. T. Dimov et al., eds., World Scientific, River Edge, NJ, 203–214.

J. C. Ward (1964), Turbulent flow in porous media, *J. Hydr. Div. ASCE* **90**, 1–12.

J. Warren and P. Root (1963), The behavior of naturally fractured reservoirs, *Soc. Pet. Eng. J.* **3**, 245–255.

H. G. Weinstein, J. E. Chappelear, and J. S. Nolen (1986), Second comparative solution project: A three-phase coning study, *J. Pet. Tech.*, March, 345–353.

J. J. Westerink and D. Shea (1989), Consistent higher degree Petrov–Galerkin methods for the solution of the transient convection-diffusion equation, *Internat. J. Numer. Methods Engrg.* **13**, 839–941.

M. F. Wheeler (1995), Environmental Studies: Mathematical, Computational, and Statistical Analysis, The IMA Volumes in Mathematics and its Applications, Vol. 79, Springer-Verlag, Berlin, New York.

S. Whitaker (1966), The equations of motion in porous media, *Chem. Eng. Sci.* **21**, 291–300.

S. Whitaker (1986), Flow in porous media I: A theoretical derivation of Darcy's law, *Transp. Porous Media* **1**, 3–25.

C. D. White and R. N. Horne (1987), Computing absolute transmissibility in the presence of fine-scale heterogeneity, SPE 16011, in The SPE Symp. on Reservoir Simulation, San Antonio, TX.

C. H. Whitson (1982), Effect of physical properties estimation on equation of state predictions, SPE 11200, in The 57th Annual Fall Technical Conference and Exhibition of the Society of Petroleum Engineers of AIME, New Orleans, LA.

E. Wichert and K. Aziz (1972), Calculate Z's for sour gases, *Hydrocarbon Processing*, May, 51–119.

P. A. Winsor (1954), Solvent Properties of Amphiphilic Compounds, Butterworths, London.

D. Wreath, G. A. Pope, and K. S. Sepehrnoori (1990), Dependence of polymer apparent viscosity on the permeable media and flow conditions, *In Situ* **14**, 263–284.

D. Yang (1992), A characteristic mixed method with dynamic finite element space for convection-dominated diffusion problems, *J. Comput. Appl. Math.* **43**, 343–353.

L. C. Young and R. E. Stephenson (1983), A generalized compositional approach for reservoir simulation, *Soc. Pet. Eng. J.* **23**, 727–742.

E. F. de Zabala, J. M. Vislocky, E. Rubin, and C. J. Radke (1982), A chemical theory for linear alkaline flooding, *Soc. Pet. Eng. J.* **12**, 245–258.

O. C. Zienkiewicz and J. Zhu (1987), A simple error estimator and adaptive procedure for practical engineering analysis, *Internat. J. Numer. Methods Engrg.* **24**, 337–357.

D. Zudkevitch and J. Joffe (1970), Correlation and prediction of vapor-liquid equilibria with the Redlich–Kwong equation of state, *American Institute of Chemical Engineers J.* **16**, 112–199.

Index